Lecture Notes in Computer Science 6812

Commenced Publication in 1973
Founding and Former Series Editors:
Gerhard Goos, Juris Hartmanis, and Jan van Leeuwen

W0044155

Udaya Parampalli   Philip Hawkes (Eds.)

Information Security and Privacy

16th Australasian Conference, ACISP 2011
Melbourne, Australia, July 11-13, 2011
Proceedings

 Springer

Volume Editors

Udaya Parampalli
The University of Melbourne
Department of Computer Science and Software Engineering
Melbourne, VIC 3010, Australia
E-mail: udaya@csse.unimelb.edu.au

Philip Hawkes
Qualcomm Incorporated
Suite 301, Level 3, 77 King Street
Sydney, NSW 2000, Australia
E-mail: phawkes@qualcomm.com

ISSN 0302-9743   e-ISSN 1611-3349
ISBN 978-3-642-22496-6   ISBN 978-3-642-22497-3 (eBook)
DOI 10.1007/978-3-642-22497-3
Springer Heidelberg Dordrecht London New York

Library of Congress Control Number: 2011931295

CR Subject Classification (1998): E.3, K.6.5, D.4.6, C.2, J.1, G.2.1

LNCS Sublibrary: SL 4 – Security and Cryptology

Typesetting: Camera-ready by author, data conversion by Scientific Publishing Services, Chennai, India

Printed on acid-free paper

Springer is part of Springer Science+Business Media (www.springer.com)

Preface

The annual Australasian Conference on Information Security and Privacy (ACISP) is the premier Australian academic conference in its field, showcasing research from around the globe on a range of topics. The 16th conference in this series—ACISP 2011—was held during July 11–13, 2011, at RMIT University in Melbourne, Australia.

There were 103 paper submissions for the conference. These submissions were reviewed by the Program Committee and a number of other individuals, whose names can be found overleaf. The Program Committee then selected 24 papers for presentation at the conference. These papers are contained in these proceedings. Theoretical research features prominently in these papers.

This year the Program Committee introduced a practice of accepting some submissions for presentation as poster papers, for the first time in ACISP history. The purpose of this practice is to allow ACISP to include more practically oriented research in our program. Ten submissions were selected as poster papers. Extended abstracts for these poster papers have been included in these proceedings.

The conference program included two invited lectures by Claude Carlet of Universities of Paris 8 and 13 and CNRS, and Nick Ellsmore of Stratsec (a BAE Systems Company). Prof. Carlet spoke about "Differentially Uniform Functions" and his paper is included in the proceedings. We would like to express our gratitude to Claude and Nick for contributing their knowledge and insight, and thus expanding the horizons of the conference delegates.

We would like to thank the authors of all submissions for offering their research for presentation at ACISP 2011. We extend our sincere thanks to the Program Committee and other reviewers for the high-quality reviews and in-depth discussions. The Program Committee made use of the EasyChair electronic submission and reviewing software written by Andrei Voronkov and maintained by the University of Manchester, UK. We would like to express our thanks to Springer, particularly Alfred Hofmann, for continuing to support the ACISP conference series and for helping in the production of the conference proceedings.

We also thank the Organizing Committee, led by the ACISP 2011 General Chair Serdar Boztaş, with key contributions from Leanne O'Doherty and Keith Tull, for their involvement in the conference. Finally, we would like to thank Qualcomm Incorporated, The University of Melbourne and the ISI-Informatics Research Group at RMIT University for their support, and the School of Mathematical and Geospatial Sciences at RMIT University for hosting the conference.

July 2011

Udaya Parampalli
Philip Hawkes

Organization

General Chair

Serdar Boztaş   RMIT University, Australia

Program Co-chairs

Udaya Parampalli   University of Melbourne, Australia
Philip Hawkes   Qualcomm Incorporated, Australia

Program Committee

Michel Abdalla	École Normale Supérieure, France
Magnus Almgren	Chalmers University of Technology, Sweden
Tuomas Aura	Microsoft Research, USA
Joonsang Baek	Institute for Infocomm Research, Singapore
Feng Bao	Institute for Infocomm Research, Singapore
Lynn Batten	Deakin University, Australia
Alex Biryukov	University of Luxembourg, Luxembourg
Colin Boyd	Queensland University of Technology, Australia
Joo Yeon Cho	Nokia A/S, Denmark
Sherman Chow	University of Waterloo, Canada
Carlos Cid	Royal Holloway, University of London, UK
Andrew Clark	Queensland University of Technology, Australia
Nicolas Courtois	University College London, UK
Yvo Desmedt	University College London, UK
Christophe Doche	Macquarie University, Australia
Pooya Farshim	Royal Holloway, University of London, UK
Praveen Gauravaram	Technical University of Denmark, Denmark
Peter Gutmann	University of Auckland, New Zealand
Kwangjo Kim	KAIST, Korea
Xuejia Lai	Shanghai Jiao Tong University, China
Mark Manulis	TU Darmstadt, Germany
Keith Martin	Royal Holloway, University of London, UK
Atefeh Mashatan	École Polytechnique Fédérale de Lausanne, Switzerland
Mitsuru Matsui	Mitsubishi Electric, Japan
Krystian Matusiewicz	Macquarie University, Australia

Chris Mitchell	Royal Holloway, University of London, UK
Atsuko Miyaji	JAIST, Japan
Yi Mu	University of Wollongong, Australia
Rei Safavi Naini	University of Calgary, Canada
Juan Gonzalez Nieto	Queensland University of Technology, Australia
Claudio Orlandi	Aarhus University, Denmark
C. Pandu Rangan	IIT, Madras, India
Vincent Rijmen	KU Leuven, Belgium and TU Graz, Austria
Bimal Roy	Indian Statistical Institute, India
Palash Sarkar	Indian Statistical Institute, India
Jennifer Seberry	University of Wollongong, Australia
Leonie Simpson	Queensland University of Technology, Australia
Damien Stehle	École Normale Supérieure de Lyon, France
Ron Stenfield	Macquarie University, Australia
Douglas Stinson	University of Waterloo, Canada
Willy Susilo	University of Wollongong, Australia
Vijay Varadharajan	Macquarie University, Australia
Maria Isabel Gonzalez Vasco	Universidad Rey Juan Carlos, Spain
Damien Vergnaud	École Normale Supérieure, France
Huaxiong Wang	Nanyang Technological University, Singapore
Kan Yasuda	NTT, Japan
Yuliang Zheng	University of North Carolina at Charlotte, USA

External Reviewers

Ejaz Ahmed	Angelo De Caro	Erland Jonsson
Toru Akishita	Yi Deng	Kiyoto Kawauchi
Martin Albrecht	Sharmila Deva Selvi	Przemyslaw Kubiak
Kazumaro Aoki	Sun Dongdong	Yee Wei Law
Frederik Armknecht	Ming Duan	Gregor Leander
Man Ho Au	Domingo Gomez	Gaëtan Leurent
Jean-Philippe Aumasson	Zheng Gong	Allison Lewko
Manuel Barbosa	Fuchun Guo	Tingting Lin
Asll Bay	Jian Guo	Joseph Liu
Rishiraj Bhattacharyya	Jinguang Han	Zhiqiang Liu
Andrey Bogdanov	Guillaume Hanrot	Yiyuan Luo
Jens-Matthias Bohli	Islam Hegazy	Vadim Lyubashevsky
Richard Brinkman	Javier Herranz	Florian Mendel
Debrup Chakraborty	Jason Hinek	Theodosis Mourouzis
Sanjit Chatterjee	Deukjo Hong	Sascha Müller
Kai-Yuen Cheong	Kathy Horadam	Mridul Nandi
Sherman S.M. Chow	Jinguang Huang	Kris Narayan
Cheng-Kang Chu	Xinyi Huang	Ta Toan Khoa Nguyen
Lizzie Coles-Kemp	Daniel Hulme	Abderrahmane Nitaj
Paolo D'Arco	Sebastiaan Indesteege	Mehrdad Nojoumian

Tatsuaki Okamoto
Kazumasa Omote
Khaled Ouafi
Sumit Pandey
Serdar Pehlivanoglu
Bertram Poettering
Elizabeth A. Quaglia
Kenneth Radke
Somindu Ramanna
Asha Rao
Reza Rezaeian Farashahi
Sondre Roenjom
Yasuyuki Sakai
Shoji Sakurai
Subhabrata Samajder
Santanu Sarkar

Yu Sasaki
Desmond Schmidt
Haya Shulman
Martijn Stam
Adriana Suarez Corona
Dongdong Sun
Li Sun
Suriadi Suriadi
Colleen Swanson
Christophe Tartary
Sui Guan Teo
Subhashini Venugopalan
Frederik Vercauteren
Eric Vetillard
Jorge Villar
Sree Vivek

Yongtao Wang
Lei Wei
Puwen Wei
Andrew White
Shuang Wu
Wei Wu
Yanjiang Yang
Huihui Yap
Po-Wah Yau
Kazuki Yoneyama
Yu Yong
Fangguo Zhang
Liangfeng Zhang
Wei Zhang
Huafei Zhu
Angela Zottarel

Table of Contents

Invited Talks

On Known and New Differentially Uniform Functions 1
Claude Carlet

Symmetric Key Cryptography

New Impossible Differential Attacks of Reduced-Round Camellia-192
and Camellia-256 16
Jiazhe Chen, Keting Jia, Hongbo Yu, and Xiaoyun Wang

Results on the Immunity of Boolean Functions against Probabilistic
Algebraic Attacks 34
Meicheng Liu, Dongdai Lin, and Dingyi Pei

Finding More Boolean Functions with Maximum Algebraic Immunity
Based on Univariate Polynomial Representation 47
Yusong Du and Fangguo Zhang

Improving the Algorithm 2 in Multidimensional Linear Cryptanalysis . . . 61
Phuong Ha Nguyen, Hongjun Wu, and Huaxiong Wang

State Convergence in the Initialisation of Stream Ciphers 75
*Sui-Guan Teo, Ali Al-Hamdan, Harry Bartlett, Leonie Simpson,
Kenneth Koon-Ho Wong, and Ed Dawson*

On Maximum Differential Probability of Generalized Feistel 89
Kazuhiko Minematsu, Tomoyasu Suzaki, and Maki Shigeri

Double SP-Functions: Enhanced Generalized Feistel Networks:
Extended Abstract 106
Andrey Bogdanov and Kyoji Shibutani

Algebraic Techniques in Differential Cryptanalysis Revisited 120
Meiqin Wang, Yue Sun, Nicky Mouha, and Bart Preneel

Hash Functions

Faster and Smoother – VSH Revisited 142
Juraj Šarinay

Cryptanalysis of the Compression Function of SIMD 157
Hongbo Yu and Xiaoyun Wang

Protocols

Electronic Cash with Anonymous User Suspension 172
   Man Ho Au, Willy Susilo, and Yi Mu

T-Robust Scalable Group Key Exchange Protocol with $O(\log n)$
complexity 189
   Tetsuya Hatano, Atsuko Miyaji, and Takashi Sato

Application-Binding Protocol in the User Centric Smart Card
Ownership Model 208
   Raja Naeem Akram, Konstantinos Markantonakis, and Keith Mayes

Access Control and Security

Security in Depth through Smart Space Cascades 226
   Benjamin W. Long

GeoEnc: Geometric Area Based Keys and Policies in Functional
Encryption Systems 241
   Mingwu Zhang and Tsuyoshi Takagi

An Efficient Rational Secret Sharing Scheme Based on the Chinese
Remainder Theorem 259
   Yun Zhang, Christophe Tartary, and Huaxiong Wang

DMIPS - Defensive Mechanism against IP Spoofing 276
   Shashank Lagishetty, Pruthvi Sabbu, and Kannan Srinathan

Public Key Cryptography

Provably Secure Key Assignment Schemes from Factoring 292
   Eduarda S.V. Freire and Kenneth G. Paterson

Efficient CCA-Secure CDH Based KEM Balanced between Ciphertext
and Key 310
   Yamin Liu, Bao Li, Xianhui Lu, and Dingding Jia

Generic Construction of Strongly Secure Timed-Release Public-Key
Encryption 319
   Atsushi Fujioka, Yoshiaki Okamoto, and Taiichi Saito

Identity-Based Server-Aided Decryption 337
   Joseph K. Liu, Cheng Kang Chu, and Jianying Zhou

A Generic Variant of NIST's KAS2 Key Agreement Protocol 353
   Sanjit Chatterjee, Alfred Menezes, and Berkant Ustaoglu

A Single Key Pair is Adequate for the Zheng Signcryption   371
  Jia Fan, Yuliang Zheng, and Xiaohu Tang

Towards Public Key Encryption Scheme Supporting Equality Test with
Fine-Grained Authorization   389
  Qiang Tang

Posters

Lattice-Based Completely Non-malleable PKE in the Standard Model
(Poster)   407
  Reza Sepahi, Ron Steinfeld, and Josef Pieprzyk

Compliance or Security, What Cost? (Poster)   412
  Craig Wright

Preimage Attacks on Full-ARIRANG (Poster)   417
  Chiaki Ohtahara, Keita Okada, Yu Sasaki, and Takeshi Shimoyama

Finding Collisions for Reduced *Luffa*-256 v2 (Poster)   423
  Bart Preneel, Hirotaka Yoshida, and Dai Watanabe

Improved Security Analysis of Fugue-256 (Poster)   428
  *Praveen Gauravaram, Lars R. Knudsen, Nasour Bagheri, and
  Lei Wei*

Improved Meet-in-the-Middle Cryptanalysis of KTANTAN (Poster)   433
  *Lei Wei, Christian Rechberger, Jian Guo, Hongjun Wu,
  Huaxiong Wang, and San Ling*

Toward Dynamic Attribute-Based Signcryption (Poster)   439
  Keita Emura, Atsuko Miyaji, and Mohammad Shahriar Rahman

A Verifiable Distributed Oblivious Transfer Protocol (Poster)   444
  Christian L.F. Corniaux and Hossein Ghodosi

Impracticality of Efficient PVSS in Real Life Security Standard
(Poster)   451
  Kun Peng

Electromagnetic Analysis Enhancement with Signal Processing
Techniques (Poster)   456
  Hongying Liu, Yukiyasu Tsunoo, and Satoshi Goto

Erratum

Compliance or Security, What Cost? (Poster) E1
  Craig Wright

Author Index   463

On Known and New Differentially Uniform Functions

Claude Carlet

LAGA, Universities of Paris 8 and Paris 13 and CNRS
Dept. of Math, Univ. of Paris 8, 2 rue de la liberté, 93526 Saint-Denis Cedex, France
claude.carlet@inria.fr

Abstract. We give a survey on the constructions of APN and differentially 4-uniform functions suitable for designing S-boxes for block ciphers. We recall why the search for more of such functions is necessary. We propose a way of designing functions which can possibly be APN or differentially 4-uniform and be bijective. We illustrate it with an example of a differentially 4-uniform (n, n)-permutation for n odd, based on the power function x^3 over the second order Galois extension of $\mathbb{F}_{2^{n+1}}$, and related to the Dickson polynomial D_3 over this field. These permutations have optimal algebraic degree and their nonlinearity happens to be rather good (but worse than that of the multiplicative inverse functions).

Keywords: Block cipher, vectorial Boolean function, S-box.

1 Introduction

Block ciphers use substitution boxes (in brief, S-boxes) to bring the confusion (a requirement already mentioned by C. Shannon) into the systems, which is necessary to withstand known (and hopefully future) attacks. Given two positive integers n and m, the functions from $\mathbb{F}_2^n$ to $\mathbb{F}_2^m$, often called (n, m)-*functions* or (if the values n and m are omitted) vectorial Boolean functions, are used as substitution boxes and play a central role in the robustness of block ciphers. The main attacks (differential attacks, linear attacks and higher order differential attacks) result in design criteria on the whole ciphers and on the particular S-boxes used in each round. These functions must be Almost Perfect Nonlinear (APN) or differentially 4-uniform (see definitions below) to allow resistance to the differential attack [1]; they must have high nonlinearity to resist the linear attack [31] and an algebraic degree at least 4 to resist the higher order differential attack [26] (which is described by Knudsen when the degree is 2 but a degree 3 seems insufficient for a reasonable resistance). Of course, in practice, since a cryptosystem must resist all these attacks, having for instance an APN function with bad nonlinearity or with low algebraic degree is less interesting than having a differentially 4-uniform function with high nonlinearity and not too low algebraic degree. Moreover, we like the S-boxes to be permutations (if the cipher is a Substitution-Permutation Network as in the AES, then this is a mandatory condition) or at least balanced (that is, uniformly distributed, with $m \leq n$; in

U. Parampalli and P. Hawkes (Eds.): ACISP 2011, LNCS 6812, pp. 1–15, 2011.

a Feistel cipher, attacks exist when the S-box is not balanced, which concretely oblige to use an expansion box - like in the DES - which makes the cipher more complex and slower). We need also the S-boxes to be efficiently computable, which in software is easier if n is a power of 2 (which is also more convenient for the design of the whole cipher), concretely if $n = 4$, 8 or 16, since it allows decomposing optimally the computation of the output in $\mathbb{F}_{2^n}$ into computations in subfields. In hardware, n does not need to be a power of 2, but we like in general the cryptosystem to be efficient in both hardware and software.

Such (n, m)-function F being given, the Boolean functions $f_1, \ldots, f_m$ defined by $F(x) = (f_1(x), \ldots, f_m(x))$, are called the *coordinate functions* of F. The design criteria on S-boxes result in necessary properties of the coordinate functions, but not only them. The linear combinations of the coordinate functions with non all-zero coefficients are called the *component functions* of F and need to satisfy the same properties as the coordinate functions (the attacks work on two linearly equivalent vectorial functions with the same complexity).

The *Walsh transform* of an (n, m)-function F maps any ordered pair $(u, v) \in \mathbb{F}_2^n \times \mathbb{F}_2^m$ to the sum (calculated in $\mathbb{Z}$):

$$W_F(u, v) := \sum_{x \in \mathbb{F}_2^n} (-1)^{v \cdot F(x) + u \cdot x},$$

where the same symbol "." is used to denote inner products in $\mathbb{F}_2^n$ and $\mathbb{F}_2^m$. For $v \neq 0$, it calculates the correlation between any component function $v \cdot F$ and any linear function $u \cdot x$. The Walsh transform satisfies Parseval's relation: $\sum_{u \in \mathbb{F}_2^n} (W_F(u, v))^2 = 2^{2n}$, for every v. The *Walsh spectrum* of F is the multi-set of all the values of the Walsh transform of F, for $u \in \mathbb{F}_2^n$ and $v \in \mathbb{F}_2^{m*}$ (where $\mathbb{F}_2^{m*} = \mathbb{F}_2^m \setminus \{0\}$); its *extended Walsh spectrum* is the multi-set of their absolute values, and its *Walsh support* is the set of those ordered pairs (u, v) such that $W_F(u, v) \neq 0$.

Any (n, m)-function F can be represented by its *algebraic normal form* (ANF):

$$F(x) = \sum_{I \subseteq \{1, \cdots, n\}} a_I \left(\prod_{i \in I} x_i \right); \; a_I \in \mathbb{F}_2^m \tag{1}$$

(this sum being calculated in $\mathbb{F}_2^m$) which exists and is unique.

The *algebraic degree* of the function (and more generally of a function defined over an affine subspace of $\mathbb{F}_2^n$) is by definition the global degree of its ANF (resp. the minimum algebraic degree of the (n, m)-functions which extend the function). A function is affine if its algebraic degree is at most 1. It is called quadratic if its algebraic degree is at most 2. The generalized Reed-Muller code of order 1 (respectively 2) is the set of (n, m)-functions of degrees at most 1 (respectively 2).

But S-boxes are rarely defined through their ANF. The known functions achieving the necessary features needed for an S-box (see below) are defined through a second representation which exists (uniquely) when $m = n$ (note this is necessarily the case when the S-box is used in an SPN) or more generally when

m divides n: we endow $\mathbb{F}_2^n$ with the structure of the field $\mathbb{F}_{2^n}$; any (n, n)-function F then admits a unique *univariate polynomial representation* over $\mathbb{F}_{2^n}$, of degree at most $2^n - 1$:

$$F(x) = \sum_{j=0}^{2^n-1} b_j x^j \ , \quad b_j \in \mathbb{F}_{2^n} \ . \tag{2}$$

The component functions are then the functions $tr_n(vF(x))$, $v \neq 0$, where $tr_n(x) = x + x^2 + x^{2^2} + \cdots + x^{2^{n-1}}$ is the trace function from $\mathbb{F}_{2^n}$ to $\mathbb{F}_2$. We denote by $w_2(j)$ the Hamming weight of the binary expansion $\sum_{s=0}^{n-1} j_s 2^s$ of j, i.e. $w_2(j) = \sum_{s=0}^{n-1} j_s$ and call it the *2-weight* of j. Then, the function F has algebraic degree $\max_{j/b_j \neq 0} w_2(j)$. This comes from the fact that, decomposing x over a basis $(\beta_1, \cdots, \beta_n)$, we have: $x = \sum_{i=1}^n x_i \beta_i$ with $x_i \in \mathbb{F}_2$ and $x^j = \prod_{s=0}^{n-1} (\sum_{i=1}^n x_i \beta_i^{2^s})^{j_s}$.

If m is a divisor of n, then any (n, m)-function F can be viewed as a function from $\mathbb{F}_{2^n}$ to itself, since $\mathbb{F}_{2^m}$ is a sub-field of $\mathbb{F}_{2^n}$.

As shown by Rivain and Prouff [36], to allow counter measures to side channel attacks which do not reduce too much the speed of the cipher, the univariate representation of the S-boxes involved in a block cipher should be calculable with a sufficiently small number of nonlinear multiplications (recall that multiplying an element by a constant is a linear function as well as multiplying the element by itself, and that multiplying the element by its square, for instance, is a nonlinear multiplication). This condition is clearly not contradicory with the need of efficiency of the S-box in terms of computability but it may represent a limitation for ensuring the other properties listed above.

An (n, m)-function F is *balanced* (that is, $|F^{-1}(z)| = 2^{n-m}$ for every $z \in \mathbb{F}_{2^m}$) if and only if its component functions are balanced (i.e. have Hamming weight 2^{n-1}) [28].

The *nonlinearity* $nl(F)$ of an (n, m)-function F is the minimum Hamming distance between all the component functions of F and all affine functions on n variables and quantifies the level of resistance of the S-box to the linear attack [31]. We have:

$$nl(F) = 2^{n-1} - \frac{1}{2} \max_{v \in \mathbb{F}_2^{m*}; \ u \in \mathbb{F}_2^n} |W_F(u, v)| \ . \tag{3}$$

Two main upper bounds are known on the nonlinearity:
1. the *covering radius bound*:

$$nl(F) \leq 2^{n-1} - 2^{n/2-1}$$

which can be directly derived from Parseval's relation and from (3); it is tight if and only if n is even and $m \leq n/2$, as proved by Nyberg [32]. The functions achieving it with equality are those such that $W_F(u, v) = \pm 2^{n/2}$ for every $v \neq 0$ and u; they are called *bent*. Since bent functions exist only for $m \leq n/2$ which seems too small with respect to n, they are not used as S-boxes (a second reason for this is that they are not balanced). But they can be used to build S-boxes (see [17]).

2. the *Sidelnikov-Chabaud-Vaudenay bound*[1] (in brief, SCV bound) [37,22], valid for $m \geq n - 1$:

$$nl(F) \leq 2^{n-1} - \frac{1}{2}\sqrt{3 \times 2^n - 2 - 2\frac{(2^n - 1)(2^{n-1} - 1)}{2^m - 1}}$$

which equals the covering radius bound when $m = n - 1$ and is strictly better when $m \geq n$. As proved in [22], the SCV bound is tight if and only if $m = n$ (it simplifies then to $nl(F) \leq 2^{n-1} - 2^{\frac{n-1}{2}}$) and if n is odd (the functions achieving it with equality are those such that $W_F(u, v) \in \{0, \pm 2^{\frac{n+1}{2}}\}$ for every $v \neq 0$ and u; they are called *almost bent*, in brief, AB). Note that the SCV bound confirms for $m \geq n$ only that no bent function exists for $m > n/2$ and obtaining a bound better than the covering radius bound for $n/2 < m < n$ is an open problem.

An (n, m) function is bent if and only if all its *derivatives* $D_a F(x) = F(x) + F(x + a)$, $a \in \mathbb{F}_2^n{}^*$, are balanced. Bent functions are also called *perfect nonlinear* (in brief, PN) for this reason. They allow optimal resistance to both the linear attack and the differential attack. But as we already mentioned, they do not exist for $n = m$; in this case and for n odd, it is the AB functions which oppose optimal resistance. For n even, the best possible nonlinearity is unknown; we know it lies between $2^{n-1} - 2^{n/2}$ (value provably achieved by the nonlinearity of a few functions, see below) and $2^{n-1} - 2^{n/2-1}$. As proved in [22], any AB function is *almost perfect nonlinear* (in brief, APN). An (n, n)-function (where n can be odd or even) is APN if all its derivatives $D_a F$, $a \in \mathbb{F}_2^n{}^*$, are 2-to-1 (i.e. every element of $\mathbb{F}_2^n$ has 0 or 2 pre-images by $D_a F$). Such APN (n, n)-functions, whose notion has been studied by Nyberg in [35], contribute to an optimal resistance to the differential attack [1]. APN-ness does not imply back AB-ness, except for quadratic functions when n is odd or more generally for functions having Walsh spectra divisible by $2^{\frac{n+1}{2}}$ [16].

More generally, F is called *differentially δ-uniform* if, for every nonzero a and every b, the equation $D_a F(x) = b$ has at most δ solutions (that is, any derivative is at most δ-to-1).

These notions, as well as the nonlinearity, are invariant under affine equivalence; two functions are called *affine equivalent* if one is equal to the other, composed on the left and on the right by affine permutations. More generally, these notions are preserved by *extended affine equivalence* (in brief, EA-equivalence); two functions are called EA-equivalent if one is affine equivalent to the other, added with an affine function. Still more generally, these notions are CCZ-invariant; two functions are called *CCZ-equivalent* if their graphs $\{(x, y) \in \mathbb{F}_2^n \times \mathbb{F}_2^n \mid y = F(x)\}$ and $\{(x, y) \in \mathbb{F}_2^n \times \mathbb{F}_2^n \mid y = G(x)\}$ are affine equivalent, that is, if there exists an affine automorphism $L = (L_1, L_2)$ of $\mathbb{F}_2^n \times \mathbb{F}_2^n$ such that $y = F(x) \Leftrightarrow L_2(x, y) = G(L_1(x, y))$. Denoting $F_1(x) = L_1(x, F(x))$ and $F_2(x) = L_2(x, F(x))$, we have then $G = F_2 \circ F_1^{-1}$. A permutation and its inverse are CCZ-equivalent. It is shown in [15] that CCZ-equivalence is more general than EA-equivalence extended by allowing replacing the permutations

---

[1] Sidelnikov proved it first in the framework of sequences - hence, for power functions - and Chabaud-Vaudenay proved it for all functions.

by their inverses. The algebraic degree is EA-invariant (when it is strictly larger than 1), but not CCZ-invariant.

A lot of work has been done during the last ten years to find new APN or differentially 4-uniform functions. However few have been found, and none possesses all the desired features. The research in the domain of S-boxes for block ciphers progresses slowly. A reason why so few infinite classes of good vectorial functions are know is that, contrarily to Boolean functions (e.g. bent functions or more generally Boolean functions with good nonlinearity, resilient functions, ...), no secondary construction (that is, construction of good functions in some number of variables from known functions in another - smaller or equal - number of variables) is known.

Only one sporadic example of an APN permutation in even number of variables is known [23]. This 6 variable function, found by Dillon and Wolfe, is complex to compute. There exist also few differentially 4-uniform permutations. The most famous example is the multiplicative inverse function (used with 8 input bits in the AES). It has very good characteristics, except that it has a peculiarity representing a potential risk that we shall detail below: its graph has low degree annihilators. Another differentially 4-uniform function found in [5] is interesting but has some drawbacks (see below). In fact, finding infinite classes of APN or differentially 4-uniform permutations in even numbers of input variables (hopefully equal to powers of two), with high (or at least, not low) algebraic degrees and whose graphs have no low degree annihilator is an open problem which does not seem easily solvable. And finding such S-boxes with a large nonlinearity is still more open.

In this paper, after a survey of the constructions of APN and differentially 4-uniform functions, we propose in Section 3 a new construction of a differentially 4-uniform permutation over $\mathbb{F}_2^n$ (n odd) based on the APN power function x^3 over $\mathbb{F}_{2^{2n+2}}$ and related to the Dickson polynomial of index 3. We study the algebraic degree of this function and its nonlinearity. We also generalize the function to the case n even. We hope that the new way of constructing differentially 4-uniform functions presented in this section can lead to functions gathering all the necessary features.

2   The Know AB, APN and Differentially 4-Uniform Functions and Their Respective Drawbacks

2.1   AB and APN Functions

• The most numerous interesting known functions are *power functions* $x \mapsto x^d$ on the field $\mathbb{F}_{2^n}$. There are two reasons for that: (1) a power function is APN, that is, its derivatives $D_a F$, $a \neq 0$, are all at most δ-to-1 if and only if one of them (any of them) is at most δ-to-1; hence the density of APN (resp. differentially 4-uniform) functions is higher for power functions than for general (n, n)-functions (2) the notion of AB power function x^d corresponds in sequence theory to the fact that the decimation by d of an m-sequence has optimal crosscorrelation with the original m-sequence; so much work had been already done for power functions in the framework of the research of good sequences for telecommunications when the notion of AB function appeared in 1994.

We list below the exponents of the known APN or AB power functions. Of course, the functions x^d and $x^{2^j d}$ are affine equivalent for every j, so we shall give only one value of d for each cyclotomic coset of 2 mod $2^n - 1$; also, power APN functions are permutations when n is odd as proved by Dobbertin (see his proof reported in [19]), and if a function is AB (resp. APN) and bijective, its inverse is AB (resp. APN); so we shall also omit $1/d$ when d is co-prime with $2^n - 1$:

- $d = 2^i + 1$ with $\gcd(i, n) = 1$; these *Gold functions* are AB for every odd n and APN for every even n, in which case they have best known nonlinearity as well: $2^{n-1} - 2^{n/2}$. But they have algebraic degree 2 and cannot be used as S-boxes.
- $d = 2^{2i} - 2^i + 1$ with $\gcd(i, n) = 1$; these *Kasami functions* have the same properties as the Gold functions and have larger degree (precisely $i + 1$ if $i < n/2$). They have best known nonlinearity $2^{n-1} - 2^{n/2}$ for n even as well. For n odd, they are related to quadratic functions by the fact that $d = \frac{2^{3i}+1}{2^i+1}$ and $2^i + 1$ is co-prime with $2^n - 1$, which means that the Kasami functions have the form $F(x) = Q_2 \circ Q_1^{-1}(x)$ where Q_1 and Q_2 are quadratic permutations; this has some similarity with a function CCZ-equivalent to a quadratic function. Maybe this could be used in an extended higher order differential attack.
- $d = 2^{(n-1)/2} + 3$ *(Welch function)* is an AB permutation for every odd n.
- $\begin{matrix} d = 2^{(n-1)/2} + 2^{(n-1)/4} - 1; & n \equiv 1 \pmod 4 \\ d = 2^{(n-1)/2} + 2^{(3n-1)/4} - 1; & n \equiv 3 \pmod 4 \end{matrix}$ *(Niho functions)*: idem.
- $d = 2^n - 2$, n odd *(inverse function)* is an APN involution for every odd n; it has nonlinearity the highest even number bounded above by $2^{n-1} - 2^{n/2}$.
- $d = 2^{\frac{4n}{5}} + 2^{\frac{3n}{5}} + 2^{\frac{2n}{5}} + 2^{\frac{n}{5}} - 1$ *((Dobbertin function)* is an APN function for every n divisible by 5. Note that n divisible by 5 is not favorable to efficiency in software. Also, its nonlinearity is not very good

All these APN functions are not balanced when they are in even dimension (moreover, Gold functions are quadratic and Kasami functions may be related to quadratic functions); this is why the differentially 4-uniform inverse function was preferred for the AES (see below).

- *Non-power functions:*

  - some non-quadratic AB functions (for n odd) and APN functions (for n even), new up to EA-equivalence have been found in [15] by applying CCZ-equivalence to Gold functions; but a risk exists that the higher order differential attacks could be extended to functions CCZ-equivalent to quadratic functions.
  - several infinite classes of quadratic AB and APN functions CCZ-inequivalent to power functions have been found in [2,3,12,13]. These functions being quadratic cannot be used as S-boxes.

- a classification under CCZ-equivalence of all APN functions up to dimension five and a (non-exhaustive) list of CCZ-inequivalent functions in dimension 6 have been given in [7]. APN functions in dimensions 6, 7 and 8 have been obtained in [8]. One of the functions in dimension 6 is CCZ-inequivalent to power functions and to quadratic functions, as proved by Edel and Pott in [24]. This function is:

$$x^3 + \alpha^{17}(x^{17} + x^{18} + x^{20} + x^{24}) + tr_2(x^{21}) + tr_3(\alpha^{18}x^9)$$
$$+\alpha^{14}\, tr_6\,(\alpha^{52}x^3 + \alpha^6 x^5 + \alpha^{19}x^7 + \alpha^{28}x^{11} + \alpha^2 x^{13}).$$

  It equals the sum of a quadratic APN function and a cubic Boolean function (it has been searched this way).
- an example of APN permutation in 6 variables has been given by J. Dillon at conference Fq 9 (the problem of finding an example of APN permutation in even number of variables had been open for ten years). But it is CCZ-equivalent to a quadratic function and its expression is complex.
  Open problem: do there exist infinite classes of APN permutations in even numbers of variables?

Differentially 4-uniform functions

- The multiplicative inverse function is differentially 4-uniform when n is even [35] (and this function is used as the S-box of the AES with $n = 8$). It has optimal nonlinearity $2^{n-1} - 2^{n/2}$. But it is the worst possible with respect to algebraic attacks (which are not yet efficient but which represent a threat) since denoting $y = x^{2^n-2} = \frac{1}{x}$ (with $\frac{1}{0} = 0$) we have the bilinear relation $x^2 y = x$ and other relations of the same kind having more generally global algebraic degree 2.
- Differentially 4-uniform functions can be obtained from APN functions by adding a Boolean function, or composing (on the right or on the left) by 2-to-1 affine functions but this does not seem to be a good way of designing S-boxes.
- The Gold functions x^{2^i+1} such that $gcd(i, n) = 2$ are straightforwardly differentially 4-uniform, but these functions are quadratic; the Kasami functions $x^{2^{2i}-2^i+1}$ such that n is divisible by 2 but not by 4 and $gcd(i, n) = 2$ are also differentially 4-uniform [6].
- Several other infinite classes of differentially 4-uniform quadratic functions have been found but clearly cannot be chosen as S-boxes.
- The functions $x^{2^{n-1}-1} + ax^5$ (n odd, $a \in \mathbb{F}_{2^n}$) and $x^{2^{n/2}+2^{n/4}+1}$ (n divisible by 4) are differentially 4-uniform, as shown in [6,5], but the first one, which is close to the inverse function, is never bijective as proved by Leander in a personal communication and its nonlinearity has not been studied; the second one, which is interesting because it has a very good (best known) nonlinearity $2^{n-1} - 2^{n/2}$ and is a power function over $\mathbb{F}_{2^n}$ (which simplifies its computation), has algebraic degree 3 which is insufficient for a concrete use; moreover, it is a permutation only when the number of variables is divisible by 4 but not by 8.

- Non-quadratic differentially 4-uniform functions can be obtained by concatenating the outputs to a bent function and to another function [17]:
  - The function $(x, y) \rightarrow (xy, (x^3 + w)(y^3 + w'))$, where w, w' and $\frac{w}{w'}$ belong to $\mathbb{F}_{2^{n/2}} \setminus \{x^3, x \in \mathbb{F}_{2^{n/2}}\}$, with $n/2$ even.
  - The function $(x, y) \rightarrow (xy, x^3(y^2 + y + 1) + y^3)$, with $n/2$ odd.
  - The function $F : X \in \mathbb{F}_{2^n} \rightarrow (X^{2^{n/2}+1}, (X^{2^{n/2}+1})^3 + wX^3 + (wX^3)^{2^{n/2}})$, which has algebraic degree 4). These functions (which have the interest of being already decomposed over $\mathbb{F}_{2^{n/2}}$ without any extra work) are not permutations and have not very good nonlinearities.

So, except for the multiplicative inverse function (which has however a potential weakness) and maybe for the Kasami functions (but these functions may be too close in some sense to quadratic functions), the known APN functions have all drawbacks. Hence, the research of more APN functions, hopefully having all the desired features, must continue.

3 A Way of Constructing Differentially Uniform Permutations

Recall that, given any integer n, every element u of $\mathbb{F}_{2^n}^*$ can be expressed uniquely in the form $h + \frac{1}{h}$ where $h \in \mathbb{F}_{2^{2n}}^*$, since the equation $h + \frac{1}{h} = u$ being equivalent to $\left(\frac{h}{u}\right)^2 + \frac{h}{u} = \frac{1}{u^2}$, has two solutions inverses of each other and differing by u, because $tr_{2n}\left(\frac{1}{u^2}\right) = 0$. This allows for every positive integer d to define the so-called Dickson polynomial $D_d(X)$ over $\mathbb{F}_{2^n}$, such that $D_d\left(h + \frac{1}{h}\right) = h^d + \frac{1}{h^d}$ for every $h \in \mathbb{F}_{2^{2n}}$ (D_d is a permutation polynomial if d is co-prime with $2^{2n} - 1$). We have $D_0(X) = 0$, $D_1(X) = X$ and $D_{d+2}(X) = XD_{d+1}(X) + D_d(X)$ for every d (which implies that $D_d(\mathbb{F}_{2^n}) \subseteq \mathbb{F}_{2^n}$). It is a simple matter to check that D_3 is APN and D_5 is differentially 4-uniform. But $D_3(X) = X^3 + X$ differs by a linear function from a Gold function, and so is not new, and D_5 having degree 5 (as univariate polynomial), its differential 4-uniformity is trivial.

Remark 1. More generally, for every $v \in \mathbb{F}_{2^n}^*$, every element u of $\mathbb{F}_{2^n}^*$ can be expressed uniquely in the form $h + \frac{v}{h}$ where $h \subset \mathbb{F}_{2^{2n}}^*$ which allows defining $D_d(X, v)$ equal by definition to $h^d + \left(\frac{v}{h}\right)^d$. In [25], the fact that, for some d, function $v \mapsto D_d(a, v)$ (called reversed Dickson polynomial) is a permutation polynomial on $\mathbb{F}_{2^n}$ is related to the fact that X^d is APN on $\mathbb{F}_{2^n}$ (resp. on $\mathbb{F}_{2^{2n}}$).

We show now a way of designing functions which can be APN or differentially 4-uniform and which can also be permutations (we illustrate it with an example). The idea is, instead of using the field structure of $\mathbb{F}_{2^n}$, to use that of $\mathbb{F}_{2^{n+1}}$. This leads to identifying an affine hyperplane in this field and a function which maps this hyperplane to a hyperplane as well. This simple idea may lead to more numerous easily computable APN or differentially 4-uniform functions. Let us illustrate it with a precise construction.

Let $N = n+1$ and $q = 2^N$. We identify any binary vector $x \in \mathbb{F}_2^n$ as an element of the linear hyperplane $H = \{x \in \mathbb{F}_q \mid tr_N(x) = 0\}$, where $tr_N(x) = \sum_{i=0}^{N-1} x^{2^i}$ is the trace function from $\mathbb{F}_q$ to $\mathbb{F}_2$. This identification is possible (through the choice of a basis of H) because $\mathbb{F}_2^n$ and H are both n-dimensional $\mathbb{F}_2$-vectorspaces. Let α be a fixed element in $\mathbb{F}_q \setminus H$. Then $x+\alpha$ ranges over $\mathbb{F}_q \setminus H$ when x ranges over H. Let tr_{2N} denote the trace function from $\mathbb{F}_{q^2}$ to $\mathbb{F}_2$ and $Tr_N^{2N}(x) = x+x^q$ the trace function from $\mathbb{F}_{q^2}$ to $\mathbb{F}_q$. We need the known lemma below (we give a proof, to be self-contained).

Lemma 1. *Let $U = \{x^{q-1}, x \in \mathbb{F}_{q^2}^*\}$ be the $(q+1)$-th order multiplicative subgroup of $\mathbb{F}_{q^2}^*$. The image of $U \setminus \{1\}$ by Tr_N^{2N} equals the set $\{\frac{1}{u}; u \in \mathbb{F}_q, tr_N(u) = 1\}$ and every element of this set is exactly the image of two conjugate elements h and $h^q = \frac{1}{h}$ of $U \setminus \{1\}$ by Tr_N^{2N}.*

Proof. For every $h \in U \setminus \{1\}$, we have $Tr_N^{2N}(h) \neq 0$ since $\ker Tr_N^{2N} = \mathbb{F}_q$ and $U \cap \mathbb{F}_q = \{1\}$.
By definition of U, we have $h^{q+1} = 1$ and therefore $Tr_N^{2N}(h) = h + \frac{1}{h}$ and thus, if $h \neq 1$: $tr_N\left(\frac{1}{Tr_N^{2N}(h)}\right) = tr_N(\frac{h}{h^2+1}) = tr_N(\frac{1}{h+1} + \frac{1}{h^2+1}) = \frac{1}{h+1} + \frac{1}{h^q+1} = \frac{1}{h+1} + \frac{1}{\frac{1}{h}+1} = 1$.
For every $u \in \mathbb{F}_q^*$ such that $tr_N\left(\frac{1}{u}\right) = 1$, the elements $h \in \mathbb{F}_{q^2}^*$ such that $h + \frac{1}{h} = u$, that is such that $\left(\frac{h}{u}\right)^2 + \frac{h}{u} = \frac{1}{u^2}$ do not belong to $\mathbb{F}_q$ since $tr_N\left(\frac{1}{u^2}\right) = tr_N\left(\frac{1}{u}\right) \neq 0$ and since $\frac{h}{u} \in \mathbb{F}_q$ would imply $tr_N\left(\left(\frac{h}{u}\right)^2 + \frac{h}{u}\right) = 0$, a contradiction. Moreover if h is a solution then h^q is also a solution; hence h and h^q are the two distinct solutions of the equation $h^2 + uh + 1 = 0$; their product h^{q+1} equals 1 and their sum $h + h^q$ equals u. Hence h belongs to $U \setminus \{1\}$ and satisfies $Tr_N^{2N}(h) = u$. $\square$

Remark 2. A simpler way of proving the surjectivity of Tr_N^{2N} from $U \setminus \{1\}$ to $\{\frac{1}{u}; u \in \mathbb{F}_q, tr_N(u) = 1\}$ and the second part of Lemma 1 is to observe that since two elements of U have the same trace Tr_N^{2N} if and only if they are conjugate, the image of $U \setminus \{1\}$ has $q/2$ elements, but the way we prove it above gives a little more insight. Note that the polynomial $z^2 + Tr_N^{2N}(h)z + 1 = (z+h)(z+h^q)$ is the minimal polynomial of h over $\mathbb{F}_q$.

We can now introduce our construction:

Definition 1. *Let n be any positive integer and $N = n+1, q = 2^N$. Let d be any integer co-prime with $q+1$. Let $\alpha \in \mathbb{F}_q$ be such that $tr_N(\alpha) = 1$. We denote by F_d the function from $\mathbb{F}_2^n$ to itself defined as follows:*

- *We identify the input x with an element of $H = \{u \in \mathbb{F}_q, tr_N(u) = 0\}$;*
- *Let h and $h^q = \frac{1}{h}$ be the elements of $U \setminus \{1\}$ such that $Tr_N^{2N}(h) = Tr_N^{2N}\left(\frac{1}{h}\right) = h + \frac{1}{h} = \frac{1}{x+\alpha}$ (these elements exist and are unique up to conjugacy, according to Lemma 1);*
- *We define $F_d(x) = \frac{1}{Tr_N^{2N}(h^d)} + \alpha = \frac{1}{h^d + \frac{1}{h^d}} + \alpha = \frac{1}{D_d\left(\frac{1}{x+\alpha}\right)} + \alpha \in H$ (viewed as an element of $\mathbb{F}_2^n$).*

For every d, function F_d is well defined because, for every $h \in U \setminus \{1\}$, h^d belongs to $U \setminus \{1\}$ and $h \mapsto h^d$ commutes with $h \mapsto h^q$.

Note that 3 is coprime with $q+1$ if and only if N is even (that is, n is odd) and 5 is coprime with $q+1$ if and only if N is not congruent with 2 modulo 4 (that is, n is not congruent with 1 modulo 4).

Proposition 1. *For every positive integer n and every integer d coprime with $q+1$, the function F_d defined above is a permutation of $\mathbb{F}_2^n$.*

Indeed, d being coprime with $q+1$, the function $h \mapsto h^d$ is a permutation of $U \setminus \{1\}$ and we have: $\{h, \frac{1}{h}\} \neq \{h', \frac{1}{h'}\} \Rightarrow \{h^d, \frac{1}{h^d}\} \neq \{h'^d, \frac{1}{h'^d}\}$.

Proposition 2. *For every N even, $F_3(x)$ equals $x + \frac{1}{(x+\alpha+1)} + \frac{1}{(x+\alpha+1)^2}$ and has algebraic degree $n-1$.*

Proof. We have $F_3(x) = \frac{(x+\alpha)^3}{x^2+\alpha^2+1} + \alpha = \frac{x^3+\alpha^2 x+\alpha}{x^2+\alpha^2+1} = x + \frac{x+\alpha}{(x+\alpha+1)^2} = x + \frac{1}{(x+\alpha+1)} + \frac{1}{(x+\alpha+1)^2}$. To study the algebraic degree we need to extend a well-known result in Lemma 2 below. We apply this lemma to the function $F(x) = F_3(x + \alpha + 1)$. Taking $k = n-1$, we have $i = 0$ and $\displaystyle\sum_{x \,/\, tr_N(x)=1} F(x) =$

$$\sum_{x \in \mathbb{F}_{2^N}^*} \left(\frac{tr_N(x)}{x} + \left(\frac{tr_N(x)}{x}\right)^2 \right) = 1 + 1 = 0, \text{ since for every } x \in \mathbb{F}_{2^N}^*, \text{ we have}$$

$\frac{tr_N(x)}{x} = 1 + x + \cdots + x^{2^{N-1}-1}$ and since $\sum_{x \in \mathbb{F}_{2^N}} (1 + x + \cdots + x^{2^{N-1}-1}) = 0$; and

taking $k = n-2$ and $i = 1$, we have $\displaystyle\sum_{x \,/\, tr_N(x)=1} xF(x) = \sum_{x \in \mathbb{F}_{2^N}^*} \left(tr_N(x) + \frac{tr_N(x)}{x} \right) = 1$.

This completes the proof. $\qquad\square$

Hence the algebraic degree of F_3 is optimal for an (n,n)- permutation.

Lemma 2. *Let $n \leq N$, let F be the restriction to an n-dimensional affine space E of a function F' from $\mathbb{F}_{2^N}$ to $\mathbb{F}_{2^N}$ and let k be a positive integer. Then F has algebraic degree at most k if and only if, for every integer i of 2-weight at most $n - k - 1$, we have $\sum_{x \in E} x^i F(x) = 0$.*

Proof. F has algebraic degree at most k if and only if, for every a, the Boolean function $tr_N(aF(x))$ over E has algebraic degree at most k, that is, if and only if for every Boolean function h of algebraic degree at most $n - k - 1$ over $\mathbb{F}_{2^N}$, we have $\sum_{x \in E} tr_N(aF(x))h(x) = 0$, that is, $\sum_{x \in E} tr_N(aF(x)h(x)) = 0$. This is equivalent to $\sum_{x \in E} F(x)h(x) = 0$ for every Boolean function h of algebraic degree at most $n - k - 1$ and therefore for every vectorial function h of algebraic degree at most $n - k - 1$. The set of functions x^i, $w_2(i) \leq n - k - 1$ generates all vectorial functions h of algebraic degrees at most $n - k - 1$ and this completes the proof. $\qquad\square$

Remark 3. Hence, the rather complex Definition 1 simplifies with Proposition 2; this gives a new approach: we know that what makes the inverse function not APN in even dimension is only its value at 0; so we consider its restriction to a hyperplane not containing 0, and since we need then the image to be also a hyperplane, we compose the inverse function by the linear mapping $x + x^2$ whose image is the hyperplane $\{x \in \mathbb{F}_{2^N} / tr_N(x) = 0\}$.

Remark 4. For every $N \not\equiv 2 \pmod 4$, that is, $n \not\equiv 1 \pmod 4$, $F_5(x) = \frac{(x+\alpha)^5}{x^4+x^2+\alpha^4+\alpha^2+1} + \alpha = \frac{x^5+\alpha^2 x^3+(\alpha^3+\alpha)x^2+\alpha^4 x+\alpha^3+\alpha}{x^4+x^2+\alpha^4+\alpha^2+1}$.

Proposition 3. *For every positive odd integer n, function F_3 is differentially 4-uniform.*

Proof. According to Proposition 2, function F_3 is EA-equivalent to $L \circ F$ where F is the restriction to a hyperplane of the multiplicative inverse function over $\mathbb{F}_{2^N}$, and $L(x) = x + x^2$. Since the multiplicative inverse function is APN and L is a linear 2-to-1 function, this completes the proof. □

We can see that the differentiality of F_3 is not a big deal since it is a direct consequence of the APNness of the inverse function. However, an interesting quality of this function is to be bijective. We know (see [29,30]) that there is no permutation EA-equivalent to the inverse function which is not in fact affinely equivalent to it. Here we have a different situation.

Remark 5. It can be shown similarly that function F_5 is differentially 8-uniform.

Remark 6. Function F_3 can be adapted to the case n even, by taking the function $x + \frac{1}{(x+\alpha)} + \frac{1}{(x+\alpha)^2}$ on $\{x \in \mathbb{F}_{2^N} / tr_N(x) = 0\}$ (and not the function $x + \frac{1}{(x+\alpha+1)} + \frac{1}{(x+\alpha+1)^2}$, because for N odd, $\alpha + 1$ has null trace). This function is differentially 4-uniform as well. Indeed, the inverse function over $\mathbb{F}_{2^N}$ is not APN for N even but its restriction to a hyperplane excluding 0 is APN. Hence the same proof as in Proposition 3 works.

Proposition 4. *Function F_3 and its generalization have nonlinearity at least $2^{n-1} - 2^{\frac{n}{2}+1}$ if n is even and $2^{n-1} - \lfloor 2^{\frac{n}{2}+1} \rfloor - 1$ if n is odd.*

Proof. The nonlinearity of F_3 and of its generalization F is equal to that of the restriction of the function $\frac{1}{x} + \frac{1}{x^2}$ to the affine hyperplane $\{x \in \mathbb{F}_{2^N} / tr_N(x) = 1\}$. For every nonzero $v \notin \mathbb{F}_2$ (we need to exclude the elements v which are orthogonal to the image $\{x \in \mathbb{F}_{2^N} / tr_N(x) = 0\}$ of the function to have one of its component functions with $v \cdot F$) and every $u \in \mathbb{F}_{2^N}$ we have

$$\sum_{x \in \mathbb{F}_{2^N} / tr_N(x)=1} (-1)^{tr_N\left(v\left(\frac{1}{x}+\left(\frac{1}{x}\right)^2\right)+ux\right)} =$$

$$\frac{1}{2}\left(\sum_{x \in \mathbb{F}_{2^N}} (-1)^{tr_N\left(v\left(\frac{1}{x}+\left(\frac{1}{x}\right)^2\right)+ux\right)} - \sum_{x \in \mathbb{F}_{2^N}} (-1)^{tr_N\left(v\left(\frac{1}{x}+\left(\frac{1}{x}\right)^2\right)+(u+1)x\right)}\right) =$$

$$\frac{1}{2}\left(\sum_{x\in\mathbb{F}_{2^N}}(-1)^{tr_N\left((v+v^{2^{N-1}})\frac{1}{x}+ux\right)}-\sum_{x\in\mathbb{F}_{2^N}}(-1)^{tr_N\left((v+v^{2^{N-1}})\frac{1}{x}+(u+1)x\right)}\right).$$

We know according to Lachaud and Wolfmann [27] that the set of values of each of these two sums equals the set of all integers $s \equiv 0 \pmod 4$ in the range $[-2^{n/2+1}+1; 2^{n/2+1}+1]$, this completes the proof, using (3).   □

The table below compares the values of the nonlinearities of F_3, computed for small n, with the values given by our bound.

n	N	$bound$	$nl(F_3)$
7	8	48	50
9	10	224	226
11	12	960	962

The nonlinearity of F_3 is less than that of the inverse function but it is still interesting. *Note that its values are different from the nonlinearities of the known APN functions, so F_3 is CCZ-inequivalent to them.*

Remark 7. Recall that with the inverse function $y = x^{2^n-2}$, $x \in \mathbb{F}_{2^n}$, we had $x^2 y = x$. Here we do not seem to have bilinear relations but we have quadratic ones: $(x+a)^2 y = 1 + x + a + x^3 + a^2 x$.

Remark 8. According to the calculations above, we have

$$\sum_{x\in\mathbb{F}_{2^N}\,/\,tr_N(x)=1}(-1)^{tr_N\left(v\left(x+\frac{1}{x}+\left(\frac{1}{x}\right)^2\right)\right)}=$$

$$\frac{1}{2}\left(\sum_{x\in\mathbb{F}_{2^N}}(-1)^{tr_N\left((v+v^{2^{N-1}})\frac{1}{x}+vx\right)}-\sum_{x\in\mathbb{F}_{2^N}}(-1)^{tr_N\left((v+v^{2^{N-1}})\frac{1}{x}+(v+1)x\right)}\right).$$

Since F_3 is a permutation for n odd, we deduce the following property of Kloosterman sums: for every even N and every $v \in \mathbb{F}_{2^N} \setminus \mathbb{F}_2$, we have
$$\sum_{x\in\mathbb{F}_{2^N}}(-1)^{tr_N\left((v+v^{2^{N-1}})\frac{1}{x}+vx\right)}=\sum_{x\in\mathbb{F}_{2^N}}(-1)^{tr_N\left((v+v^{2^{N-1}})\frac{1}{x}+(v+1)x\right)}.$$

Remark 9. We studied other functions:

1. Under the hypotheses of Definition 1, let $\beta \in \mathbb{F}_{q^2}$ be such that $Tr_N^{2N}(\beta) = 1$ and let H' be a linear hyperplane of $\mathbb{F}_q$ (we can take $H' = H$ or any other hyperplane). We can then define the function G_d as follows: for every $x \in H$, let h and h^q be the elements of $U \setminus \{1\}$ such that $Tr_q^{q^2}(h) = Tr_q^{q^2}(h^q) = \frac{1}{x+a}$, as above. Note that $\beta + \frac{h^d}{Tr_q^{q^2}(h^d)}$ belongs to $\mathbb{F}_q$ since $Tr_q^{q^2}\left(\frac{h^d}{Tr_q^{q^2}(h^d)}\right) = 1$ and $\mathbb{F}_q = \{z \in \mathbb{F}_{q^2} \,|\, Tr_q^{q^2}(z) = 0\}$. We define $G_d(x) = \beta + \frac{h^d}{Tr_q^{q^2}(h^d)}$ (identified with an element of $\mathbb{F}_2^n$) if $\beta + \frac{h^d}{Tr_q^{q^2}(h^d)} \in H'$, and $G_d(x) = \gamma + \beta + \frac{h^d}{Tr_q^{q^2}(h^d)} = \beta + \frac{h^{dq}}{Tr_q^{q^2}(h)}$

otherwise, where $\gamma \in \mathbb{F}_q \setminus H'$. This function is well defined since the output does not depend on the choice between h and h^q. But it is linearly equivalent to F_d: taking $H' = H$, it equals $\beta + \frac{h^d}{Tr_q^{q^2}(h^d)} + \gamma \, tr_q \left(\beta + \frac{h^d}{Tr_q^{q^2}(h^d)} \right) = \frac{1}{Tr_q^{q^2}(h^d)} + \gamma \, tr_q \left(\frac{1}{Tr_q^{q^2}(h^d)} \right) + cst.$ So it equals F_d composed with a linear function, which is necessarily a permutation since otherwise, G_d could not be differentially 4-uniform for $d = 3$.

2. Let H_d be defined as follows: $H_d(x) = \frac{(\beta+x)^{d(q-1)}}{Tr_N^{2N}((\beta+x)^{d(q-1)})} + \beta$. It is well defined and bijective since it equals the function $x \in \mathbb{F}_q \mapsto (\beta+x)^{q-1} \in U \setminus \{1\}$ which is bijective since $\beta \notin \mathbb{F}_q$, composed with the function $h \in U \setminus \{1\} \mapsto h^d \in U \setminus \{1\}$, the function $h \in U \setminus \{1\} \mapsto \frac{h}{Tr_N^{2N}(h)} \in \{z \in \mathbb{F}_{q^2} \mid Tr_N^{2N}(z) = 1\}$ and the function $z \in \mathbb{F}_{q^2} \mid Tr_N^{2N}(z) = 1 \mapsto \beta + z \in \mathbb{F}_q$. But the resulting function does not seem to be differentially δ-uniform for interesting values of δ as computer investigation suggests.

Acknowledgement. We thank Lilya Budaghyan for a nice observation on how proving the differentiality of the function. We also thank Tomas Roche for his computer investigations and Gregor Leander for a useful information.

References

1. Biham, E., Shamir, A.: Differential Cryptanalysis of DES-like Cryptosystems. Journal of Cryptology 4(1), 3–72 (1991)
2. Bracken, C., Byrne, E., Markin, N., McGuire, G.: New families of quadratic almost perfect nonlinear trinomials and multinomials. Finite Fields and their Applications 14, 703–714 (2008)
3. Bracken, C., Byrne, E., Markin, N., McGuire, G.: A few more quadratic APN functions. arXiv:0804.4799v1 (2007)
4. Bracken, C., Byrne, E., McGuire, G., Nebe, G.: On the equivalence of quadratic APN functions. To appear in Designs, Codes and Cryptography (2011)
5. Bracken, C., Leander, G.: A highly nonlinear differentially 4 uniform power mapping that permutes fields of even degree. Finite Fields and Their Applications 16(4), 231–242 (2010)
6. Bracken, C., Leander, G.: New families of functions with differential uniformity of 4. In: Proceedings of the conference BFCA 2008, Copenhagen (2008) (to appear)
7. Brinkmann, M., Leander, G.: On the classification of APN functions up to dimension five. Designs, Codes and Cryptography 49(1-3), 273–288 (2008); Revised and extended version of a paper with the same title in the Proceedings of the Workshop on Coding and Cryptography WCC 2007, pp. 39-48 (2007)
8. Browning, K., Dillon, J.F., Kibler, R.E., McQuistan, M.: APN polynomials and related codes. Special volume of Journal of Combinatorics, Information and System Sciences, honoring the 75-th birthday of Prof. D.K.Ray-Chaudhuri 34, 135–159 (2009)
9. Budaghyan, L.: The simplest method for constructing APN polynomials EA-inequivalent to power functions. In: Carlet, C., Sunar, B. (eds.) WAIFI 2007. LNCS, vol. 4547, pp. 177–188. Springer, Heidelberg (2007)

10. Budaghyan, L., Carlet, C.: Classes of Quadratic APN Trinomials and Hexanomials and Related Structures. IEEE Trans. Inform. Theory 54(5), 2354–2357 (2008)
11. Budaghyan, L., Carlet, C.: On CCZ-equivalence and its use in secondary constructions of bent functions. In: Proceedings of WCC 2009 (2009)
12. Budaghyan, L., Carlet, C., Leander, G.: Two classes of quadratic APN binomials inequivalent to power functions. IEEE Trans. Inform. Theory 54(9), 4218–4229 (2008)
13. Budaghyan, L., Carlet, C., Leander, G.: Constructing new APN functions from known ones. Finite Fields and Applications 15(2), 150–159 (2009)
14. Budaghyan, L., Carlet, C., Leander, G.: On a construction of quadratic APN functions. In: Proceedings of ITW workshop, Taormina, Italy, October 11-16 (2009)
15. Budaghyan, L., Carlet, C., Pott, A.: New Classes of Almost Bent and Almost Perfect Nonlinear Polynomials. In: Proceedings of the Workshop on Coding and Cryptography 2005, Bergen, pp. 306–315 (2005); A completed version has been published in IEEE Trans. Inform. Theory 52(3), 1141–1152 (March 2006)
16. Canteaut, A., Charpin, P., Dobbertin, H.: Binary m-sequences with three-valued crosscorrelation: A proof of Welch's conjecture. IEEE Trans. Inform. Theory 46(1), 4–8 (2000)
17. Carlet, C.: Relating three nonlinearity parameters of vectorial functions and building APN functions from bent functions. Designs, Codes and Cryptography 59(1-3), 89–109 (2010); post-proceedings of WCC 2009
18. Carlet, C.: Boolean Functions for Cryptography and Error Correcting Codes. In: Crama, Y., Hammer, P. (eds.) Boolean Models and Methods in Mathematics, Computer Science, and Engineering, pp. 257–397. Cambridge University Press, Cambridge (2010); Preliminary version available at
   http://www-rocq.inria.fr/codes/Claude.Carlet/pubs.html
19. Carlet, C.: Vectorial Boolean Functions for Cryptography. In: Crama, Y., Hammer, P. (eds.) Boolean Models and Methods in Mathematics, Computer Science, and Engineering, pp. 398–469. Cambridge University Press, Cambridge (2010); Preliminary version available at
   http://www-rocq.inria.fr/codes/Claude.Carlet/pubs.html
20. Carlet, C., Charpin, P., Zinoviev, V.: Codes, bent functions and permutations suitable for DES-like cryptosystems. Designs, Codes and Cryptography 15(2), 125–156 (1998)
21. Carlet, C., Ding, C.: Nonlinearities of S-boxes. Finite Fields and its Applications 13(1), 121–135 (2007)
22. Chabaud, F., Vaudenay, S.: Links between Differential and Linear Cryptanalysis. In: De Santis, A. (ed.) EUROCRYPT 1994. LNCS, vol. 950, pp. 356–365. Springer, Heidelberg (1995)
23. Dillon, J.F.: APN polynomials: an update. In: Conference Finite Fields and Applications Fq9, Dublin, Ireland (July 2009)
24. Edel, Y., Pott, A.: A new almost perfect nonlinear function which is not quadratic. Advances in Mathematics of Communications 3(1), 59–81 (2009)
25. de Hou, X., Mullen, G.L., Sellers, J.A., Yucas, J.L.: Sellers and J. L. Yucas. Reversed Dickson polynomials over finite fields. Finite Fields and Their Applications 15(6), 748–773 (2009)
26. Knudsen, L.: Truncated and higher order differentials. In: Preneel, B. (ed.) FSE 1994. LNCS, vol. 1008, pp. 196–211. Springer, Heidelberg (1995)
27. Lachaud, G., Wolfmann, J.: The Weights of the Orthogonals of the Extended Quadratic Binary Goppa Codes. IEEE Trans. Inform. Theory 36, 686–692 (1990)

28. Lidl, R., Niederreiter, H.: Finite Fields. Encyclopedia of Mathematics and its Applications, vol. 20. Addison-Wesley, Reading (1983)
29. Li, Y., Wang, M.: On EA-equivalence of certain permutations to power mappings. Designs, Codes and Cryptography 58(3), 259–269 (2010)
30. Li, Y., Wang, M.: On permutation polynomials EA-equivalent to the inverse function over $GF(2^n)$. IACR ePrint Archive 2010/573
31. Matsui, M.: Linear cryptanalysis method for DES cipher. In: Helleseth, T. (ed.) EUROCRYPT 1993. LNCS, vol. 765, pp. 386–397. Springer, Heidelberg (1994)
32. Nyberg, K.: Perfect non-linear S-boxes. In: Davies, D.W. (ed.) EUROCRYPT 1991. LNCS, vol. 547, pp. 378–386. Springer, Heidelberg (1991)
33. Nyberg, K.: On the construction of highly nonlinear permutations. In: Rueppel, R.A. (ed.) EUROCRYPT 1992. LNCS, vol. 658, pp. 92–98. Springer, Heidelberg (1993)
34. Nyberg, K.: New bent mappings suitable for fast implementation. In: Anderson, R. (ed.) FSE 1993. LNCS, vol. 809, pp. 179–184. Springer, Heidelberg (1994)
35. Nyberg, K.: Differentially uniform mappings for cryptography. In: Helleseth, T. (ed.) EUROCRYPT 1993. LNCS, vol. 765, pp. 55–64. Springer, Heidelberg (1994)
36. Rivain, M., Prouff, E.: Provably Secure Higher-Order Masking of AES. In: Mangard, S., Standaert, F.-X. (eds.) CHES 2010. LNCS, vol. 6225, pp. 413–427. Springer, Heidelberg (2010)
37. Sidelnikov, V.M.: On the mutual correlation of sequences. Soviet Math. Dokl. 12, 197–201 (1971)

New Impossible Differential Attacks of Reduced-Round Camellia-192 and Camellia-256*

Jiazhe Chen[1,2], Keting Jia[3], Hongbo Yu[4], and Xiaoyun Wang[1,2,3,**]

[1] Key Laboratory of Cryptologic Technology and Information Security,
Ministry of Education, Shandong University, Jinan 250100, China
[2] School of Mathematics, Shandong University, Jinan 250100, China
jiazhechen@mail.sdu.edu.cn
[3] Institute for Advanced Study, Tsinghua University, Beijing 100084, China
[4] Department of Computer Science and Technology,
Tsinghua University, Beijing 100084, China
{ktjia,yuhongbo,xiaoyunwang}@mail.tsinghua.edu.cn

Abstract. Camellia, which is a block cipher selected as a standard by ISO/IEC, is one of the most widely used block ciphers. In this paper, we propose several 6-round impossible differentials of Camellia with FL/FL^{-1} layers in the middle of them. With the impossible differentials and a well-organized precomputed table, impossible differential attacks on 10-round Camellia-192 and 11-round Camellia-256 are given, and the time complexities are $2^{175.3}$ and $2^{206.8}$ respectively. In addition, an impossible differential attack on 15-round Camellia-256 without FL/FL^{-1} layers and whitening is also be given, which needs about $2^{236.1}$ encryptions. To the best of our knowledge, these are the best cryptanalytic results of Camellia-192/-256 with FL/FL^{-1} layers and Camellia-256 without FL/FL^{-1} layers to date.

Keywords: Camellia Block Cipher, Cryptanalysis, Impossible Differential, Impossible Differential Attack.

1 Introduction

Block cipher Camellia is proposed by NTT and Mitsubishi in 2000 [1]. Its block size is 128 bits and it supports 128-, 192- and 256-bit key sizes with 18, 24 and 24 rounds respectively. Camellia was selected as an e-government recommended cipher by CRYPTREC [5] and recommended in NESSIE [15] block cipher portfolio. Then it was selected as an international standard by ISO/IEC 18033-3 [8].

The structure of Camellia is Feistel structure with FL/FL^{-1} layers inserted every 6 rounds. The FL and FL^{-1} functions are keyed linear functions which are designed to provide non-regularity across rounds [1]. In the past years, Camellia

---

* Supported by 973 Project (No.2007CB807902), the National Natural Science Foundation of China (Grant No.60931160442) and Graduate Independent Innovation Foundation of Shandong University (No. 11140070613183).
** Corresponding author.

U. Parampalli and P. Hawkes (Eds.): ACISP 2011, LNCS 6812, pp. 16–33, 2011.

has attracted the attention of the cryptanalytic community. The square-type attacks are efficient to attack Camellia, which can be used to analyze 9-round Camellia-128 and 10-round Camellia-256 [11]. Furthermore, Hatano *et al.* used the higher order differential attack to analyze the last 11 rounds Camellia-256, with complexity $2^{255.6}$ [7].

There are a number of results on simple versions of Camellia which exclude the FL/FL^{-1} layers and whitening in recent years [6,10,13,14,16,17,18,19]. Among them, the impossible differential attacks [4] are most efficient [13,14,17,18]. Since the existence of FL/FL^{-1} layers will probably destroy the impossibility, none of the impossible differentials in these attacks includes the FL/FL^{-1} layers. For the attacks of Camellia-256 without FL/FL^{-1} layers and whitening, the 14-round attack in [13] was pointed out to be incorrect by [20]. Later Mala *et al.* [14] pointed out a flaw in [20] and showed that the time complexities of the 12-round Camellia-128 and 16-round Camellia-256 attacks were more than exhaustive search. As a result, the best analysis of Camellia-256 without FL/FL^{-1} layers and whitening dated back to [12], which was a 13-round attack with complexity $2^{211.7}$.

Our Contribution. In this paper, we present 6-round impossible differentials with FL/FL^{-1} layers in the middle, which turn out to be first impossible differentials with FL/FL^{-1} layers. Due to one of these impossible differentials and a precomputation table that is carefully constructed, we propose impossible differential attacks on 10-round Camellia-192 and 11-round (round 1~11) Camellia-256 with complexity $2^{175.3}$ and $2^{206.8}$ respectively. Then by carefully using the subkey relations and one of the 8-round impossible differentials without FL/FL^{-1} layers proposed in [18], we also present an impossible differential attack on 15-round Camellia-256 without FL/FL^{-1} layers and whitening, and the complexity is about $2^{236.1}$ encryptions.

The rest of this paper is organized as follows. We give some notations and a brief description of Camellia in Section 2. Some properties of Camellia and 6-round impossible differentials with FL/FL^{-1} layers are given in Section 3. Section 4 describes the impossible differential attacks on reduced-round Camellia with FL/FL^{-1} layers and whitening. The impossible differential attack on 15-round Camellia-256 without FL/FL^{-1} layers and whitening is illustrated in Section 5. Finally, we conclude the paper in Section 6.

2   Preliminaries

Some notions used in this paper and a simple description of the Camellia algorithm are given in this section.

2.1   Notations

L^{r-1}, L'^{r-1} : the left half of the 128-bit r-th round input
R^{r-1}, R'^{r-1} : the right half of the 128-bit r-th round input
ΔL^{r-1} : the difference of L^{r-1} and L'^{r-1}

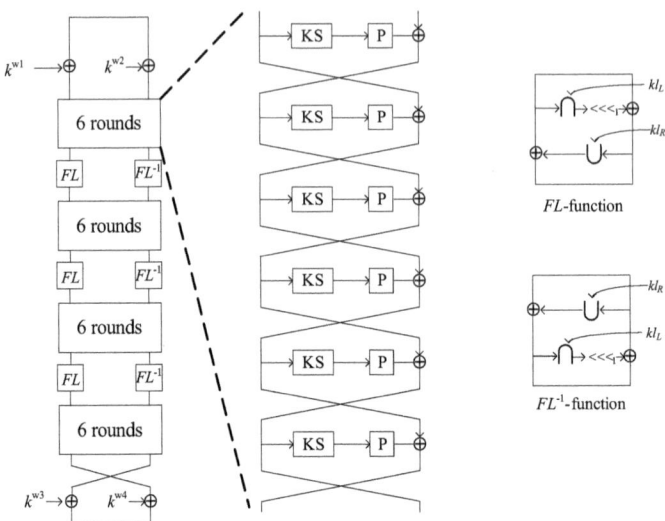

Fig. 1. Camellia-192/-256

ΔR^{r-1} : the difference of R^{r-1} and R'^{r-1}

S^r, S'^r: the output value of the S-box layer of the r-th round

ΔS^r: the output difference of the S-box layer of the r-th round

A_i: the i-th byte of a 64-bit value A $(i = 1, ..., 8)$

$B \lll j$: left rotation of B by j bits

$X_{L(64)}$: the left half of a 128-bit word X

$X_{R(64)}$: the right half of a 128-bit word X

$Y_{L(32)}$: the left half of a 64-bit word Y

$Y_{R(32)}$: the right half of a 64-bit word Y

$||$: the cascade of two words

$\overline{x}$: the bitwise complement of x

$\oplus, \cap, \cup$: bitwise exclusive-OR(XOR), AND, OR

2.2 The Camellia Algorithm

Camellia [1] is a 128-bit block cipher with Feistel structure. It has 18 rounds for 128-bit key, and 24 rounds for 192-/256-bit key. We give the encryption procedure of Camellia-192/-256 as follows, see Fig. 1.

Encryption Procedure. The input of the encryption procedure is a 128-bit plaintext M, and 64-bit subkeys k^{wi} $(i = 1, ..., 4)$, k^r $(r = 1, ..., 24)$ and kl^j $(j = 1, ..., 6)$. First M is XORed with k^{w1} and k^{w2} to get two 64-bit intermediate values L^0 and R^0: $L^0||R^0 = M \oplus (k^{w1}||k^{w2})$. Then the following operations are carried out for $i = 1$ to 24, except for $r = 6, 12$ and 18:

$$L^r = R^{r-1} \oplus F(L^{r-1}, k^r), R^r = L^{r-1}.$$

For $r = 6$, 12 and 18, do the following:

$$L^{*r} = R^{r-1} \oplus F(L^{r-1}, k^r), R^{*r} = L^{r-1}.$$
$$L^r = FL(L^{*r}, kl^{2r/6-1}), R^r = FL^{-1}(R^{*r}, kl^{2r/6}).$$

Finally the 128-bit ciphertext C is computed as: $C = (R^{24}||L^{24}) \oplus (k^{w3}||k^{w4})$.

The FL function is defined as: $(X_{L(32)}||X_{R(32)}, kl_{L(32)}||kl_{R(32)}) \mapsto (Y_{L(32)}||Y_{R(32)})$, where:

$$Y_{R(32)} = ((X_{L(32)} \cap kl_{L(32)}) \lll 1) \oplus X_{R(32)},$$
$$Y_{L(32)} = (Y_{R(32)} \cup kl_{R(32)}) \oplus X_{L(32)}.$$

The FL^{-1} function is the inverse of FL function, and FL and FL^{-1} are linear as long as the keys are fixed [2].

The round function F is composed of the key-addition layer, S-box layer S and linear transformation P. In the key-addition layer, the input of the round function is XORed with the subkey. There are four 8×8 S-boxes S_1, S_2, S_3, S_4 used in the S-box layer, and each S-box is used twice. Finally, the linear transformation $P : (\{0,1\}^8)^8 \rightarrow (\{0,1\}^8)^8$ maps $(z_1, ..., z_8) \rightarrow (y_1, ..., y_8)$. P function and its inverse function P^{-1} are:

$y_1 = z_1 \oplus z_3 \oplus z_4 \oplus z_6 \oplus z_7 \oplus z_8$
$y_2 = z_1 \oplus z_2 \oplus z_4 \oplus z_5 \oplus z_7 \oplus z_8$
$y_3 = z_1 \oplus z_2 \oplus z_3 \oplus z_5 \oplus z_6 \oplus z_8$
$y_4 = z_2 \oplus z_3 \oplus z_4 \oplus z_5 \oplus z_6 \oplus z_7$
$y_5 = z_1 \oplus z_2 \oplus z_6 \oplus z_7 \oplus z_8$
$y_6 = z_2 \oplus z_3 \oplus z_5 \oplus z_7 \oplus z_8$
$y_7 = z_3 \oplus z_4 \oplus z_5 \oplus z_6 \oplus z_8$
$y_8 = z_1 \oplus z_4 \oplus z_5 \oplus z_6 \oplus z_7$

$z_1 = y_2 \oplus y_3 \oplus y_4 \oplus y_6 \oplus y_7 \oplus y_8$
$z_2 = y_1 \oplus y_3 \oplus y_4 \oplus y_5 \oplus y_7 \oplus y_8$
$z_3 = y_1 \oplus y_2 \oplus y_4 \oplus y_5 \oplus y_6 \oplus y_8$
$z_4 = y_1 \oplus y_2 \oplus y_3 \oplus y_5 \oplus y_6 \oplus y_7$
$z_5 = y_1 \oplus y_2 \oplus y_5 \oplus y_7 \oplus y_8$
$z_6 = y_2 \oplus y_3 \oplus y_5 \oplus y_6 \oplus y_8$
$z_7 = y_3 \oplus y_4 \oplus y_5 \oplus y_6 \oplus y_7$
$z_8 = y_1 \oplus y_4 \oplus y_6 \oplus y_7 \oplus y_8$

Key Schedule. For Camellia-256, the 256-bit main key $K = K_L||K_R$, where K_L and K_R are 128 bits. And for Camellia-192, the 192-bit main key $K = K_L||K_{RL(64)}$ and $K_{RR(64)} = \overline{K_{RL(64)}}$. Using K_L and K_R, the key schedule algorithm first calculate K_A and K_B, which is described in Fig. 2. Where F is the round function of Camellia and C_i ($1 \leq i \leq 6$) are constants used as the keys. Then the subkeys k^{wi} ($i = 1, ..., 4$), k^r ($r = 1, ..., 24$) and kl^j ($j = 1, ..., 6$) are derived from rotating K_L, K_R, K_A or K_B. For details of Camellia, we refer to [1].

It can be known from Fig. 2 that, if K_B and K_R are known, K_A is known. Therefore, one can get K_L using the relation between K_L and K_A described in [14], Section 3.2. So once K_B and K_R are known, K can be computed.

3 Properties and 6-Round Impossible Differentials of Camellia with FL/FL^{-1} Functions

In this section, we first give some useful properties of Camellia and then propose several impossible differentials.

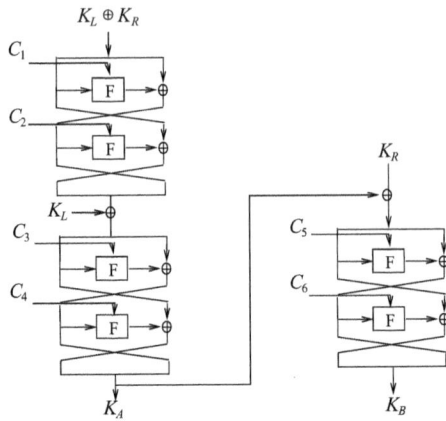

Fig. 2. The Calculation of K_A and K_B

Property 1. *For a 3-round Camellia structure, if the input difference is of the form* $\Delta L^i = (0, a, 0, 0, 0, 0, 0, 0)$, $\Delta R^i = (0, 0, 0, 0, 0, 0, 0, 0)$, *then:*
1. $\Delta L^{i+1} = (0, b, b, b, b, b, 0, 0)$, $\Delta S^{i+2} = (0, b_2, b_3, b_4, b_5, b_6, 0, 0)$,
$\Delta L^{i+2} = \Delta R^{i+3} = P(a, b_2, b_3 \oplus a, b_4 \oplus a, b_5 \oplus a, b_6 \oplus a, 0, 0)$,
2. $\Delta S_l^{i+3} = (P^{-1}(\Delta L^{i+3}))_l$, *for* $l = 1, 3, 4, ..., 8$,
where $a, b, b_2, b_3, b_4, b_5, b_6$ *are non-zero bytes.*

Property 2. *For a 3-round Camellia structure, the necessary conditions of* $\Delta L^{i+3} = (0, 0, 0, 0, 0, 0, 0, 0)$ *and* $\Delta R^{i+3} = (0, a, 0, 0, 0, 0, 0, 0)$ *are:*
1. $\Delta L^{i+1} = (0, b, b, b, b, b, 0, 0)$, $\Delta S^{i+2} = (0, b_2, b_3, b_4, b_5, b_6, 0, 0)$,
$\Delta L^i = P(a, b_2, b_3 \oplus a, b_4 \oplus a, b_5 \oplus a, b_6 \oplus a, 0, 0)$, *and*
2. $\Delta S_l^{i+1} = (P^{-1}(\Delta R^i))_l$, *for* $l = 1, 3, 4, ..., 8$,
where $a, b, b_2, b_3, b_4, b_5, b_6$ *are non-zero bytes.*

To better describe the properties, we also illustrate them in Fig. 3. Property 1.1 and 2.1 are trivial, which have been used in most previous impossible attacks of Camellia. Property 1.2 and 2.2 are also simple, but to the best of our knowledge, none of the previous attacks used them. The proofs of the properties are similar and the proof of Property 1 is given as an example.

Proof. Apparently, ΔS^{i+1} is of the form $(0, b, 0, 0, 0, 0, 0, 0)$, where b an is unknown non-zero byte. And $\Delta L^{i+1} = (0, b, b, b, b, b, 0, 0)$ as P function is linear. After the key-addition layer and S-box layer, it can be obtained that $\Delta S^{i+2} = (0, b_2, b_3, b_4, b_5, b_6, 0, 0)$, where b_2, b_3, b_4, b_5 and b_6 are unknown non-zero bytes.
     Since $\Delta L^{i+2} = P(\Delta S^{i+2}) \oplus \Delta L^i$ and $P^{-1}(\Delta L^i) = (a, 0, a, a, a, a, 0, 0)$,

$$\Delta L^{i+2} = P(a, b_2, b_3 \oplus a, b_4 \oplus a, b_5 \oplus a, b_6 \oplus a, 0, 0).$$

Finally, because $\Delta S^{i+3} = P^{-1}(\Delta L^{i+1} \oplus \Delta L^{i+3})$, $P^{-1}(\Delta L^{i+1}) = (0, b, 0, 0, 0, 0, 0, 0)$ and P^{-1} function is linear,

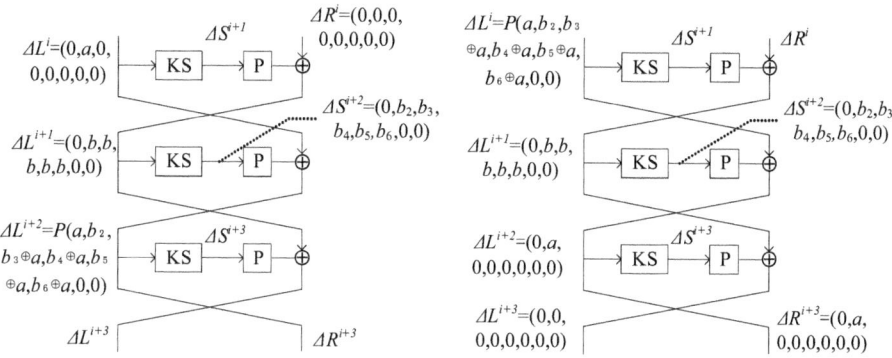

Fig. 3. Properties of 3-round Camellia

$$\Delta S_l^{i+3} = (P^{-1}(\Delta L^{i+3}))_l, \text{ for } l = 1, 3, 4, ..., 8. \qquad \square$$

Property 3. *(from [9]) Let x, x', k be 32-bit values, and $\Delta x = x \oplus x'$, then the differential properties of AND and OR operations are:*

$$(x \cap k) \oplus (x' \cap k) = (x \oplus x') \cap k = \Delta x \cap k,$$
$$(x \cup k) \oplus (x' \cup k) = (x \oplus k \oplus (x \cap k)) \oplus (x' \oplus k \oplus (x' \cap k)) = \Delta x \oplus (\Delta x \cap k).$$

Property 4. *Let $M = (m_1, m_2, m_3, m_4, m_5, m_6, m_7, m_8)$ be the input difference of FL function, and $N = (n_1, n_2, n_3, n_4, n_5, n_6, n_7, n_8)$ be the the output difference of FL, where n_l, m_l $(l = 1, ..., 8)$ are arbitrary 8-bit values. Then if $n_i = 0$ $(i \in \{5, 6, 7, 8\})$, $n_{i-4} = m_{i-4}$.*

Proof. Let us denote the subkey used for AND operation as k_L and the subkey used for OR operation as k_R. By Property 3, the following equations must hold:

$$((M_{L(32)} \cap k_L) \lll 1) \oplus M_{R(32)} = N_{R(32)}$$
$$M_{L(32)} \oplus N_{R(32)} \oplus (N_{R(32)} \cap k_R) = N_{L(32)} \qquad (1)$$

Then if $n_i = 0$ $(i \in \{5, 6, 7, 8\})$, it can be deduced from Equation (1) that $n_{i-4} = m_{i-4}$. $\qquad \square$

Impossible Differential. Now we demonstrate that the 6-round differential in Fig. 4 is impossible. The input difference is

$$((0, 0, 0, 0, 0, 0, 0, 0); (0, a, 0, 0, 0, 0, 0, 0)),$$

where a is arbitrary non-zero byte. The output difference of the first round is

$$((0, a, 0, 0, 0, 0, 0, 0); (0, 0, 0, 0, 0, 0, 0, 0)).$$

Then by Property 1, the output differences of the second round and third round are

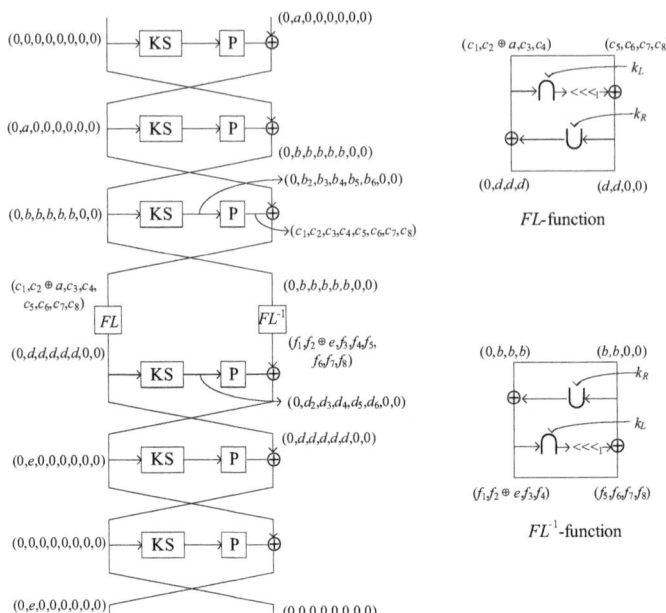

Fig. 4. 6-round Impossible Differential with the FL/FL^{-1} Layer in the Middle

$$((0, b, b, b, b, b, 0, 0); (0, a, 0, 0, 0, 0, 0, 0)) \text{ and}$$
$$((c_1, c_2 \oplus a, c_3, c_4, c_5, c_6, c_7, c_8); (0, b, b, b, b, b, 0, 0))$$

with probability 1, as long as

$$(c_1, c_2, c_3, c_4, c_5, c_6, c_7, c_8) = P(0, b_2, b_3, b_4, b_5, b_6, 0, 0),$$

where $b, b_2, b_3, b_4, b_5, b_6$ are unknown non-zero bytes, $(0, b_2, b_3, b_4, b_5, b_6, 0, 0)$ is evolved from $(0, b, b, b, b, b, 0, 0)$ after the S-box layer and c_l ($l = 1, .., 8$) are unknown bytes.

Similarly, in the backward direction, we know that for arbitrary non-zero byte e, if the output difference of the sixth round is

$$((0, e, 0, 0, 0, 0, 0, 0); (0, 0, 0, 0, 0, 0, 0, 0)),$$

then the input difference of the fourth round is

$$((0, d, d, d, d, d, 0, 0); (f_1, f_2 \oplus e, f_3, f_4, f_5, f_6, f_7, f_8)),$$

where d is an unknown non-zero byte and f_l ($l = 1, .., 8$) are unknown bytes.

Now the input and output differences of the FL function are determined. It can be deduced from Property 4 that $c_3 = d$ and $c_4 = d$, which means $c_3 = c_4$. But this implies $b_4 = 0$ as

$$c_3 = b_2 \oplus b_3 \oplus b_5 \oplus b_6,$$
$$c_4 = b_2 \oplus b_3 \oplus b_4 \oplus b_5 \oplus b_6,$$

which contradict $b_4 \neq 0$ (By the input and output difference of FL^{-1} function, we can also deduce another contradiction that $d_4 = 0 \nLeftrightarrow d_4 \neq 0$). As a result, the differential

$$((0,0,0,0,0,0,0,0); (0,a,0,0,0,0,0,0)) \xrightarrow{6-round} ((0,e,0,0,0,0,0,0); (0,0,0,0,0,0,0,0))$$

is impossible.

Actually, there are three more 6-round impossible differentials with FL/FL^{-1} layers in the middle, which are:

$$((0,0,0,0,0,0,0,0); (a,0,0,0,0,0,0,0)) \xrightarrow{6-round}_{\nrightarrow} ((e,0,0,0,0,0,0,0); (0,0,0,0,0,0,0,0))$$

$$((0,0,0,0,0,0,0,0); (0,0,a,0,0,0,0,0)) \xrightarrow{6-round}_{\nrightarrow} ((0,0,e,0,0,0,0,0); (0,0,0,0,0,0,0,0))$$

$$((0,0,0,0,0,0,0,0); (0,0,0,a,0,0,0,0)) \xrightarrow{6-round}_{\nrightarrow} ((0,0,0,e,0,0,0,0); (0,0,0,0,0,0,0,0))$$

4 Impossible Differential Attacks on Camellia with FL/FL^{-1} Functions and Whitening

In this section, we present impossible differential attacks on 11-round Camellia-256 and 10-round Camellia-192 using the impossible differential proposed in Section 3.

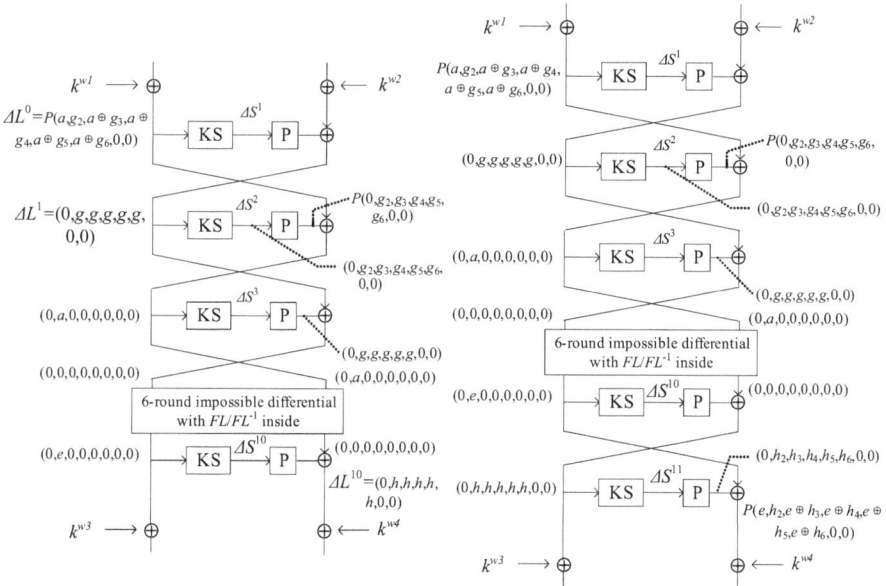

Fig. 5. Impossible Differential Attacks on 10-Round Camellia-192 and 11-Round Camellia-256 with Whitening and FL/FL^{-1}

4.1   Impossible Differential Attack on 11-Round Camellia-256

We add 3 rounds on the top and 2 rounds on the bottom of the 6-round impossible differential to analyze 11-round Camellia-256, see Fig. 5 in the right. In order to deal with the whitening keys, we first denote equivalent subkeys $k^a = k^{w1} \oplus k^1$, $k^b = k^{w2} \oplus k^2$, $k^c = k^{w1} \oplus k^3$, $k^d = k^{w4} \oplus k^{10}$ and $k^e = k^{w3} \oplus k^{11}$. Then the cipher acts as there are no whitening keys except that the round functions use the equivalent subkeys instead of the original ones. The attack aims to discard the wrong equivalent subkeys by means of the impossible differential, i.e., if there is a plaintext/ciphertext pair meets the differential in Fig. 5, then the corresponding subkey must be wrong. After finding the correct equivalent subkeys, the main key will be recovered by the key schedule. Before introducing the attack procedure, we first set up a precomputation table which is used to reduce the time complexity of the attack.

Precomputation. A precomputation hash table H for rounds $2 \sim 3$ is set up here, which contains all possible pairs that can follow the differential in rounds $2 \sim 3$ and their corresponding subkeys k^b, k_2^c. This table can also be used for rounds $10 \sim 11$, as in the backward direction, the differences are the same as that of rounds $2 \sim 3$. The table is constructed as follows:

For every $(L^1, g, k^b, L_2^2, a, k_2^c)$, sieve the ones satisfying $S_2(L_2^2 \oplus k_2^c) \oplus S_2(L_2^2 \oplus a \oplus k_2^c) = g$, where g and a are non-zero bytes. There are 2^{160} $(L^1, g, k^b, L_2^2, a, k_2^c)$s, and 2^{152} of which remain after the sieve. Then compute $L'^1 = L^1 \oplus (0, g, g, g, g, g, 0, 0)$, $T = F(L^1, k^b)$, $T' = F(L'^1, k^b)$, $\Delta T = T \oplus T'$ and insert (k^b, k_2^c) into the row indexed by $(L^1, g, \Delta T \oplus (0, a, 0, 0, 0, 0, 0, 0), L_2^2 \oplus T_2)$. As we aim to recover the equivalent subkeys, the values L^1, L^2 and R^1 are not related to the whitening keys. Consequently, for each pair, once we get the value $(L^1, \Delta L_2^1, \Delta R^1, R_2^1)$, we can access the corresponding row in H to get the equivalent subkeys such that the pair satisfies the differential.

Because there are only 2^{40} ΔTs which lead to 2^{48} $\Delta T \oplus (0, a, 0, 0, 0, 0, 0, 0)$s, there are 2^{128} rows in H and each row contains 2^{24} 72-bit equivalent subkeys (k^b, k_2^c). As a result, the memory complexity of the table is about $2^{155.2}$ bytes and the time complexity of the precomputation is less than 2^{161} one round encryptions.

Data Collection. Choose 2^n structures of plaintexts, and each structure contains plaintexts with the following form:

$$(P(y_1, y_2, y_3, y_4, y_5, y_6, \alpha, \beta); (x_1, x_2, x_3, x_4, x_5, x_6, x_7, x_8)),$$

where y_i $(i = 1, ..., 6)$ and x_j $(j = 1, ..., 8)$ take all possible values and α, β are fixed in each structure. As a result, there are 2^{112} plaintexts in each structure and we can get $2^n \times 2^{112 \times 2 - 1} = 2^{n+223}$ plaintext pairs totally. For each of the pairs, $(P^{-1}(\Delta L^0))_7 = 0$, $(P^{-1}(\Delta L^0))_8 = 0$.

Ask for the encryptions of the plaintexts in each structure to get the corresponding ciphertexts, and keep the pairs whose ciphertext differences satisfy the following form by birthday paradox:

$$((0, h, h, h, h, h, 0, 0); P(e, h_2, e \oplus h_3, e \oplus h_4, e \oplus h_5, e \oplus h_6, 0, 0)),$$

where e, h, h_2, h_3, h_4, h_5 and h_6 are non-zero bytes. So there are $2^{n+223-72} = 2^{n+151}$ pairs remaining.

Key Recovery. In the key recovery procedure, we use Property 2 and the precomputation table H to discard the wrong keys.

1. For $l = 1, 3, ..., 8$, guess k_l^a and keep the pairs that satisfy the equation $\triangle S_l^1 = (P^{-1}(\triangle R^0))_l$. Next guess k_2^a, so (L^1, L'^1) can be computed. For each of the remaining pairs, do Step 2.
2. Initialize a table Γ of 2^{144} all possible values (k^b, k_2^c, k^e, k_2^d), for each of the remaining pairs, access the row $(L^1, \triangle L_2^1, \triangle R^1, R_2^1)$ and the row $(L^{10}, \triangle L_2^{10}, \triangle L^{11}, L_2^{11})$ in table H. Then combine the values in the two rows to get (k^b, k_2^c, k^e, k_2^d), and remove the corresponding value from Γ.
3. If Γ is not empty, output the 208-bit value $(k^a, k^b, k_2^c, k_2^d, k^e)$, jump out of the iteration and go to step 4 to recover the main key; otherwise go to Step 1 and try another guess.
4. The following equations are deduced from Table 3 in [1]:

$$k^a = (K_L \lll 0)_{L(64)} \oplus (K_B \lll 0)_{L(64)}, \tag{2}$$
$$k^b = (K_L \lll 0)_{R(64)} \oplus (K_B \lll 0)_{R(64)}, \tag{3}$$
$$k_2^c = ((K_L \lll 0)_{L(64)} \oplus (K_R \lll 15)_{L(64)})_2, \tag{4}$$
$$k^e = (K_B \lll 111)_{L(64)} \oplus (K_A \lll 45)_{L(64)}, \tag{5}$$
$$k_2^d = ((K_B \lll 111)_{R(64)} \oplus (K_L \lll 45)_{R(64)})_2. \tag{6}$$

We guess every possible value of K_L. For each guess, K_B can be calculated by Equations (2) and (3), then sieve this (K_L, K_B) pair by Equation (6). For each of the (K_L, K_B)s that satisfy Equation (6), further compute 64 bits of K_A by Equation (5). Then guess the other 64-bits of K_A, by the key schedule of Camellia-256, K_R can be fully determined by K_B and K_A. Equation (4) will further reduce the keys by a factor of 2^8. So we get about $2^{192} \times 2^{-8} \times 2^{-8} = 2^{176}$ (K_L, K_R)s and the correct $K = K_L || K_R$ can be obtained by trial encryption.

Complexity. In Step 1, it is expected that about $2^{n+151} \times 2^{-8 \times 7} = 2^{n+95}$ pairs will be kept for each guess of k^a. Step 2 removes 2^{48} wrong (k^b, k_2^c, k^e, k_2^d)s for each pair remained after Step 1, so a proportion of $\frac{2^{48}}{2^{144}} = 2^{-96}$ of wrong (k^b, k_2^c, k_2^d, k^e)s are removed for each pair. Consequently, the number of remaining wrong 208-bit values $(k^a, k^b, k_2^c, k_2^d, k^e)$ after analyzing all the pairs is $\sigma = 2^{64} \times 2^{144} \times (1 - 2^{-96})^{2^{n+95}}$. In order to let $\sigma \ll 0$, we choose $n = 9$. Then the data complexity is 2^{121} chosen plaintexts, and the complexity of choosing proper pairs that meet the required form of ciphertext differences by birthday paradox is less than 2^{121} one-round encryptions.

The complexity of Step 1 is about $2 \times (\sum_{i=1}^{7} 2^{160-8(i-1)} \times 2^{8i}) \times \frac{1}{8} + 2 \times 2^{64} \times 2^{104} \approx 2^{170}$ one round encryptions, equivalent to $2^{166.5}$ encryptions. There are 2^{24} values in H, so in Step 2, 2×2^{24} memory access to H and 2^{48} memory access to

Γ are needed for each pair, which result in $2^{64} \times 2^{104} \times 2^{48} = 2^{216}$ memory access. As one memory access is equivalent to one XOR operation and there are 52 XOR operations in one round Camellia, the complexity of Step 2 is about $2^{216} \times \frac{1}{52} \times \frac{1}{11} \approx 2^{206.8}$ 11-round encryptions. The complexity of Step 4 is about 2^{184} 2-round encryptions, so the time complexity is dominated by Step 2, which about $2^{206.8}$ encryptions. And the memory complexity is dominated by storing the 2^{160} pairs after the data collection phase, which is about 2^{166} bytes.

4.2   Impossible Differential Attack on 10-Round Camellia-192

We remove one round from the bottom of the 11-round attack, and give an attack on 10-round Camellia-192, see Fig. 5 in the left. In this attack, the table H is also set up in the precomputation phase. The choice of plaintexts is the same as the 11-round attack, and the ciphertext pairs are sieved by the difference:

$$(0, e, 0, 0, 0, 0, 0, 0; 0, h, h, h, h, h, 0, 0),$$

where e and h are non-zero values. After the sieve, about 2^{120} pairs remain.

Denote equivalent subkeys $k^a = k^{w1} \oplus k^1$, $k^b = k^{w2} \oplus k^2$, $k^c = k^{w1} \oplus k^3$ and $k^d = k^{w3} \oplus k^{10}$. The key recovery phase is as follows:

1. Guess k_2^d and check whether $\Delta S_2^{10} = \Delta L_2^{10}$, discard the pairs that do not satisfy this condition. Then for $l = 1, 3, ..., 8$, guess k_l^a and discard the pairs that do not satisfy the equation $\Delta S_l^1 = (P^{-1}(\Delta R^0))_l$. Next guess k_2^a, so (L^1, L'^1) can be computed. For each of the remaining pairs, do Step 2.
2. Initialize a table Γ' of 2^{72} all possible values (k^b, k_2^c), for each of the remaining pairs, access the row $(L^1, \Delta L_2^1, \Delta R^1, R_2^1)$ in table H. For each value in the row, remove the corresponding value from Γ'.
3. If Γ' is not empty, output the 144-bit value (k^a, k^b, k_2^c, k_2^d), jump out of the iteration and go to step 4; otherwise go to Step 1 and continue the iteration.
4. The main key can be recovered when (k^a, k^b, k_2^c, k_2^d) is obtained by the similar method of the 11-round attack, except that there are only four equations that can be used:

$$k^a = (K_L \lll 0)_{L(64)} \oplus (K_B \lll 0)_{L(64)}, \tag{7}$$

$$k^b = (K_L \lll 0)_{R(64)} \oplus (K_B \lll 0)_{R(64)}, \tag{8}$$

$$k_2^c = ((K_L \lll 0)_{L(64)} \oplus (K_R \lll 15)_{L(64)})_2, \tag{9}$$

$$k_2^d = ((K_B \lll 111)_{L(64)} \oplus (K_L \lll 45)_{R(64)})_2. \tag{10}$$

Again, we guess every possible value of K_L. For each guess, K_B can be calculated by Equations (7) and (8), then sieve this (K_L, K_B) pair by Equation (10). For the (K_L, K_B) that satisfy Equation (10), compute 8 bits of K_R by Equation (9), and further guess the rest unknown 56 bits of K_R. Furthermore, we test whether the (K_L, K_R, K_B) can pass the key schedule of Camellia-192. About $2^{184} \times 2^{-8} \times 2^{-128} = 2^{48}$ keys will remain, and the correct $K = K_L || K_R$ can be obtained by trial encryption.

In this attack, the time complexity is dominated by Step 4, which is about 2^{176} 6-round encryptions, equivalent to about $2^{175.3}$ 10-round encryptions. The memory complexity is $2^{155.2}$ bytes, which is dominated by the precomputation. If we do not take the pre-/post- whitening key into account, then we deal with the original subkeys instead of the equivalent ones and Step 4 is unnecessary. So the complexity would be determined by Step 2, which is about 2^{144} encryptions.

5   Impossible Differential Cryptanalysis of 15-Round Camellia-256 without FL/FL^{-1} Layers and Whitening

In this section, we give an improved impossible differential attack on Camellia-256 by using the 8-round impossible differential without FL/FL^{-1} layers in Fig. 7, which was proposed in [18]. By adding 4 rounds on the top and 3 rounds on the bottom, we can attack 15-round Camellia-256 without FL/FL^{-1} layers and whitening, see Fig. 6. We give in Table 1 the corresponding positions of k^1, k^2 and k^{15} in K_B, and the corresponding positions of k^3, k^{14}, k_2^4 and k_2^{13} in K_R. It is obvious that there are close relations among the subkeys, i.e., there are common bits in some of the subkeys, which can be used to reduced the complexity of the attack. From Table 1 we know that the subkey bits used in the 15-th round are also involved in the subkeys of the first and second round. As the first

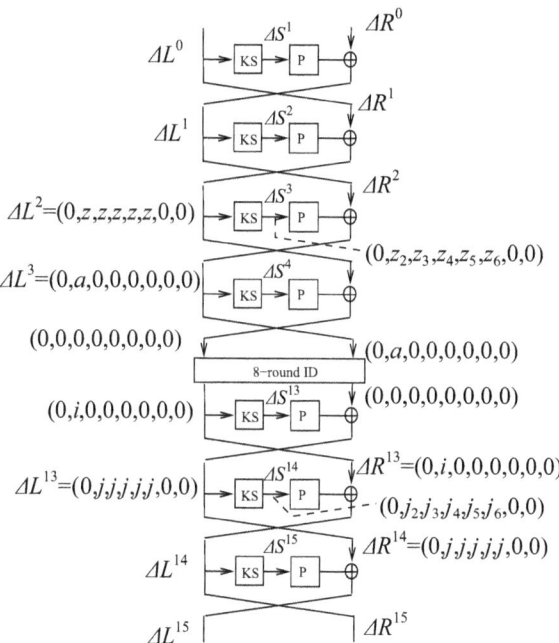

Fig. 6. Impossible Differential Attack on 15-Round Camellia-256 without FL/FL^{-1} Layers and Whitening

two rounds contribute most to the time complexity, we set up a precomputation table below to store the required pairs in advance.

Precomputation. First, we set up a hash table Γ_1 for the first two rounds of the 15-round model. Property 2 implies that $\Delta L^1 = P(a, z_2, z_3 \oplus a, z_4 \oplus a, z_5 \oplus a, z_6 \oplus a, 0, 0)$, so we choose all the 2^{48} ΔL^1s with the required form. Furthermore, all possible ΔL^0, $L^0 \oplus k^1$ and $(R^0 \oplus k^2)_2$ are chosen. For each of the values, we compute $T = P(S(L^0 \oplus k^1))$, $T' = P(S(L^0 \oplus k^1 \oplus \Delta L^0))$, and get the value $(R^0 \oplus k^2)_l$ ($l = 1, 3, ..., 8$) from ΔL^1_l, $(P^{-1}(\Delta L^0))_l$, T, and the corresponding differential table of S-box. Insert $L^0 \oplus k^1$ and $R^0 \oplus k^2$ into the row indexed by $\Delta R^0 = T \oplus T' \oplus \Delta L^1$, ΔL^0, bits $61 \sim 64$ of $L^0 \oplus k^1$, and bits $1 \sim 4, 13 \sim 60$ of $R^0 \oplus k^2$. As 2^{184} $(\Delta L^0, \Delta L^1, L^0 \oplus k^1, (R^0 \oplus k^2)_2)$s are totally chosen and we get 2^{184} $(\Delta L^0, \Delta L^1, L^0 \oplus k^1, R^0 \oplus k^2)$s whose $(\Delta L^1, P^{-1}(\Delta L^0))$s are possible input/output pairs of the S-box layer in the second round. Furthermore, there are 2^{184} rows in the table, so we get about one $(L^0 \oplus k^1, R^0 \oplus k^2)$ in each row on average.

The complexity of the precomputation is about 2×2^{184} 2-round encryptions, equivalent to $2^{182.1}$ 15-round encryptions. The table requires about $2^{184} \times 16 = 2^{188}$ bytes of memory.

Data Collection. For $2^{122.5}$ known plaintexts, ask for the encryptions and insert the ciphertexts into a hash table indexed by the 7-th and 8-th bytes of $P^{-1}(\Delta R^{15})$. That is because by Property 1, the right half of ciphertexts must have the form

$$\Delta R^{15} = \Delta L^{14} = P(i, j_2, j_3 \oplus i, j_4 \oplus i, j_5 \oplus i, j_6 \oplus i, 0, 0).$$

By birthday paradox, we can get $2^{244} \times 2^{-16} = 2^{228}$ pairs that the 7-th and 8-th bytes of $P^{-1}(\Delta R^{15})$ are 0.

Key Recovery. In this phase, our attack is benefit from the subkey relation and the precomputation. Note that once we know $61 \sim 64$ of k^1 and bits $1 \sim 4, 13 \sim 60$ of k^2 (which are also included in the 15-th round), we can access the corresponding $(L^0 \oplus k^1, R^0 \oplus k^2)$ from Γ_1. The procedure is demonstrated as follows.

1. For $l = 1, 3, ..., 8$, guess k^{15}_l ($K_B : 61 \sim 68, 73 \sim 124$) and remove the pairs that do not satisfy $\Delta S^{15}_l = (P^{-1}(\Delta L^{15}))_l$. About $2^{228-7\times8} = 2^{172}$ pairs will be kept. From Table 1, bits $61 \sim 64$ of k^1 and bits $1 \sim 4, 13 \sim 60$ of k^2 are known.
2. For each of the remaining pairs, compute bits $61 \sim 64$ of $L^0 \oplus k^1$ and bits $1 \sim 4, 13 \sim 60$ of $R^0 \oplus k^2$, then access the value in the corresponding row in Γ_1. Insert $(\Delta L^0, \Delta R^0, L^0, R^0)$ to a hash table Γ_2 indexed by bits $1 \sim 60$ of k^1 ($K_B : 1 \sim 60$) and bits $5 \sim 12, 61 \sim 64$ of k^2 ($K_B : 69 \sim 76, 124 \sim 128$). As a result, Γ_2 has 2^{72} rows with about 2^{100} $(\Delta L^0, \Delta R^0, L^0, R^0)$s in each.
3. Guess bits $1 \sim 60$ of k^1 and bits $5 \sim 12, 61 \sim 64$ of k^2, access the corresponding row in Γ_2, and compute 2^{100} (L^2, L'^2)s and (R^2, R'^2)s by two-round

encryptions as the whole K_B is known. The memory of table Γ_2 will be freed once we finish using it.

4. From Property 1, it is clear that if a pair follows the path in Fig. 6, it has to satisfy $\Delta S_l^{14} = (P^{-1}(\Delta L^{14}))_1 \oplus (P^{-1}(\Delta L^{14}))_l$ $(l = 3, ..., 6)$ and $\Delta S_2^{14} = (P^{-1}(\Delta L^{14}))_2$. Then:

   (a) Further guess k_2^{14} $(K_R : 5 \sim 12)$, partially decrypt round 15 and round 14 to discard the pairs which do not satisfy $\Delta S_2^{14} = (P^{-1}(\Delta L^{14}))_2$. After this procedure, the number of remaining pairs is $2^{100-8} = 2^{92}$.

   (b) For $l = 3, ..., 6$, guess k_l^{14} $(k_R : 13 \sim 44)$ and keep the pairs which satisfy $\Delta S_l^{14} = (P^{-1}(\Delta L^{14}))_1 \oplus (P^{-1}(\Delta L^{14}))_l$. There are $2^{92-8\times4} = 2^{60}$ pairs being kept.

5. (a) Guess bits $45 \sim 47$ of k_R, now k_2^3, k_3^3, and k_4^3 $(K_R : 24 \sim 47)$ are known. Detect if $\Delta S_2^3 = (P^{-1}(\Delta R^2))_2$, $\Delta S_3^3 = (P^{-1}(\Delta R^2))_1 \oplus (P^{-1}(\Delta R^2))_3$, and $\Delta S_4^3 = (P^{-1}(\Delta R^2))_1 \oplus (P^{-1}(\Delta R^2))_4$. The number of remaining pairs is $2^{60-8\times3} = 2^{36}$.

   (b) For $l = 5, 6$, guess k_l^3 $(k_R : 48 \sim 63)$ and keep the pairs that satisfy $\Delta S_l^3 = (P^{-1}(\Delta R^2))_1 \oplus (P^{-1}(\Delta R^2))_l$. There are $2^{36-8\times2} = 2^{20}$ pairs being kept.

6. Guess k_1^{14} $(K_R : 125 \sim 128, 1 \sim 4)$ (now the whole k^{14} is known) and k_2^{13} $(K_R : 69 \sim 76)$, keep the pairs that meet $\Delta S_2^{13} = \Delta L_2^{13}$. The number of remaining of pairs will be $2^{20-8} = 2^{12}$.

7. Guess the rest 8 bits of k^3 $(K_R : 64 \sim 68, 77 \sim 79)$, now the whole k^3 $(K_R : 16 \sim 79)$ is known. We further guess k_2^4 $(K_R : 88 \sim 95)$ and check if there is a pair satisfy $\Delta S_2^4 = \Delta L_2^4$. If there is a pair satisfy this, then discard the key guess. Otherwise for every 219-bit key guess, exhaustively search the rest 37 bits of K_R to calculate K_A, use the relation of K_A and K_L to recover K_L, and test the resulting (K_L, K_R) by trial encryption.

Complexity. The data complexity is $2^{122.5}$ known plaintexts. In the data collecting phase, the computation of the 7-th and 8-th bytes of $P^{-1}(\Delta R^{15})$ is less

Table 1. Corresponding Bit Positions of the Subkeys in K_B and K_R

Subkey bytes	Bit positions in K_B	Subkey bytes	Bit positions in K_B	Subkey bytes	Bit positions in K_R	Subkey bytes	Bit positions in K_R
k_1^1	$1 \sim 8$	k_5^2	$97 \sim 104$	k_1^3	$16 \sim 23$	k_1^{14}	$125 \sim 4$
k_2^1	$9 \sim 16$	k_6^2	$105 \sim 112$	k_2^3	$24 \sim 31$	k_2^{14}	$5 \sim 12$
k_3^1	$17 \sim 24$	k_7^2	$113 \sim 120$	k_3^3	$32 \sim 39$	k_3^{14}	$13 \sim 20$
k_4^1	$25 \sim 32$	k_8^2	$121 \sim 128$	k_4^3	$40 \sim 47$	k_4^{14}	$21 \sim 28$
k_5^1	$33 \sim 40$	k_1^{15}	$61 \sim 68$	k_5^3	$48 \sim 55$	k_5^{14}	$29 \sim 36$
k_6^1	$41 \sim 48$	k_2^{15}	$69 \sim 76$	k_6^3	$56 \sim 63$	k_6^{14}	$37 \sim 44$
k_7^1	$49 \sim 56$	k_3^{15}	$77 \sim 84$	k_7^3	$64 \sim 71$	k_7^{14}	$45 \sim 52$
k_8^1	$57 \sim 64$	k_4^{15}	$85 \sim 92$	k_8^3	$72 \sim 79$	k_8^{14}	$53 \sim 60$
k_1^2	$65 \sim 72$	k_5^{15}	$93 \sim 100$	k_2^4	$88 \sim 95$	k_2^{13}	$69 \sim 76$
k_2^2	$73 \sim 80$	k_6^{15}	$101 \sim 108$				
k_3^2	$81 \sim 88$	k_7^{15}	$109 \sim 116$				
k_4^2	$89 \sim 96$	k_8^{15}	$117 \sim 124$				

than 2/8 one round encryption, so the complexity of computing the 7-th and 8-th bytes of $P^{-1}(\Delta R^{15})$ for all the ciphertexts is at most $2^{122.5} \times \frac{1}{4} \times \frac{1}{15} \approx 2^{116.6}$ 15-round encryptions. Below we elaborate the complexity of each step in the key-recovery phase.

1. The complexity is about $7 \times 2 \times 2^8 \times 2^{228} = 2^{240}$ one round encryptions, which is about $2^{236.1}$ encryptions.
2. This step needs about $2 \times 2^{56} \times 2^{172} = 2^{229}$ memory access and $2^{72} \times 2^{100} \times 16 = 2^{176}$ bytes of memory.
3. The complexity of this step is about $2 \times 2^{128} \times 2^{100} = 2^{229}$ two round encryptions.
4. (a) The complexity of this step is about $2 \times 2^{136} \times 2^{100} = 2^{237}$ one round encryptions, equivalent to $2^{233.1}$ encryptions.
   (b) The complexity of the each operation in this step is about one round encryption, so the complexity of is about: $2 \times \sum_{i=0}^{3}(2^{144+8\times i} \times 2^{92-8\times i} \times \frac{1}{15}) \approx 2^{235.1}$.
5. (a) The complexity of this step is about $2 \times \frac{1}{15} \times 2^{171} \times (2^{60} + 2^{52} + 2^{44}) \approx 2^{228.1}$.
   (b) The complexity of this step is about $2 \times \frac{1}{15} \times (2^{179} \times 2^{36} + 2^{187} \times 2^{28}) \approx 2^{213.1}$.
6. This step requires $2 \times 2^{203} \times 2^{20} \times \frac{1}{15} \approx 2^{220.1}$ encryptions.
7. In step 7, we expect $2^{219} \times (1 - 2^{-8})^{2^{12}} \approx 2^{196.6}$ of the key guess remained. So about $2^{196.6+37} = 2^{233.6}$ trail encryptions are required to recover the whole key. The complexity of this step is thus $2 \times 2^{219} \times [1 + (1 - 2^{-8}) + \ldots + (1 - 2^{-8})^{2^{12}}] \times \frac{1}{15} + 2^{233.6} \approx 2^{233.6}$.

As a result, the time complexity is dominated by Step 1, which is about $2^{236.1}$ 15-round encryptions.

Table 2. Summary of the Attacks on Camellia

Cipher	#Rounds	FL/FL^{-1}	Attack Type	Data	Time	Source
Camellia-128	8	×	Truncated DC	$2^{83.6}$CP	$2^{55.6}$	[10]
	9	✓	Square Attack	2^{48}CP	2^{122}	[11]
	9	×	Collision Attack	$2^{113.6}$CP	2^{121}	[19]
	9	×	Square Attack	2^{66}CP	$2^{84.8}$	[6]
	11	×	Impossible DC	2^{118}CP	2^{126}MA	[12]
	12	×	Impossible DC	$2^{116.3}$CP	$2^{116.6}$	[14]
Camellia-192	12	×	Impossible DC	2^{120} CP	2^{181}	[18]
	10	✓	Impossible DC	2^{121}CP	$2^{175.3}$	this paper
Camellia-256	10	✓	Square Attack	2^{48}CP	2^{210}	[11]
	last 11 rounds	✓	Higher Order DC	2^{93}CC	$2^{255.6}$	[7]
	11	✓	Impossible DC	2^{121}CP	$2^{206.8}$	this paper
	12	×	Linear Attack	2^{119}KP	2^{247}	[16]
	12	×	Square Attack	2^{66}CP	$2^{249.6}$	[6]
	13	×	Impossible DC	2^{120}CP	$2^{211.7}$	[12]
	15	×	Impossible DC	$2^{122.5}$KP	$2^{236.1}$	this paper

KP: known plaintext; CP: chosen plaintext; CC: chosen ciphertext; DC: differential attack

6   Conclusion

In this paper, we present several 6-round impossible differentials with FL/FL^{-1} layers in the middle, which lead to impossible differential attacks on 10-round Camellia-192 and 11-round Camellia-256. Then an impossible differential cryptanalysis of 15-round Camellia-256 without FL/FL^{-1} layers and whitening is given by carefully using the subkey relation and a precomputation table. A summary of the previous attacks and our analysis of Camellia is given in Table 2.

Acknowledgement

We are grateful to the anonymous reviewers for their valuable comments on this paper.

References

1. Aoki, K., Ichikawa, T., Kanda, M., Matsui, M., Moriai, S., Nakajima, J., Tokita, T.: Camellia: a 128-bit block cipher Suitable for Multiple Platforms-Design and Analysis. In: SAC 2000. LNCS, vol. 2012, pp. 39–56. Springer, Heidelberg (2001)
2. Aoki, K., Ichikawa, T., Kanda, M., Matsui, M., Moriai, S., Nakajima, J., Tokita, T.: Specification of Camellia-a 128-bit block cipher. version 2.0 (2001), http://info.isl.ntt.co.jp/crypt/eng/camellia/specifications.html
3. Biham, E., Shamir, A.: Differential cryptanalysis of the Data Encryption Standard. Springer, Heidelberg (1993)
4. Biham, E., Biryukov, A., Shamir, A.: Cryptanalysis of Skipjack reduced to 31 rounds using impossible differentials. In: Stern, J. (ed.) EUROCRYPT 1999. LNCS, vol. 1592, pp. 12–23. Springer, Heidelberg (1999)
5. CRYPTREC-Cryptography Research and Evaluation Committees, report, Archive (2002), http://www.cryptrec.go.jp/english/index.html
6. Duo, L., Li, C., Feng, K.: Square like attack on camellia. In: Qing, S., Imai, H., Wang, G. (eds.) ICICS 2007. LNCS, vol. 4861, pp. 269–283. Springer, Heidelberg (2007)
7. Hatano, Y., Sekine, H., Kaneko, T.: Higher Order Differential Attack of Camellia (II). In: Nyberg, K., Heys, H.M. (eds.) SAC 2002. LNCS, vol. 2595, pp. 129–146. Springer, Heidelberg (2003)
8. International Standardization of Organization (ISO), International Standard-ISO/IEC 18033-3, Information technology-Security techniques-Encryption algorithms -Part 3: Block ciphers (2005)
9. Kühn, U.: Improved cryptanalysis of MISTY1. In: Daemen, J., Rijmen, V. (eds.) FSE 2002. LNCS, vol. 2365, pp. 61–75. Springer, Heidelberg (2002)
10. Lee, S., Hong, S.H., Lee, S.-J., Lim, J.-I., Yoon, S.H.: Truncated differential cryptanalysis of Camellia. In: Kim, K.-c. (ed.) ICISC 2001. LNCS, vol. 2288, pp. 32–38. Springer, Heidelberg (2002)
11. Lei, D., Chao, L., Feng, K.: New observation on Camellia. In: Preneel, B., Tavares, S. (eds.) SAC 2005. LNCS, vol. 3897, pp. 51–64. Springer, Heidelberg (2006)
12. Lu, J.: Cryptanalysis of block ciphers. PhD Thesis, Department of Mathematics, Royal Holloway. University of London, England (2008)

13. Lu, J., Kim, J.-S., Keller, N., Dunkelman, O.: Improving the efficiency of impossible differential cryptanalysis of reduced Camellia and MISTY1. In: Malkin, T.G. (ed.) CT-RSA 2008. LNCS, vol. 4964, pp. 370–386. Springer, Heidelberg (2008)
14. Mala, H., Shakiba, M., Dakhilalian, M., Bagherikaram, G.: New results on impossible differential cryptanalysis of reduced–round Camellia–128. In: Jacobson Jr., M.J., Rijmen, V., Safavi-Naini, R. (eds.) SAC 2009. LNCS, vol. 5867, pp. 281–294. Springer, Heidelberg (2009)
15. NESSIE-New European Schemes for Signatures, Integrity, and Encryption, final report of European project IST-1999-12324. Archive (1999), https://www.cosic.esat.kuleuven.be/nessie/Bookv015.pdf
16. Shirai, T.: Differential, linear, boomerang and rectangle Cryptanalysis of Reduced-Round Camellia. In: Proceedings of the Third NESSIE Workshop, Munich, Germany, (November 6-7, 2002)
17. Sugita, M., Kobara, K., Imai, H.: Security of reduced version of the block cipher Camellia against truncated and impossible differential cryptanalysis. In: Boyd, C. (ed.) ASIACRYPT 2001. LNCS, vol. 2248, pp. 193–207. Springer, Heidelberg (2001)
18. Wu, W., Zhang, W., Feng, D.: Impossible differential cryptanalysis of Reduced-Round ARIA and Camellia. Journal of Computer Science and Technology 22(3), 449–456 (2007)
19. Wu, W., Feng, D., Chen, H.: Collision attack and pseudorandomness of reduced-round Camellia. In: Handschuh, H., Hasan, M.A. (eds.) SAC 2004. LNCS, vol. 3357, pp. 252–266. Springer, Heidelberg (2004)
20. Wu, W., Zhang, L., Zhang, W.: Improved impossible differential cryptanalysis of reduced-round Camellia. In: Avanzi, R., Keliher, L., Sica, F. (eds.) SAC 2008. LNCS, vol. 5381, pp. 442–456. Springer, Heidelberg (2009)

A 8-Round Impossible Differential without FL/FL^{-1} Layer

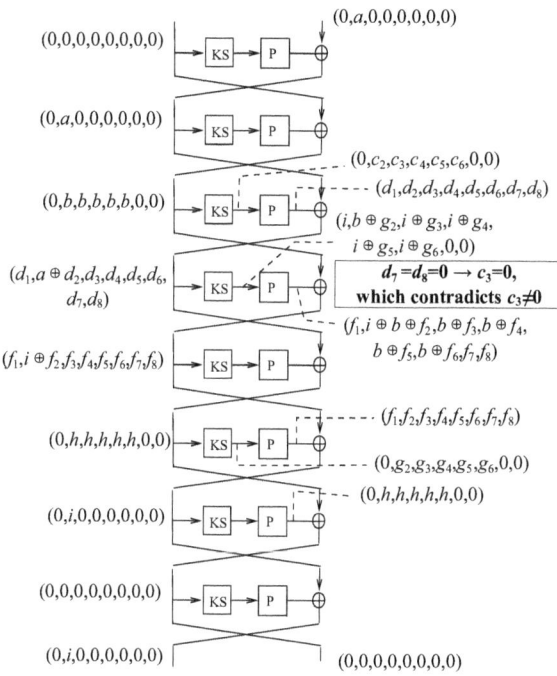

Fig. 7. 8-round Impossible Differential without FL/FL^{-1} Layer (from [18])

Results on the Immunity of Boolean Functions against Probabilistic Algebraic Attacks*

Meicheng Liu[1], Dongdai Lin[1], and Dingyi Pei[2]

[1] The State Key Laboratory of Information Security, Institute of Software,
Chinese Academy of Sciences, Beijing 100190, China
[2] College of Mathematics and Information Sciences, Guangzhou University,
Guangzhou 510006, China
meicheng.liu@gmail.com, ddlin@is.iscas.ac.cn, gztcdpei@scut.edu.cn

Abstract. In this paper, we study the immunity of Boolean functions against probabilistic algebraic attacks. We first show that there are functions, using as filters in a linear feedback shift register based nonlinear filter generator, such that probabilistic algebraic attacks outperform deterministic ones. Then we introduce two notions, algebraic immunity distance and k-error algebraic immunity, to measure the ability of Boolean functions resistant to probabilistic algebraic attacks. We analyze both lower and upper bounds on algebraic immunity distance, and also present the relations among algebraic immunity distance, k-error algebraic immunity, algebraic immunity and high order nonlinearity.

Keywords: Boolean functions, algebraic attacks, algebraic immunity, algebraic immunity distance, k-error algebraic immunity, high order nonlinearity.

1 Introduction

Algebraic attacks, which cleverly use over-defined systems of multi-variable non-linear equations to recover the secret key, have been regarded as a great threat against stream ciphers based on linear feedback shift register (LFSR). A new cryptographic property known as algebraic immunity (AI) is used to scale the ability of Boolean functions to resist algebraic attacks. The AI of a function is defined by the minimum degree of nonzero annihilators of the function or its complement. There are a number of literatures, e.g. [2,5,9,10,11,14,15,16,23], investigating algebraic immunity but few referring to the immunity of Boolean functions against probabilistic algebraic attacks. As long ago as 2003, Courtois and Meier [7] described the probabilistic scenario of algebraic attacks:

S4. There exists a nonzero function g of low degree such that gf can be approximated by a function of low degree with probability $1 - \varepsilon$.

---

* This work was in part supported by the National 973 Program of China under Grant 2011CB302400, the National Natural Science Foundation of China under Grants 10971246 and 60970152, the Grand Project of Institute of Software under Grant YOCX285056 and the CAS Special Grant for Postgraduate Research, Innovation and Practice.

U. Parampalli and P. Hawkes (Eds.): ACISP 2011, LNCS 6812, pp. 34–46, 2011.

In 2005, Braeken and Preneel [3] generalized S4 to the two scenarios:

S4a. There exists a nonzero function g of low degree such that $gf = g$ on $\{x \mid f(x) = 0\}$ with probability $1 - \varepsilon$.

S4b. There exists a nonzero function g of low degree such that $gf = 0$ on $\{x \mid f(x) = 1\}$ with probability $1 - \varepsilon$.

The probability for the scenario S4a is equal to $p = 1 - \frac{d(gf,g)}{2^n - wt(f)}$, and equal to $p = 1 - \frac{d(gf,0)}{wt(f)}$ for the scenario S4b. Taking g equal to a function of degree r and with Hamming weight 2^{n-r}, the minimum between $d(gf, f)$ and $d(gf, 0)$ is smaller than or equal to the average 2^{n-r-1}. Then $p \geq 1 - 2^{-r}$ for a balanced function. In order to compute the probability for applying scenarios S4a and S4b, by means of the Walsh spectrum the authors derived a formula for measuring the distance, denoted by X in their paper, between a given function and a function with annihilator equal to an indicator of a flat of small co-dimension (i.e. a product of small number of affine functions). The authors also gave an example of a function on 6 variables for which probabilistic algebraic attacks perform better than classical algebraic attacks if the length of the LFSR is less than or equal to 18.

Later, Pometun [22] introduced the notion of the high order partial nonlinearity as a measure of the ability of Boolean functions resistant to probabilistic algebraic attacks, and showed that the r-th order partial nonlinearity is always less than or equal to the r-th order nonlinearity for a balanced function. The author also constructed a class of vulnerable functions against probabilistic algebraic attacks:

$$F(x) = x_1 x_2 + x_1 x_2 \cdots x_n + f(x)(x_1 x_2 + 1).$$

Since $x_1 x_2 F(x) = x_1 x_2 + x_1 x_2 \cdots x_n$, we know that $x_1 x_2 F(x) = x_1 x_2$ holds with very high probability. Nevertheless, the function $F + 1$ admits annihilators of degree 3, such as $x_1 x_2 (x_3 + 1)$, which implies that the function is also vulnerable to classical algebraic attacks.

Recently, Pasalic [21] claimed that from time complexity point of view deterministic algebraic attacks are in general more efficient than probabilistic ones for practical sizes L (e.g. $L = 256$) of LFSR in the context of their application to certain LFSR-based stream ciphers under an assumption that the minimum distance of the code derived by shortening Reed-Muller code which depends on the filter function meets the Gilbert-Varshamov (GV) bound. Nevertheless, one should still verify whether the structure of the function itself allows a low-degree approximation that is satisfied with high probability. This raises the question of whether there exist Boolean functions using as filters in an LFSR-based nonlinear filter generator for which probabilistic algebraic attacks outperform deterministic ones for practical sizes L.

In this paper, we consider this question and further research the immunity of Boolean functions against probabilistic algebraic attacks. At the beginning, we discuss the complexities and validity of probabilistic algebraic attacks, and show two examples of functions that probabilistic algebraic attacks outperform deterministic ones for practical sizes L. Furthermore, we introduce the notions of algebraic immunity distance and k-error algebraic immunity. The algebraic immunity distance of r-th order is the minimum distance between a given function and all functions with AI at most r. It is always hold that the algebraic immunity distance is less than or equal to the aforementioned distance X. For the case $r = 1$, the algebraic immunity distance is the minimum of the distances X of the function and its complement. The algebraic immunity distance is similar to but different from high order partial nonlinearity. For a balanced function, the former is half of the latter. In this context, some upper bounds on algebraic immunity distance can be derived from previous results. A lower bound on algebraic immunity distance of functions with designated AI was implied by Carlet [4]. The bound is confirmed to be tight for balanced function in this paper. We also present other new bounds, including both upper and lower bounds, on algebraic immunity distance. The bounds reveal the relations among algebraic immunity distance, algebraic immunity and high order nonlinearity. The notion of k-error algebraic immunity is a dual concept of algebraic immunity distance. The former relates to minimum degree for applying PAA with high probability while the latter relates to maximum probability for applying PAA with small degree. Several properties of k-error algebraic immunity is then obtained by its relation with algebraic immunity distance. Lastly a sufficient and necessary condition for Boolean functions to achieve designated k-error algebraic immunity is described.

2   Preliminary

Let $\mathbb{F}_2$ be the binary field. An n-variable Boolean function is a mapping from $\mathbb{F}_2^n$ into $\mathbb{F}_2$. Denote by $\mathbf{B}_n$ the set of all n-variable Boolean functions. An n-variable Boolean function can be uniquely represented as a truth table of length 2^n, $f = [f(0,0,\cdots,0), f(1,0,\cdots,0),\cdots, f(1,1,\cdots,1)]$. Denote $1_f = \{x \mid f(x) = 1\}$. The number of ones in the truth table of f is called the Hamming weight of f, denoted by $\mathrm{wt}(f)$. If $\mathrm{wt}(f) = 2^{n-1}$, then f is called balanced. The number of $x \in \mathbb{F}_2^n$ at which $f(x) \neq g(x)$ is called the Hamming distance between f and g, denoted by $\mathrm{d}(f,g)$. It is well known that $\mathrm{d}(f,g) = \mathrm{wt}(f+g)$.

An n-variable Boolean function can also be uniquely represented as a multivariate polynomial over $\mathbb{F}_2$: $f(x) = \sum_{c \in \mathbb{F}_2^n} a_c x^c$, $x^c = x_1^{c_1} x_2^{c_2} \cdots x_n^{c_n}$, $a_c \in \mathbb{F}_2$, called algebraic normal form (ANF). The algebraic degree of f, denoted by $\deg(f)$, is defined as $\max\{\mathrm{wt}(c) \mid a_c \neq 0\}$.

The minimum distance between f and Boolean functions with degree at most r is called r-th order nonlinearity of f, denoted by $\mathrm{nl}_r(f)$. That is $\mathrm{nl}_r(f) = \min\{\mathrm{d}(f,g) \mid \deg(g) \leq r\}$. It is called nonlinearity, denoted by $\mathrm{nl}(f)$, if $r = 1$.

Definition 1. *[20] The algebraic immunity of the function f, denoted by $\mathrm{AI}(f)$, is defined as*

$$\mathrm{AI}(f) = \min\{\deg(g) \mid gf = 0 \ or \ g(f+1) = 0, \ g \neq 0\}.$$

3   Probabilistic Algebraic Attacks

This section mainly focuses on the complexities of probabilistic algebraic attacks on an LFSR-based nonlinear filter generator.

3.1   Time Complexity

Let p be the probability for S4a or S4b. Then an overdetermined system of nonlinear equations with degree r is obtained where each equation holds with probability p. In the affine case probabilistic algebraic attacks relate to the (fast) correlation attacks [3], so we always consider the nonlinear case here. One can use the linearization algorithm to solve the system, where $R = \sum_{i=0}^{r} \binom{L}{i}$ equations are used and hold with probability p^R. Then the time complexity of probabilistic algebraic attacks (PAA) is $p^{-R} R^w$ ($w \approx 2.807$ is the exponent of the Gaussian reduction), compared with T^w of classical algebraic attacks (AA) ($T = \sum_{i=0}^{\mathrm{AI}(f)} \binom{L}{i}$), and $2ED \log_2 D$ [12] of fast algebraic attacks (FAA) ($E = \sum_{i=0}^{e} \binom{L}{i}$ and $D = \sum_{i=0}^{d} \binom{L}{i}$ with $e = \deg(g) < d = \deg(gf)$).

Our work shows that there are functions for which probabilistic algebraic attacks outperform deterministic ones for practical sizes L of LFSR. An example is the function

$$F(x) = f(x_1, \cdots, x_{n-1}) + x_n$$

where $f(x) = 1$ if $\mathrm{wt}(x) < t - 1$ ($t \leq n/2$) and otherwise $f(x) = 0$. First, the function F is balanced and has algebraic immunity t [16, Lemma 3.5]. Second, the function $f+1$ admits no annihilator of degree less than $n-t+1$ [16, Theorem 3.1], so any nonzero multiple h of F has degree equal to or more than $n - t + 1$, then $d \geq n-t+1$ for any $e \geq 1$. Third, taking $g = x_1 \cdots x_{r-1}(x_n + 1)$ gives $\mathrm{wt}(gF) = \sum_{i=0}^{t-r-1} \binom{n-r}{i}$ (since gF takes values 1 at and only at $(1, \cdots, 1, x_r, \cdots, x_{n-1}, 0)$ with $\mathrm{wt}(x_r, \cdots, x_{n-1}) \leq t - r - 1$), and therefore $p = 1 - \sum_{i=0}^{t-r-1} \binom{n-r}{i}/2^{n-1}$ for S4b. Fixing t, the probability p becomes closer to 1 as n increases. For reasonable n and t, probabilistic algebraic attacks will outperform deterministic ones, even including fast algebraic attacks (e.g., see Table 1). For instance, when $L = 256$, $n = 29$ and $t = 8$, we can calculate $\mathrm{wt}(gF) = 101584 = 2^{16.6}$ and $p = 1 - 2^{-11.4} = 0.9996$ for $r = 2$, then the time complexity of the PAA is $2^{60.1}$ while the AA requires $2^{136.4}$ operations and the FAA runs in $2^{120.6}$ or larger. Nevertheless, the function F has nonlinearity equal to $\mathrm{nl}(F) = 2 \sum_{i=0}^{t-2} \binom{n-1}{i}$ according to [16, Lemma 3.5], which is low.

Table 1. Time complexities of PAA ($r = 2$), AA and FAA for $L = 256$

(n,t)	(18,4)	(20,5)	(22,6)	(24,7)	(26,7)	(28,8)	(29,8)	(30,9)	(32,10)
PAA	$2^{48.3}$	$2^{57.7}$	$2^{72.7}$	$2^{93.7}$	$2^{60.4}$	$2^{71.7}$	$2^{60.1}$	$2^{86.3}$	$2^{104.1}$
AA	$2^{76.9}$	$2^{92.8}$	2^{108}	$2^{122.4}$	$2^{122.4}$	$2^{136.4}$	$2^{136.4}$	$2^{149.8}$	$2^{162.8}$
FAA $\geq$	$2^{94.6}$	$2^{98.5}$	$2^{102.4}$	$2^{106.2}$	$2^{113.6}$	$2^{117.1}$	$2^{120.6}$	$2^{120.6}$	2^{124}

Another example with a little higher nonlinearity is the function

$$F(x) = \begin{cases} 1, & \text{if } \operatorname{wt}(x) < t \\ (x_1 x_2 \cdots x_r + 1)b(x_{r+1}, \cdots, x_n), & \text{if } t \leq \operatorname{wt}(x) \leq n - t \\ 0, & \text{if } \operatorname{wt}(x) > n - t \end{cases}$$

where b, if any, is a function such that $\operatorname{wt}(F) = 2^{n-1}$. The function F has algebraic immunity equal to or more than t [16, Corollary 4.2] and admits the function $g = x_1 x_2 \cdots x_r$ such that $\operatorname{wt}(gF) = \sum_{i=0}^{t-r-1} \binom{n-r}{i}$. Again, for reasonable n and t probabilistic algebraic attacks will outperform deterministic ones. Despite of low nonlinearities resulting in vulnerabilities to the affine case of the PAA, the quadratic version of the PAA might performs better than the affine one. For example, for $L = 256$, $n = 29$, $t = 8$ and $r = 2$, the quadratic one requires $2^{60.1}$ operations while the affine one runs in $p^{-(L+1)}(L+1)^w \approx 2^{129.3}$ with $p \approx 0.75$.

3.2   Data Complexity[1]

Assume that there are N equations each of which holds with probability p, and that every R equations of these equations are independent. Then $\binom{N}{R}$ systems are established. Now, we need p^{-R} systems of equations to mount PAA. Note that $\binom{N}{R} \geq (\frac{N}{R})^R \geq p^{-R}$ for $N \geq p^{-1}R$. Then the amount of keystream used in the PAA is at most $p^{-1}R$, which is much smaller than T of the AA and $D + E$ of the FAA.

3.3   Validity

Without loss of generality, we suppose that f coincides with S4b. Denote by $\Pr[A]$ the probability of an event A and let

$$\delta_f(g) = \Pr[g(x) = 0 \mid x \in 1_f] - \Pr[g(x) = 0],$$

where $1_f = \{x \mid f(x) = 1\}$. If $\delta_f(g) \approx 0$, then solving the equation systems of $g(x) = 0$ ($x \in 1_f$) is almost equivalent to solve the equation systems of $g(x) = 0$, and therefore probabilistic algebraic attacks cannot be applied. Hence, probabilistic algebraic attacks make necessary that $\delta_f(g) \not\approx 0$. The value $\delta_f(g)$ reflects the validity of probabilistic algebraic attacks on f using the function g. The smaller $\delta_f(g)$ is, the worse probabilistic algebraic attacks behave; but not vice versa.

Theorem 1. *Let f be an n-variable balanced Boolean function. Then*

$$\max\{|\delta_f(g)| \mid \deg(g) \leq r\} = \frac{2^{n-1} - \operatorname{nl}_r(f)}{2^n}.$$

---

[1] This section was suggested by Frederik Armknecht and an anonymous referee.

Proof. Since the function f is balanced, we have

$$\delta_f(g) = (1 - \frac{\mathrm{wt}(gf)}{2^{n-1}}) - (1 - \frac{\mathrm{wt}(g)}{2^n}) = \frac{\mathrm{wt}(g) - 2\mathrm{wt}(gf)}{2^n}.$$

Because $\mathrm{wt}(g) + \mathrm{wt}(f) = 2\,\mathrm{wt}(gf) + \mathrm{d}(f,g)$, we obtain

$$\delta_f(g) = -\frac{2^{n-1} - \mathrm{d}(f,g)}{2^n}.$$

Therefore

$$\max\{|\delta_f(g)| \mid \deg(g) \le r\} = \frac{2^{n-1} - \mathrm{nl}_r(f)}{2^n}.$$

Note that $\mathrm{nl}_r(f+1) = \mathrm{nl}_r(f)$. Theorem 1 also applies to $f+1$. Therefore the scenarios S4a and S4b have both been considered in Theorem 1. The theorem shows that if the function f is balanced and of good r-th order nonlinearity, then f may be robust against r-th order probabilistic algebraic attacks to some extent.

4 Algebraic Immunity Distance

Hereinafter, S4a and S4b are included into the scenario:

S4′. There exists a nonzero function g of low algebraic immunity such that $f = g$ with probability $1 - \varepsilon$.

In this section, we consider the set of the n-variable Boolean functions with algebraic immunity $\le r$, and discuss the minimum distance between a given function and that set.

Denote by $\mathcal{AI}_r = \{f \in \mathbf{B}_n \mid \mathrm{AI}(f) \le r\}$ the set of the n-variable Boolean functions with algebraic immunity $\le r$. By convention $\mathcal{AI}_0 = \{0, 1\}$.

Proposition 2. *Let* $r \ge 1$. *Then* $\mathcal{AI}_r = \{gh + c \mid g, h \in \mathbf{B}_n, c \in \mathbb{F}_2, 1 \le \deg(g) \le r\}$.

Proof. Denote by $\mathcal{A}$ the right part of the equality. Let us prove $\mathcal{A} \subset \mathcal{AI}_r$ first. For $f \in \mathcal{A}$, there exist $g, h \in \mathbf{B}_n$ with $1 \le \deg(g) \le r$ such that $f = gh + c$. Then $(g+1)(f+c) = (g+1)gh = 0$ where $g+1 \ne 0$, so $\mathrm{AI}(f) \le r$. Hence $f \in \mathcal{AI}_r$. Next we check that $\mathcal{AI}_r \subset \mathcal{A}$. It is clear that $\mathcal{AI}_0 \subset \mathcal{A}$. For $f \in \mathcal{AI}_r \backslash \mathcal{AI}_0$, there exists a nonzero function g with degree $\le r$ such that $g(f+c) = 0$ for $c \in \mathbb{F}_2$. Then $f = (g+1)(f+c) + c$. Since $f \notin \mathcal{AI}_0$, we have $g \ne 1$. Therefore $1 \le \deg(g) \le r$. Then $f \in \mathcal{A}$. Hence $\mathcal{AI}_r = \mathcal{A}$.

It is significant to study the set $\mathcal{AI}_r$, since its complement $\mathbf{B}_n \backslash \mathcal{AI}_r$ contains all the functions with algebraic immunity $\ge r+1$. Some results for the case $r = 1$ was presented by Tu and Deng [24].

Now we introduce the notion of algebraic immunity distance.

Definition 2. *The minimum distance between the function f and Boolean functions with algebraic immunity $\leq r$ is called the r-th order algebraic immunity distance, denoted by $\mathrm{dai}_r(f)$, i.e.,*

$$\mathrm{dai}_r(f) = \mathrm{d}(f, \mathcal{AI}_r) = \min\{\mathrm{d}(f,g) \mid \mathrm{AI}(g) \leq r\}.$$

Remark 1. *The algebraic immunity distance is always less than or equal to the distance X [3] between a given function and functions with annihilator equal to a product of small number of affine functions. For the case $r = 1$, the algebraic immunity distance is the minimum between the distance X of the function and that of its complement.*

Remark 2. *The r-th order algebraic immunity distance is similar to but not the same as the r-th order partial nonlinearity [22] which is given by*

$$\mathrm{nlp}_r(f) = \min\{2^n \Pr[g \neq f | f = c] \mid c \in \mathbb{F}_2, 1 \leq \deg(g) \leq r\}.$$

The algebraic immunity distance implicitly reflects the maximum probability for applying probabilistic algebraic attacks while the partial nonlinearity explicitly describes it. The latter relates to the Hamming weight and is therefore difficult to be analyzed.

Proposition 3. $\mathrm{dai}_r(f) = \min\{\mathrm{d}(gf, 0), \mathrm{d}(gf, g) \mid 1 \leq \deg(g) \leq r\}.$

Proof. By Proposition 2, we have $\mathcal{AI}_r = \{gh, gh + 1 \mid 1 \leq \deg(g) \leq r\}$. Then

$$\begin{aligned}
\mathrm{dai}_r(f) &= \mathrm{d}(f, \mathcal{AI}_r) \\
&= \min\{\mathrm{d}(f, gh), \mathrm{d}(f, gh + 1) \mid 1 \leq \deg(g) \leq r, g, h \in \mathbf{B}_n\} \\
&\leq \min\{\mathrm{d}(f, gf), \mathrm{d}(f + 1, g(f + 1)) \mid 1 \leq \deg(g) \leq r\}. \quad (1)
\end{aligned}$$

It is clear that $\mathrm{d}(f, gh) = \mathrm{wt}(f + gh) \geq \mathrm{wt}((g + 1)(f + gh)) = \mathrm{wt}(gf + f) = \mathrm{d}(f, gf)$. Similarly, $\mathrm{d}(f + 1, gh) \geq \mathrm{d}(f + 1, g(f + 1))$. Therefore

$$\mathrm{dai}_r(f) \geq \min\{\mathrm{d}(f, gf), \mathrm{d}(f + 1, g(f + 1)) \mid 1 \leq \deg(g) \leq r\}. \quad (2)$$

By (1) and (2) it follows that

$$\begin{aligned}
\mathrm{dai}_r(f) &= \min\{\mathrm{d}(f, gf), \mathrm{d}(f + 1, g(f + 1)) \mid 1 \leq \deg(g) \leq r\} \\
&= \min\{\mathrm{d}((g + 1)f, 0), \mathrm{d}((g + 1)f, g + 1) \mid 1 \leq \deg(g) \leq r\} \\
&= \min\{\mathrm{d}(gf, 0), \mathrm{d}(gf, g) \mid 1 \leq \deg(g) \leq r\}.
\end{aligned}$$

Remark 3. *The above theorem shows that $\mathrm{dai}_r(f) = \frac{1}{2}\mathrm{nlp}_r(f)$ for balanced function f.*

4.1   Bounds on Algebraic Immunity Distance

Braeken and Preneel [3] proved that $X \leq 2^{n-r-1}$ by Eq.(3) and Eq.(4) in their paper. Then $\mathrm{dai}_r(f) \leq 2^{n-r-1}$ since $\mathrm{dai}_r(f) \leq X$. This can also be explained by the fact that taking g equal to a function with degree r and with Hamming weight 2^{n-r} gives that $\mathrm{dai}_r(f) \leq \min\{\mathrm{d}(gf,0), \mathrm{d}(gf,g)\} \leq \frac{1}{2}[\mathrm{d}(gf,0) + \mathrm{d}(gf,g)] = \frac{1}{2}\mathrm{wt}(g) = 2^{n-r-1}$. In [22], Pometun observed that for balanced f, $\mathrm{nlp}_r(f) \leq \mathrm{nl}_r(f)$. By Remark 3 it follows that $\mathrm{dai}_r(f) \leq \frac{1}{2}\mathrm{nl}_r(f)$ for a balanced function. Next we present some new results on both upper and lower bounds on algebraic immunity distance.

Theorem 4. *Let* $f \in \mathbf{B}_n$. *If* $\mathrm{nl}_r(f) = \min\{\mathrm{d}(g,f) \mid 1 \leq \deg(g) \leq r\}$, *then* $\mathrm{dai}_r(f) \leq \frac{1}{2}\mathrm{nl}_r(f)$.

Proof. Since $\mathrm{d}(g,f) = \mathrm{wt}((g+1)f) + \mathrm{wt}(g(f+1))$, we have

$$\mathrm{dai}_r(f) = \min\{\mathrm{wt}(gf), \mathrm{wt}(g(f+1)) \mid 1 \leq \deg(g) \leq r\}$$
$$= \min\{\mathrm{wt}((g+1)f), \mathrm{wt}(g(f+1)) \mid 1 \leq \deg(g) \leq r\}$$
$$\leq \min\{\frac{\mathrm{d}(g,f)}{2} \mid 1 \leq \deg(g) \leq r\}$$
$$= \frac{\mathrm{nl}_r(f)}{2}.$$

Note that for a balanced function it always holds that $\mathrm{nl}_r(f) = \min\{\mathrm{d}(g,f) \mid 1 \leq \deg(g) \leq r\}$. Therefore the result of [22] is a special case of Theorem 4.

It is well known that any Boolean function h with $\mathrm{AI} > r$ has Hamming weight $\sum_{i=0}^r \binom{n}{i} \leq \mathrm{wt}(h) \leq 2^n - \sum_{i=0}^r \binom{n}{i}$ [10]. In other words, if $\min\{\mathrm{wt}(h), \mathrm{wt}(h+1)\} < \sum_{i=0}^r \binom{n}{i}$ then the function h has $\mathrm{AI} \leq r$. Therefore Theorem 5 follows.

Theorem 5. *Let* $f \in \mathbf{B}_n$ *and* $\mathrm{wt}_{\min}(f) = \min\{\mathrm{wt}(f), \mathrm{wt}(f+1)\}$. *Then*

$$\mathrm{dai}_r(f) \leq \mathrm{wt}_{\min}(f) - \sum_{i=0}^r \binom{n}{i} + 1.$$

Proof. Without loss of generality, we assume that $\mathrm{wt}_{\min}(f) = \mathrm{wt}(f)$. Let h be a function such that $1_h \subset 1_f$ and $\mathrm{wt}(h) = \sum_{i=0}^r \binom{n}{i} - 1$. Then $\mathrm{AI}(h) \leq r$ and $\mathrm{d}(f,h) = \mathrm{wt}(f) - \mathrm{wt}(h) = \mathrm{wt}(f) - \sum_{i=0}^r \binom{n}{i} + 1$, showing that $\mathrm{dai}_r(f) \leq \mathrm{wt}_{\min}(f) - \sum_{i=0}^r \binom{n}{i} + 1$.

Corollary 6. *Let* $n > 1$ *be an odd integer and* $f \in \mathbf{B}_n$. *Then* $\mathrm{dai}_{\frac{n-1}{2}}(f) \leq 1$.

Now we discuss the lower bounds on algebraic immunity distance. The result of Carlet [4] implies a lower bound on algebraic immunity distance of functions with designated AI.

Lemma 7. *[4, Proposition 5] Let* $f, g \in \mathbf{B}_n$ *and* $\deg(g) = r$. *Then* $\mathrm{wt}(gf) \geq \sum_{i=0}^{\mathrm{AI}(f)-r-1} \binom{n-r}{i}$.

The two classes of balanced functions constructed in Section 3.1 both admit a function g with degree r such that $\mathrm{wt}(gf) = \sum_{i=0}^{\mathrm{AI}(f)-r-1} \binom{n-r}{i}$. Note that Lemma 7 also applies to $f + 1$. Then Theorem 8 follows.

Theorem 8. *Let* $f \in \mathbf{B}_n$. *Then* $\mathrm{dai}_r(f) \geq \sum_{i=0}^{\mathrm{AI}(f)-r-1} \binom{n-r}{i}$ *and there exist balanced functions achieving the bound.*

Based on computation experiments, Pometun [22] stated that if the second order partial nonlinearity of Boolean function f equals $\mathrm{nlp}_2(f) = 2$, then its algebraic immunity $\mathrm{AI}(f) \leq 3$. (In fact, this can be explained by the fact that if $\mathrm{dai}_r(f) \leq 1$ then $\mathrm{AI}(f) \leq r+1$.) Further Pometun conjectured that there exists a connection between algebraic immunity and partial nonlinearity. Note that $\mathrm{nlp}_r(f) = 2\,\mathrm{dai}_r(f)$ for a balanced function. From cryptographic viewpoint it seems that the above results is the answer to the problem.

Theorem 9. *Let* $f \in \mathbf{B}_n$ *and* $\mathrm{wt}_{\max}(f) = \max\{\mathrm{wt}(f), \mathrm{wt}(f+1)\}$. *Then*

$$\mathrm{dai}_r(f) \geq 2^{n-r-1} + \frac{1}{2}\,\mathrm{nl}_r(f) - \frac{1}{2}\,\mathrm{wt}_{\max}(f).$$

Proof. Let g be a function of degree d, $1 \leq d \leq r$. Then $\mathrm{wt}(g) \geq 2^{n-d} \geq 2^{n-r}$. Since $\mathrm{wt}(gf) = \frac{1}{2}[\mathrm{wt}(g) + \mathrm{wt}(f) - d(g, f)]$, we have

$$\mathrm{wt}(gf) = \frac{1}{2}[\mathrm{wt}(g) + d(g+1, f) - \mathrm{wt}(f+1)]$$
$$\geq \frac{1}{2}[2^{n-r} + \mathrm{nl}_r(f) - \mathrm{wt}(f+1)]. \qquad (3)$$

Taking $f + 1$ in place of f, we know

$$\mathrm{wt}(g(f+1)) \geq \frac{1}{2}[2^{n-r} + \mathrm{nl}_r(f) - \mathrm{wt}(f)]. \qquad (4)$$

By Proposition 3 and from (3) and (4) we obtain

$$\mathrm{dai}_r(f) \geq 2^{n-r-1} + \frac{1}{2}\,\mathrm{nl}_r(f) - \frac{1}{2}\,\mathrm{wt}_{\max}(f).$$

For large r, the bound of Theorem 9 may be negative. However, we only need consider small r in practice, for example, $r = 2$. Theorem 9 states that balanced functions is optimal among the functions of the same r-th order nonlinearity. This coincides with the viewpoint that balancedness is an important property in cryptography for Boolean functions. The theorem also shows that if a balanced function has high r-th order nonlinearity, then the r-th order algebraic immunity distance of the function is not bad. Again, it states that a balanced function of good r-th order nonlinearity can avoid the scenario S4′ to some extent.

From Theorem 4 and Theorem 9, we obtain that $\mathrm{dai}_1(f) = \frac{1}{2}\,\mathrm{nl}(f)$ for balanced function, showing that in the affine case there is no better approximation for probabilistic algebraic attacks than the correlation attacks. This was also observed by Braeken and Preneel [3].

Corollary 10. *Let* $f \in \mathbf{B}_n$. *If* $\mathrm{nl}_k(f) > \mathrm{wt}_{\max}(f) - 2^{n-k}$, *then* $\mathrm{dai}_r(f) > 2^{n-r-1} - 2^{n-k-1}$ *for* $1 \leq r \leq k$ *and* $\mathrm{AI}(f) > k$.

Since Bent functions have Hamming weight $2^{n-1} \pm 2^{\frac{n}{2}-1}$ and nonlinearity $2^{n-1} - 2^{\frac{n}{2}-1}$, by Corollary 10 we obtain the result proven by Tu and Deng [24]: Bent functions with $n \geq 4$ have algebraic immunity greater than 1.

Corollary 11. *Let $f \in \mathbf{B}_n$. If $\mathrm{AI}(f) \leq k$, then $\mathrm{nl}_k(f) \leq \mathrm{wt}_{\max}(f) - 2^{n-k}$ and therefore $\mathrm{nl}_k(f) \leq 2^{n-1} - 2^{n-k-1}$.*

Proof. The first half part is clear. Then

$$\mathrm{nl}_k(f) \leq \min\{\mathrm{wt}_{\max}(f) - 2^{n-k}, \mathrm{wt}_{\min}(f)\}$$

$$\leq \frac{1}{2}[\mathrm{wt}_{\max}(f) - 2^{n-k} + \mathrm{wt}_{\min}(f)] = 2^{n-1} - 2^{n-k-1}.$$

For instance, $\mathrm{nl}(f) \leq 2^{n-2}$ when $\mathrm{AI}(f) \leq 1$ [24], and $\mathrm{nl}_2(f) \leq 3 \cdot 2^{n-3}$ when $\mathrm{AI}(f) \leq 2$.

Some results on the lower bound on the high order nonlinearity of Boolean functions with designated algebraic immunity was presented by Carlet [4], Mesnager [19] and Lobanov [17]. To the best of our knowledge, this is the first time that a new upper bound is obtained.

5 k-Error Algebraic Immunity

If a function has low algebraic immunity distance of small order, then the function would be vulnerable to probabilistic algebraic attacks. The lower bound on algebraic immunity distance gives an upper bound on the probability for applying the attacks. A high probability relates to sufficiently small distance. Fixing the distance, there is a lower bound on the degree for applying the attacks. In this section, we consider this lower bound, i.e., the minimum AI of the functions having a small Hamming distance to a given function. This leads to the notion of k-error algebraic immunity.

Definition 3. *Let $k \geq 0$ be an integer and $f \in \mathbf{B}_n$. The k-error algebraic immunity of the function f is defined as*

$$\mathrm{AI}^k(f) = \min\{\mathrm{AI}(f + \varepsilon) \mid \mathrm{wt}(\varepsilon) \leq k, \varepsilon \in \mathbf{B}_n\}.$$

The new notion of k-error algebraic immunity generalizes the notion of algebraic immunity which is exactly $\mathrm{AI}^0(f)$, and also generalizes the notion of the extended algebraic immunity (EAI) proposed by Zhang [25] which is a special case of $\mathrm{AI}^1(f)$ (since for extended algebraic immunity the function ε only takes over the two functions 0 and $(x_1+1)(x_2+1)\cdots(x_n+1)$). In brief, we have $\mathrm{AI}^0(f) = \mathrm{AI}(f)$ and $\mathrm{AI}^1(f) \leq \mathrm{EAI}(f)$.

By Definition 2 and Definition 3, we can obtain the following result.

Corollary 12. *Let $f \in \mathbf{B}_n$. Then $\mathrm{dai}_r(f) = \min\{k \mid \mathrm{AI}^k(f) \leq r\}$.*

Proof. Let $k_{\min} = \min\{k \mid \mathrm{AI}^k(f) \leq r\}$. It is clear that $\mathrm{dai}_r(f) \leq k_{\min}$. On the other hand, there is a function $h \in \mathcal{AI}_r$ such that $\mathrm{d}(f, h) = \mathrm{dai}_r(f)$. Then $\mathrm{AI}^{\mathrm{dai}_r(f)}(f) \leq \mathrm{AI}(h) \leq r$, showing that $k_{\min} \leq \mathrm{dai}_r(f)$. Hence $k_{\min} = \mathrm{dai}_r(f)$.

The duality between algebraic immunity distance and k-error algebraic immunity indicates that $\mathrm{AI}^{\mathrm{dai}_r(f)}(f) \leq r$ and $\mathrm{dai}_{\mathrm{AI}^k(f)}(f) \leq k$.

Corollary 13. *Let* $f \in \mathbf{B}_n$ *and* $k < \sum_{i=0}^r \binom{n-\mathrm{AI}(f)+r+1}{i}$. *Then* $\mathrm{AI}^k(f) \geq \mathrm{AI}(f) - r$.

Proof. Let $r_0 = \mathrm{AI}(f) - r - 1$. We know $\min\{k \mid \mathrm{AI}^k(f) \leq r_0\} = \mathrm{dai}_{r_0}(f)$ by Corollary 12. Furthermore, it holds that $\mathrm{dai}_{r_0}(f) \geq \sum_{i=0}^r \binom{n-r_0}{i}$ by Theorem 8. This states that if $k < \sum_{i=0}^r \binom{n-r_0}{i}$ then $\mathrm{AI}^k(f) \geq r_0 + 1 = \mathrm{AI}(f) - r$.

Taking $r = 1$ gives $\sum_{i=0}^r \binom{n-\mathrm{AI}(f)+r+1}{i} = n - \mathrm{AI}(f) + 3$ and therefore $\mathrm{AI}^{n-\mathrm{AI}(f)+2}(f) \geq \mathrm{AI}(f) - 1$, which implies the result of [25]: $\mathrm{EAI}(f) \geq \mathrm{AI}(f) - 1$. In particular, if f is a function with AI $\lceil \frac{n}{2} \rceil$, then $\mathrm{AI}^{\lfloor \frac{n}{2} \rfloor + 2}(f) \geq \lceil \frac{n}{2} \rceil - 1$.

Next we will discuss the sufficient and necessary condition for Boolean functions to achieve high possible k-error algebraic immunity.

Let g be an annihilator of f with algebraic degree $< d$. Let

$$g(x) = \sum_{c \in \mathbb{F}_2^n, \mathrm{wt}(c) < d} g_c x^c, \ g_c \in \mathbb{F}_2.$$

We have $g(b) = 0$ for $b \in 1_f$. Then

$$\sum_{c \in \mathbb{F}_2^n, \mathrm{wt}(c) < d} b^c g_c = 0, \ \text{for } b \in 1_f. \tag{5}$$

The above equations on g_c's are homogeneous linear. Denote the coefficient matrix of the equations by $V(f, d)$, which is a $\mathrm{wt}(f) \times \sum_{i=0}^{d-1} \binom{n}{i}$ matrix. Then f has no annihilator of algebraic degree $< d$ if and only if the rank of the matrix $V(f, d)$ equals the number of g_c's which is $\sum_{i=0}^{d-1} \binom{n}{i}$, i.e., $V(f, d)$ has full column rank (see also [23]). In the view of this result we can affirm the following theorem.

Proposition 14. *Let* $f \in \mathbf{B}_n$. *Then* $\mathrm{AI}^k(f) \geq d$ *if and only if all the matrices obtained by removing* k *rows from* $V(f, d)$ *and all the matrices obtained by removing* k *rows from* $V(f+1, d)$ *have full column rank.*

6   Conclusion

As described in [3,22], probabilistic algebraic attacks works more effectively than deterministic algebraic attacks with their applications to the nonlinear filter generator if the filter function has very low algebraic immunity distance of small order. Two classes of vulnerable functions are demonstrated in this paper, but both of them do not have good nonlinearities. We leave as an open problem whether there are algebraic immunity functions with good nonlinearity vulnerable to probabilistic algebraic attacks. Another problem is the practical applications of probabilistic algebraic attacks.

The algebraic immunity distance and k-error algebraic immunity of Boolean functions relate to their resistances to probabilistic algebraic attacks. The results of Section 4 imply the lower bound on algebraic immunity distance of r-th order of a balanced function

$$\max\{2^{n-r-1} + \frac{1}{2}\,\mathrm{nl}_r(f) - 2^{n-2}, \sum_{i=0}^{\mathrm{AI}(f)-r-1} \binom{n-r}{i}\},\tag{6}$$

and the upper bound

$$\min\{2^{n-r-1}, \frac{1}{2}\,\mathrm{nl}_r(f), 2^{n-1} - \sum_{i=0}^{r} \binom{n}{i} + 1\}.\tag{7}$$

The lower bound shows that Boolean functions with good high order nonlinearity and good algebraic immunity have algebraic immunity distance not too bad. The upper bound gives the minimum value of the probability for applying probabilistic algebraic attacks. However, it is not yet clear how to find the best approximation for the attacks.

Acknowledgement. The authors thank the anonymous referees of ACISP 2011 and SCC 2010 for their valuable comments on this paper. Meicheng Liu is grateful to Frederik Armknecht for helpful conversations on probabilistic algebraic attacks, and also for his careful reading of the manuscript and useful suggestions.

References

1. Armknecht, F.: Improving fast algebraic attacks. In: Roy, B., Meier, W. (eds.) FSE 2004. LNCS, vol. 3017, pp. 65–82. Springer, Heidelberg (2004)
2. Armknecht, F., Carlet, C., Gaborit, P., Künzli, S., Meier, W., Ruatta, O.: Efficient computation of algebraic immunity for algebraic and fast algebraic attacks. In: Vaudenay, S. (ed.) EUROCRYPT 2006. LNCS, vol. 4004, pp. 147–164. Springer, Heidelberg (2006)
3. Braeken, A., Preneel, B.: Probabilistic algebraic attacks. In: Smart, N.P. (ed.) Cryptography and Coding 2005. LNCS, vol. 3796, pp. 290–303. Springer, Heidelberg (2005)
4. Carlet, C.: On the higher order nonlinearities of algebraic immune functions. In: Dwork, C. (ed.) CRYPTO 2006. LNCS, vol. 4117, pp. 584–601. Springer, Heidelberg (2006)
5. Carlet, C., Feng, K.: An infinite class of balanced functions with optimal algebraic immunity, good immunity to fast algebraic attacks and good nonlinearity. In: Pieprzyk, J. (ed.) ASIACRYPT 2008. LNCS, vol. 5350, pp. 425–440. Springer, Heidelberg (2008)
6. Cohen, G., Honkala, I., Litsyn, S., Lobstein, A.: Covering codes. North-Holland, Amsterdam (1997)
7. Courtois, N., Meier, W.: Algebraic attacks on stream ciphers with linear feedback. In: Biham, E. (ed.) EUROCRYPT 2003. LNCS, vol. 2656, pp. 345–359. Springer, Heidelberg (2003)

8. Courtois, N.T.: Fast algebraic attacks on stream ciphers with linear feedback. In: Boneh, D. (ed.) CRYPTO 2003. LNCS, vol. 2729, pp. 176–194. Springer, Heidelberg (2003)
9. Dalai, D.K., Maitra, S., Sarkar, S.: Basic theory in construction of Boolean functions with maximum possible annihilator immunity. Designs, Codes and Cryptography 40(1), 41–58 (2006)
10. Dalai, D.K., Gupta, K.C., Maitra, S.: Results on algebraic immunity for cryptographically significant boolean functions. In: Canteaut, A., Viswanathan, K. (eds.) INDOCRYPT 2004. LNCS, vol. 3348, pp. 92–106. Springer, Heidelberg (2004)
11. Du, Y., Pei, D.: Construction of Boolean functions with maximum algebraic immunity and count of their annihilators at lowest degree. Sci. China Inf. Sci, 53(4), 780–787 (2010)
12. Hawkes, P., Rose, G.: Rewriting variables: The complexity of fast algebraic attacks on stream ciphers. In: Franklin, M. (ed.) CRYPTO 2004. LNCS, vol. 3152, pp. 390–406. Springer, Heidelberg (2004)
13. Li, N., Qu, L., Qi, W., et al.: On the construction of Boolean Functions with optimal algebraic immunity. IEEE Trans. Inform. Theory 54(3), 1330–1334 (2008)
14. Li, N., Qi, W.: Boolean functions of an odd number of variables with maximum algebraic immunity. Sci. China Ser. F-Inf. Sci. 50(3), 307–317 (2007)
15. Liu, M., Pei, D., Du, Y.: Identification and construction of Boolean functions with maximum algebraic immunity. Sci. China. Inf. Sci, 53(7), 1379–1396 (2010)
16. Liu, M., Du, Y., Pei, D., Lin, D.: On designated-weight Boolean functions with highest algebraic immunity. Sci. China. Math, 53(11), 2847–2854 (2010)
17. Lobanov, M.: Tight bounds between algebraic immunity and nonlinearities of high orders., http://eprint.iacr.org/2007/444
18. MacWilliams, F.J., Sloane, N.J.A.: The theory of error correcting codes. North-Holland, New York (1977)
19. Mesnager, S.: Improving the Lower Bound on the Higher Order Nonlinearity of Boolean Functions With Prescribed Algebraic Immunity. IEEE Transactions on Information Theory 54(8), 3656–3662 (2008)
20. Meier, W., Pasalic, E., Carlet, C.: Algebraic attacks and decomposition of boolean functions. In: Cachin, C., Camenisch, J.L. (eds.) EUROCRYPT 2004. LNCS, vol. 3027, pp. 474–491. Springer, Heidelberg (2004)
21. Pasalic, E.: Probabilistic versus deterministic algebraic cryptanalysis – a performance comparison. IEEE Transactions on Information Theory 55(11), 5233–5240 (2009)
22. Pometun, S.: Study of Probabilistic Scenarios of Algebraic Attacks on Stream Ciphers. Journal of Automation and Information Sciences 41(2), 67–80 (2009), http://eprint.iacr.org/2007/448
23. Qu, L., Feng, G., Li, C.: On the Boolean functions with maximum possible algebraic immunity: construction and a lower bound of the count., http://eprint.iacr.org/2005/449
24. Tu, Z., Deng, Y.: Algebraic Immunity Hierarchy of Boolean Functions. ChinaCrypt (2007), http://eprint.iacr.org/2007/259
25. Zhang, X., Pieprzyk, J., Zheng, Y.: On algebraic immunity and annihilators. In: Rhee, M.S., Lee, B. (eds.) ICISC 2006. LNCS, vol. 4296, pp. 65–80. Springer, Heidelberg (2006)

Finding More Boolean Functions with Maximum Algebraic Immunity Based on Univariate Polynomial Representation*

Yusong Du and Fangguo Zhang

School of Information Science and Technology, Sun Yat-sen University,
510006 Guangzhou, China
yusongdu@hotmail.com, isszhfg@mail.sysu.edu.cn

Abstract. Algebraic immunity is an important cryptographic property for Boolean functions against algebraic attacks. Constructions of Boolean functions with the maximum algebraic immunity (MAI Boolean functions) by using univariate polynomial representation of Boolean functions over finite fields have received more and more attention. In this paper, how to obtain more MAI Boolean functions from a known MAI Boolean function under univariate polynomial representation is further investigated. The sufficient condition of Boolean functions having the maximum algebraic immunity obtained by changing a known MAI Boolean function under univariate polynomial representation is given. With this condition, more balanced MAI Boolean functions under univariate polynomial representation can be obtained. The algebraic degree and the nonlinearity of these Boolean functions are analyzed.

Keywords: stream ciphers, algebraic attacks, Boolean functions, algebraic immunity, nonlinearity.

1   Introduction

In order to resist algebraic attacks, Boolean functions used in stream ciphers should have large *algebraic immunity* (AI) [1,2]. Construction of Boolean functions with the *maximum algebraic immunity* (MAI Boolean functions) is an important problem [3]. Nowadays, there have been many constructions of MAI Boolean functions. However, many of constructed functions were not proven to have good nonlinearity.

In 2008, Carlet and Feng exploited the univariate polynomial representation of Boolean functions over finite fields and constructed a class of balanced MAI Boolean functions with good nonlinearity [4], which is called Carlet-Feng functions. From then on, MAI Boolean functions under univariate (bivariate) polynomial representation received more and more attention and the nonlinearity of balanced MAI Boolean functions was improved further [5,6,7].

* This work is supported by National Natural Science Foundation of China (Grant No. 61070168, Grant No. 10971246, Grant No. 10871222) and Research Fund for the Doctoral Program of Higher Education of China (20094410110001).

U. Parampalli and P. Hawkes (Eds.): ACISP 2011, LNCS 6812, pp. 47–60, 2011.
© Springer-Verlag Berlin Heidelberg 2011

More recently, P.Rizomiliotis discussed the resistance of Boolean functions against (fast) algebraic attacks and provided a sufficient and necessary condition of Boolean function having the maximum AI under univariate polynomial representation [8]. Before long, X.Zeng et al. exploited the sufficient and necessary condition and provided more constructions of MAI Boolean functions under univariate polynomial representation [9].

Inspired by two papers above, in this paper, we would like to consider a typical method of finding MAI Boolean functions. That is to obtain new MAI Boolean functions by changing the majority Boolean function. The idea of this method firstly appeared in [10]. It was then realized by N.Li and W.Qi in [11,12] for Boolean functions in odd number of variables. It was further generalized in [13] for Boolean functions in any number of variables. With this method all the MAI Boolean functions can be obtained theoretically. However, the disadvantage of this method is mainly the poor cryptographic properties (except AI) of the majority Boolean function, which may result the failure to prove newly-constructed MAI Boolean functions having good cryptographic properties. This motivate us to consider replacing the majority Boolean functions with a new MAI Boolean function under univariate polynomial representation possessing good cryptographic properties. We hope that some newly-constructed Boolean functions in this way can be proven to have good cryptographic properties.

Recall that finding an odd n-variable MAI Boolean function is equivalent to finding an invertible submatrix in a given $2^{n-1} \times 2^{n-1}$ matrix W [11]. We would like to know if there exists similar relation under univariate polynomial representation. If it exists, what does the given matrix W like and how to find invertible submatrixes efficiently? And how about Boolean functions in even number of variables? In this paper we manage to find the answers.

The rest of the paper is organized as follows. Section 2 provides some preliminaries and recalls the sufficient and necessary condition given by P.Rizomiliotis. Section 3 gives the sufficient condition of Boolean functions having the maximum AI obtained by changing a known MAI Boolean function under univariate polynomial representation. Section 4 provides some concrete methods of finding more MAI Boolean functions in odd number of variables and further discusses the case when Boolean functions have even number of variables. Section 5 analyzes the algebraic degree and the nonlinearity.

2   Preliminaries

Let n always be a positive integer in this paper. An n-variable Boolean function may be viewed as a mapping from $\mathbb{F}_2^n$ to $\mathbb{F}_2$. We denote by $\mathbb{B}_n$ the set of all the n-variable Boolean functions.

Any n-variable Boolean function has a unique representation as a multivariate polynomial over $\mathbb{F}_2$, called the *algebraic normal form*(ANF),

$$f(x_1, x_2, \cdots, x_n) = a_0 + \sum_{1 \le i \le n} a_i x_i + \sum_{1 \le i < j \le n} a_{ij} x_i x_j + \cdots + a_{12\cdots n} x_1 x_2 \cdots x_n,$$

where $a_0, a_i, a_{ij}, \ldots, a_{12\cdots n}$ belong to $\mathbb{F}_2$. The algebraic degree of Boolean function f, denoted by $\deg(f)$, is the degree of this polynomial, i.e., the number of variables in the highest order term with nonzero coefficient. A boolean function is *affine* if there exists no term of degree strictly greater than 1 in the ANF.

For the simplicity, we omit $*$ and replace $f*g$ with fg to denote the polynomial multiplication over $\mathbb{F}_2$. A Boolean function $g \in \mathbb{B}_n$ is called an *annihilator* of $f \in \mathbb{B}_n$ if $fg = 0$. The lowest algebraic degree of all the nonzero annihilators of f and $1+f$ is called *algebraic immunity* of f or $1+f$ [1,2], denoted by $\mathcal{AI}(f)$. It has been proved that $\mathcal{AI}_n(f) \leq \lceil \frac{n}{2} \rceil$ for a given $f \in \mathbb{B}_n$ [14]. A Boolean function $f \in \mathbb{B}_n$ has the *maximum algebraic immunity* (MAI) if $\mathcal{AI}_n(f) = \lceil \frac{n}{2} \rceil$.

For $f \in \mathbb{B}_n$, the set of $x = (x_1, x_2, \cdots, x_n) \in \mathbb{F}_2^n$ for which $f(x) = 1$ (resp. $f(x) = 0$) is called the on-set (resp. off-set), denoted by 1_f (resp. 0_f). The Hamming weight of f is the cardinality of 1_f, denoted by $\mathrm{wt}(f)$. f is called balanced if $\mathrm{wt}(f) = 2^{n-1}$. If $f \in \mathbb{B}_n$ is an MAI Boolean function then f is balanced when n is odd and $\sum_{i=0}^{\frac{n}{2}-1} \binom{n}{i} \leq \mathrm{wt}(f) \leq \sum_{i=0}^{\frac{n}{2}} \binom{n}{i}$ when n is even [14].

The Hamming distance of $f \in \mathbb{B}_n$ from $g \in \mathbb{B}_n$ is the Hamming weight of $f+g$. The nonlinearity of an n-variable Boolean function f is its minimum Hamming distance from all the n-variable affine functions. The nonlinearity of f can be described through its Walsh transform: $\mathrm{nl}(f) = 2^{n-1} - \frac{1}{2}\max_{\omega \in \mathbb{F}_2^n}|W_f(\omega)|$, where $W_f(\omega) = \sum_{x \in \mathbb{F}_2^n}(-1)^{f(x)+\omega \cdot x}$ and $\omega \cdot x \in \mathbb{F}_2$ is the usual inner product over $\mathbb{F}_2^n$.

By identifying the finite field $\mathbb{F}_{2^n}$ with the vector space $\mathbb{F}_2^n$, an n-variable Boolean function f can be written as a univariate polynomial over $\mathbb{F}_{2^n}$:

$$f(x) = \sum_{i=0}^{2^n-1} f_i x^i,$$

where $f_0, f_{2^n-1} \in \mathbb{F}_2$ and $f_{2i} = (f_i)^2 \in \mathbb{F}_{2^n}$, $1 \leq i \leq 2^n - 2$. The algebraic degree $\deg(f)$ (not the degree of the polynomial over $\mathbb{F}_{2^n}$) is given by the largest integer $s = \mathrm{wt}_2(k)$, such that $f_k \neq 0$, where $\mathrm{wt}_2(k)$ is the number of nonzero coefficients in the binary representation of k.

A cyclotomic coset C_d modulo $2^n - 1$ can be written as

$$C_d = \{d, d \cdot 2, \cdots, d \cdot 2^{n_d-1}\}$$

where n_d is the smallest integer such that $d = d \cdot 2^{n_d} (\mathrm{mod} 2^n - 1)$ and d is the coset leader of C_d.

Denote by $\Gamma(n)$ the set of all the coset leader modulo $2^n - 1$, and by $r_d(x)$ the minimal polynomial of α^d over $\mathbb{F}_2$. Denote by $R_d(x)$ the product of the minimal polynomial over $\mathbb{F}_2$ of all the elements $\alpha^{-i} \in \mathbb{F}_{2^n}$, where $\mathrm{wt}_2(i) = d$, i.e.,

$$R_d(x) = \prod_{i \in \Gamma(n), \mathrm{wt}_2(i)=d} r_{-i}(x),$$

for $1 \leq d \leq n - 1$. Let $R_n(x) = x + 1$, $R_0(x) = x$ and

$$R_{d_1,d_2}(x) = \prod_{i=d_1}^{d_2} R_i(x), \quad 0 \leq d_1 \leq d_2 \leq n.$$

Define the $\sum_{i=0}^{d_1} \binom{n}{i} \times \sum_{i=0}^{d_2} \binom{n}{i}$ matrix $\mathbf{R}_{d_1+1,d_2}$, for $d_1 < d_2$, as follows. The rth row of $\mathbf{R}_{d_1+1,d_2}$ consists of the coefficients of the polynomial $x^r R_{d_1+1,d_2}(x)$ for $0 \le r \le \sum_{i=0}^{d_1} \binom{n}{i} - 1$, appended with zeros. Namely, matrix $\mathbf{R}_{d_1+1,d_2}$ produces a linear cyclic code $\mathcal{C}(N,k)$, where $N = \sum_{i=0}^{d_2} \binom{n}{i}$ and $k = \sum_{i=0}^{d_1} \binom{n}{i}$. Clearly, $\mathbf{R}_{d_1+1,d_2}$ has full rank $\sum_{i=0}^{d_1} \binom{n}{i}$, i.e., $\operatorname{rank}(\mathbf{R}_{d_1+1,d_2}) = \sum_{i=0}^{d_1} \binom{n}{i}$.

Let $\mathcal{A} \subseteq \mathbb{F}_{2^n}^*$. We denote by $\mathbf{R}_{d+1,n-1}^{(\mathcal{A})}$ the submatrix of $\mathbf{R}_{d+1,n-1}$, such that the jth column of $\mathbf{R}_{d+1,n-1}$, for $0 \le j \le 2^n - 2$, belongs to $\mathbf{R}_{d+1,n-1}^{(\mathcal{A})}$, if $\alpha^j \in \mathcal{A}$. Particularly, for $f \in \mathbb{B}_n$, we have submatrix $\mathbf{R}_{d+1,n-1}^{(1_f)}$ and submatrix $\mathbf{R}_{d+1,n-1}^{(0_f)}$ of $\mathbf{R}_{d+1,n-1}$.

Based on notations above, P.Rizomiliotis gave a sufficient and necessary condition of Boolean function having the maximum AI under univariate polynomial representation.

Lemma 1. *[8] Let $f \in \mathbb{B}_n$ and $d = \lceil \frac{n}{2} \rceil - 1$. $\mathcal{AI}(f) = d + 1$ if and only if*

$$\operatorname{rank}(\mathbf{R}_{d+1,n-1}^{(1_f)}) = \sum_{i=0}^{d} \binom{n}{i}$$

and

$$\operatorname{rank}([\gamma_{1_f}(d) \quad \mathbf{R}_{t+1,n-1}^{(0_f)}]) = \sum_{i=0}^{d} \binom{n}{i}$$

where $\gamma_{1_f}(d) = \mathbf{R}_{d+1,n-1}^{(1_f)} \cdot \mathbf{1}_{\operatorname{wt}(f)}^T$ and $\mathbf{1}_{\operatorname{wt}(f)}^T$ is the transpose of the all ones vector with length $\operatorname{wt}(f)$.

Let n be odd and $f \in \mathbb{B}_n$. It is well-known that $\mathcal{AI}(f) = \frac{n+1}{2}$ if and only if f is balanced and has not nonzero annihilators of degree less than $\frac{n+1}{2}$ [3].

Lemma 2. *Let n be odd, $f \in \mathbb{B}_n$ and $d = \frac{n-1}{2}$. $\mathcal{AI}(f) = d + 1$ if and only if $\operatorname{rank}(\mathbf{R}_{d+1,n-1}^{(1_f)})$ is an invertible square matrix.*

3   Deciding Boolean Functions Having Maximum AI under Univariate Polynomial Representation

In this section, we give the decision condition of Boolean functions having the maximum AI obtained by changing a known MAI Boolean function under univariate polynomial representation. We begin with the definition of the majority function[15].

Lemma 3. *Let $d = \lceil \frac{n}{2} \rceil - 1$ and $x \in \mathbb{F}_2^n$. $F_n \in \mathbb{B}_n$ satisfies*

$$F_n(x) = \begin{cases} 1 & \operatorname{wt}(x) \le d \\ 0 & \operatorname{wt}(x) > d \end{cases},$$

where $\operatorname{wt}(x)$ is the number of nonzero components of vector $x \in \mathbb{F}_2^n$. Then $\mathcal{AI}(F_n) = d + 1 = \lceil \frac{n}{2} \rceil$.

Boolean function F_n is called the *majority function* and has the maximum AI. A typical idea is to obtain a new MAI function by changing the majority function F_n [10]. When n is odd, this idea can be converted to the problem of finding out an invertible submatrix in a given $2^{n-1} \times 2^{n-1}$ matrix [11]. When n is even, it can be converted to the problem of finding out two submatrixes with full column rank in a $\sum_{i=d+1}^{n} \binom{n}{i} \times \sum_{i=0}^{d} \binom{n}{i}$ matrix with $d = \lceil \frac{n}{2} \rceil - 1$ [13].

Now we consider replace the majority function with a new MAI Boolean function under univariate polynomial representation. The new function comes from a known class of MAI Boolean functions under univariate polynomial representation.

Lemma 4. *[8] Let $f \in \mathbb{B}_n$ and α be a primitive element of the finite field $\mathbb{F}_{2^n}$. f is considered as a univariate polynomial over $\mathbb{F}_{2^n}$ and satisfies*

$$1_f = \{1, \alpha, \alpha^2, \cdots, \alpha^{D_n-1}\} \cup S,$$

where $S \subset \{\alpha^{D_n}, \alpha^{D_n+1}, \cdots, \alpha^{\hat{D}_n-1}\}$, $D_n = \sum_{i=0}^{\lceil \frac{n}{2} \rceil - 1} \binom{n}{i}$ and $\hat{D}_n = 2^n - D_n$. Then $\mathcal{AI}(f) = \lceil \frac{n}{2} \rceil$.

For the convenience of the description of this paper, in Lemma 4, the definition of S and $\hat{D}_n$ has been changed, but the function f in Lemma 4 is essentially same as the function f given by Definition 2 in [8]. Based on Lemma 4, we can define the majority function under univariate polynomial representation.

Definition 1. *Let α be a primitive element of the finite field $\mathbb{F}_{2^n}$. Boolean function $F_n \in \mathbb{B}_n$ is called the majority function in n variables under univariate polynomial representation if its on-set is exactly equal to $\{1, \alpha, \alpha^2, \cdots, \alpha^{D_n-1}\}$ where $D_n = \sum_{i=0}^{\lceil \frac{n}{2} \rceil - 1} \binom{n}{i}$.*

In Lemma 4, if n is odd, then $D_n = \hat{D}_n = 2^{n-1}$ and $F_n \in \mathbb{B}_n$ is the unique function defined as Lemma 4. For the simplicity, in the following content of this paper, we call $F_n \in \mathbb{B}_n$ the majority function and we always let α be a primitive element of the finite field $\mathbb{F}_{2^n}$, $D_n = \sum_{i=0}^{\lceil \frac{n}{2} \rceil - 1} \binom{n}{i}$ and $\hat{D}_n = 2^n - D_n$.

Definition 2. *For $1 \leq j_1 < j_2 < \cdots < j_t \leq D_n$ and $1 \leq i_1 < i_2 < \cdots < i_s \leq \hat{D}_n - 1$, $f = F_n(i_1, i_2, \cdots, i_s; j_1, j_2, \cdots, j_t) = F_n(\mathcal{A}; \mathcal{B}) \in \mathbb{B}_n$ is defined as*

$$f(x) = \begin{cases} F_n(x) + 1 & \text{if } x \in \mathcal{A} \cup \mathcal{B} \\ F_n(x) & \text{else} \end{cases}$$

where

$$\mathcal{A} = \{\alpha^{j_1-1}, \alpha^{j_2-1}, \cdots, \alpha^{j_t-1}\} \subseteq 1_{F_n},$$
$$\mathcal{B} = \{\alpha^{D_n+i_1-1}, \alpha^{D_n+i_2-1}, \cdots, \alpha^{D_n+i_s-1}\} \subset 0_{F_n},$$

$1 \leq |\mathcal{A}| = t \leq D_n$ and $1 \leq |\mathcal{B}| = s \leq \hat{D}_n - 1$.

Boolean function $f = F_n(\mathcal{A}; \mathcal{B})$ is obtained by changing the majority function according to $\mathcal{A}$ and $\mathcal{B}$. We want to know what $\mathcal{A}$ and $\mathcal{B}$ should be when $\mathcal{AI}(f) = \lceil \frac{n}{2} \rceil$. The following corollary gives an answer directly from Lemma 2 and the definition of matrix $\mathbf{R}_{d+1,n-1}$ with $d = \lceil \frac{n}{2} \rceil - 1$.

Corollary 1. *Let n be odd. For any integers i $(1 \leq i < 2^{n-1})$, if*

$$\mathcal{A}_i = \{\alpha^i, \alpha^{i+1}, \cdots, \alpha^{D_n-1}\}$$

and

$$\mathcal{B}_i = \{\alpha^{D_n-1+i}, \alpha^{D_n+i}, \cdots, \alpha^{2^n-2}\},$$

then $\mathcal{AI}(F_n(\mathcal{A}_i; \mathcal{B}_i)) = \frac{n+1}{2}$.

Proof. Let $d = \frac{n-1}{2}$. According to the definition of matrix $\mathbf{R}_{d+1,n-1}$, the first i rows of $\mathbf{R}_{d+1,n-1}^{(1_{F_n} \backslash \mathcal{A}_i)}$ form an upper triangular matrix, which is invertible, and all the entries on the rest of rows of $\mathbf{R}_{d+1,n-1}^{(1_{F_n} \backslash \mathcal{A}_i)}$ are zero. Similarly, the last $2^{n-1} - i$ rows of $\mathbf{R}_{d+1,n-1}^{(\mathcal{B}_i)}$ form a lower triangular matrix, which is also invertible, and all the entries on the rest of rows of $\mathbf{R}_{d+1,n-1}^{(\mathcal{B}_i)}$ are zero. Therefore, $\mathbf{R}_{d+1,n-1}^{((1_{F_n} \backslash \mathcal{A}_i) \cup \mathcal{B}_i)}$ is an invertible matrix and $\mathcal{AI}(F_n(\mathcal{A}_i; \mathcal{B}_i)) = \frac{n+1}{2}$ from Lemma 2.   □

Definition 3. *Let $f \in \mathbb{B}_n$ and $d = \lceil \frac{n}{2} \rceil - 1$. The matrix $V(1_f)$ is defined to be $(\mathbf{R}_{d+1,n-1}^{(1_f)})^T$ and the matrix $V(0_f)$ is defined to be $(\mathbf{R}_{d+1,n-1}^{(0_f)})^T$ where T is the transpose of the matrix. Particularly, the $(\hat{D}_n - 1) \times D_n$ matrix W_n is defined to be*

$$W_n = V(0_{F_n}) \cdot V(1_{F_n})^{-1},$$

for the majority function $F_n \in \mathbb{B}_n$.

With the notations in Definition 2, we can write $\alpha^{j_{t+1}-1}, \alpha^{j_{t+2}-1}, \cdots, \alpha^{j_{D_n}-1}$ as the rest of elements of 1_{F_n} after excluding all the elements of $\mathcal{A}$. Similarly, $0, \alpha^{D_n+i_{s+1}-1}, \alpha^{D_n+i_{s+2}-1}, \cdots, \alpha^{D_n+i_{\hat{D}_n-1}-1}$ are the rest of elements of 0_{F_n} after excluding all the elements of $\mathcal{B}$.

We denote by $W_n(\mathcal{A}; \mathcal{B})$ the $s \times t$ submatrix with all the entries on rows $i_1, i_2, \cdots, i_s$ and columns $j_1, j_2, \cdots, j_t$ of W_n, and by $W_n^*(\mathcal{A}; \mathcal{B})$ the $(\hat{D}_n - s - 1) \times (D_n - t)$ submatrix with all the entries on rows $i_{s+1}, i_{s+2}, \cdots, i_{\hat{D}_n-1}$ and columns $j_{t+1}, j_{t+2}, \cdots, j_{D_n}$ of W_n. We call $W_n^*(\mathcal{A}; \mathcal{B})$ the *complementary matrix* of $W_n(\mathcal{A}; \mathcal{B})$.

According to Definition 3, if $f \in \mathbb{B}_n$ satisfies $1_f = (1_{F_n} \backslash \mathcal{A}) \cup \mathcal{B}$, then $V(1_f)$ and $V(0_f)$ can be obtained by swapping t rows of $V(1_{F_n})$ with s rows of $V(0_{F_n})$. Since $W_n = V(0_{F_n}) \cdot V(1_{F_n})^{-1}$, from some linear algebra knowledge (or see Lemma 6 in [13] directly), it is not hard to prove that $\operatorname{rank} V(1_f) = D_n$ if and only if $W_n(\mathcal{A}; \mathcal{B})$ have full column rank and $\operatorname{rank} V(0_f) = D_n$ if and only if $W_n^*(\mathcal{A}; \mathcal{B})$ have full column rank.

When n is even, $\operatorname{rank}([\gamma_{1_f}(d) \ \mathbf{R}_{d+1,n-1}^{(0_f)}]) = D_n$ if $V(0_f)$ have full column rank D_n. Therefore, from Lemma 1 $\mathcal{AI}(f) = \lceil \frac{n}{2} \rceil$ if both $W_n(\mathcal{A}; \mathcal{B})$ and $W_n^*(\mathcal{A}; \mathcal{B})$ have full column rank.

Theorem 1. *Let n be even. $\mathcal{AI}(F_n(\mathcal{A}; \mathcal{B})) = \frac{n}{2}$ if both $W_n(\mathcal{A}; \mathcal{B})$ and $W_n^*(\mathcal{A}; \mathcal{B})$ have full column rank.*

When n is odd, if $|\mathcal{B}| = |\mathcal{A}|$, then $F_n(\mathcal{A}; \mathcal{B})$ is balanced. From Lemma 2 we have the following result.

Theorem 2. *Let n be odd. $\mathcal{AI}(F_n(\mathcal{A};\mathcal{B})) = \frac{n+1}{2}$ if and only if $W_n(\mathcal{A};\mathcal{B})$ is an invertible square matrix.*

Theorem 1 and Theorem 2 mean that finding MAI Boolean functions in n variables under univariate polynomial representation can be converted to finding submatrixes with full column rank in $(\hat{D}_n - 1) \times D_n$ matrix W_n. It is interesting to study the properties of W_n.

Theorem 3. *Let $d = \lceil \frac{n}{2} \rceil - 1$. The ith column of W_n equals to the coefficient list (written as a column vector) of polynomial $q_{i-1}(x) = x^{i-1} \cdot b(x)$ mod $(R_{d+1,n-1}(x))$ where $1 \le i \le D_n$ and $b(x) \cdot x^{D_n} \equiv 1$ mod $(R_{d+1,n-1}(x))$.*

Proof. According to the definition of $\mathbf{R}_{d+1,n-1}$, the ith row of $\mathbf{R}_{d+1,n-1}$ equals to the coefficient list of polynomial $x^{i-1}R_{d+1,n-1}(x)$. since

$$\begin{pmatrix} V(1_{F_n}) \\ V(0_{F_n}) \end{pmatrix} = (\mathbf{R}_{d+1,n-1})^T,$$

we have

$$\begin{pmatrix} V(1_{F_n}) \\ V(0_{F_n}) \end{pmatrix} \cdot V(1_{F_n})^{-1} = \begin{pmatrix} I_n \\ W_n \end{pmatrix} = ((V(1_{F_n})^{-1})^T \cdot \mathbf{R}_{d+1,n-1})^T,$$

where I_n is the identity matrix. Thus the ith column of $\begin{pmatrix} I_n \\ W_n \end{pmatrix}$ is a codeword of the linear cyclic code generated by $(\mathbf{R}_{d+1,n-1})^T$, which equals to the coefficient list (written as a column vector) of the polynomial denoted by $C_i(x) = x^{i-1} + q_{i-1}(x) \cdot x^{D_n}$ with $C_i(x) \equiv 0$ mod $(R_{d+1,n-1}(x))$ where $\deg(q_{i-1}(x)) < 2^n - D_n - 1$. Therefore the ith column of W_n equals to the coefficient list of polynomial $q_{i-1}(x)$, then equals to the coefficient list of polynomial $q_{i-1}(x) = x^{i-1} \cdot b(x)$ mod $(R_{d+1,n-1}(x))$ where $b(x) \cdot x^{D_n} \equiv 1$ mod $(R_{d+1,n-1}(x))$ since $\gcd(x^{D_n}, R_{d+1,n-1}(x)) = 1$. □

Corollary 2. *For $1 \le i \le \hat{D}_n - 1$ and $1 \le j \le D_n$, the entry on row i and column j of W_n is denoted by $W_n^{(i,j)}$. Then $W_n^{(i,j)}$ satisfies the following recursive relation:*

$$\begin{cases} W_n^{(i,1)} = a_{i-1} \\ W_n^{(i,j)} = W_n^{(i-1,j-1)} + W_n^{(\hat{D}_n-1,j-1)} \cdot a_{i-1} & \text{for } j \ge 2 \end{cases}$$

where $a_{i-1} \in \mathbb{F}_2$ is the coefficient of the term x^{i-1} in the $q_0(x)$ defined in Theorem 3 and $W_n^{(0,j)} = 0$ for $1 \le j \le D_n$.

Example 1. Let $n = 5$. $d = \lceil \frac{n}{2} \rceil - 1 = 2$. It can be verified that

$$q_0(x) = b(x) = 1 + x^4 + x^5 + x^6 + x^7 + x^8 + x^{10} + x^{12} + x^{13} + x^{14}.$$

Then the first column of W_5 is $(1,0,0,0,1,1,1,1,1,0,1,0,1,1,1)^T$, and the second column of W_5 is

$$(1,1,0,0,1,0,0,0,0,1,1,1,1,0,0)^T$$
$$= (0,1,0,0,0,1,1,1,1,1,0,1,0,1,1)^T \oplus (1,0,0,0,1,1,1,1,1,0,1,0,1,1,1)^T,$$

and the third column of W_5 is $(0,1,1,0,0,1,0,0,0,0,1,1,1,1,0)^T$, and so on. Finally, we have

$$W_5 = \begin{pmatrix}
1\,1\,0\,0\,1\,0\,0\,0\,0\,1\,1\,1\,1\,0\,0\,0 \\
0\,1\,1\,0\,0\,1\,0\,0\,0\,0\,1\,1\,1\,1\,0\,0 \\
0\,0\,1\,1\,0\,0\,1\,0\,0\,0\,0\,1\,1\,1\,1\,0 \\
0\,0\,0\,1\,1\,0\,0\,1\,0\,0\,0\,0\,1\,1\,1\,1 \\
1\,1\,0\,0\,0\,1\,0\,0\,1\,1\,1\,1\,1\,1\,1\,1 \\
1\,0\,1\,0\,1\,0\,1\,0\,0\,0\,0\,0\,0\,1\,1\,1 \\
1\,0\,0\,1\,1\,1\,0\,1\,0\,1\,1\,1\,1\,0\,1\,1 \\
1\,0\,0\,0\,0\,1\,1\,0\,1\,1\,0\,0\,0\,1\,0\,1 \\
1\,0\,0\,0\,1\,0\,1\,1\,0\,0\,0\,1\,1\,0\,1\,0 \\
0\,1\,0\,0\,0\,1\,0\,1\,1\,0\,0\,0\,1\,1\,0\,1 \\
1\,1\,1\,0\,1\,0\,1\,0\,1\,0\,1\,1\,1\,1\,1\,0 \\
0\,1\,1\,1\,0\,1\,0\,1\,0\,1\,0\,1\,1\,1\,1\,1 \\
1\,1\,1\,1\,0\,0\,1\,0\,1\,1\,0\,1\,0\,1\,1\,1 \\
1\,0\,1\,1\,0\,0\,0\,1\,0\,0\,0\,1\,0\,0\,1\,1 \\
1\,0\,0\,1\,0\,0\,0\,0\,1\,1\,1\,1\,0\,0\,0\,1
\end{pmatrix}.$$

4   Finding MAI Boolean Functions under Univariate Polynomial Representation

In this section, we give some concrete methods of finding MAI Boolean functions in odd number of variables and further discuss the case when Boolean functions have even number of variables.

When n is odd, from Corollary 2, we can find the nonzero entries in W_n and obtain new MAI Boolean functions directly.

Corollary 3. *Let n be odd. For $1 \le i \le D_n - 1$ and $1 \le j \le D_n$, $f \in \mathbb{B}_n$ is defined as*

$$f(x) = \begin{cases} F_n(x) + 1 & \text{if } x \in \{\alpha^{D_n+i-1}, \alpha^{j-1}\} \\ F_n(x) & \text{else} \end{cases}$$

Then, $\mathcal{AI}(f) = \lceil \frac{n}{2} \rceil$ if and only if the entry on row i and column j of W_n is nonzero.

Section 3 shows that finding an MAI Boolean function in odd n variables under univariate polynomial representation is equivalent to finding an invertible square matrix . Although it is hard to find out all the invertible matrixes, finding some special invertible matrixes in W_n may be possible.

In the following two algorithms, Algorithm 1 aims at finding some upper triangular submatrix of W_n, while algorithm 2 can give an upper triangular submatrix as well as a lower triangular submatrix. The invertible submatrixes found by Example 2 and Example 3 are emphasized with bold font in W_5, which has been shown in Example 1.

Algorithm 1. *The algorithm can be divided into 3 steps.*

1. *Find integers $j\,(1 \le j < D_n)$ and $s \ge 2$ such that the coefficient of x^l in $q_{j-1}(x) \bmod (R_{d+1,n-1})$ equals to 0 for $D_n - s \le l < D_n - 1$ and 1 for $l = D_n - s - 1$.*
2. *Let $i_1 = D_n - s, i_2 = D_n - s + 1, \cdots, i_s = D_n - 1$ and $j_1 = j, j_2 = j + 1, \cdots, j_s = j + s - 1$.*
3. *Output an MAI Boolean function $f = F_n(i_1, \cdots, i_s; j_1, \cdots, j_s)$.*

Example 2. Let $n = 5$. $D_5 = 16$

1. $q_5(x) \bmod (R_{d+1,n-1}) = x + x^4 + x^6 + x^7 + x^9 + x^{11}$. The coefficients of terms x^{12}, x^{13}, x^{14} are zeros and the coefficients of x^{11} is 1. Thus, we find out two integers $j = 5 + 1 = 6$ and $s = 4$ satisfying the given conditions.
2. Let $i_1 = D_5 - s = 12, i_2 = D_5 - s + 1 = 13, i_3 = D_5 - s + 2 = 14, i_4 = D_5 - 1 = 15$ and $j_1 = j = 6, j_2 = j + 1 = 7, j_3 = j + 2 = 8, j_4 = j + s - 1 = 9$.
3. Output an MAI Boolean function $f = F_n(12, 13, 14, 15; 6, 7, 8, 9)$.

Algorithm 2. *The algorithm can be divided into 6 steps and outputs two MAI Boolean functions.*

1. *Find integers j $(1 \le j < D_n)$ and $s \ge 3$ such that the coefficient of x^l in $q_{j-1}(x) \bmod (R_{d+1,n-1})$ equals to 0 for $D_n - s \le l < D_n - 1$ and 1 for $l = D_n - s - 1$.*
2. *Find k $(1 \le k < D_n - 1)$ and $2 \le s' < s$ such that the coefficient of x^l in $q_{j-1}(x) \bmod (R_{d+1,n-1})$ equals to 0 for $k \le l < k + s' - 2$ and 1 for $k - 1$ and $l = k + s' - 2$.*
3. *Let $i_1 = k, i_2 = k + 1, i_{s'} = k + s' - 1$ and $j_1 = j, j_2 = j + 1, j_{j'} = j + s' - 1$.*
4. *Output an MAI Boolean function $f = F_n(i_1, \cdots, i_{s'}; j_1, \cdots, j_{s'})$.*
5. *Let $i_1 = k + s', i_2 = k + s' + 1, i_{s'} = k + 2s' - 1$.*
6. *Output an MAI Boolean function $f = F_n(i_1, \cdots, i_{s'}; j_1, \cdots, j_{s'})$.*

Example 3. Let $n = 5$. $D_5 = 16$

1. $q_4(x) \bmod (R_{d+1,n-1}) = 1 + x^3 + x^5 + x^6 + x^8 + x^{10}$. The coefficients of terms $x^{11}, x^{12}, x^{13}, x^{14}$ are zeros and the coefficients of x^{10} is 1. Thus, we find out two integers $j = 4 + 1 = 5$ and $s = 5$ satisfying the given conditions.
2. Note that the coefficients of terms x, x^2 are zeros and the coefficients of x^0 and x^3 are 1. Thus, $k = 1$ and $s' = 3$ satisfy the given conditions.
3. Let $i_1 = k = 1, i_2 = k + 1 = 2, i_3 = k + s' - 1 = 3$ and $j_1 = j = 5, j_2 = j + 1 = 6, j_s = j + 3 - 1 = 7$.
4. Output an MAI Boolean function $f = F_n(1, 2, 3; 5, 6, 7)$.
5. Let $i_1 = k + s' = 4, i_2 = k + s' + 1 = 5, i_3 = k + 2s' - 1 = 6$.
6. Output an MAI Boolean function $f = F_n(4, 5, 6; 5, 6, 7)$.

Now we consider finding MAI Boolean functions in even number of variables. From Theorem 1, we need to guarantee two submatrixes in W_n having full column rank at same time, which is difficult, if we want to obtain a new MAI Boolean function in even n variables from the majority function F_n. Our idea in this paper is to restrict choice of $\mathcal{B}$ firstly such that $W_n^*(\mathcal{A}; \mathcal{B})$ always has full column rank, then find $W_n(\mathcal{A}; \mathcal{B})$ with full column rank.

Lemma 5. *Let n be even. $\mathcal{AI}(F_n(\mathcal{A}; \mathcal{B})) = \frac{n}{2}$ if $W_n(\mathcal{A}; \mathcal{B})$ has full column rank and $\mathcal{B} \subset \{\alpha^{D_n}, \alpha^{D_n+1}, \cdots, \alpha^{\hat{D}_n-2}\}$.*

Proof. Let $d = \frac{n}{2} - 1$, from the definition of matrix $\mathbf{R}_{d+1,n-1}$, the last D_n columns of $\mathbf{R}_{d+1,n-1}$ form a lower triangular matrix, which is invertible. Then, from the definition of matrix W_n, the last D_n rows of W_n form a upper triangular matrix, which is invertible. Therefore, $W_n(\mathcal{A}; \mathcal{B})$ is a submatrix with full column rank taken only from the first $\hat{D}_n - D_n - 1$ rows of W_n and its complementary submatrix $W_n^*(\mathcal{A}; \mathcal{B})$ always has full column rank. This means that $\mathcal{AI}(f) = \lceil \frac{n}{2} \rceil$. □

Generally, in fact, excluding arbitrary D_n rows of W_n which form an invertible matrix, if $W_n(\mathcal{A}, \mathcal{B})$ is taken only from the rest of rows of W_n, its complementary $W_n^*(\mathcal{A}, \mathcal{B})$ always has full column rank. Thus, we have following result.

Lemma 6. *Let n be even and $\mathcal{S} \subset 0_{F_n} \backslash \{0\}$ such that $W_n(1_{F_n}; \mathcal{S})$ is an invertible square matrix. For $\mathcal{A} \subset 1_{F_n}$ and $\mathcal{B} \subset 0_{F_n} \backslash (\mathcal{S} \cup \{0\})$, $\mathcal{AI}(F_n(\mathcal{A}; \mathcal{B})) = \frac{n}{2}$ if $W_n(\mathcal{A}; \mathcal{B})$ has full column rank.*

With the notations in Lemma 4, let $\mathcal{S} = \{\alpha^{D_n}, \alpha^{D_n+1}, \cdots, \alpha^{\hat{D}_n-2}\}$, i.e., the on-set of the function is $\{1, \alpha, \alpha^2, \cdots, \alpha^{\hat{D}_n-2}\}$ and the off-set of the function is $\{\alpha^{\hat{D}_n-1}, \alpha^{\hat{D}_n}, \cdots, \alpha^{2^n-2}\} \cup \{0\}$, we denote by $\bar{F}_n \in \mathbb{B}_n$ this function .

Definition 4. *Let n be even. With notations in Definition 3, the $(\hat{D}_n - 1) \times D_n$ matrix $\bar{W}_n$ is defined to be*

$$\bar{W}_n = V(1_{\bar{F}_n}) \cdot V(0_{\bar{F}_n})^{-1},$$

for the Boolean function $\bar{F}_n \in \mathbb{B}_n$.

Let

$$\bar{\mathcal{A}} = \{\alpha^{i_1-1}, \alpha^{i_2-1}, \cdots, \alpha^{i_s-1}\} \subset 1_{\bar{F}_n},$$

and

$$\bar{\mathcal{B}} = \{\alpha^{\hat{D}_n+j_1-2}, \alpha^{\hat{D}_n+j_2-2}, \cdots, \alpha^{\hat{D}_n+j_t-2}\} \subset 0_{\bar{F}_n},$$

where $D_n + 1 \leq i_1 < i_2 < \cdots < i_s \leq \hat{D}_n - 1$ and $1 \leq j_1 < j_2 < \cdots < j_t \leq D_n$. Similarly, we denote by $\bar{W}_n(\bar{\mathcal{A}}; \bar{\mathcal{B}})$ the $s \times t$ submatrix with all the entries on rows $i_1, i_2, \cdots, i_s$ and columns $j_1, j_2, \cdots, j_t$ of $\bar{W}_n$.

Lemma 7. *Let $\mathcal{S} - (0_{\bar{F}_n} \backslash (\bar{\mathcal{B}} \cup \{0\})) \cup \bar{\mathcal{A}}$. $W_n(1_{F_n}; \mathcal{S})$ is an invertible square matrix if and only if $\bar{W}_n(\bar{\mathcal{A}}; \bar{\mathcal{B}})$ is an invertible square matrix.*

Proof. Let $f \in \mathbb{B}_n$ such that $0_f = \mathcal{S} \cup \{0\} = (0_{\bar{F}_n} \backslash \bar{\mathcal{B}}) \cup \bar{\mathcal{A}}$. $V(0_f)$ is an invertible matrix if and only if $\bar{W}_n(\bar{\mathcal{A}}; \bar{\mathcal{B}})$ is an invertible square matrix. Since $V(0_f) \cdot V(1_{F_n})^{-1} = W_n(1_{F_n}; \mathcal{S})$, $V(0_f)$ is an invertible matrix if and only if $W_n(1_{F_n}; \mathcal{S})$ is an invertible square matrix. Therefore, $W_n(1_{F_n}; \mathcal{S})$ is an invertible square matrix if and only if $\bar{W}_n(\bar{\mathcal{A}}; \bar{\mathcal{B}})$ is an invertible square matrix. □

From lemma 6 and Lemma 7, we have the following result.

Theorem 4. *Let n be even and $\mathcal{S} = (0_{\bar{F}_n} \backslash (\bar{\mathcal{B}} \cup \{0\})) \cup \bar{\mathcal{A}}$. For $\mathcal{A} \subset 1_{F_n}$ and $\mathcal{B} \subset 0_{F_n} \backslash (\mathcal{S} \cup \{0\})$, $\mathcal{AI}(F_n(\mathcal{A}; \mathcal{B})) = \frac{n}{2}$ if $W_n(\mathcal{A}; \mathcal{B})$ has full column rank and $\bar{W}_n(\bar{\mathcal{A}}; \bar{\mathcal{B}})$ is an invertible square matrix.*

Theorem 5. *Let n be even and $d = \frac{n}{2} - 1$. The ith column of $\bar{W}_n$ equals to the coefficient list (written as a column vector) of polynomial $q_{i-1}(x) = x^{\hat{D}_n+i-2} \bmod (R_{d+1,n-1}(x))$ where $1 \le i \le D_n$.*

Proof. Being similar to the proof of Theorem 3, we have

$$\begin{pmatrix} V(1_{\bar{F}_n}) \\ V(0_{\bar{F}_n}) \end{pmatrix} \cdot V(0_{\bar{F}_n})^{-1} = \begin{pmatrix} \bar{W}_n \\ I_n \end{pmatrix} = ((V(0_{\bar{F}_n})^{-1})^T \cdot \mathbf{R}_{d+1,n-1})^T$$

Thus the ith column of $\begin{pmatrix} \bar{W}_n \\ I_n \end{pmatrix}$ is a codeword of the linear cyclic code generated by $(\mathbf{R}_{d+1,n-1})^T$, which equals to the coefficient list (written as a column vector) of the polynomial denoted by $C_i(x) = q_{i-1}(x) + x^{i-1}x^{\hat{D}_n-1}$ with $C_i(x) \equiv 0 \bmod (R_{d+1,n-1}(x))$ where $\deg(q_{i-1}(x)) < \hat{D}_n - 1$. Therefore the ith column of W_n equals to the coefficient list of polynomial $q_{i-1}(x) = x^{\hat{D}_n+i-2} \bmod (R_{d+1,n-1}(x))$. $\square$

Based on the discussion above, when finding an MAI Boolean function in even number of variables, we can find an invertible square matrix $\bar{W}_n(\bar{A}; \bar{B})$ in $\bar{W}_n$ firstly, then restrict the choice of $\mathcal{B}$ according to $\bar{A}$ and $\bar{B}$, finally we find a submatrix $W_n(\mathcal{A}; \mathcal{B})$ with full column rank and obtain $F_n(\mathcal{A}; \mathcal{B})$ with the maximum AI.

5   Analysis of Algebraic Degree and Nonlinearity

In this section, we analyze the algebraic degree and the nonlinearity of $f = F_n(\mathcal{A}; \mathcal{B}) \in \mathbb{B}_n$ given by Definition 2.

Theorem 6. *When $f = F_n(\mathcal{A}; \mathcal{B}) \in \mathbb{B}_n$ is balanced, $\deg(f) = n - 1$ if and only if*

$$\sum_{x \in \mathcal{A} \cup \mathcal{B}} x \ne \frac{1 + \alpha^{D_n}}{1 + \alpha}$$

Proof. Let $f(x) = \sum_{i=0}^{2^n-1} f_i x^i$ be the univariate polynomial representation of f. Then $f_0 = f(0) = 0$ and $f_{2^n-1} = 0$, i.e., $\deg(f) \le n - 1$ since f is balanced. For every $i \in \{1, 2, \cdots, 2^n - 2\}$:

$$f_i = \sum_{j=0}^{2^n-2} f(\alpha^j)\alpha^{-ij}.$$

We define $F(x) = \sum_{j=0}^{2^n-2} f(\alpha^j)x^j$. Then $\deg(f) = n-1$ if and only if $F(\alpha^{-i}) \ne 0$ for some i with $\mathrm{wt}_2(i) = n-1$. Note that $i_k = 2^n - 1 - 2^k$ with $k = 0, 1, \cdots, n-1$

are all the positive integers less than $2^n - 1$ such that $\text{wt}_2(i_k) = n - 1$. Thus, $\deg(f) = n - 1$ if and only if $F(\alpha^{-(2^n-1-2^k)}) = F(\alpha^{2^k}) = \sum_{j=0}^{2^n-2} f(\alpha^j)\alpha^{j2^k} = F(\alpha)^{2^k} \neq 0$ for some integer k with $0 \leq k \leq n - 1$. Finally, $F(\alpha)^{2^k} \neq 0$ if and only if $F(\alpha) \neq 0$, for $F(\alpha)$ we have

$$F(\alpha) = \sum_{x \in 1_f} x = \sum_{x \in (1_{F_n} \setminus \mathcal{A}) \cup \mathcal{B}} x = \sum_{x \in 1_{F_n}} x + \sum_{x \in \mathcal{A} \cup \mathcal{B}} x = \sum_{x \in \mathcal{A} \cup \mathcal{B}} x + \frac{1 + \alpha^{D_n}}{1 + \alpha}.$$

This completes the proof.   □

Let $\lambda \in \mathbb{F}_{2^n}^*$. Carlet and Feng proved

$$\max_{\lambda \in \mathbb{F}_{2^n}^*} \left| \sum_{i=2^{n-1}-1}^{2^n-2} (-1)^{\text{tr}(\lambda \alpha^i)} \right| \leq 2^{\frac{n}{2}} \cdot n \ln 2 + 1$$

when considering the nonlinearity of Boolean functions constructed by them[4]. With a similar proof idea, Zeng et al. recently gave a better bound[9], i.e.,

$$\max_{\lambda \in \mathbb{F}_{2^n}^*} \left| \sum_{i=0}^{2^{n-1}-1} (-1)^{\text{tr}(\lambda \alpha^i)} \right| \leq 2^{\frac{n}{2}} \cdot c_n \ln 2 + 1$$

where $c_n = \frac{\ln 2}{3}(n - 1) + \frac{5}{6} + \frac{1}{3\sqrt{3}} + \frac{1}{6\sqrt{2}} < n \ln 2$ for large n. Based on this, for any integer k $(0 \leq k \leq 2^n - 2)$, we have

$$\max_{\lambda \in \mathbb{F}_{2^n}^*} \left| \sum_{i=k}^{2^{n-1}-1+k} (-1)^{\text{tr}(\lambda \alpha^i)} \right| \leq 2^{\frac{n}{2}} \cdot c_n \ln 2 + 1$$

and the following result.

Lemma 8. *If* $g \in \mathbb{B}_n$ *satisfies* $1_g = \{\alpha^k, \alpha^{k+1}, \cdots, \alpha^{k+2^{n-1}-1}\}$ *where* $0 \leq k \leq 2^n - 2$, *then* g *is balanced and* $\text{nl}(g) > 2^{n-1} - 2^{\frac{n}{2}} \cdot c_n \ln 2 - 1$.

Theorem 7. *Let* $f = F_n(\mathcal{A}; \mathcal{B}) \in \mathbb{B}_n$ *be balanced. For any integer* k $(1 \leq k \leq D_n)$, *if* $\mathcal{A}$ *satisfies*

$$\mathcal{S}_1 = \{1, \alpha, \cdots, \alpha^{k-1}\} \subseteq \mathcal{A} \subseteq 1_{F_n}$$

and $\mathcal{B}$ *satisfies*

$$\mathcal{S}_2 = \{\alpha^{D_n}, \alpha^{D_n+1}, \cdots, \alpha^{D_n+k-1}, \cdots, \alpha^{D_n+k+\frac{\hat{D}_n - D_n}{2}-1}\} \subseteq \mathcal{B} \subseteq 0_{F_n} \setminus \{0\},$$

then $\text{nl}(f) > 2^{n-1} - 2^{\frac{n}{2}} \cdot c_n \ln 2 - |\mathcal{A} \setminus \mathcal{S}_1| - |\mathcal{B} \setminus \mathcal{S}_2| - 1$, *where* $c_n = \frac{\ln 2}{3}(n - 1) + \frac{5}{6} + \frac{1}{3\sqrt{3}} + \frac{1}{6\sqrt{2}}$.

Proof. Let $g \in \mathbb{B}_n$ such that $1_g = \{\alpha^k, \alpha^{k+1}, \cdots, \alpha^{k+2^{n-1}-1}\}$. It is not hard to see that

$$1_f = (1_{F_n} \setminus \mathcal{A}) \cup \mathcal{B} = (1_g \setminus (\mathcal{A} \setminus \mathcal{S}_1)) \cup (\mathcal{B} \setminus \mathcal{S}_2).$$

When we identify the vector space $\mathbb{F}_2^n$ with the finite field $\mathbb{F}_{2^n}$, we can take for inner product: $\lambda \cdot x = \text{tr}(\lambda x)$ where $\text{tr}(\cdot)$ is the absolute trace function. Then, for $\lambda \in \mathbb{F}_{2^n}^*$, we have

$$W_f(\lambda) = -2 \sum_{x \in 1_f} (-1)^{\text{tr}(\lambda x)} = -2 \sum_{x \in (1_{F_n} \setminus \mathcal{A}) \cup \mathcal{B}} (-1)^{\text{tr}(\lambda x)}$$

$$= -2 \left(\sum_{x \in 1_g} (-1)^{\text{tr}(\lambda x)} - \sum_{x \in \mathcal{A} \setminus \mathcal{S}_1} (-1)^{\text{tr}(\lambda x)} + \sum_{x \in \mathcal{B} \setminus \mathcal{S}_2} (-1)^{\text{tr}(\lambda x)} \right).$$

Then

$$|W_f(\lambda)| \leq 2 \left| \sum_{x \in 1_g} (-1)^{\text{tr}(\lambda x)} \right| + 2|\mathcal{A} \setminus \mathcal{S}_1| + 2|\mathcal{B} \setminus \mathcal{S}_2|.$$

Note that $W_f(0) = 0$ since f is balanced and $\text{nl}(g) > 2^{n-1} - 2^{\frac{n}{2}} \cdot c_n \ln 2 - 1$ by Lemma 8. Therefore,

$$\text{nl}(f) \geq 2^{n-1} - \max_{\lambda \in \mathbb{F}_{2^n}^*} \left| \sum_{x \in 1_g} (-1)^{\text{tr}(\lambda x)} \right| - |\mathcal{A} \setminus \mathcal{S}_1| - |\mathcal{B} \setminus \mathcal{S}_2|$$

$$= \text{nl}(g) - |\mathcal{A} \setminus \mathcal{S}_1| - |\mathcal{B} \setminus \mathcal{S}_2|.$$

$$> 2^{n-1} - 2^{\frac{n}{2}} \cdot c_n \ln 2 - |\mathcal{A} \setminus \mathcal{S}_1| - |\mathcal{B} \setminus \mathcal{S}_2| - 1.$$

This proves the theorem. □

With ideas in Section 4, it is possible to find out $F_n(\mathcal{A}; \mathcal{B})$ with the maximum AI satisfying the conditions in Theorem 7. Generally, Theorem 7 still holds for any n-variable Boolean function g such that $1_g = (1_{F_n} \setminus \mathcal{S}_1) \cup \mathcal{S}_2$ with $\mathcal{S}_1 \subseteq 1_{F_n}$ and $\mathcal{S}_2 \subseteq 0_{F_n} \setminus \{0\}$. If such a Boolean function g has better nonlinearity, we may find MAI Boolean functions with better nonlinearity.

6   Conclusion

In this paper, how to find more MAI Boolean functions by changing the new majority function under univariate polynomial representation is further investigated. It shows that finding MAI Boolean functions in n variables under univariate polynomial representation can be converted to finding submatrixes with full column rank in the $(\hat{D}_n - 1) \times D_n$ matrix W_n. It studies some basic properties of W_n, which makes finding some special invertible submatrixes be possible. It also analyzes the algebraic degree and the nonlinearity of Boolean functions obtained in this way.

References

1. Courtois, N., Meier, W.: Algebraic attacks on stream ciphers with linear feedback. In: Biham, E. (ed.) EUROCRYPT 2003. LNCS, vol. 2729, pp. 345–359. Springer, Heidelberg (2003)
2. Meier, W., Pasalic, E., Carlet, C.: Algebraic attacks and decomposition of boolean functions. In: Cachin, C., Camenisch, J.L. (eds.) EUROCRYPT 2004. LNCS, vol. 3027, pp. 474–491. Springer, Heidelberg (2004)
3. Canteaut, A.: Open problems related to algebraic attacks on stream ciphers. In: Ytrehus, Ø. (ed.) WCC 2005. LNCS, vol. 3969, pp. 120–134. Springer, Heidelberg (2006)
4. Carlet, C., Feng, K.: An infinite class of balanced functions with optimal algebraic immunity, good immunity to fast algebraic attacks and good nonlinearity. In: Pieprzyk, J. (ed.) ASIACRYPT 2008. LNCS, vol. 5350, pp. 425–440. Springer, Heidelberg (2008)
5. Tu, Z., Deng, Y.: A Conjecture on Binary String and Its Applications on Constructing Boolean Functions of Optimal Algebraic Immunity. Cryptology ePrint Archive, Report 2009/272., http://eprint.iacr.org/2009/272.pdf
6. Tang, X., Tang, D., Zeng, X., Hu, L.: Balanced Boolean functions with (almost) optimal algebraic immunity and very high nonlinearity. Cryptology ePrint Archive, Report 2010/443., http://eprint.iacr.org/2010/443
7. Wang, Q., Peng, J., Kan, H., Xue, X.: Constructions of cryptographically significant Boolean functions using primitive polynomials. IEEE Trans. Inform. Theory 56(6), 3048–3053 (2010)
8. Rizomiliotis, P.: On the Resistance of Boolean Functions Against Algebraic Attacks Using Univariate Polynomial Representation. IEEE Trans. Inform. Theory 56(8), 4014–4024 (2010)
9. Zeng, X., Carlct, C., Shan, J., Hu, L.: Balanced Boolean Functions with Optimum Algebraic Immunity and High Nonlinearity. Cryptology ePrint Archive, Report /2010/606, http://eprint.iacr.org/2010/606
10. Qu, L., Li, C.: On the Boolean functions with maximum possible algebraic immunity: construction and a lower bound of the count. Cryptology ePrint Archive, Report 2005 /449, http://eprint.iacr.org/2005/449
11. Li, N., Qi, W.: Construction and analysis of boolean functions of $2t+1$ variables with maximum algebraic immunity. In: Lai, X., Chen, K. (eds.) ASIACRYPT 2006. LNCS, vol. 4284, pp. 84–98. Springer, Heidelberg (2006)
12. Li, N., Qi, W.: Boolean functions of an odd number of variables with maximum algebraic immunity. Sci. China Ser. F-Information Sciences 50(3), 307–317 (2007)
13. Liu, M., Pei, D., Du, Y.: Identification and construction of Boolean functions with maximum algebraic immunity. Sci. China Ser. F-Information Sciences 53(7), 1379–1396 (2010)
14. Carlet, C., Dalai, D.K., Gupta, K.C., Maitra, S.: Algebraic Immunity for Cryptographically Significant Boolean Functions: Analysis and Construction. IEEE Trans. Inform. Theory 52(7), 3105–3121 (2006)
15. Dalai, D.K., Maitra, S., Sarkar, S.: Basic Theory in Construction of Boolean Functions with Maximum Possible Annihilator Immunity. Designs, Codes and Cryptography 40(1), 41–58 (2006)

Improving the Algorithm 2 in Multidimensional Linear Cryptanalysis

Phuong Ha Nguyen, Hongjun Wu, and Huaxiong Wang

Division of Mathematical Sciences,
School of Physical and Mathematical Sciences
Nanyang Technological University, Singapore
ng0007ha@e.ntu.edu.sg, {wuhj,hxwang}@ntu.edu.sg

Abstract. In FSE'09 Hermelin *et al.* introduced the Algorithm 2 of multidimensional linear cryptanalysis. If this algorithm is m-dimensional and reveals l bits of the last round key with N plaintext-ciphertext pairs, then its time complexity is $\mathcal{O}(mN2^l)$. In this paper, we show that by applying the *Fast Fourier Transform* and *Fast Walsh Hadamard Transform* to the Algorithm 2 of multidimensional linear cryptanalysis, we can reduce the time complexity of the attack to $\mathcal{O}(N + \lambda 2^{m+l})$, where λ is $3(m + l)$ or $4m+3l$. The resulting attacks are the best known key recovery attacks on 11-round and 12-round *Serpent*. The data, time, and memory complexity of the previously best known attack on 12-round *Serpent* are reduced by factor of $2^{7.5}$, $2^{11.7}$, and $2^{7.5}$, respectively. This paper also simulates the experiments of the improved Algorithm 2 in multidimensional linear cryptanalysis on 5-round *Serpent*.

Keywords: Multidimensional linear cryptanalysis, Linear Cryptanalysis, Serpent, Fast Fourier Transform, Fast Walsh Hadamard Transform.

1 Introduction

In 1993, Matsui [13] introduced the linear cryptanalysis and two algorithms, Algorithm 1 and Algorithm 2. These two algorithms exploit block cipher's linear approximation between the plaintext P, the ciphertext C and the secret key with a certain probability. Algorithm 2 is the modified version of Algorithm 1 by relaxing the first round and/or the last round of the linear approximation of Algorithm 1. Algorithm 2 can recover multiple secret key bits instead of one bit in Algorithm 1 with the same number of samples N required by the attack. However, the time complexity of the distillation phase of Algorithm 2 is much higher than that of Algorithm 1. Without loss of generality, the Substitution Permutation Network (SPN) n-round block cipher is considered, the last round of cipher is relaxed with l-bit subkey K_n of the last round key involved in the attack and N samples given. The time complexity for recovering l-bit K_n is $\mathcal{O}(2^l N)$. Matsui [15] suggested a method to reduce the time complexity to $\mathcal{O}(N + 2^{2l})$ by using one pre-computed table T. In 2007, Collard [7] *et al.* realized that table

U. Parampalli and P. Hawkes (Eds.): ACISP 2011, LNCS 6812, pp. 61–74, 2011.
© Springer-Verlag Berlin Heidelberg 2011

T is a *circulant* matrix and applied the *Fast Fourier Transform (FFT)* to this table to reduce the time complexity to $\mathcal{O}(N + 3l2^l)$.

There are several important papers [12,2,9] which extended Matsui's algorithms, i.e., m linear approximations are used instead of one linear approximation in the attack. By combining m linear approximations together in the attack, the number of samples N is expected to be reduced. Two of the most outstanding models are the full Biryukov's model [11] and Hermelin's model [10] which is known as multidimensional linear cryptanalysis. The time complexity of the extended Algorithm 2 of these models is $\mathcal{O}(mN2^l)$. Even if we apply the algorithm in [7] to the full Biryukov's model, the time complexity of the extended Algorithm 2 is $\mathcal{O}(2^mN + 3l2^{l+m})$, which is still too high. Due to the high time complexity, the extended Algorithm 2 of these models can not be used to attack as many rounds as Algorithm 2 of Matsui or the extended Algorithm 1 does.

The extended Algorithm 2 in [10] computes the information for the attack by using a very natural algorithm which, however, it has a high computation cost, i.e., $\mathcal{O}(mN2^l)$. Our main contribution is to show that by applying the FFT and *Fast Walsh Hadamard Transform (FWHT)* to the extended Algorithm 2, the time complexity of the attack is reduced to $\mathcal{O}(N + \lambda 2^{m+l})$, where λ is $3(m+l)$ or $4m + 3l$. Firstly, we work with N pairs of samples (P, C) to extract the information for the attack only once. Secondly, we use two assistant tables to compute the information extracted from N samples. Finally, we apply FFT, $FWHT$ to these tables to derive the correct key from 2^l guessed keys. Based on the steps above, we introduce two methods, *method 1* and *method 2*, for two cases of the extended Algorithm 2 to reduce the time complexity to $\mathcal{O}(N + \lambda 2^{m+l})$, where λ is $3(m+l)$ or $4m+3l$. Applying these methods to the extended Algorithm 2, we attack 11-round *Serpent* with 2^{116} data complexity, $2^{107.5}$ time complexity, 2^{104} memory complexity. The previously best known result on 11-round *Serpent* is given in [7] with 2^{118} data complexity, $2^{114.3}$ time complexity, 2^{108} memory complexity. Therefore, the attack in this paper is the best on 11 rounds. For 12-round *Serpent*, we have two attacks. The first reduces the data and time complexity in [17] by factors of $2^{5.5}$ and $2^{19.4}$, respectively, but the memory requirements are higher by a large factor of 2^{100}. In comparison with [17], the second attack reduces the data, time, and memory complexity by $2^{7.5}$, $2^{11.7}$, and $2^{7.5}$, respectively. Hence our second attack on 12 rounds is much better than [17].

This paper is organized as follows. Section 2 introduces the background as well as some notations and definitions. Next, Section 3 gives the brief description of Algorithm 2 of the full Biryukov model and of the Hermelin model. In this section, the algorithm with time complexity $\mathcal{O}(mN2^l)$, which is used to compute the information for determining the correct key, is also explained. Then, Section 4 describes two efficient computing methods for two cases of the extended Algorithm 2. The experimental results on 5-round *Serpent* and the best cryptanalytic results against 11-round and 12-round *Serpent* are reported in Section 5. Lastly, Section 6 concludes this paper.

2 Notations and Background

The notations and definitions used in the rest of paper are introduced in this section. For the sake of convenience we follow the notations in [9].

Let $V_m = GF(2)^m$ denote the space of m-dimensional binary vectors. If $a = (a_1, \ldots, a_m), b = (b_1, \ldots, b_m)$ are two vectors in V_m, its inner product $a \cdot b$ is defined as follows: $a \cdot b = \bigoplus_{i=1}^{m} a_i b_i$. The function $f : V_m \to V_1$ is called a Boolean function. The function $f = (f_1, \ldots, f_m) : V_l \to V_m$ is called a vectorial Boolean function, where $f_i : V_l \to V_1$ is a Boolean function for $i = 1, \cdots, m$.

Let X be a random variable in V_m and $p_\eta = Pr(X = \eta)$, where $\eta \in V_m$. Then $p = (p_0, p_1, \ldots, p_{2^m-1})$ is the probability distribution of random variable X. If we associate with a vectorial Boolean function $f : V_l \to V_m$ a random variable $Y := f(X)$, where X is uniformly distributed in V_l, then the probability distribution of Y is $p(f) := (p_0(f), \ldots, p_{2^m-1}(f))$ where $p_\eta(f) = Pr(f(X) = \eta)$, for all $\eta \in V_m$.

The correlation between a binary random variable X and 0 is $\rho = Pr(X = 0) - Pr(X = 1)$. The value $\epsilon = \rho/2$ is called a bias of the random variable X. Let $g : V_m \to V_1$ be a Boolean function. Its correlation with 0 is defined as

$$\rho = 2^{-m}(\#\{\eta \in V_m | g(\eta) = 0\}) - \#\{\eta \in V_m | g(\eta) = 1\})$$
$$= 2Pr(g(X) = 0) - 1,$$

where X is uniformly distributed in V_m.

We recall important results on the *Fast Fourier Transform* [8], *Fast Walsh-Hadamard Transform* [16] and *Parseval's* theorem. Given a k-dimensional vector $\mathbf{E} = (E_1, \ldots, E_k)$ and a matrix $\mathbf{F}^{k \times k}$, we have k-dimensional vector $\mathbf{D} = \mathbf{F}\mathbf{E}^T$, where $\mathbf{E}^T$ is the transpose of $\mathbf{E}$. The matrix $\mathbf{F}$ is a *Hadamard* matrix if $\mathbf{F}(i, j) = (-1)^{ij}$ for $i, j = 0, \cdots, k - 1$. The matrix $\mathbf{F}$ is a *Fourier* matrix if $\mathbf{F}(i, j) = (e)^{2\pi\sqrt{-1}ij/k}$, for $i, j = 0, \cdots, k - 1$. If matrix $\mathbf{F}$ is either *Fourier* or *Hadamard*, then vector $\mathbf{D}$ can be computed with complexity $\mathcal{O}(k \log k)$ instead of $\mathcal{O}(k^2)$ by FFT or $FWHT$, respectively.

If $\mathbf{F}$ is a matrix then we denote $\mathbf{F}[\cdot, j], \mathbf{F}[i, \cdot]$ the j-th column and the i-th row of matrix $\mathbf{F}$, respectively.

3 Algorithm 2 of Multidimensional Linear Cryptanalysis

The m-dimensional linear cryptanalysis based on m linear approximations is introduced by Hermelin *et.al* [9]. In the attack, 2^m linear approximations which are the combinations of m linear approximation are exploited to reduce the data complexity. At first, we briefly review the multidimensional linear cryptanalysis as well as the results of [9], which are needed for our efficient computing methods. Then, we present the extended Algorithm 2 of multidimensional linear cryptanalysis of the full Biryukov's model and the Hermelin's model and we explain the high time complexity of the extended Algorithm 2 afterwards.

3.1   Construction of Multidimensional Probability Distribution

Let $f : V_l \to V_n$ be a vectorial Boolean function and binary vectors $w_i \in V_n, u_i \in V_l$ $(i = 1, \ldots, m)$ be selection patterns such that pairs (u_i, w_i) are linearly independent. Define the functions g_i as

$$g_i(\eta) := w_i \cdot f(\eta) \oplus u_i \cdot \eta, \quad \forall \eta \in V_l$$

and g_i has correlation ρ_i $(i = 1, \cdots, m)$. Then $\rho_1, \ldots, \rho_m$ are called the base-correlations and $g_1, \ldots, g_m$ are the base approximations of f. Let $g = (g_1, \ldots, g_m)$ be an m-dimensional vectorial Boolean function and denote $p = (p_0, \ldots, p_{2^m-1})$ probability distribution of g.

Lemma 1. [9] Let $g = (g_1, \ldots, g_m) : V_l \to V_m$ be a vectorial Boolean function and $p = (p_0, \ldots, p_{2^m-1})$ its probability distribution. Then

$$2^l p_\eta = 2^{-m} \sum_{a \in V_m} \sum_{b \in V_l} (-1)^{a \cdot (g(b) \oplus \eta)}, \quad \eta \in V_m.$$

Define

$$\rho(a) = 2^{-l} \sum_{b \in V_l} (-1)^{a \cdot g(b)} = Pr(a \cdot g(X) = 0) - Pr(a \cdot g(X) = 1),$$

where X is an random variable uniformly distributed in V_l. Thus, the combined approximation $a \cdot g$ has the correlation $\rho(a)$ for all $a \in V_m$.

Corollary 1. Let $g : V_l \to V_m$ be a vectorial Boolean function with probability distribution p and correlations $\rho(a)$ of the combined approximations $a \cdot g$, for all $a \in V_m$. Then for $\eta \in V_m$,

$$p_\eta = 2^{-m} \sum_{a \in V_m} (-1)^{a \cdot \eta} \rho(a). \tag{1}$$

3.2   Brief Analysis on Algorithm 2 of the Full Biryukov's Model and Hermelin's Model

Algorithm 2 of the multidimensional linear cryptanalysis is the extension of the Matsui's Algorithm 2 [13] by using 2^m linear approximations constructed from m base linear approximations. Let us consider the SPN block cipher with n rounds. Let $u \cdot P \oplus v \cdot C_{n-1} \oplus w \cdot \bar{K}$ be the $(n-1)$-round linear approximation of the block cipher, where u, v, w are the selection patterns of the plaintext P, the output C_{n-1} of $(n-1)$-th round, and the inner key $\bar{K}$, respectively. In the Matsui's Algorithm 2, this linear approximation is extended to n rounds by replacing $v \cdot C_{n-1} = z \cdot S^{-1}(K_n \oplus C_l)$, where z is a selection pattern corresponding to v on the input of the last round, S^{-1} is the inverse of the last S-box layer, K_n is the l bits out of the last round key and C_l is the l bits out of the ciphertext C involved in the attack. Let us denote $z \cdot S^{-1}(K_n \oplus C_l)$ function $f(K_n \oplus C_l)$, then we have the following linear approximation which is used in Matsui's Algorithm 2

$$u \cdot P \oplus f(K_n \oplus C_l) \oplus w \cdot \bar{K}.$$

Since the above linear approximation has correlation, the linear approximation $u \cdot P \oplus f(K_n \oplus C_l)$ also has correlation and then Matsui's Algorithm 2 [13] will determine the right key K_n based on this linear approximation. Instead of using one linear approximation, the extended Algorithm 2 uses many linear approximations. Let us assume that the extended Algorithm 2 is m-dimensional and recovers l-bit subkey K_n of the last round key. The m linear approximations used in the extended Algorithm 2 as base linear approximations are $g_i := u_i \cdot P \oplus f(K_n \oplus C_l)$ for $i = 1, \ldots, m$, where u_i are the selection patterns of plaintext. Define the vectorial boolean function $g = (g_1, \ldots, g_m)$. Then, $a \cdot g := a \cdot (u_1 \cdot P, \ldots, u_m \cdot P) \oplus f(K_n \oplus C_l)$ are combined approximations with the correlation $\rho(a)$ for $a \in V_m$.

The aim of the extend Algorithm 2 is to exploit 2^m combined approximations $a \cdot g$ in order to reduce the the data complexity N, i.e., $[10] N \sim \mathcal{O}(1/c)$, where $c = \sum_{\forall a \in V_m, a \neq 0} \rho^2(a)$ is called a capacity of the system. In practice, the correlation $\rho(a)$ of $a \cdot g$ is computed by *Piling-Up Lemma* [9]. The probability distribution $p = (p_0, \ldots, p_{2^m-1})$ of g is not uniformly distributed. The distillation phase [2,3] working on N samples (P, C) helps us to derive the correct K_n from 2^l guessed keys K. The time complexity of this phase is the major factor in the total time complexity of the attack. Let $\mathbf{p}^{2^l \times 2^m}$ be the matrix in which each row is empirical probability distributions of g for each candidate key K, i.e., $\mathbf{p}[K, \cdot] = (\mathbf{p}[K, 0], \ldots, \mathbf{p}[K, 2^{m-1}])$ ($\forall K \in V_l$). In [11], Hermelin *et al.* proved that the full Biryukov's model is equal to the convolution model and the distillation phase of the convolution model uses the same algorithm with the Hermelin's model of Algorithm 2 to compute the empirical m-dimensional probability distribution of g, i.e., $\mathbf{p}$. The algorithm in [11] is as follows:

Input: N pairs of plaintext-ciphertexts $\{(P_1, C_1), \ldots, (P_N, C_N)\}$,
         $g = (g_1, \ldots, g_m)$
Output: the matrix $\mathbf{p}$ of empirical probability distribution for all
           candidate $K_n \in V_l$
foreach $K \in V_l$, $\eta \in V_m$ **do**
  | set $\mathbf{p}[K, \eta] = 0$.
end
foreach $K \in V_l$, $t:=0, \ldots, N$ **do**
  | Calculate $\eta = g(P_t, C_t, K) = (g_1(P_t, C_t, K), \ldots, g_m(P_t, C_t, K))$;
  | $\mathbf{p}[K, \eta]$++;
end
foreach $K \in V_l$, $\eta \in V_m$ **do**
  | $\mathbf{p}[K, \eta] = \mathbf{p}[K, \eta]/N$.
end

Algorithm 1. The Hermelin's Algorithm to compute the $\mathbf{p}$ with given N samples plaintext-ciphertext

In Hermelin's Algorithm, we use only m linear approximations $g_1, \ldots, g_m$ and the distillation phase computes all values of $\mathbf{p}[K, \cdot]$ ($\forall K \in V_l$) with $mN2^l$ time complexity. Based on the *Wrong Key Hypothesis* [10], if K is the wrong key, its $\mathbf{p}[K, \cdot]$ is uniformly distributed and the right key K's $\mathbf{p}[K, \cdot]$ is non uniformly distributed. The χ^2 or LLR statistic [10] is applied to determine what kind of probability distribution for all $\mathbf{p}[K, \cdot]$ ($\forall K \in V_l$). Then, the correct key K is determined by the information received from the chosen statistic.

Although the algorithm used to compute all $\mathbf{p}[K, \cdot]$ ($\forall K \in V_l$) directly from N samples (P, C) is very elegant and can be easily implemented, its complexity is much higher in comparison to the extended Algorithm 1 [11,14] or to Algorithm 2 of Matsui. In cryptanalysis, it is the bottleneck in terms of time complexity. Therefore the extended Algorithm 2 is not comparable to Algorithm 2 of Matsui in terms of the number of rounds attacked or time complexity.

4   Efficient Computation of Distillation Phase of Extended Algorithm 2 of Matsui

In this section, we describe two efficient computing methods, *method 1* and *method 2*, for two cases in multidimensional linear cryptanalysis. These methods use Lemma 1, Corollary 1, the *FFT*, *FWHT* and two pre-computed tables T, E to avoid repeatedly working on N samples and 2^l guessed keys K_n in order to reduce the time complexity from $\mathcal{O}(mN2^l)$ to $\mathcal{O}(N + \lambda 2^{m+l})$, where λ is a positive number which is specified for each case.

4.1   Case 1 and *Method 1*

In the first case, we have N pairs of samples (P, C) and m base linear approximations $g_i := u_i \cdot P \oplus f(K_n \oplus C_l)$ for $i = 1, \ldots, m$, where C_l is l bits out of ciphertext C involved in attack. All m base approximation g_i share one function $f(\cdot)$ and m selection patterns of plaintext $u_1, \ldots, u_m$ are linearly independent. Suppose $g = (g_1, \ldots, g_m)$, we then need to compute the empirical probability distribution matrix $\mathbf{p}$ of g.

Lemma 2. *For the first case, there exists a method (method 1) which computes the empirical probability distribution $\mathbf{p}$ of g with $\mathcal{O}(N + (4m + 3l)2^{m+l})$ time complexity.*

Due to the word count limit of this paper, the proof and description of *method 1* are presented in Appendix A.

4.2   Case 2 and *Method 2*

In this case, the first and the last round keys of cipher are considered in the attack. We have N pairs of samples (P, C) and m base linear approximations $g_i := f_1^i(K_1 \oplus P_{l_1}) \oplus f_2(K_n \oplus C_{l_2})$ for $i = 1, \ldots, m$, where K_1 is l_1 bits of the first round key, P_{l_1} is l_1 bits of the plaintext P, K_n is l_2 bits of the last round key, C_{l_2} is l_2 bits of the ciphertext C, f_1^i, f_2 ($i = 1, \ldots, m$) are the boolean functions in [13,7]. The functions $f_1^i(\cdot)$ are constructed in the same way of $f_2(\cdot)$ for $i = 1, \ldots, m$. Suppose $g = (g_1, \ldots, g_m)$, we then need to compute the empirical probability distribution matrix $\mathbf{p}$ of g.

Lemma 3. *For the second case, there exists a method (method 2) which computes the empirical probability distribution* **p** *of* g *with* $\mathcal{O}(N + 3(l_1 + l_2 + m)2^{l_1+l_2+m})$ *time complexity.*

Due to the word count limit of this paper, the proof and description of *method 2* are presented in Appendix B.

5 Results on Cryptanalysis of Serpent

In this section, the experimental results of 5-round *Serpent* are reported first. *Serpent* [1] is the SPN block cipher which is one of the five AES finalists. *Serpent* has 128-bit block size and supports 128-bit, 192-bit and 256-bit keys. Based on the experimental results, we manage to confirm some facts claimed in [7] for the extended Algorithm 2. Then, we describe our attacks against 11-round and 12-round *Serpent*. The cryptanalytic results show that the multidimensional linear cryptanalysis can reduce the number of samples N and the time complexity in comparison to other methods, i.e., the linear cryptanalysis of Matsui and the differential-linear cryptanalysis on *Serpent* [17].

5.1 Experimental Results on 5-Round Serpent

In this subsection, the experiment on 5-round Serpent aims to confirm the results in [3,10] by using the extended Algorithm 2 with *method 1*. We call the extended Algorithm 2 with *method 1* or *method 2* improved extended Algorithm 2. For simplicity, the χ^2 statistic is applied to 4-, 7-, 10-dimensional linear cryptanalysis with $N = 2^{20}, 2^{21}, \ldots, 2^{26}$ and $l = 12$. The capacities are $2^{-23}, 2^{-22}, 2^{-21}$ for 4-, 7-, 10-dimensional linear cryptanalysis, respectively. If the correct l-bit key K_n has rank r out of 2^l possible guessed keys, then the attack obtains an $(l - \log_2 r)$-bit advantage over exhaustive search [3]. We test fifty keys and compute the advantage [2,3] for each key. Then, we construct the comparison table of average advantages for 4-, 7-, and 10-dimensional linear cryptanalysis for each set of N pairs. Table 1 displays the average of advantages for each set of N samples and 4-, 7-, 10-dimensional linear cryptanalysis.

The results confirm the facts that the extended Algorithm 2 with χ^2 statistic does not work as well as expected and when $m = 4$ the average of advantages is the highest. Furthermore, the experiments tell us that, given capacity c and if $N = 2^2/c$,then the advantage ≥ 1 of certain key has more than 50% success

Table 1. The average of advantages on 5-round *Serpent*

$\log_2(N)$	20	21	22	23	24	25	26
m=4	1.30	1.45	1.60	1.64	1.65	1.65	1.55
m=7	1.46	1.50	1.39	1.19	1.29	1.35	1.37
m=10	1.21	1.19	1.30	1.25	1.38	1.55	1.52

probability. According to the experimental results in [3], if the combination of LLR statistic and maximum KL is used instead of χ^2 statistic, then the multidimensional linear cryptanalysis works better, i.e., the average of advantages is much higher for each set of N samples.

5.2   Cryptanalysis of 11-Round and 12-Round Reduced Serpent

Before going to the cryptanalysis of 11-round and 12-round *Serpent*, we recall the fact that both methods do not require any decryption or encryption operation as that in Algorithm 2 of Matsui or in the differential-linear cryptanalysis [17]. We only need to extract the subset of bits involved in the plaintext P and the ciphertext C to compute the table E. The linear cryptanalysis can reach 11-round reduced *Serpent* by Algorithm 2 of Matsui based on the 9-round linear approximation [1,7,5]. The improved extended Algorithm 2 uses the 9-round linear approximation in [7,6,4] which is called *9-round linear approximation of Collard*. The 10-round linear approximation is constructed by adding one more round before the above 9-round linear approximation. The 11-round linear approximation, which is constructed from this 10-round linear approximation with the last round S_4 relaxed, is as follows:

$$S_2 \longrightarrow \underbrace{\underbrace{S_3 \longrightarrow \ldots \longrightarrow S_3}_{\text{9-round linear approximation of Collard}} \longrightarrow \underbrace{S_4}_{\text{round relaxed}}}_{\text{10-round}} .$$

In S_2 at the first round, there are 15 active S-boxes. The output masks of these active S-boxes described in hexadecimal number are:

S-box	0	3	5	6	9	10	11	13	15	16	17	24	28	29	30
output-mask	e	d	7	c	3	6	4	b	4	b	2	1	1	e	8

Table 2. Comparison of the attacks against reduced-round *Serpent*

Round		complexity		
		data	time	memory
11	Lin.cryptanalysis [5]	2^{118} KP	2^{166} En	2^{121}
	Lin.cryptanalysis [5]	2^{118} KP	$2^{173.5}$ En	2^{97}
	Lin.cryptanalysis [7]	2^{118} KP	$2^{114.3}$ En	2^{108}
	MultiDim.cryptanalysis [this paper](*method 1*)	2^{116} KP	$2^{107.5}$ En	2^{104}
	MultiDim.cryptanalysis [this paper](*method 1*)	2^{118} KP	$2^{109.5}$ En	2^{100}
12	Differential-Lin.cryptanalysis [17]	$2^{123.5}$ CP	$2^{249.4}$ En	$2^{128.5}$
	MultiDim.cryptanalysis [this paper](*method 2*)	2^{118} KP	$2^{228.8}$ En	2^{228}
	MultiDim.cryptanalysis [this paper](*method 1*)	2^{116} KP	$2^{237.5}$ En	2^{121}

En - Encryptions, KP - Known Plaintexts, CP - Chosen Plaintexts.

Based on the observation on active S-boxes of S_2, we choose 56 linearly independent selection patterns of plaintext $u_1, \ldots, u_{56}$, i.e., we have 56-dimensional linear cryptanalysis attack.

If S_4 at the 11-th round has 12 active S-boxes, then in the 11-th round, the number of key bits l is 48 and the capacity c of 56-dimensional system is computed as follows:

$$c = \sum_{a \neq 0, a \in V_{56}} \rho^2(a)$$

$$= (2^{-57})^2 \sum_{m=0}^{11} C_m^{11} 8^m 2^{11-m} 4^4 (2^{-4m} 2^{-2(11-m)} 2^{-8}) \tag{2}$$

$$= 2^{-114}.$$

Let the number of samples required N be 2^{116}. If *method 1* is used, then the total time complexity is $(N/352 + 2^{111.6}/352)$ which is equal to $2^{107.5}$ 11-round *Serpent* encryptions and the memory will be 2^{104}.

If S_4 at the 11-round has 11 active S-boxes, then in the 11-th round, l is 44 and the capacity is 2^{-116}. Let N be 2^{118}, then the total time complexity is $(N/352 + 2^{107.5}/352)$ which is equal to $2^{109.5}$ 11-round *Serpent* encryptions and the memory is 2^{100}.

We attack the 12-round *Serpent* by using the above 11-round linear approximation and relaxing 12-th round S_4, which has 11 active S-boxes.

$$S_1 \longrightarrow S_2 \longrightarrow \underbrace{S_3 \longrightarrow \ldots \longrightarrow S_3}_{\text{9-round linear approximation of Collard}} \longrightarrow \underbrace{S_4}_{\text{round relaxed}} \ .$$
$$\underbrace{\qquad\qquad\qquad\qquad\qquad\qquad\qquad\qquad\qquad\qquad\qquad\qquad}_{\text{11-round}}$$

The capacity is 2^{-116}. Let $N = 2^{118}$, $l_1 = 128$ bits and $l_2 = 44$ bits. Based on *method 2*, the total time complexity is $(N/384 + 2^{237.4}/384)$ which is equal to $2^{228.8}$ 12-round *Serpent* encryptions and the memory is 2^{228}.

We can attack 12-round *Serpent* by *method 1* by relaxing the first and the 12-th rounds. We search for all 2^{128} possible keys in S_1 at the first round and use *method 1* for the other 11 rounds similarly to the case $N = 2^{116}$. Then the time and memory complexity are $2^{237.5}, 2^{121}$. Table 2 shows that the improved extended Algorithm 2 is better than all the previously known algorithms, i.e., linear cryptanalysis and differential-linear cryptanalysis, on 11-round and 12-round *Serpent*.

6   Conclusion

We studied two methods to reduce the time complexity in distillation phase of extended Algorithm 2 from $\mathcal{O}(mN2^l)$ to $\mathcal{O}(N + \lambda 2^{m+l})$, where m is the number of dimension of the attack, N is the number of samples needed, l is the number of key bits in the first and/or in the last rounds. These methods are introduced when we combine the Corollary 1, the Lemma 1, exploiting the repeated structure of data and key and using 2 assistant pre-computed tables T, E. Applying $FFT, FWHT$ to the tables T, E, we have two efficient computing methods to determine the correct key out of 2^l guessed keys .

We have simulated the experiments on 5-round *Serpent* to check the improved extended Algorithm 2 and to confirm claims in [3,10]. Based on the results of the experiments, we develop attack on 11-round, 12-round *Serpent* by using the improved extended Algorithm 2. These results are the best among those reported so far. On 12-round *Serpent*, we can reduce the data complexity, time and memory complexity of the previously best known attack by factors of $2^{7.5}$, $2^{11.7}$ and $2^{7.5}$, respectively. The result of 11-round *Serpent* is even better than the best currently reported, i.e., data complexity, time complexity, and memory complexity are reduced by factor of $2^2, 2^7$, and 2^4, respectively. The improved extended Algorithm 2 is competitive to the extended Algorithm 1 and Algorithm 2 of Matsui in terms of data complexity, time complexity and the number of rounds attacked. The extended Algorithm 2 usually involves many active S-boxes in the outer round(s) because it uses many linear approximations. Intuitively, it is easy to reach the limit of complexity in terms of time complexity and memory complexity. It implies that the bound of number of rounds attacked in improved extended Algorithm 2 is close to that in Algorithm 2 of Matsui somehow.

Acknowledgements

This work was supported in part by the Singapore National Research Foundation under Research Grant NRF-CRP2- 2007-03. The first author is supported by the Singapore International Graduate (SINGA) Scholarship.

References

1. Biham, E., Dunkelman, O., Keller, N.: Linear cryptanalysis of reduced round serpent. In: Matsui, M. (ed.) FSE 2001. LNCS, vol. 2355, pp. 16–27. Springer, Heidelberg (2002)
2. Biryukov, A., De Cannière, C., Quisquater, M.: On multiple linear approximations. In: Franklin, M. (ed.) CRYPTO 2004. LNCS, vol. 3152, pp. 1–22. Springer, Heidelberg (2004)
3. Cho, J.Y., Hermelin, M., Nyberg, K.: A new technique for multidimensional linear cryptanalysis with applications on reduced round serpent. In: Lee, P.J., Cheon, J.H. (eds.) ICISC 2008. LNCS, vol. 5461, pp. 383–398. Springer, Heidelberg (2009)
4. Collard, B.: Private Communication (2010)
5. Collard, B., Standaert, F.-X., Quisquater, J.-J.: Improved and multiple linear cryptanalysis of reduced round serpent. In: Pei, D., Yung, M., Lin, D., Wu, C. (eds.) Inscrypt 2007. LNCS, vol. 4990, pp. 51–65. Springer, Heidelberg (2008)
6. Collard, B., Standaert, F.-X., Quisquater, J.-J.: Improved and Multiple Linear Cryptanalysis of Reduced Round Serpent - Description of the Linear Approximations (2007) (unpublished manuscript)

7. Collard, B., Standaert, F.-X., Quisquater, J.-J.: Improving the time complexity of matsui's linear cryptanalysis. In: Nam, K.-H., Rhee, G. (eds.) ICISC 2007. LNCS, vol. 4817, pp. 77–88. Springer, Heidelberg (2007)
8. Cormen, T.H., Stein, C., Rivest, R.L., Leiserson, C.E.: Introduction to Algorithms. Hill Higher Education. McGraw-Hill Higher Education, New York (2001)
9. Hermelin, M., Cho, J.Y., Nyberg, K.: Multidimensional linear cryptanalysis of reduced round serpent. In: Mu, Y., Susilo, W., Seberry, J. (eds.) ACISP 2008. LNCS, vol. 5107, pp. 203–215. Springer, Heidelberg (2008)
10. Hermelin, M., Cho, J.Y., Nyberg, K.: Multidimensional extension of matsui's algorithm 2. In: Dunkelman, O. (ed.) FSE 2009. LNCS, vol. 5665, pp. 209–227. Springer, Heidelberg (2009)
11. Hermelin, M., Nyberg, K.: Dependent linear approximations: The algorithm of biryukov and others revisited. In: Pieprzyk, J. (ed.) CT-RSA 2010. LNCS, vol. 5985, pp. 318–333. Springer, Heidelberg (2010)
12. Kaliski Jr., B.S., Robshaw, M.J.B.: Linear cryptanalysis using multiple approximations. In: Desmedt, Y.G. (ed.) CRYPTO 1994. LNCS, vol. 839, pp. 26–39. Springer, Heidelberg (1994)
13. Matsui, M.: Linear cryptanalysis method for DES cipher. In: Helleseth, T. (ed.) EUROCRYPT 1993. LNCS, vol. 765, pp. 386–397. Springer, Heidelberg (1994)
14. Nguyen, P.H., Wei, L., Wang, H., Ling, S.: On multidimensional linear cryptanalysis. In: Steinfeld, R., Hawkes, P. (eds.) ACISP 2010. LNCS, vol. 6168, pp. 37–52. Springer, Heidelberg (2010)
15. Matsui, M.: The first experimental cryptanalysis of the data encryption standard. In: Desmedt, Y.G. (ed.) CRYPTO 1994. LNCS, vol. 839, pp. 1–11. Springer, Heidelberg (1994)
16. Rao Yarlagadda, R.K., Hershey, J.E.: Hadamard Matrix Analysis and Synthesis: with Applications to Communications and Signal/image Processing. Kluwer Academic Publishers, Norwell (1997)
17. Dunkelman, O., Indesteege, S., Keller, N.: A differential-linear attack on 12-round serpent. In: Chowdhury, D.R., Rijmen, V., Das, A. (eds.) INDOCRYPT 2008. LNCS, vol. 5365, pp. 308–321. Springer, Heidelberg (2008)

Appendix

A   Proof and Description of Lemma 2

Proof. The Lemma 1 and the corollary 1 show that the probability distribution p of g can be calculated from correlations $\rho(a)$ of combined approximations $a \cdot g$ for $a \in V_m$. The $a \cdot g$ combined approximations are defined as follows:

$$a \cdot g := \begin{cases} 0 & \text{if } a \text{ is a zero vector} \\ a \cdot (u_1 P, \ldots, u_m P) \oplus f(K_n \oplus C_l) & a \in V_m, a \neq 0 . \end{cases}$$

Let $\mathbf{r}^{2^m \times 2^l}[a, K]$ be the matrix whose elements $\mathbf{r}[a, K]$ are the correlations of linear approximations $a \cdot g$ when $K_n = K$. We have 2 following cases:

1. a is zero vector. Since $a \cdot g$ is equal to 0, $\mathbf{r}[0, K] = 1$, $(\forall K \in V_l)$.

2. a is not zero vector. *method 1* below will compute $\mathbf{r}[a, K]$ ($K \in V_l$).

Since our aim is to reduce as much as possible the number of times of working with N samples for computing $\mathbf{r}[a, K]$, we will need 2 assistant tables $T^{2^m \times 2^l}$ and $E^{2^m \times 2^l}$.

In practice, the empirical $\mathbf{r}[a, K]$ is computed as follows [7]:

$$\mathbf{r}[a, K] = \frac{\#\{(P_t, C_t) : a \cdot g(P_t, C_t, K) = 0\} - \#\{(P_t, C_t) : a \cdot g(P_t, C_t, K) = 1\}}{N},$$

$$t = 1, \ldots, N.$$

We go through all $a \in V_m, (P_t, C_t)$ for $t = 1, \ldots, N$ to compute the value of table T. The elements of this table are computed as follows:

$$a(u_1 P_t, \ldots, u_m P_t) = \begin{cases} 0 \to T(a, (C_l)_t) + + \\ 1 \to T(a, (C_l)_t) - - \end{cases},$$

where $(C_l)_t$ is the l bits of ciphertext C_t involved in function $f(\cdot)$, $\forall a \in V_m$ and (P_t, C_t) ($t = 1, \ldots, N$).

Let $T[a, v] = T[a, v]/N$ ($a \in V_m, v \in V_l$), then

$$\mathbf{r}[a, K] = \sum_{v=0}^{2^l - 1} (-1)^{f(K \oplus v)} T[a, v].$$

Let $\mathbf{S}^{2^l \times 2^l}$ be a matrix with $\mathbf{S}[i, j] = (-1)^{f(i \oplus j)}$ ($\forall i, j \in V_l$). Hence,

$$\mathbf{r}[a, \cdot] = \mathbf{S} T[a, \cdot]. \tag{3}$$

Based on the proposition in [7], matrix $\mathbf{S}$ is a *circulant* matrix. Therefore, vector $\mathbf{r}[a, \cdot]$ is computed with complexity $3l2^l$ and we only need 2^l memory units to store the first column $((-1)^{f(0)}, \ldots, (-1)^{f(2^l - 1)})^T$ of matrix $\mathbf{S}$ to calculate $\mathbf{r}[a, \cdot]$.

The time complexity is very high if we directly compute T from N samples, i.e., $\mathcal{O}(2^m N)$. We show a way to compute T more efficiently by assistant table E. We construct $\mathbf{E}^{2^m \times 2^l}$ as follows:

$$E[h_1, h_2] = \#\{(P_t, C_t), t = 1, \ldots, N : (u_1 \cdot P_t, \ldots, u_m \cdot P_t) = h_1 \text{ and } (C_l)_t = h_2\},$$
$$h_1 \in V_m, h_2 \in V_l.$$

The time complexity for constructing $\mathbf{E}$ is mN and the memory complexity is 2^{m+l}. If we do m computing $u_1 \cdot P_t, \ldots, u_m \cdot P_t$ for each sample (P_t, C_t) at the same time, then the time complexity for constructing E is N.

For $\forall v \in V_l, \forall a \in V_m$, We have

$$T[a, v] = \sum_{h_1=0}^{2^m - 1} (-1)^{a h_1} E[h_1, v]. \tag{4}$$

Let $\mathbf{H}^{2^m \times 2^m}$ be a matrix with $H[i,j] = (-1)^{ij}$ ($\forall i,j \in V_m$). Then,

$$T[\cdot, v] = \mathbf{H}E[\cdot, v]. \tag{5}$$

Since matrix $\mathbf{H}$ is a *Hadamard* matrix, $T[\cdot, v]$ can be computed using *FWHT* algorithm with time complexity $m2^m$ and memory complexity 2^m.

In summary, the steps taken in *method 1* are as follows:

1. Compute $\mathbf{E}^{2^m \times 2^l}$ with N time complexity and 2^{m+l} memory complexity.
2. For $\forall v \in V_l$, compute $T[\cdot, v] = (T[0, v], \ldots, T[2^m - 1, v])$ using *FWHT* algorithm in (5). Time complexity is $2^l m 2^m (= m 2^{l+m})$ and memory complexity is 2^{m+l}.
3. For $\forall a \in V_m, a \neq 0$, compute $\mathbf{r}[a, \cdot]$ by the algorithm of *circulant* matrix [7] in (3). Time complexity is $2^m 3 l 2^l (= 3l 2^{l+m})$ and memory complexity is 2^l.
4. If a is a zero vector, then $\mathbf{r}[a, \cdot]$ is vector 1.
5. We need 2^{m+l} memory units to store all $\mathbf{r}[a, K]$ ($\forall K \in V_l, \forall a \in V_m$).
6. If we need to compute $\mathbf{p}[K, \cdot]$ ($\forall K \in V_l$), then time complexity is $3m 2^m 2^l (= 3m 2^{l+m})$ and 2^{m+l} memory units needed for all $\mathbf{p}$.

The total time complexity is $\mathcal{O}(N + 2^{l+m}(4m + 3l))$ and memory complexity is $\mathcal{O}(2^{m+l})$.

B   Proof and Description of Lemma 3

Proof. The combined approximations $a \cdot g$ are defined as follows:

$$a \cdot g = \begin{cases} a \cdot f_1(K_1 \oplus P_{l_1}) \oplus f_2(K_n \oplus C_{l_2}) \\ \quad \text{if } a \text{ is not a zero vector} \\ 0 \text{ if } a \text{ is a zero vector} \end{cases},$$

where $f_1(K_1 \oplus P_{l_1}) = (f_1^1(K_1 \oplus P_{l_1}), \ldots, f_1^m(K_1 \oplus P_{l_1}))$, $\forall a \in V_m$.

Let $\mathbf{r}^{2^m \times 2^{l_1} \times 2^{l_2}}$ be the matrix containing all the correlations of $a \cdot g$ with $a \in V_m, K_1 \in V_{l_1}, K_n \in V_{l_2}$. We have the following 2 cases:

1. If a is vector 0, then $\mathbf{r}[a, k_1, k_2] = 1$ ($\forall k_1 \in V_{l_1}, \forall k_2 \in V_{l_2}$).
2. If a is not vector 0, then we can compute $\mathbf{r}[a, k_1, k_2]$ ($\forall k_1 \in V_{l_1}, \forall k_2 \in V_{l_2}$) by *method 2* below.

We need 2 assistant tables $T^{2^m \times 2^{l_1} \times 2^{l_2}}$ and $E^{2^{l_1} \times 2^{l_2}}$.
Let

$$\mathbf{r}[a, k_1, \cdot] = (\mathbf{r}[a, k_1, 0], \ldots, \mathbf{r}[a, k_1, 2^{l_2} - 1]),$$
$$T[a, k_1, \cdot] = (T[a, k_1, 0], \ldots, T[a, k_1, 2^{l_2} - 1]).$$

For $\forall a \in V_m$, (P_t, C_t) $(t = 1, \ldots, N)$, $\forall k_1 \in V_{l_1}$, the elements of table T are calculated as follows:

$$af_1(k_1 \oplus (P_{l_1})_t) = \begin{cases} 0 \to T[a, k_1, (C_{l_2})_t] + + \\ 1 \to T[a, k_1, (C_{l_2})_t] - - \end{cases} ,$$

where $(P_{l_1})_t$ and $(C_{l_2})_t$ are l_1 bits of P_t and l_2 bits of C_t involved in the attack, respectively.

Let $T[a, k_1, k_2] = T[a, k_1, k_2]/N$.

With the same argument in *method 1*, we have

$$\mathbf{r}[a, k_1, k_2] = \sum_{v=0}^{2^{l_2}-1} (-1)^{f_2(k_2 \oplus v)} T[a, k_1, v].$$

Let $\mathbf{S}_2^{2^{l_2} \times 2^{l_2}}$ be a matrix with $\mathbf{S}_2[i, j] = (-1)^{f_2(i \oplus j)}$ $(\forall i, j \in V_{l_2})$. Hence,

$$\mathbf{r}[a, k_1, \cdot] = \mathbf{S}_2 T[a, k_1, \cdot]. \tag{6}$$

According to [7], matrix $\mathbf{S}_2$ is a *circulant* matrix. Hence, vector $\mathbf{r}[a, k_1, \cdot]$ is computed with time complexity $3l_2 2^{l_2}$ and 2^{l_2} memory units.

In order to efficiently compute $T[a, k_1, v]$, we construct the table $\mathbf{E}^{2^{l_1} \times 2^{l_2}}$ as follows:

$$E[h_1, h_2] = \sharp\{(P_t, C_t), t = 1, \ldots, N : (P_{l_1})_t = h_1, (C_{l_2})_t = h_2\}, \forall h_1 \in V_{l_1}, \forall h_2 \in V_{l_2}.$$

The time complexity to construct $\mathbf{E}$ is N and the memory complexity is $2^{l_1+l_2}$. Then,

$$T[a, k_1, v] = \sum_{u=0}^{2^{l_1}-1} (-1)^{af_1(k_1 \oplus u)} E[u, v]. \tag{7}$$

Let $\mathbf{S}_a^{2^{l_1} \times 2^{l_1}}$ be a matrix with $\mathbf{S}_a[i, j] = (-1)^{af_1(i \oplus j)}$ $(\forall i, j \in V_{l_1})$. Then,

$$T[a, \cdot, k_2] = \mathbf{S}_a E[\cdot, v], \tag{8}$$

where $E[., v]$ is the column of matrix $\mathbf{E}$.

Based on [7], matrix S_a is a *circulant* matrix. Hence, the vector $T[a, \cdot, k_2]$ is computed in $3l_1 2^{l_1}$ computations and the memory needed is 2^{l_1}.

In summary, the steps taken in *method 2* are as follows:

1. Construct table $\mathbf{E}$. Time complexity is N and the memory needed for $\mathbf{E}$ is $2^{l_1+l_2}$.
2. For $\forall a \in V_m, \forall v \in V_{l_2}$, we compute all $T[a, \cdot, k_2]$ by (8). Time complexity is $2^m 2^{l_2} 3l_1 2^{l_1}(= 3l_1 2^{l_1+l_2+m})$. We need $2^{m+l_1+l_2}$ memory units for all $T[a, k_1, v]$ $(\forall a \in V_m, \forall k_1 \in V_{l_1}, \forall v \in V_{l_2})$.
3. For $\forall a \in V_m, a \neq 0, \forall k_1 \in V_{l_1}$ compute all $\mathbf{r}[a, k_1, \cdot]$ by (6). The time complexity is $2^m 2^{l_1} 3l_2 2^{l_2}(= 3l_2 2^{l_1+l_2+m})$. If a is a vector zero, the $\mathbf{r}[a, k_1, \cdot]$ is the vector 1 $(\forall k_1 \in V_{l_1})$. We need $2^{l_1+l_2+m}$ memory for $\mathbf{r}$.
4. If we need to compute $\mathbf{p}$, then time complexity is $3m 2^m 2^{l_1+l_2}(= 3m 2^{l_1+l_2+m})$ and memory complexity is $2^{m+l_1+l_2}$.

Consequently, the total time complexity is $\mathcal{O}(N + 3(l_1 + l_2 + m)2^{l_1+l_2+m})$ and the memory complexity is $\mathcal{O}(2^{m+l_1+l_2})$.

State Convergence in the Initialisation of Stream Ciphers

Sui-Guan Teo[1], Ali Al-Hamdan[1], Harry Bartlett[1,2], Leonie Simpson[1,2],
Kenneth Koon-Ho Wong[1], and Ed Dawson[1]

[1] Information Security Institute,
Queensland University of Technology
126 Margaret Street, Brisbane Qld 4001, Australia
{sg.teo,kk.wong,e.dawson}@qut.edu.au, a.alhamdan@student.qut.edu.au
[2] Faculty of Science and Technology,
Queensland University of Technology
GPO Box 2434, Brisbane Qld 4001, Australia
{h.bartlett,lr.simpson}@qut.edu.au

Abstract. An initialisation process is a key component in modern
stream cipher design. A well-designed initialisation process should ensure
that each key-IV pair generates a different keystream. In this paper, we
analyse two ciphers, A5/1 and Mixer, for which this does not happen
due to state convergence. We show how the state convergence problem
occurs and estimate the effective key-space in each case.

Keywords: Stream cipher, initialisation, state convergence, A5/1, Mixer.

1 Introduction

Modern stream cipher applications use a secret key and an initialisation vector (IV) to form an initial internal state before keystream generation begins.
A good initialisation process should ensure that each key-IV pair generates a
distinct keystream. This is possible for recent proposals where the state size is
large enough but may not be the case for older designs. This paper analyses two
stream ciphers, A5/1 and Mixer, where state convergence occurs during initialisation, resulting in different key-IV pairs producing the same keystream. We
show how state convergence occurs in each case and demonstrate that increasing the number of iterations in the initialisation process effectively decreases the
security provided.

2 Background and Notation

Keystream generators for stream ciphers operate by maintaining an internal state
and applying update and output functions to the state. In many cases, the state
space is provided by a combination of linear and/or nonlinear feedback shift
registers (LFSR/NLFSR respectively). In this paper, we consider two ciphers

U. Parampalli and P. Hawkes (Eds.): ACISP 2011, LNCS 6812, pp. 75–88, 2011.
© Springer-Verlag Berlin Heidelberg 2011

based on binary shift registers, where each register stage holds one bit. We use the notation $R_t[i]$ to denote the contents of stage i of register R at time t where $i = 0, 1, \ldots, r - 1$, for an r-bit register. The state S of a stream cipher is of size s bits. For the two ciphers examined in this paper, s is the sum of the component register lengths.

Modern keystream generators take two inputs: a secret key and an IV, of size l and j bits respectively. A stream cipher with an l-bit key and a j-bit IV has a keyspace of 2^l bits and an IV-space of 2^j bits. Let $k_0, k_1, \ldots k_{l-1}$ represent the l-bit key and $v_0, v_1, \ldots v_{j-1}$ represent the j-bit IV. Before keystream generation commences, a key-IV pair is used to form an internal state value. This process is referred to as initialisation and can be considered as a mapping from binary vectors of length $l + j$ to those of length s.

The purpose of the initialisation process is to diffuse the key-IV pair across the entire state and make mathematical relationships between the key-IV pair and the keystream hard to establish. The initialisation process is often performed in three phases: key-loading, IV-loading and the diffusion phase. In the key-loading and IV-loading phase, the secret key and IV are transferred to the stream cipher's state. When both the secret key and IV have been transferred, the stream cipher is in its *"loaded state"*. Following this, the diffusion phase begins. This is generally the most complex phase and it is important, as using the loaded state directly to begin keystream generation could make the stream cipher vulnerable to correlation attacks. The diffusion phase consists of a number of iterations, denoted α in this paper, of the initialisation state-update function. Each iteration of the initialisation state-update function can be considered as a function which maps the state space to itself. This mapping should be one-to-one and nonlinear in nature. After the initialisation process is complete, the keystream generator is said to be in its *initial state*.

To prevent time-memory-data tradeoff attacks, modern stream ciphers have internal states which are at least the size of the key-IV pair. That is, $2^s \geq 2^{l+j}$. Since the state space is at least the size of the space spanned by all key-IV pairs, it is reasonable to expect that the initialisation process will be one-to-one, that is, each distinct key-IV pair should map to a distinct state at the end of initialisation.

Some initialisation processes are not one-to-one. Considering the state-update function in the forwards direction, for a given S_t, there is a single S_{t+1}. However, when considering the reverse direction, for a given value of S_{t+1}, there may be multiple values for S_t. That is, multiple states converge during one iteration of the initialisation state update function. If this state convergence occurs at any point during the initialisation process, the same initial state will be attained at the end of initialisation. Multiple distinct key-IV pairs will then generate the same keystream. This could leave the stream cipher vulnerable to ciphertext-only attacks [4] or time-memory-data tradeoff attacks [1].

One key factor to consider when designing the initialisation process is the number of iterations of the initialisation state-update function to be performed. A small number can be performed quickly, which may be desirable in applications where rekeying is frequent. However, an initialisation function with very few iterations may not provide sufficient diffusion and could leave the cipher vulnerable to attacks, including correlation and algebraic attacks. Many iterations might provide resistance to attacks, but the time taken to re-key could make it unsuitable for real-time applications. Therefore, it is important to balance security and performance when designing stream cipher initialisation functions. Where state convergence occurs during initialisation, increasing the number of iterations can reduce both the rekeying efficiency and the security provided by the cipher. State convergence results in multiple key-IV pairs producing the same keystream at the end of initialisation. A user who encrypts multiple messages and chooses a different key-IV pair cannot now be sure that this will result in a distinct keystream for each message. Clearly, this is not desirable.

3   Case Studies

3.1   A5/1 Stream Cipher

Description of A5/1. A5/1 [3] is a bit based stream cipher based on three LFSRs, denoted A, B and C, with lengths of 19, 22 and 23 bits respectively, giving a state size of 64 bits. Each LFSR has a primitive feedback polynomial. A single 64-bit secret key is used for each conversation and a 22-bit frame number is used as the IV. The three registers are regularly clocked during loading of the key and IV (frame number), while a majority clocking mechanism is used for the diffusion phase and for keystream generation. This is the only nonlinear operation performed.

To implement the majority clocking scheme, each register has a clocking tap: stages $A_t[8]$, $B_t[10]$ and $C_t[10]$. The contents of these stages determine which registers will be clocked at the next iteration: Those registers for which the clock control bits agree with the majority value are clocked. For example, if $A_t[8] = 0$, $B_t[10] = 1$ and $C_t[10] = 0$, then the majority value is 0 and registers A and C are clocked. Thus, either two or three registers are clocked at each step. Figure 1 shows the components of the A5/1 keystream generator, including the feedback taps and the clocking tap for each register.

Initialisation Process. Prior to loading, all stages of the three registers are set to zero. Each register is autonomous during key and IV loading. Each register is regularly clocked 64 times and each key bit, k_i, is XORed with the register feedback to form the new value of stage 0. Following this, each register is regularly clocked 22 times as the IV is loaded in the same manner [2].

The diffusion phase involves performing 100 iterations of the initialisation state update function using the majority clocking scheme. At the end of this phase an initial state is obtained.

Previous Work. Few previous analyses of A5/1 focussed specifically on the effect of state convergence during initialisation. Two papers that deal with this topic as part of a broader analysis are Golić [6] (based on [5]) and Biryukov, Shamir and Wagner [2].

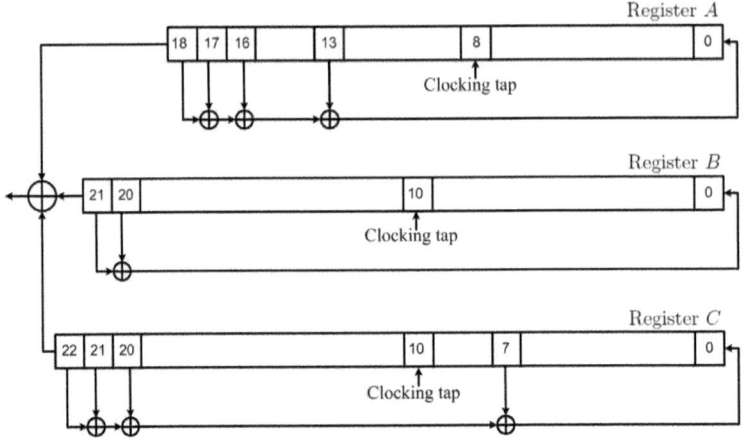

Fig. 1. A5/1 Stream cipher

Golić [6] considered the inverse mapping for the majority clocking function and identified some states with no pre-image and which therefore cannot be reached from any loaded state in a single iteration. He demonstrated that these states comprise $\frac{3}{8}$ of the loaded states of the system. Thus, the usable state space shrinks by a factor of $\frac{5}{8}$ (from 2^{64} to $5 \times 2^{61} \approx 2^{63.32}$) at the first iteration of the diffusion phase. Golić also identified some states with unique pre-images and others with up to four pre-image states. Figure 2 presents a graphical summary of the six cases identified by Golić. In this figure, (R_i, R_j, R_k) is any permutation of the set $\{A, B, C\}$ of registers and the shaded stage in each register is its clocking tap. The symbol **x** represents either 0 or 1, while **#** represents the complement of **x**; a blank square represents a bit which can take either value.

The proportion of loaded states for each case in Figure 2 is presented in Table 1, along with the corresponding number of pre-images. Note that the case identified as (i) cannot be clocked back to any valid state. That is, states of this form cannot be reached after the first iteration of the initialisation state update function.

Biryukov, Shamir and Wagner [2] also provide convergence estimates when exploring the efficiency of their attack. They report that, of 10^8 randomly chosen states, only about 15% can be clocked back 100 iterations. That is, 85% of states could not be reached by a 100 iteration forward clocking process.

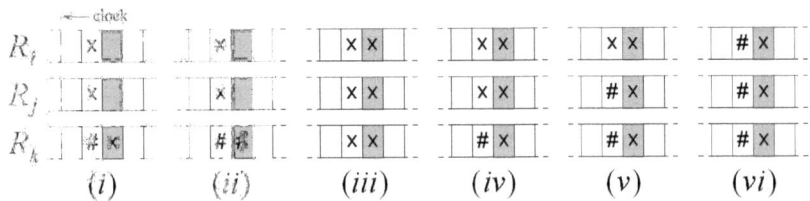

Fig. 2. A5/1 pre-image cases

Table 1. Proportions of states in each of Golić's cases

Case	(i)	(ii)	(iii)	(iv)	(v)	(vi)
Proportion of states	$\frac{3}{8}$	$\frac{3}{8}$	$\frac{1}{32}$	$\frac{3}{32}$	$\frac{3}{32}$	$\frac{1}{32}$
Number of pre-images	0	1	1	2	3	4

Our analysis. As the total size of key and IV for A5/1 $(64 + 22 = 86$ bits) exceeds the 64 bit state size, a degree of compression occurs during the loading phases of initialisation. In fact, as the state-update function is linear during the loading phases, it can be shown that there are 2^{22} key-IV pairs corresponding to each possible loaded state.

Nonlinear operations in the state-update function are introduced during the diffusion phase via majority clocking. However, this also introduces state convergence. (This convergence continues into the keystream generation stage but this is beyond the scope of this paper.) This effect was reported by Golić [5,6] and quantified to some extent by Biryukov, Shamir and Wagner [2]. Our analysis supports and extends these results.

Golić's results demonstrate that the majority clocking process is not one-to-one and can result in state convergence in one iteration. We extend Golić's logic to identify the states which cannot be reached after each of the first six iterations of the diffusion phase. We show that state convergence continues with each iteration, though not uniformly at each iteration, contrary to Golić's assumptions [6]. Some of the inaccessible states we identified for multiple iterations are presented in Figure 3.

We now sketch the reasoning used to identify states that are inaccessible after two iterations. We use the term "downstream" to refer to the stages in Figure 2 and 3 that are to the left of the clocking stages. By reversing the logic of the majority clocking process, the following conditions apply when we invert an iteration:

1. A state obtained by clocking a pair of registers must have the contents of the stages immediately downstream of the clocking bit in these registers identical in value to one another, and different in value from the clocking bit of the third register.
2. A state obtained by clocking all three registers must have the contents of the stages immediately downstream of the clocking tap identical in all three registers.

For Figure 2, we note that condition 1 applies to case (ii), condition 2 applies to case (iii), both conditions apply to cases (iv) to (vi), but neither applies to case (i). In cases (iv), (v) and (vi), condition 1 applies to different numbers of the three possible pairs of registers.

Applying this logic to the pattern labelled "2 steps" in Figure 3 shows that such a state can arise only by clocking a combination of registers that includes

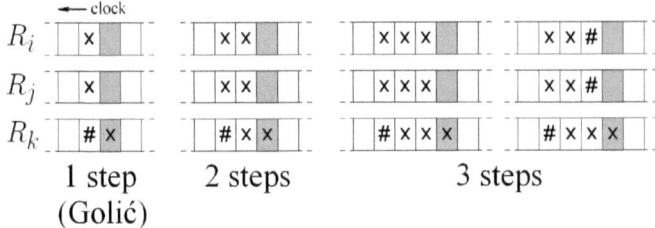

Fig. 3. Inaccessible states for various numbers of iterations (steps)

register R_k. But this implies that any previous state belongs to case (i) of Figure 2 (possibly with additional values specified among the clocking bits). Since case (i) cannot be reached by the first clocking step, this "2 steps" state cannot be reached at the subsequent clocking step. (Note: it can, however, be reached by the first clocking step, since case (i) is a valid loaded state.)

We now show that this pattern is the only inaccessible pattern at this step. Any state which is inaccessible after two iterations must clock back only to states that were inaccessible after the first step. So all such states must be contained in the image space (under clocking) of case (i) above. This image space can be found by completing the unspecified values in case (i) in all possible ways and applying the clocking rule to each (see Figure 4(a)). When this is done, we find that many of the image states are accessible, as they have multiple pre-images, some of which are accessible (see Figure 4(b) for an example). If we discard these states and retain those which can clock back only to case (i), we find that the pattern presented above is indeed the only new inaccessible pattern at the second step.

A similar process can be followed to identify inaccessible patterns after α iterations. There is a branching tree of patterns for these inaccessible states: as well as the two "3 step" patterns presented in Figure 3, there are five distinct patterns at the fourth iteration, 17 at the fifth iteration and many more at each

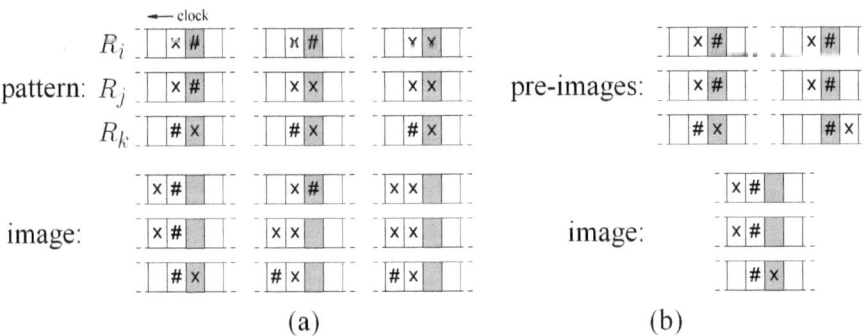

Fig. 4. Determining inaccessible states at the second step:(a) Results of clocking case (i) forwards. (b) Possible pre-images for one of these results

subsequent iteration. Table 2 presents the cumulative proportion of inaccessible states (out of all possible loaded states) after each of the first six iterations, together with the corresponding proportion and number of accessible states.

Table 2. Proportion of available states after α iterations

α (number of iterations)	1	2	3	4	5	6
new proportion inaccessible	$\frac{3}{8}$	$\frac{3}{64}$	$\frac{9}{512}$	$\frac{57}{4096}$	$\frac{423}{32768}$	$\frac{6453}{524288}$
cumulative proportion inaccessible	0.375	0.422	0.439	0.453	0.466	0.479
proportion accessible	0.625	0.578	0.561	0.547	0.534	0.521
number of accessible states	$2^{63.322}$	$2^{63.209}$	$2^{63.165}$	$2^{63.129}$	$2^{63.094}$	$2^{63.061}$

The number and complexity of the patterns obtained so far indicates that obtaining a general expression for the number of accessible states after a given number of iterations is not a simple task for large values. Extrapolating from the known values in Table 2 provides an approximation. Using an exponential extrapolation based on 2–6 iterations, we obtain an approximation of the proportion of accessible states after 100 iterations of around 5% of the number of loaded states.

Another approach to determining the extent of state convergence over the entire diffusion phase is to perform exhaustive experimental evaluation of a scaled-down version with three LFSRs and a majority clocking arrangement, but only a 15-bit internal state. (LFSR lengths of 4, 5 and 6 bits were used.) All possible loaded states were used and the number of distinct states remaining after each iteration was recorded. Results for small numbers of iterations align very closely with those reported in Table 2, while the proportion of distinct states observed after 100 iterations was found to be $\frac{6278}{32768} = 19.2\%$ of the original number. This is similar to Biryukov, Shamir and Wagner's [2] results for random sampling with A5/1 itself.

Summary. State convergence occurs during the diffusion phase of the A5/1 initialisation process (and also during keystream generation) as a result of the majority clocking operation. Increasing the number of iterations in the diffusion phase results in a further reduction of the total number of distinct initial states, decreasing both security and efficiency.

The total number of distinct internal states of A5/1 is reduced to approximately half of the loaded value after six iterations. This is equivalent to a loss of around one bit of its internal state. After 100 iterations, the total number of distinct states is potentially reduced to 15%–20% of the number of loaded states (i.e. effective key space of $2^{61.26}$–$2^{61.68}$).

3.2   Mixer

Description of Mixer. Mixer is a bit-based stream cipher proposed by Kanso [7] which uses a 128-bit key and a 64-bit IV. The keystream generator is based on two shift registers, denoted A and B, of lengths 128-bits and 89-bits respectively, giving a total state size of 217 bits. Figure 5 illustrates the components of Mixer and their interaction during both initialisation (includes both solid and dotted lines) and keystream generation (solid lines are used only). A is a regularly clocked LFSR and B is an irregularly clocked NLFSR which is controlled by A as follows. An integer function, F_{INT}, takes the contents of w stages of A as input and outputs an integer $c(b)$: the number of times B is to be clocked. The Mixer specification does not fix the value for w or specify the tap positions for F_{INT}, but recommends that $w \in \{2, 3, \ldots 7\}$ be used for efficiency reasons.

Although the feedback function of B is nonlinear, it can be approximated by a linear function with very high probability. We take this approach. A nonlinear Boolean function, $g(x)$, takes inputs from five stages of A to determine whether the output of B will be used or discarded.

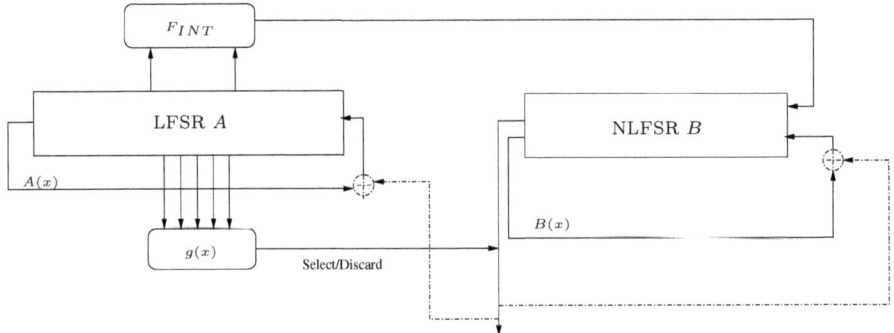

Fig. 5. Mixer state update functions

Initialisation Process. During the loading phase of the initialisation function, the key is loaded into A such that $A[i] - k_i$, for $0 \leq i \leq 127$ and the IV is loaded into B such that $B[j] = v_j$, for $0 \leq j \leq 63$. The remaining stages of B are filled with ones. The diffusion phase involves performing 200 iterations of the initialisation state update function. Each iteration is performed as follows:

1. Clock register A once.
2. For the updated state A_{t+1}, calculate:
   (a) The integer value $c_{t+1}(b)$ using F_{INT}.
   (b) The output of the nonlinear Boolean function $g_{t+1}(x)$
3. Clock register B $c_{t+1}(b)$ times.
4. If $g_{t+1}(x) = 0$ then this iteration is complete.
5. If $g_{t+1}(x) = 1$ then XOR the output bit of B (after $c_{t+1}(b)$ clocks) with the contents of both register stages $A[127]$ and $B[88]$.

We refer to the XOR operation in Step 5 as the *mixing* operation. This is the only operation in the initialisation process where the contents of the two registers are directly combined. The output bit from B which is XORed is referred to as *the mixing bit*, and denoted m. During initialisation no keystream is produced. After 200 iterations of the above process, Mixer is in an initial state and is ready to begin keystream generation.

Our Analysis. The total key-IV space of Mixer ($128 + 64$ bits) indicates the potential for 2^{192} distinct initial states. However, this does not occur. Our analysis of the initialisation process begins with the observation that the state update function is not one-to-one. In this section we examine the state convergence during one iteration of the initialisation state update function, and across multiple iterations of the initialisation process.

The Mixer initialisation state update function requires calculation of $g(x)$, as the update of $A[127]$ and $B[88]$ with the mixing value m is conditional on the value of $g(x)$. The possibilities for the state transitions from S_t to S_{t+1} are:

1. $g(x) = 0$. No mixing operation occurs, regardless of the value of m.
2. $g(x) = 1$ and $m = 0$. The mixing operation occurs but the contents of $A[127]$ and $B[88]$ remain unchanged after the mixing operation. That is, the outcome is the same as when $g(x) = 0$.
3. $g(x) = 1$ and $m = 1$: The mixing operation occurs, and the contents of $A[127]$ and $B[88]$ are complemented.

A is an LFSR with a primitive feedback function and $g(x)$ is a balanced nonlinear Boolean function. If A was autonomous, then the probability that $g(x) = 1$ would be very close to 0.5. After the first iteration the feedback from B complicates this. However, assuming this probability is still very close to 0.5 and considering the four possible combinations of $g(x)$ and m values, effective mixing occurs with a probability of 0.25.

Consider inverting the initialisation state update function. That is, given S_{t+1} we want to obtain S_t. Recall that A is a regularly clocked LFSR, which controls the clocking of B. The value of $g_{t+1}(x)$ is readily calculated. The possibilities for the state transitions from S_{t+1} to S_t are conditional on $g_{t+1}(x)$ and m:

1. $g_{t+1}(x) = 0$. No mixing occurred. In this case, we use A_{t+1} to calculate $c_{t+1}(b)$, and clock A back once and B back $c_{t+1}(b)$ times.
2. $g_{t+1}(x) = 1$. Mixing has occurred, but the effect depends on the value of m:
   (a) If $m = 0$ then again use A_{t+1} to calculate $c_{t+1}(b)$, and clock A back once and B back $c_{t+1}(b)$ times.
   (b) If $m = 1$ then complement both $A_{t+1}[127]$ and $B_{t+1}[88]$, and then use A_{t+1} to calculate $c_{t+1}(b)$, and clock A back once and B back $c_{t+1}(b)$ times.

The difficulty in inverting the state update function lies with computing the value of m. We cannot obtain this directly from B_{t+1} as it is discarded from B after the mixing operation. Therefore, given $g(x) = 1$ we consider two possibilities (m equals 0 or 1). Thus there are two possible previous states. Figure 6 shows the format of two states at time t which converge to the same state at time $t + 1$. Note that x' and m' represent the complements of x and m respectively. The contents of the other register stages must be the same.

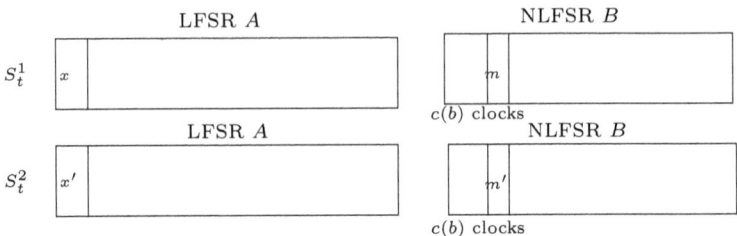

Fig. 6. States which converge to the same next state

For the first iteration of the diffusion phase 50% of all loaded states have $g(x) = 0$. Each of these produces a distinct next state. For the other 50% $g(x) = 1$ and these states can be grouped into pairs that converge to the same next state. Thus, after the first iteration of the state update function, the number of distinct states is only 75% of the number of loaded states.

At the next iteration, we consider firstly those states for which $g_1(x) = 0$. Applying the argument above, after the second iteration the number of distinct states is 75% of the size of this group. For the states where $g_1(x) = 1$, the pairing argument may not hold (some of the relevant states may have been eliminated in the previous iteration) so the number of remaining states may be more than 75%.

Combining these results gives upper and lower bounds on the number of distinct states after two iterations of 62.5% and 56.25% of the number of loaded states, respectively. Continuing these arguments for α iterations gives upper and lower bounds on the proportion of states remaining as $n_{upper} = \frac{N}{2}(1 + 2^{-\alpha})$ and $n_{lower} = N \times 0.75^{\alpha}$ where N is the number of loaded states.

As an alternative approach to estimating the degree of state convergence, we ran some computer simulations for a reduced-round diffusion phase. We set $w = 2$ and took inputs to F_{INT} from $A[70]$ and $A[71]$. In our experiments, 100 loaded states were randomly generated. For each loaded state, α iterations of the Mixer initialisation process were performed, for $\alpha = 1, 2, \ldots, 30$. We refer to the initial state resulting from this process as the *target* initial state. For each value of α and for each target obtained, the state was clocked back α times and all loaded states which generate the same target were recovered.

Data corresponding to $\alpha = 5, 10, 15, 20, 25$ and 30 have been collated to form Table 3. For each value of α, the table includes:

- The total number (Total) of loaded states found for all 100 target states.
- The minimum number (Min) and maximum number (Max) of loaded states found for any target.
- The mean and standard deviation (S.D) of the number of loaded states for each target state.

Table 3. Number of Mixer loaded states for 100 random targets

α	Total	Min	Max	Mean	S.D
5	766	1	32	7.66	6.47
10	3327	2	256	33.27	35.072
15	8120	2	1024	81.2	96.522
20	14239	4	1152	142.39	149.068
25	20328	4	1344	203.28	211.736
30	23180	4	1848	231.8	242.39

The table clearly shows that as α increases, the number of loaded states corresponding to a target also increases. That is, the number of loaded states which converge to a particular initial state increases with α. Also, it is clear that the rate of state convergence is not uniform across all key-IV pairs which form the loaded states.

From our experiments, we plotted a graph of the mean number of loaded states per target, n, against α. Two versions of this experiment were run: one in which candidate loaded states must conform to the specifications (with $B[64], \ldots, B[88] = 1, 1, \ldots, 1$) and another without this restriction. These are labelled Format check and No Format check respectively in Figure 7. For reference, the figure also includes the graphs of two other curves: $n = 1.25^{\alpha}$ and $n = 1.5^{\alpha}$.

Our experimental sample size of 100 trials represents a very small fraction of the 2^{192} possible loaded states. This, coupled with the non-uniform rate of convergence, may have affected the accuracy of our estimate of the number of loaded states converging to each target after α iterations.

Summary. State convergence during the diffusion phase is largely due to the mixing operation. This operation results in convergence at each iteration, reducing the number of distinct states by a factor of between 0.75 and 1.0. Increasing the number of iterations in the diffusion phase results in a further reduction in the number of distinct states. Both theoretical and experimental results support this. Further analysis on the effect larger w values or different tap positions has on state convergence remains future work.

4   Discussion

Traditional stream cipher designs used a state space of equal size to that of the secret key. For applications that make use of an IV as well as a key, the state space of these ciphers is less than the key-IV space, so it is clear that compression will occur. That is, multiple key-IV pairs will produce the same keystream. For some ciphers, this is further compounded by state convergence, reducing the effective key size. This was the case for A5/1 where both compression and further convergence occur.

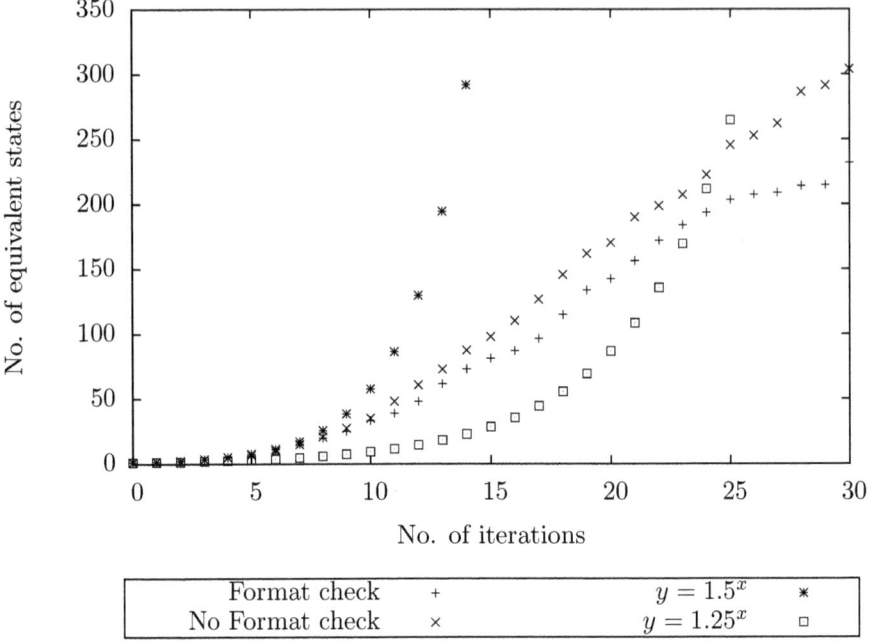

Fig. 7. Mean number of loaded states per target for various α

Modern stream ciphers have much larger state space. This permits a designer to avoid the compression issue associated with traditional designs as each key-IV pair can map to a distinct loaded state. However, problems with state convergence may still occur if the initialisation process is not carefully considered. It is crucial that the state-update function during initialisation is one-to-one. This is exactly the problem for Mixer. The compression problem experienced with A5/1 is avoided but the convergence problem remains.

Where state convergence does occur, it might not occur uniformly across all possible keys. For example, it is possible that only a single key-IV pair generates a particular initial state and associated keystream, while another initial state could have been generated by many key-IV pairs. Thus, not all keystreams are equally likely. This observation has implications for the effectiveness of time-memory-data tradeoff (TMDT) attacks.

During the pre-computation phase of a TMDT attack, for a given IV an attacker selects a few keys and generates a length of keystream corresponding to each key. This key-keystream pair is stored in a lookup table. During the real-time phase, the attacker compares a segment of keystream they have obtained with the entries in the lookup table. If there is a match, the attacker assumes the key corresponding to the matching segment is the correct key. If the initialisation process was one-to-one, the attacker would be able to use this secret key with the other IVs to correctly decrypt other messages. However, if the initialisation process was not one-to-one, it is possible for the key the attacker obtains is not the correct key but one that also produces the same initial state when used with the given IV. For an alternative IV, the two keys may not result in the same initial state, resulting in the incorrect decryption of other messages. That is, if the initialisation process is one-to-one, the TMDT attack is a deterministic attack. However, if the initialisation process is not one-to-one, the TMDT attack may be a probabilistic attack.

The overall security provided by a stream cipher with state convergence problems is also directly related to the number of distinct initial states that can be obtained as a result of the initialisation process. That is, we need to consider the total number of key-IV pairs and the total number of distinct initial states. If the state convergence is such that the total number of distinct initial states is less than the total number of key-IV pairs, then it is possible that the same secret key with different IVs will produce the same keystream. If the total number of distinct initial states is less than the total number of keys, then clearly for any given IV there will be multiple keys that produce the same keystream, so the effective keyspace is reduced.

5   Conclusion

A common belief in symmetric key cryptography is that increasing the number of iterations of a nonlinear process increases the security provided by the cipher. This is accompanied by a corresponding decrease in efficiency. For some applications, an appropriate tradeoff can be identified. However where the non-linear function is not one-to-one, as in the case of A5/1 and Mixer, increasing the number of iterations decreases the efficiency of the rekeying process with no corresponding increase in security.

Stream cipher proposals usually include both design specifications and an analysis section outlining resistance against common attacks. The focus of the security analysis is generally only on the cipher's keystream generation function. Less attention is paid to the analysis of the initialisation process. We recommend that stream cipher designers consider carefully the design of the initialisation process, and perform sufficient analysis to ensure that state update function is one-to-one so that state convergence does not occur.

Acknowledgments. The authors would like to thank the anonymous reviewers for their helpful comments. Computational resources and services used in this work were provided by the HPC and Research Support Unit, Queensland University of Technology.

References

1. Biryukov, A., Shamir, A.: Cryptanalytic time/Memory/Data tradeoffs for stream ciphers. In: Okamoto, T. (ed.) ASIACRYPT 2000. LNCS, vol. 1976, pp. 1–13. Springer, Heidelberg (2000)

2. Biryukov, A., Shamir, A., Wagner, D.: Real time cryptanalysis of A5/1 on a PC. In: Schneier, B. (ed.) FSE 2000. LNCS, vol. 1978, pp. 1–18. Springer, Heidelberg (2001)
3. Briceno, M., Goldberg, I., Wagner, D.: A Pedagogical Implementation of A5/1 (1999), http://cryptome.org/jya/a51-pi.htm
4. Dawson, E., Nielsen, L.: Automated Cryptanalysis of XOR Plaintext Strings. Cryptologia 20(2), 165–181 (1996)
5. Golić, J.D.: Cryptanalysis of alleged A5 stream cipher. In: Fumy, W. (ed.) EUROCRYPT 1997. LNCS, vol. 1233, pp. 239–255. Springer, Heidelberg (1997)
6. Golić, J.D.: Cryptanalysis of Three Mutually Clock-Controlled Stop/Go Shift Registers. IEEE Transactions on Information Theory 46(3), 1081–1090 (2002)
7. Kanso, A.A.: Mixer — A new stream cipher. Journal of Discrete Mathematical Sciences and Cryptography 11(2), 159–179 (2008)

On Maximum Differential Probability of Generalized Feistel

Kazuhiko Minematsu[1], Tomoyasu Suzaki[1], and Maki Shigeri[2]

[1] NEC Corporation. 1753, Shimonumabe, Nakahara, Kawasaki 211-8666, Japan
[2] NEC Software Hokuriku, Ltd. 1, Anyoji, Hakusan, Ishikawa 920-2141, Japan
{k-minematsu@ah,t-suzaki@pd,m-shigeri@pb}.jp.nec.com

Abstract. The maximum differential probability (MDP) is an important security measure for blockciphers. We investigate MDP of Type-2 generalized Feistel structure (Type-2 GFS), one of the most popular cipher architectures. Previously MDP of Type-2 GFS has been studied for partition number (number of sub-blocks) $k = 2$ by Aoki and Ohta, and $k = 4$ by Kim et al. These studies are based on ad-hoc case analysis and it seems rather difficult to analyze larger k by hand. In this paper, we abstract the idea of previous studies and generalize it for any k, and implement it using computers. We investigate Type-2 GFS of $k = 4, 6, 8$ and 10 with $k+1$ rounds, and obtain $O(p^k)$ bound for all cases, when the round function is invertible and its MDP is p. The bound for $k = 4$ is improved from Kim et al. and those for larger k are new. We also investigate an improvement of Type-2 GFS proposed by Suzaki and Minematsu, and obtain similar bounds as Type-2.

Keywords: blockcipher, generalized Feistel, differential probability.

1 Introduction

Generalized Feistel Structure (GFS) is a top-level blockcipher scheme based on Feistel permutation. While classical Feistel cipher partitions an input into two sub-blocks, GFS partitions it into k sub-blocks. k is called the partition number. Zhang et al. [2] proposed such a generalization and defined Type-1, 2, and 3 GFSs. Among them, Type-2 has been received much attention as its implementation can be quite small yet provides large parallereizability and high throughput, which is desirable for emerging ultra-small devices such as RFID. In fact there are some modern blockciphers based on Type-2, e.g., CLEFIA [3] ($k = 4$) and HIGHT [7] ($k = 8$). When we build a blockcipher based on Type-2 GFS, we must evaluate its strength against Differential Cryptanalysis (DC) [9]. Formally, this requires us to find (an upper bound of) the cipher's maximum differential probability (MDP)[1]. For an N-bit blockcipher, if maximum differential probability (MDP) of the cipher is sufficiently close to 2^{-N}, the cipher is immune to DC. Such a cipher is called "provably secure against DC"[15].

---

[1] It is also called Maximum Expected Differential Probability, MEDP. See Sect. 2.

U. Parampalli and P. Hawkes (Eds.): ACISP 2011, LNCS 6812, pp. 89–105, 2011.
© Springer-Verlag Berlin Heidelberg 2011

The first result was provided by Nyberg and Knudsen [12]. They studied MDP of the classical Feistel cipher, corresponding to Type-2 with $k = 2$. Later, Aoki and Ohta [13] optimized their result and proved MDP bound p^2 for 3-round Feistel with invertible round functions of MDP p. As MDP bound becomes difficult to compute as k grows, it is common to evaluate the maximum differential characteristic probability [16] for $k > 2$ [14][4] for a substitute of MDP. This is much simpler, but generally does not provide a rigorous MDP bound. Recently, Kim et al. proved that 5-round 4-partition Type-2 GFS with invertible round function has MDP bound $p^4 + 2p^5$ [10]. Their analysis consists of many probability calculations found by a divide-and-conquer, (possibly) by hand. Nyberg [8] also studied a variant of 4-partition Type-2 GFS and proved a similar bound. These heuristic approaches seem difficult to extend for larger k.

In this paper, we study the ad-hoc analyses of previous studies and distill a formal procedure that basically works for any k. In particular our algorithm is deeply inspired by Kim et al.'s work [10]. We implement our algorithm on computers, and provide MDP of Type-2 for k up to 10. To our knowledge, this is the first attempt to derive MDP of GFS for any k. Here, the situation is quite contrastive to the Substitution-Permutation Network (SPN). It has been extensively studied and a succinct MDP bound is known for any k (e.g. [5][6])[2]. Basically our algorithm is not a surprising one, however, its implementation needed much cares as the real program must handle various cases which were not appeared at Kim et al.'s analysis. For k-partition Type-2 GFS with invertible round function, we obtained a bound $O(p^k)$ with $k+1$-round, for all $k = 4, 6, 8, 10$. In particular, our bound for $k = 4$ is $p^4 + p^5$, thus slightly improves Kim et al.'s result (for exact bounds of other k, see Sect. 4). Note that we have not studied the tightness of these bounds (it is generally difficult to see as our postulate is only the MDP's upper bound of round function). At the same time, our bounds are at least close to the lowest possible if p itself is too.

We also investigate a generalization/improvement of Type-2 GFS, recently proposed by Suzaki and Minematsu [14], which we call Type-2i (i for improvement) GFS. They proposed to use a sub-block shuffle different from the cyclic shift employed by Type-2, and proved a better diffusion property for $k \geq 6$. They presented various optimally-diffusive shuffles found via exhaustive search or built from the De Bruijn graph. Although differential characteristic probability of Type-2i have been investigated [14][4], no result is known on its MDP. Using our bounding algorithm we investigated Type-2i with $k = 6, 8$. We were interested in knowing whether Type-2i has better MDP than Type-2; say if k-round Type-2i could have $O(p^k)$ bound. However, as far as we investigated, the MDPs are almost the same as Type-2 or even slight worse, especially in proving MDP $O(p^k)$ for k-partition. This phenomenon is probably a consequence of a limit of our proof method, and it is open if we can improve or not.

Finally, we have to mention that the impact of our result on current blockciphers is rather minor, as Type-2 with large k is not widely deployed, and there

---

[2] Here, k denotes the number of sub-blocks in a round where one sub-block is given to a unit cryptographic permutation (typically S-box).

is no real proposal based on Type-2i (it is quite new). Instead, we expect our study helps understanding Type-2 and Type-2i and boosts the spread of Type-2/2i based ciphers.

2 Preliminaries

2.1 Type-2 GFS and Its Generalization

Our targets are Type-2 GFS and its further generalization [14], which we call Type-2i. As Type-2i includes Type-2, we first explain Type-2i. For even integer $k \geq 2$, we consider a blockcipher over $\mathcal{M} = (\{0,1\}^\kappa)^k$ for some κ. Let π be a permutation over $\mathcal{M}$, in particular a *shuffle* of k sub-blocks. The single round of k-partition Type-2i GFS using shuffle π is,

$$
\begin{aligned}
&(x_0, x_1, \ldots, x_{k-1}) \\
&\quad \to \pi\left(x_0, F(K_0, x_0) \oplus x_1, x_2, F(K_1, x_2) \oplus x_3, \ldots, F(K_{(k-2)/2}, x_{k-2}) \oplus x_{k-1}\right),
\end{aligned}
\tag{1}
$$

where $x_i \in \{0,1\}^\kappa$ and $F(K, *)$ is a keyed function with key K. If above round is iterated for r times the resulting cipher is denoted by $\Phi_k^r[\pi]$ (here κ is not important hence omitted). For the decryption, we perform an inversion of Eq. (1) using the inverse, π^{-1}. In particular, if π is the right cyclic shift[3], i.e.,

$$
\pi_{\mathrm{cyc}}(x_0, \ldots, x_{k-1}) \stackrel{\mathrm{def}}{=} (x_{k-1}, x_0, x_1, \ldots, x_{k-2}),
\tag{2}
$$

the resulting cipher corresponds to Type-2 GFS. For convenience we also use the index representation for π. Hence π_{cyc} can be written as $\pi_{\mathrm{cyc}}(0, 1, \ldots, k-1) = (k-1, 0, 1, \ldots, k-2)$, or we simply use $(k-1, 0, 1, \ldots, k-2)$ to mean π_{cyc}.

The Conditions for Shuffles. To make $\Phi_k^r[\pi]$ cipher secure, π should diffuse an input difference to all outputs. For instance, identity shuffle makes no sense since an input difference never diffuses no matter how many rounds are iterated, indicating a simple distinguisher from random permutation. Formally, π must have finite DRmax, a measure of goodness-of-diffusion introduced by [14]. The definition of DRmax is related to the diameter of directed graphs induced by the shuffle[4], and reflects the needed rounds for full-diffusion. For example, DRmax of π_{cyc} with k-partition is k. Suzaki and Minematsu proved that, for $k \geq 6$ there are shuffles with smaller DRmax than π_{cyc}, by presenting shuffles with minimum DRmax up to $k = 16$ and a construction assuring DRmax $= 2\log_2 k$ for any k being power of two. See [14] for exact presentation of shuffles.

---

[3] One may think of using the left cyclic shift instead. From the property of cyclic shift the results of this paper are the same for both definitions.

[4] An idea similar to Massey [1].

2.2   Maximum Differential Probability

For any keyed function $F(K, *)$ having binary domain and range we define

$$\mathrm{DP}^{F(K,*)}(\alpha \to \beta) \stackrel{\text{def}}{=} \Pr[F(K, X) \oplus F(K, X \oplus \alpha) = \beta], \text{ and} \qquad (3)$$

$$\mathrm{DP}_{\max}^{F(K,*)} \stackrel{\text{def}}{=} \max_{\alpha \neq 0, \beta} \mathrm{DP}^{F(K,*)}(\alpha \to \beta), \qquad (4)$$

where probability is defined by K and X, which is independent and uniform. For simplicity we omit K and write as $F(X)$, assuming a fixed probability distribution of K. Theoretically, the lower bound of MDP for N-bit keyed function (permutation) is $(1/2^N)$ $(1/(2^N - 1))$, achieved by random N-bit function (permutation).

Caveat. If $K \in \mathcal{K}$ is uniform $\mathrm{DP}^{F(K,*)}$ is the average of probability $\mathrm{DP}_{\max}^{F(\eta,*)}$, where probability is defined solely by X, for all $K = \eta$. In some cases $\mathrm{DP}_{\max}^{F(K,*)}$ is called the Maximum Expected Differential Probability (MEDP) if we need to distinguish it from $\mathrm{DP}_{\max}^{F(\eta,*)}$. We use the word "MDP" to mean MEDP.

Our purpose is to derive an upper bound of $\mathrm{DP}_{\max}^{\Phi_k^r[\pi]}$, for various π (shuffle), even k (partition number), and r (round). In a similar manner to [13] and [10], we assume the following for $\Phi_k^r[\pi]$;

Assumption 1

 - For r-round, k-partition blockcipher there are $(k/2)r$ keys of F (where one round has $k/2$ keys, as shown by Eq. (1)) and they are independently sampled from a fixed probability distribution.
 - F is invertible, i.e., a permutation over $\{0,1\}^\kappa$ for any $K = \eta$.
 - $\mathrm{DP}_{\max}^F \leq p$ for some $1/(2^\kappa - 1) \leq p \leq 1$.

With these assumptions, for given $k \geq 2$ we primarily try to find the smallest r such that $\mathrm{DP}_{\max}^{\Phi_k^r[\pi]} \leq O(p^k)$. This goal is rational, since we generally can not expect to have p^{k+1} as MDP (as our proof is independent of the real value of p). We call such r the *limit round*. As well as previous studies, the following lemma [15] plays a crucial role in our analysis :

Lemma 1. *For any keyed function F, we have*

$$\sum_\beta \mathrm{DP}^F(\alpha \to \beta) = 1 \text{ for any } \alpha, \text{ and}$$

$$\sum_\alpha \mathrm{DP}^F(\alpha \to \beta) = 1 \text{ for any } \beta, \text{ if } F \text{ is invertible.}$$

3   MDP Bounding Algorithm

3.1   Overview

With Assumption 1, previous results are $\mathrm{DP}_{\max}^{\Phi_2^3[\pi_{\mathrm{cyc}}]} \leq p^2$ [13] and $\mathrm{DP}_{\max}^{\Phi_4^5[\pi_{\mathrm{cyc}}]} \leq p^4 + 2p^5$ [10]. We further extend these results, i.e., derive $\mathrm{DP}_{\max}^{\Phi_k^r[\pi]}$ for various

k, r, and π. Our algorithm can basically work for any π, but for simplicity we focus on the even-odd ones [14], which means that, any i-th input sub-block for even i is mapped to an output j-th sub-block for odd j, and vice versa. This includes Type-2 and Type-2i of Suzaki-Minematsu. According to [14] the smallest DRmax is always achieved by even-odd shuffles. In order to keep the notational compatibility with Kim et al., we index the input sub-blocks from 1 to k throughout Sect. 3. The number of F functions involved in $\Phi_k^r[\pi]$ is $s = (k/2) \cdot r$, and we indexed them from top left to bottom right, such as $F_1, \ldots, F_s$ (See Fig. 1).

We explain the procedure for deriving a bound of $\mathrm{DP}_{\max}^{\Phi_k^r[\pi]}(\alpha \rightarrow \beta)$, where $\alpha = (\alpha_1, \ldots, \alpha_k)$ and $\beta = (\beta_1, \ldots, \beta_k)$, are input and output differential. The overall MDP bound is derived as the maximum for all pairs of (α, β). The output differential of F_i is denoted by δ_i except for the last two rounds (i.e. for $(k/2)(r-2) < i \leq s$), as the output differentials of the last two rounds are determined by other variables, since π is even-odd. We define $\delta = (\delta_1, \ldots, \delta_q)$, where $q = (k/2) \cdot (r-2)$. See Fig. 1 for example.

Our algorithm has three search parameters: n, nd, and ub. The algorithm first tries to prove $\mathrm{DP}_{\max}^{\Phi_k^r[\pi]} \leq p^n$. If this is not possible it tries $\mathrm{DP}_{\max}^{\Phi_k^r[\pi]} \leq \sum_{i=n,\ldots,nd} c_i \cdot p^i$ for non-negative integers $c_{n+1}, \ldots, c_{nd}$ and $c_n \geq 1$. The parameter ub determines the size of window search over δ.

3.2 Ordered Sum

For given $\Phi_k^r[\pi]$, the input/output differential of each F_i is represented as a sum (XOR) of α, β, and δ, which we call *clause*. For $i = 1, \ldots, s$ the differential equation for F_i is denoted by E_i, and its input (output) side clause is denoted by $\mathsf{E}_{i,0}$ ($\mathsf{E}_{i,1}$). See Fig. 1 for the case of $\Phi_4^5[\pi_{\mathrm{cyc}}]$. Moreover, we write the set of δ appeared in $\mathsf{E}_{i,j}$ as $\mathcal{H}_{i,j}$ and define $\overline{\mathcal{H}}_i = \mathcal{H}_{i,0} \cup \mathcal{H}_{i,1}$. For example, in Fig. 1 we have $\mathsf{E}_{7,0} = \alpha_2 \oplus \delta_1 \oplus \delta_6$, and $\mathsf{E}_{7,1} = \alpha_3 \oplus \beta_3 \oplus \delta_4$, and $\mathcal{H}_{7,0} = \{\delta_1, \delta_6\}$, $\mathcal{H}_{7,1} = \{\delta_4\}$, and $\overline{\mathcal{H}}_7 = \{\delta_1, \delta_4, \delta_6\}$.

The list of all differential equations are written as $\mathcal{E} = (\mathsf{E}_1, \ldots, \mathsf{E}_s)$. From Markov cipher theory [16], the MDP is written as

$$\mathrm{DP}_{\max}^{\Phi_k^r[\pi]} = \max_{\alpha, \beta \in (\{0,1\}^\kappa)^k \setminus \{(0\ldots,0)\}} \sum_{\delta_i \in \{0,1\}^\kappa : i=1,\ldots,s} \prod_j \mathrm{DP}(\mathsf{E}_j), \tag{5}$$

where $\mathrm{DP}(\mathsf{E}_j)$ denotes $\mathrm{DP}^F(\mathsf{E}_j)$. Note that $\mathrm{DP}^F(\mathsf{E}_j) = \mathrm{DP}^F(\mathsf{E}_{j,0} \rightarrow \mathsf{E}_{j,1})$ is a function of α, β, and δ. As the number of possible (α, β) is huge, we collect them into $(2^k - 1)^2$ cases[5], where each case assigns $\alpha_i = 0$ or $\alpha_i \neq 0$, and $\beta_i = 0$ or $\beta_i \neq 0$, for all i (here all-zero pattern is omitted). We also utilize a tool called ordered sum [10] for reducing the complexity w.r.t. δ. Since original presentation of Kim et al. is slight informal, we here present a formal one. The proof is clear from Lemma 1.

---

[5] Generally some input-output differential patterns may have its dual and we could exploit it to reduce the complexity, as performed by Kim et al.

Proposition 1. *Let $\mathcal{E}' = (E_{i_1}, \ldots, E_{i_M}) \subseteq \mathcal{E}$ be a list of M differential equations. We assume $\bigcup_{j=1}^{M} \overline{\mathcal{H}}_{i_j} = \{\delta_{h_1}, \ldots, \delta_{h_M}\}$. If $\overline{\mathcal{H}}_{i_j} \setminus \bigcup_{j'=1}^{j-1} \overline{\mathcal{H}}_{i_{j'}} = \{\delta_{h_j}\}$ (where $\bigcup_{j'=1}^{0} \overline{\mathcal{H}}_{i_{j'}}$ is defined as empty) and $\{\delta_{h_j}\} \notin \mathcal{H}_{i_j,0} \cap \mathcal{H}_{i_j,1}$, for $1 \leq j \leq M$, we have*

$$\sum_{\delta_{h_1}, \ldots, \delta_{h_M}} DP(E_{i_1}) \cdot DP(E_{i_2}) \cdot \cdots \cdot DP(E_{i_M})$$

$$= \sum_{\delta_{h_1}} DP(E_{i_1}) \sum_{\delta_{h_2}} DP(E_{i_2}) \cdots \sum_{\delta_{h_M}} DP(E_{i_M}) = 1.$$

Here, the condition $\{\delta_{h_j}\} \notin \mathcal{H}_{i_j,0} \cap \mathcal{H}_{i_j,1}$ is needed to make sure that each equation E_{i_j} in the summation $\sum_{\delta_{h_j}} DP(E_{i_j})$ contains δ_{h_j} in its left or right hand side, but not both. If both sides contain δ_{h_j}, the ordered sum is not guaranteed to be 1. Such a case does not happen in the initial stage of our algorithm. However, it can happen later as the algorithm may rewrite variables of $\mathcal{E}$.

Example 1. Let $E_1 = [\alpha_1 \to \delta_1]$ and $E_2 = [\beta_1 \oplus \delta_1 \to \delta_2]$. Then we can take an ordered sum for E_1 and E_2 since

$$\sum_{\delta_1, \delta_2} DP(E_1) \cdot DP(E_2) = \sum_{\delta_1} DP(\alpha_1 \to \delta_1) \sum_{\delta_2} DP(\beta_1 \oplus \delta_1 \to \delta_2) \qquad (6)$$

$$= \sum_{\delta_1} DP(\alpha_1 \to \delta_1) = 1 \qquad (7)$$

holds for any (including 0) α_1 and β_1. However, if E_2 is replaced with $[\beta_1 \oplus \delta_1 \oplus \delta_2 \to \delta_2]$, the second equality does not hold true (as $\{\delta_2\} \in \mathcal{H}_{2,0} \cap \mathcal{H}_{2,1}$), thus ordered sum can not be taken.

3.3   Details of Bounding Algorithm

Constraints, Labels, and States. The algorithm handles many linear constraints on clauses, such as $\alpha_1 \oplus \delta_1 = 0$ or $\beta_1 \neq 0$. The set of these constraints are defined by two clause sets, $\mathcal{C}_0, \mathcal{C}_1$. If we have $V \in \mathcal{C}_0$ ($V \in \mathcal{C}_1$) for clause V, it means $V = 0$ ($V \neq 0$) is given as a linear constraint. The algorithm also uses *label* and *state*, each assigned for a clause and an equation in $\mathcal{E}$. A label is one of **nz**, **zr**, and **any**. For clause V, if $V = 0$ ($V \neq 0$) is deduced from $\mathcal{C}_0$ and $\mathcal{C}_1$ we write $V = \mathbf{zr}$ ($V = \mathbf{nz}$). Otherwise we write $V = \mathbf{any}$. Especially, the label of $E_{i,j}$ is denoted by $L_{i,j} \in \{\mathbf{zr}, \mathbf{nz}, \mathbf{any}\}$. For instance, if $\mathcal{C}_0 = \{\alpha_1, \alpha_2\}$ and $E_{1,1} = \alpha_1 \oplus \alpha_2$ we have $L_{1,1} = \mathbf{zr}$. Moreover if $\mathcal{C}_1 = \{\delta_1, \delta_2\}$ and $E_{2,1} = \delta_1 \oplus \delta_2$ we have $L_{2,1} = \mathbf{any}$. We define the label set $\mathcal{L} = (L_1, \ldots, L_s)$, where $L_i = (L_{i,0}, L_{i,1})$.

The state set is defined as $\mathcal{S} = (S_1, \ldots, S_s)$, where S_i represents a status on the probability of E_i. Possible states are 0 (probability is zero), 1 (probability is one), p (probability is at most p), and u (probability is unknown). Hence $S_i \in \{0, 1, p, u\}$.

Update. The working structure of the algorithm is $\Pi \overset{\text{def}}{=} (\mathcal{E}, \mathcal{L}, \mathcal{S}, \mathcal{C}_0, \mathcal{C}_1)$. Following Lemma 1, the algorithm first updates Π considering their relationships in a similar manner to Kim et al. For example, let us assume $E_1 = [\alpha_1 \to \delta_1]$ and $E_2 = [\delta_2 \oplus \alpha_2 \to \alpha_3 \oplus \delta_3]$ and $\mathcal{C}_0 = \{\alpha_3 \oplus \delta_3\}$, $\mathcal{C}_1 = \{\alpha_1\}$. Now, since $\alpha_1 \in \mathcal{C}_1$ and

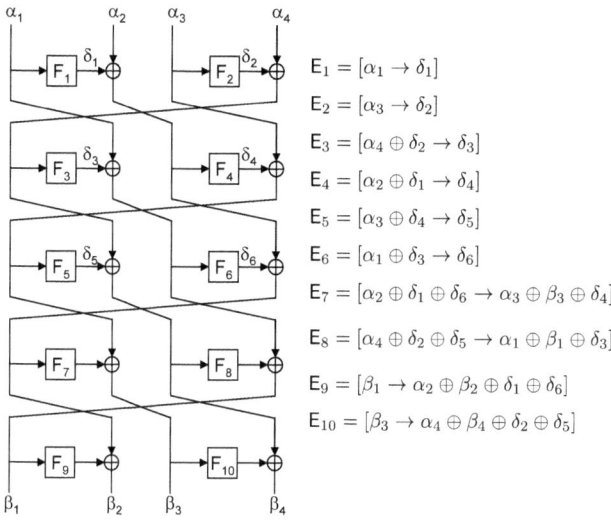

$E_1 = [\alpha_1 \rightarrow \delta_1]$

$E_2 = [\alpha_3 \rightarrow \delta_2]$

$E_3 = [\alpha_4 \oplus \delta_2 \rightarrow \delta_3]$

$E_4 = [\alpha_2 \oplus \delta_1 \rightarrow \delta_4]$

$E_5 = [\alpha_3 \oplus \delta_4 \rightarrow \delta_5]$

$E_6 = [\alpha_1 \oplus \delta_3 \rightarrow \delta_6]$

$E_7 = [\alpha_2 \oplus \delta_1 \oplus \delta_6 \rightarrow \alpha_3 \oplus \beta_3 \oplus \delta_4]$

$E_8 = [\alpha_4 \oplus \delta_2 \oplus \delta_5 \rightarrow \alpha_1 \oplus \beta_1 \oplus \delta_3]$

$E_9 = [\beta_1 \rightarrow \alpha_2 \oplus \beta_2 \oplus \delta_1 \oplus \delta_6]$

$E_{10} = [\beta_3 \rightarrow \alpha_4 \oplus \beta_4 \oplus \delta_2 \oplus \delta_5]$

Fig. 1. Type-2 GFS with $k = 4$, $r = 5$

each F is invertible, δ_1 must be non-zero. Then we make $(\mathsf{L}_{1,0}, \mathsf{L}_{1,1}) = (\mathbf{nz}, \mathbf{nz})$ and add $\{\delta_1\}$ to $\mathcal{C}_1$, and make $(\mathsf{L}_{2,0}, \mathsf{L}_{2,1}) = (\mathbf{zr}, \mathbf{zr})$ and add $\{\delta_2 \oplus \alpha_2\}$ to $\mathcal{C}_0$ in a similar manner. As we obtain $(\mathsf{L}_{1,0}, \mathsf{L}_{1,1}) = (\mathbf{nz}, \mathbf{nz})$ and $(\mathsf{L}_{2,0}, \mathsf{L}_{2,1}) = (\mathbf{zr}, \mathbf{zr})$, we make $\mathsf{S}_1 = p$ and $\mathsf{S}_2 = 1$. If we observe an inconsistency, for instance $\alpha_1 \in \mathcal{C}_0$ and $\delta_1 \in \mathcal{C}_1$ with $E_1 = [\alpha_1 \rightarrow \delta_1]$, we make $(\mathsf{L}_{1,0}, \mathsf{L}_{1,1}) = (\mathbf{zr}, \mathbf{nz})$ and thus $\mathsf{S}_1 = 0$. This implies the overall differential probability is zero.

In addition, we update $\mathcal{E}$ so that the number of δ_i appeared is minimized using the knowledge of $\mathcal{C}_0$. For example if $\delta_1 \oplus \alpha_1 \oplus \beta_2 \in \mathcal{C}_0$, we substitute δ_1 by $\alpha_1 \oplus \beta_2$ for every equations of $\mathcal{E}$ containing δ_1. Of course, if $\delta_i = 0$ we remove it from $\mathcal{E}$. The update of $\mathcal{E}$ can be done via a triangulation of $\mathcal{C}_0$, considering it as a binary matrix. The procedure is trivial, hence we skip the details. We define $\mathcal{U} \subseteq \{\delta_1, \ldots, \delta_q\}$ as the set of remaining δ_is after update of $\mathcal{E}$. As a result, the members of $\mathcal{U}$ are independent of $\mathcal{C}_0$'s constraints, while those of $\overline{\mathcal{U}} \stackrel{\text{def}}{=} \{\delta_1, \ldots, \delta_q\} \setminus \mathcal{U}$ are linearly dependent on α, β, and other members of $\overline{\mathcal{U}}$. The exact update procedure is shown as Algorithm 2 of Appendix A.

Bound Derivation. Given an initial constraint $\alpha, \beta \in \{\mathbf{nz}, \mathbf{zr}\}^k \setminus \{(\mathbf{zr}, \ldots, \mathbf{zr})\}$, the algorithm first sets $\mathcal{C}_0$ and $\mathcal{C}_1$ as $\mathcal{C}_0 = \{\alpha_i : \alpha_i = \mathbf{zr}\} \cup \{\beta_i : \beta_i = \mathbf{zr}\}$ and $\mathcal{C}_1 = \{\alpha_i : \alpha_i = \mathbf{nz}\} \cup \{\beta_i : \beta_i = \mathbf{nz}\}$. Then the algorithm iteratively updates Π and $\mathcal{C}_0$, $\mathcal{C}_1$ as described above, until there is no change of $\mathcal{C}_0$ and $\mathcal{C}_1$. Next, the algorithm checks if it is possible to build an ordered sum[6] for all $\delta_i \in \mathcal{U}$ *without* using

---

[6] Practically, an ordered sum can be smaller than 1 as $\delta_i \in \mathcal{U}$ can have some impossible values implied by $\mathcal{C}_1$, e.g., if $\{\delta_1 \oplus \alpha_1, \alpha_1\} \subseteq \mathcal{C}_1$ the summation for $\delta_1 = \alpha_1$ is redundant. For upper-bound derivation, it is sufficient to assume that a possible ordered sum has always probability 1.

(at least) n equations whose states are p, which implies $\mathrm{DP}^{\Phi_k^r[\pi]}_{\max}(\alpha \to \beta) \le p^n$. If this is possible, we say we can "extract p^n". This is done by a simple exhaustive search using $\mathcal{E}$ and $\mathcal{S}$. If we cannot extract p^n, we give additional constraints, represented as $(\delta_{i_1}, \ldots, \delta_{i_h}) = (v_{i_1}, \ldots, v_{i_h})$, with $(v_{i_1}, \ldots, v_{i_h}) \in \{\mathrm{zr}, \mathrm{nz}\}^{i_h}$ for $\{\delta_{i_1}, \ldots, \delta_{i_h}\} \subseteq \mathcal{U}$ with $\delta_{i_j} = \mathrm{any}$, for all $h = 1, \ldots, ub$. Here, $(\delta_{i_1}, \ldots, \delta_{i_h})$ is called a δ-combination and $(v_{i_1}, \ldots, v_{i_h})$ is called a v-assignment. Hence if $|\{\delta_i \in \mathcal{U} : \delta_i = \mathrm{any}\}| = \rho$, we have $\sum_{h=1}^{ub} \binom{\rho}{h}$ δ-combinations. For each δ-combination, $(\delta_{i_1}, \ldots, \delta_{i_h})$, we have 2^h v-assignments.

We choose a δ-combination, and for each v-assignment we temporarily update Π and find maximum of $m \in \{n, \ldots, nd\}$ such that we can extract p^m. If all 2^h v-assignments are successful (i.e. we can extract p^n or more), the sum of the results for all v-assignments is $\mathrm{DP}^{\Phi_k^r[\pi]}_{\max}(\alpha \to \beta)$. In this case the bound will be the form of $\sum_{h=n}^{nd} c_n p^n$ for non-negative integer c_n with $\sum_{i=n}^{nd} c_i = 2^h$. We try all δ-combinations until we success. If all δ-combinations fail we declare the total failure and quit. A case of 4-partition 5-round Type-2 is shown by Example 2.

Example 2. $\Phi_4^5[\pi_{\mathrm{cyc}}]$ with $\alpha = (\mathrm{nz}, \mathrm{nz}, \mathrm{nz}, \mathrm{nz})$, $\beta = (\mathrm{nz}, \mathrm{nz}, \mathrm{zr}, \mathrm{nz})$ (a sub-case of Case 9-3 of [10]). We assume $n = 4$, $nd = 5$, and $ub = 1$. Then we have

$\mathsf{E}_1 = [\alpha_1 \to \delta_1],$	$\mathsf{L}_1 = (\mathrm{nz}, \mathrm{nz}),$	$\mathsf{S}_1 = p,$
$\mathsf{E}_2 = [\alpha_3 \to \delta_2],$	$\mathsf{L}_2 = (\mathrm{nz}, \mathrm{nz}),$	$\mathsf{S}_2 = p,$
$\mathsf{E}_3 = [\alpha_4 \oplus \delta_2 \to \delta_3],$	$\mathsf{L}_3 = (\mathrm{any}, \mathrm{any}),$	$\mathsf{S}_3 = u,$
$\mathsf{E}_4 = [\alpha_2 \oplus \delta_1 \to \delta_4],$	$\mathsf{L}_4 = (\mathrm{any}, \mathrm{any}),$	$\mathsf{S}_4 = u,$
$\mathsf{E}_5 = [\alpha_3 \oplus \delta_4 \to \delta_5],$	$\mathsf{L}_5 = (\mathrm{any}, \mathrm{any}),$	$\mathsf{S}_5 = u,$
$\mathsf{E}_6 = [\alpha_1 \oplus \delta_3 \to \delta_6],$	$\mathsf{L}_6 = (\mathrm{any}, \mathrm{any}),$	$\mathsf{S}_6 = u,$
$\mathsf{E}_7 = [\alpha_2 \oplus \delta_1 \oplus \delta_6 \to \alpha_3 \oplus \beta_3 \oplus \delta_4],$	$\mathsf{L}_7 = (\mathrm{any}, \mathrm{any}),$	$\mathsf{S}_7 = u,$
$\mathsf{E}_8 = [\alpha_4 \oplus \delta_2 \oplus \delta_5 \to \alpha_1 \oplus \beta_1 \oplus \delta_3],$	$\mathsf{L}_8 = (\mathrm{nz}, \mathrm{nz}),$	$\mathsf{S}_8 = p,$
$\mathsf{E}_9 = [\beta_1 \to \alpha_2 \oplus \beta_2 \oplus \delta_1 \oplus \delta_6],$	$\mathsf{L}_9 = (\mathrm{nz}, \mathrm{nz}),$	$\mathsf{S}_9 = p,$
$\mathsf{E}_{10} = [\beta_3 \to \alpha_4 \oplus \beta_4 \oplus \delta_2 \oplus \delta_5],$	$\mathsf{L}_{10} = (\mathrm{zr}, \mathrm{zr}),$	$\mathsf{S}_{10} = 1,$

where $\mathcal{C}_0 = \{\beta_3, \alpha_4 \oplus \beta_4 \oplus \delta_2 \oplus \delta_5\}$ and $\mathcal{C}_1 = \{\alpha_1, \alpha_2, \alpha_3, \alpha_4, \beta_1, \beta_2, \beta_4, \delta_1, \delta_2, \alpha_2 \oplus \beta_2 \oplus \delta_1 \oplus \delta_6, \alpha_4 \oplus \delta_2 \oplus \delta_5, \alpha_1 \oplus \beta_1 \oplus \delta_3\}$.

As $\delta_2 = \alpha_4 \oplus \beta_4 \oplus \delta_5$, we rewrite $\mathsf{E}_2, \mathsf{E}_3, \mathsf{E}_8, \mathsf{E}_{10}$ and try to build an ordered sum with $\mathcal{U} - \{\delta_1, \delta_3, \delta_4, \delta_5, \delta_6\}$. However we cannot extract p^4 since there are 4 equations with state p and remaining equations have $|\overline{\mathcal{H}}_i| = 2$ (an ordered sum must contain at least one equation with $|\overline{\mathcal{H}}_i| = 1$). Then we consider δ_3 (whose label is any) as a δ-combination. If $\delta_3 = \mathrm{nz}$ we add δ_3 to $\mathcal{C}_1$ and S_3 is updated to p. Then we can extract p^4 as

$$\sum_{\delta_1, \delta_3(=\mathrm{nz}), \delta_4, \delta_5, \delta_6} \mathrm{DP}(\mathsf{E}_1) \cdot \cdots \cdot \mathrm{DP}(\mathsf{E}_{10})$$

$$\le p^4 \sum_{\delta_1, \delta_3(=\mathrm{nz}), \delta_4, \delta_5, \delta_6} \mathrm{DP}(\mathsf{E}_1) \cdot \mathrm{DP}(\mathsf{E}_4) \cdot \mathrm{DP}(\mathsf{E}_5) \cdot \mathrm{DP}(\mathsf{E}_6) \cdot \mathrm{DP}(\mathsf{E}_7)$$

$$\le p^4 \sum_{\delta_1} \mathrm{DP}(\mathsf{E}_1) \cdot \sum_{\delta_4} \mathrm{DP}(\mathsf{E}_4) \cdot \sum_{\delta_5} \mathrm{DP}(\mathsf{E}_5) \cdot \sum_{\delta_3} \mathrm{DP}(\mathsf{E}_6) \cdot \sum_{\delta_6} \mathrm{DP}(\mathsf{E}_7) \le p^4. \quad (8)$$

If $\delta_3 = \mathbf{zr}$ we obtain $\delta_5 = \beta_4$ and add δ_3 and $\delta_5 \oplus \beta_4$ to $\mathcal{C}_0$ and update states as $S_3 = 1$, $S_5 = S_6 = S_7 = p$. Then $\mathcal{U}$ is changed to $\{\delta_1, \delta_4, \delta_6\}$ and we can extract p^5 with a similar computation as Eq. (8). Summing up both cases, we obtain

$$\mathrm{DP}_{\max}^{\Phi_4^5[\pi_{\mathrm{cyc}}]}(\alpha \to \beta) \leq \sum_{\substack{\delta_1, \delta_3(=\mathrm{nz}), \\ \delta_4, \delta_5, \delta_6}} \prod_{i=1,\ldots,10} \mathrm{DP}(\mathsf{E}_i) + \sum_{\substack{\delta_1, \delta_3(=\mathrm{zr}), \\ \delta_4, \delta_5, \delta_6}} \prod_{j=1,\ldots,10} \mathrm{DP}(\mathsf{E}_j) \quad (9)$$

$$\leq p^4 + p^5. \quad (10)$$

The basic flow of the algorithm is depicted at Fig. 2. The pseudocode is in Appendix A. The algorithm here generates one polynomial of p for each (α, β). We need some post-processing for taking the maximum among them. Hence, it may be the case the total MDP bound we obtain is of the form $\max\{\mathrm{poly}_1(p), \mathrm{poly}_2(p), \ldots, \}$, say, if $\mathrm{poly}_1(p) = p^4$ and $\mathrm{poly}_2(p) = 10p^5$. This is in fact the case, see Table 3.

3.4 Optimized Version

The algorithm described above immediately terminates if it finds a δ-combination that provides $\sum_{i=n,\ldots,nd} c_i \cdot p^i$ with non-negative integer c_i. Hence the result may have the leading coefficient, c_n, being larger than 1. We therefore implemented an "optimized version" that minimizes c_n with an exhaustive search for all δ-combinations. It produces multiple candidate polynomials of p for each (α, β), where $\mathrm{DP}_{\max}^{\Phi_k^r[\pi]}(\alpha \to \beta)$ is the minimum of them (see Appendix B).

4 Experimental Results

4.1 Results for Type-2

We implemented the described algorithm and performed it using eight 2GHz, dual-core computers. For Type-2, the experiments are done[7] for $k = 4, 6, 8$ and 10. We first ran our algorithm and saw if unoptimized (w.r.t the leading coefficient, see Sect. 3.4) bound was obtained. If this was successful, we then tried to optimize the bound. The Type-2 uses the cyclic shift as defined by Eq. (2). For completeness, see Table 1. Unoptimized bounds were relatively easy to derive. In contrast the optimized ones needed much time: for $(k, r, n) = (8, 9, 8)$ we spent about three weeks for the optimized bound, $p^8 + 26p^9$, while unoptimized one was derived within a few days. For $k = 10$ and $r = 11$ we could only derive unoptimized bound. For $k \geq 12$, our algorithm seems infeasible for now.

The results are shown by Table 2. During the experiment, nd was set to $n+1$ as larger values did not provide a noticeable improvement, and ub was set from 1 to 5. In Table 2, "suc" indicates that the obtained bound is $c_n p^n + c_{n+1} p^{n+1} + \cdots + c_{nd} p^{nd}$ with some $c_n \geq 1$. "fail(x)" indicates the failure, where x is the number of (α, β) pairs that fails (corresponding to line 30 of Algorithm 1 in Appendix A).

---

[7] We did not test for $k = 2$ as the bound of Aoki-Ohta is already p^2.

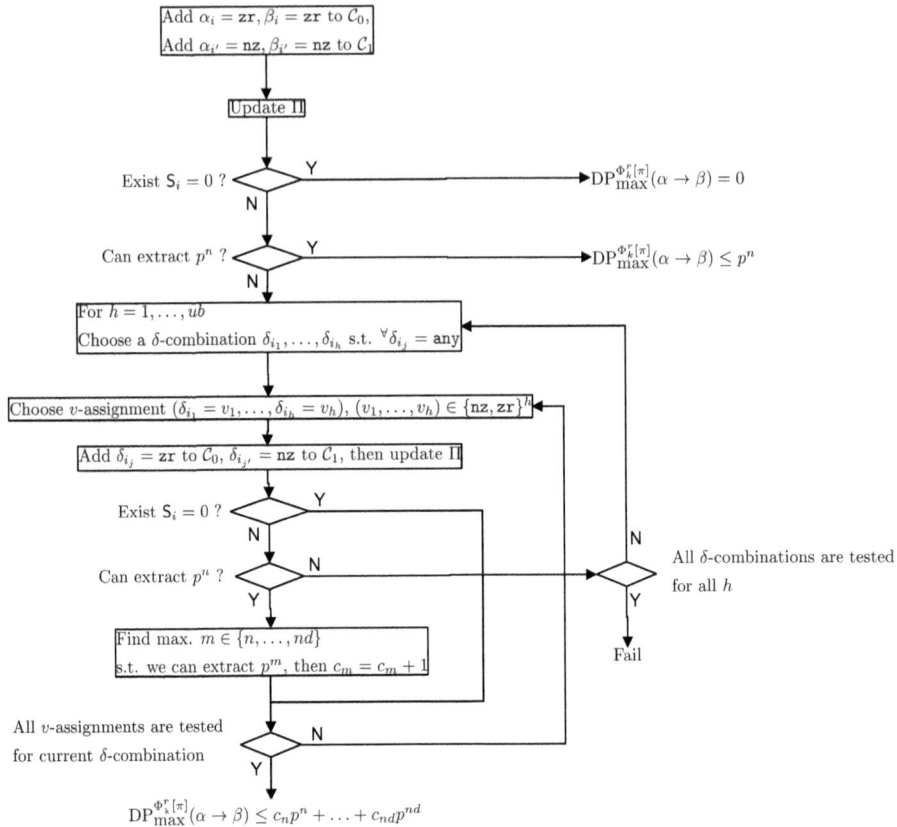

Fig. 2. The flow of our algorithm

For $k = 4$ and $r = 5$, we derived $p^4 + p^5$, which improves the previous bound of $p^4 + 2p^5$. This is due to our finer case analysis: Kim et al. divided (α, β) pairs into 24 cases, whereas we divided it into $(2^4 - 1)^2 = 225$ cases. For instance, Kim et al. considered $\alpha = (\mathbf{nz}, \mathbf{nz}, \mathbf{nz}, \mathbf{nz})$ with $\beta_1 = \mathbf{nz}$ and $\beta_3 = \mathbf{zr}$ as an initial constraint and derived $p^4 + 2p^5$ for this case, whereas we analyze $\alpha = (\mathbf{nz}, \mathbf{nz}, \mathbf{nz}, \mathbf{nz})$ with each of 15 cases of β, and derive p^4 or $p^4 + p^5$ for each case. One of the case is given at Example 2. Our result implies that the limit round (see Sect. 2.2) for Type-2 is $k + 1$, where we verified it for $k \leq 10$. It would be an interesting open problem to prove it for any k. More generally we conjecture the following:

Conjecture 1. Under Assumption 1, $\mathrm{DP}_{\max}^{\Phi_k^r[\pi_{\mathrm{cyc}}]} = O(p^{r-1})$, for $1 \leq r \leq k + 1$.

Note that it is generally impossible to have $O(p^k)$ for k rounds of Type-2. If input difference $\alpha = (\mathbf{zr}, \ldots, \mathbf{zr}, \mathbf{nz})$ is given to $\Phi_k^k[\pi_{\mathrm{cyc}}]$, we always have $\beta_k = \alpha_k$ with probability 1 (this is quite easy to verify. See Fig. 2 of [14] for instance). In this sense the limit rounds we derived is tight.

Table 1. Investigated shuffles of Type-2 (left) and Type-2i (right). For compatibility with [14] the block index is from 0 to $k-1$.

k	shuffle
4	$\{3,0,1,2\}$
6	$\{5,0,1,2,3,4\}$
8	$\{7,0,1,2,3,4,5,6\}$
10	$\{9,0,1,2,3,4,5,6,7,8\}$

k	name	shuffle
6	Opt (6)	$\{3,0,1,4,5,2\}$
8	Opt (8-1)	$\{3,0,1,4,7,2,5,6\}$
8	Opt (8-2)	$\{3,0,7,4,5,6,1,2\}$

Table 2. Derived MDP bounds of Type-2 GFS

k	r	n	result	bound
4	5	4	suc	$p^4 + p^5$ (improved from [10])
6	5	4	suc	p^4
6	5	5	fail (33)	
6	6	5	suc	$p^5 + p^6$
6	6	6	fail (66)	
6	7	6	suc	$p^6 + 10p^7$
8	7	6	suc	p^6
8	7	7	fail (176)	
8	8	7	suc	$p^7 + 3p^8$
8	8	8	fail (360)	
8	9	8	suc	$p^8 + 26p^9$
10	11	10	suc	$14p^{10} + 2p^{11}$ (unoptimized)

4.2 Results for Type-2i

We also investigated shuffles provided by [14] for $k = 6$ and 8. According to [14] there are one shuffle with $\mathrm{DRmax}(\pi) = 5$ (while $\mathrm{DRmax}(\pi_{\mathrm{cyc}}) = 6$) and two shuffles with $\mathrm{DRmax}(\pi) = 6$ (while $\mathrm{DRmax}(\pi_{\mathrm{cyc}}) = 8$), except isomorphic ones. They are called "optimum block shuffles". See Table 1.

Since these shuffles provide a faster diffusion than the cyclic shift, one may expect some improved results from Type-2, say $O(p^6)$ bound with $k = 6$ and $r = 6$. However the result does not follow this expectation, as shown by Table 3. Interestingly, the obtained bounds are roughly the same as Type-2 as long as $r \leq k$ or in a sense better, for a smaller number of failures. For instance, if $(k, r, n) = (8, 8, 8)$ we have 360 failures for Type-2, 240 for Opt (8-1), and only 16 for Opt (8-2). In contrast, the result is quite unpredictable for $r \geq k + 1$ with $n = k$. For $k = 6$ we need 8 rounds to have $O(p^6)$ with Opt (6). For $k = 8$ we need 9 rounds to have $O(p^8)$ for Opt (8-2), which is the same as Type-2. While

Table 3. Derived MDP bounds for Type-2i

k	r	n	shuffle	result	bound
8	7	6	Opt (8-1)	suc	$p^6 + 3p^7$
			Opt (8-2)	suc	$p^6 + 3p^7$
8	7	7	Opt (8-1)	fail (24)	
			Opt (8-2)	fail (6)	
8	8	7	Opt (8-1)	suc	$p^7 + 7p^8$
			Opt (8-2)	suc	$p^7 + 7p^8$
8	8	8	Opt (8-1)	fail (240)	
			Opt (8-2)	fail (16)	
8	9	8	Opt (8-1)	fail (53)	
			Opt (8-2)	suc	$\max\{p^8 + 31p^9, 2p^8\}$
8	10	8	Opt (8-1)	fail (16)	
8	11	8	Opt (8-1)	fail (4)	
8	12	8	Opt (8-1)	suc	$13p^8 + 19p^9$ (unopt.)

k	r	n	shuffle	result	bound
6	5	4	Opt (6)	suc	p^4
6	5	5	Opt (6)	fail (6)	
6	6	5	Opt (6)	suc	$p^5 + p^6$
6	6	6	Opt (6)	fail (36)	
6	7	6	Opt (6)	fail (3)	
6	8	6	Opt (6)	suc	$p^6 + 9p^7$

Table 4. An intermediate outputs of failure case with Opt (6) and $(k, r, n) = (6, 7, 6)$. $\alpha = (\mathbf{zr}, \mathbf{nz}, \mathbf{zr}, \mathbf{nz}, \mathbf{zr}, \mathbf{zr})$ and $\beta = (\mathbf{zr}, \mathbf{nz}, \mathbf{zr}, \mathbf{zr}, \mathbf{zr}, \mathbf{nz})$. E_is with $\mathsf{L}_{i,0} = \mathsf{L}_{i,1} = \mathbf{zr}$ (thus $\mathsf{S}_i = 1$) are omitted.

E_i			
round 1	E_1:	E_2:	E_3:
round 2	E_4: $\alpha_4 \to \delta_{12}$	E_5: $\alpha_2 \to \delta_{11}$	E_6:
round 3	E_7: $\delta_{11} \to \delta_9$	E_8: $\delta_{12} \to \delta_8$	E_9:
round 4	E_{10}: $\delta_8 \to \delta_{11}$	E_{11}: $\alpha_4 + \delta_9 \to \delta_5$	E_{12}: $\alpha_2 \to \delta_4$
round 5	E_{13}: $\delta_5 \to \beta_6 + \delta_8$	E_{14}:	E_{15}: $\delta_4 + \delta_{12} \to \alpha_4 + \delta_9$
round 6	E_{16}: $\alpha_2 \to \delta_5$	E_{17}: $\beta_6 \to \delta_4 + \delta_{12}$	E_{18}:
round 7	E_{19}:	E_{20}:	E_{21}:

$\mathsf{S}_i(\mathcal{H}_i)$			
round 1	S_1: 1	S_2: 1	S_3: 1
round 2	S_4: $p(\delta_{12})$	S_5: $p(\delta_{11})$	S_6: 1
round 3	S_7: $p(\delta_9, \delta_{11})$	S_8: $p(\delta_{12}, \delta_8)$	S_9: 1
round 4	S_{10}: $p(\delta_8, \delta_{11})$	S_{11}: $p(\delta_5, \delta_9)$	S_{12}: $p(\delta_4)$
round 5	S_{13}: $p(\delta_5, \delta_8)$	S_{14}: 1	S_{15}: $p(\delta_4, \delta_9, \delta_{12})$
round 6	S_{16}: $p(\delta_5)$	S_{17}: $p(\delta_4, \delta_{12})$	S_{18}: 1
round 7	S_{19}: 1	S_{20}: 1	S_{21}: 1

$$\mathcal{C}_0 = \{\delta_1 + \delta_9 + \alpha_4, \delta_2, \delta_3 + \delta_8 + \beta_6, \delta_6 + \delta_{11}, \delta_7, \delta_{10}, \delta_{13}, \delta_{14}, \delta_{15}\}$$
$$\mathcal{C}_1 = \{\alpha_2, \alpha_4, \beta_2, \beta_6, \delta_1, \delta_3, \delta_4, \delta_5, \delta_6, \delta_8, \delta_9, \delta_{11}, \delta_{12}, \delta_4 + \delta_{12}, \alpha_4 + \delta_9\}$$
$$\mathcal{U} = \{\delta_4, \delta_5, \delta_8, \delta_9, \delta_{11}, \delta_{12}\}$$

Opt (8-1) needs 12 rounds to have $O(p^8)$. That is, Conjecture 1 seems to hold with these shuffles for $r \leq k$, but it is unclear for $r = k + 1$; the limit round may depend on the shuffle.

An Analysis of Anomalous Failure Case. To see why Type-2i fails to provide $O(p^k)$ when $r = k + 1$, let us show one failure case (among three) of $(k, r, n) = (6, 7, 6)$ with Opt (6). Table. 4 is built from intermediate outputs of our algorithm for $\alpha = (\mathbf{zr}, \mathbf{nz}, \mathbf{zr}, \mathbf{nz}, \mathbf{zr}, \mathbf{zr})$ and $\beta = (\mathbf{zr}, \mathbf{nz}, \mathbf{zr}, \mathbf{zr}, \mathbf{zr}, \mathbf{nz})$. We need to take an ordered sum using $\mathcal{U} = \{\delta_4, \delta_5, \delta_8, \delta_9, \delta_{11}, \delta_{12}\}$ (here δ_i is indexed in the reverse-order: from bottom right to top left). However there are 11 S_is not being 1, hence we could extract p^5 at best. In addition, all $\delta_i \in \mathcal{U}$ are non-zero ($\mathbf{nz}$) thus there is no possible δ-combination. Therefore the algorithm declares failure and quits.

We still have a chance to improve the result by introducing some constraints other than δ-combination (though we did not find one). Another plausible approach is taking the structure of F into account (not only MDP being p). In any case, we consider this failure as a consequence of the limit of our approach.

5   Conclusion

In this paper, we have derived MDP bounds of Type-2 GFS and its generalization w.r.t the block shuffle, called Type-2i. For Type-2, the bounds of $k = 2, 4$-partition have been known, while we have shown bounds for $k = 6, 8, 10$. For Type-2i, we derived bounds for $k = 6, 8$, which is the first result on Type-2i, as far as we know. In deriving bounds, we utilized the idea of Kim et al.'s analysis for 4-partition Type-2, and distilled a formal procedure that basically works for any k with any block shuffle and implemented it on computers. Although MDP of SPN structure has been much studied and the bound is known for any k, this is the first attempt to derive MDP of GFS for any k. Our result would provide a fresh insight on the Type-2 and Type-2i GFSs and help build real ciphers based on them.

A future direction is reducing the complexity: for $k > 10$ the complexity of the current algorithm is not practical. Reducing the complexity will enable a more comprehensive investigation. In addition, it would be interesting to study if we can improve our bounds on Type-2i, which can fill the gap between differential characteristic probability evaluation (such as [4]) and ours.

Acknowledgments

We would like to thank Hiroyasu Kubo, Hirokatsu Nakagawa, Teruo Saito, Manabu Kurohashi, Daisuke Ikemura and Takeshi Kawabata for discussions and software implementation. We are deeply grateful for anonymous reviewers providing many useful comments.

References

1. Massey, J.: On the Optimality of SAFER+ Diffusion. In: Second AES Candidate Conference. National Institute of Standards and Technology (1999)
2. Zheng, Y., Matsumoto, T., Imai, H.: On the construction of block ciphers provably secure and not relying on any unproved hypotheses. In: Brassard, G. (ed.) CRYPTO 1989. LNCS, vol. 435, pp. 461–480. Springer, Heidelberg (1990)

3. Shirai, T., Shibutani, K., Akishita, T., Moriai, S., Iwata, T.: The 128-bit Block-cipher CLEFIA. In: Biryukov, A. (ed.) FSE 2007. LNCS, vol. 4593, pp. 181–195. Springer, Heidelberg (2007)

4. Shibutani, K.: On the Diffusion Properties of Generalized Feistel Structures. In: Biryukov, A., Gong, G., Stinson, D.R. (eds.) SAC 2010. LNCS, vol. 6544, pp. 211–228. Springer, Heidelberg (2011)

5. Park, S., Sung, S., Lee, S., Lim, J.: Improving the upper bound on the maximum differential and the maximum linear hull probability for SPN structures and AES. In: Johansson, T. (ed.) FSE 2003. LNCS, vol. 2887, pp. 247–260. Springer, Heidelberg (2003)

6. Hong, S., Lee, S., Lim, J., Sung, J., Cheon, D., Cho, I.: Provable security against differential and linear cryptanalysis for the SPN structure. In: Schneier, B. (ed.) FSE 2000. LNCS, vol. 1978, p. 273. Springer, Heidelberg (2001)

7. Hong, D., Sung, J., Hong, S., Lim, J., Lee, S., Koo, B., Lee, C., Chang, D., Lee, J., Jeong, K., Kim, H., Kim, J., Chee, S.: HIGHT: A new block cipher suitable for low-resource device. In: Goubin, L., Matsui, M. (eds.) CHES 2006. LNCS, vol. 4249, pp. 46–59. Springer, Heidelberg (2006)

8. Nyberg, K.: Generalized Feistel Networks. In: Kim, K.-c., Matsumoto, T. (eds.) ASIACRYPT 1996. LNCS, vol. 1163, pp. 90–104. Springer, Heidelberg (1996)

9. Biham, E., Shamir, A.: Differential cryptanalysis of DES-like cryptosystems. In: Menezes, A., Vanstone, S.A. (eds.) CRYPTO 1990. LNCS, vol. 537, pp. 2–21. Springer, Heidelberg (1991)

10. Kim, J., Lee, C., Sung, J., Hong, S., Lee, S., Lim, J.: Seven New Block Cipher Structures with Provable Security against Differential Cryptanalysis. IEICE Trans. Fundamentals E91-A(10) (2008)

11. Corporation, S.: The 128-bit Blockcipher CLEFIA Security and Performance Evaluations. Revision 1.0 (June 1, 2007)

12. Nyberg, K., Knudsen, L.R.: Provable security against differential cryptanalysis. In: Brickell, E.F. (ed.) CRYPTO 1992. LNCS, vol. 740, pp. 566–574. Springer, Heidelberg (1993)

13. Aoki, K., Ohta, K.: Strict Evaluation of the Maximum Average of Differential Probability and the Maximum Average of Linear Probability. IEICE Trans. Fundamentals E80-A(1), 2–8 (1997)

14. Suzaki, T., Minematsu, K.: Improving the generalized feistel. In: Hong, S., Iwata, T. (eds.) FSE 2010. LNCS, vol. 6147, pp. 19–39. Springer, Heidelberg (2010)

15. Matsui, M.: New Structure of Block Ciphers With Provable Security against Differential and Linear Cryptanalysis. In: Gollmann, D. (ed.) FSE 1996. LNCS, vol. 1039. Springer, Heidelberg (1996)

16. Lai, X.: On the Design and Security of Block Ciphers. Hartung-Gorre (1992)

A   Pseudocode of MDP Derivation Algorithm

The following is the pseudocode of (unoptimized version of) our algorithm. As it reflects the real program there are some minor differences from the description of the body texts. In particular the update function does not change $\mathcal{E}$, instead OrdSum function does. The label of a clause is determined via triangulation of $\mathcal{C}_0$, represented as rank($\mathcal{C}_0 \cup \{V\}$) step in Algorithm 3. Here, rank denotes the rank computation of the matrix corresponding $\mathcal{C}_0 \cup \{V\}$, where each element corresponds to a binary row vector.

Algorithm 1. MDP derivation for $(\alpha_1, \ldots, \alpha_k) \in \{\mathbf{nz}, \mathbf{zr}\}^k$, $(\beta_1, \ldots, \beta_k) \in \{\mathbf{nz}, \mathbf{zr}\}^k$

1: **procedure** INITIALIZATION
2:     $\mathcal{C}_0 \leftarrow \emptyset, \mathcal{C}_1 \leftarrow \emptyset$
3:     $\mathcal{L} \leftarrow (\mathbf{any}, \ldots, \mathbf{any}), \mathcal{S} \leftarrow (u, \ldots, u)$.
4:     Add $\{\alpha_i : \alpha_i = \mathbf{nz}\}_{i=1,\ldots,k}$ and $\{\beta_i : \beta_i = \mathbf{nz}\}_{i=1,\ldots,k}$ to $\mathcal{C}_1$
5:     Add $\{\alpha_i : \alpha_i = \mathbf{zr}\}_{i=1,\ldots,k}$ and $\{\beta_i : \beta_i = \mathbf{zr}\}_{i=1,\ldots,k}$ to $\mathcal{C}_0$
6: **end procedure**
7: $(t, \Pi) \leftarrow$ UPDATE(Π)
8: **if** $t =$ IMP **then return** "DP$_{\max}^{\Phi_k^r[\pi]}(\alpha \rightarrow \beta) = 0$"
9: **else if** ORDSUM(n, Π)=Suc **then return** "DP$_{\max}^{\Phi_k^r[\pi]}(\alpha \rightarrow \beta) \leq p^n$"
10: **else**
11:     **for** $h = 1, \ldots, ub$ **do**
12:         **for all** $i_1, \ldots, i_h$, s.t. LABELING$(\delta_{i_j}, \mathcal{C}_0, \mathcal{C}_1) = \mathbf{any}$ **do**
13:             $c_n, \ldots, c_{nd} \leftarrow 0, w \leftarrow$ Suc
14:             **for all** $(v_1, \ldots, v_h) \in \{\mathbf{nz}, \mathbf{zr}\}^h$ **do**
15:                 $\widetilde{\mathcal{C}}_0 \leftarrow \mathcal{C}_0, \widetilde{\mathcal{C}}_1 \leftarrow \mathcal{C}_1$,
16:                 add $\{\delta_{i_j} : v_j = \mathbf{nz}\}$ to $\widetilde{\mathcal{C}}_1$ and $\{\delta_{i_{j'}} : v_{j'} = \mathbf{zr}\}$ to $\widetilde{\mathcal{C}}_0$, $\widetilde{\Pi} \leftarrow (\mathcal{E}, \mathcal{L}, \mathcal{S}, \widetilde{\mathcal{C}}_0, \widetilde{\mathcal{C}}_1)$
17:                 $(t, \widetilde{\Pi}) \leftarrow$ UPDATE$(\widetilde{\Pi})$
18:                 **if** $t =$ OK **then**
19:                     **if** ORDSUM$(n, \widetilde{\Pi}) =$ Fail **then** $w \leftarrow$ Fail, **break**
20:                     **else**
21:                         $s^* \leftarrow \max\{s : s \in \{n, \ldots, nd\}, \text{ORDSUM}(s, \widetilde{\Pi}) = \text{Suc}\}$
22:                         $c_{s^*} \leftarrow c_{s^*} + 1$
23:                     **end if**
24:                 **end if**
25:             **end for**
26:             **if** $w =$ Suc **then return** "DP$_{\max}^{\Phi_k^r[\pi]}(\alpha \rightarrow \beta) \leq \sum_{i=n,\ldots,nd} c_i \cdot p^i$"
27:             **end if**
28:         **end for**
29:     **end for**
30:     **return** "Fail (DP$_{\max}^{\Phi_k^r[\pi]}(\alpha \rightarrow \beta)$ is not derived)"
31: **end if**

Algorithm 2. Update

```
1: function UPDATE(Π)
2:     while γ = 1 do
3:         γ ← 0
4:         for all L_{i,j} = any do L_{i,j} ←LABELING(E_{i,j}, C_0, C_1)
5:         end for
6:         for all i = 1, ..., s, S_i = u do
7:             if L_i = (zr, nz) or (nz, zr) then S_i ← 0, return ("IMP", Π)
8:             else if L_i = (zr, zr) then S_i ← 1
9:             else if L_i = (nz, nz) then S_i ← p
10:            else if L_i = (nz, any) then S_i ← p, L_{i,1} ← nz, add E_{i,1} to C_1, γ ← 1
11:            else if L_i = (any, nz) then S_i ← p, L_{i,0} ← nz, add E_{i,0} to C_1, γ ← 1
12:            else if L_i = (zr, any) then S_i ← 1, L_{i,1} ← zr, add E_{i,1} to C_0, γ ← 1
13:            else if L_i = (any, zr) then S_i ← 1, L_{i,0} ← zr, add E_{i,0} to C_0, γ ← 1
14:            end if
15:        end for
16:    end while
17:    return ("OK", Π)
18: end function
```

Algorithm 3. Determine label of clause V

```
1: function LABELING(V, C_0, C_1)
2:     if rank(C_0 ∪ {V}) = rank(C_0) then return zr
3:     end if
4:     for all V' ∈ C_1 do
5:         if rank(C_0 ∪ {V ⊕ V'}) = rank(C_0) then return nz
6:         end if
7:     end for
8:     return any
9: end function
```

Algorithm 4. Ordered sum

```
1: function ORDSUM(n, Π)
2:     Determine U using triangulation of C_0
3:     Obtain linear expressions for δ_i ∈ Ū, rewrite E' ← E
4:     E' ← E' \ {E_i : S_i = 1}
5:     ν ← |U|, {δ_{u_1}, ..., δ_{u_ν}} ← U:
6:     for all E_{i_1}, ..., E_{i_n} s.t. S_{i_1} = ··· = S_{i_n} = p do
7:         Ẽ ← E' \ {E_{i_1}, ..., E_{i_n}}
8:         for all {E_{i_1}, ..., E_{i_ν}} ∈ Ẽ s.t. ⋃_{h=1}^{ν} H̄_{i_h}) = U do
9:             if H̄_{i_j} \ ⋃_{h=1}^{j-1} H̄_{i_h} = {δ_{u_j}} and {δ_{u_j}} ∉ H_{i_j,0} ∩ H_{i_j,1}, for j = 1, ..., ν
      then
10:                 return "Suc"
11:            end if
12:        end for
13:    end for
14:    return "Fail"
15: end function
```

B   A Sample of Program Output

The following is an output sample of (optimized version of) our program applied to Type-2 with $(k, r, n) = (6, 7, 6)$. The optimized version produces multiple polynomials for each (α, β), hence we manually do some post-processing to derive overall MDP.

```
k=6, r=7, n=6, nd=7, asmpt_lb=0, asmpt_ub=4, a_ub=0x5c000000, a_lb=0xfc000000,
b_ub=0x04000000, b_lb=0xfc000000
detail=ON
Fi
R 1 F 1: a1->d15     F 2: a3->d14     F 3: a5->d13
R 2 F 4: a2+d15->d12     F 5: a4+d14->d11     F 6: a6+d13->d10
R 3 F 7: a3+d12->d9 F 8: a5+d11->d8 F 9: a1+d10->d7
R 4 F10: a4+d9+d14->d6  F11: a6+d8+d13->d5  F12: a2+d7+d15->d4
R 5 F13: a5+d6+d11->d3  F14: a1+d5+d10->d2  F15: a3+d4+d12->d1
R 6 F16: a6+d3+d8+d13->a1+b1+d5+d10 F17: a2+d2+d7+d15->a3+b3+d4+d12
    F18: a4+d1+d9+d14->a5+b5+d6+d11
R 7 F19: b1->a2+b2+d2+d7+d15     F20: b3->a4+b4+d1+d9+d14     F21: b5->a6+b6+d3+d8+d13


********************

a_lp = 0xFC000000(=0b111111)
b_lp = 0xFC000000(=0b111111)
DP=p^6+p^7+p^6+p^6  number=2     di=(d1,d4)
DP=p^6+p^7+p^7+p^7+p^6+p^7+p^7+p^7  number=3     di=(d1,d2,d12)
DP=p^6+p^7+p^7+p^7+p^7+p^7+0+p^7+p^7+p^7+p^7+p^7+p^7+0+0+0     number=4     di=(d4,d7,d9,d10)
a_lp = 0xFC000000(=0b111111)
b_lp = 0xF8000000(=0b111110)
DP=p^6+p^6  number=1     di=(d1)
DP=p^6+p^7+p^7+p^7  number=2     di=(d1,d11)
DP=p^6+p^7+p^7+0+p^7+p^7+p^7+0  number=3     di=(d4,d9,d12)
DP=p^6+p^7+p^7+0+p^7+0+p^7+0+p^7+0+p^7+0+p^7+0+0  number=4     di=(d4,d7,d9,d12)
a_lp = 0xFC000000(=0b111111)
b_lp = 0xF4000000(=0b111101)
DP=p^6+p^7+p^7+p^7  number=2     di=(d1,d6)
DP=p^6+p^7+p^7+0+p^7+p^7+p^7+0  number=3     di=(d4,d9,d12)
DP=p^6+p^7+p^7+0+p^7+0+p^7+0+p^7+0+p^7+0+0+p^7+0+0  number=4     di=(d4,d7,d9,d12)
a_lp = 0xFC000000(=0b111111)
b_lp = 0xF0000000(=0b111100)
DP=p^6+p^7  number=1     di=(d7)
DP=p^6+p^7+p^7+0     number=2     di=(d7,d10)
DP=p^6+p^7+p^7+0+p^7+p^7+0+p^7  number=3     di=(d1,d6,d8)
DP=p^6+p^7+p^7+0+p^7+0+p^7+0+p^7+0+0+p^7+p^7+0+0+0  number=4     di=(d6,d8,d9,d11)
a_lp = 0xFC000000(=0b111111)
b_lp = 0xEC000000(=0b111011)
DP=p^6+p^6  number=1     di=(d2)
DP=p^6+p^7+p^7+p^7  number=2     di=(d1,d4)
DP=p^6+p^7+p^7+0+p^7+p^7+p^7+0  number=3     di=(d4,d8,d11)
DP=p^6+p^7+p^7+0+p^7+0+p^7+0+p^7+0+0+p^7+p^7+0+0+0  number=4     di=(d5,d7,d8,d10)
a_lp = 0xFC000000(=0b111111)
b_lp = 0xE8000000(=0b111010)
DP=p^6  number=0
```

Double SP-Functions: Enhanced Generalized Feistel Networks[*]

Extended Abstract

Andrey Bogdanov[1] and Kyoji Shibutani[2]

[1] Katholieke Universiteit Leuven, ESAT/COSIC and IBBT, Belgium
andrey.bogdanov@esat.kuleuven.be
[2] Sony Corporation, Japan
kyoji.shibutani@jp.sony.com

Abstract. This work deals with the security and efficiency of type-I and type-II generalized Feistel networks (GFNs) with 4 lines. We propose to instantiate the GFNs with double SP-functions (substitution-permutation layer followed by another substitution-permutation layer) instead of single SP-functions (one substitution-permutation layer). We provide tight lower bounds on the number of differentially and linearly active functions and S-boxes in such ciphers. Based on these bounds, we show that the instantiation with double SP-functions using MDS diffusion has a proportion of differentially and linearly active S-boxes by up to 33% and 50% higher than that with single SP-functions for type-I and type-II GFNs, respectively. This opens up the possibility of designing more efficient block ciphers based on GFN structure. Note that type-I and type-II GFNs are the only non-contracting GFNs with 4 lines under a reasonable definition of a GFN.

Keywords: block cipher, generalized Feistel network, type-I GFN, type-II GFN, double SP-functions, active S-boxes, trail probability, substitution-permutation network.

1 Introduction

In this paper, we will demonstrate that instantiating the GFNs with double SP-functions (two subsequent substitution-permutation layers) is significantly more efficient with respect to differential and linear cryptanalysis than using single SP-functions for this purpose in terms of the proportion of active S-boxes.

1.1 Background

Generalized Feistel networks. Type-I and type-II [12] GFNs are block ciphers with specific internal structures as shown in Figure 1. In this paper, we focus on GFNs with 4 equally wide lines: One round of both type-I and type-II

---

[*] A part of this work has been presented at the Seventh International Workshop on Coding and Cryptography in April 2011, Paris, France.

U. Parampalli and P. Hawkes (Eds.): ACISP 2011, LNCS 6812, pp. 106–119, 2011.
© Springer-Verlag Berlin Heidelberg 2011

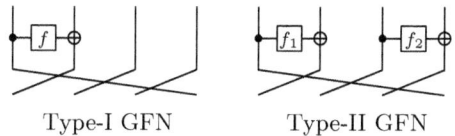

Type-I GFN   Type-II GFN

Fig. 1. Round transforms of type-I and type-II GFNs with 4 lines

GFNs divides the round input α into 4 equally long parts: $\alpha = [\alpha_1, \alpha_2, \alpha_3, \alpha_4]$. A round of type-I GFN outputs $[\alpha_2 \oplus f(\alpha_1), \alpha_3, \alpha_4, \alpha_1]$ for some keyed nonlinear function f. A round of type-II GFN outputs $[\alpha_2 \oplus f_1(\alpha_1), \alpha_3, \alpha_4 \oplus f_2(\alpha_3), \alpha_1]$ for keyed nonlinear functions f_1 and f_2. Note that the inverse of type-I GFN is type-III GFN with some appropriate input and output shuffling.

The functions of the round transforms often exhibit the *substitution-permutation* (SP) structure (subkey addition followed by a layer of m S-boxes s_i and a linear diffusion map M). If two SP-functions are applied one after another, we speak about *double SP-functions*. See Figure 2 for both single and double SP-functions.

Upper-bounding trail probability for Feistel networks. Apart from the concrete block cipher and hash function proposals based on Feistel networks instantiated with SP-type functions, thorough security analysis of more generic designs has been performed in the literature with respect to differential and linear cryptanalysis. Most cryptanalysis is aimed at the derivation of upper bounds on the probability of differential and linear trails by lower-bounding the number of differentially and linearly active S-boxes [3] — a tool that turned out to be highly useful for the security evaluation of various designs.

For balanced Feistel networks (BFNs), the work [4] proves the minimum number of active S-boxes in BFNs with SP-functions when the diffusion matrix is the same in all rounds (*single-round diffusion*). The papers [8,9] deal with the difference cancellation effect for such BFNs and introduces the diffusion switching mechanism which relies on using several distinct diffusion matrices over multiple rounds (*multiple-round diffusion*). The lower bounds on the number of active S-boxes for BFN with SP-functions and multiple-round diffusion are proven in [7]. Those for BFNs with SPS-functions and single-round diffusion are analyzed in [1]. Note that using distinct diffusion matrices in different rounds (as required by the multiple-round diffusion) reduces the efficiency.

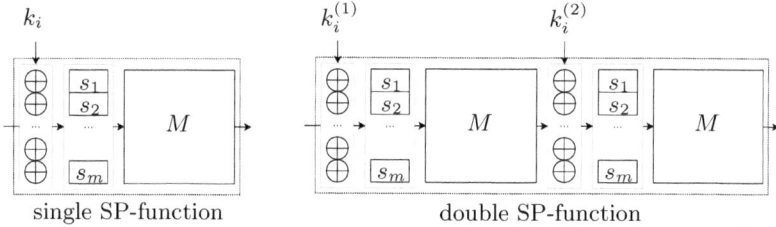

single SP-function   double SP-function

Fig. 2. Single and double SP-functions

For GFNs, lower bounds on the number of active S-boxes are obtained for type-I and type-II GFNs with SP-functions and single-round diffusion in [11] and [5], respectively. Bounds for unbalanced Feistel networks with contracting multiple-round diffusion are derived in [2]. Rough lower bounds for type-I and type-II with single SP-functions and multiple-round diffusion were proven by [6]. The work [6] also provides some numeric analysis for two specific cases of type-I and type-II GFNs with single SP-functions and multiple-round diffusion.

In this work, we will focus on 4-line type-I and type-II GFNs with double SP-functions and single-round diffusion (i.e. using the same diffusion matrix in all diffusion layers of the cipher) and obtain tight lower bounds on the number of active S-boxes in such constructions.

Efficiency metrics. A metric has to be defined to enable an efficiency comparison between different designs. Since a single metric capturing all implementation types and details is unlikely to exist, one might consider any comparison not based on concrete implementation figures illustrative. However, there are indeed metrics that adequately reflect at least some important efficiency properties. Inspired by [7], we use the proportion of active S-boxes in all S-boxes for this purpose with respect to differential and linear cryptanalysis, nonlinear operations being often most costly to implement:

Definition 1 (Efficiency metrics, E_m and E). *The efficiency metric E_m is defined as $E_m = \lim_{r \to \infty} \frac{A_{m,r}}{S_{m,r}}$, where $A_{m,r}$ is the number of active S-boxes over r rounds and $S_{m,r}$ is the total number of S-box computations over r rounds. The efficiency metric E is defined as $E = \lim_{m \to \infty} E_m$.*

As above (Figures 1 and 2), m corresponds to the block size of a cipher (for 4-line GFNs, the block size is $4m$ components in a block, m components in a line). The reason for E_m being asymptotic in the number of rounds is technical: One can operate with security results without having to extend them to an arbitrary number of rounds. Sometimes, for clarity, it is desirable to compare just two efficiency numbers, which is possible for large blocks (e.g. for hash functions or wide-block encryption) and justifies the usage of E as an efficiency metric in such cases. The metrics make most sense for tight bounds and iterative trails.

1.2 Contributions and Outline

GFNs with double SP-functions. We propose to instantiate the type-I and type-II GFNs with invertible double SP-functions (substitution-permutation layer followed by another substitution-permutation layer) instead of single SP-functions (one substitution-permutation layer). The intuition behind this specific choice of functions is as follows:

- Due to the second S-box layer, double SP-functions allow, on the one hand, to limit the analysis to the differential and linear activity patterns of functions and, on the other hand, to have effectively a higher number of active S-boxes.
- The second diffusion layer of a double SP-function constrains the differential effect (many differential trails contributing to the same differential) which might be present for SPS-functions.

- Having an odd number of SP-layers does not enable to prove tight bounds on the number of active S-boxes by working with the number of active functions only. An even number of SP-layers is similar to the case of double SP-functions.
- The invertibility prevents a function from absorbing differences: If a nonzero difference enters a bijective function the output difference will also be nonzero.

Under some reasonable restrictions, among the 4-line GFNs, the type-I and type-II structures are the only two unique non-contracting GFNs. All the other 4-line GFNs exhibit a differential effect, since at least one line is XOR-updated more than once before being used as an input to a function there [2]. This effectively reduces the proportion of active S-boxes for contracting GFNs, which is not the case for type-I and type-II GFNs.

Truncated trails and lower bounds on active functions. We use a string-based technique to show tight lower bounds on the number of differentially and linearly active functions for the GFNs. We demonstrate an equivalence between truncated differential and linear trails as well as imposed structural constraints which allows to deal with differential and linear cryptanalysis simultaneously.

We obtain that, for 4-line type-I and type-II GFNs with invertible functions, at least a half of their functions over 14 and 6 rounds, respectively, are active (Section 3). Note that this is not necessarily the case for GFNs with more than 4 lines: type-II GFNs with 8 lines do not seem to provide a proportion of more than 0.35 active functions [10].

Table 1. Efficiency E 4-line GFNs with single and double invertible SP-functions using MDS diffusion matrices with respect to differential and linear cryptanalysis, see also Figures 1 and 2. For more comparison see Section 4 and Table 2

	type-I GFN			type-II GFN		
	E	E_4	E_8	E	E_4	E_8
single SP	0.188 [11]	0.250 [11]	0.219 [11]	0.167 [5]	0.229 [5]	0.198 [5]
double SP (this paper)	0.250	0.313	0.281	0.250	0.313	0.281
advantage of double SP	33.3%	25.0%	28.3%	50.0%	36.7%	41.9%

Improved efficiency of GFNs. For double SP-functions, the lower bound on the number of active functions for type-I and type-II GFNs directly translates to the lower bound on the number of active S-boxes. Based on the demonstrated bounds, we show that the instantiation with double SP-functions provides a proportion E of differentially and linearly active S-boxes by up to 33% and 50% higher than that with single SP-functions using MDS diffusion for type-I and type-II GFNs, respectively. In other words, GFNs with double SP-functions outperform GFNs with single SP-functions in terms of differential and linear efficiency by a considerable margin. This opens up the possibility of designing significantly more efficient block ciphers based on GFN structures. A brief comparison is provided in Table 1. For more comparative analysis see Section 4.

2   Equivalence of Differential and Linear Truncated Trails

Here we analyze constraints on the truncated differential and linear trails of type-I and type-II GFNs. We demonstrate an equivalence between differential and linear truncated trails for the GFNs with respect to these constraints. This allows to study truncated differential and linear trails simultaneously by treating them as bit strings.

2.1   Truncated Differential Trails and Constraints

A differential trail for an iterative block cipher is a sequence of input and output differences for the consecutive rounds of the cipher. Let

$$\Delta x_i, \Delta x_{i+1}, \Delta x_{i+2}, \Delta x_{i+3}$$

be the input difference to a type-I or type-II GFN with 4 lines. Then a differential trail over t functions is the sequence of $t + 4$ differences

$$\Delta x_i, \Delta x_{i+1} \ldots, \Delta x_{i+t+2}, \Delta x_{i+t+3}.$$

Let the bit value d_{i+j} be defined as:

$$d_{i+j} = \begin{cases} 0, & \text{if } \Delta x_{i+j} = 0 \\ 1, & \text{if } \Delta x_{i+j} \neq 0 \end{cases} \text{ for } j \in \{0, \ldots, t+3\}.$$

Then the string of $t + 4$ bits

$$d_i, d_{i+1}, \ldots, d_{i+t+3} \tag{1}$$

is called a *truncated differential trail* over t functions illustrated in Figure 3.

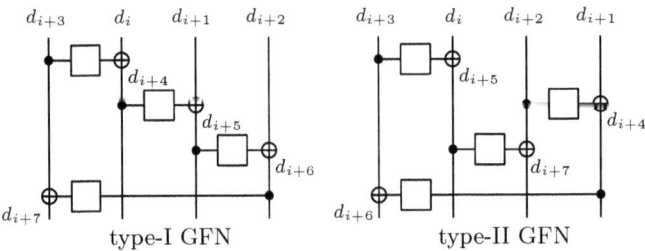

Fig. 3. Truncated differential trails of type-I (4 rounds) and type-II (2 rounds) GFNs with 4 lines. $d_{i+j} \in \{0, 1\}$, where $d_{i+j} = 1$ indicates that the line is differentially active

Due to the properties of XOR used to update lines and the invertibility of the functions, the propagation of differences through type-I and type-II GFNs with 4 lines obeys the following rules:

Property 1 (Differential zero rule for GFN-I). If two of d_i, d_{i+3}, d_{i+4} are zero, then all of them are zero, where $i = 0, 1, 2, \ldots$

Property 2 (Differential nonzero rule for GFN-I). If d_i, d_{i+3}, d_{i+4} are not all zero, at least two of them are nonzero, where $i = 0, 1, 2, \ldots$

Property 3 (Differential zero rule for GFN-II). If two of d_i, d_{i+3}, d_{i+5} are zero, then all of them are zero. Similarly, if two of d_{i+1}, d_{i+2}, d_{i+4} are zero, then all of them are zero, where $i = 0, 2, 4, \ldots$

Property 4 (Differential nonzero rule for GFN-II). If d_i, d_{i+3}, d_{i+5} are not all zero, at least two of them are nonzero. Similarly, if d_{i+1}, d_{i+2}, d_{i+4} are not all zero, at least two of them are nonzero, where $i = 0, 2, 4, \ldots$

2.2   Truncated Linear Trails and Constraints

A linear trail for an iterative block cipher is a sequence of input and output selection patterns for the consecutive rounds of the cipher. Let

$$\Gamma x_i, \Gamma x_{i+1}, \Gamma x_{i+2}, \Gamma x_{i+3}$$

be the input selection pattern for a type-I or type-II GFN with 4 lines. Then a linear trail over t functions is the sequence of $t + 4$ selection patterns

$$\Gamma x_i, \Gamma x_{i+1} \ldots, \Gamma x_{i+t+2}, \Gamma x_{i+t+3}.$$

Similarly to truncated differential trails, let the bit value l_{i+j} be defined as:

$$l_{i+j} = \begin{cases} 0, & \text{if } \Gamma x_{i+j} = 0 \\ 1, & \text{if } \Gamma x_{i+j} \neq 0 \end{cases} \text{ for } j \in \{0, \ldots, t + 3\}.$$

Then the string of $t + 4$ bits

$$l_i, l_{i+1}, \ldots, l_{i+t+3} \tag{2}$$

is called a *truncated linear trail* over t functions illustrated in Figure 4.

Like for differential trails, the propagation of selection patterns through type-I and type-II GFNs with 4 lines with invertible functions is due to the following rules:

Property 5 (Linear zero rule for GFN-I). If two of l_i, l_{i+1}, l_{i+4} are zero, then all of them are zero, where $i = 0, 1, 2, \ldots$

Property 6 (Linear nonzero rule for GFN-I). If l_i, l_{i+1}, l_{i+4} are not all zero, at least two of them are nonzero, where $i = 0, 1, 2, \ldots$

Property 7 (Linear zero rule for GFN-II). If two of l_i, l_{i+2}, d_{i+5} are zero, then all of them are zero. Similarly, if two of l_{i+1}, l_{i+3}, l_{i+4} are zero, then all of them are zero, where $i = 0, 2, 4, \ldots$

Property 8 (Linear nonzero rule for GFN-II). If l_i, l_{i+2}, l_{i+5} are not all zero, at least two of them are nonzero. Similarly, if l_{i+1}, l_{i+3}, l_{i+4} are not all zero, at least two of them are nonzero, where $i = 0, 2, 4, \ldots$

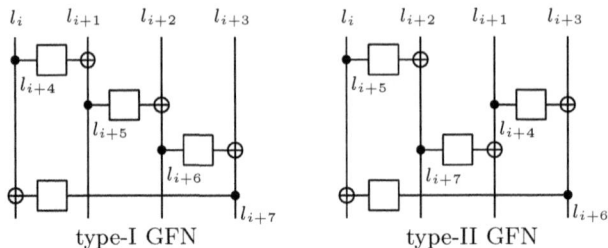

Fig. 4. Truncated linear trails of type-I (4 rounds) and type-II (2 rounds) GFNs with 4 lines. $l_{i+j} \in \{0,1\}$, where $l_{i+j} = 1$ indicates that the line is linearly active

2.3   Active Functions and Equivalence for Type-I GFNs

With respect to differential cryptanalysis, we look for a tight lower bound on the number of differentially active functions among t consecutive functions of type-I GFN. In other words, for some positive number λ_d, our aim is to prove

$$\sum_{j=3}^{t+2} d_{i+j} \geq \lambda_d, \qquad (3)$$

see Figure 3. At the same time, a tight lower bound λ_l on the number of linearly active functions among t functions means that (cf. Figure 4)

$$\sum_{j=1}^{t} l_{i+j} \geq \lambda_l. \qquad (4)$$

Direct manipulations with the indexes of d_{i+j} and l_{i+j} yield

Proposition 1. *Under the change of variables $d_{i+j} \mapsto l_{i+t+3-j}$, $j \in \{0, \dots, t + 3\}$ the following holds for a 4-line type-I GFN with invertible functions:*

- *truncated differential trail (1) translates to truncated linear trail (2),*
- *Property 1 translates to Property 5,*
- *Property 2 translates to Property 6, and*
- *if inequality (3) holds for $\lambda_d = \lambda$, then inequality (4) holds for $\lambda_l = \lambda$, and vice versa.*

2.4   Active Functions and Equivalence for Type-II GFNs

Similarly to type-I GFNs, we explore tight lower bounds λ_d and λ_l on the number of differentially and linearly active functions among t functions of type-II GFN, i.e.:

$$\sum_{j=2}^{t+3} d_{i+j} \geq \lambda_d \qquad (5)$$

and

$$\sum_{j=2}^{t+3} l_{i+j} \geq \lambda_l, \tag{6}$$

respectively (see Figures 3 and 4). Also here, we obtain the following

Proposition 2. *Under the change of variables $d_{i+j} \mapsto l_{i+t+3-j}$, $j \in \{0, \ldots, t + 3\}$ the following holds for a 4-line type-II GFN with invertible functions:*

- *truncated differential trail (1) translates to truncated linear trail (2),*
- *Property 3 translates to Property 7,*
- *Property 4 translates to Property 8, and*
- *if inequality (5) holds for $\lambda_d = \lambda$, then inequality (6) holds for $\lambda_l = \lambda$, and vice versa.*

Propositions 1 and 2 say that once we have a proof that the minimum number of differentially active functions among t consecutive functions of type-I and type-II GFNs with 4 lines is λ, we automatically obtain a proof that the minimum number of linearly active functions among t functions of the cipher is also λ.

3   Bounds for Active Functions

We focus on differentially active functions in this section, since one automatically obtains a proof for the minimum number of linearly active functions from a proof for the minimum number of differentially active functions as shown in the previous section (Propositions 1 and 2).

3.1   Some Truncated Differential Trails

Let *function i* of type-I or type-II GFN indicate the function whose output XOR-updates line number i. We refer to the XOR connecting to the i-th function's output as *the XOR of function i*. Then, if at least two of three lines connecting to the XOR of function i are non-active, the XOR is called *all-zero XOR*. Also, if at least one of the three lines connecting to the XOR of function i is active, the XOR is called *nonzero XOR*. These notions are related to Properties 1 to 4. For instance, when the XOR of function i of type-I GFN is all-zero, d_i, d_{i+3} and d_{i+4} are all zero due to Property 1. Also, when the XOR of function i of type-II GFN is nonzero, at least two of d_i, d_{i+3}, d_{i+5} are nonzero due to Property 4. Using these notions, the following truncated differential trails (treated as bit strings[1]) are derived (the proof is provided in the full version):

GFN-I-1 (consecutive all-zero XORs). If the XORs of functions i and $i + 1$ of type-I GFN are both all-zero, the forward and backward difference propagations will follow the truncated differential trail:

$$d_{i-7}d_{i-6}...d_i d_{i+1}...d_{i+8}d_{i+9} = 11 * 10110010001111.$$

---

[1] In the bit strings, $*$ denotes 0 or 1.

GFN-I-2 (no consecutive all-zero XORs). If the XOR of function i of type-I GFN is all-zero and there are no consecutive all-zero XORs, the forward and backward difference propagations will follow the truncated differential trail:

$$d_{i-4}d_{i-3}d_{i-2}d_{i-1}d_id_{i+1}d_{i+2}d_{i+3}d_{i+4}d_{i+5} = 11 * 1011001.$$

GFN-II-1 (consecutive all zero XORs in even-numbered functions). If the XORs of functions i and $i+2$ of type-II GFN are both all-zero, the forward and backward difference propagations will follow the truncated differential trail:

$$d_{i-6}d_{i-5}...d_id_{i+1}...d_{i+9}d_{i+10} = 1 * 111001001010111.$$

GFN-II-2 (consecutive all-zero XORs in odd-numbered functions). If the XORs of functions $i+1$ and $i+3$ of type-II GFN are both all-zero, the forward and backward difference propagations will follow the truncated differential trail:

$$d_{i-5}d_{i-4}...d_id_{i+1}...d_{i+10}d_{i+11} = 111011000010111 * 1.$$

GFN-II-3 (no consecutive all-zero XORs in even- and odd-numbered functions). If the XORs of functions i of type-II GFN is all-zero and there is no consecutive all-zero XORs in even- and odd-numbered functions, the forward and backward difference propagations will follow the truncated differential trail:

$$d_{i-3}d_{i-2}d_{i-1}d_id_{i+1}d_{i+2}d_{i+3}d_{i+4}d_{i+5}d_{i+6}d_{i+7} = 1110110 * 0 * 1.$$

GFN-II-4 (no consecutive all-zero XORs in even and odd numbered functions). If the XORs of functions $i+1$ of type-II GFN is all-zero and there is no consecutive all-zero XORs in even and odd numbered functions, the forward and backward difference propagations will follow the truncated differential trail:

$$d_{i-4}d_{i-3}d_{i-2}d_{i-1}d_id_{i+1}d_{i+2}d_{i+3}d_{i+4}d_{i+5}d_{i+6} = 1 * 1110010 * 1.$$

We employ the above bit strings to demonstrate the minimum number of differentially active functions of type-I and type-II GFNs in the following subsections.

3.2   Differentially Active Functions of Type-I GFNs

Using the truncated differential trails of Subsection 3.1, one derives the following statements (proofs are omitted here due to space limitations but can be found in the full version of the paper):

Lemma 1. *For 4-line type-I GFNs with invertible functions, every nontrivial differential trail over 14 rounds with at most 4 all-zero XORs has at least 7 active functions.*

Lemma 2. *For 4-line type-I GFNs with invertible functions, every nontrivial differential trail over 14 rounds with consecutive all-zero XORs has at most 4 all-zero XORs.*

Lemma 3. *For 4-line type-I GFNs with invertible functions, every nontrivial differential trail over 14 rounds without consecutive all-zero XORs has at most 4 all-zero XORs.*

Lemmata 1 to 3 yield

Proposition 3 (Active functions for type-I GFNs). *The 4-line type-I GFN with invertible functions provides at least 7 differentially active functions over 14 consecutive rounds for each non-trivial input difference.*

3.3   Differentially Active Functions in Type-II GFNs

Lemma 4. *For 4-line type-II GFNs with invertible functions, every differential trail over 6 rounds with at most 3 all zero XORs has at least 6 active functions.*

Lemma 5. *For 4-line type-II GFNs with invertible functions, every nontrivial differential trail over 6 rounds with consecutive all-zero XORs in even numbered functions has at most 3 all-zero XORs.*

Lemma 6. *For 4-line type-II GFNs with invertible functions, every nontrivial differential trail over 6 rounds, with consecutive all-zero XORs in odd numbered functions has at most 3 all-zero XORs.*

Lemma 7. *For 4-line type-II GFNs with invertible functions, every nontrivial differential trail over 6 rounds, without consecutive all-zero XORs in both even numbered rounds and odd numbered rounds has at most 3 all-zero XORs.*

Again, Lemmata 4 to 7 yield

Proposition 4 (Active functions for type-II GFNs). *The 4-line type-II GFN with invertible functions provides at least 6 differentially active functions over 6 rounds for each non-trivial input difference.*

4   Comparative Efficiency of GFNs

4.1   Converting Active Functions to Active S-Boxes

If $u \in \mathbb{F}_2^{nm}$ is represented by a bundle $u = (u_1, \ldots, u_m)$ of m elements in $\mathbb{F}_2^n$, $u_i \in \mathbb{F}_2^n$, $w(u)$ denotes the bundle weight of u, that is, the number of nonzero $\mathbb{F}_2^n$-components in u, $w(u) = \#\{u_i : u_i \neq 0, 1 \leq i \leq m\}$. The *branch number* of a linear map $M : \mathbb{F}_2^{nm} \to \mathbb{F}_2^{nm}$, is then defined as $\mathcal{B}(M) = \min_{u \neq 0}\{w(u) + w(M \cdot u)\}$. Higher values of $\mathcal{B}(M)$ indicate stronger diffusion, since a higher number of n-bit elements at inputs and outputs of an SP-function are nonzero then. The maximum value of $\mathcal{B}(M)$ is $m+1$. Linear transforms M with the highest branch number can be built from the generator matrices of maximum distance separable codes and are called *MDS*. Note that, for GFNs utilizing SP-functions with the diffusion matrix M, $\mathcal{B}(M)$ and $\mathcal{B}(^tM)$ imply the diffusion property for differential and linear attacks, respectively [4, 5], where ^tM is the transpose matrix of M.

When a type-I or type-II GFN is instantiated with double SP-functions, the minimum number of differentially and linearly active functions directly translates to a lower bound on the number of differentially and linearly active S-boxes, unlike the Feistel constructions with single SP-functions for which quite involving techniques are usually necessary at this point. We formulate this formally as

Proposition 5 (Active functions to active S-boxes). *Let $\mathcal{B}$ be the branch number of the diffusion matrix M or its transpose ^tM. Whenever a function is active (differentially or linearly) in type-I or type-II GFNs with double SP-functions, it provides at least $\mathcal{B}$ (differentially or linearly) active S-boxes.*

Combining Proposition 5 with Propositions 3 and 4 gives the minimum number of differentially active S-boxes for type-I and type-II GFNs. Then the equivalence between differential and linear cryptanalysis (Propositions 1 and 2) yields the minimum number of linearly active S-boxes. Thus, one directly obtains:

Theorem 1 (Active S-boxes for type-I GFNs). *For each nontrivial differential or linear trail, every $14R$, $R \geq 1$, rounds of 4-line type-I GFN with double SP-functions provide at least $7\mathcal{B}R$ active S-boxes (differentially or linearly), where $\mathcal{B}$ is the branch number of the diffusion matrix or its transpose in the SP-functions.*

Theorem 2 (Active S-boxes for type-II GFNs). *For each nontrivial differential or linear trail, every $6R$, $R \geq 1$, rounds of 4-line type-II GFN with double SP-functions provide at least $6\mathcal{B}R$ active S-boxes (differentially or linearly), where $\mathcal{B}$ is the branch number of the diffusion matrix or its transpose in the SP-functions.*

Theorems 1 and 2 can be seen as the main results of this paper. One can see that their bounds are actually tight. The lower bounds on the number of active functions translate to upper bounds on the differential and linear trail probabilities in a standard way: If p and q are the maximum linear and differential probabilities of the S-boxes, the probability of a $14R$-round nontrivial linear and differential trail will be upper-bounded by $p^{7\mathcal{B}R}$ and $q^{7\mathcal{B}R}$, respectively, for type-I GFNs. For type-II GFNs, the probability of a $6R$-round nontrivial differential and linear trail will be upper-bounded by $p^{6\mathcal{B}R}$ and $q^{6\mathcal{B}R}$, respectively.

4.2   GFNs: Double SP-Functions vs Single SP-Functions

Now we can compare type-I and type-II GFNs with single and double SP-functions with respect to the efficiency metrics E and E_m. The usefulness of these metrics is not limited to reflecting the time performance of some software implementations. We also expect it to indicate efficiency regarding such crucial parameters as energy and area consumption of a design in hardware. Note that the double SP-functions might require an additional memory buffer between the first SP-function and the second SP-function in certain implementations.

Recall that m is the number of components in each of the 4 lines of the cipher constructions under consideration. We perform comparison for MDS diffusion matrix M, i.e. for $\mathcal{B}(M) = m + 1$. The results are given in Table 2 and Figure 5.

Table 2. Efficiency metrics E_m and E for 4-line type-I and type-II GFNs with MDS diffusion: single SP-functions vs double SP-functions, see also Figure 5

	r	$A_{m,r}$	$S_{m,r}$	E_m	E
GFN-I, single SP [11]	$16R$	$[3(m+1)+1]R$	$16mR$	$\frac{3m+4}{16m}$	$3/16$
GFN-II, single SP [5]	$6R$	$[2(m+1)+2]R$	$12mR$	$\frac{2m+3}{12m}$	$1/6$
GFN-I, double SP (Th. 1)	$14R$	$[7(m+1)]R$	$28mR$	$\frac{m+1}{4m}$	$1/4$
GFN-II, double SP (Th. 2)	$6R$	$[6(m+1)]R$	$24mR$	$\frac{m+1}{4m}$	$1/4$

$A_{m,r}$ = number of active S-boxes in r rounds for m components in a line
$S_{m,r}$ = number of all S-boxes in r rounds for m components in a line
E_m = $\lim_{r \to \infty} \frac{A_{m,r}}{S_{m,r}}$, Definition 1
E = $\lim_{m \to \infty} E_m$, Definition 1

As one can see from Figure 5, type-I and type-II GFNs perform consistently better with double SP-functions than with single SP-functions with respect to E_m for all block sizes. For short blocks ($m = 2$), the advantage is at least 20% for type-I GFN and at least 28% for type-II GFN. For longer blocks ($m = 32$), E_m becomes close to E and the advantage amounts to about 33% and 50%, respectively. These results show that the instantiation with the double SP-functions can more than halve the required number of rounds compared to that with the single SP-function. This implies that the double SP construction is still more efficient than the single SP construction, even if a single round computation of the double SP is twice as slow as that of the single SP. Therefore, our results are not a tradeoff between the number of S-boxes in a single-round and the required number of rounds. See also Table 1.

Furthermore, we compare with the bounds for 4-line type-I and type-II GFNs using multiple-round diffusion with optimal diffusion matrices proven in [6]. E_m given by Theorems 1 to 4 of [6] for type-I and type-II GFNs is $(m+1)/6m$ for both differentially and linearly active S-boxes (i.e. $E_4 = 0.208$, $E_8 = 0.188$, and $E = 0.167$). A comparison of these values to Tables 1 and 2 yields that type-I and type-II GFNs with double SP-functions and single-round diffusion are much more efficient than the respective constructions with single SP-functions and

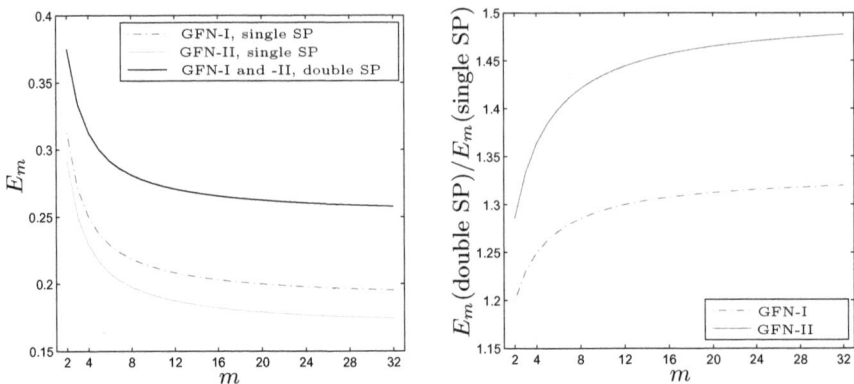

Fig. 5. Efficiency metric E_m for type-I and type-II GFNs with 4 lines: Absolute values of E_m (on the left) and normalized advantage of double SP-functions over single SP-functions (on the right)

multiple-round diffusion. However, the proven bounds of [6] do not appear to be tight. That is why we do not include this consideration into Table 2 and Figure 5.

5   Conclusions

In this paper, we have proposed to instantiate the functions of 4-line type-I and type-II GFNs with invertible double SP-functions having single-round diffusion (the same matrix in all rounds). We proved that at least a half of their functions are differentially and linearly active. This result allows us to show that every 14 rounds of type-I GFNs add at least $7\mathcal{B}$ active S-boxes and every 6 rounds type-II GFNs add at least $6\mathcal{B}$ active S-boxes, where $\mathcal{B}$ is the branch number of the underlying linear diffusion matrix.

We demonstrate that 4-line type-I and type-II GFNs are consistently more efficient with respect to differential and linear cryptanalysis when instantiated with double SP-functions than when standard single SP-functions are employed (advantage of up to 33% and 50% for type-I and type-II GFNs, respectively). This opens up the possibility of building considerably more efficient cryptographic primitives based upon GFNs.

Acknowledgements. Andrey Bogdanov is a postdoctoral fellow of the Fund for Scientific Research - Flanders (FWO). He was also partially supported by the Research Fund K.U.Leuven grant "A mathematical theory for the design of symmetric primitives". This work is sponsored in part by the IAP Programme P6/26 BCRYPT of the Belgian State (Belgian Science Policy). We would like to thank the reviewers of FSE 2011 and ACISP 2011 for their insightful comments.

References

1. Bogdanov, A.: On the Differential and Linear Efficiency of Balanced Feistel Networks. Inf. Process. Lett. 110(20), 861–866 (2010)
2. Bogdanov, A.: On Unbalanced Feistel Networks with Contracting MDS Diffusion. Des. Codes Cryptography. Special issue: Coding and Cryptography 2009 (2010)
3. Daemen, J., Rijmen, V.: The Design of Rijndael: AES – The Advanced Encryption Standard. Springer, Heidelberg (2002)
4. Kanda, M.: Practical Security Evaluation against Differential and Linear Cryptanalyses for Feistel Ciphers with SPN Round Function. In: Stinson, D.R., Tavares, S. (eds.) SAC 2000. LNCS, vol. 2012, pp. 324–338. Springer, Heidelberg (2001)
5. Shibutani, K.: On the Diffusion of Generalized Feistel Structures Regarding Differential and Linear Cryptanalysis. In: Biryukov, A., Gong, G., Stinson, D.R. (eds.) SAC 2010. LNCS, vol. 6544, pp. 211–228. Springer, Heidelberg (2011)
6. Shirai, T., Araki, K.: On Generalized Feistel Structures Using the Diffusion Switching Mechanism. IEICE Transactions 91-A(8), 2120–2129 (2008)
7. Shirai, T., Preneel, B.: On Feistel Ciphers Using Optimal Diffusion Mappings Across Multiple Rounds. In: Lee, P.J. (ed.) ASIACRYPT 2004. LNCS, vol. 3329, pp. 1–15. Springer, Heidelberg (2004)
8. Shirai, T., Shibutani, K.: Improving Immunity of Feistel Ciphers against Differential Cryptanalysis by Using Multiple MDS Matrices. In: Roy, B., Meier, W. (eds.) FSE 2004. LNCS, vol. 3017, pp. 260–278. Springer, Heidelberg (2004)
9. Shirai, T., Shibutani, K.: On Feistel Structures Using a Diffusion Switching Mechanism. In: Robshaw, M.J.B. (ed.) FSE 2006. LNCS, vol. 4047, pp. 41–56. Springer, Heidelberg (2006)
10. Suzaki, T., Minematsu, K.: Improving the Generalized Feistel. In: Hong, S., Iwata, T. (eds.) FSE 2010. LNCS, vol. 6147, pp. 19–39. Springer, Heidelberg (2010)
11. Wu, W., Zhang, W., Lin, D.: Security on Generalized Feistel Scheme with SP Round Function. I. J. Network Security 3(3), 215–224 (2006)
12. Zheng, Y., Matsumoto, T., Imai, H.: On the Construction of Block Ciphers Provably Secure and Not Relying on Any Unproved Hypotheses. In: Brassard, G. (ed.) CRYPTO 1989. LNCS, vol. 435, pp. 461–480. Springer, Heidelberg (1990)

Algebraic Techniques in Differential Cryptanalysis Revisited[*]

Meiqin Wang[1,2,3,**], Yue Sun[1], Nicky Mouha[2,3,***], and Bart Preneel[2,3]

[1] School of Mathematics, Shandong University, Jinan 250100, China
[2] Department of Electrical Engineering ESAT/SCD-COSIC,
Katholieke Universiteit Leuven, Kasteelpark Arenberg 10, B-3001 Heverlee, Belgium
[3] Interdisciplinary Institute for BroadBand Technology (IBBT), Belgium
mqwang@sdu.edu.cn

Abstract. At FSE 2009, Albrecht *et al.* proposed a new cryptanalytic method that combines algebraic and differential cryptanalysis. They introduced three new attacks, namely Attack A, Attack B and Attack C. For Attack A, they explain that the time complexity is difficult to determine. The goal of Attacks B and C is to filter out wrong pairs and then recover the key. In this paper, we show that Attack C does not provide an advantage over differential cryptanalysis for typical block ciphers, because it cannot be used to filter out any wrong pairs that satisfy the ciphertext differences. Furthermore, we explain why Attack B provides no advantage over differential cryptanalysis for PRESENT. We verify our results for PRESENT experimentally, using both PolyBoRi and MiniSat. Our work helps to understand which equations are important in the differential-algebraic attack. Based on our findings, we present two new differential-algebraic attacks. Using the first method, our attack on 15-round PRESENT-80 requires 2^{59} chosen plaintexts and has a worst-case time complexity of $2^{73.79}$ equivalent encryptions. Our new attack on 14-round PRESENT-128 requires 2^{55} chosen plaintexts and has a worst-case time complexity of $2^{112.83}$ equivalent encryptions. Although these attacks have a higher time complexity than the differential attacks, their data complexity is lower.

Keywords: Differential-Algebraic Attack, Block Cipher, PRESENT.

---

[*] This work was supported in part by the Research Council K.U.Leuven: GOA TENSE, the IAP Program P6/26 BCRYPT of the Belgian State (Belgian Science Policy), and in part by the European Commission through the ICT program under contract ICT-2007-216676 ECRYPT II.

[**] This author is supported by 973 Project (No.2007CB807902), NSFC Projects (No.61070244 and No.60931160442), Outstanding Young Scientists Foundation Grant of Shandong Province (No.BS2009DX030), IIFSDU Project (No. 2009TS087).

[***] This author is funded by a research grant of the Institute for the Promotion of Innovation through Science and Technology in Flanders (IWT-Vlaanderen).

U. Parampalli and P. Hawkes (Eds.): ACISP 2011, LNCS 6812, pp. 120–141, 2011.
© Springer-Verlag Berlin Heidelberg 2011

1   Introduction

Differential cryptanalysis [6, 7] is one of classic cryptanalytic methods for block ciphers. Resistance against differential cryptanalysis is a typical design criterion for new block ciphers. Algebraic cryptanalysis is a general method to attack ciphers. It has been widely used to cryptanalyze many primitives such as stream ciphers [13, 16], multivariate cryptosystems [19] and in particular block ciphers [14, 15, 17, 22]. The basic idea of algebraic cryptanalysis is to express the block cipher as a large multivariate polynomial system of equations. The secret key of the cipher is the solution of this system of equations. If the system is very sparse, overdefined or structured, it may be solved faster than a generic non-linear system of equations. By solving the system of equations for the block cipher, the key can be recovered with only a few plaintext-ciphertext pairs.

There are several methods to solve these systems of equations, such as computing a Gröbner basis or using a SAT solver. To compute a Gröbner basis, PolyBoRi [11] can be used. MiniSat [18] is a fast SAT solver. The advantage of computing a Gröbner basis is that useful equations can be generated, but this computation is typically slower than using a SAT solver and can more easily run out of memory.

However, the feasibility of algebraic cryptanalysis against block ciphers still remains a source of speculation. The main problem is that the size of the corresponding algebraic system is so large (thousands of variables and equations) that it seems infeasible to correctly predict the complexity of solving such polynomial systems. Therefore, algebraic cryptanalysis has so far had limited success in targeting modern block ciphers.

Recently, some works combining statistical cryptanalysis and algebraic cryptanalysis were presented [2–4, 20, 26]. Specifically, the combination of differential cryptanalysis and algebraic cryptanalysis appears to offer an advantage in reducing the data complexity. In [2, 3], Albrecht et $al.$ propose new differential-algebraic cryptanalytic methods, which they refer to as Attack A, Attack B and Attack C. In order to describe them, let p denote the probability of the r-round differential characteristic for an N-round block cipher.

In Attack A, the system of equations consists of the equations of the plaintext bits, ciphertext bits, and subkey bits, the equations of the key schedule, and the linear equations resulting from the differential characteristic and the filter equations of the last $(N - r)$ rounds (i.e. the equations that must hold if the output difference after round r holds). Attack A recovers the key by solving this system of equations for each of the about $1/p$ plaintext-ciphertext pairs.

In Attack B, the same system of equations is used. The longest time to find that the system of equations is inconsistent, is measured. If this time is exceeded, a right pair is found with a high probability.

In Attack C, the system of equations only consists of the filter equations after r rounds for an r-round differential and the key schedule algorithm after r rounds. The conditions resulting from the differential characteristic and the conditions from the plaintext to the corresponding ciphertext are omitted in Attack C. The goal of Attack C in [3] is to filter out wrong pairs by solving the system of equations using tools such as PolyBoRi or MiniSat, and use the remaining right pair to recover the subkey bits.

In differential cryptanalysis, the filtering process can only filter out the wrong pairs according to the difference values of the ciphertext pairs. That is, after the filtering process, a lot of wrong pairs may still remain, which may increase the time complexity to recover the key in the differential attack. However, in Attack B and Attack C, Albrecht *et al.* claim that the right pairs can be identified with a good probability if the equations after the r-th round of the differential characteristic are inconsistent. They claim that with their technique, the time complexity will be lower than in the standard differential attack. Their work received a lot of attention in the cryptographic community [5, 8, 12, 21, 23], because it gives hope for the combination of a statistical attack and an algebraic attack.

In this paper, we will revisit the differential-algebraic attack given by Albrecht *et al.*, which they applied to PRESENT [9]. We find that Albrecht's method cannot filter out most of the wrong pairs satisfying the ciphertexts differences. However, we will show that wrong pairs that do not satisfy the ciphertext differences, can easily be filtered out without the algebraic method. Using [3, 4], it is not possible to filter out more wrong pairs than using differential cryptanalysis.

Firstly, we show that Attack C typically cannot be used to filter out wrong pairs that do not satisfy the difference values of the ciphertexts to improve the differential cryptanalysis. Secondly, we verify using PolyBoRi and MiniSat2 that Attack B does not improve the current differential results for the PRESENT block cipher. The reason is that there are too few usable equations in the system of equations to derive an inconsistency for the wrong pairs or to find a solution for the right pairs. Based on our findings, we introduce two new methods that can more reliably use the right pairs to solve the right key within an acceptable time. For wrong pairs, no solution will be produced. One method is to fix certain key bits in the system of equations. This will allow an inconsistency to be derived faster. Another method is to use more than one plaintext-ciphertext pair to construct the system of equations.

We apply our attack methods to a reduced-round PRESENT block cipher. With the first method, we attack 15-round PRESENT-80 with 2^{59} chosen plaintexts and $2^{73.79}$ equivalent encryptions in the worst case. The 2R-differential attack on 15-round PRESENT-80 has a data complexity of more than 2^{59} and a time complexity of less than 2^{62} memory accesses. Therefore, the time complexity of the differential-algebraic attack for PRESENT-80 is much larger than that of the differential attack, but the data complexity is lower and the key does not have to be the same for every pair. If the number of chosen plaintext pairs that the attacker can obtain is limited, the algebraic-differential attack might be the only feasible attack. Note, however, that more rounds can be attacked in the case of PRESENT-80 using differential cryptanalysis (16 rounds instead of 15 rounds). We also provide a new attack on 14-round PRESENT-128 with a data complexity of 2^{55} chosen plaintexts and a worst-case time complexity of $2^{112.83}$ equivalent encryptions.

With our second method, the time complexity will be larger than with the first method for 15-round PRESENT-80. It is an open question whether the second method can offer an improvement for other block ciphers.

Our work also points out which equations are important in the differential-algebraic attack. With pure algebraic cryptanalysis, a 5-round PRESENT block cipher [15, 22] can be attacked. Compared to this result, our differential-algebraic attack can attack more rounds, but the data complexity will be higher than that for the pure algebraic attack.

This paper is organized as follows. Section 2 describes Albrecht's differential-algebraic attack. In Sect. 3, we show why Attack C cannot filter out more wrong pairs than differential cryptanalysis for most block ciphers. We verify using Poly-BoRi and MiniSat2 that Attack B cannot improve the differential cryptanalysis of the PRESENT block cipher. In Sect. 4, we present two methods that can be used to successfully solve the right key with the right pairs. Our attack methods are then applied to a reduced-round PRESENT block cipher. We conclude the paper in Sect. 5.

2   Description of Albrecht's Differential-Algebraic Attack

In [2, 3], Albrecht *et al.* proposed three types of attacks that combine algebraic techniques with differential cryptanalysis. They are referred to as Attack A, Attack B and Attack C. We now describe these three types of attacks.

Attack A. For an r-round differential characteristic $\Delta = (\delta_0, \delta_1, \ldots, \delta_r)$, the probability of the differential characteristic is denoted by p. For a pair of plaintexts (P', P''), where $P' \oplus P'' = \delta_0$, and the corresponding ciphertexts (C', C''), two systems of equations F' and F'' are constructed under the same encryption key K. With the differential characteristic, the following linear equations are constructed:

$$X'_{i,j} \oplus X''_{i,j} = \Delta X_{i,j} \rightarrow \Delta Y_{i,j} = Y'_{i,j} \oplus Y''_{i,j} \ ,$$

where $X'_{i,j}$ and $X''_{i,j}$ are the j-th bit of the input to the S-box layer in round i for the systems F' and F'' respectively. The corresponding output bits are $Y'_{i,j}$ and $Y''_{i,j}$. The values resulting from the differential characteristic are $\Delta X_{i,j}$ and $\Delta Y_{i,j}$. The linear expressions corresponding to bits of active S-boxes hold with some non-negligible probability. For the non-active S-boxes, the following linear relations also hold with non-negligible probability:

$$X'_{i,j} \oplus X''_{i,j} = 0 = Y'_{i,j} \oplus Y''_{i,j} \ .$$

If the r-round differential characteristic is used to recover the key for N rounds, the differences from the $(r+1)$-th round to the N-th round can be derived from the output difference of the r-th round. Theses differences after the r-th round are described by equations. Attack A combines the two systems of equations F' and F'', the above linear relations resulting from the differential characteristic and the equations from the difference values after round r to produce the system of equations $\overline{F}$ that holds with probability p. If about $1/p$ systems corresponding to $1/p$ pairs of plaintext-ciphertext can be solved, a right pair is expected to be found which can then be used to obtain the right key. However, the time complexity to solve the system about $1/p$ times may be very high.

Attack B. Attack B uses the same system equations as Attack A to filter out the wrong pairs. In a differential attack, the ciphertext difference values are commonly used to filter out wrong pairs. However, in Attack B, by measuring the time t it maximally takes to find that the system is inconsistent, it is assumed that a right pair has been identified with high probability if a time t has elapsed without finding an inconsistency. More specifically, Attack B assumes that $\Delta Y_{1,j}$ holds with a high probability after time t has elapsed. With the remaining pairs, the subkey bits involved in the active S-boxes in the first round can be recovered. An alternative form of Attack B is to recover key bits from the last round. It is assumed that if time t passes for a given plaintext-ciphertext pair, a right pair has been found. In this case, some subkey bits in the last rounds will be fixed, and then it is checked whether time t still passes without contradiction. The time to find an inconsistency or a reduced-round PRESENT block cipher was measured in Appendix C of [3].

Attack C. In Attack C, the differential is used instead of the differential characteristic as in Attack B. If the r-round differential $\delta_0 \rightarrow \delta_r$ is used to recover the key for N rounds, the system of equations only consists of the equations resulting from the round functions from round $(r + 1)$ to round N, the relations for the difference values from the $(r + 1)$-th round to the N-th round, and the equations of key schedule from the $(r + 1)$-th round to the N-th round. In this system of equations, there are no equations to restrict the relations between the plaintext and the corresponding ciphertext, and there are no equations for the difference values from the first round to the r-th round. By solving the system of equations and waiting for a fixed time t, a contradiction can be found in the system of equations. If one tested pair did not produce a contradiction after a fixed time, it is assumed to be a right pair satisfying the differential. Then with the right pair, the partial information for the subkey bits can be recovered. Appendix D in [3] measured the time to find an inconsistency for a reduced-round PRESENT block cipher. Based on this measured time, attacks on 16-round PRESENT-80, 17, 18 and 19 rounds of PRESENT-128 block cipher were given in [2, 3].

3   Inapplicability of Albrecht *et al.*'s Attacks

3.1   Inapplicability of Attack C

In this section, we will show that Attack C typically cannot be used to filter out the wrong pairs satisfying the difference values of the ciphertexts. Therefore, the right pairs cannot be identified and the key cannot be recovered. Moreover, Attack C can not filter out more wrong pairs than differential cryptanalysis to improve the differential cryptanalysis. As in the previous description, the system of equations in Attack C consists of the equations resulting from the round functions from round $(r+1)$ to round N, the relations resulting from the difference values from the $(r+1)$-th round to the N-th round, and the equations of key schedule from the $(r+1)$-th round to the N-th round. Let C'_i and C''_i be the i-th bit of ciphertext pair C' and C'' respectively, and ΔC_i is the i-th bit of the difference value of ciphertext pair C' and C''. We then classify these equations into three groups, Group A, Group B and Group C.

Group A. The linear equations resulting from the difference values of ciphertexts corresponding to the non-active S-boxes in the last round are

$$\Delta C_i = C'_i \oplus C''_i = 0 \ ,$$

where the i-th bit position corresponds to an output bit of any non-active S-box.

Group B. The equations resulting from the difference values of ciphertexts corresponding to the active S-boxes in the last round are

$$(\Delta C_{i_1} \| \Delta C_{i_2} \| \cdots \| \Delta C_{i_a}) = (C'_{i_1} \| C'_{i_2} \| \cdots \| C'_{i_a}) \oplus (C''_{i_1} \| C''_{i_2} \| \cdots \| C''_{i_a})$$
$$= \delta_N, \delta_N \in \Gamma_N \ ,$$

where $i_1, i_2, \ldots, i_a$ correspond to output bits of the active S-boxes, and Γ_N is the set of the ciphertext difference values.

Group C. The remaining equations are the equations resulting from the round functions from round $(r+1)$ to round N, the relations resulting from the difference values from the $(r+1)$-th round to the $(N-1)$-th round, and the equations of key schedule from the $(r+1)$-th round to the N-th round.

If a plaintext-ciphertext pair satisfies all the equations in Group A, Group B and Group C, it must be a right pair for the given differential. In the differential attack, the wrong pairs that do not satisfy the equations in Group A and Group B are easy to filter out using a look-up table combined with a time-memory trade-off. Because the equations in Group C involve unknown subkey bits, they cannot easily be used to filter out the remaining wrong ciphertext pairs after the filtering process with the ciphertext differences. In Attack C, Albrecht *et al.* wish to measure the maximum time t to identify a pair as a wrong pair with all the equations in Group A, B and C. In fact, the equations in Group A and Group B can easily be used to find a contradiction because they are only related to the ciphertext difference values. For a typical block cipher, it is impossible to find contradictions for the equations in Group C. To understand why this is the case, we claim the following.

Claim 1. *If there is a wrong ciphertext pair that satisfies all the equations in Group A and Group B but does not satisfy the equations in Group C, it is impossible for a typical block cipher to find a contradiction for the equations in Group C.*

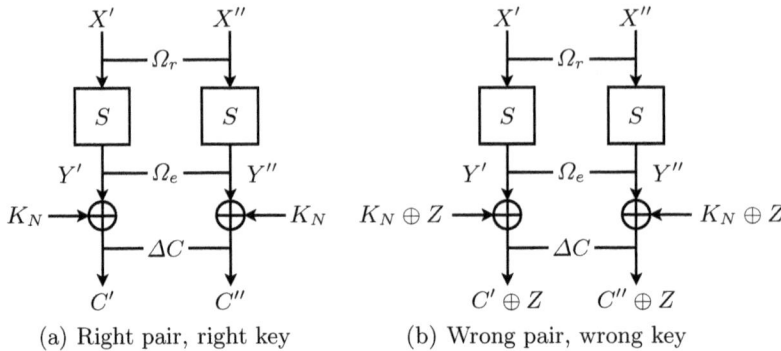

(a) Right pair, right key   (b) Wrong pair, wrong key

Fig. 1. It is not possible to detect that $(C' \oplus Z, C'' \oplus Z)$ is a wrong pair (see Claim 1)

Proof. We consider a block cipher based on a substitution-permutation network (SPN). For other structures (Feistel, Generalized Feistel,...), a similar proof can be given. We assume that the difference value of the ciphertext pair satisfies the equations in Group A and Group B, but does not satisfy the equations in Group C. First, we will prove Claim 1 for a 1R-attack and extend the proof to an sR-attack[1] ($s = 1, 2, 3, \ldots$).

In a 1R-attack, the wrong ciphertext pair satisfies the output difference values of all non-active and active S-boxes in the last round, but does not satisfy the input difference of some active S-boxes in the last round. In most SPN block ciphers, after the S-box layer in the last round, the whitening subkeys will be XORed.

Let us introduce the shortened notation

$$X_i' \leftarrow X_{i,j_1}' || X_{i,j_2}' || \ldots || X_{i,j_m}' \ ,$$

where $X_{i,j}'$ is the j-th bit of the input to the S-box layer in round i. We can then describe the round function for the last round as follows:

$$Y_N' = S[X_N'] \ , \quad C_N' = Y_N' \oplus K_N \ ,$$
$$Y_N'' = S[X_N''] \ , \quad C_N'' = Y_N'' \odot K_N \ ,$$

where X_N' and X_N'' are the inputs of the S-box layer S in the last round for the system F' and F'' respectively, and Y_N' and Y_N'' are the corresponding outputs. The values C_N' and C_N'' are the ciphertext bits, and K_N' is the whitening subkey in the last round.

We now consider Fig. 1. Under the right key, the wrong ciphertext pair $(C' \oplus Z, C'' \oplus Z)$ will result in the output difference of the S-box Ω_e and the input difference of the S-box Ω_w, however, the right pair (C', C'') will result

---

[1] An sR-attack means that the r-round differential is used to recover the key for $(r+s)$ rounds of the block cipher. We require in this paper that $s \ll N$, which is the case for typical differential attacks.

in the output difference and the input difference for the S-box as Ω_e and Ω_r respectively. As the subkey bits in the above equations are unknown variables, we will solve the following system of equations,

$$X'_N \oplus X''_N = \Omega_r.$$

We can obtain

$$S^{-1}[Y'_N] \oplus S^{-1}[Y''_N] = \Omega_r,$$

where S^{-1} denotes the inverse S-boxes Layer. Then we have

$$S^{-1}[C'_N \oplus K_N] \oplus S^{-1}[C''_N \oplus K_N] = \Omega_r \ .$$

Because the right pair always can produce the difference from $\Omega_r \mapsto \Omega_e$ for the active S-boxes, there is at least one pair of input values (X'_r, X''_r) and the corresponding output values (Y'_r, Y''_r) satisfying the following equations:

$$X'_r \oplus X''_r = \Omega_r, \ \ Y' \oplus Y'' = \Omega_e \ .$$

We have

$$S^{-1}[Y'_r] \oplus S^{-1}[Y''_r] = X'_r \oplus X''_r = \Omega_r.$$

For the wrong pair $(C' \oplus Z, C'' \oplus Z)$, let the whitening subkey in the last round satisfy the following equations:

$$C'_N \oplus Z \oplus K_N = Y'_r \ , \ \ C''_N \oplus Z \oplus K_N = Y''_r \ .$$

The resulting wrong whitening subkey $K_N \oplus Z$ in the last round can make the wrong pair $(C' \oplus Z, C'' \oplus Z)$ produce the right input difference Ω_r, so the wrong pair $(C' \oplus Z, C'' \oplus Z)$ cannot be filtered out with the system of equations in the last round.

The proof for 1R-attack is helpful to understand the idea. The analysis of the sR-attack works in a similar way. As stated by Biham and Shamir [7] (and similarly by Selçuk [24]):

"Each surviving pair suggests several possible values for [the subkey] bits. Right pairs always suggest the correct value for [the subkey] bits (along with several wrong values), while wrong pairs suggest random values [for the subkey bits]."

This statement is true for typical block ciphers. Therefore, any remaining wrong pair must produce some solutions for the subkey satisfying the difference values in the last s-round. The solution may be the right subkey or the wrong subkey. Thus, it is impossible for most block ciphers to produce a contradiction for the sR-attack in the above s-round equations.

The equations for the key schedule may lead to a contradiction in Group C for the derived subkey value for the last s rounds, but the number of the subkey bits involved in the last s rounds is usually not large enough to produce a contradiction, assuming the key schedule is random. However, assume that the equations for the key schedule result in a contradiction for the subkey values of the last s rounds. Then, this contradiction holds for all values of the subkeys. That is, the contradiction is independent of the subkey values. The contradiction must be a contradiction on the difference of the ciphertext pair: a contradiction on the values of the ciphertext pair cannot appear because the ciphertext is calculated as $C = Y_N \oplus K_N$. Therefore, this contradiction can be included into Group A or Group B. Because the differential cryptanalysis attack uses the equations of Group A and Group B to filter the ciphertext values, an inconsistency in the key schedule does not improve the differential attack.   □

In order to verify Claim 1, we tested the filtering time for different values of N and r of the PRESENT block cipher. In our tests, we constructed wrong ciphertext pairs that only satisfy the equations in Group A and Group B, but do not satisfy the equations in Group C when evaluated on the correct key. We used the source code provided by Albrecht [1] to apply Attack C with PolyBoRi-0.6 and MiniSat2. We performed a Gröbner basis computation to generate the filtering equations from the $(r + 1)$-th round to the $(r + 4)$-th round for the differential characteristic ($2 \leq r \leq 14$) for PRESENT-80. These filtering equations can speed up the procedure of producing the contradiction.

However, there is no contradiction for any ciphertext pair with PolyBoRi-0.6 after six hours of computation. MiniSat2 always obtained the wrong solution for the key. In Table 1, we list these test results. For the wrong pairs under the right key, the wrong solution can be obtained within t seconds. We tested 20 wrong pairs for different values of r and N, and list one example of a wrong pair (P', P'') and the corresponding right key K. Due to space limitations, we only present the difference values for the wrong pair in the last row of Table 1 and the differential characteristic for the right pair in Table 2. In Table 2, the output difference for the wrong pair of the r-th ($r = 12$) round is not equal to the output difference of the characteristic, but the output difference of the 13-th round is equal to the output difference of the characteristic. Therefore, this is a wrong pair.

At the same time, we construct the wrong ciphertext pairs for PRESENT-80 which do not satisfy the equations in any Group, the contradiction can be produced quickly and the filtering time is listed in Table 3. In addition, we construct some wrong ciphertext pairs that only satisfy the equation in Group A, the time to produce the contradiction is listed in Table 4. Moreover, we use a look-up table combined with a time-memory trade-off in differential cryptanalysis to filter out these pairs. As a result, our filter is more efficient than Attack C.

The computer we used is an IBM X3950 M2 with a CPU clock frequency of 2.4 GHz and 64GB RAM. From Tables 3 and 4, our test time with PolyBoRi approaches the corresponding time in Appendix D of [3], but our tested time with MiniSat2 is greater. The main reason is that our CPU is not same as Albrecht's. However, we can deduce that the wrong pairs Albrecht *et al.* used are wrong pairs that do not satisfy the equations in Group A or Group B, so they did not filter out wrong pairs that do satisfy the equations in Group A and Group B. Furthermore, even if Attack C is used as a filter for wrong pairs that do not satisfy the equations in Group A and Group B, its efficiency is much lower than the filter used in differential cryptanalysis. This shows that Attack C does not provide an advantage over differential cryptanalysis for most block ciphers.

Using Group A and Group B in a Differential Attack. We now clarify in more detail how the equations of Group A and Group B can be used in a differential attack. We consider two types of differential attacks:

(a) By generating a table of all possible ciphertext differences (corresponding to all solutions to the equations of Group A and Group B), wrong pairs can easily be filtered out. Because key counters will be used for the subkey bits corresponding to the active S-boxes, the number of output differences is less than the number of key counters required. Therefore, the table of all possible ciphertext differences provides only a relatively small overhead.

(b) In the filtering process, for each pair of ciphertexts (C', C''), a table is made of all possible input differences for the last round. This table does not depend on the value of the subkey bits in the last round. If we do not find a valid input difference for a particular pair of ciphertexts, this pair is identified as a wrong pair (i.e. it does not satisfy the equations of Group A and Group B). In this way, it is only necessary to make table of all input differences, and not all ciphertext differences. Typically, the table of all input differences should be small. For the remaining pairs, subkey bits in the last round will be guessed (instead of using key counters), to filter out pairs. For a wrong key, no pairs will remain, but the right pair will remain for the right key.

Note that (b) is in fact a time-memory trade-off applied to (a). In both (a) and (b), if output differences are invalid for some active S-boxes, they can be filtered using smaller tables. Then, the table that is described in (a) and (b) will be used to filter out the remaining pairs. In the next paragraph, we describe in detail how (a) can be used for a 2R attack on PRESENT. To construct a filter for a 3R and 4R attack on PRESENT, (b) can be used.

Relation to the Work of [4]. The equation system that Albrecht *et al.* set up in [4], is similar to the system of [3], except that the ciphertext bits (C'_i and C''_i)

Table 1. Attack C's Filtering Test for Wrong Pairs with MiniSat2

N	r	P'	P''	K	$t(s)$
8-10	7	$8b29917c174f21b7$	$8c29917c174f26b7$	$2b8bc6ad5d4b869101c2$	12.20-12.77
9-11	8	$d549bf122a09edfa$	$d249bf122a09eafa$	$5d05c98dce5da5894fc5$	12.26-12.92
10-12	9	$f5fc5a0d3979d9d3$	$f2fc5a0d3979ded3$	$f53e4ecaf9ce361ee6d7$	12.11-13.03
11-13	10	$50d752ee7f6017d7$	$57d752ee7f6010d7$	$afc238c99ce160d8254b$	12.22-12.73
12-14	11	$155fdec5b70e8b3a$	$125fdec5b70e8c3a$	$b544c98fce9474d53925$	12.33-12.92
13-15	12	$504ad07e763a8289$	$574ad07e763a8589$	$a7ece17b6ab73269d7e9$	12.01-12.71

N: the round number we attack; r: the round number of the differential; K: right key; (P', P''): one example of wrong pairs; t: the wrong solution obtained within t seconds.

are variables instead of fixed values. This equation system is used to compute a Gröbner basis for PRESENT up to degree $D = 3$ using PolyBoRi. Polynomials that contain non-ciphertext variables are removed.

The resulting equations are used as a first filter for the ciphertext pairs. The probability p_1 that a random ciphertext pair passes the first filter, is estimated by Albrecht *et al.* as $p_1 \approx 2^{-50.669}$ for a 2R-attack on PRESENT-80 and PRESENT-128. Afterwards, [4] uses Attack C to filter out the remaining pairs. They estimate the total filtering probability $p_2 \approx 2^{-51.669}$ for PRESENT-80 and $p_2 \approx 2^{-51.361}$ for PRESENT-128.

For a 2R-attack on PRESENT, it is straightforward to write a fast program to compute the total number of ciphertext differences. We find that $11664 \approx 2^{13.51}$ ciphertext differences are possible, and store them in a small table. This results in the accurate filtering probability of $p_a = 2^{13.51}/2^{64} = 2^{-50.49}$ for both PRESENT-80 and PRESENT-128. When we derive the probability of p_1 ourselves, using the equations in [4, Fig. 2], we find that $p_1 = p_2 = p_a = 2^{-50.49}$. This confirms our result, and shows that the calculation of p_1 and p_2 in [4] is not correct. The accurate filtering probability p_a is slightly lower than the probability of the rough filter used by Wang [25].

By storing the output differences in a small table, we can easily filter out the wrong ciphertext pairs without using the algebraic method. Furthermore, we calculate that the reinterpretation of Attack C in [4] as a technique to filter ciphertext differences, does not result in a better filter. Therefore, Attack C does not provide an advantage over differential cryptanalysis in the case of a 2R-attack on PRESENT.

For a 3R-attack and a 4R-attack on PRESENT, we used a look-up table combined with a time-memory trade-off to filter out 1000 randomly generated wrong pairs. We note that although the filtering probability of our filter and Attack C is same, our filter is much faster than Attack C.

3.2 Inapplicability of Attack B to PRESENT

Attack B involves two other types of equations, besides the equations in Group A, Group B and Group C in Attack C. The first type of equations is the linear equations derived from the difference values from round 1 to round r, and the second type of equations is the round functions and the key schedule algorithm from round 1 to round r. In this way, the restriction from the plaintext to the corresponding ciphertext was added. Although we cannot show that Attack B does not provide an advantage over differential cryptanalysis for any block cipher, we make the following two observations for Attack B:

Observation 1. If N approaches the maximum number of rounds that can be attacked with a pure algebraic attack, the linear equations for the inner rounds and the round functions restricting the relation between the plaintext and the ciphertext are all usable to solve the system of equations. There are three possible subcases:

1. If the key size is much larger than the block size, for a wrong pair, the probability that a solution can be found for the key in the system of equations is non-negligible. In this way, there is a non-negligible probability that a contradiction for the wrong pairs cannot be produced. Attack B will likely fail.
2. If the key size is smaller than the block size, for a wrong pair, the probability that no solution can be found for the key in the system of equations is high. In this way, the contradiction for the wrong pairs can be produced and the right solution for the right pair can be found with a high probability. Attack B is likely to succeed.
3. If the key size approaches the block size, Attack B can either succeed or fail.

Observation 2. If N is much larger than the maximum number of rounds that can be attacked with a pure algebraic attack, the linear equations for the inner rounds and the round functions and the key schedule algorithm for the inner rounds are not crucial to solve the system of equations. Only the equations for the outer rounds are relevant. We consider two subcases.

1. If there are few active S-boxes in the outer rounds, the restriction conditions are so few that a contradiction will be produced with low probability. Attack B will likely fail.
2. If there are many active S-boxes in the outer rounds, there are enough restriction conditions to derive a contradiction with high probability. Attack B is then likely to succeed.

Table 2. Difference Values for Wrong Pair and Right Pair in Attack C

R		Δ_{wrong}	Δ_{right}	R		Δ_{wrong}	Δ_{right}
I		$x_2=7, x_{14}=7$	$x_2=1, x_{14}=1$				
R1	S	$x_2=1, x_{14}=1$	$x_2=1, x_{14}=1$	R8	S	$x_0=9, x_2=9$	$x_8=9, x_{10}=9$
R1	P	$x_0=4, x_3=4$	$x_0=4, x_3=4$	R8	P	$x_0=5, x_{12}=5$	$x_2=5, x_{14}=5$
R2	S	$x_0=5, x_3=5$	$x_0=5, x_3=5$	R9	S	$x_0=1, x_{12}=1$	$x_2=1, x_{14}=1$
R2	P	$x_0=9, x_8=9$	$x_0=9, x_8=9$	R9	P	$x_0=1, x_3=1$	$x_0=4, x_3=4$
R3	S	$x_0=4, x_8=4$	$x_0=4, x_8=4$	R10	S	$x_0=3, x_3=3$	$x_0=5, x_3=5$
R3	P	$x_8=1, x_{10}=1$	$x_8=1, x_{10}=1$	R10	P	$x_0=9, x_4=9$	$x_0=9, x_8=9$
R4	S	$x_8=3, x_{10}=3$	$x_8=9, x_{11}=9$	R11	S	$x_0=4, x_4=4$	$x_0=4, x_8=4$
R4	P	$x_2=5, x_6=5$	$x_2=5, x_{14}=5$	R11	P	$x_8=1, x_9=1$	$x_8=1, x_{10}=1$
R5	S	$x_2=1, x_6=1$	$x_2=1, x_{14}=1$	R12	S	$x_8=9, x_9=9$	$x_8=9, x_{10}=9$
R5	P	$x_0=4, x_1=4$	$x_0=4, x_3=4$	R12	P	$\mathbf{x_2=3, x_{14}=3}$	$\mathbf{x_2=5, x_{14}=5}$
R6	S	$x_0=5, x_1=5$	$x_0=5, x_3=5$	R13	S	$\mathbf{x_2=1, x_{14}=1}$	$\mathbf{x_2=1, x_{14}=1}$
R6	P	$x_0=3, x_8=3$	$x_0=9, x_8=9$	R13	P	$\mathbf{x_0=4, x_3=4}$	$\mathbf{x_0=4, x_3=4}$
R7	S	$x_0=1, x_8=1$	$x_0=4, x_8=4$				
R7	P	$x_0=1, x_2=1$	$x_8=1, x_{10}=1$				

Rj: output difference after round j (S: after S-box layer, P: after permutation layer); Δ_{wrong}: differential value for wrong pair; Δ_{right}: differential value for right pair.

Table 3. Filter Time for Wrong Pairs Not Satisfying Equations in any Group

N	r	♯trails	PolyBoRi	MiniSat2	N	r	♯trails	PolyBoRi	MiniSat2
9	8	20	3.51-3.85	4.06-4.64	13	12	20	4.99-5.34	4.96-5.25
10	8	20	4.89-5.23	7.57-8.44	14	12	20	6.67-6.83	8.86-9.26
11	8	20	7.89-8.41	11.29-12.34	15	12	20	9.69-10.20	12.80-13.15
10	9	20	3.92-4.27	4.55-4.79	14	13	20	5.66-5.78	5.07-5.37
11	9	20	5.32-5.66	8.40-8.66	15	13	20	7.02-7.50	9.08-9.38
12	9	20	6.24-6.59	12.19-12.45	16	13	20	7.99-8.51	12.91-13.58
11	10	20	4.28-4.67	4.73-4.99	15	14	20	6.06-6.18	5.24-5.52
12	10	20	4.75-5.09	8.35-8.59	16	14	20	6.50-6.95	9.04-9.47
13	10	20	6.93-7.05	12.32-12.59	17	14	20	8.48-8.88	13.17-13.77
12	11	20	4.66-5.02	4.87-5.12					
13	11	20	6.09-6.42	8.69-8.97					
14	11	20	7.41-10.17	12.42-12.75					

♯trails: the number of wrong pairs we test;
PolyBoRi: the filtering time in seconds with PolyBori;
MiniSat2: the filtering time in seconds with Minisat2.

Table 4. Filter Time for Wrong Pairs Only Satisfying Equations in Group A

N	r	♯trails	PolyBoRi	MiniSat2
10	8	20	5.07-5.55	8.09-8.53
11	9	20	6.33-6.68	7.34-7.81
12	10	20	6.02-6.45	7.53-8.12

In order to verify our observations for a small number of rounds, we apply Attack B to PRESENT-80 with for $N = 4$, $r = 3$. The block size and the key size for PRESENT-80 are 64 and 80, respectively. We have tested 10 wrong pairs satisfying the filter conditions in Group A and Group B, but not satisfying the conditions in Group C. We found that among 10 wrong pairs, only one wrong pair was filtered out within 1500 seconds. The reason is that the key size is larger than the block size.

As N and r increase, we ran several tests and list the results in Table 5. We identify different differential characteristics for the PRESENT-80 block cipher. For any value of r we tested, the characteristics have two active S-boxes from round 1 to round r. There will be two active S-boxes in round $(r + 1)$ and 6, 7 or 8 active S-boxes in round $(r + 2)$. Round $r + 3$ has at least 12 active S-boxes and round $(r + 4)$ has 16 active S-boxes. We use MiniSat2 to filter out the wrong pairs. For $N = r$, $N = r + 1$ or $N = r + 2$, no wrong pairs were filtered out. For $N = r + 3$, very few wrong pairs were filtered out. Although for $N = r + 4$, more wrong pairs were filtered out compared to $N = r + 3$, lots of wrong pairs still remain. The reason is that there are more active S-boxes in round $(r + 4)$ than in round $(r + 3)$. This result is consistent with Table 10.8 of [2], where $N = r + 4$ is used as well.

Further experiments are listed in Table 5. In Table 5, the plaintext pairs are all wrong pairs and we cannot filter them out within 1500 seconds. Even if wrong pairs can be filtered out after 1500 seconds, the time complexity of Attack B would become much higher than differential cryptanalysis. Due to space limitations, we only present the difference values for the pair in the last row of Table 5 and the characteristics for the right pair in Table 6. For the pair in Table 6, the output difference of the r-th ($r = 14$) round is same as that of the characteristics, but the difference values from round 2 to round 10 are different from that of the characteristic. Therefore, this pair is a wrong pair. We also confirmed experimentally that Attack B cannot filter out wrong pairs that do not satisfy the output difference for the first round.

Observation 2 can be derived from the following statements:

1. SAT solvers use a tree-structured search algorithm, where branching is performed by heuristic guesses based on non-algebraic criteria. In order to reduce the search time, we must minimize both the average search depth and the dependencies of the unknown variables. In this way, those equations should be identified that tend to result in an inconsistency sooner.
2. In the system of equations in Attack B, the equations that lead to inconsistencies the soonest, are the equations related to the difference values, the round functions in the outer rounds such as the previous few rounds and the later few rounds. In contrast, the equations related to the difference values and the round functions in the inner rounds do not easily lead to inconsistencies. Therefore, the equations in the inner rounds can be removed in order to reduce the solving time.
3. Since the equations for the difference value in the outer rounds are very important for the solving process, we must obtain enough such equations to ensure there are enough restrictions for the dependent unknown subkey bits. If there are fewer active S-boxes in the outer rounds, there are not enough restrictions on the involved unknown subkey bits to obtain the right solution or filter out the wrong solutions. In other words, if there are more active S-boxes in the outer rounds, the solving process or the filtering process will be more efficient.

Table 5. Attack B's Filtering Test for Wrong Pairs Satisfying Ciphertext Difference Values with MiniSat2 (Timeout $t = 1500$ s)

N	r	P'	P''	K
5-7	4	$67279b1efdb93674$	$60279b1efdb93174$	$9ad864e12a6ecc872280$
6-8	5	$cdc43299824183d4$	$cac43299824184d4$	$70be32f5dd35396cdbfd$
7-9	6	$bc887a5de0597dd6$	$bb887a5de0597ad6$	$716d96982927070b6da$
8-10	7	$c53f11ab7329e7cf$	$c23f11ab7329e0cf$	$78bf3977acaffded898a$
9-11	8	$6d736a36a28d4f93$	$6a736a36a28d4893$	$5e7f5234d2063c5dd11d$
10-12	9	$94bd4ffd6585072e$	$93bd4ffd6585002e$	$1e00538c107f7abc4a73$
11,12,13	10	$f02f740d8d4b6d37$	$f72f740d8d4b6a37$	$df76f9fdaf4ead07d9a2$
12,13,14	11	$85f4ab19cf1dd9ac$	$82f4ab19cf1ddeac$	$5d0de0769a874e36d362$
13,14,15	12	$ca8b8755e65217af$	$cd8b8755e65210af$	$2d0d71c7a40d3084ac3a$
15,16,17	14	$934c64486fa9ed41$	$944c64486fa9ea41$	$8b1c1828ec601df09214$

Table 6. Difference Values for Wrong Pair and Right Pair in Attack B

R		Δ_{wrong}	Δ_{right}	R		Δ_{wrong}	Δ_{right}
I		$x_2 = 7, x_{14} = 7$	$x_2 = 7, x_{14} = 7$				
R1	S	$x_2 = 1, x_{14} = 1$	$x_2 = 1, x_{14} = 1$	R8	S	$x_8 = 5, x_{10} = 5$	$x_8 = 9, x_{10} = 9$
R1	P	$x_0 = 4, x_3 = 4$	$x_0 = 4, x_3 = 4$	R8	P	$x_2 = 5, x_{10} = 5$	$x_2 = 5, x_{14} = 5$
R2	S	$x_0 = 9, x_3 = 9$	$x_0 = 5, x_3 = 5$	R9	S	$x_2 = 1, x_{10} = 1$	$x_2 = 1, x_{14} = 1$
R2	P	$x_0 = 9, x_{12} = 9$	$x_0 = 9, x_8 = 9$	R9	P	$x_0 = 4, x_2 = 4$	$x_0 = 4, x_3 = 4$
R3	S	$x_0 = 4, x_{12} = 4$	$x_0 = 4, x_8 = 4$	R10	S	$x_0 = 5, x_2 = 5$	$x_0 = 5, x_3 = 5$
R3	P	$x_8 = 1, x_{11} = 1$	$x_8 = 1, x_{10} = 1$	R10	P	$x_0 = 5, x_8 = 5$	$x_0 = 9, x_8 = 9$
R4	S	$x_8 = 9, x_{11} = 9$	$x_8 = 9, x_{10} = 9$	R11	S	$x_0 = 4, x_8 = 4$	$x_0 = 4, x_8 = 4$
R4	P	$x_2 = 9, x_{14} = 9$	$x_2 = 5, x_{14} = 5$	R11	P	$x_8 = 1, x_{10} = 1$	$x_8 = 1, x_{10} = 1$
R5	S	$x_2 = 4, x_{14} = 4$	$x_2 = 1, x_{14} = 1$	R12	S	$x_8 = 9, x_{10} = 9$	$x_8 = 9, x_{10} = 9$
R5	P	$x_8 = 4, x_{11} = 4$	$x_0 = 4, x_3 = 4$	R12	P	$x_2 = 5, x_{14} = 5$	$x_2 = 5, x_{14} = 5$
R6	S	$x_8 = 5, x_{11} = 5$	$x_0 = 5, x_3 = 5$	R13	S	$x_2 = 1, x_{14} = 1$	$x_2 = 1, x_{14} = 1$
R6	P	$x_2 = 9, x_{10} = 9$	$x_0 = 9, x_8 = 9$	R13	P	$x_0 = 4, x_3 = 4$	$x_0 = 4, x_3 = 4$
R7	S	$x_2 = 4, x_{10} = 4$	$x_0 = 4, x_8 = 4$	R14	S	$x_2 = 4, x_{10} = 4$	$x_0 = 4, x_8 = 4$
R7	P	$x_8 = 4, x_{10} = 4$	$x_8 = 1, x_{10} = 1$	R14	P	$x_0 = 9, x_8 = 9$	$x_0 = 9, x_8 = 9$

Rj: output difference after round j (S: after S-box layer,
P: after permutation layer); Δ_{wrong}: differential value for wrong pair;
Δ_{right}: differential value for right pair.

It is noted that if there are more active S-boxes in the outer rounds, the filtering process will be efficient, but it is not favorable to filter out the wrong ciphertext pairs directly according to the difference value of the ciphertexts. This will further increase the time complexity.

To overcome these problems, we propose the following two methods for the differential-algebraic attack. The first method is to fix certain key bits to ensure with a high probability that the right key can be recovered from the right pair. The second method has the same goal, but adds some extra equations. We will describe these two attacks in Sect. 4.

4   New Differential-Algebraic Attacks

In Sect. 3, we showed that neither Attack C nor Attack B can improve the differential cryptanalysis of the PRESENT block cipher. We also explained why Attack C does not provide an improvement for most block ciphers. The reason is that the attacks cannot filter out the wrong pairs satisfying the ciphertext difference values to identify the right pair. We present two methods that can find the right solution in acceptable time t, based on the system of equations constructed in Attack B. For the right pair, we can solve the right key within time t. If a pair cannot be filtered within time t, we discard it and consider another pair.

Attack 1 Based on Fixing Certain Key Bits. According to the key schedule algorithm and the outer rounds of the characteristic, fix the key bits related to the active S-boxes in the top rounds or the bottom rounds. In this way, inconsistencies can be found sooner. As we showed in Sect. 3.2, Attack B cannot be used to filter out most wrong pairs. Therefore, our attack fixes key bits in all tested pairs. The idea of fixing key bits was already proposed in [3]. The difference with Attack 1 is that we recover the entire key, and not only subkey bits from the last rounds.

Attack 2 Based on Multiple Pairs. Because the equations for the difference values in the outer rounds lead to inconsistencies sooner, appending more such equations will be helpful to find the inconsistency. Using multiple plaintext-ciphertext pairs to construct more equations of outer rounds will make the solving process or the filtering process more efficient. For example, if two plaintext-ciphertext pairs are used to perform the attack, the number of such equations will double. This means that if we use two right pairs to solve the system of equations, the right key can be found. However, if there is at least one wrong pair involved in the two pairs, the key cannot be found. In addition, if we use three plaintext-ciphertext pairs, the efficiency can be improved further. However, as the number of pairs increase, the number of combinations of pairs grows exponentially and the time complexity increases. So the number of pairs to construct the system of equations should not be too high.

Our experiments show that some wrong pairs can be filtered out quickly, but others cannot. However, if most of the wrong pairs cannot be filtered out, the attack becomes infeasible. So we attack the PRESENT block cipher with the above approaches and try to solve the right key with the right pairs.

4.1   Attack 1 for the PRESENT Block Cipher

We now apply Attack 1 to the PRESENT block cipher. The results are listed in Table 7. If we use $r = 13$ to attack $N = 15$ rounds of PRESENT-80, the probability of the characteristic is 2^{-58} (using the last 13 rounds of the 14-round characteristic of [25]). The filtering probability according to the difference value for the ciphertext pair is $2^{-50.49}$ (as calculated at the end of Sect. 3.1). The CPU clock frequency is 2.4 GHz. From Table 7, we find that it takes at most 523.16 s to find an inconsistency. The table also shows that we should guess at least 34 key bits, so the time complexity will be $2^{34} \cdot 2^{58-50.49} \cdot 2.4 \cdot 10^9 \cdot 523.16 = 2^{34} \cdot 2^{7.51} \cdot 2^{31.16} \cdot 2^{9.03} = 2^{81.70}$ CPU cycles. We assume that a single encryption costs at least 16 CPU cycles per round[2]. Therefore, the time complexity for our attack ($2^{73.79}$ equivalent encryptions) is better than exhaustive search (2^{80}).[3] The data complexity is 2^{59} chosen plaintexts. For the 2R-differential attack, the data complexity must be higher than 2^{59} chosen plaintexts, because then one right plaintext-ciphertext pair is not sufficient to recover the key with a high success probability. However, the time complexity of the 15-round 2R-differential

---

[2] The bitsliced implementation of PRESENT by Albrecht achieves 16.5 cycles per round [2].

[3] We used 20 trials to obtain time t. Although more trials may result in a longer time t, we expect that our attack will still be much faster than exhaustive search.

Table 7. Time to Solve Right Key under Some Fixed Key Bits with MiniSat2

K_s	N	r	♯trails	N_k	t(s)	K_s	N	r	♯trails	N_k	t(s)
80	10	10	20	32	45.18-285.20	80	14-17	14	20	36	63.47-120.08
80	11	10	20	32	64.45-564.87	128	10	10	20	79	43.75-288.63
80	12	10	20	32	61.88-591.56	128	11	10	20	78	63.38-821.45
80	13	10	20	32	53.49-497.96	128	12	10	20	75	79.83-966.38
80	11	11	20	33	60.19-151.28	128	13	10	20	72	89.15-751.30
80	12	11	20	33	53.01-316.94	128	11	11	20	79	98.35-662.19
80	13	11	20	33	56.64-528.03	128	12	11	20	79	58.73-483.92
80	14	11	20	33	56.25-104.26	128	13	11	20	79	69.41-805.18
80	12	12	20	34	97.19-487.77	128	14	11	20	71	78.20-891.08
80	13	12	20	34	69.24-680.41	128	12	12	20	82	57.35-115.11
80	14	12	20	34	61.09-110.02	128	13	12	20	82	118.08-668.53
80	15	12	20	34	59.25-77.82	128	14	12	20	78	61.84-251.14
80	13-16	13	20	34	85.54-523.16	128	15	12	20	66	64.86-309.90

N_k: the number of fixed key bits.

attack must be lower than 2^{62} memory accesses (the time complexity given for the 16-round differential attack in [25]). Depending on the processor, one memory access requires about 2 to 10 CPU cycles. This means the complexity of the differential-algebraic attack for PRESENT-80 is much higher than that of the differential attack, but the data complexity is lower. Depending on how many chosen plaintext-ciphertext pairs the attacker can obtain, the algebraic-differential attack might however be the only feasible attack.

For PRESENT-128, we could not identify the right pairs for $r > 12$ using the method from [2]. If we use the 12-round differential characteristic with the probability 2^{-54} to attack 14-round PRESENT-128, the time complexity will be about $2^{78+54-50.49+31.16+7.97} = 2^{120.64}$ CPU cycles, or about $2^{112.83}$ equivalent encryptions. The data complexity is 2^{55} chosen plaintexts.

4.2   Attack 2 for the PRESENT Block Cipher

We respectively use two pairs and three pairs to attack the PRESENT. The test results are listed in Tables 8 and 9. For the right pairs, the right key can be solved within t seconds. We ran 10 trials for different values of r and N, and one example of right pairs $\{(P'_0, P''_0), (P'_1, P''_1)\}$ or $\{(P'_0, P''_0), (P'_1, P''_1), (P'_2, P''_2)\}$ and list the corresponding right key K. As in Attack 1, we can solve the right key from the right pairs, but the wrong pairs cannot always be filtered out. So we perform the test with the right pairs to recover the right key. We obtained the following results:

1. For $N = r + 3$ or $N = r + 4$ rounds of PRESENT-80 with the r-round differential characteristic, the right key can be solved with the two right pairs. Some test results are listed in Table 8. However, because we use two right pairs, this means that if m pairs of ciphertexts remain after filtering

Table 8. Time to Solve Right Key using Two Right Pairs with MiniSat2

K_s	N	r	P_0', P_1'	P_0'', P_1''	K	$t(s)$
80	12	9	$39121b2bffad3bbc$, $91f1a75a4f4d33e0$	$3e121b2bffad3cbc$, $96f1a75a4f4d34e0$	$4634342e33 \parallel$ $0d53e8cd71$	132.88-377.13
80	13	10	$67bb6eecd081767c$, $6f62c9bd561f718e$	$60bb6eecd081717c$, $6862c9bd561f768e$	$6fcaf3033d \parallel$ $39296c0f66$	122.00-849.89
80	14	11	$c2b3135aa3b8f3b4$, $8a43480c3122ab14$	$c5b3135aa3b8f4b4$, $8d43480c3122ac14$	$22c587b7b2 \parallel$ $607cddab90$	129.01-213.98
80	15	12	$c2b3135aa3b8f3b4$, $85c6576306a6a545$	$125fcb08afed6df3$, $82c6576306a6a245$	$155fcb08af \parallel$ $ed6af317f1$	133.64-141.75
80	13	9	$0c03406225bf97cd$, $0bbd25aea7c5b0c9$	$0b03406225bf90cd$, $0cbd25aea7c5b7c9$	$cca9deeb2c \parallel$ $0d98071ca6$	115.61-133.35
80	14	10	$9434381cb8083429$, $0b40a64e215244c6$	$9334381cb8083329$, $0c40a64e215243c6$	$ab7b47fdf8 \parallel$ $93fb87c9cd$	124.22-132.99
80	15	11	$8814d6bea07fd660$, $f02e367f419a412e$	$8f14d6bea07fd160$, $f72e367f419a462e$	$a7d16cda8d \parallel$ $b76ec42756$	130.48-144.89
80	16	12	$cbaef2f923614742$, $b37ee1f334c4207b$	$ccaef2f923614042$, $b47ee1f334c4277b$	$6b9b4087a6 \parallel$ $254f2bbef2$	189.26-280.49

according to the ciphertext difference, we must consider $\binom{m}{2}$ combinations of two pairs. However, the solving time for $\binom{m}{2}$ combinations of two pairs becomes unacceptable. If we attack 16-round PRESENT-80 with a 13-round differential characteristic with the probability 2^{-58}, we choose 2^{59} pairs of plaintexts and the filtering probability with the ciphertext difference is about $2^{-25.711}$, so the number of the remaining ciphertext pairs is about $2^{33.289}$ which will be combined to produce $2^{65.578}$ combinations of two pairs. The time complexity will be $2^{65.578} \cdot 2^{31.16} \cdot t > 2^{88}$. We have not identified the right pairs for $r = 13$, so we cannot test the time for t and it should be more than 100 seconds according to the test time for $r < 13$. Therefore, Attack 2 is slower than exhaustive search.

2. For $N = r + 2$ rounds of PRESENT-80, only few combinations of two right pairs can be used to solve the right key, so the success rate is too low.

3. For $N = r + 4$ rounds of PRESENT-128 with the r-round differential, only few combinations of two right pairs can be used to recover the right key and the success rate is also very low.

4. For $N = r+3$ rounds of PRESENT-80 and $N = r+4$ rounds of PRESENT-128 with the r-round differential, the right key can be solved with the three right pairs. The test results are listed in Table 9. However, because we use three pairs, this means that if m pairs of ciphertexts remain, there are $\binom{m}{3}$ combinations of three pairs. However, the solving time for $\binom{m}{3}$ combinations of three pairs becomes unacceptable.

From the above results, Attack 2 (using two pairs or three pairs for PRESENT) has no advantage over Attack 1 (fixing certain key bits). Maybe these attacks have some advantage for other ciphers. For example, if there would be more ac-

Table 9. Time to Solve Right Key using Three Right Pairs with MiniSat2

K_s	N	r	P'_0, P'_1, P'_2	P''_0, P''_1, P''_2	K	$t(s)$
80	11	9	$d9591ff50fc1df6d,$ $f9866c0009f3bf44,$ $0e768137f568779d$	$de591ff50fc1d86d,$ $fe866c0009f3b844,$ $09768137f568709d$	$66efab8af3 \parallel$ $74afe67553$	177.77-1402.2
80	12	10	$3a659aa3dc72107c,$ $62129df1a637b88f,$ $c566bb319010f0df$	$3d659aa3dc72177c,$ $65129df1a637bf8f,$ $c266bb319010f7df$	$2dc9fceff3 \parallel$ $174f9919c4$	240.70-578.68
80	13	11	$383663a9bc01cec5,$ $88042f67e3b59e95,$ $c842b19a415d9105$	$3f3663a9bc01c9c5,$ $8f042f67e3b59995,$ $cf42b19a415d9605$	$a0f5a7209b \parallel$ $b95180a21c$	247.53-t $(t > 2500)$
80	14	12	$2ddbc9427defb9ee,$ $2aa2624e2cb1dede,$ $4d19fefd126a29ee$	$2adbc9427defbeee,$ $2da2624e2cb1d9de,$ $4a19fefd126a2eee$	$3200679dd6 \parallel$ $3d29ae18bc$	293.21-408.40
80	12	9	$3d84126858c7435e,$ $32a6811bd0c6a32e,$ $cd66cbdb18c23c55$	$3a84126858c7445e,$ $35a6811bd0c6a42e,$ $ca66cbdb18c23b55$	$5da70ed0b5 \parallel$ $13fb14435c$	216.35-239.90
80	13	10	$e519cccfa40ce691,$ $e5aa80afcfc216a3,$ $8a179faf87127908$	$e219cccfa40ce191,$ $e2aa80afcfc211a3,$ $8d179faf87127e08$	$72ada6021d \parallel$ $d2667ab4e5$	238.47-258.13
80	14	11	$f5a33b54749b6624,$ $b2f64b6c661d6101,$ $2d106b5e6d2b4e24$	$f2a33b54749b6124,$ $b5f64b6c661d6601,$ $2a106b5e6d2b4924$	$8ab6e28d86 \parallel$ $9ef6858a87$	292.15-319.56
80	15	12	$e6005b48d2abd194,$ $41909dfa1ac196d9,$ $0e43381eb485d900$	$e1005b48d2abd694,$ $46909dfa1ac191d9,$ $0943381eb485de00$	$393d660706 \parallel$ $1dbe32c806$	271.31-340.26
128	13	9	$9d6902f268514522,$ $95d585a882e6e250,$ $2da0d2114f1805c2$	$9a6902f268514222,$ $92d585a882e6e550,$ $2aa0d2114f1802c2$	$0578224d0c9eba10 \parallel$ $bb0fd3b56d8b4834$	235.64-265.20
128	14	10	$972331fa763f86bd,$ $50d342a2a6dce17a,$ $efdfd44485f1ee81$	$902331fa763f81bd,$ $57d342a2a6dce67a,$ $e8dfd44485f1e981$	$d8ca446899016e69 \parallel$ $17641f71e11d09f5$	235.16-291.02
128	15	11	$76971713b1f0d438,$ $aed2ee07ad11dc6d,$ $e609bfed79d4143b$	$71971713b1f0d338,$ $a9d2ee07ad11db6d,$ $e109bfed79d4133b$	$9e3328405c865b25 \parallel$ $2201229c273fd1dd$	285.00-303.82
128	16	12	$eb449a907d31f33e,$ $84363465aaddb304,$ $e3a2e5866f5814a9$	$ec449a907d31f43c,$ $83363465aaddb404,$ $e4a2e5866f5813a9$	$73fdf364db99c472 \parallel$ $bb7a8e563b20a1f2$	316.21-414.30

tive S-boxes involved in the outer rounds in PRESENT, maybe we could obtain the right key using two right pairs with a high success probability.

5   Conclusion

The cryptanalytic method combining differential cryptanalysis and algebraic cryptanalysis has been a focus topic in the field of the cryptanalysis of symmetric ciphers. At FSE 2009, Albrecht *et al.* propose new differential-algebraic attacks, which they claim improves the results of the differential cryptanalysis. In this paper, we revisited Albrecht's cryptanalytic method and identified that the time complexity to identify the right pairs is not correct. Firstly, we showed that Attack C cannot be used to filter out the wrong pairs satisfying the difference value of the ciphertexts for most block ciphers to improve the differential cryptanalysis. We identified some important properties for Attack B and showed that Attack B does not provide an advantage over differential cryptanalysis for PRESENT. Faugère *et al.* presented a similar attack for DES, however, they could only attack 8-round DES with a 5-round differential characteristic. Their attack for DES is accordant with our Observation 1 in Sect. 3.2 because the key size for DES is smaller than the block size.

In this paper, we introduce two new methods to perform a differential-algebraic attack. The first method is to fix certain key bits to solve the system of equations and the second method is to use multiple pairs to construct the system of equations. This method is more efficient for the PRESENT block cipher and its data complexity is better than that of the differential attack, but the time complexity is worse. Although we did not significantly improve the results of the differential cryptanalysis for PRESENT, our work indicates which equations are important in the differential-algebraic attack. For the differential-algebraic attack, we obtain the following three conclusions:

1. Compared with the differential cryptanalysis, the differential-algebraic attack can reduce the data complexity, but the time complexity increases. Compared with the algebraic cryptanalysis, the differential-algebraic attack can attack more rounds because the relations resulting from the differential characteristic are very important for the solving process.
2. In order to make the solving process in the differential-algebraic attack more efficient, more active S-boxes should be involved in the outer rounds. However, more active S-boxes will reduce the filtering probability with the ciphertext difference and it will increase the time complexity. The lower bound for the number of the active S-boxes should be used to ensure the system of equations can be solved reliably. The detailed analysis of this case can be seen as future work.
3. If the methods to solve systems of equations can be improved, and if the computational power available increases, we expect that differential-algebraic attacks will gain in importance.

Acknowledgments. The authors would like thank the anonymous reviewers for their detailed comments and suggestions.

References

1. Albrecht, M.: Tools for the algebraic cryptanalysis of cryptographic primitives., http://www.ecrypt.eu.org/tools/tools-for-algebraic-cryptanalysis
2. Albrecht, M.: Algorithmic Algebraic Techniques and their Application to Block Cipher Cryptanalysis. PhD thesis, Royal Holloway, University of London (2010)

3. Albrecht, M., Cid, C.: Algebraic Techniques in Differential Cryptanalysis. In: Dunkelman, O. (ed.) FSE 2009. LNCS, vol. 5665, pp. 193–208. Springer, Heidelberg (2009)

4. Albrecht, M., Cid, C., Dullien, T., Faugère, J.-C., Perret, L.: Algebraic precomputations in differential and integral cryptanalysis. In: INSCRYPT 2010, p. 18 (2010) (to appear)

5. Bard, G.V.: Algebraic Cryptanalysis. Security and Cryptology, vol. XXXIV. Springer, Heidelberg (2009)

6. Biham, E., Shamir, A.: Differential Cryptanalysis of DES-like Cryptosystems. J. Cryptology 4(1), 3–72 (1991)

7. Biham, E., Shamir, A.: Differential Cryptanalysis of the Full 16-Round DES. In: Brickell, E.F. (ed.) CRYPTO 1992. LNCS, vol. 740, pp. 487–496. Springer, Heidelberg (1993)

8. Blondeau, C., Gérard, B.: Multiple Differential Cryptanalysis: Theory and Practice. In: Joux, A. (ed.) FSE 2011. LNCS, vol. 6733, pp. 35–54. Springer, Heidelberg (2011)

9. Bogdanov, A., Knudsen, L.R., Leander, G., Paar, C., Poschmann, A., Robshaw, M.J.B., Seurin, Y., Vikkelsoe, C.: PRESENT: An Ultra-Lightweight Block Cipher. In: Paillier, P., Verbauwhede, I. (eds.) CHES 2007. LNCS, vol. 4727, pp. 450–466. Springer, Heidelberg (2007)

10. Boneh, D. (ed.): CRYPTO 2003. LNCS, vol. 2729. Springer, Heidelberg (2003)

11. Brickenstein, M., Dreyer, A.: PolyBoRi: A framework for Gröbner-basis computations with Boolean polynomials. J. Symb. Comput. 44(9), 1326–1345 (2009)

12. Cho, J.Y.: Linear Cryptanalysis of Reduced-Round PRESENT. In: Pieprzyk, J. (ed.) CT-RSA 2010. LNCS, vol. 5985, pp. 302–317. Springer, Heidelberg (2010)

13. Courtois, N.: Fast Algebraic Attacks on Stream Ciphers with Linear Feedback. In: Boneh (ed.) [10], pp. 176–194

14. Courtois, N., Bard, G.V.: Algebraic Cryptanalysis of the Data Encryption Standard. In: Galbraith, S.D. (ed.) Cryptography and Coding 2007. LNCS, vol. 4887, pp. 152–169. Springer, Heidelberg (2007)

15. Courtois, N., Debraize, B.: Specific S-Box Criteria in Algebraic Attacks on Block Ciphers with Several Known Plaintexts. In: Lucks, S., Sadeghi, A.-R., Wolf, C. (eds.) WEWoRC 2007. LNCS, vol. 4945, pp. 100–113. Springer, Heidelberg (2008)

16. Courtois, N., Meier, W.: Algebraic Attacks on Stream Ciphers with Linear Feedback. In: Biham, E. (ed.) EUROCRYPT 2003. LNCS, vol. 2656, pp. 345–359. Springer, Heidelberg (2003)

17. Courtois, N., Pieprzyk, J.: Cryptanalysis of Block Ciphers with Overdefined Systems of Equations. In: Zheng, Y. (ed.) ASIACRYPT 2002. LNCS, vol. 2501, pp. 267–287. Springer, Heidelberg (2002)

18. Eén, N., Sörensson, N.: An Extensible SAT-solver. In: Giunchiglia, E., Tacchella, A. (eds.) SAT 2003. LNCS, vol. 2919, pp. 502–518. Springer, Heidelberg (2004)

19. Faugère, J.-C., Joux, A.: Algebraic Cryptanalysis of Hidden Field Equation (HFE) Cryptosystems Using Gröbner Bases. In: Boneh (ed.) [10], pp. 44–60

20. Faugère, J.-C., Perret, L., Spaenlehauer, P.-J.: Algebraic-Differential Cryptanalysis of DES. In: Western European Workshop on Research in Cryptology - WEWoRC 2009, pp. 1–5 (2009)

21. Gong, Z., Hartel, P., Nikova, S., Zhu, B.: Towards Secure and Practical MACs for Body Sensor Networks. In: Roy, B.K., Sendrier, N. (eds.) INDOCRYPT 2009. LNCS, vol. 5922, pp. 182–198. Springer, Heidelberg (2009)

22. Nakahara, J., Sepehrdad, P., Zhang, B., Wang, M.: Linear (Hull) and Algebraic Cryptanalysis of the Block Cipher PRESENT. In: Garay, J.A., Miyaji, A., Otsuka, A. (eds.) CANS 2009. LNCS, vol. 5888, pp. 58–75. Springer, Heidelberg (2009)
23. Özen, O., Varıcı, K., Tezcan, C., Kocair, Ç.: Lightweight Block Ciphers Revisited: Cryptanalysis of Reduced Round PRESENT and HIGHT. In: Boyd, C., Nieto, J.M.G. (eds.) ACISP 2009. LNCS, vol. 5594, pp. 90–107. Springer, Heidelberg (2009)
24. Selçuk, A.A.: On Probability of Success in Linear and Differential Cryptanalysis. J. Cryptology 21(1), 131–147 (2008)
25. Wang, M.: Differential Cryptanalysis of Reduced-Round PRESENT. In: Vaudenay, S. (ed.) AFRICACRYPT 2008. LNCS, vol. 5023, pp. 40–49. Springer, Heidelberg (2008)
26. Wang, M., Wang, X., Hui, L.C.: Differential-algebraic cryptanalysis of reduced-round of Serpent-256. SCIENCE CHINA Information Sciences 53(3), 546–556 (2010)

Faster and Smoother – VSH Revisited

Juraj Šarinay*

EPFL IC LACAL, Station 14, CH-1015 Lausanne, Switzerland
juraj.sarinay@epfl.ch

Abstract. We reconsider the provably collision resistant Very Smooth Hash and propose a small change in the design aiming to improve both performance and security. While the original proofs of security based on hardness of factoring or discrete logarithms are preserved, we can base the security on the k-sum problem studied by Wagner and more recently by Minder & Sinclair. The new approach allows to output shorter digests and brings the speed of Fast VSH closer to the range of "classical" hash functions. The modified VSH is likely to remain secure even if factoring and discrete logarithms are easy, while this would have a devastating effect on the original versions. This observation leads us to propose a variant that operates modulo a power of two to increase the speed even more. A function that offers an equivalent of 128-bit collision resistance runs at 68.5 MB/s on a 2.4 GHz Intel Core 2 CPU, more than a third of the speed of SHA-256.

Keywords: hash functions, generalized birthday problem, knapsacks, provable security.

1 Introduction

A hash function is a mapping that on input a string of arbitrary length outputs a fixed-length digest. Such functions are among the most basic building blocks in cryptology. In the recent years there have been several attempts to design provably secure hash functions. Several of them follow the same general idea and hash by computing sums in finite Abelian groups.

Very Smooth Hash. The function designed by Contini et al. [7] is a provably collision resistant hash function based on arithmetic in multiplicative groups modulo an integer. If the modulus is of a proper form, collision resistance of the function can be proved under an assumption heuristically linked to well-known problems, such as integer factoring and discrete logarithm. The function is relatively practical, but still considerably less convenient than the established hash functions. For example, in order to reach collision resistance that corresponds to 1024-bit RSA security, one needs to compute a digest that is 1516 bits long. The function was reported to be about 25 times slower than SHA-1. It is our goal to improve the multiplicative VSH in terms of both efficiency and security.

---

* Supported by a grant of the Swiss National Science Foundation, 200021-116712.

U. Parampalli and P. Hawkes (Eds.): ACISP 2011, LNCS 6812, pp. 142–156, 2011.

The generalized VSH-DL variants proposed in [16] are not targeted in this paper. While the functions output shorter digests, their performance is much worse than in case of the multiplicative versions.

Knapsack Based Functions. From a high level point of view, VSH can be seen as a multiplicative knapsack. Compression function families based on 0-1 knapsacks in groups were introduced by Impagliazzo & Naor. Such functions are parametrized by k randomly selected elements of a finite group G denoted by $a_1, \ldots, a_k$. The function maps a k-bit string $b_1 \ldots b_k$ to $\sum_{i \in T} a_i$ where $T \subseteq \{1, \ldots k\}$ is the set of indices i such that $b_i = 1$. If $k > \lg |G|$,[1] the functions compress. Under certain assumptions it was shown that the families were universal and one way. Examples were given in groups $(\mathbb{Z}_n, +)$ and $(\mathbb{Z}_n^*, \times)$ and the security of some variants related to hardness of discrete logarithm and factoring [13,14].

A similar additive function was proposed by Damgård in [9]. It is again a 0-1 knapsack, where a_i are positive integers and the group operation is addition. In a concrete example, k equaled 256 and the a_i were random integers with bit length under 120. The function was broken by Camion & Patarin using techniques that turn out to be central to our paper [6]. Joux et al. successfully applied lattice reduction to the class functions [15].

Following Ajtai's discovery of the one-way function related to worst case assumptions on lattices [1], Goldreich et al. showed that the function is actually collision resistant [12]. It is a 0-1 knapsack on k random elements of the additive group $\mathbb{Z}_q^n$ for prime $q = O(n^c)$ and $n \lg q < k \leq \frac{q}{2n^4}$.

Micciancio generalized the concept to certain rings in [20,21]. Parametrize the function by k elements of a ring R. For $T \subseteq R$, define a compression function $T^k \to R$ as $\sum_{i=1}^n x_i \cdot a_i$. For appropriate choices of R, the family is pre-image and collision resistant. Properties of the family are connected to worst-case assumption on special classes of lattices. For comprehensive treatment of the family, see also [24,18]. A practical hash function SWIFFT following the principles was proposed in [17,19] and its modification SWIFFTX was submitted as a SHA-3 candidate [2]. The generalized knapsacks considerably lower the key length and the number of operations needed to hash a bit. If $b = \lg |T|$, then kb bits can be hashed in k ring multiplications and $k - 1$ ring additions and only k ring elements are needed to specify a particular function.

Incremental Hashing. Bellare and Micciancio proposed a family of incremental hash functions in [5]. The functions are designed to map strings of arbitrary length to group elements. Fix an integer b, assume that the message is composed of kb bits. Denote the i-th b-bit chunk of the input by x_i. Define a function $f : \mathbb{N} \times \{0, 1\}^b \to G$ and map x_i to $f(i, x_i)$.[2] Obtain k group elements, output their product in G. The authors define a *balance problem* in groups and use it as a security assumption in proofs. The groups considered include $(\mathbb{Z}_n, +)$, $(\mathbb{Z}_n^*, \times)$,

---

[1] lg stands for base 2 logarithm.

[2] The f is considered a random oracle in proofs and instantiated by a real-world hash function such as SHA.

$(\mathbb{Z}_q^n, +)$, $(\mathbb{Z}_2^n, +)$. The functions defined in these groups were named AdHASH, MuHASH, LtHASH and XHASH, respectively. In case of multiplicative groups, security follows from hardness of discrete logarithm. The authors observe that the construction appears more secure than what the proof suggests, there seems to be no attack even if discrete logarithms were easy to compute.

The (Extended) k-tree Algorithm. All the above functions map an input string of length kb to k group elements that are then added in G. This can be seen as if we had k lists of 2^b group elements each and selected a single element from every list. This view is valid also for 0-1 knapsacks where $b = 1$, simply think of the lists as containing the identity element of G in addition to a_i. The problem of inverting such a function was considered by Wagner, it is known as the generalized birthday problem [28,27]. In many groups it can be solved by the *tree algorithm* in time $O\left(k2^{\frac{S}{1+\lg k}}\right)$ where $S = \lg |G|$, provided all the lists contain enough elements. The idea is a generalization of the attack of Camion and Patarin on Damgård's function. Minder and Sinclair extended the algorithm to cases where the length of the lists is restricted [22].

Recent Provably Secure Hash Functions. The fastest known attack on SWIFFT applies the extended tree algorithm while ignoring the ring structure and the relation to lattices [19].

Finiasz et al. proposed a code based function FSB that is essentially a knapsack in the group $(\mathbb{Z}_2^n, +)$ [3]. Joux et al. broke the function applying Wagner's algorithm [8]. Later variants of FSB [4,10,11] consider Wagner's attack in security analysis. This paper is in part inspired by the known applications of the tree algorithm to hash functions.

Main Contributions. We propose two new variants of VSH and interpret them as knapsacks or k-sums. This brings the function closer to SWIFFT or FSB and allows us to adapt the cryptanalytic methods previously used on the two compression functions. We quantify security of the new VSH variants applying the known results on complexity of the extended k-tree algorithm. This suggests that our minor modification may improve security by many orders of magnitude compared to the original VSH variants. We point out evidence that the functions remain secure even if factoring and discrete logarithms are easy. Practicality of the new functions is demonstrated on an implementation.

2 Very Smooth Hash Algorithm

We recall two variants of VSH from [7] that are the starting points for our improvements. Denote the i-th prime number by p_i. In addition let $p_0 = -1$.

Fast VSH. Let n be an S-bit integer, fix a small integer $b > 0$. Let k be the maximal integer such that $\prod_{i=1}^{k} p_{i2^b} < n$. Have an l-bit message m, denote the r-th b-bit chunk of m by $m[r]$ with $0 \leq m[r] < 2^b$. The Fast VSH algorithm proceeds as follows:

1. Let $x_0 = 1$.
2. Let $\mathcal{L} = \lceil \frac{l}{bk} \rceil$. Pad the message with zero bits up to an integral multiple of bk.
3. Append a bk-bit binary representation of l to the message, denote the new chunks $m[\mathcal{L}k+1]$ to $m[(\mathcal{L}+1)k]$.
4. For $j = 0, 1, \ldots, \mathcal{L}$ in succession compute

$$x_{j+1} = x_j^2 \times \prod_{i=1}^{k} p_{(i-1)2^b + m[jk+i]+1} \quad \mod n \ .$$

5. Return $x_{\mathcal{L}+1}$.

Step 4 maps the S bits of x_j and kb fresh message bits to a new S-bit value x_{j+1}. The x_j can be viewed as a chaining variable, Fast VSH employs a variant of the Merkle-Damgård transform [9]. This special chaining mode is made use of in the original security proof that links collision resistance to the hardness of the following problem:

Definition 1 (VSSR: Very Smooth number nontrivial modular Square Root). *Let n be the product of two primes of approximately the same size and let $k' \leq (\log n)^c$. Given n, find x such that $x^2 \equiv \prod_{i=0}^{k'} p_i^{e_i} \mod n$ and at least one of the $e_0, e_1, \ldots, e_{k'}$ is odd.*

The (Fast) VSH security proof establishes that any collision in Fast VSH leads to a solution to VSSR with $k' = k2^b$.

Fast VSH-DL. If the modulus is replaced by an S-bit prime number $p = 2q+1$ for prime q, one obtains the Discrete Log variant of Fast VSH. This function maps $\mathcal{L}bk$ bits to S bits for fixed $\mathcal{L} \leq S - 2$. It uses the same iteration as Fast VSH, but because the length is fixed, no length needs to be appended. To hash $\mathcal{L}bk$ bits proceed as follows:

1. Let $x_0 = 1$.
2. For $j = 0, 1, \ldots, \mathcal{L}$ in succession compute

$$x_{j+1} = x_j^2 \times \prod_{i=1}^{k} p_{(i-1)2^b + m[jk+i]+1} \quad \mod p \ .$$

3. Return $x_{\mathcal{L}+1}$.

The compression function can be extended to process inputs of arbitrary length by applying the Merkle-Damgård transform. The function is proved collision resistant assuming the following problem is hard:

Definition 2 (VSDL: Very Smooth number Discrete Log). *Let p, q be prime numbers with $p = 2q + 1$ and let $k' \leq (\log p)^c$. Given p, find integers $e_1, e_2, \ldots, e_{k'}$ such that $2^{e_1} \equiv \prod_{i=2}^{k'} p_i^{e_i} \mod p$ with $|e_i| < q$ for $i = 1, 2, \ldots, k'$, and at least one of $e_1, e_2, \ldots, e_{k'}$ is non-zero.*

2.1   Security

Security of Fast VSH is related to the hardness of factoring the modulus n. The designers base the assessment of collision resistance of Fast VSH on the following assumption:

Computational VSSR Assumption. Solving VSSR is as hard as factoring a hard to factor S'-bit modulus, where S' is the least positive integer such that

$$L'[2^{S'}] \geq \frac{L'[n]}{k2^b} , \tag{1}$$

where the function

$$L'[n] = e^{1.923(\log n)^{1/3}(\log\log n)^{2/3}} \tag{2}$$

approximates the running time of Number Field Sieve factoring the integer n.

Heuristically, given $k2^b$ solutions to VSSR, one can use linear algebra to find $x \not\equiv \pm y \mod n$ such that $x^2 \equiv y^2 \mod n$ and obtain factorization of the modulus. If NFS is assumed to be the fastest method for factoring integers of the form of n, the VSSR problem is no easier than a fraction $\frac{1}{k2^b}$ of the cost of NFS. The above assumption can directly be used to derive a provable lower bound on collision resistance for a particular Fast VSH variant.

In contrast, no computational VSDL assumption was made in the proposal. Therefore no computational lower bound on collision resistance of Fast VSH-DL can immediately be derived.

Generating Collisions. Collisions in Fast VSH are trivial to create if the factorization of the modulus n is known. This allows to compute $\varphi(n) = (p-1)(q-1)$. The hash function computes a modular multi-exponentiation $\prod_{i=1}^{k2^b} p_i^{e_i} \mod n$. The product does not change if an integral multiple of $\varphi(n)$ is added to any of the exponents. Creating messages that lead to appropriate differences in e_i is rather easy. Note that the collisions created in this way are quite long, measuring some Skb bits. Creating short collisions appears to be a much harder problem, even if p and q are known.

Similarly, the ability to compute discrete logarithms modulo p leads to an immediate straightforward way to create collisions in Fast VSH-DL.

3   A Variant without Modular Squaring

We now proceed to modify Fast VSH and Fast VSH-DL such that the collision attacks from the previous section are no longer possible. The changes we make preserve the original security proofs.

Note that if $kb > S$, the operation

$$\prod_{i=1}^{k} p_{(i-1)2^b + m[jk+i]+1} \mod n$$

in Step 4 of Fast VSH compresses its input. It can therefore be extended to a hash function in the usual ways, such as the plain Merkle-Damgård mode. Our new Fast VSH variant will impose the condition $kb > S$ and build a compression function only.

Faster VSH. Have an S-bit modulus n, let k, b be integers such that $kb > S$ and $k2^b < \log(n)^c$. Define a compression function H from kb bits to S bits that outputs

$$H(m) = \prod_{i=1}^{k} p_{(i-1)2^b + m[i] + 1} \quad \mod n .$$

The function computes a modular product of precisely k out of the primes $p_1, \ldots, p_{k2^b}$. If $n = pq$, we can prove collision resistance of the compression function based on the VSSR problem:

Theorem 1. *A collision in Faster VSH solves VSSR for $k' = k2^b$.*

Proof. Suppose there are two messages $m \neq m'$ such that $H(m) = H(m')$ and

$$H(m) = \prod_{i=1}^{k} s_i \quad \mod n \tag{3}$$

$$H(m') = \prod_{i=1}^{k} t_i \quad \mod n . \tag{4}$$

Denote the hash value $H(m)$ by x. From (3) and (4) it follows that

$$x^2 \equiv \prod_{i=1}^{k} t_i \prod_{i=1}^{k} s_i \quad \mod n .$$

Because $m \neq m'$, at least one of the exponents on the right hand side equals one, the above expression solves VSSR. □

If n is replaced by the prime $p = 2q + 1$ for prime q, we are in the Fast VSH-DL setting. In a direct analogy of the original proof we can link the security of modified Fast VSH-DL to VSDL.

Theorem 2. *A collision in Faster VSH-DL solves VSDL for $k' = k2^b$.*

The modular squaring is no more crucial for the security proof. Our modified function is secure under the original assumptions, the proofs can be easily adapted to the new setting.

Note however, that our modification limits all the exponents in the prime products to one. The factoring attack on Fast VSH and the discrete log attack on Fast VSH-DL do not extend to the modified variants.

Performance. The change proposed in this section slows the functions down. Only $bk - S$ fresh bits are processed per iteration due to the use of the ordinary Merkle-Damgård mode. In contrast, the original Fast VSH processed bk bits per iteration. The performance of the modified variant is approximately a fraction $1 - \frac{S}{bk}$ of the original. The greater the compression ratio of the new function, the less significant the slowdown.

The proposed modification to Fast VSH does not appear to be beneficial at all. The mere fact that the original collision-finding attacks do not work is not worth the performance loss. The security level implied by security proofs is the same after all.

As we will show in the next section, the modification allows a radical change in security assessment and in the end permits faster and/or more secure hashing.

4   The k-Sum Problem and the Tree Algorithm

The following equivalent view of the two new VSH variants will be useful in security analysis. Split the primes into k lists $L_1, \ldots, L_k$, such that L_i contains the 2^b primes $p_{(i-1)2^b+1}, \ldots, p_{i2^b}$.

Let $f_i : \{0,1\}^b \to L_i$ be a function that interprets the bit string $m[i]$ as an integer and returns the element on position $m[i]$ from the list L_i. The new compression functions can be rewritten as follows:

$$H(m) = f_1(m[1]) * f_2(m[2]) * \cdots * f_k(m[k]) \tag{5}$$

where $m[i]$ is the i-th b-bit chunk of the input m. In the above description, $*$ stands for modular multiplication. For the rest of this section, let $*$ simply denote a group operation in a finite Abelian group G such that $\lg |G| \approx S$.

To measure the security of compression functions of the above type consider the following problem:

Definition 3 (The k-sum problem). *Given a group G, an element $y \in G$ and disjoint lists of group elements $L_1, \ldots, L_k$ find $g_i \in L_i$ such that*

$$g_1 * g_2 * \ldots * g_k = y .$$

Pre-image resistance of our two new compression functions is easily seen to be equivalent to an instance of the above problem. We show that collisions also naturally correspond to an instance of a k-sum problem.

Given the k lists L_i, form new lists L_i' containing all the elements gh^{-1} for $g, h \in L_i$. Size of L_i' is approximately 2^{2b}. A collision in H corresponds to a solution to the k-sum problem with the lists L_i' and target value 1. Note that a solution to the new k-sum problem leads in turn to a pair of colliding messages. A solution is necessarily of the form $g_1 h_1^{-1} * g_2 h_2^{-1} * \ldots * g_k h_k^{-1}$. If $m[i] \in \{0,1\}^b$ is the unique value such that $f_i(m[i]) = g_i$ and $m'[i] \in \{0,1\}^b$ is such that $f_i(m'[i]) = h_i$, then the concatenation of $m[i]$ collides with the concatenation of $m'[i]$. This leads to a collision if $m \neq m'$, or equivalently if the solution to the

k-list problem does not select 1 in all of the lists.[3] In general, collision search for H is as hard as pre-image search for a function (over the same group) that compresses twice as much as H.

Known Lower Bounds. For certain groups, hardness of the k-sum problem follows from more "usual" assumptions. Wagner shows that the problem in a cyclic group is no easier than discrete logarithms. Impagliazzo & Naor prove that a 0-1 knapsack in any group G is hard whenever there is a homomorphism onto G that is hard to invert [14]. Their theorem easily extends to k-sums.

The connection of k-sums to homomorphisms was also pointed out in [5]. The discrete logarithm is a special case. Squaring in the multiplicative group of quadratic residues modulo $n = pq$ is another example of an onto homomorphism. Solving the k-sum problem in the group is therefore at least as hard as factoring n.

Wagner's Tree Algorithm. The tree algorithm by Wagner solves the k-sum problem in about $k2^{\frac{S}{1+\lg k}}$ group operations, provided all the lists contain at least $2^{\frac{S}{1+\lg k}}$ elements. To simplify the exposition, we deliberately neglect other operations the algorithm performs, such as sorting and comparisons. The algorithm needs a sequence of groups K_j normal in G for $j = 0, \ldots 1 + \lg k$, such that $K_0 \subseteq K_1 \subseteq K_2 \subseteq \ldots \subseteq K_{1+\lg k}$ and $|K_j| \approx |G|^{\frac{j}{1+\lg k}}$. In addition, the algorithm makes use of homomorphisms $\rho_j : G \rightarrow G/K_j$ with $K_j = \mathrm{Ker}\, \rho_j$.

Without loss of generality one can assume that the target value y is the identity of G. For a different y, simply multiply all elements in one of the lists by y^{-1} and solve for 1. The tree algorithm successively merges pairs of the lists. Only "useful" entries that fall in the subgroups K_j are kept. For details of the algorithm, the reader is referred to [27].

It is not necessary that the K_j form a sequence of normal subgroups of G. A "weaker" chain of subsets may be sufficient, as demonstrated in Wagner's paper for groups $(\mathbb{Z}_M, +)$ where a chain of intervals in used. We will further assume that the structure of G allows the tree algorithm to run with k lists for any k. This is a strong assumption, the group structure will often limit the applicability and/or exact performance of the algorithm. Our goal is to estimate the cost of the tree algorithm from below. The assumption well fits the purpose and greatly simplifies further analysis.

Extended Tree Algorithm. Wagner's analysis assumes the lists are long enough. If it is not the case, one can combine the short lists into (fewer) longer lists and invoke the ordinary tree algorithm once the lists are long enough. The extended tree algorithm by Minder and Sinclair[4] [22] builds lists of maximal length 2^u where

$$u = \frac{S - b2^p}{\lg k - p} \tag{6}$$

---

[3] We might remove the element 1 from all the lists and limit ourselves to colliding messages that differ in all b-bit blocks.

[4] The algorithm was described for XOR as the group operation, but can be generalized to other groups.

such that p is the least integer satisfying

$$S \leq (\lg k - p + 1)b2^p . \tag{7}$$

We will use the two above expressions to measure the security of Fast VSH variants. To simplify the exposition, we will measure the complexity of a particular instance of the k-sum problem exclusively by the maximum expected length of the lists produced. The workload will usually be slightly higher, but a lower bound on the cost is sufficient for our purposes.

What If Factoring and DL Are Easy? Interestingly, the k-sum problems over multiplicative groups appear to remain hard even if factoring and discrete logarithms are easy. The multiplicative group $\mathbb{Z}_n^*$ for $n = pq$ is isomorphic to $(\mathbb{Z}_{p-1}, +) \times (\mathbb{Z}_{q-1}, +)$ and the multiplicative group $\mathbb{Z}_p^*$ is isomorphic to $(\mathbb{Z}_{p-1}, +)$. Suppose we can factor and compute discrete logarithms efficiently. This allows us to compute the isomorphisms from the multiplicative to the additive group. While a group isomorphism transposes a k-sum problem instance to a setting where we may be able to compute faster, the lengths of the lists are preserved. Even in $(\mathbb{Z}_m, +)$ no faster approach than the (extended) tree algorithm is known.

For all the multiplicative functions we consider, the cost of factoring and/or discrete logarithms is negligible compared to the cost of the tree algorithm. We may therefore simply consider the steps to come for free in our analysis, even if a run of the tree algorithm is preceded by DL computations.

4.1   Security of Faster VSH

Table 1 captures the collision resistance level implied by the computational VSSR assumption ("factoring hardness") and the new security estimates based on the extended k-tree algorithm for two variants of Faster VSH. The parameters originate from [7]. The modulus is assumed to be a product of two large primes. The factoring hardness measure is the complexity of factoring a hard to factor integer of said bit length. Columns 5 and 6 are computed using (6) and (7).

Note that the two security measures are in somewhat incompatible units. The exponent in column 5 means "bit security", the number in column 4 is "RSA security".

Hardness of factoring a 1024-bit RSA modulus is considered equivalent to roughly "80-bit" security. If Faster VSH with $S = 1516$ is assessed as a k-sum problem, it provides collision resistance of at least 335 bits. For these parameters,

Table 1. Security estimates for variants of Faster VSH

k	S	b	factoring coll	coll	pre
256	1516	8	1024 bits	$2^{334.7}$	$2^{502.0}$
1024	2874	8	2048 bits	$2^{472.4}$	$2^{616.7}$

the removal of squarings and the use of Merkle-Damgård mode slows down the original Fast VSH by approximately 74%. This is only a moderate slowdown given the great increase in security.

The slowdown is only approx. 35% for the other variant with $S = 2874$, while the gap between the hardness of factoring an 2048-bit modulus and 472-bit security is even greater.

Collision resistance of 335 or 472 bits is more than enough for many years to come. We can therefore aim for lower security levels and tweak the parameters, in particular the digest length, to gain performance. Precise parameters for practical instances of the functions as well as speed measurements are given in Section 6.

5   A Variant without Modular Reduction

The lower bound on the complexity of the extended tree algorithm only depends on the number of lists k, the length of the lists 2^b and the output length S. While the structure of the particular group does affect the practicality of the algorithm to an extent, it almost certainly makes the job harder than what our estimates suggest.

Because we choose not to rely on the group structure, we propose to replace the modulus in Faster VSH by a power of two. No costly modular reductions are needed in the computation of the compression function, because reducing modulo a power of two can be done for free. This will lead to a considerable speed-up while maintaining the k-list security. Note that the prime $2 = p_1$ cannot be used in this setting, because it does not belong to the multiplicative group of integers modulo 2^S. The lists are to be filled with small primes starting with $p_2 = 3$. To hash bk input bits, compute the following product:

$$H(m) = \prod_{i=1}^{k} p_{(i-1)2^b+m[i]+2} \mod 2^S .$$

Any hash value will be an odd number, i.e. the least significant bit is always 1. Therefore only $S-1$ leftmost bits of the S-bit modular product should be output. Call the function *Smoother VSH*.[5]

Technically, the provable connection between the security of the function and the VSSR assumption (or a variant of VSDL) is preserved. This provides little confidence in security, because factoring the number 2^S is trivial and so are discrete logarithms in (the large cyclic subgroup of) $\mathbb{Z}_{2^S}^*$.

The security assessment based on the extended tree algorithm remains valid. Recall that it does not rely on the modulus, but only on the values k, b and S.

6   Experimental Results

We propose seven parameter sets with varying security, speed, and memory requirements. All the parameter choices were tailored to meet one of the three

---

[5] Now we are running Very Smooth Hash modulo a smooth number.

collision resistance levels 2^{128}, 2^{192} and 2^{256}. The length of the modulus is always an integral multiple of word size (64 bits in our case). The value b has a significant impact on performance. Small b results in many multiplications, but keeps the memory requirements down. With larger b one saves on multiplications, but needs much more memory. The ideal value is best determined empirically. The value $b = 8$ used in all our variants is the most convenient choice also from an implementation point of view.

The pre-image and collision resistance are measured as list lengths in the extended tree algorithm. For comparison, in the case of Faster VSH modulo a product of two primes, we include "factoring" security levels implied by the Computational VSSR assumption. The one bit difference in output length caused by the even modulus can be neglected, we consider the two variants equivalent for the purposes of the tree algorithm.

The modified functions are very simple and quite efficient in software. Our C implementation uses GNU MP 5.0.1 to perform arithmetic on large integers. The Faster VSH code uses Montgomery arithmetic to improve performance [23].

The speed was measured on a 2.40 GHz Intel Core 2 CPU running a 64-bit system. Table 2 displays speed measurements for the variants running in the Merkle-Damgård mode. The table also captures the total size of the lists of small prime numbers, represented by three bytes each. All the primes used in the proposed variants are at most 21 bits long, therefore the product of any three primes fits in 64 bits. Our implementation processes them in triples to save on multiplications.

The values in columns 3 and 4 were computed using (6) and (7), column 5 follows from (1) and (2).

Use of the modulus 2^S makes Smoother VSH approximately twice as fast as in the case of a random S-bit modulus M. For comparison, the speed of SHA-256 on the same platform is approximately 160 MB/s.[6] Our fastest variant with 128-bit collision resistance runs at more than third of that speed.

Note that none of the seven variants of Faster VSH would be considered sufficiently secure based on the original computational VSSR assumption, possibly with the exception of the last one that is as secure as a 1013-bit RSA modulus.

Table 2. Proposed parameters for Faster VSH and Smoother VSH

k	S	coll	pre	factoring coll	memory	Smoother VSH speed	Faster VSH speed
128	640	2^{128}	2^{192}	375	96 kB	45.5 MB/s	24.9 MB/s
256	768	2^{128}	2^{170}	452	192 kB	63.4 MB/s	33.2 MB/s
512	896	2^{128}	2^{160}	528	384 kB	68.5 MB/s	35.8 MB/s
192	960	2^{192}	2^{288}	603	144 kB	34.0 MB/s	18.1 MB/s
384	1152	2^{192}	2^{256}	727	288 kB	48.0 MB/s	24.3 MB/s
256	1280	2^{256}	2^{384}	839	192 kB	27.9 MB/s	14.6 MB/s
512	1536	2^{256}	2^{341}	1013	384 kB	39.8 MB/s	19.8 MB/s

---

[6] OpenSSL benchmark.

7   Choice of the List Elements

The analysis in Section 4 assumes the k lists L_i contain random elements of the group G. This is not at all the case for our functions. We deliberately ignore this and expect the tree algorithm to behave as if the elements were random. Similar reasoning was used in cryptanalysis of SWIFFT and FSB, where the lists are far from random [19,10,4].

Small prime numbers used in VSH have been known to have two positive effects. Because a k-list function is defined by $k2^b$ group elements, if full S bits were used for every single entry, the memory requirements would soon become prohibitive. For the very same reason the function should not be implemented in its equivalent additive representation.

Another advantage of the use of small primes is speed. Multiplication of the S-bit modulus by a small (say 21-bit) prime is much easier an operation than a full $S \times S$ bit multiplication. The small prime numbers are absolutely necessary for VSH to remain practical.

7.1   Minimal Distance of Colliding Inputs

The list elements used in VSH are independent in a rather strong sense. Suppose there is a pair of colliding inputs $m \neq m'$ under Faster VSH such that

$$H(m) = \prod_{i=1}^{k} s_i \quad \mathrm{mod}\ n$$

$$H(m') = \prod_{i=1}^{k} t_i \quad \mathrm{mod}\ n$$

and the two inputs differ in precisely $l \leq k$ of the b-bit chunks that select a particular element from a list. Given $H(m) = H(m')$, there is a congruence of two products of primes:

$$\prod_{i=1}^{l} s_i' = \prod_{i=1}^{l} t_i' \quad \mathrm{mod}\ n\ .$$

At least one of the products must exceed n. Therefore if d is the maximal bit length of any prime in our lists, then

$$l \geq \frac{S}{d}\ .$$

More precisely, if the first $k2^b$ primes are used to fill the lists, the largest prime is approximately $k2^b \ln(k2^b)$. Its bit length is then

$$d \approx b + \lg(bk \ln 2 + k \ln k)\ .$$

Any pair of colliding messages must differ in at least

$$l \gtrsim \frac{S}{b + \lg(bk \ln 2 + k \ln k)}$$

of the b-bit input chunks.

As an example, if $k = 128$ and $S = 640$, then $l \approx 35$. For the variant with $k = 512$ and $S = 1536$, the minimal number of different 8-bit chunks for colliding messages is $l \approx 75$.

Although we do not have a more direct link to collision resistance, we believe this property may support confidence in the functions. Small changes in input *never* lead to a collision. This property was in some form present in the original VSH as well, but (to our knowledge) it was never explicitly described. It is however considerably easier to quantify in our new setting.

A similar property was derived by Zémor & Tillich for hash functions based on (Cayley) graphs [25,26].

8   On Provable Security

The main advantage of Fast VSH was the (heuristic) connection to well-known hard algorithmic problems. The modifications we propose in Sections 3 and 5 drop this feature, security is no longer supported by hardness of factoring or discrete logarithms.

Although our first modification to VSH preserved the proof, we ignored the security level implied by the VSSR assumption and measured it in a different way. This new viewpoint renders the proof and the VSSR assumption rather useless and provides some justification for dropping the VSSR and VSDL altogether with the smooth modulus in Section 5.

Note that in case of VSDL and the corresponding variant, there never has been a hardness estimate leading to meaningful bounds on collision resistance. All that was known is that discrete logs imply collisions. Our modification unifies the DL and factoring variants and allows the two to be proved secure based on a single new assumption.

The modified VSH shares several features with the functions SWIFFT and FSB. The two are provably secure compression functions also building on the hardness of the k-sum problem in commutative groups. The main feature shared is the method for measuring security. All the three function families build on their own hardness assumptions, originating from number theory (VSH), lattice theory (SWIFFT) or coding theory (FSB). If the k-sum problem were better understood, the various assumptions could be replaced by a single universal k-sum assumption. The functions would then all share a single security proof.

9   Conclusions

We proposed modifications to Very Smooth Hash that allow a radical change in security assessment. There was no need to apply the extended tree algorithm to the original multiplicative variants of VSH before, because the known attacks were considerably faster.

Our new variants of Faster VSH are designed to prevent factoring and DL attacks. The tree algorithm becomes a useful tool to measure security. We have shown that (independent of factoring or discrete logarithms) there is a deep combinatorial problem behind Faster VSH, the same problem that supports several other k-list hash functions.

We designed a new multiplicative compression function Smoother VSH that relies exclusively on the hardness of the k-sum problem. The function has still much longer output compared to "classical" hash functions of comparable security, but its speed has become reasonable. There is room for improvement in terms of implementation.

The extreme simplicity and clear structure of the new functions can be considered advantages. On the other hand, more insight in the complexity of multiplicative k-sum problems is desirable.

Acknowledgements. The author would like to thank Arjen Lenstra, Ron Steinfeld, Scott Contini, Dimitar Jetchev and the anonymous reviewers for useful comments on the text.

References

1. Ajtai, M.: Generating hard instances of lattice problems (extended abstract). In: STOC, pp. 99–108 (1996)
2. Arbitman, Y., Dogon, G., Lyubashevsky, V., Micciancio, D., Peikert, C., Rosen, A.: SWIFFTX: A Proposal for the SHA-3 Standard, Submission to NIST (2008)
3. Augot, D., Finiasz, M., Sendrier, N.: A Fast Provably Secure Cryptographic Hash Function (2003)
4. Augot, D., Finiasz, M., Sendrier, N.: A family of fast syndrome based cryptographic hash functions. In: Dawson, E., Vaudenay, S. (eds.) Mycrypt 2005. LNCS, vol. 3715, pp. 64–83. Springer, Heidelberg (2005)
5. Bellare, M., Micciancio, D.: A new paradigm for collision-free hashing: Incrementality at reduced cost. In: Fumy, W. (ed.) EUROCRYPT 1997. LNCS, vol. 1233, pp. 163–192. Springer, Heidelberg (1997)
6. Camion, P., Patarin, J.: The knapsack hash function proposed at crypto'89 can be broken. In: Davies, D.W. (ed.) EUROCRYPT 1991. LNCS, vol. 547, pp. 39–53. Springer, Heidelberg (1991)
7. Contini, S., Lenstra, A.K., Steinfeld, R.: VSH, an efficient and provable collision-resistant hash function. In: Vaudenay, S. (ed.) EUROCRYPT 2006. LNCS, vol. 4004, pp. 165–182. Springer, Heidelberg (2006)
8. Coron, J.-S., Joux, A.: Cryptanalysis of a provably secure cryptographic hash function. Cryptology ePrint Archive, Report 2004/013 (2004)
9. Damgård, I.: A design principle for hash functions. In: Brassard, G. (ed.) CRYPTO 1989. LNCS, vol. 435, pp. 416–427. Springer, Heidelberg (1990)
10. Finiasz, M., Gaborit, P., Sendrier, N.: Improved fast syndrome based cryptographic hash functions. In: ECRYPT Hash Function Workshop 2007 (2007)
11. Finiasz, M.: Syndrome based collision resistant hashing. In: Buchmann, J., Ding, J. (eds.) PQCrypto 2008. LNCS, vol. 5299, pp. 137–147. Springer, Heidelberg (2008)

12. Goldreich, O., Goldwasser, S., Halevi, S.: Collision-free hashing from lattice problems. Electronic Colloquium on Computational Complexity (ECCC) 3(42) (1996)
13. Impagliazzo, R., Naor, M.: Efficient cryptographic schemes provably as secure as subset sum. In: FOCS, pp. 236–241. IEEE, Los Alamitos (1989)
14. Impagliazzo, R., Naor, M.: Efficient cryptographic schemes provably as secure as subset sum. J. Cryptology 9(4), 199–216 (1996)
15. Joux, A., Granboulan, L.: A practical attack against knapsack based hash functions (extended abstract). In: De Santis, A. (ed.) EUROCRYPT 1994. LNCS, vol. 950, pp. 58–66. Springer, Heidelberg (1995)
16. Lenstra, A.K., Page, D., Stam, M.: Discrete logarithm variants of VSH. In: Nguyên, P.Q. (ed.) VIETCRYPT 2006. LNCS, vol. 4341, pp. 229–242. Springer, Heidelberg (2006)
17. Lyubashevsky, V., Micciancio, D., Peikert, C., Rosen, A.: Provably Secure FFT Hashing. In: 2nd NIST Cryptographic Hash Function Workshop (2006)
18. Lyubashevsky, V., Micciancio, D.: Generalized compact knapsacks are collision resistant. In: Bugliesi, M., Preneel, B., Sassone, V., Wegener, I. (eds.) ICALP 2006. LNCS, vol. 4052, pp. 144–155. Springer, Heidelberg (2006)
19. Lyubashevsky, V., Micciancio, D., Peikert, C., Rosen, A.: SWIFFT: A modest proposal for FFT hashing. In: Nyberg, K. (ed.) FSE 2008. LNCS, vol. 5086, pp. 54–72. Springer, Heidelberg (2008)
20. Micciancio, D.: Generalized compact knapsacks, cyclic lattices, and efficient one-way functions from worst-case complexity assumptions. In: FOCS, pp. 356–365. IEEE Computer Society, Los Alamitos (2002)
21. Micciancio, D.: Generalized compact knapsacks, cyclic lattices, and efficient one-way functions. Computational Complexity 16(4), 365–411 (2007)
22. Minder, L., Sinclair, A.: The extended k-tree algorithm. In: Mathieu, C. (ed.) SODA, pp. 586–595. SIAM, Philadelphia (2009)
23. Montgomery, P.L.: Modular multiplication without trial division. Mathematics of Computation 44, 519 (1985)
24. Peikert, C., Rosen, A.: Efficient collision-resistant hashing from worst-case assumptions on cyclic lattices. In: Halevi, S., Rabin, T. (eds.) TCC 2006. LNCS, vol. 3876, pp. 145–166. Springer, Heidelberg (2006)
25. Tillich, J.-P., Zémor, G.: Group-theoretic hash functions. In: Cohen, G.D., Litsyn, S., Lobstein, A., Zémor, G. (eds.) Algebraic Coding 1993. LNCS, vol. 781, pp. 90–110. Springer, Heidelberg (1994)
26. Tillich, J.-P., Zémor, G.: Hashing with SL_2. In: Desmedt, Y. (ed.) CRYPTO 1994. LNCS, vol. 839, pp. 40–49. Springer, Heidelberg (1994)
27. Wagner, D.: A generalized birthday problem. Long version, http://www.eecs.berkeley.edu/~daw/papers/genbday-long.ps
28. Wagner, D.: A generalized birthday problem. In: Yung, M. (ed.) CRYPTO 2002. LNCS, vol. 2442, pp. 288–303. Springer, Heidelberg (2002)

Cryptanalysis of the Compression Function of SIMD

Hongbo Yu[1] and Xiaoyun Wang[2,*]

[1] Department of Computer Science and Technology, Tsinghua University,
Beijing 100084, China
yuhongbo@tsinghua.edu.cn
[2] Tsinghua University and Shandong University, China
xiaoyunwang@tsinghua.edu.cn, xywang@sdu.edu.cn

Abstract. SIMD is one of the second round candidates of the SHA-3 competition hosted by NIST. In this paper, we present the first attack for the compression function of the reduced SIMD-256 and the full SIMD-512 (the tweaked version) using the modular difference method. For SIMD-256, we give a free-start near collision attack on the compression function reduced to 20 steps with complexity 2^{116}. And for SIMD-512, we give a free-start near collision attack on the 24-step compression function with complexity 2^{235}. Furthermore, we give a distinguisher attack for the full compression function of SIMD-512 with complexity 2^{475}. Our attacks are also applicable for the final compression function of SIMD.

Keywords: SIMD, SHA-3 Candidate, near collision, distinguishing attack.

1 Introduction

Hash functions play a fundamental role in modern Cryptography. Due to the collision attacks on the series general hash functions [7,8,1], NIST hosted the SHA-3 hash function competition to select a new cryptographic hash function as the standard [5]. Until November 2008, NIST accepted 51 out of 64 submissions as the first round. In July 2009, NIST announced 14 second round candidates.

The hash function SIMD is one of the second round candidates, and it is designed by Leurent et al[4]. SIMD is a wide-pipe design based on the MD iterative structure. In Indocrypt 2009 [3], Mendel et al give a distinguisher attack on the SIMD-512 compression function with complexity $5.2^{425.8}$ using a differential distinguisher. Because Mendel et al's attack, the designers found some bad properties of Feistel structure in SIMD, and they tweaked the SIMD by changing rotation constants and permutations for diffusion between parallel Feistels. In 2010 [6], Nikolic et al give the distinguishing attack for the compression function

* Supported by the National Natural Science Foundation of China(NSFC Grant No.60803125), 973 Project(No.2007CB807902), 863 Project(No.2006AA01Z420) and the Tsinghua University Initiative Scientific Research Program (No.2009THZ01002).

U. Parampalli and P. Hawkes (Eds.): ACISP 2011, LNCS 6812, pp. 157–171, 2011.

of 12-step SIMD-512 using the rotation distinguisher [2]. In this paper, we give
series cryptanalysis results for the compression functions of the SIMD-256 and
SIMD-512:

1. For SIMD-256 reduced to 20 steps, we give a free-start near collision attack
   with complexity 2^{116}. So far, this is the first analysis result for the SIMD-256.
2. For SIMD-512 reduced to 24 steps, we give a free-start near collision attack
   with complexity 2^{235}. And for the full SIMD-512 compression function, we
   give a distinguishing attack with complexity 2^{475} using a difference distinguisher.

This paper is organized as follows. In section 2, we define some notations and
give a brief description of SIMD. In section 3, we give the free-start near collision
attack on 20-step SIMD-256. In section 4, we give a near collision attack on the
24-step SIMD-512 and a distinguishing attack for the full SIMD-512. Finally we
conclude the paper in section 5.

2   Notations and Description of SIMD

The following notations can be used in this paper.

2.1   Notations

1. $+$ and $-$ denote addition and subtraction modular 2^{32}.

2. $x_{i,j}$ is the j-th bit of x_i, where x_i is a 32-bit word and $x_{i,32}$ is the most
   significant bit.

3. $x_i[j]$ and $x_i[-j]$ (where x is a 32-bit word) are the resultant values of changing only the j-th bit of the word x_i from 0 to 1 and 1 to 0 respectively.

4. $x_i[j_1, j_2, ..., j_l]$ is the value resulting by changing the j_1-th, j_2-th, ...and j_l-th
   bits of x_i. Again the $+$ sign means that the bit is changed from 0 to 1, and
   the $-$ sign means that the bit is changed from 1 to 0.

5. $\lll$ denote left-rotation by n-bit.

6. h_i denote the chaining values in step i of SIMD-256 (or SIMD-512).

2.2   Description of SIMD

SIMD is an iterative hash function that follow the Merkle-Damgård design. The
SIMD family hash function is based on two functions SIMD-256 and SIMD-512.
The SIMD-n with $n \leq 256$ is defined as a truncation of SIMD-256, and SIMD-n
with $256 \leq n \leq 512$ is defined as a truncation of SIMD-512. Each function SIMD-
n takes as input a message of arbitrary size, and outputs a digest of n bits. The

input message is padded and then divided into k 512-bit (resp. 1024-bit) blocks for SIMD-256 (resp. SIMD-512).

The compression function of SIMD-256 (resp. SIMD-512) takes a 512-bit (resp. 1024-bit) chaining value and a 512-bit (resp. 1024-bit) message and output another 512-bit (resp. 1024-bit) chaining value. Each 512-bit (resp. 1024-bit) block is first expanded into 4096 bits (resp. 8192 bits). The compression functions of SIMD consist of 4 rounds, and each includes 8 steps. The feed-forward consists of 4 additional steps with the IV as the message input. Each step has 4 (resp. 8) parallel Feistel ladders, and they interact together because of the permutations $p(i)$'s. At each step, a new value is computed in each Feistel ladder, and this new value is sent to another Feistel ladder at the following step. In the rest of this paper, we count the steps and bits starting from 1. In step i, the $j - th$ feistel ladder given are as follows:

$$a_i^j = (d_{i-1}^j + w_i^j + \Phi(a_{i-1}^j, b_{i-1}^j, c_{i-1}^j)) \lll s_i + (a_{i-1}^{p^{(i)}(j)}) \lll r_i$$
$$b_i^j = a_{i-1}^j \lll r_i$$
$$c_i^j = b_{i-1}^j$$
$$d_i^j = c_{i-1}^j$$

For SIMD-256, the permutation used at step i is $p^{(i-1) \mod 3}$, and it is defined in the following:

j	0	1	2	3
$p^{(0)}(j)$	1	0	3	2
$p^{(1)}(j)$	2	3	0	1
$p^{(2)}(j)$	3	2	1	0

For SIMD-512, $p_i = p^{(i-1) \mod 7}$, and the seven permutations are defined:

j	0	1	2	3	4	5	6	7
$p^{(0)}(j)$	1	0	3	2	5	4	7	6
$p^{(1)}(j)$	6	7	4	5	2	3	0	1
$p^{(2)}(j)$	2	3	0	1	6	7	4	5
$p^{(3)}(j)$	3	2	1	0	7	6	5	4
$p^{(4)}(j)$	5	4	7	6	1	0	3	2
$p^{(5)}(j)$	7	6	5	4	3	2	1	0
$p^{(6)}(j)$	4	5	6	7	0	1	2	3

In this paper, we omit to describe the message expansion algorithm because our attack is independent with the message expansion.

The Boolean functions Φ used in the first 4 steps of each round is the chosen function IF and last 4 steps is the majority function MAJ. The feed-forward steps use the Boolean function IF. The rotation constants r_i and s_i can refer to the original paper [4], and they also are shown in our detail differential paths.

3   The Free-Start Near Collision Attack on the Reduced SIMD-256

In this section, we use the modular difference method to find a differential path with high probability. The modular differential method was presented in Eurocrypt 2005 by Wang *et al* [7], and it is a precise difference that uses integer modular subtraction in conjunction with exclusive-or as a measure of difference. There are four steps in attacking a hash function using the modular difference method. The first step is to select an appropriate message or initial value difference, which determines the success probability of the attack. The key step of the modular difference attack is to select a feasible differential path according to the selected message difference or initial value difference. This difficult step requires intelligent analysis, sophisticated technique, lots of patience and good luck. The third phase is to derive the sufficient conditions that guarantee the feasibility of the differential path. In the process of searching for differential paths, the chaining variable conditions can be determined. A feasible differential path implies that all the chaining variable conditions deduced from the path do not contradict each other. The last step is the message/IV modification which forces the modified messages/IV to satisfy additional sufficient conditions.

3.1   Constructing the Specific Differential Path for 20-Step SIMD-256

In this attack, we introduce the difference only in the IV. We keep the message difference to be zero because the expanded message has a minimal distance of 520 (resp. 1032) for SIMD-256 (resp. SIMD-512). Before the search of the differential path, we observe that the difference propagation of SIMD is slower in the backward direction than that in the forward direction. Our basic attack strategy is first to introduce one bit difference in the intermediate chaining value and trace this difference in the forward and backward direction using the modular difference method to get a difference path with high probability. Because the IV is introduced in the four feed-forward steps as messages, we then adjust the differential path slightly so that the differences in IV can be used to cancel some difference of the the feed-forward steps.

For SIMD-256, we introduce 1-bit initial difference in the last ladder of the Step 16, and go 16 steps in the backward direction to obtain the initial difference, then we trace the 1-bit difference in the forward direction for 4 steps. Furthermore, we go forward another 4 steps under the specific IV difference for the feed-forward steps. In this way, we can get a 20-step differential path which is shown in Table 3.2. The sufficient conditions for the differential path is given in Table 3. It's easy to compute the probability of the differential path of Table 3.2 which holds with probability 2^{-186} for the selected IV difference.

3.2   Message/IV Modification

In order to get a free-start near collision, we need to carry out the message or IV modification technique to fulfill the conditions in the IV and the chaining vales

Table 1. The conditions distribution of the Step 0 to 9 in Table 5

Chaining variables	conditions Number
d_0, d_1, d_2, d_3	120
d_4, d_5, d_6, d_7	146
a_8, b_8, c_8, d_8	320

of the first round. In this section, we use the message/IV modification to fulfill the conditions in IV and a_1, a_2, a_3, and a_4. We denote the chaining values x_i in the j-th ladder as x_i^j.

1. From the Table 3, there are 29 conditions in IV (a_0, b_0, c_0 and d_0 of the four ladders). These conditions can be satisfied by choosing the IV values freely.
2. There are 18 conditions in a_1^j, $j = 0, 1, 2, 3$, and these conditions can be satisfied by modifying the corresponding d_0^j. From the step update formula

$$a_1^j = (d_0^j + w_0^j + IF(a_0^j, b_0^j, c_0^j)) \lll 23 + a_0^{p^{(j)}} \lll 3,$$

if the i-th bit in a_1^j doesn't satisfied, it's enough to set $d_0^j = d_0^j \oplus 2^{(i-24)} \mod 32$.
3. There are 12 conditions in a_2^j, $j = 0, 1, 2, 3$. From the trace

$$a_{2,i}^j \rightarrow d_{1,(i-17)}^j \mod 32 \rightarrow c_{0,(i-17)}^j \mod 32,$$

the i-th bit of a_2^j can be modified by negating the $(i-17) mod 32$ bit of c_0^j. In the same way, we need to set the additional condition $a_{0,(i-17)mod32}^j = 1$ so that the change in c_0^j can not impact the output of a_1^j.
4. There are 10 conditions in a_3^j, $j = 0, 1, 2, 3$. From the trace

$$a_{3,i}^j \rightarrow d_{2,(i-27)}^j \mod 32 \rightarrow c_{1,(i-27)}^j \mod 32 \rightarrow b_{0,(i-27)}^j \mod 32,$$

the i-th bit of a_3^j can be modified by negating the $(i-27) \mod 32$ bit of b_0^j. We have to set the additional conditions $a_{0,(i-27) \mod 32}^j = 0$ and $a_{1,(i-27) \mod 32}^j = 1$ so that the change in b_0^j can not impact the output of a_1^j and a_2^j.
5. There are 10 conditions in a_4^j, $j = 0, 1, 2, 3$. From the trace

$$a_{4,i}^j \rightarrow d_{3,(i-3)}^j \mod 32 \rightarrow c_{2,(i-3)}^j \mod 32 \rightarrow b_{1,(i-3)}^j \mod 32 \rightarrow a_{0,(i-3)}^j \mod 32,$$

the i-th bit of a_4^j can be modified by negating the bit $a_{0,(i-3) \mod 32}^j$. The additional conditions $b_{0,(i-3) \mod 32}^j = c_{0,(i-3) \mod 32}^j$ and $a_{1,(i-3) \mod 32}^j = 0$, and $a_{2,(i-3) \mod 32}^j = 1$ are needed so that the change in b_0^j can not impact the output of a_1^j and a_2^j and a_3^j. Furthermore, the change of $a_{0,(i-3) \mod 32}^j$ will cause the change in $a_1^{p^{(j)}}$, we have to adjust the value of $d_0^{p^{(j)}}$ to cancel this change.

Table 2. The differential path for 20-step SIMD-256

step	r	s	h_i^0	h_i^1	h_i^2	h_i^3	pr
0			a_0^0 $b_0^0[-13]$ c_0^0 d_0^0	a_0^1 $b_0^1[7,13]$ $c_0^1[-21,26]$ d_0^1	$a_0^2[7,-3]$ $b_0^2[-1]$ $c_0^2[-32]$ d_0^2	a_0^3 $b_0^3[-19,20,23,24,-25]$ c_0^3 $d_0^3[15]$	2^{-14}
1	3	23	a_1^0 b_1^0 $c_1^0[-13]$ d_1^0	a_1^1 b_1^1 $c_1^1[7,13]$ $d_1^1[-21,26]$	a_1^2 $b_1^2[-6,10]$ $c_1^2[-1]$ $d_1^2[-32]$	$a_1^3[15]$ b_1^3 $c_1^3[-19,20,23,24,-25]$ d_1^3	2^{-15}
2	23	17	a_2^0 b_2^0 c_2^0 $d_2^0[-13]$	$a_2^1[-11,-12,-13,-14,15]$ b_2^1 c_2^1 $d_2^1[7,13]$	$a_2^2[-17]$ b_2^2 $c_2^2[-6,10]$ $d_2^2[-1]$	a_2^3 $b_2^3[6]$ c_2^3 $d_2^3[-19,20,23,24,-25]$	2^{-18}
3	17	27	$a_3^0[-8]$ b_3^0 c_3^0 d_3^0	a_3^1 $b_3^1[-28,-29,-30,-31,32]$ c_3^1 d_3^1	a_3^2 $b_3^2[-2]$ c_3^2 $d_3^2[-6,10]$	$a_3^3[14,-18]$ b_3^3 $c_3^3[6]$ d_3^3	2^{-13}
4	27	3	a_4^0 $b_4^0[-3]$ c_4^0 d_4^0	a_4^1 b_4^1 $c_4^1[-28,-29,-30,-31,32]$ d_4^1	a_4^2 b_4^2 $c_4^2[-2]$ d_4^2	a_4^3 $b_4^3[9,-13]$ c_4^3 $d_4^3[6]$	2^{-10}
5	3	23	a_5^0 b_5^0 $c_5^0[-3]$ d_5^0	a_5^1 b_5^1 c_5^1 $d_5^1[-28,-29,-30,-31,32]$	a_5^2 b_5^2 c_5^2 $d_5^2[-2]$	$a_5^3[29]$ b_5^3 $c_5^3[9,-13]$ d_5^3	2^{-10}
6	23	17	a_6^0 b_6^0 c_6^0 $d_6^0[-3]$	$a_6^1[-13,-14,15]$ b_6^1 c_6^1 d_6^1	$a_6^2[-19]$ b_6^2 c_6^2 d_6^2	a_6^3 $b_6^3[20]$ c_6^3 $d_6^3[9,-13]$	2^{-8}
7	17	27	a_7^0 b_7^0 c_7^0 d_7^0	a_7^1 $b_7^1[-30,-31,32]$ c_7^1 d_7^1	a_7^2 $b_7^2[-4]$ c_7^2 d_7^2	$a_7^3[-8]$ b_7^3 $c_7^3[20]$ d_7^3	2^{-6}
8	27	3	a_8^0 b_8^0 c_8^0 d_8^0	a_8^1 b_8^1 $c_8^1[-30,-31,32]$ d_8^1	a_8^2 b_8^2 $c_8^2[-4]$ d_8^2	a_8^3 $b_8^3[-3]$ c_8^3 $d_8^3[20]$	2^{-6}
9	28	19	a_9^0 b_9^0 c_9^0 d_9^0	a_9^1 b_9^1 c_9^1 $d_9^1[-30,-31,32]$	a_9^2 b_9^2 c_9^2 $d_9^2[-4]$	$a_9^3[7]$ b_9^3 $c_9^3[-3]$ d_9^3	2^{-6}
10	19	22	a_{10}^0 b_{10}^0 c_{10}^0 d_{10}^0	$a_{10}^1[20]$ b_{10}^1 c_{10}^1 d_{10}^1	a_{10}^2 b_{10}^2 c_{10}^2 d_{10}^2	a_{10}^3 $b_{10}^3[26]$ c_{10}^3 $d_{10}^3[-3]$	2^{-3}
11	22	7	a_{11}^0 b_{11}^0 c_{11}^0 d_{11}^0	a_{11}^1 $b_{11}^1[10]$ c_{11}^1 d_{11}^1	a_{11}^2 b_{11}^2 c_{11}^2 d_{11}^2	a_{11}^3 b_{11}^3 $c_{11}^3[26]$ d_{11}^3	2^{-2}
12	7	28	a_{12}^0 b_{12}^0 c_{12}^0 d_{12}^0	a_{12}^1 b_{12}^1 $c_{12}^1[10]$ d_{12}^1	a_{12}^2 b_{12}^2 c_{12}^2 d_{12}^2	a_{12}^3 b_{12}^3 c_{12}^3 $d_{12}^3[26]$	2^{-2}
13	28	19	a_{13}^0 b_{13}^0 c_{13}^0 d_{13}^0	a_{13}^1 b_{13}^1 c_{13}^1 $d_{13}^1[10]$	a_{13}^2 b_{13}^2 c_{13}^2 d_{13}^2	$a_{13}^3[13]$ b_{13}^3 c_{13}^3 d_{13}^3	2^{-2}
14	19	22	a_{14}^0 b_{14}^0 c_{14}^0 d_{14}^0	a_{14}^1 b_{14}^1 c_{14}^1 d_{14}^1	a_{14}^2 b_{14}^2 c_{14}^2 d_{14}^2	a_{14}^3 $b_{14}^3[32]$ c_{14}^3 d_{14}^3	2^{-1}
15	22	7	a_{15}^0 b_{15}^0 c_{15}^0 d_{15}^0	a_{15}^1 b_{15}^1 c_{15}^1 d_{15}^1	a_{15}^2 b_{15}^2 c_{15}^2 d_{15}^2	a_{15}^3 b_{15}^3 $c_{15}^3[32]$ d_{15}^3	2^{-1}

Table 2. (*continued*)

16	7	28	a_{16}^0 b_{16}^0 c_{16}^0 d_{16}^0	a_{16}^1 b_{16}^1 c_{16}^1 d_{16}^1	a_{16}^2 b_{16}^2 c_{16}^2 d_{16}^2	a_{16}^3 b_{16}^3 c_{16}^3 $d_{16}^3[32]$	2^{-1}
17	29	9	a_{17}^0 b_{17}^0 c_{17}^0 d_{17}^0	a_{17}^1 b_{17}^1 c_{17}^1 d_{17}^1	a_{17}^2 b_{17}^2 c_{17}^2 d_{17}^2	$a_{17}^3[-9]$ b_{17}^3 c_{17}^3 d_{17}^3	2^{-1}
18	9	15	$a_{18}^0[-18]$ b_{18}^0 c_{18}^0 d_{18}^0	a_{18}^1 b_{18}^1 c_{18}^1 d_{18}^1	a_{18}^2 b_{18}^2 c_{18}^2 d_{18}^2	a_{18}^3 $b_{18}^3[-18]$ c_{18}^3 d_{18}^3	2^{-2}
19	15	5	a_{19}^0 $b_{19}^0[-1]$ c_{19}^0 d_{19}^0	$a_{19}^1[1,-2]$ b_{19}^1 c_{19}^1 d_{19}^1	a_{19}^2 b_{19}^2 c_{19}^2 d_{19}^2	a_{19}^3 b_{19}^3 $c_{19}^3[-18]$ d_{19}^3	2^{-4}
20	5	29	a_{20}^0 b_{20}^0 $c_{20}^0[-1]$ d_{20}^0	a_{20}^1 $b_{20}^1[6,-7]$ c_{20}^1 d_{20}^1	a_{20}^2 b_{20}^2 c_{20}^2 d_{20}^2	$a_{20}^3[-6]$ b_{20}^3 c_{20}^3 $d_{20}^3[-18]$	2^{-5}
21	4	13	$a_{21}^0[-10]$ b_{21}^0 c_{21}^0 $d_{21}^0[-1]$	a_{21}^1 b_{21}^1 $c_{21}^1[6,-7]$ d_{21}^1	$a_{21}^2[-16,20]$ b_{21}^2 c_{21}^2 d_{21}^2	$a_{21}^3[-31]$ $b_{21}^3[-10]$ c_{21}^3 d_{21}^3	2^{-8}
22	13	10	$a_{22}^0[-11,-23]$ $b_{22}^0[-23]$ c_{22}^0 d_{22}^0	a_{22}^1 b_{22}^1 c_{22}^1 $d_{22}^1[6,-7]$	$a_{22}^2[-11,-12]$ $b_{22}^2[1,-29]$ c_{22}^2 d_{22}^2	a_{22}^3 $b_{22}^3[-12]$ $c_{22}^3[-10]$ d_{22}^3	2^{-11}
23	10	25	$a_{23}^0[-21,-22]$ $b_{23}^0[-1,-21]$ $c_{23}^0[-23]$ d_{23}^0	$a_{23}^1[-14,19,-31]$ b_{23}^1 c_{23}^1 d_{23}^1	$a_{23}^2[-1,-21,-25]$ $b_{23}^2[-21,-22]$ $c_{23}^2[1,-29]$ d_{23}^2	a_{23}^3 b_{23}^3 $c_{23}^3[-12]$ $d_{23}^3[-10]$	2^{-16}
24	25	4	a_{24}^0 $b_{24}^0[-14,-15]$ $c_{24}^0[-1,-21]$ $d_{24}^0[-23]$	$a_{24}^1[-14,-18,-26]$ $b_{24}^1[-7,12,-24]$ c_{24}^1 $d_{24}^1[-7]$	$a_{24}^2[-7,12,-24]$ $b_{24}^2[-14,-18,-26]$ $c_{24}^2[-21,-22]$ $d_{24}^2[1,-29]$	$a_{24}^3[-16,19]$ b_{24}^3 c_{24}^3 $d_{24}^3[-12]$	2^{-21}

In fact, we can also modify the conditions in a_5 and a_6, even a_7 and a_8, but it's more expensive and need to set many pre-conditions. It's worth to note that the conditions $a_{2,11}^1$, $a_{3,8}^0$, $a_{3,14}^3$ and $a_{3,18}^3$ can not be modified using the IV modification, because the additional conditions needed to modify these bits are contradict the fix conditions in Table 3. We can modify these four conditions by the corresponding message. Due to the message expansion, we can not select the message completely arbitrary. So we use the IV modification as much as possible instead of message modification .

After the message/IV modification for IV and $a_1 \sim a_4$, the differential path in Table 3.2 holds with probability 2^{-116}. This way, we can find a 25-bit free-start near collision for the 20-step SIMD-256 with complexity 2^{116} which is lower than the birthday attack.

Table 3. The sufficient conditions for the differential path in Table 3.2

a_0	$a_{0,13}^0 = 0$	$a_{0,7}^1 = 0,\ a_{0,10}^1 = 0,\ a_{0,13}^1 = 0,\ a_{0,21}^1 = 1,\ a_{0,26}^1 = 1$	$a_{0,1}^2 = 0,\ a_{0,3}^2 = 1,\ a_{0,7}^2 = 0,\ a_{0,32}^2 = 1$	$a_{0,19}^3 = 1,\ a_{0,20}^3 = 0,\ a_{0,23}^3 = 0,\ a_{0,24}^3 = 1,\ a_{0,25}^3 = 0$
b_0	$b_{0,13}^0 = 1$	$b_{0,7}^1 = 0,\ b_{0,13}^1 = 0$	$b_{0,1}^2 = 1$	$b_{0,15}^3 = a_{0,12}^3,\ b_{0,19}^3 = 1,\ b_{0,20}^3 = 0,\ b_{0,23}^3 = 0,\ b_{0,24}^3 = 0,\ b_{0,25}^3 = 1$
c_0		$c_{0,21}^1 = 1,\ c_{0,26}^1 = 0$	$c_{0,3}^2 = b_{0,3}^2,\ c_{0,7}^3 = b_{0,7}^3$	
d_0				$d_{0,15}^3 = 0$
a_1	$a_{1,13}^0 = 1$	$a_{1,7}^1 = 1,\ a_{1,13}^1 = 1,\ a_{1,20}^1 = a_{0,8}^1,\ a_{1,21}^1 = a_{0,9}^1,\ a_{1,22}^1 = 1,\ a_{1,23}^1 = a_{0,11}^1,\ a_{1,24}^1 = a_{0,12}^1$	$a_{1,6}^2 = 0,\ a_{1,10}^2 = 0,\ a_{1,1}^2 = 1,\ a_{1,26}^2 = a_{0,14}^2$	$a_{1,15}^3 = 0,\ a_{1,19}^3 = 1,\ a_{1,20}^3 = 1,\ a_{1,23}^3 = 1,\ a_{1,24}^3 = 1,\ a_{1,25}^3 = 1$
a_2	$a_{2,23}^0 = a_{1,17}^0$	$a_{2,11}^1 = 1,\ a_{2,12}^1 = 1,\ a_{2,13}^1 = 1,\ a_{2,14}^1 = 1,\ a_{2,15}^1 = 0$	$a_{2,6}^2 = 1,\ a_{2,10}^2 = 1,\ a_{2,17}^2 = 1$	$a_{2,1}^3 = a_{1,27}^3,\ a_{2,6}^3 = 0,\ a_{2,29}^3 = a_{1,23}^3$
a_3	$a_{3,8}^0 = 1$	$a_{3,28}^1 = 0,\ a_{3,29}^1 = 0,\ a_{3,30}^1 = 0,\ a_{3,31}^1 = 0,\ a_{3,32}^1 = 1$	$a_{3,2}^2 = 0$	$a_{3,6}^3 = 1,\ a_{3,14}^3 = 0,\ a_{3,18}^3 = 1$
a_4	$a_{4,3}^0 = a_{2,18}^0$	$a_{4,28}^1 = a_{3,1}^1,\ a_{4,29}^1 = a_{3,2}^1,\ a_{4,30}^1 = a_{3,3}^1,\ a_{4,31}^1 = b_{3,4}^1,\ a_{4,32}^1 = a_{3,5}^1$	$a_{4,2}^2 = a_{3,7}^2$	$a_{4,9}^3 = a_{3,14}^3,\ a_{4,13}^3 = a_{3,18}^3,\ a_{4,26}^3 = a_{3,2}^3$
a_5	$a_{5,3}^0 = a_{4,32}^0$	$a_{5,22}^1 = a_{4,10}^1,\ a_{5,23}^1 = a_{4,11}^1,\ a_{5,24}^1 = a_{4,12}^1$	$a_{5,28}^2 = a_{4,16}^2$	$a_{5,29}^3 = 0,\ a_{5,9}^3 = a_{4,6}^3,\ a_{5,13}^3 = a_{4,10}^3$
a_6		$a_{6,13}^1 = 1,\ a_{6,14}^1 = 1,\ a_{6,15}^1 = 0$	$a_{6,19}^2 = 1$	$a_{6,20}^3 = a_{4,17}^3,\ a_{6,23}^3 = a_{5,17}^3$
a_7		$a_{7,30}^1 = a_{5,7}^1,\ a_{7,31}^1 = a_{5,8}^1,\ a_{7,32}^1 = a_{5,9}^1 \oplus 1$	$a_{7,4}^2 = a_{5,13}^2$	$a_{7,8}^3 = 1,\ a_{7,20}^3 = a_{6,3}^3$
a_8		$a_{8,30}^1 = 1,\ a_{8,31}^1 = 1,\ a_{8,32}^1 = 1$	$a_{8,4}^2 = 1$	$a_{8,3}^3 = 0,\ a_{8,11}^3 = a_{7,12}^3$
a_9		$a_{9,1}^1 = a_{8,24}^1$		$a_{9,3}^3 = 1,\ a_{9,7}^3 = 0$
a_{10}		$a_{10,20}^1 = 0$		$a_{10,26}^3 = 0$
a_{11}		$a_{11,10}^1 = 0$		$a_{11,26}^3 = 1$
a_{12}		$a_{12,10}^1 = a_{11,3}^1$		$a_{12,17}^3 = a_{11,6}^3$
a_{13}				$a_{13,13}^3 = 0$
a_{14}				$a_{14,32}^3 = a_{12,4}^3$
a_{15}				$a_{15,32}^3 = a_{14,10}^3$
a_{16}				$a_{16,12}^3 = a_{15,2}^3$
a_{17}	$a_{17,9}^0 = a_{16,21}^0$			$a_{17,9}^3 = 1$
a_{18}	$a_{18,18}^0 = 1$	$a_{18,18}^1 = a_{17,24}^1,\ a_{18,19}^1 = a_{17,25}^1$		$a_{18,18}^3 = 0$
a_{19}	$a_{19,1}^0 = 0$	$a_{19,1}^1 = 0,\ a_{19,2}^1 = 1$		$a_{19,1}^3 = a_{18,23}^3,\ a_{19,18}^3 = 1$
a_{20}	$a_{20,1}^0 = 1,\ a_{20,6}^0 = a_{19,5}^0,\ a_{20,19}^0 = 1$	$a_{20,6}^1 = 0,\ a_{20,7}^1 = 1$	$a_{20,12}^2 = a_{19,11}^2,\ a_{20,16}^2 = a_{19,15}^2$	$a_{20,6}^3 = 0,\ a_{20,27}^3 = a_{19,26}^3$
a_{21}	$a_{21,8}^0 = 1,\ a_{21,10}^0 = 1,\ a_{21,30}^0 = a_{20,7}^0,$	$a_{21,6}^1 = 1,\ a_{21,7}^1 = 0$	$a_{21,8}^2 = 1,\ a_{21,16}^2 = 1,\ a_{21,20}^2 = 0,\ a_{21,30}^2 = a_{20,7}^2,\ a_{21,31}^2 = a_{20,8}^2$	$a_{21,31}^3 = 1,\ a_{21,10}^3 = 0$
a_{22}	$a_{22,11}^0 = 1,\ a_{22,12}^0 = a_{21,9}^0,\ a_{22,23}^0 = 1$	$a_{22,4}^1 = a_{21,1}^1,\ a_{22,9}^1 = a_{21,6}^1,\ a_{22,21}^1 = a_{21,18}^1$	$a_{22,1}^2 = 0,\ a_{22,11}^2 = 1,\ a_{22,12}^2 = 1,\ a_{22,15}^2 = a_{21,12}^2,\ a_{22,23}^2 = 1,\ a_{22,29}^2 = 0,$	$a_{22,10}^3 = 0,\ a_{22,12}^3 = 1$
a_{23}	$a_{23,1}^0 = 0,\ a_{23,21}^0 = 1,\ a_{23,22}^0 = 1,\ a_{23,23}^0 = 1$	$a_{23,14}^1 = 1,\ a_{23,19}^1 = 0,\ a_{21,31}^1 = 1$	$a_{23,1}^2 = 1,\ a_{23,21}^2 = 1,\ a_{23,25}^2 = 1,\ a_{23,29}^2 = 1$	$a_{23,12}^3 = 1$
a_{24}		$a_{24,14}^1 = 1,\ a_{24,18}^1 = 1,\ a_{24,26}^1 = 1$	$a_{24,7}^2 = 1,\ a_{24,12}^2 = 0,\ a_{24,24}^2 = 1$	$a_{24,16}^3 = 1,\ a_{24,19}^3 = 0$

4   Free-Start Near Collision and Distinguishing Attack on SIMD-512

4.1   Free-Start Near Collision Attack for the Compression Function of 24-Step SIMD-512

In this section, we will show that finding a free-start near collision for the 24-step SIMD-512 can be done with less effort than the birthday attack using our differential path. Similar to the Section 3, we introduce the 1-bit difference in chaining value of the 20-th step, and trace the difference in the forward and backward direction using the modular difference method. We get a near-collision differential path for 24-step SIMD-512 in Table 4 with probability about 2^{-364}. We utilize the message/IV modification technique to fulfill the 129 conditions in IV and $a_1 \sim a_4$. After the message/IV modification, the complexity to find a 51-bit free-start near collision for 24-step SIMD-512 is about 2^{235} operations.

4.2   A Differential Distinguisher for the Compression Function of Full SIMD-512

Our strategy to find a differential path for the full SIMD-512 is a little different from that of the 24-step near collision differential path. We start from 1-bit difference in the Step 24 and trace this difference in backward direction. We introduce the long difference bit carries from the Step 8 down to Step 3, and shrink the difference bit carries from the Step 2 so that the Hamming weight of the difference in IV as lower as possible. The differential distinguisher for the full SIMD-512 compression function is shown in Table 5, and its most expensive part focuses on Steps 3 to 9.

It's worth to note that the bit carry in a_3^7 of step 3 in Table 5 cross the 32-th bit. The bit difference $d_2^7[4]$ cause the bit difference $a_3^7[-31, -32, -1, 2]$. Let $t_3^7 = (d_2^7 + w_2^7 + \Phi(a_2^7, b_2^7, c_2^7)) \lll 27$. In order to cause the bit carry across the 32-bit, we need set the additional conditions $t_{3,31}^7 = 1$ and $t_{3,32}^7 = 1$.

Let $h_i^j = (a_i^j, b_i^j, c_i^j, d_i^j)$ denote the 128-bit outputs of the $j - th$ ladder in Step i, and $h_i = (h_i^0, h_i^1, ..., h_i^7)$ is the 1024-bit output of the Step i of SIMD-512 compression function. Let $x_i = (x_i^0, x_i^1, ..., x_i^7)$ be a 256-bit value, and x can be a, b, c, d and w. There are 941 sufficient conditions in Table 5. We divided those conditions into two parts. The first part is from Step 0 to 9 and the second part is from Step 10 to 36. The number of conditions in the first part is 586 and that in the second part is 355. The conditions distribution of the first part is as follows.

So we start from the chaining values $h_8 = (a_8, b_8, c_8, d_8)$ and go both in the backward and forward direction to compute the IV and output h_{36}. The 320 conditions in a_8, b_8, c_8 and d_8 can be fulfilled by choosing the chaining variables h_8 randomly. And the 466 conditions in d_4, d_5, d_6 and d_7 can be modified by the message words w_8, w_7, w_6 and w_5.

The detail of our distinguishing algorithm for the full compression function of SIMD-512 is as follows.

Table 4. The near-collision differential path for 24-step SIMD-512

step	r	s	h_i^0	h_i^1	h_i^2	h_i^3	h_i^4	h_i^5	h_i^6	h_i^7	pr
0			$a_0^0[6]$	a_0^1	$a_0^2[-17]$	$a_0^3[24]$	$a_0^4[2,-11]$	a_0^5	$a_0^6[32]$	$a_0^7[-18]$	2^{-34}
			$b_0^0[-8]$	$b_0^1[-1,\ldots,-7,8,15]$	b_0^2	$b_0^3[-15]$	b_0^4	$b_0^5[-16]$	$b_0^6[-28]$	$b_0^7[-18,27]$	
			$c_0^0[28]$	c_0^1	c_0^2	$c_0^3[-7]$	$c_0^4[21]$	$c_0^5[-14]$	c_0^6	c_0^7	
			d_0^0	$d_0^1[-18]$	$d_0^2[-4]$	$d_0^3[29]$	d_0^4	$d_0^5[-10,23]$	$d_0^6[-31]$	$d_0^7[6,12]$	
1	3	23	a_1^0	a_1^1	a_1^2	a_1^3	a_1^4	$a_1^5[-1]$	$a_1^6[-22]$	$a_1^7[-29,30]$	2^{-29}
			$b_1^0[9]$	b_1^1	$b_1^2[-20]$	$b_1^3[27]$	$b_1^4[5,-14]$	b_1^5	$b_1^6[3]$	$b_1^7[-21]$	
			$c_1^0[-8]$	$c_1^1[-1,\ldots-7,8,15]$	c_1^2	$c_1^3[-15]$	c_1^4	$c_1^5[-16]$	$c_1^6[-28]$	$c_1^7[-18,27]$	
			$d_1^0[28]$	d_1^1	d_1^2	$d_1^3[-7]$	$d_1^4[21]$	$d_1^5[-14]$	d_1^6	d_1^7	
2	23	17	a_2^0	$a_2^1[25]$	a_2^2	$a_2^3[-25]$	$a_2^4[6]$	$a_2^5[-31]$	a_2^6	a_2^7	2^{-30}
			b_2^0	b_2^1	b_2^2	b_2^3	b_2^4	$b_2^5[-24]$	$b_2^6[-13]$	$b_2^7[-20,21]$	
			$c_2^0[9]^0$	c_2^1	$c_2^2[-20]$	$c_2^3[27]$	$c_2^4[5,-14]$	c_2^5	$c_2^6[3]$	$c_2^7[-21]$	
			$d_2^0[-8]$	$d_2^1[-1,\ldots,-7,8,15]$	d_2^2	$d_2^3[-15]$	d_2^4	$d_2^5[-16]$	$d_2^6[-28]$	$d_2^7[-18,27]$	
3	17	27	$a_3^0[-3]$	$a_3^1[28]$	a_3^2	a_3^3	a_3^4	$a_3^5[-11]$	a_3^6	$a_3^7[-13,22]$	2^{-19}
			b_3^0	$b_3^1[10]$	b_3^2	$b_3^3[-10]$	$b_3^4[23]$	$b_3^5[-16]$	b_3^6	b_3^7	
			c_3^0	c_3^1	c_3^2	c_3^3	c_3^4	$c_3^5[-24]$	$c_3^6[-13]$	$c_3^7[-20,21]$	
			$d_3^0[9]$	d_3^1	$d_3^2[-20]$	$d_3^3[27]$	$d_3^4[5,-14]$	d_3^5	$d_3^6[3]$	$d_3^7[-21]$	
4	27	3	$a_4^0[-12,-13,-14,15]$	a_4^1	a_4^2	a_4^3	a_4^4	a_4^5	a_4^6	a_4^7	2^{-17}
			$b_4^0[-30]$	$b_4^1[23]$	b_4^2	b_4^3	b_4^4	$b_4^5[-6]$	b_4^6	$b_4^7[-8,17]$	
			c_4^0	$c_4^1[10]$	c_4^2	$c_4^3[-10]$	$c_4^4[23]$	$c_4^5[-16]$	c_4^6	c_4^7	
			d_4^0	d_4^1	d_4^2	d_4^3	d_4^4	$d_4^5[-24]$	$d_4^6[-13]$	$d_4^7[-20,21]$	
5	3	23	a_5^0	a_5^1	a_5^2	a_5^3	a_5^4	$a_5^5[4,-5]$	$a_5^6[11]$	a_5^7	2^{-16}
			$b_5^0[-15,-16,-17,18]$	b_5^1	b_5^2	b_5^3	b_5^4	b_5^5	b_5^6	b_5^7	
			$c_5^0[-30]$	$c_5^1[23]$	c_5^2	c_5^3	c_5^4	$c_5^5[-6]$	c_5^6	$c_5^7[-8,17]$	
			d_5^0	$d_5^1[10]$	d_5^2	$d_5^3[-10]$	$d_5^4[23]$	$d_5^5[-16]$	d_5^6	d_5^7	
6	23	17	a_6^0	a_6^1	a_6^2	$a_6^3[-27]$	$a_6^4[8]$	$a_6^5[-1]$	a_6^6	a_6^7	2^{-15}
			b_6^0	b_6^1	b_6^2	b_6^3	b_6^4	b_6^5	$b_6^6[27,-28]$	$b_6^7[2]$	
			$c_6^0[-15,-16,-17,18]$	c_6^1	c_6^2	c_6^3	c_6^4	c_6^5	c_6^6	c_6^7	
			$d_6^0[-30]$	$d_6^1[23]$	d_6^2	d_6^3	d_6^4	$d_6^5[-6]$	d_6^6	$d_6^7[-8,17]$	
7	17	27	a_7^0	a_7^1	a_7^2	a_7^3	a_7^4	$a_7^5[-1]$	a_7^6	$a_7^7[-3]$	2^{-12}
			b_7^0	b_7^1	b_7^2	$b_7^3[-12]$	$b_7^4[25]$	$b_7^5[-18]$	b_7^6	b_7^7	
			c_7^0	c_7^1	c_7^2	c_7^3	c_7^4	c_7^5	$c_7^6[27,-28]$	$c_7^7[2]$	
			$d_7^0[-15,-16,-17,18]$	d_7^1	d_7^2	d_7^3	d_7^4	d_7^5	d_7^6	d_7^7	
8	27	3	$a_8^0[-18,-19,-20,21]$	a_8^1	a_8^2	a_8^3	a_8^4	a_8^5	a_8^6	a_8^7	2^{-12}
			b_8^0	b_8^1	b_8^2	b_8^3	b_8^4	$b_8^5[-28]$	b_8^6	$b_8^7[-30]$	
			c_8^0	c_8^1	c_8^2	$c_8^3[-12]$	$c_8^4[25]$	$c_8^5[-18]$	c_8^6	c_8^7	
			d_8^0	d_8^1	d_8^2	d_8^3	d_8^4	$d_8^5[27,-28]$	d_8^6	$d_8^7[2]$	
9	28	19	a_9^0	a_9^1	a_9^2	a_9^3	a_9^4	a_9^5	a_9^6	$a_9^7[21]$	2^{-10}
			$b_9^0[-14,-15,-16,17]$	b_9^1	b_9^2	b_9^3	b_9^4	b_9^5	b_9^6	b_9^7	
			c_9^0	c_9^1	c_9^2	c_9^3	c_9^4	$c_9^5[-28]$	c_9^6	$c_9^7[-30]$	
			d_9^0	d_9^1	d_9^2	$d_9^3[-12]$	$d_9^4[25]$	$d_9^5[-18]$	d_9^6	d_9^7	
10	19	22	a_{10}^0	a_{10}^1	a_{10}^2	$a_{10}^3[0,-2]$	$a_{10}^4[15]$	a_{10}^5	a_{10}^6	a_{10}^7	2^{-9}
			b_{10}^0	b_{10}^1	b_{10}^2	b_{10}^3	b_{10}^4	b_{10}^5	b_{10}^6	$b_{10}^7[8]$	
			$c_{10}^0[-14,-15,-16,17]$	c_{10}^1	c_{10}^2	c_{10}^3	c_{10}^4	c_{10}^5	c_{10}^6	c_{10}^7	
			d_{10}^0	d_{10}^1	d_{10}^2	d_{10}^3	d_{10}^4	$d_{10}^5[-28]$	d_{10}^6	$d_{10}^7[-30]$	
11	22	7	a_{11}^0	a_{11}^1	a_{11}^2	a_{11}^3	a_{11}^4	$a_{11}^5[-3]$	a_{11}^6	a_{11}^7	2^{-8}
			b_{11}^0	b_{11}^1	b_{11}^2	$b_{11}^3[-24]$	$b_{11}^4[5]$	b_{11}^5	b_{11}^6	b_{11}^7	
			c_{11}^0	c_{11}^1	c_{11}^2	c_{11}^3	c_{11}^4	c_{11}^5	c_{11}^6	$c_{11}^7[8]$	
			$d_{11}^0[-14,-15,-16,17]$	d_{11}^1	d_{11}^2	d_{11}^3	d_{11}^4	d_{11}^5	d_{11}^6	d_{11}^7	
12	7	28	a_{12}^0	a_{12}^1	a_{12}^2	a_{12}^3	a_{12}^4	a_{12}^5	a_{12}^6	a_{12}^7	2^{-4}
			b_{12}^0	b_{12}^1	b_{12}^2	b_{12}^3	b_{12}^4	$b_{12}^5[-10]$	b_{12}^6	b_{12}^7	
			c_{12}^0	c_{12}^1	c_{12}^2	$c_{12}^3[-24]$	$c_{12}^4[5]$	c_{12}^5	c_{12}^6	c_{12}^7	
			d_{12}^0	d_{12}^1	d_{12}^2	d_{12}^3	d_{12}^4	d_{12}^5	d_{12}^6	$d_{12}^7[8]$	
13	28	19	a_{13}^0	a_{13}^1	a_{13}^2	a_{13}^3	a_{13}^4	a_{13}^5	a_{13}^6	$a_{13}^7[27]$	2^{-4}
			b_{13}^0	b_{13}^1	b_{13}^2	b_{13}^3	b_{13}^4	b_{13}^5	b_{13}^6	b_{13}^7	
			c_{13}^0	c_{13}^1	c_{13}^2	c_{13}^3	c_{13}^4	$c_{13}^5[-10]$	c_{13}^6	c_{13}^7	
			d_{13}^0	d_{13}^1	d_{13}^2	$d_{13}^3[-24]$	$d_{13}^4[5]$	d_{13}^5	d_{13}^6	d_{13}^7	
14	19	22	a_{14}^0	a_{14}^1	a_{14}^2	a_{14}^3	$a_{14}^4[27]$	a_{14}^5	a_{14}^6	a_{14}^7	2^{-3}
			b_{14}^0	b_{14}^1	b_{14}^2	b_{14}^3	b_{14}^4	b_{14}^5	b_{14}^6	$b_{14}^7[14]$	
			c_{14}^0	c_{14}^1	c_{14}^2	c_{14}^3	c_{14}^4	c_{14}^5	c_{14}^6	c_{14}^7	
			d_{14}^0	d_{14}^1	d_{14}^2	d_{14}^3	d_{14}^4	$d_{14}^5[-10]$	d_{14}^6	d_{14}^7	
15	22	7	a_{15}^0	a_{15}^1	a_{15}^2	a_{15}^3	a_{15}^4	a_{15}^5	a_{15}^6	a_{15}^7	2^{-2}
			b_{15}^0	b_{15}^1	b_{15}^2	b_{15}^3	$b_{15}^4[17]$	b_{15}^5	b_{15}^6	b_{15}^7	
			c_{15}^0	c_{15}^1	c_{15}^2	c_{15}^3	c_{15}^4	c_{15}^5	c_{15}^6	$c_{15}^7[14]$	
			d_{15}^0	d_{15}^1	d_{15}^2	d_{15}^3	d_{15}^4	d_{15}^5	d_{15}^6	d_{15}^7	

Table 4. (*continued*)

step	r	s	h^0	h^1	h^2	h^3	h^4	h^5	h^6	h^7	Pr
16	7	28	a_{16}^0	a_{16}^1	a_{16}^2	a_{16}^3	a_{16}^4	a_{16}^5	a_{16}^6	a_{16}^7	2^{-2}
			b_{16}^0	b_{16}^1	b_{16}^2	b_{16}^3	b_{16}^4	b_{16}^5	b_{16}^6	b_{16}^7	
			c_{16}^0	c_{16}^1	c_{16}^2	c_{16}^3	$c_{16}^4[17]$	c_{16}^5	c_{16}^6	c_{16}^7	
			d_{16}^0	d_{16}^1	d_{16}^2	d_{16}^3	d_{16}^4	d_{16}^5	d_{16}^6	$d_{16}^7[14]$	
17	29	9	a_{17}^0	a_{17}^1	a_{17}^2	a_{17}^3	a_{17}^4	a_{17}^5	a_{17}^6	$a_{17}^7[23]$	2^{-2}
			b_{17}^0	b_{17}^1	b_{17}^2	b_{17}^3	b_{17}^4	b_{17}^5	b_{17}^6	b_{17}^7	
			c_{17}^0	c_{17}^1	c_{17}^2	c_{17}^3	c_{17}^4	c_{17}^5	c_{17}^6	c_{17}^7	
			d_{17}^0	d_{17}^1	d_{17}^2	d_{17}^3	$d_{17}^4[17]$	d_{17}^5	d_{17}^6	d_{17}^7	
18	9	15	a_{18}^0	a_{18}^1	a_{18}^2	a_{18}^3	a_{18}^4	a_{18}^5	a_{18}^6	a_{18}^7	2^{-1}
			b_{18}^0	b_{18}^1	b_{18}^2	b_{18}^3	b_{18}^4	b_{18}^5	b_{18}^6	$b_{18}^7[32]$	
			c_{18}^0	c_{18}^1	c_{18}^2	c_{18}^3	c_{18}^4	$c_{18}^5 6$	c_{18}^6	c_{18}^7	
			d_{18}^0	d_{18}^1	d_{18}^2	d_{18}^3	d_{18}^4	d_{18}^5	d_{18}^6	d_{18}^7	
19	15	5	a_{19}^0	a_{19}^1	a_{19}^2	a_{19}^3	a_{19}^4	a_{19}^5	a_{19}^6	a_{19}^7	2^{-1}
			b_{19}^0	b_{19}^1	b_{19}^2	b_{19}^3	b_{19}^4	b_{19}^5	b_{19}^6	b_{19}^7	
			c_{19}^0	c_{19}^1	c_{19}^2	c_{19}^3	c_{19}^4	c_{19}^5	c_{19}^6	$c_{19}^7[32]$	
			d_{19}^0	d_{19}^1	d_{19}^2	d_{19}^3	d_{19}^4	d_{19}^5	d_{19}^6	d_{19}^7	
20	5	29	a_{20}^0	a_{20}^1	a_{20}^2	a_{20}^3	a_{20}^4	a_{20}^5	a_{20}^6	a_{20}^7	2^{-1}
			b_{20}^0	b_{20}^1	b_{20}^2	b_{20}^3	b_{20}^4	b_{20}^5	b_{20}^6	b_{20}^7	
			c_{20}^0	c_{20}^1	c_{20}^2	c_{20}^3	c_{20}^4	c_{20}^5	c_{20}^6	c_{20}^7	
			d_{20}^0	d_{20}^1	d_{20}^2	d_{20}^3	d_{20}^4	d_{20}^5	d_{20}^6	$d_{20}^7[32]$	
21	29	9	a_{21}^0	a_{21}^1	a_{21}^2	a_{21}^3	a_{21}^4	a_{21}^5	a_{21}^6	$a_{21}^7[9]$	2^{-1}
			b_{21}^0	b_{21}^1	b_{21}^2	b_{21}^3	b_{21}^4	b_{21}^5	b_{21}^6	b_{21}^7	
			c_{21}^0	c_{21}^1	c_{21}^2	c_{21}^3	c_{21}^4	c_{21}^5	c_{21}^6	c_{21}^7	
			d_{21}^0	d_{21}^1	d_{21}^2	d_{21}^3	d_{21}^4	d_{21}^5	d_{21}^6	d_{21}^7	
22	9	15	a_{22}^0	a_{22}^1	a_{22}^2	a_{22}^3	a_{22}^4	a_{22}^5	$a_{22}^6[18]$	a_{22}^7	2^{-2}
			b_{22}^0	b_{22}^1	b_{22}^2	b_{22}^3	b_{22}^4	b_{22}^5	$b_{22}^6[18]$	$b_{22}^7[18]$	
			c_{22}^0	c_{22}^1	c_{22}^2	c_{22}^3	c_{22}^4	c_{22}^5	c_{22}^6	c_{22}^7	
			d_{22}^0	d_{22}^1	d_{22}^2	d_{22}^3	d_{22}^4	d_{22}^5	d_{22}^6	d_{22}^7	
23	15	5	$a_{23}^0[-1,-2,3]$	a_{23}^1	a_{23}^2	a_{23}^3	a_{23}^4	a_{23}^5	a_{23}^6	a_{23}^7	2^{-5}
			b_{23}^0	b_{23}^1	b_{23}^2	b_{23}^3	b_{23}^4	$b_{23}^5[1]$	b_{23}^6	b_{23}^7	
			c_{23}^0	c_{23}^1	c_{23}^2	c_{23}^3	c_{23}^4	c_{23}^5	c_{23}^6	$c_{23}^7[18]$	
			d_{23}^0	d_{23}^1	d_{23}^2	d_{23}^3	d_{23}^4	d_{23}^5	d_{23}^6	d_{23}^7	
24	5	29	a_{24}^0	a_{24}^1	$a_{24}^2[6]$	a_{24}^3	a_{24}^4	a_{24}^5	a_{24}^6	a_{24}^7	2^{-6}
			$b_{24}^0[-6,-7,8]$	b_{24}^1	b_{24}^2	b_{24}^3	b_{24}^4	b_{24}^5	b_{24}^6	b_{24}^7	
			c_{24}^0	c_{24}^1	c_{24}^2	c_{24}^3	c_{24}^4	$c_{24}^5[1]$	c_{24}^6	c_{24}^7	
			d_{24}^0	d_{24}^1	d_{24}^2	d_{24}^3	d_{24}^4	d_{24}^5	d_{24}^6	$d_{24}^7[18]$	
25	4	13	$a_{25}^0[10]$	$a_{25}^1[10]$	$a_{25}^2[-30]$	$a_{25}^3[5]$	$a_{25}^4[15,-24]$	$a_{25}^5[13]$	a_{25}^6	a_{25}^7	2^{-11}
			b_{25}^0	$b_{25}^1[10]$	b_{25}^2	b_{25}^3	b_{25}^4	b_{25}^5	b_{25}^6	b_{25}^7	
			$c_{25}^0[-6,-7,8]$	c_{25}^1	c_{25}^2	c_{25}^3	c_{25}^4	c_{25}^5	c_{25}^6	c_{25}^7	
			d_{25}^0	d_{25}^1	d_{25}^2	d_{25}^3	d_{25}^4	$d_{25}^5[1]$	d_{25}^6	d_{25}^7	
26	13	10	$a_{26}^0[-5,11,25,28]$	a_{26}^1	$a_{26}^2[25]$	a_{26}^3	$a_{26}^4[23]$	$a_{26}^5[-26]$	$a_{26}^6[-6,11,18]$	$a_{26}^7[5,-11,-28]$	2^{-23}
			$b_{26}^0[23]$	$b_{26}^1[-11]$	b_{26}^2	$b_{26}^3[18]$	$b_{26}^4[-5,28]$	b_{26}^5	$b_{26}^6[26]$	b_{26}^7	
			c_{26}^0	$c_{26}^1[10]$	c_{26}^2	c_{26}^3	c_{26}^4	c_{26}^5	c_{26}^6	c_{26}^7	
			$d_{26}^0[-6,-7,8]$	d_{26}^1	d_{26}^2	d_{26}^3	d_{26}^4	d_{26}^5	d_{26}^6	d_{26}^7	
27	10	25	$a_{27}^0[-6,15,31]$	$a_{27}^1[-16,21,28]$	$a_{27}^2[-4]$	$a_{27}^3[1,-32]$	$a_{27}^4[3,14]$	$a_{27}^5[-7]$	$a_{27}^6[3,6,-15,21]$	a_{27}^7	2^{-36}
			b_{27}^0	$b_{27}^1[3,6,-15,21]$	b_{27}^2	$b_{27}^3[3]$	$b_{27}^4[1]$	$b_{27}^5[-4]$	$b_{27}^6[-16,21,28]$	$b_{27}^7[-6,15,-21]$	
			$c_{27}^0[-6,-7,8]$	$c_{27}^1[23]$	$c_{27}^2[-11]$	$c_{27}^3[18]$	$c_{27}^4[-5,28]$	c_{27}^5	$c_{27}^6[26]$	c_{27}^7	
			d_{27}^0	d_{27}^1	$d_{27}^2[10]$	d_{27}^3	d_{27}^4	d_{27}^5	d_{27}^6	d_{27}^7	
28	25	4	$a_{28}^0[7,28]$	a_{28}^1	$a_{28}^2[-9,15,28,31]$	$a_{28}^3[1]$	$a_{28}^4[8,24,-31]$	$a_{28}^5[-9,21,27]$	$a_{28}^6[-3,-29]$	$a_{28}^7[16]$	2^{-49}
			$b_{28}^0[8,24,-31]$	$b_{28}^1[-9,14,21]$	$b_{28}^2[-29]$	$b_{28}^3[-25,26]$	$b_{28}^4[7,28]$	$b_{28}^5[-32]$	$b_{28}^6[-8,14,28,31]$	b_{28}^7	
			c_{28}^0	$c_{28}^1[3,6,-15,21]$	c_{28}^2	$c_{28}^3[3]$	$c_{28}^4[1]$	$c_{28}^5[-4]$	$c_{28}^6[-16,21,28]$	$c_{28}^7[-6,15,-21]$	
			d_{28}^0	$d_{28}^1[23]$	$d_{28}^2[-11]$	$d_{28}^3[18]$	$d_{28}^4[-5,28]$	d_{28}^5	$d_{28}^6[26]$	d_{28}^7	

1. Select a 1024-bit chaining values h_8 and a 1024-bit message M randomly, and let $h_8' = h_8 \oplus \Delta h_8$ where Δh_8 is the difference in Step 8. Modify h_8 to satisfy the 146 conditions. Compute the chaining values h_7 to h_4 and h_7' to h_4' in the backward directions.
2. Modify the conditions in d_7, d_6, d_5 and d_4 by h_8 and the expanded message w_8, w_7, w_6 and w_5 respectively. Update the message M according to the 1024-bit expanded message w_5, w_6, w_7 and w_8 and compute the new expanded message W.
3. Compute the chaining values h_3 to h_0 and h_3' to h_0' in the backward directions. If Δh_0 is equal to the fixed difference ΔIV in Table 5, go to Step 4; Otherwise, return Step 1.

Table 5. The differential path for the full SIMD-512

step	r	s	h^0	h^1	h^2	h^3	h^4	h^5	h^6	h^7	p_r
0			a_0^0	$a_0^1[23,-24]$	$a_0^2[-3,26,-27]$	$a_0^3[-6]$	$a_0^4[11,20,-21,29]$	a_0^5	$a_0^6[-4,-23,32]$	a_0^7	2^{-50}
			b_0^0	$b_0^1[1,2,3,-4,-15]$	b_0^2	b_0^3	b_0^4	$b_0^5[-7,-8,9]$	$b_0^6[10,-25]$	$b_0^7[4,7,-11,-13,24,-27]$	
			$c_0[3]^0$	c_0^1	$c_0^2[-17,18]$	$c_0^3[-8,9,26]$	$c_0^4[-3,-5,18]$	$c_0^5[4,5,-17,32]$	c_0^6	c_0^7	
			d_0^0	d_0^1	$d_0^2[8]$	$d_0^3[-4,15]$	d_0^4	$d_0^5[-23]$	d_0^6	$d_0^7[3,-12,16,30]$	
1	3	23	a_1^0	a_1^1	$a_1^2[-18,-19,...,22,31]$	$a_1^3[-27,-31]$	$a_1^4[11]$	a_1^5	a_1^6	$a_1^7[21]$	2^{-55}
			b_1^0	$b_1^1[26,-27]$	$b_1^2[-6,29,-30]$	$b_1^3[-9]$	$b_1^4[14,23,-24,32]$	b_1^5	$b_1^6[3,-7,-26]$	b_1^7	
			c_1^0	$c_1^1[1,2,3,-4,-15]$	c_1^2	c_1^3	c_1^4	$c_1^5[-7,-8,9]$	$c_1^6[10,-25]$	$c_1^7[4,7,-11,-13,24,-27]$	
			$d_1[3]^0$	d_1^1	$d_1^2[-17,18]$	$d_1^3[-8,9,26]$	$d_1^4[-3,-5,18]$	$d_1^5[4,5,-17,32]$	d_1^6	d_1^7	
2	23	17	$a_2^0[-20,-21,22]$	$a_2^1[19,-21]$	a_2^2	$a_2^3[11,25]$	$a_2^4[3,20,21,22,23,-24]$	$a_2^5[-2,-17,21]$	a_2^6	a_2^7	2^{-56}
			b_2^0	b_2^1	$b_2^2[-9,-10,...,13,22]$	$b_2^3[-18,-22]$	$b_2^4[2]$	b_2^5	b_2^6	$b_2^7[12]$	
			c_2^0	$c_2^1[26,-27]$	$c_2^2[-6,29,-30]$	$c_2^3[-9]$	$c_2^4[14,23,-24,32]$	c_2^5	$c_2^6[3,-7,-26]$	c_2^7	
			d_2^0	$d_2^1[1,2,3,-4,-15]$	d_2^2	d_2^3	d_2^4	$d_2^5[-7,-8,9]$	$d_2^6[10,-25]$	$d_2^7[4,7,-11,-13,24,-27]$	
3	17	27	$a_3^0[17,18,...,-27]$	$a_3^1[-14,-15,...,19]$	a_3^2	a_3^3	$a_3^4[-17]$	$a_3^5[2]$	a_3^6	$a_3^7[8,-9,22,23,...,-30,-31,-32,-1,2]$	2^{-75}
			$b_3^0[-5,-6,7]$	$b_3^1[4,-6]$	b_3^2	$b_3^3[10,28]$	$b_3^4[5,6,7,8,-9,20]$	$b_3^5[-2,6,-19]$	b_3^6	b_3^7	
			c_3^0	c_3^1	$c_3^2[-9,-10,...,13,22]$	$c_3^3[-18,-22]$	$c_3^4[2]$	c_3^5	c_3^6	$c_3^7[12]$	
			d_3^0	$d_3^1[26,-27]$	$d_3^2[-6,29,-30]$	$d_3^3[-9]$	$d_3^4[14,23,-24,32]$	d_3^5	$d_3^6[3,-7,-26]$	d_3^7	
4	27	3	a_4^0	$a_4^1[-22,-29]$	$a_4^2[16,-32],$	a_4^3	a_4^4	a_4^5	$a_4^6[-6,-7,-8,-9]$	a_4^7	2^{-68}
			$b_4^0[12,13,...,-22]$	$b_4^1[-9,-10,...,14]$	b_4^2	b_4^3	$b_4^4[-12]$	$b_4^5[29]$	b_4^6	$b_4^7[3,-4,17,18,...,-25,-26,...,29]$	
			$c_4^0[-5,-6,7]$	$c_4^1[4,-6]$	c_4^2	$c_4^3[10,28]$	$c_4^4[5,6,7,8,-9,20]$	$c_4^5[-2,6,-19]$	c_4^6	c_4^7	
			d_4^0	d_4^1	$d_4^2[-9,-10,...,13,22]$	$d_4^3[-18,-22]$	$d_4^4[2]$	d_4^5	d_4^6	$d_4^7[12]$	
5	3	23	a_5^0	a_5^1	$a_5^2[13,-32]$	$a_5^3[-14]$	a_5^4	a_5^5	$a_5^6[-30,32]$	$a_5^7[-14,-15]$	2^{-65}
			b_5^0	$b_5^1[-25,32]$	$b_5^2[-3,19]$	b_5^3	b_5^4	b_5^5	$b_5^6[0,10,-11,-12]$	b_5^7	
			$c_5^0[12,13,...,-22]$	$c_5^1[-9,-10,...,14]$	c_5^2	c_5^3	$c_5^4[-12]$	$c_5^5[29]$	c_5^6	$c_5^7[3,-4,17,18,...,-25,-26,...,29]$	
			$d_5^0[-5,-6,7]$	$d_5^1[4,-6]$	d_5^2	$d_5^3[10,28]$	$d_5^4[5,6,7,8,-9,20]$	$d_5^5[-2,6,-19]$	d_5^6	d_5^7	
6	23	17	$a_6^0[3,-4,22]$	a_6^1	a_6^2	$a_6^3[13,-27,-28,...,32]$	$a_6^4[22,23,...,-31]$	$a_6^5[-19]$	$a_6^6[-29]$	a_6^7	2^{-71}
			b_6^0	b_6^1	$b_6^2[4,-23]$	$b_6^3[-5]$	b_6^4	b_6^5	$b_6^6[-21,23]$	$b_6^7[-5,-6]$	
			c_6^0	$c_6^1[-25,32]$	$c_6^2[-3,19]$	c_6^3	c_6^4	c_6^5	$c_6^6[-9,-10,-11,-12]$	c_6^7	
			$d_6^0[12,13,...,-22]$	$d_6^1[-9,-10,...,14]$	d_6^2	d_6^3	$d_6^4[-12]$	$d_6^5[29]$	d_6^6	$d_6^7[3,-4,17,18,...,-25,-26,...,29]$	
7	17	27	$a_7^0[8,9,-10]$	a_7^1	a_7^2	$a_7^3[-27]^3$	a_7^4	$a_7^5[-24,-25,...,28]$	a_7^6	$a_7^7[-21,-22,...,25]$	2^{-51}
			$b_7^0[7,20,-21]$	b_7^1	b_7^2	$b_7^3[-12,-13,...17,30]$	$b_7^4[7,8,...,-16]$	$b_7^5[-4]$	$b_7^6[-14]$	b_7^7	
			c_7^0	c_7^1	$c_7^2[4,-23]$	$c_7^3[-5]$	c_7^4	c_7^5	$c_7^6[-21,23]$	$c_7^7[-5,-6]$	
			d_7^0	$d_7^1[-25,32]$	$d_7^2[-3,19]$	d_7^3	d_7^4	d_7^5	$d_7^6[-9,-10,-11,-12]$	d_7^7	
8	27	3	$a_8^0[12]$	$a_8^1[-28]$	$a_8^2[-6]$	a_8^3	a_8^4	$a_8^5[28]$	$a_8^6[12]$	a_8^7	2^{-47}
			$b_8^0[3,4,-5]$	b_8^1	b_8^2	$b_8^3[-22]$	b_8^4	$b_8^5[-19,-20,...,23]$	b_8^6	$b_8^7[-16,-17,...,20]$	
			$c_8^0[7,20,-21]$	c_8^1	c_8^2	$c_8^3[-12,-13,...17,30]$	$c_8^4[7,8,...,-16]$	$c_8^5[-4]$	$c_8^6[-14]$	c_8^7	
			d_8^0	d_8^1	$d_8^2[4,-23]$	$d_8^3[-5]$	d_8^4	d_8^5	$d_8^6[-21,23]$	$d_8^7[-5,-6]$	
9	28	19	a_9^0	$a_9^1[-15]$	$a_9^2[-10,23]$	a_9^3	a_9^4	a_9^5	$a_9^6[10]$	$a_9^7[7,-26]$	2^{-48}
			$b_9^0[8]$	$b_9^1[-24]$	$b_9^2[-2]$	b_9^3	b_9^4	$b_9^5[24]$	$b_9^6[8]$	b_9^7	
			$c_9^0[3,4,-5]$	c_9^1	c_9^2	$c_9^3[-22]$	c_9^4	$c_9^5[-19,-20,...,23]$	c_9^6	$c_9^7[-16,-17,...,20]$	
			$d_9^0[7,20,-21]$	d_9^1	d_9^2	$d_9^3[-12,-13,...17,30]$	$d_9^4[7,8,...,-16]$	$d_9^5[-4]$	$d_9^6[-14]$	d_9^7	
10	19	22	a_{10}^0	a_{10}^1	a_{10}^2	$a_{10}^3[20]$	a_{10}^4	a_{10}^5	$a_{10}^6[-4]$	a_{10}^7	2^{-27}
			b_{10}^0	$b_{10}^1[-2]$	$b_{10}^2[10,-29]$	b_{10}^3	b_{10}^4	b_{10}^5	$b_{10}^6[29]$	$b_{10}^7[-13,26]$	
			$c_{10}^0[8]$	$c_{10}^1[-24]$	$c_{10}^2[-2]$	c_{10}^3	c_{10}^4	$c_{10}^5[24]$	$c_{10}^6[8]$	c_{10}^7	
			$d_{10}^0[3,4,-5]$	d_{10}^1	d_{10}^2	$d_{10}^3[-22]$	d_{10}^4	$d_{10}^5[-19,-20,...23]$	d_{10}^6	$d_{10}^7[-16,-17,...,20]$	

Table 5. (*continued*)

step	r	s	h^0_i	h^1_i	h^2_i	h^3_i	h^4_i	h^5_i	h^6_i	h^7_i	p_r
11	22	7	a^0_{11} b^0_{11} $c^0_{11}[-2]$ $d^0_{11}[8]$	a^1_{11} b^1_{11} $c^1_{11}[-2]$ $d^1_{11}[-24]$	a^2_{11} b^2_{11} $c^2_{11}[10,-29]$ $d^2_{11}[-2]$	$a^3_{11}[-29]$ $b^3_{11}[10]$ c^3_{11} d^3_{11}	a^4_{11} b^4_{11} c^4_{11} d^4_{11}	a^5_{11} b^5_{11} c^5_{11} $d^5_{11}[24]$	a^6_{11} $b^6_{11}[-26]$ $c^6_{11}[29]$ $d^6_{11}[8]$	$a^7_{11}[23]$[7] b^7_{11} $c^7_{11}[-13,26]$ d^7_{11}	2^{-15}
12	7	28	$a^0_{12}[4]$ b^0_{12} c^0_{12} d^0_{12}	$a^1_{12}[-20]$ b^1_{12} c^1_{12} $d^1_{12}[-2]$	a^2_{12} b^2_{12} c^2_{12} $d^2_{12}[10,-29]$	a^3_{12} $b^3_{12}[-4]$ $c^3_{12}[10]$ d^3_{12}	a^4_{12} b^4_{12} c^4_{12} d^4_{12}	$a^5_{12}[-20,-21,22]$ b^5_{12} c^5_{12} d^5_{12}	a^6_{12} b^6_{12} $c^6_{12}[-26]$ $d^6_{12}[29]$[6]	a^7_{12} $b^7_{12}[30]$ c^7_{12} $d^7_{12}[-13,26]$	2^{-15}
13	28	19	a^0_{13} $b^0_{13}[32]$ c^0_{13} d^0_{13}	$a^1_{13}[-21]$ $b^1_{13}[-16]$ c^1_{13} d^1_{13}	$a^2_{13}[-29,-30,31]$ b^2_{13} c^2_{13} d^2_{13}	a^3_{13} b^3_{13} $c^3_{13}[-4]$ $d^3_{13}[10]$	a^4_{13} b^4_{13} c^4_{13} d^4_{13}	a^5_{13} $b^5_{13}[-16,-17,18]$ c^5_{13} d^5_{13}	a^6_{13} b^6_{13} c^6_{13} $d^6_{13}[-26]$	$a^7_{13}[13]$ b^7_{13} $c^7_{13}[30]$ d^7_{13}	2^{-14}
14	19	22	a^0_{14} b^0_{14} $c^0_{14}[32]$ d^0_{14}	$a^1_{14}[21]$ $b^1_{14}[-8]$ $c^1_{14}[-16]$ d^1_{14}	a^2_{14} $b^2_{14}[-16,-17,18]$ c^2_{14} d^2_{14}	a^3_{14} b^3_{14} c^3_{14} $d^3_{14}[-4]$	a^4_{14} b^4_{14} c^4_{14} d^4_{14}	a^5_{14} b^5_{14} $c^5_{14}[-16,-17,18]$ d^5_{14}	a^6_{14} b^6_{14} c^6_{14} d^6_{14}	a^7_{14} $b^7_{14}[32]$ c^7_{14} $d^7_{14}[30]$	2^{-13}
15	22	7	a^0_{15} b^0_{15} c^0_{15} $d^0_{15}[32]$	a^1_{15} b^1_{15} $c^1_{15}[-8]$ $d^1_{15}[-16]$	a^2_{15} $b^2_{15}[11]$ $c^2_{15}[-16,-17,18]$ d^2_{15}	$a^3_{15}[6]$ b^3_{15} c^3_{15} d^3_{15}	a^4_{15} b^4_{15} c^4_{15} d^4_{15}	a^5_{15} b^5_{15} c^5_{15} $d^5_{15}[-16,-17,18]$	a^6_{15} b^6_{15} c^6_{15} d^6_{15}	$a^7_{15}[5]$ b^7_{15} $c^7_{15}[32]$ d^7_{15}	2^{-12}
16	7	28	$a^0_{16}[-28]$ b^0_{16} c^0_{16} d^0_{16}	a^1_{16} b^1_{16} c^1_{16} $d^1_{16}[-8]$	a^2_{16} b^2_{16} $c^2_{16}[11]$ $d^2_{16}[16]$	a^3_{16} b^3_{16} c^3_{16} d^3_{16}	a^4_{16} b^4_{16} c^4_{16} d^4_{16}	$a^5_{16}[12]$ b^5_{16} c^5_{16} d^5_{16}	a^6_{16} b^6_{16} c^6_{16} d^6_{16}	a^7_{16} $b^7_{16}[12]$ c^7_{16} $d^7_{16}[32]$	2^{-9}
17	29	9	a^0_{17} $b^0_{17}[-25]$ c^0_{17} d^0_{17}	$a^1_{17}[-17]$ b^1_{17} c^1_{17} d^1_{17}	a^2_{17} b^2_{17} c^2_{17} $d^2_{17}[11]$	a^3_{17} b^3_{17} c^3_{17} d^3_{17}	a^4_{17} b^4_{17} c^4_{17} d^4_{17}	a^5_{17} $b^5_{17}[9]$ c^5_{17} d^5_{17}	a^6_{17} b^6_{17} c^6_{17} d^6_{17}	a^7_{17} b^7_{17} $c^7_{17}[12]$ d^7_{17}	2^{-5}
18	9	15	a^0_{18} b^0_{18} $c^0_{18}[-25]$ d^0_{18}	a^1_{18} $b^1_{18}[-26]$ c^1_{18} d^1_{18}	a^2_{18} b^2_{18} c^2_{18} d^2_{18}	a^3_{18} b^3_{18} c^3_{18} d^3_{18}	a^4_{18} b^4_{18} c^4_{18} d^4_{18}	a^5_{18} b^5_{18} $c^5_{18}[9]$ d^5_{18}	a^6_{18} b^6_{18} c^6_{18} d^6_{18}	a^7_{18} b^7_{18} c^7_{18} $d^7_{18}[12]$	2^{-4}
19	15	5	a^0_{19} b^0_{19} c^0_{19} $d^0_{19}[-25]$	a^1_{19} b^1_{19} $c^1_{19}[-26]$ d^1_{19}	a^2_{19} b^2_{19} c^2_{19} d^2_{19}	a^3_{19} b^3_{19} c^3_{19} d^3_{19}	a^4_{19} b^4_{19} c^4_{19} d^4_{19}	a^5_{19} b^5_{19} c^5_{19} $d^5_{19}[9]$	a^6_{19} b^6_{19} c^6_{19} d^6_{19}	$a^7_{19}[17]$ b^7_{19} c^7_{19} d^7_{19}	2^{-4}
20	5	29	a^0_{20} b^0_{20} c^0_{20} d^0_{20}	a^1_{20} b^1_{20} c^1_{20} $d^1_{20}[-26]$	a^2_{20} b^2_{20} c^2_{20} d^2_{20}	a^3_{20} b^3_{20} c^3_{20} d^3_{20}	a^4_{20} b^4_{20} c^4_{20} d^4_{20}	$a^5_{20}[6]$ b^5_{20} c^5_{20} d^5_{20}	a^6_{20} b^6_{20} c^6_{20} d^6_{20}	a^7_{20} $b^7_{20}[22]$ c^7_{20} d^7_{20}	2^{-3}
21	29	9	a^0_{21} b^0_{21} c^0_{21} d^0_{21}	a^1_{21} b^1_{21} c^1_{21} d^1_{21}	a^2_{21} b^2_{21} c^2_{21} d^2_{21}	a^3_{21} b^3_{21} c^3_{21} d^3_{21}	a^4_{21} b^4_{21} c^4_{21} d^4_{21}	a^5_{21} $b^5_{21}[3]$ c^5_{21} d^5_{21}	a^6_{21} b^6_{21} c^6_{21} d^6_{21}	a^7_{21} b^7_{21} $c^7_{21}[22]$ d^7_{21}	2^{-2}
22	9	15	a^0_{22} b^0_{22} c^0_{22} d^0_{22}	a^1_{22} b^1_{22} c^1_{22} d^1_{22}	a^2_{22} b^2_{22} c^2_{22} d^2_{22}	a^3_{22} b^3_{22} c^3_{22} d^3_{22}	a^4_{22} b^4_{22} c^4_{22} d^4_{22}	a^5_{22} b^5_{22} $c^5_{22}[3]$ d^5_{22}	a^6_{22} b^6_{22} c^6_{22} d^6_{22}	a^7_{22} b^7_{22} c^7_{22} $d^7_{22}[22]$[7]	2^{-2}
23	15	5	a^0_{23} b^0_{23} c^0_{23} d^0_{23}	a^1_{23} b^1_{23} c^1_{23} d^1_{23}	a^2_{23} b^2_{23} c^2_{23} d^2_{23}	a^3_{23} b^3_{23} c^3_{23} d^3_{23}	a^4_{23} b^4_{23} c^4_{23} d^4_{23}	a^5_{23} b^5_{23} c^5_{23} $d^5_{23}[3]$	a^6_{23} b^6_{23} c^6_{23} d^6_{23}	$a^7_{23}[27]$ b^7_{23} c^7_{23} d^7_{23}	2^{-2}
24	5	29	a^0_{24} b^0_{24} c^0_{24} d^0_{24}	a^1_{24} b^1_{24} c^1_{24} d^1_{24}	a^2_{24} b^2_{24} c^2_{24} d^2_{24}	a^3_{24} b^3_{24} c^3_{24} d^3_{24}	a^4_{24} b^4_{24} c^4_{24} d^4_{24}	a^5_{24} b^5_{24} c^5_{24} d^5_{24}	a^6_{24} b^6_{24} c^6_{24} d^6_{24}	a^7_{24} $b^7_{24}[32]$ c^7_{24} d^7_{24}	2^{-1}
25	4	13	a^0_{25} b^0_{25} c^0_{25} d^0_{25}	a^1_{25} b^1_{25} c^1_{25} d^1_{25}	a^2_{25} b^2_{25} c^2_{25} d^2_{25}	a^3_{25} b^3_{25} c^3_{25} d^3_{25}	a^4_{25} b^4_{25} c^4_{25} d^4_{25}	a^5_{25} b^5_{25} c^5_{25} d^5_{25}	a^6_{25} b^6_{25} c^6_{25} d^6_{25}	a^7_{25} b^7_{25} $c^7_{25}[32]$ d^7_{25}	2^{-1}
26	13	10	a^0_{26} b^0_{26} c^0_{26} d^0_{26}	a^1_{26} b^1_{26} c^1_{26} d^1_{26}	a^2_{26} b^2_{26} c^2_{26} d^2_{26}	a^3_{26} b^3_{26} c^3_{26} d^3_{26}	a^4_{26} b^4_{26} c^4_{26} d^4_{26}	a^5_{26} b^5_{26} c^5_{26} d^5_{26}	a^6_{26} b^6_{26} c^6_{26} d^6_{26}	a^7_{26} b^7_{26} c^7_{26} $d^7_{26}[32]$	2^{-1}
27	10	25	a^0_{27} b^0_{27} c^0_{27} d^0_{27}	a^1_{27} b^1_{27} c^1_{27} d^1_{27}	a^2_{27} b^2_{27} c^2_{27} d^2_{27}	a^3_{27} b^3_{27} c^3_{27} d^3_{27}	a^4_{27} b^4_{27} c^4_{27} d^4_{27}	a^5_{27} b^5_{27} c^5_{27} d^5_{27}	a^6_{27} b^6_{27} c^6_{27} d^6_{27}	$a^7_{27}[25]$ b^7_{27} c^7_{27} d^7_{27}	2^{-1}
28	25	4	a^0_{28} b^0_{28} c^0_{28} d^0_{28}	a^1_{28} b^1_{28} c^1_{28} d^1_{28}	a^2_{28} b^2_{28} c^2_{28} d^2_{28}	$a^3_{28}[18]$ b^3_{28} c^3_{28} d^3_{28}	a^4_{28} b^4_{28} c^4_{28} d^4_{28}	a^5_{28} b^5_{28} c^5_{28} d^5_{28}	a^6_{28} b^6_{28} c^6_{28} d^6_{28}	a^7_{28} $b^7_{28}[18]$ c^7_{28} d^7_{28}	2^{-2}
29	4	13	a^0_{29} b^0_{29} c^0_{29} d^0_{29}	a^1_{29} b^1_{29} c^1_{29} d^1_{29}	$a^2_{29}[22]$ b^2_{29} c^2_{29} d^2_{29}	a^3_{29} $b^3_{29}[22]$ c^3_{29} d^3_{29}	a^4_{29} b^4_{29} c^4_{29} d^4_{29}	a^5_{29} b^5_{29} c^5_{29} d^5_{29}	a^6_{29} b^6_{29} c^6_{29} d^6_{29}	a^7_{29} b^7_{29} $c^7_{29}[18]$ d^7_{29}	2^{-3}

Table 5. (*continued*)

step	r	s	h_i^0	h_i^1	h_i^2	h_i^3	h_i^4	h_i^5	h_i^6	h_i^7	p_r
30	13	10	a_{30}^0 b_{30}^0 c_{30}^0 d_{30}^0	a_{30}^1 b_{30}^1 c_{30}^1 d_{30}^1	a_{30}^2 $b_{30}^2[3]$ c_{30}^2 d_{30}^2	a_{30}^3 b_{30}^3 $c_{30}^3[22]$ d_{30}^3	$a_{30}^4[3]$ b_{30}^4 c_{30}^4 d_{30}^4	a_{30}^5 b_{30}^5 c_{30}^5 d_{30}^5	a_{30}^6 b_{30}^6 c_{30}^6 d_{30}^6	a_{30}^7 b_{30}^7 c_{30}^7 $d_{30}^7[18]$	2^{-4}
31	10	25	a_{31}^0 b_{31}^0 c_{31}^0 d_{31}^0	a_{31}^1 b_{31}^1 c_{31}^1 d_{31}^1	a_{31}^2 b_{31}^2 $c_{31}^2[3]$ d_{31}^2	a_{31}^3 b_{31}^3 c_{31}^3 $d_{31}^3[22]$	a_{31}^4 $b_{31}^4[13]$ c_{31}^4 d_{31}^4	a_{31}^5 b_{31}^5 c_{31}^5 d_{31}^5	$a_{31}^6[13]$ b_{31}^6 c_{31}^6 d_{31}^6	$a_{31}^7[-11,12]$ b_{31}^7 c_{31}^7 d_{31}^7	2^{-6}
32	25	4	a_{32}^0 b_{32}^0 c_{32}^0 d_{32}^0	a_{32}^1 b_{32}^1 c_{32}^1 d_{32}^1	a_{32}^2 b_{32}^2 c_{32}^2 $d_{32}^2[3]$	$a_{32}^3[26]$ b_{32}^3 $c_{32}^3[13]$ d_{32}^3	$a_{32}^4[6]$ b_{32}^4 c_{32}^4 d_{32}^4	a_{32}^5 $b_{32}^5[6]$ c_{32}^5 d_{32}^5	a_{32}^6 b_{32}^6 c_{32}^6 d_{32}^6	a_{32}^7 $b_{32}^7[-4,5]$ c_{32}^7 d_{32}^7	2^{-8}
33	4	13	$a_{33}^0[10]$ b_{33}^0 c_{33}^0 d_{33}^0	$a_{33}^1[-4,8]$ b_{33}^1 c_{33}^1 d_{33}^1	$a_{33}^2[-7]$ b_{33}^2 c_{33}^2 d_{33}^2	$a_{33}^3[-19]$ $b_{33}^3[30]$ c_{33}^3 d_{33}^3	$a_{33}^4[-1,10,24]$ $b_{33}^4[8]$ c_{33}^4 $d_{33}^4[13]$	a_{33}^5 $b_{33}^5[10]$ $c_{33}^5[6]$ d_{33}^5	$a_{33}^6[-4,13,-17,30]$ b_{33}^6 c_{33}^6 d_{33}^6	a_{33}^7 $b_{33}^7[-4,5]$ c_{33}^7 d_{33}^7	2^{-19}
34	13	10	$a_{34}^0[-17,25,-30]$ $b_{34}^0[23]$ c_{34}^0 d_{34}^0	a_{34}^1 $b_{34}^1[-17,21]$ c_{34}^1 d_{34}^1	$a_{34}^2[5,-14,23]$ $b_{34}^2[-20]$ c_{34}^2 d_{34}^2	$a_{34}^3[23,-32]$ $b_{34}^3[-32]$ $c_{34}^3[30]$ d_{34}^3	$a_{34}^4[17,-20]$ $b_{34}^4[5,-14,23]$ $c_{34}^4[8]$ d_{34}^4	$a_{34}^5[-3,-17,20,21]$ $b_{34}^5[11,-17,26,-30]$ $c_{34}^5[10]$ d_{34}^5	a_{34}^6 b_{34}^6 c_{34}^6 $d_{34}^6[6]$	$a_{34}^7[2,-5,17,21,-22]$ b_{34}^7 c_{34}^7 $d_{34}^7[-4,5]$	2^{-37}
35	10	25	$a_{35}^0[1,-10,28]$ b_{35}^0 $c_{35}^0[23]$ d_{35}^0	$a_{35}^1[27,-30]$ $b_{35}^1[3,-8,-27]$ $c_{35}^1[-17,21]$ d_{35}^1	$a_{35}^2[10,-13,-27,30,31]$ b_{35}^2 $c_{35}^2[-20]$ d_{35}^2	$a_{35}^3[1,12,-15,19,27,-31]$ $b_{35}^3[1,15,-24]$ $c_{35}^3[-32]$ $d_{35}^3[30]$	$a_{35}^4[11,-28]$ $b_{35}^4[1,-10]$ $c_{35}^4[5,-14,23]$ $d_{35}^4[8]$	$a_{35}^5[3,-8,25,-27,29,30]$ $b_{35}^5[27,-30]$ c_{35}^5 $d_{35}^5[10]$	$a_{35}^6[31]$ $b_{35}^6[-13,-27,30,31]$ $c_{35}^6[11,-17,26,-30]$ d_{35}^6	$a_{35}^7[1,-24,29]$ $b_{35}^7[12,-15,27,31,-32]$ c_{35}^7 d_{35}^7	2^{-59}
36	25	4	$a_{36}^0[20,-23]$ $b_{36}^0[-3,21,26]$ c_{36}^0 $d_{36}^0[23]$	$a_{36}^1[-3,26]$ $b_{36}^1[20,-23]$ $c_{36}^1[3,-8,-27]$ $d_{36}^1[-17,21]$	$a_{36}^2[5,-8,13,20,-24,26]$ $b_{36}^2[3,-6,-20,23,24]$ c_{36}^2 $d_{36}^2[-20]$	$a_{36}^3[2,3,-6,-8,19,20,23,24]$ $b_{36}^3[5,-8,12,20,-24,26]$ $c_{36}^3[1,15,-24]$ $d_{36}^3[-32]$	$a_{36}^4[-1,12,-20,22,23,28]$ $b_{36}^4[4,-21]$ $c_{36}^4[1,-10]$ $d_{36}^4[5,-14,23]$	$a_{36}^5[4,14,-21,-27]$ $b_{36}^5[-1,18,-20,22,23,28]$ $c_{36}^5[27,-30]$ d_{36}^5	$a_{36}^6[-17,22,26]$ $b_{36}^6[24]$ $c_{36}^6[-13,-27,30,31]$ $d_{36}^6[11,-17,26,-30]$	$a_{36}^7[2,7,20,24]$ $b_{36}^7[-17,22,26]$ $c_{36}^7[12,-15,27,31,-32]$ d_{36}^7	2^{-86}

4. Compute h_9 to h_{36} and h_9' to h_{36}' in the forward direction using h_8, h_8' and the expanded message W. If the differences Δh_{36} is equal to the fixed output difference in Table 5, stop; Otherwise, go back to Step 1.

By running the algorithm above, we can find a pair (M, IV) and (M, IV') which has the fixed input and output differences in Table 5 with about 2^{475} SIMD-512 compression function computations. By using the more sophisticated message/IV modification techniques, the complexity can be improved further. But for a random function with output length n-bit, to find a plain pair (P, P') which satisfy the fixed input and output difference has the probability 2^{-n}. Furthermore, our differential distinguisher is applicable for both the compression function and the final compression function of SIMD-512.

5   Conclusions

In this paper, we find some differential paths using the modular difference method for the reduced and full SIMD compression functions. Based on our differential path, we give the free-start near collision and distinguisher attack for the SIMD. Our attack does not contract with any security claims of the designers.

References

1. Biham, E., Chen, R., Joux, A., Carribault, P., Lemuet, C., Jalby, W.: Collisions of SHA-0 and reduced SHA-1. In: Cramer, R. (ed.) EUROCRYPT 2005. LNCS, vol. 3494, pp. 36–57. Springer, Heidelberg (2005)
2. Khovratovich, D., Nikolic, I.: Rotational Cryptanalysis of ARX. In: Hong, S., Iwata, T. (eds.) FSE 2010. LNCS, vol. 6147, pp. 333–346. Springer, Heidelberg (2010)
3. Mendel, F., Nad, T.: A distinguisher for the compression function of SIMD-512. In: Roy, B., Sendrier, N. (eds.) INDOCRYPT 2009. LNCS, vol. 5922, pp. 219–232. Springer, Heidelberg (2009)
4. Leurent, G., Bouillaguet, C., Fouque, P.A.: SIMD Is a Message Digest, Submission to NIST(round 2) (2009)
5. National Institute of Standards and Technoloy: Annoucing Request for Candidate Algorithm Nominations for a New Cryptographic Hash Algorithm (SHA-3) Family., http://nist.gov
6. Nikolić, I., Pieprzyk, J., et al.: Rotational Cryptanalysis of (Modified) Versions of BMW and SIMD, http://ehash.iaik.tugraz.at/wiki/SIMD
7. Wang, X.Y., Yu, H.B.: How to break MD5 and other hash functions. In: Cramer, R. (ed.) EUROCRYPT 2005. LNCS, vol. 3494, pp. 19–35. Springer, Heidelberg (2005)
8. Wang, X.Y., Yin, Y.L., Yu, H.B.: Finding collisions in the full SHA-1. In: Shoup, V. (ed.) CRYPTO 2005. LNCS, vol. 3621, pp. 17–36. Springer, Heidelberg (2005)

Electronic Cash with Anonymous User Suspension

Man Ho Au, Willy Susilo, and Yi Mu

Centre for Computer and Information Security Research
School of Computer Science and Software Engineering
University of Wollongong, Australia
{aau,wsusilo,ymu}@uow.edu.au

Abstract. Electronic cash (E-cash) is the digital counterpart of cash payment. They allow users to spend anonymously unless they "double spend" their electronic coins. However, it is not possible to prevent users from misbehaving under some other subjective definitions of misbehavior, such as money laundering. One solution is to incorporate a trusted third party (TTP), which, upon complaint, uses its power to deanonymize the suspected user. This solution, known as fair e-cash, is not fully satisfactory since additional measure has to be taken to stop misbehaving users from further abusing the system after they have been identified. We present a e-cash system with anonymous user suspension, EC-AUS, which features an suspension manager (SM) that is capable of suspending the underlying user that participates in any suspicious transaction. Suspended users cannot participate in any transaction. The suspension is anonymous in the sense that no party, not even SM, can tell the identities of the suspended users nor link their past transactions. If they are found innocent later, their suspension can be revoked easily.

1 Introduction

E-cash was introduced by David Chaum [18] as an electronic counterpart of physical money. Extensive research [19,28,24,20,7,17,23,14] has been done on the subject since then. In an e-cash scheme, a user withdraws an electronic coin from the bank and the user can spend it to any merchant, who will deposit the coin back to the bank.

A secure and practical e-cash should possess three essential properties, namely, *anonymity*, *balance* and *exculpability*. *Anonymity* (also referred to as privacy), is a distinctive feature of cash payments offers a customer. It means that payments do not leak the customers' whereabouts, spending patterns or personal preferences. *Balance* means that no collusion of users and merchants together can deposit more than they withdraw *without* being detected. Finally, *exculpability* refers to the fact that honest spenders cannot be accused to have double-spent.

Too much privacy may cause problems in the regulatory levels since there is no way misbehaving users can be identified, let alone being punished. Spending the same electronic coin twice, also known as *double-spending*, is a prominent

U. Parampalli and P. Hawkes (Eds.): ACISP 2011, LNCS 6812, pp. 172–188, 2011.
© Springer-Verlag Berlin Heidelberg 2011

example of misbehavior. Existing e-cash schemes tackle this dilemma by incorporating mechanisms such that spending an electronic coin twice provides sufficient information for everyone to compute the user's identity.

Unfortunately, misbehavior cannot always be represented by mathematical relationships such as spending the same electronic coin twice. For instance, it is hard to define mathematically transactions for money laundering, illegal goods purchasing and blackmailing. Fair e-cash [15] addresses the issue by introducing an administrative party, called Open Authority (OA), which is capable of outputting the identity of a user participating in a transaction. This solution, however, does not stop the user from further abusing the system. The user can still spend all his other electronic coins after his identity is revealed. This gives the opportunity for the misbehaving user to transfer his money to some other accounts. In order to stop this, OA will have to open identities of all the transactions to check the flow of the money. The problem can be tackled using the technique of traceable signatures [27] in which the administrative party discloses some secret information, also known as tracing information, of a particular user, which, enables everyone to test if a spending belongs to that specific user. This property is sometimes known as coin traceability [10]. The problem is, once the tracing information is disclosed, there is no way to restore the user's privacy even if he/she is found innocent later.

We think it is important to equip e-cash systems with anonymous user suspension in which users can be suspended without sacrificing their privacy. Suspended users are simply stopped from accessing the system, while their identities remain hidden. Law-enforcing agent can thus suspend users that participate in dubious transactions, investigate the case, and un-suspend the suspect if he/she is found innocent.

Our Contributions. We propose an *electronic cash with anonymous user suspension* (EC-AUS). We formalize the security model for such a system and prove that our construction is secure under this model. Furthermore, we also evaluate the performance of our system.

Paper Outline. In Section 2, we present preliminary information on the various cryptographic tools used in our construction. In Section 3, we formalize the syntax and security properties for EC-AUS. We present our construction and analyze the algorithmic complexity in Section 4. We discuss extensions and several other issues in Section 5 and conclude the paper in Section 6.

Related Work. Our EC-AUS is constructed based on the blacklisting technique from blacklistable anonymous authentication systems [29,8]. Their idea can be summarized as follow. For each authentication, a user with secret key x provides the server with a unique value, called ticket t, which is b^x in some cyclic group $\mathbb{G}$ for a random nonce b. The server provides the user with a blacklist $\{(t_1, b_1), (t_2, b_2), \ldots, (t_n, b_n)\}$. In order to authenticate, the user proves to the server, in zero-knowledge, that $t_i \neq b_i^x$ for $i = 1$ to n and $t = b^x$. This assures the server that the authenticating user is not on the blacklist. If the server would like to blacklist this user later, the entry (t, b) is appended to the blacklist. If the Decisional Diffie-Hellman (DDH) Problem is hard in $\mathbb{G}$, the ticket t is unlinkable and thus user anonymity is preserved.

2   Preliminaries

In this section we define some notations and review cryptographic tools that we use as building blocks in our EC-AUS construction.

Notations. $|S|$ represents the cardinality of a set S. If S is a non-empty set, $a \in_R S$ means that a is drawn uniformly at random from S. If n is a positive integer, we write $[n]$ to mean the set $\{1, 2, \ldots, n\}$. If $s_1, s_2 \in \{0,1\}^*$, then $s_1 \| s_2 \in \{0,1\}^*$ is the concatenation of binary strings s_1 and s_2. We say that a function $\mathsf{negl}(\lambda)$ is a negligible function [3], if for all polynomials $f(\lambda)$, for all sufficiently large λ, $\mathsf{negl}(\lambda) < 1/f(\lambda)$.

Bilinear Map. A pairing is a bilinear mapping from a pair of group elements to a group element. Specifically, let $\mathbb{G}_1, \mathbb{G}_2$ be cyclic groups of prime order p. A function $\hat{e} : \mathbb{G}_1 \times \mathbb{G}_1 \to \mathbb{G}_2$ is said to be a pairing if it satisfies the following properties:

- (Bilinearity.) $\hat{e}(u^x, v^y) = \hat{e}(u, v)^{xy}$ for all $u, v \in \mathbb{G}_1$ and $x, y \in \mathbb{Z}_p$.
- (Non-Degeneracy.) $\hat{e}(g, g) \neq 1_{\mathbb{G}_2}$, the identity element of $\mathbb{G}_2$.
- (Efficient Computability.) $\hat{e}(u, v)$ is efficiently computable for all u, v.
- (Unique Representation.) All elements in $\mathbb{G}_1$, $\mathbb{G}_2$ have unique binary representation.

Proof of Knowledge. In a *Zero-Knowledge Proof of Knowledge (ZKPoK)* protocol [25], a prover convinces a verifier that some statement is true, while the verifier learns nothing except the validity of the statement. Σ-protocols are a special type of three-move ZKPoK protocols, which can be converted into non-interactive *Signature Proof of Knowledge (SPK)* schemes or simply signature schemes [26] that are secure in the *Random Oracle (RO)* Model [4]. Σ-protocols can be transformed to 4-move perfect zero-knowledge ZKPoK protocols [21]. They can also be transformed to 3-move concurrent zero-knowledge protocol in the auxiliary string model using trapdoor commitment schemes [22].

We follow the notation introduced in [13]. For instance, $PK\{(x) : y = g^x\}$ denotes a Σ-protocol that proves the knowledge of $x \in \mathbb{Z}_p$ such that $y = g^x$ for some $y \in \mathbb{G}$. The values inside the parenthesis on the left of the colon denotes variables whose knowledge is to be proven, while values on the right of the colon except those inside the parenthesis denote publicly known value. We use $SPK\{(x) : y = g^x\}(M)$ to denote the transformation of the above Σ-protocol into signature of knowledge, which is secure in the random oracle model due to Fiat-Shamir heuristic. We employ several existing Σ-protocols as building blocks in our construction of EC-AUS. In particular, the ZKPoK of Knowledge and Inequalities of Discrete Logarithms due to Camenisch and Shoup [12].

BBS+ Signature. We briefly review the signature scheme proposed in [1], which is based on the schemes of [11] and [6]. This signature scheme also serves as building blocks in a number of cryptographic systems [2,29,9] and is referred to as BBS+ signature or credential signature.

Let $g_0, g_1, g_2, \ldots, g_\ell, g_{\ell+1} \in \mathbb{G}_1$ be generators of $\mathbb{G}_1$. Let $\hat{e}$ be a bilinear map as discussed. Let $w = g_0^\gamma$ for some $\gamma \in_R \mathbb{Z}_p$. The public key of the signature scheme is $(g_0, \ldots, g_\ell, w, \hat{e})$, and the signing key is (γ).

A signature on messages $(m_1, \ldots, m_\ell)$ is a tuple (A, e, z), where e, z are random values in $\mathbb{Z}_p$ chosen by the signer such that $A = (g_0 g_1^{m_1} \cdots g_\ell^{m_\ell} g_{\ell+1}^z)^{\frac{1}{\gamma+e}}$. Such a signature can be verified by checking if

$$\hat{e}(A, wg_0^e) \overset{?}{=} \hat{e}(g_0 g_1^{m_1} \cdots g_\ell^{m_\ell} g_{\ell+1}^z, g_0).$$

It was proved in [1] that BBS+ is unforgeable under adaptively chosen message attack if the q-SDH assumption holds, where q is the number of signature queries, and that they also proposed a ZKPoK protocol which allows one to prove possession of message-signature pairs.

3   Security Definition

We present the syntax of EC-AUS, followed by the security properties that any EC-AUS construction must satisfy.

3.1   Syntax

The entities in EC-AUS are the *Suspension Manager (SM)*, *Bank (B)*, a set of *Merchants (M)* and a set of *users (U)*. EC-AUS consists of the following protocols/algorithms:

- $(bpk, bsk) \leftarrow BSetup(1^\lambda)$. This algorithm is executed by the bank B to set up the system. On input of one or more security parameters (say, 1^λ), the algorithm outputs a pair consisting of public key bpk and private key bsk. B keeps bsk private and publishes bpk to the public. bpk is an implicit input to all the algorithms described below.
- $(pk, sk) \leftarrow KeyGen$. This algorithm is executed by the user or merchant to generate her key pairs. We assume there exists some kind of public key infrastructure that ensures the public key pk is properly certified and is a unique identifier for the user or merchant.
- $\{SUL \leftarrow SSetup\}$. SM maintains a suspended user list SUL which is available to all entities in the system and is empty initially.
- $AccEstablish(B(bsk, pk_U), U(pk_U, sk_U))$. This protocol is executed between B and a legitimate user U with public key pk_U to establish an account. Upon successful completion of the protocol, the user obtains an account secret cred, which she keeps private to herself, and is thereby eligible for conducting transactions in the system.

- Withdraw($B(\text{bsk}, \text{SUL}, \text{pk}_U), U(\text{cred}, \text{SUL}, \text{sk}_U)$). This protocol is executed between B and a legitimate user U to withdraw an electronic coin. Upon successful completion of the protocol, the user obtains an electronic coin cn, which she keeps private to herself.
- Spend($M(\text{pk}_M, \text{sk}_M, \text{SUL}), U(\text{cred}, \text{pk}_M, \text{cn}, \text{SUL})$). This protocol is executed between a merchant M with public key pk_M and a legitimate user U to spend an electronic coin. Upon successful completion of the protocol, M accepts the coin and obtains a transcript trans.
- Deposit($B(\text{bsk}, \text{pk}_M), M(\text{trans}, \text{sk}_M)$). This protocol is executed between B and a merchant M for the later to deposit an electronic coin. Upon successful completion of the protocol, B either accepts the request or outputs pk^*, along with trans_1, trans_2, Π which serves as a proof that the party with public key pk^* has spent an electronic coin twice in transactions with transcripts trans_1, trans_2.
- $0/1 \leftarrow$ VerGuilt($\text{trans}_1, \text{trans}_2, \text{pk}^*, \Pi$). Everyone can execute this algorithm to check if the party with public key pk^* indeed spent an electronic coin twice in transactions whose transcripts are trans_1 and trans_2.
- *Suspension*. This is a suite of three algorithms: $\varpi \leftarrow$ Extract(trans), SUL $\leftarrow$ Add(SUL', ϖ) and SUL' $\leftarrow$ Remove(SUL, ϖ). These algorithms are executed by SM to suspend or un-suspend a user. On input of a Spend protocol transcript trans, Extract extracts and returns a *ticket* ϖ from the transcript. The suspended user list SUL is a collection of tickets. On input of a SUL and a ticket, Add returns a new SUL that contains all the tickets in the input SUL' in addition to the input ticket. On the other hand, on input of SUL' and a ticket, Remove returns a new SUL that contains all the tickets in it, except the one(s) equivalent to the input ticket.

  When we say that a user Alice is suspended, we mean that there exists a Spend transaction between Alice and a merchant M with transcript trans such that the SM has invoked Add(SUL, Extract(trans)) and no Remove($\cdot$, Extract(trans)) has been invoked afterwards. If Alice is suspended, she cannot conduct Withdraw or Spend. We would like to stress that SM learns nothing about the identity of Alice, nor link any of Alice's past action. All SM does is to suspended an anonymous user that has participated in a spend that results in transcript trans.

3.2   Security Requirements

We first describe various security properties that an EC-AUS construction must possess. Their formal definitions will be given in Appendix A.

- Balance. The bank B is assured that no collusion of users and merchants can deposit more than they withdraw *without being identified*. Consequently, any double spender in the system will be identified.
- Suspension-Correctness. B is assured to accept Withdraw, while Ms are assured to accept Spend, only from Us who are not suspended. On the other hand, honest users that are not currently suspended by SM can always conduct the above transaction with honest B or Ms.

- Anonymity. All that B, M and SM collude together can infer about the identity of a spender is whether that user is suspended at the time of protocol execution, and whether she is in possession of a valid electronic coin.
- Exculpability. An honest user will not be falsely accused of having spent an electronic coin twice. That is, B cannot output $(\text{trans}_1, \text{trans}_2, \text{pk}^*, \Pi)$ such that $1 \leftarrow \text{VerGuilt}(\text{trans}_1, \text{trans}_2, \text{pk}^*, \Pi)$ even if B colludes with M and SM.

The trust placed on various parties regarding the security requirements are summarized in Table 1. The table is interpreted as follows. If Party A is to be assured Security Requirement B, he/she needs to trust the party with tick mark. For instance, users, bank and merchants need to trust that SM is honest for suspension-correctness to hold. Indeed, that is the only trust placed in our system. For instance, an honest user is guaranteed anonymity and exculpability even if the bank, merchant, suspension manager are malicious.

Table 1. Trust Relationship of various parties

Party A	Security Requirements B	Bank	Suspension Manager	Merchant	User
Bank	Balance	N/A	×	×	×
User/Bank/Merchant	Suspension-Correctness	×	✓	×	×
User	Anonymity	×	×	×	×
User	Exculpability	×	×	×	×

4 Our System

4.1 High Level Description

We provide a high level description of EC-AUS, which combines the technique of the e-cash scheme due to [2,10] and the anonymous blacklisting technique from [29].

The Setup. Let $\mathbb{G}$ be a cyclic group and g, h, h_0, h_1 are generators of $\mathbb{G}$. User and Merchant are equipped with key pairs of the form (g^x, x) where g is a generator of $\mathbb{G}$. The bank chooses a signature scheme and assume the key pair is (pk_{Sig}, sk_{Sig}). The public key of the bank is pk_{Sig}. The secret key is sk_{Sig}. The suspension manager makes available an empty list, SUL.

Account Creation. User U with public key g^x creates an account with the bank B by submitting a value h^x to the bank, along with a proof-of-correctness.

Withdrawing an E-Coin. U first needs to show B he/she is not suspended. Both parties first obtain the current $\text{SUL} = \{(t_1, b_1), (t_2, b_2), \ldots, (t_n, b_n)\}$ from suspension manager SM. U proves to B that using his secret key x, none of the relationships $t_i = b_i^x$ hold. B only issues U with an electronic coin if U is not suspended. An electronic coin for U is simply a signature $\sigma_{x,y}$ from B on values (x, y), where y is a random number unknown to B. $\sigma_{x,y}$ is issued in a "blind" way such that B learns nothing about x and y.

Spending an E-Coin. U needs to prove to M that he/she is not suspended before M would accept payment from U. Both parties first obtain the current $SUL = \{(t_1, b_1), (t_2, b_2), \ldots, (t_n, b_n)\}$ from suspension manager SM. U and M agree on a unique transaction identifier R and a random value b. U then computes $S = h_0{}^y$, $T = h^x h_1^{Ry}$ and $t = b^x$ and proves the following facts.

1. U knows $\sigma_{x,y}$ which is a valid signature from B on values x, y.
2. S, T, t are formed correctly with respective to x and y.
3. $t_i \neq b_i^x$ for $i = 1$ to n.

Depositing an E-Coin. M submits (S, T, t, R) to B, along with the transcript trans of the spend operation. After checking the transcript, B checks if S is in its database. If yes, it is a coin that has been spent before. If not, it stores (S, T, t) in its database and credits M.

Dealing with Double-Spending. If B (S, T', t', R') is in its database, The public key of the double-spender can be computed as $(\frac{T^{R'}}{T^R})^{\frac{1}{R'-R}}$. Indeed, due to the soundness of the proof in the spend protocol, $T = h^x h_1^{Ry}$ and $T' = h^x h_1^{R'y}$. Thus $(\frac{T^{R'}}{T^R})^{\frac{1}{R'-R}} = ((h^x)^{R'-R})^{\frac{1}{R'-R}} = h^x = u$.

Suspension. To suspend a user, SM appends the value (b, t), in the protocol transcript of a spend operation, to SUL. Note that SM does not know the identity of the user being suspended; he just suspend the user that engage in this transaction. To un-suspend the user, SM removes that entry from SUL.

4.2 Construction Details

We now present our cryptographic construction of EC-AUS.

Parameters. Let λ be a sufficiently large security parameter. Let $(\mathbb{G}_1, \mathbb{G}_2)$ be a bilinear group pair such that $|\mathbb{G}_1| = |\mathbb{G}_2| = p$ for some prime p of λ bits. Also, let $\mathbb{G}$ be a group of order p where DDH Assumption holds. Let $g, g_0, g_1, g_2, g_3 \in \mathbb{G}_1$, $h, h_0, h_1 \in \mathbb{G}$ be generators of $\mathbb{G}_1$ and $\mathbb{G}$ respectively such that the relative discrete logarithm of the generators are unknown.[1] Let $H_0 : \{0,1\}^* \to \mathbb{G}$ and $H : \{0,1\}^* \to \mathbb{Z}_p$ be secure cryptographic hash functions, both of which will be modeled as random oracles.

BSetup, SSetup, KeyGen. The bank B randomly chooses $\gamma \in_R \mathbb{Z}_p$ and computes $w = g_0^\gamma$. The bank secret key is $\mathsf{bsk} = (\gamma)$ and the public key is $\mathsf{bpk} = (w)$. The user U (resp. merchant M) randomly chooses $x \in_R \mathbb{Z}_p$ and computes $u = h^x$. The secret key is $\mathsf{sk} = (x)$ and the public key is $\mathsf{pk} = (u)$. The suspension manager SM initializes the suspended user list SUL.

AccEstablish. User U sends her public key $\mathsf{pk}_U = (u)$ to B, along with the following the zero-knowledge proof-of-knowledge $PK\{(x) : u = h^x\}$ to open an account in the bank. U stores the account secret $\mathsf{cred} = (x)$.

---

[1] This can be done by setting the generators to be the output of a cryptographic hash function of some publicly known seeds. It is important for the users to verify this. For instance, knowledge of the discrete logarithm of h_1 to base h_0 would allow the bank to break the anonymity of the system.

Withdraw

1. U and B retrieve the current SUL from SM and parse SUL as $\{(t_1, b_1), \ldots, (t_n, b_n)\}$.
2. U initializes the request, claims to be the user with public key u who has already registered an account.
3. B sends a random challenge $m \in_R \mathbb{Z}_p$ to U.
4. U sends a pair (C, Π_1) to B, where $C = g_1^x g_2^y g_3^{z'} \in \mathbb{G}_1$ is a commitment of $x, y \in_R \mathbb{Z}_p$ using randomness z' and Π_1 is a signature proof of knowledge of

$$SPK_1 \left\{ (x, y, z') : C = g_1^x g_2^y g_3^{z'} \wedge u = h^x \left(\bigwedge_{i \in [n]} t_i \neq b_i^x \right) \right\} (m) \quad (1)$$

on challenge m, which proves that C is correctly formed.
5. The B returns failure if the verification of Π_1 returns invalid. Otherwise B sends U a tuple (A, e, z''), where $e, z'' \in_R \mathbb{Z}_p$ and $A = (g_0 C g_3^{z''})^{\frac{1}{e+\gamma}} \in \mathbb{G}_1$.
6. U computes $z = z' + z''$. She returns failure if $\hat{e}(A, wg_0^e) \neq \hat{e}(g_0 g_1^x g_2^y g_3^z, g_0)$. Otherwise she stores $cn = (A, e, x, y, z)$ as her electronic coin.

Note that (A, e, z) is a BBS+ signature on values (x, y).

Spend. During an execution of this protocol between a user U and the merchant M, U's private input is her electronic coin $cn = (A, e, x, y, z)$. Let R be the string that uniquely identifies this transaction. In particular, R includes the public key $\mathsf{pk_M}$ of M, the version of SUL used and a random nonce nonce. When the protocol terminates, M outputs success or failure, indicating whether the payment is accepted. Both parties retrieve the current SUL from SM and parse SUL as $\{(t_1, b_1), \ldots, (t_n, b_n)\}$.

1. *(Challenge.)* M sends a random challenge $m \in_R \mathbb{Z}_p$ to U.
2. *(Suspension Check.)* U returns failure if $t_i = b_i^x$ for some i (indicating that she is suspended). She proceeds otherwise.
3. *(Proof Generation.)* U returns to M a tuple (S, T, t, Π_2), where $S = h_0^y$, $T = u h_1^{yR}$, $t = b^x$ where $b = H_0(R)$. S is called the serial number of the coin while T is called a double-spending equation. The pair (S, T) allows the bank to identify the double spender. t is the ticket associated with the transaction which allows suspension. Finally, Π_2 is a signature proof of knowledge of:

$$SPK_2 \left\{ (A, e, x, y, z) : \begin{array}{l} \hat{e}(A, wg_0^e) = \hat{e}(g_0 g_1^x g_2^y g_3^z, g_0) \wedge \\ S = h_0^y \wedge T = h^x (h_1^R)^y \wedge \\ t = b^x \wedge \left(\bigwedge_{i \in [n]} t_i \neq b_i^x \right) \end{array} \right\} (m) \quad (2)$$

on the challenge m.

4. *(Proof Verification.)* M returns `failure` if the verification of Π_2 returns `invalid`. Otherwise it returns `success`.

Deposit. The merchant M submits the tuple (S, T, t, R, Π_2) to B. B first verifies if R contains a fresh nonce `nonce`, the public key $\mathsf{pk_M}$ of M and obtains version of SUL used in this transaction. B then verifies Π_2. It runs through its database of spent coin, which is a list of tuples $(S_i, T_i, t_i, R_i, \Pi_{2,i})$. If S is not equal to any of the S_i, B credits M and appends (S, T, t, R, Π_2) to the list.

If R contains a reused nonce `nonce`, B outputs $\mathsf{pk_M}$.

Otherwise, suppose there exists an entry $(S_j, T_j, t_j, R_j, \Pi_{2,j})$ for some index j in the list such that $S = S_j$, B computes $u^* = (\frac{T^{R_j}}{T_j^R})^{\frac{1}{R_j - R}}$ and outputs $\mathsf{pk^*} = u^*$, $\mathsf{trans_1} = (S_j, T_j, t_j, R_j, \Pi_{2,j})$, $\mathsf{trans_2} = (S, T, t, R, \Pi_2)$ and $\Pi = (\mathsf{trans_1}, \mathsf{trans_2})$, indicating that u^* is the public key of the double spender.

VerGuilt. Since the computation of the identity of the double spender does not require `bsk`, everyone can verify the correctness of the bank's computation based on the two given transcripts.

Suspension. The three algorithms Extract, Add, Remove are all very simple and efficient. $\mathsf{Extract}(\langle S, T, t, R, \Pi_2 \rangle)$ returns the ticket $(t, b = H_0(R))$ in the input transcript. Of course, SM should also verify Π_2 to ensure that the transcript is valid. $\mathsf{Add}(\mathsf{SUL}, (t, b))$ returns $\mathsf{SUL'}$, which is the same as the input SUL, with the input ticket (t, b) appended to it. $\mathsf{Remove}(\mathsf{SUL}, (t, b))$ returns $\mathsf{SUL'}$, which is the same as the input SUL, with all entries equal to the input ticket (t, b) dropped.

Formal security analysis of our construction is presented in Appendix A.

4.3   Efficiency Analysis

We analyze the efficiency of our construction in terms of both time and space/communication complexities. Both complexities are linear in the size of SUL for Withdraw and Spend protocols. Below we analyze the most expensive operation, Spend, in our system.

Assume SUL contains n tickets. A proof Π_2 of SPK_2 consists of 2 $\mathbb{G}_1$ elements, n $\mathbb{G}$ elements and $2n + 10$ $\mathbb{Z}_p$ elements. The total communication complexity for a Spend protocol is thus $n + 1$ ℓ-bit strings, 5 $\mathbb{G}_1$ elements, n $\mathbb{G}$ elements and $2n + 10$ $\mathbb{Z}_p$ elements.

A breakdown of time complexity of the Spend protocol into the number of pairing operations and *multi-exponentiations (multi-EXPs)*[2] in various groups is shown in Table 2. Operations such as $\mathbb{G}$ addition and hashing have been omitted as computing them takes relatively insignificant time. Some preprocessing is possible at the user's side. In fact, all but $2n$ multi-EXPs in $\mathbb{G}$ can be precomputed by the user.

---

[2] A multi-EXP computes the product of exponentiations faster than performing the exponentiations separately. We assume that one multi-EXP operation multiplies up to 3 exponentiations.

Table 2. Number of operations during a Spend protocol with a SUL of size n

Operation	User w/o Preproc.	User w/ Preproc.	Merchant
$\mathbb{G}_1$ multi-EXP	6	0	3
$\mathbb{G}$ multi-EXP	$3n + 6$	$2n$	$2n + 3$
Pairing	1	0	1

5   Discussions

5.1   Incorporating Tracing Authority and Open Authority

Introduction of Open Authority (OA). It is relatively straightforward to introduce an Open Authority (OA) which is capable of revealing the public key of the user of any Spend transaction. One could simply require all users to verifiably encrypt [12] their public key g^x into ciphertext C_x under the public key of the OA in the Spend transaction. This allows the OA to decrypt C_x and obtains the public key of the user participating in the Spend transaction.

Introduction of Tracing Authority (TA). We can also introduce TA in EC-AUS based on the idea of traceable signature [27]. Each user is issued a traceable signature signing key $k_{x,trace}$ from TA such that $k_{x,trace}$ is bind to the user public key g^x. All users are required to create a traceable signature σ_x using his key $k_{x,trace}$ in the Spend transaction. If the TA would like to trace the spending of a particular user, he/she reveal the tracing information of $k_{x,trace}$ so that every one can link the traceable signature σ_x from user with key $k_{x,trace}$ and thus link all his past actions.

Three levels of anonymity revocation. The trio of TA, OA and SM provide a balance between users' privacy and accountability. For instance, when a suspicious transaction is identified, the law-enforcing agent can at once request a suspension from SM on the underlying user. Since SM reveals least information on the user, the threshold of issue could be fairy low. After preliminary investigations, law-enforcing agent could request TA to release the tracing information so that all transaction regarding the suspect can be linked and provide more information for the law-enforcing agent to make further investigation. Finally, he/she could request OA to reveal the identity of the spender for prosecution.

5.2   Managing the Size of SUL and the Bank's Database

Our system does not scale well with the size of the SUL. Thus, we assume that suspended users are eventually un-suspended if they are found innocent or their identity are revealed by the OA for prosecution. This would help keeping the size of SUL a minimum. Using practical parameters, modern computer handles a multi-EXP at around 2 ms. Realistically, EC-AUS would support SUL of size up to several thousands.

Another issue is the requirement that the bank has to keep record of all the electronic coins deposited. One solution is to limit the lifetime of the user account as well as the coins. Users are required to establish a new account and have their electronic coins re-issued at the end of the period. The account and coins are made valid only for a specific period of time, say, a month, several months or a year which offers a trade-off between database size and the frequency of the re-issue. Thus, the bank only needs to record spent coins of the current time period.

6   Conclusion

We presented EC-AUS, an electronic cash system with anonymous user suspension. Since suspended users remain anonymous, misbehavior can be judged subjectively and imposed with less cation. We also discuss how to limit the size of the bank's storage. We believe the ability to suspend users while maintaining their anonymity is a worthwhile endeavor. We left it as an open problem of constructing schemes whose complexities is independent to the size of SUL.

References

1. Au, M.H., Susilo, W., Mu, Y.: Constant-Size Dynamic k-TAA. In: Prisco, R.D., Yung, M. (eds.) SCN 2006. LNCS, vol. 4116, pp. 111–125. Springer, Heidelberg (2006)
2. Au, M.H., Susilo, W., Mu, Y.: Practical Compact E-Cash. In: Pieprzyk, J., Ghodosi, H., Dawson, E. (eds.) ACISP 2007. LNCS, vol. 4586, pp. 431–445. Springer, Heidelberg (2007)
3. Bellare, M.: A Note on Negligible Functions. J. Cryptology 15(4), 271–284 (2002)
4. Bellare, M., Rogaway, P.: Random Oracles are Practical: A Paradigm for Designing Efficient Protocols. In: ACM Conference on Computer and Communications Security, pp. 62–73 (1993)
5. Boneh, D., Boyen, X.: Short Signatures without Random Oracles. In: Cachin, C., Camenisch, J.L. (eds.) EUROCRYPT 2004. LNCS, vol. 3027, pp. 56–73. Springer, Heidelberg (2004)
6. Boneh, D., Boyen, X., Shacham, H.: Short Group Signatures. In: Franklin, M. (ed.) CRYPTO 2004. LNCS, vol. 3152, pp. 41–55. Springer, Heidelberg (2004)
7. Brands, S.: Untraceable Off-line Cash in Wallets with Observers (Extended Abstract). In: Stinson, D.R. (ed.) CRYPTO 1993. LNCS, vol. 773, pp. 302–318. Springer, Heidelberg (1994)
8. Brickell, E., Li, J.: Enhanced Privacy ID: A Direct Anonymous Attestation Scheme with Enhanced Revocation Capabilities. In: WPES, pp. 21–30 (2007)
9. Camenisch, J., Dubovitskaya, M., Neven, G.: Oblivious Transfer with Access Control. In: Al-Shaer, E., Jha, S., Keromytis, A.D. (eds.) ACM Conference on Computer and Communications Security, pp. 131–140. ACM, New York (2009)
10. Camenisch, J., Hohenberger, S., Lysyanskaya, A.: Compact E-Cash. In: Cramer, R. (ed.) EUROCRYPT 2005. LNCS, vol. 3494, pp. 302–321. Springer, Heidelberg (2005)
11. Camenisch, J., Lysyanskaya, A.: A Signature Scheme with Efficient Protocols. In: Cimato, S., Galdi, C., Persiano, G. (eds.) SCN 2002. LNCS, vol. 2576, pp. 268–289. Springer, Heidelberg (2003)

12. Camenisch, J., Shoup, V.: Practical Verifiable Encryption and Decryption of Discrete Logarithms. In: Boneh, D. (ed.) CRYPTO 2003. LNCS, vol. 2729, pp. 126–144. Springer, Heidelberg (2003)

13. Camenisch, J., Stadler, M.: Efficient Group Signature Schemes for Large Groups (Extended Abstract). In: Kaliski Jr., B.S. (ed.) CRYPTO 1997. LNCS, vol. 1294, pp. 410–424. Springer, Heidelberg (1997)

14. Canard, S., Gouget, A.: Divisible E-Cash Systems can be Truly Anonymous. In: Naor, M. (ed.) EUROCRYPT 2007. LNCS, vol. 4515, pp. 482–497. Springer, Heidelberg (2007)

15. Canard, S., Traoré, J.: On Fair E-cash Systems Based on Group Signature Schemes. In: Safavi-Naini, R., Seberry, J. (eds.) ACISP 2003. LNCS, vol. 2727, pp. 237–248. Springer, Heidelberg (2003)

16. Canetti, R.: Universally Composable Security: A New Paradigm for Cryptographic Protocols. Cryptology ePrint Archive, Report 2000/067 (2000), http://eprint.iacr.org/

17. Chan, A.H., Frankel, Y., Tsiounis, Y.: Easy Come - Easy Go Divisible Cash. In: Nyberg, K. (ed.) EUROCRYPT 1998. LNCS, vol. 1403, pp. 561–575. Springer, Heidelberg (1998)

18. Chaum, D.: Blind Signatures for Untraceable Payments. In: Advances in Cryptology: Proceedings of CRYPTO 1982, pp. 199–203. Plenum, New York (1983)

19. Chaum, D., Fiat, A., Naor, M.: Untraceable Electronic Cash. In: Goldwasser, S. (ed.) CRYPTO 1988. LNCS, vol. 403, pp. 319–327. Springer, Heidelberg (1990)

20. Chaum, D., Pedersen, T.P.: Transferred Cash Grows in Size. In: Rueppel, R.A. (ed.) EUROCRYPT 1992. LNCS, vol. 658, pp. 390–407. Springer, Heidelberg (1993)

21. Cramer, R., Damgård, I., MacKenzie, P.D.: Efficient Zero-Knowledge Proofs of Knowledge without Intractability Assumptions. In: Imai, H., Zheng, Y. (eds.) PKC 2000. LNCS, vol. 1751, pp. 354–373. Springer, Heidelberg (2000)

22. Damgård, I.: Efficient Concurrent Zero-Knowledge in the Auxiliary String Model. In: Preneel, B. (ed.) EUROCRYPT 2000. LNCS, vol. 1807, pp. 418–430. Springer, Heidelberg (2000)

23. Eng, T., Okamoto, T.: Single-Term Divisible Electronic Coins. In: De Santis, A. (ed.) EUROCRYPT 1994. LNCS, vol. 950, pp. 306–319. Springer, Heidelberg (1995)

24. Franklin, M.K., Yung, M.: Secure and Efficient Off-Line Digital Money (Extended Abstract). In: Lingas, A., Carlsson, S., Karlsson, R. (eds.) ICALP 1993. LNCS, vol. 700, pp. 265–276. Springer, Heidelberg (1993)

25. Goldwasser, S., Micali, S., Rackoff, C.: The Knowledge Complexity of Interactive Proof-Systems (Extended Abstract). In: STOC, pp. 291–304 (1985)

26. Goldwasser, S., Micali, S., Rivest, R.L.: A digital signature scheme secure against adaptive chosen-message attacks. SIAM J. Comput. 17(2), 281–308 (1988)

27. Kiayias, A., Tsiounis, Y., Yung, M.: Traceable Signatures. In: Cachin, C., Camenisch, J. (eds.) EUROCRYPT 2004. LNCS, vol. 3027, pp. 571–589. Springer, Heidelberg (2004)

28. Okamoto, T., Ohta, K.: Universal Electronic Cash. In: Feigenbaum, J. (ed.) CRYPTO 1991. LNCS, vol. 576, pp. 324–337. Springer, Heidelberg (1992)

29. Tsang, P.P., Au, M.H., Kapadia, A., Smith, S.W.: Blacklistable Anonymous Credentials: Blocking Misbehaving Users without TTPs. In: ACMCCS 2007, pp. 72–81 (2007)

A   Formal Security Analysis

A.1   Security Model

We use a simulation-based approach to define security of EC-AUS formally. We would like to remark that the definition we give do not entail all formalities necessary to fit into the universal composability framework [16]; our goal here is to prove security of our construction. Our model is static in the sense that the adversary could not corrupt honest users and merchants during the execution of the system.

We summarize the ideas of the model. The players in the system are the suspension manager SM, the bank B, a set of users Us and a set of merchants Ms. In the real world there are a number of players who communicate via cryptographic protocols. Then there is an adversary $\mathcal{A}$, who controls the dishonest players in the system. We define an entity called environment, $\mathcal{E}$, who provides the inputs to the players and receives their outputs. $\mathcal{E}$ also interacts freely with the adversary $\mathcal{A}$.

In the ideal world, we have the same players. However, they do not communicate directly. Rather, there exists a trusted party $\mathcal{T}$ who is responsible for all handling operations for all players. Specifically, $\mathcal{T}$ computes the outputs of the players from their inputs, that is, applies the functionality that the cryptographic protocols are supposed to realize. The environment $\mathcal{E}$ again provides the inputs to, and receives the outputs from, the players, and interacts arbitrarily with $\mathcal{A}$ who controls the dishonest players.

The ideal world. First we define the ideal world specification of EC-AUS. Communication between a player and the trusted party $\mathcal{T}$ is not anonymous. Ideal world EC-AUS supports the following operations. These operations are scheduled according to the environment $\mathcal{E}$'s wish. Each call to the operation is assigned a unique identifier tid. We also describe the behavior of $\mathcal{T}$ based on the inputs of the ideal world players for the following operations. The SUL in the ideal world is a list of tid of Spend operation.

- $\text{tid}_0 \leftarrow$ SSetup/BSetup/KeyGen($\mathcal{HP}, \mathcal{AP}$). The system begin when $\mathcal{E}$ invokes this operation which specified the set of honest players $\mathcal{HP}$ and dishonest players $\mathcal{AP}$. This must be the first operation in the schedule and can only be called once.

- $\text{tid}_A \leftarrow$ AccEstablish(i). $\mathcal{E}$ instructs user U_i to establish an account with bank B. U_i sends a request to $\mathcal{T}$, $\mathcal{T}$ checks U_i has never established an account before and informs B that U_i would like to establish an account. B returns accept/reject to $\mathcal{T}$ and $\mathcal{T}$ forward it to U_i. Both U_i and B output $(\text{tid}_A, \text{accept/reject})$ to $\mathcal{E}$ individually.

- $\text{tid}_W \leftarrow$ Withdraw(i). $\mathcal{E}$ instructs user U_i to withdraw an electronic coin from B. U_i sends a request to $\mathcal{T}$, $\mathcal{T}$ requests the current version of SUL from SM and forward SUL to U_i, along with a check result that indicate if U_i is suspended or not. U_i replies to $\mathcal{T}$ if he/she chooses to proceed or not. $\mathcal{T}$ requests the same version of SUL from SM, check if U_i has ever participated in the Spend specified in this SUL and forwards SUL to B, a bit indicating if U_i is suspended and the request that U_i would like to withdraw an electronic coin. B returns `accept/reject` to $\mathcal{T}$ and $\mathcal{T}$ forwards it to U_i. If B returns `accept`, $\mathcal{T}$ stores $\mathtt{tid}_W$ as U_i's un-spent coin. Both U_i and B output $(\mathtt{tid}_W, \mathtt{accept/reject})$ to $\mathcal{E}$ individually.

- $\mathtt{tid}_S \leftarrow$ Spend$(i, \mathtt{tid}_W, j)$. $\mathcal{E}$ instructs user U_i to spend the electronic coin he/she obtains in transaction $\mathtt{tid}_W$ to merchant M_j. U_i sends a request to $\mathcal{T}$, $\mathcal{T}$ requests the current version of SUL from SM and forward SUL to U_i, along with a check result that indicate if U_i is suspended or not as well as whether $\mathtt{tid}_W$ corresponds to an un-spent coin of U_i. U_i replies to $\mathcal{T}$ if he/she chooses to proceed or not. $\mathcal{T}$ requests the same version of SUL from SM, check if U_i is suspended and forwards SUL to M_j, the request that an anonymous user that would like to spend a coin to M_j, and whether this user is suspended or not and whether U_i is having a valid coin (valid means $\mathtt{tid}_W$ corresponds to a Withdraw that U_i participated in, it might be a spent-coin though). M_j returns `accept/reject` to $\mathcal{T}$ and $\mathcal{T}$ forwards it to U_i. If M_j returns `accept`, $\mathcal{T}$ marked $\mathtt{tid}_W$ as U_i's spent coin. Both U_i and M_j output $(\mathtt{tid}_S, \mathtt{accept/reject})$ to $\mathcal{E}$ individually.

- $\mathtt{tid}_D \leftarrow$ Deposit$(j, \mathtt{tid}_S)$. $\mathcal{E}$ instructs user M_j to deposit the electronic coin he/she obtains in transaction $\mathtt{tid}_S$. M_j sends a request to $\mathcal{T}$, $\mathcal{T}$ requests the version of SUL used during Spend of $\mathtt{tid}_S$ from SM and forward SUL to B, along with a check result that indicate if $\mathtt{tid}_S$ corresponds to a Spend$i, \mathtt{tid}_W, j$ that results in M_j outputting `accept` and that U_i is not suspended based on SUL. Next $\mathcal{T}$ also informs B if $\mathtt{tid}_W$ corresponds to a deposited-coin. If yes, $\mathcal{T}$ gives B an identity, U_i or M_j, indicating if the coin is spent twice by U_i or deposited twice by M_j. If not, $\mathcal{T}$ marks $\mathtt{tid}_W$ as a deposited coin from M_j. Both B and M_j output $(\mathtt{tid}_D, \mathtt{accept/reject}(U_i/M_j))$ to $\mathcal{E}$ individually.

- $\mathtt{tid}_V \leftarrow$ VerGuilt$(P, \mathtt{tid}_D, U_i/M_j)$. $\mathcal{E}$ instructs any player P to query if B outputs a correct double-spender in transaction $\mathtt{tid}_D$. $\mathcal{T}$ replies with a bit, indicating if the correct double-spender is outputted in $\mathtt{tid}_D$.

- $\mathtt{tid}_{Sus} \leftarrow Suspend(\mathtt{tid}_\S)$. $\mathcal{E}$ instructs SM to add the Spend identified by $\mathtt{tid}_S$ to SUL.

- $\mathtt{tid}_{Un-Sus} \leftarrow Un-Suspend(\mathtt{tid}_\S)$. $\mathcal{E}$ instructs SM to removes the entry $\mathtt{tid}_S$ from SUL.

Ideal world EC-AUS provides all the desired security properties. Firstly, all Spend transaction are anonymous. $\mathcal{T}$ only informs and M a certain anonymous user would like to spend an e-coin. Thus, anonymity and exculpability is guaranteed. Secondly, $\mathcal{T}$ verifies if validity of the user during Withdraw, Spend and Depositand thus balance is assured. Finally, $\mathcal{T}$ consults SM for SUL and checks if the underlying user is suspended for the B and M and thus suspension-correctness is attained.

Next, we define a cryptographic EC-AUS which also supports the above eight types of transaction. Since there is no trusted party $\mathcal{T}$, the functionalities are realized through cryptographic means. Below we highlight the difference.

- $\text{tid}_0 \leftarrow$ SSetup/BSetup/KeyGen($\mathcal{HP}, \mathcal{AP}$). SM, B, Us and Ms invokes the respective algorithms SSetup, BSetup and KeyGen.
- $\text{tid}_A \leftarrow$ AccEstablish(i). U_i and B obtains the current version of SUL from SM individually and engage in the AccEstablish protocol.
- $\text{tid}_W \leftarrow$ Withdraw(i). U_i and B obtains the current version of SUL from SM individually and engage in the Withdraw protocol.
- $\text{tid}_S \leftarrow$ Spend(i, tid_W, j). U_i and M obtains the current version of SUL from SM individually and engage in the Spend protocol.
- $\text{tid}_D \leftarrow$ Deposit(j, tid_S). M_j and B engage in the Deposit protocol in which B obtains from SM the version of SUL used in the Spend protocol identified by tid_S.
- $\text{tid}_V \leftarrow$ VerGuilt($P, \text{tid}_D, U_i/M_j$). P interacts with B who proves to P the identity of the double-spender is correctly computted.

Informally speaking, a cryptographic system is secure if for every real world adversary $\mathcal{A}$ and every environment $\mathcal{E}$, there exists an ideal world adversary $\mathcal{S}$ controlling the same players in the ideal world as $\mathcal{A}$ does in the real world such that, $\mathcal{E}$ cannot tell whether it is running in the real world interacting with $\mathcal{A}$ or it is running in the ideal world interacting with $\mathcal{S}$ which has blackbox access to $\mathcal{A}$. The rationale is that since by default the ideal world EC-AUS is secure, and the real world EC-AUS is indistinguishable to the ideal world EC-AUS, the real world EC-AUS is also secure. Formally, we define it in Definition 1.

Definition 1 (Security). *Let* $\mathbf{Real}_{\mathcal{E},\mathcal{A}}(\lambda)$ *(resp.* $\mathbf{Ideal}_{\mathcal{E},\mathcal{S}_A}(\lambda)$ *) be the probability that* $\mathcal{E}$ *outputs 1 when run in the real world (resp. ideal world) with adversary* $\mathcal{A}$ *(resp.* $\mathcal{S}$ *having blackbox access to* $\mathcal{A}$*). A EC-AUS construction is secure if*

$$|\mathbf{Real}_{\mathcal{E},\mathcal{A}}(\lambda) - \mathbf{Ideal}_{\mathcal{E},\mathcal{S}_A}(\lambda)| = \mathsf{negl}(\lambda)$$

for every PPT algorithms $\mathcal{E}$*,* $\mathcal{A}$*.*

A.2   Security Analysis

The security of our EC-AUS construction depends on the following two assumptions:

Definition 2 (DDH). *The Decisional Diffie-Hellman (DDH) problem in group* $\mathbb{G}$ *is defined as follows: On input of a quadruple* $(g, g^a, g^b, g^c) \in \mathbb{G}^4$*, output 1 if* $c = ab$ *and 0 otherwise. We say that the DDH assumption holds if no probabilistic polynomial time (PPT) algorithm has non-negligible advantage over random guessing in solving the DDH problem.*

Definition 3 (q-SDH). *The q-Strong Diffie-Hellman (q-SDH) problem in* $\mathbb{G}$ *is defined as follows: On input of a* $(q+1)$*-tuple* $(g, g^x, g^{x^2}, \ldots, g^{x^q}) \in \mathbb{G}$*, output* a pair $(A, e) \in \mathbb{G} \times \mathbb{Z}_p$ such that $A^{(x+e)} = g$ where $|\mathbb{G}| = p$. We say that the q-SDH assumption holds if no PPT algorithm has non-negligible advantage in solving the q-SDH problem.

The q-SDH assumption was introduced by Boneh and Boyen [5] when they proposed a new short signature. They derived a lower bound on any generic algorithm that solves the q-SDH problem.

Regarding the security of EC-AUS, we have the following theorem.

Theorem 1. *If the q-SDH assumption holds in $\mathbb{G}_1$ and the DDH assumption holds in $\mathbb{G}$, our construction of* EC-AUS *satisfies Definition 1 in the random oracle model.*

Proof of Theorem 1 is done by showing the indistinguishability between adversary actions in the real world and the ideal world. The idea of the proof is that, given a real world adversary $\mathcal{A}$, we show how to construct an ideal world adversary $\mathcal{S_A}$[3] such that no environment $\mathcal{E}$ can distinguish whether it is interacting with $\mathcal{A}$ or $\mathcal{S}$. The proof is divided into three cases according to the subset of players controlled by $\mathcal{A}$. In the first case, $\mathcal{A}$ controls the SM, a subset of merchants and users. This covers the security requirement of balance. In the second case, $\mathcal{A}$ controls SM, the bank, a subset of merchants and users. This covers the security requirement of exculpability and anonymity. In the third case, $\mathcal{A}$ control a subset of merchants and users. This covers the security requirement of suspension-correctness. We would like to remark that the three cases are orthogonal because on one hand, $\mathcal{S}$ has to represent all honest players to $\mathcal{A}$, while on the other hand $\mathcal{S}$ has to represent all dishonest players to $\mathcal{E}$. Thus, an adversary $\mathcal{A}$ controlling fewer parties does not necessarily makes the construction of $\mathcal{S}$ easier. We complete the proof with the following three lemmas.

Lemma 1. *For any environment $\mathcal{E}$ and real world adversaries $\mathcal{A}$ controlling the* SM, *some subsets of merchants and users, there exists an ideal world simulator $\mathcal{S_A}$ such that*

$$|\mathbf{Real}_{\mathcal{E},\mathcal{A}}(\lambda) - \mathbf{Ideal}_{\mathcal{E},\mathcal{S_A}}(\lambda)| = \mathsf{negl}(\lambda)$$

Proof. (Sketch) We construct $\mathcal{S}$ as follow. On one hand, $\mathcal{S}$ represents the honest merchants Ms, users Us and the bank B to $\mathcal{A}$ while on the other hand, $\mathcal{S}$ represents the dishonest Us, Ms and SM to $\mathcal{T}$ as well as $\mathcal{E}$ based on the actions from $\mathcal{A}$. $\mathcal{S}$ forwards all the messages between between $\mathcal{E}$ and $\mathcal{A}$. Next, for all AccEstablish involving a dishonest user, $\mathcal{S}$, playing the role of B, extracts the secret x from $\mathcal{A}$ and uses x as an index for the underlying dishonest user. It then represents that dishonest user to $\mathcal{T}$ and initiates an AccEstablish request. For all Withdraw involving a dishonest user, $\mathcal{S}$ extracts the values (x, y, z') from Eq.1. For all Spend events involving a dishonest user, $\mathcal{S}$, playing the role of an honest merchant, runs through its list of (x, y, z) extracted and locate the user by testing if $S = h_0^y$ and $t = b^x$ and locate the tid of the corresponding Withdraw

---
[3] The subscript $\mathcal{A}$ is used to emphasis that $\mathcal{S}$ is given blackbox access to $\mathcal{A}$.

event in the ideal world. It then represents that dishonest user to $\mathcal{T}$ and initiates a Spend request.

$\mathcal{S}$'s behavior in the view of $\mathcal{E}$ is exactly the same as $\mathcal{A}$ would provide, except in the case when $\mathcal{S}$ cannot extract the values from $\mathcal{A}$, or the extracted values do not matches with previously extracted one. This represents $\mathcal{A}$ is able to break the soundness of the various zero-knowledge proves, which happens with negligible probability under the q-SDH Assumption.

Lemma 2. *For any environment $\mathcal{E}$ and any real world adversaries $\mathcal{A}$ controlling SM, the bank, some subsets of merchants and users, there exists an ideal world simulator $\mathcal{S}_{\mathcal{A}}$ such that*

$$|\mathbf{Real}_{\mathcal{E},\mathcal{A}}(\lambda) - \mathbf{Ideal}_{\mathcal{E},\mathcal{S}_{\mathcal{A}}}(\lambda)| = \mathsf{negl}(\lambda)$$

Proof. (Sketch) Construction of such $\mathcal{S}$ is straightforward. $\mathcal{S}$ forwards all the messages between $\mathcal{E}$ and $\mathcal{A}$. For all events when $\mathcal{S}$ has to represent an honest user, $\mathcal{S}$ employs the zero-knowledge simulator to simulate the proofs using a random and different (x, y). $\mathcal{S}$'s behavior in the view of $\mathcal{E}$ is exactly the same as $\mathcal{A}$ would provide, except in the case when $\mathcal{A}$ is able to break the zero-knowledgeness. This happens with negligible probability under the DDH Assumption.

Lemma 3. *For any environment $\mathcal{E}$ and any real world adversaries $\mathcal{A}$ controlling some subsets of merchants and users, there exists an ideal world simulator $\mathcal{S}_{\mathcal{A}}$ such that*

$$|\mathbf{Real}_{\mathcal{E},\mathcal{A}}(\lambda) - \mathbf{Ideal}_{\mathcal{E},\mathcal{S}_{\mathcal{A}}}(\lambda)| = \mathsf{negl}(\lambda)$$

Proof. (Sketch) Construction of such $\mathcal{S}$ is straightforward. Again, $\mathcal{S}$ forwards all the messages between $\mathcal{E}$ and $\mathcal{A}$. For all events when $\mathcal{S}$ has to deal with dishonest user, $\mathcal{S}$ extracts the user secret x from the zero-knowledge proofs. $\mathcal{S}$'s behavior in the view of $\mathcal{E}$ is exactly the same as $\mathcal{A}$ would provide, except in the case when $\mathcal{S}$ cannot extract the values from $\mathcal{A}$, or the extracted values do not matches with previously extracted one. This represents $\mathcal{A}$ is able to break the soundness of the various zero-knowledge proves, which happens with negligible probability under the q-SDH Assumption.

T-Robust Scalable Group Key Exchange Protocol with $O(\log n)$ Complexity

Tetsuya Hatano, Atsuko Miyaji*, and Takashi Sato

Japan Advanced Institute of Science and Technology
miyaji@jaist.ac.jp

Abstract. Group key exchange (GKE) allows a large group of n parties to share a common secret key over insecure channels. The goal of this paper is to present T-robust scalable GKE with communicational and computational complexity $O(\log n)$ for the size of n parties. As a result, our GKE not only has a resistance to party failures resulting from party crashes, run-down batteries, and network failures, but also satisfies scalability: each party does not need to have the same environment such as computational resources, batteries, etc. The previous schemes in this area focus on Burmester-Desmedt GKE with complexity $O(n)$ (BDI) and without scalability. As a result, the previous robust GKEs, proposed by Jarecki, Kim and Tsudik (JKT), need computational complexity $O(n)$ without scalability although it allows any T-party fault in any position.

We, by focusing the well-known Burmester-Desmedt GKE with complexity $O(\log n)$ (BDII), propose a new robust GKE with scalability, called CH-GKE. CH-GKE can *reduce the communicational and computational complexity and allow parties be in different environments*. Then, we extend CH-GKE to increase the number of faults and present T-robust scalable efficient GKE by a novel combination of CH-GKE and JKT. Our T-robust scalable GKE can work in flexible settings between fault tolerance and efficiency, such as communicational and computational complexity.

Keywords: group key exchange, robustness, scalability.

1 Introduction

A group key exchange protocol (GKE) allows a large group of n parties to share a common secret key over an insecure channel, and thus, parties in the group can encrypt and decrypt messages among group members. Secure communication within a large group has become an integral part of many applications. For example, ad hoc wireless networks are deployed in many areas such as homes, schools, disaster areas, etc., where a network is susceptible to attacks ranging from passive eavesdropping to active interference. Besides ad hoc networks, another environment where ad hoc groups are popular is in the context of new emerging social networks such as Facebook and LinkedIn.

Widely-known GKEs based on the DH-key exchange protocol, such as BDI [4] and BDII [5], can work with constant rounds. The important difference in efficiency between BDI and BDII is that BDI needs communicational complexity $O(n)$ while BDII

* This study is partly supported by Grant-in-Aid for Scientific Research (A), 21240001.

U. Parampalli and P. Hawkes (Eds.): ACISP 2011, LNCS 6812, pp. 189–207, 2011.

works with only communicational complexity $O(\log n)$. Another important difference in practicality between BDI and BDII is scalability. BDI assumes that all parties work in the same environment. On the other hand, parties in BDII can work in different environments. For example, some parties may have large computational resources, but others may have low resources; while some parties may have almost unlimited electrical power, others may run on small batteries. On the other hand, both schemes are not robust: if some parties fail during protocol execution, then some other parties cannot share a common secret key. Therefore, the protocol must be re-started from scratch whenever a player fails, which increases the computational, communicational, and round complexity, since total complexity of protocols is multiplied by the number of faults. This is why we need a constant-round GKE that is robust to some parties' failures.

The first robust GKE was proposed by [1], which needs round complexity $O(n)$. Subsequently, the constant-round robust GKE was proposed by [6], which needs communicational complexity $O(n^2)$. The efficient robust GKE, called JKT in this paper, was proposed by [9]. JKT works with constant-round complexity and both communicational[1] and computational complexity $O(n + T)$, and tolerates up to T-party failures. JKT has a useful feature of *flexible trade-off* between complexity and fault tolerance. The feature is practical because complexity of GKE can be arranged according to the reliability of network. R-TDH1 [3] achieves the full robustness for a tree-based GKE [10], however, it does not have a feature of flexible trade-off between complexity and fault tolerance.

In this paper, we focus on JKT, which is T-robust GKE and satisfies a feature of flexible trade-off between complexity and fault tolerance. JKT is constructed by adding robustness to BDI, and thus, it inherits all features described above from BDI: it can achieve neither $O(\log n + T)$ complexity nor scalability. We present T-robust GKE among n parties, which can achieve efficient communicational and computational complexity as well as scalability up to T-party failures. From the feature of scalability, our T-robust GKE can work in different environments of parties with large resources and/or low resources. Our T-robust GKE can work with both computational and communicational complexity $O(\log n + T)$. This is the first result that can work with $O(\log n)$ complexity according to the size of parties n. Let us explain how we construct the efficient T-robust GKE. In order to achieve such efficiency, we investigate adding robustness to BDII and construct a new scalable constant-round robust GKE, which is secure in the standard model under the Square Decision Diffie-Hellman assumption. The proposed GKE, called *CH-GKE* in this paper, inherits efficiency and scalability described above from BDII. For example, CH-GKE works with 2 round complexity with communicational and computational complexity $O(\log n)$, which tolerates up to $\frac{n}{2}$ party failures. Then, we generalize the construction of CH-GKE to increase the number of party failures. Finally, we combine both CH-GKE and JKT, and propose T-robust GKE with $O(\log n)$ complexity, where it tolerates up to any T-party failures in any positions.

This paper is organized as follows. Section 2 summarizes computational assumptions, security assumptions and definitions of GKE, together with notations. Section 3

---

[1] Communicational complexity can be measured from the point of view of maximum number of sent or received messages. In JKT, maximum number of sent messages is $O(T)$ but that of received messages is $O(n + T)$.

reviews the previous GKEs related with our scheme BDI, JKT, and BDII. Section 4 presents CH-GKE and its generalization. Then, *T*-robust GKE is presented in Section 5. Section 6 compares our *T*-robust GKE with previous GKEs.

2 Preliminary

This section summarizes notations, assumptions, and the basic security notions used in this paper.

2.1 The Security Assumptions, and Model of GKE

Let G be a cyclic group of prime order p and let k be a security parameter.

Definition 1. A DDH (Decision Diffie Hellman) parameter generator $I G_{DDH}$ is a probabilistic polynomial time (PPT) algorithm that, on input 1^k, outputs a cyclic group G of prime order p. The *DDH problem* with respect to $I G_{DDH}$ is: given g, g^a, g^b, $y \in G$ to decide whether $y = g^{ab}$ or not.

Definition 2. A Square-DDH (Square-Decision Diffie Hellman) parameter generator $I G_{Square-DDH}$ is a probabilistic polynomial time (PPT) algorithm that, on input 1^k, outputs a cyclic group G of prime order p. The *Square-DDH problem* with respect to $I G_{Square-DDH}$ is: given g, g^x, $y \in G$ to decide whether $y = g^{x^2}$ or not.

Definition 3. Let $P_1, P_2, \dots, P_n$ be interactive polynomial-time Turing Machines with history tapes that take part in a Protocol Π. Protocol Π is a *Group Key Exchange* (GKE) if each member P_i computes the same key $K = K_i$ when all group members follow the protocol as specified. We call P_i a party and the n parties a group.

Definition 4 ([11]). Let Π be a GKE protocol with n parties, let k be a security parameter, and let $\mathcal{P} = \{P_1, \cdots, P_n\}$ be a set of n parties, where n is bounded above by a polynomial in k. We assume that parties do not deviate from the protocol. An active adversary[2] is given access to the following Send, Execute, Reveal, Corrupt and Test oracles, where all except Test oracle are queried several times, and Test oracle is asked only once at any time during the adversary's execution and on a fresh instance[3]: Send(P, i, m) sends message m to instance Π_P^i, and outputs the reply generated by this instance; Execute$(P_1, i_1, \cdots, P_n, i_n)$ executes the protocol between unused i_j-th instances of party P_j, $\{\Pi_{P_j}^{i_j}\}_{1 \le j \le n}$, and outputs the transcript of the execution; Reveal(P, i) outputs a session key sk_P^i for a terminated instance Π_P^i; Corrupt(P) outputs the long-term secret key of a party P; Test(P, i) chooses a bit $b \in \{0, 1\}$ uniformly at random and outputs sk_P^i or a random session key if $b = 1$ or $b = 0$, respectively.

---

[2] We follow the security model in [11] except deleting Corrupt and Send oracles, and those related definitions. Because our protocol does not give a long-term secret key to any party P.
[3] Π_P^i is a fresh instance unless the following is true: $\mathcal{A}$, at some point, queried Reveal (P, i) or Reveal (P', j) with $P' \in pid_P^i$, where pid_P^i denotes the party identity for Π_P^i.

Finally, the adversary outputs a guess bit b'. Then, Succ, the event in which $\mathcal{A}$ wins the game for a protocol Π, occurs if $b = b'$ where b is the hidden bit used by the Test oracle. The advantage of $\mathcal{A}$ is defined as $\mathsf{Adv}_\Pi(k) = |\mathrm{Prob}[\mathsf{Succ}] - 1/2|$. We say Π is a secure group key exchange protocol against an active adversary, if, for any PPT active adversary $\mathcal{A}$, $\mathsf{Adv}_\Pi^{\mathsf{KE}}$ is negligible (in k).

In the passive adversary model, Send oracle is ignored. We focus solely on the passive case since the Katz-Yung compiler [11] or a variant of [7] transforms any GKE secure against a passive adversary into one secure against outside active adversaries.

2.2 Notation and Assumptions on GKE

We make some assumptions necessary to compute the computational complexity. The GKE we will build consists of multiplications on $\mathbf{G}$, scalar multiplications on $\mathbf{G}$, and inversions on $\mathbf{G}$, whose computational complexity are denoted by M, EM, and I, respectively.

This paper focuses on GKE which can be robust against some number of node faults, while keeping both communicational and computational complexity per party down. Let us first make some observations on GKE. In this paper, when we evaluate the communicational complexity per party, it is from the point of view of the party with the maximum *sent* and *received* data. We distinguish between point-to-point and broadcast communication, while we do not distinguish between multicast and broadcast communication. We use $\mathbf{p}$ (resp. $\mathbf{b}$) to denote messages in $\mathbf{G}$ through point-to-point (resp. broadcast) communication, both of which are investigated in two cases of sent and received messages. The computational complexity is measured by the number of M, EM, and I.

We also use the phrase of "*auxiliary elements*", introduced in [8]. In some GKEs, some parties compute data, which help other parties compute a shared key. That is, those parties cannot compute a shared key without auxiliary elements". Actually, failures happen for those parties who need auxiliary elements to compute a shared key but those are not sent to them. In order to achieve a GKE robust against party faults, we will discuss how we provide auxiliary elements in spite of fault parties.

3   Background

This section summarizes previous GKEs: BDI [4], its fault-tolerant version [9], and BDII [5]. In BDI, parties are arranged in a ring (See Figure 3.1). When n parties P_1, $P_2, \cdots, P_n$ wish to generate a session key, they proceed as follows (the indices are taken modulo n so that party P_0 is P_n and party P_{n+1} is P_1).

Protocol 1 (BDI[4])

1. *Each P_i computes $z_i = g^{r_i}$ for a secretly chosen $r_i \in \mathbb{Z}_p^*$ and sends it to P_{i-1} and P_{i+1}.*
2. *Each P_i computes $x_{[i-1,\,i+1]} = \left(\frac{z_{i+1}}{z_{i-1}}\right)^{r_i} = g^{r_i r_{i+1} - r_{i-1} r_i}$ and broadcasts.*
3. *Each P_i computes a shared key $K = (z_{i-1})^{n r_i} \cdot x_{[i-1,\,i+1]}^{n-1} \cdot x_{[i,\,i+2]}^{n-2} \cdots$*
   $x_{[i-3,\,i-1]} = g^{r_1 r_2 + r_2 r_3 + \cdots + r_n r_1}$.

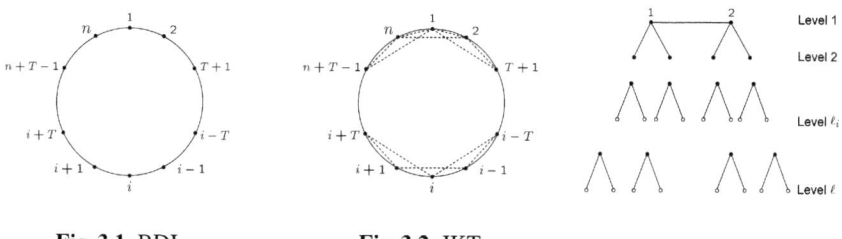

Fig. 3.1. BDI   **Fig. 3.2.** JKT   **Fig. 3.3.** BDII

The original BDI is not robust because if any message in the 2nd round is not delivered, then all parties abort, since all parties need auxiliary elements broadcasted in the 2nd round to compute a shared key. BDI is modified to achieve robustness in [9], which is called JKT in this paper. JKT uses Hamilton cycle or Hamilton path to compute a shared key. The maximum failures as well as the key computation in the Hamilton-cycle JKT are different from ones in the Hamilton-path JKT when both compute the same amount of auxiliary elements in the 2nd round. Here we present both Hamilton-cycle and Hamilton-path JKT in the case of sending the same amount of auxiliary elements and discuss each differences (see Figure 3.2).

Protocol 2 (JKT[9])

1. *Each P_i computes $z_i = g^{r_i}$ for a secretly chosen $r_i \in \mathbb{Z}_p^*$ and broadcasts.*
2. *Let* ActiveList$_1$ *be the list of indices of all parties who complete the 1st round. Each P_i computes $x_{[k,\,i]} = \left(\frac{z_i}{z_k}\right)^{r_i} = g^{r_i^2 - r_k r_i}$ for T nearest neighbors to the right and T nearest neighbors to the left among parties $k \in$ ActiveList$_1$ and broadcasts.*
3. *Let* ActiveList$_2$ *be the list of indices of all parties who complete the 2nd round. Each P_i sorts the parties in* ActiveList$_2$ *in the same order. We assume that the alive parties constructs a Hamilton cycle or Hamilton path taken twice: $\{P_{a_1}, P_{a_2}, \cdots, P_{a_m}\}$ or $\{P_{a_1}, \cdots, P_{a_{m-1}}, P_{a_m}, P_{a_{m-1}}, \cdots, P_{a_2}\}$, respectively. In the case of Hamilton cycle or path, each P_{a_i} computes a shared key $K = z_{a_{i-1}}^{m \cdot r_{a_i}} \cdot X_{a_i}^{m-1} \cdot X_{a_{i+1}}^{m-2} \cdots X_{a_{i-2}} = g^{r_{a_1} r_{a_2} + r_{a_2} r_{a_3} + \cdots + r_{a_m} r_{a_1}}$ or $K = z_{a_{i-1}}^{(2m-2) \cdot r_{a_i}} \cdot X_{a_i}^{2m-3} \cdot X_{a_{i+1}}^{2m-4} \cdots X_{a_{i-2}} = g^{2(r_{a_1} r_{a_2} + r_{a_2} r_{a_3} + \cdots + r_{a_{m-1}} r_{a_m})}$, respectively. Here, $X_{a_i} = x_{[a_{i-1},\,a_i]} \cdot (x_{[a_{i+1},\,a_i]})^{-1} = g^{r_{a_i} r_{a_{i+1}} - r_{a_{i-1}} r_{a_i}}$.*

Let us discuss how many auxiliary elements in the 2nd round is necessary to achieve *T*-robust GKE. *T*-robust GKE means that GKE toterates all patterns of party faults up to *T*. In the case of Hamilton cycle, $2(T + 1)$ auxiliary elements are necessary for *T*-robust GKE. In the case of Hamilton path, $2T$ auxiliary elements are enough to achieve *T*-robust GKE. The detailed comparisons among two types of JKT and our scheme will be shown in Section 6. The security of JKT is given in the theorem below.

Theorem 1 ([9]). Assuming the Square-DDH over $\mathbb{G}$ is hard, JKT is a secure group GKE protocol.

Another typical GKE is BDII, proposed by the same authors as BDI. Both BDI and BDII have different features. BDI is fully contributory, but requires $O(n)$ computational and message complexity for any party. In fact, all parties are arranged symmetrically, and thus all parties need to have the same computational resources. On the other hand,

BDII is not contributory, but can work with $O(\log n)$ computational and message complexity for any party. Furthermore, parties are not arranged symmetrically, and thus BDII can adapt to the situation of parties with different computational resources.

Up to now, no fault-tolerant version of BDII has been proposed. In order to achieve robust GKE with $O(\log n)$ message size, we focus on BDII in this paper. In BDII, parties are arranged in a binary tree (See Figure 3.3). Therefore, all but the leaves of the tree, each has one parent and two children. We denote the parent, the left child, and the right child by $\texttt{parent}(i)$, $\texttt{l.child}(i)$, and $\texttt{r.child}(i)$, respectively, and denote the set of ancestors of a party P_i by $\texttt{ancestor}(i)$. Parties P_1 and P_2 are parents to each other, that is, P_1 (resp. P_2) is the parent of P_2 (resp. P_1). Such a relation is used to compute $x_{\texttt{l.child}(i)}$ and $x_{\texttt{r.child}(i)}$ for $i = 1$ or 2. However, for a party P_i, either P_1 or P_2 is included in $\texttt{ancestor}(i)$.

When n parties $P_1, P_2, \cdots, P_n$ wish to generate a session key, they proceed as follows.

Protocol 3 (BDII[5])

1. *Each P_i computes $z_i = g^{r_i}$ for a secretly chosen $r_i \in \mathbb{Z}_p^*$ and sends it to its neighbors.*
2. *Each P_i computes both* $x_{\texttt{l.child}(i)} = \left(\dfrac{z_{\texttt{parent}(i)}}{z_{\texttt{l.child}(i)}} \right)^{r_i} = g^{r_{par(i)}r_i - r_i r_{l.child(i)}}$ *and*
   $x_{\texttt{r.child}(i)} = \left(\dfrac{z_{\texttt{parent}(i)}}{z_{\texttt{r.child}(i)}} \right)^{r_i}$
   $= g^{r_{par(i)}r_i - r_i r_{r.child(i)}}$ *and multicasts these to its left and right descendants, respectively.*
3. *Each P_i computes a shared key $K = (z_{\texttt{parent}(i)})^{r_i} \cdot \Pi_{j \in \texttt{ancestor}(i)} x_j = g^{r_1 r_2}$.*

Theorem 2 ([7]). Assuming the DDH over $\mathbb{G}$ is hard, BDII is a secure group GKE protocol.

4   Robust GKE with $O(\log n)$ complexity

This section presents our GKE with robustness with $O(\log n)$ complexity, called *Cross-Help GKE (CH-GKE)*. We start with intuition on how to make BDII robust, then presents CH-GKE and its generalization.

4.1   Intuition

Our robust GKE is constructed over BDII in Section 3. Let us discuss why BDII is not robust in detail. In BDII, a party P_i in level ℓ_i makes two auxiliary elements for two children in level $\ell_i + 1$. In the key-sharing phase, a party P_i computes a shared key by using all auxiliary elements sent by ancestors in the path from the parent of P_i to the root P_1 or P_2 (see Figure 3.3). Note that the path has been determined uniquely in BDII. This is why if a party P_i in level ℓ_i fails after the 1st round and cannot send any auxiliary element to descendants, no descendant can compute a shared key. However, unlike BDI, any ancestor of P_i as well as any party who is not in the same path from the failed P_i to the root can compute a shared key.

Before showing our strategy, let us start with a primitive construction of robust GKE. In BDII, an auxiliary element computed by the parent of P_i is necessary for P_i to compute a shared key. Suppose that we add a few extra edges to the graph of BDII and two or more parties compute auxiliary elements for P_i, then one of the alive auxiliary elements enables P_i to compute a shared key. For example, suppose that we add 1 more edge to P_i from parties in level $\ell_i - 1$ and add 2 more edges to parties in level $\ell_i + 1$ from P_i. Then, if either path from level $\ell_i - 1$ to P_i is alive, 4 parties in level $\ell_i + 1$ as well as P_i can compute a shared key. However, in order to let the 2×4 path available, P_i computes and multicasts 2×4 auxiliary elements for 4 parties in level $\ell_i + 1$. Thus, if both I_i edges from parties in level $\ell_i - 1$ to P_i and O_i edges from P_i to parties in level $\ell_i + 1$ exist[4], then the number of auxiliary elements is $I_i \times O_i$. That is, computational and communicational complexity increases multiplicatively according to the number of edges coming in and coming out. Our scheme can reduce the multiplicative cost to the additive cost.

Our scheme, CH-GKE, achieves efficient robust GKE, by realizing a *cross-help* idea with additive computational and communicational complexity. In order to realize such efficient cross-help, we introduce the following ideas.

1. *The division and restoration of auxiliary elements*
   In order to achieve robustness, paths to enable key-sharing need to be increased, which also increases the number of auxiliary elements. For example, for $I_i \times O_i$ paths, $I_i \times O_i$ auxiliary elements are required. In order to reduce computational and communicational complexity, we generalize a technique used in [9]: divide $I_i \times O_i$ auxiliary elements into $I_i + O_i$ parts; divided parts are computed and multicasted; then, in the key-sharing phase, a necessary auxiliary element is restored from two parts of I_i and O_i.

2. *A new relation between P_1 (resp. P_2) and descendants of P_2 (resp. P_1)*
   Parties P_1 and P_2 are sisters of each other for those who are descendants of P_1 and P_2. That is, P_1 (resp. P_2) is the aunt for parties P_i who are P_2's (resp. P_1's) children. See Figure 4.1.

Let us show an overview of CH-GKE briefly. We denote a child, a parent, and ancestors in the same notation as in Section 3. In addition to these, the sister of i who has the same parent as i, the left child and the right child of the sister, and the sister of parent are denoted by $\mathtt{sister}(i)$, $\mathtt{l.niece}(i)$ and $\mathtt{r.niece}(i)$, and $\mathtt{aunt}(i)$, respectively.

The relations among neighbors of party P_i are shown in Figure 4.2. In the basic construction of CH-GKE, the cross-help is done by sisters: parent and aunt of P_i in level $\ell_i (i \geq 2)$ make P_i's auxiliary elements, and P_i makes auxiliary elements for 4 parties such as children and nieces in level $\ell_i + 1$. Then, P_i computes and multicasts $2 + 4 = 6$ auxiliary elements. In the general construction, the cross-help is generalized in such a way that I_i parties in level $\ell_i - 1$ make P_i's auxiliary elements, and P_i in level ℓ_i makes auxiliary elements for O_i parties in level $\ell_i + 1$. Then, P_i computes $I_i + O_i$ auxiliary elements and multicasts them.

4.2   Cross-Help GKE (CH-GKE)

This section presents the basic version of CH-GKE, which tolerates all patterns of party fault if either sister is alive in a binary tree. Parties P_1 and P_2 in level 1 can cross-help each other, and thus, CH-GKE tolerates even if either party fails after the 1st round.

---

[4] I_i edges are input edges to P_i. O_i edges are output edges from P_i.

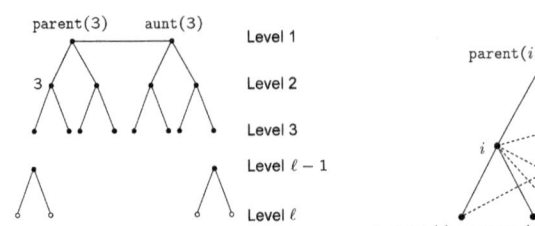

Fig. 4.1. Party Tree of CH-GKE

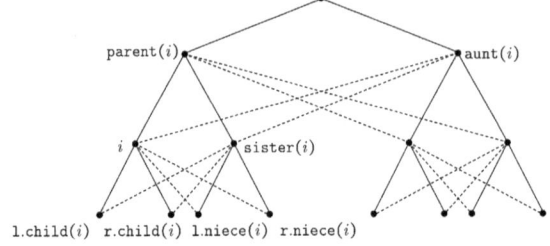

Fig. 4.2. CH-GKE

Figure 4.2 shows the relations of parties in CH-GKE, where a dotted line represents a flow of additional auxiliary elements to BDII. For example, additional auxiliary elements constructed by $\mathtt{aunt}(i)$ are sent to i and $\mathtt{sister}(i)$.

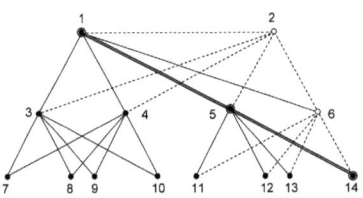

Fig. 4.3. Example of CH-GKE

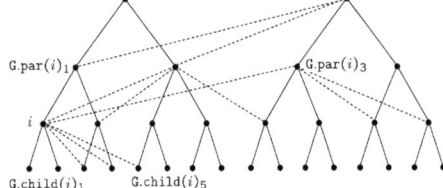

Fig. 4.4. Generalized CH-GKE

Protocol 4 (CH-GKE)

1. *Each party P_i computes $z_i = g^{r_i}$ for a (private) uniformly and randomly chosen $r_i \in \mathbb{Z}_q^*$ and sends it to its neighbors.*

2. *Let* $\mathtt{ActiveList}_1$ *be the list of indices of all parties who complete the 1st round. Then, P_i ($i \in \{1,2\}$) in level 1 computes 5 auxiliary elements $y_{i[\mathtt{aunt}(i), i]}$, $y_{i[i, \mathtt{l.child}(i)]}$, $y_{i[i, \mathtt{r.child}(i)]}$, $y_{i[i, \mathtt{l.niece}(i)]}$, and $y_{i[i, \mathtt{r.niece}(i)]}$, and multicasts them to parties in levels ≥ 2, where*

$$y_{i[\mathtt{aunt}(i), i]} = \left(\frac{z_{\mathtt{aunt}(i)}}{z_i}\right)^{r_i} = g^{r_{\mathtt{aunt}(i)} \cdot r_i - r_i^2};$$

$$y_{i[i, \mathtt{l.child}(i)]} = \left(\frac{z_i}{z_{\mathtt{l.child}(i)}}\right)^{r_i} = g^{r_i^2 - r_{\mathtt{l.child}(i)} \cdot r_i}; \quad y_{i[i, \mathtt{r.child}(i)]} = \left(\frac{z_i}{z_{\mathtt{r.child}(i)}}\right)^{r_i} = g^{r_i^2 - r_{\mathtt{r.child}(i)} \cdot r_i}$$

$$y_{i[i, \mathtt{l.niece}(i)]} = \left(\frac{z_i}{z_{\mathtt{l.niece}(i)}}\right)^{r_i} = g^{r_i^2 - r_{\mathtt{l.niece}(i)} \cdot r_i}; \quad y_{i[i, \mathtt{r.niece}(i)]} = \left(\frac{z_i}{z_{\mathtt{r.niece}(i)}}\right)^{r_i} = g^{r_i^2 - r_{\mathtt{r.niece}(i)} \cdot r_i}.$$

Let P_i be an inner-node party in level $\ell_i \geq 2$. Then, P_i computes 6 auxiliary elements $y_{i[\mathtt{parent}(i), i]}$, $y_{i[\mathtt{aunt}(i), i]}$, $y_{i[i, \mathtt{l.child}(i)]}$, $y_{i[i, \mathtt{r.child}(i)]}$, $y_{i[i, \mathtt{l.niece}(i)]}$, and

$y_{i[i, \text{r.niece}(i)]}$, and multicasts them to parties in level $\geq \ell_i + 1$, where

$$y_{i[\text{parent}(i), i]} = \left(\frac{z_{\text{parent}(i)}}{z_i}\right)^{r_i} = g^{r_{\text{parent}(i)} \cdot r_i - r_i^2}; \quad y_{i[\text{aunt}(i), i]} = \left(\frac{z_{\text{aunt}(i)}}{z_i}\right)^{r_i} = g^{r_{\text{aunt}(i)} \cdot r_i - r_i^2};$$

$$y_{i[i, \text{1.child}(i)]} = \left(\frac{z_i}{z_{1.\text{child}(i)}}\right)^{r_i} = g^{r_i^2 - r_{1.\text{child}(i)} \cdot r_i}; \quad y_{i[i, \text{r.child}(i)]} = \left(\frac{z_i}{z_{\text{r.child}(i)}}\right)^{r_i} = g^{r_i^2 - r_{\text{r.child}(i)} \cdot r_i}$$

$$y_{i[i, \text{1.niece}(i)]} = \left(\frac{z_i}{z_{1.\text{niece}(i)}}\right)^{r_i} = g^{r_i^2 - r_{1.\text{niece}(i)} \cdot r_i}; \quad y_{i[i, \text{r.niece}(i)]} = \left(\frac{z_i}{z_{\text{r.niece}(i)}}\right)^{r_i} = g^{r_i^2 - r_{\text{r.niece}(i)} \cdot r_i}.$$

3. Let $\mathtt{ActiveList_2}$ be the list of indices of all parties who complete the 2nd round. Then, P_i $(i \in \{1, 2\})$ in level 1 computes $K = z_{\text{aunt}(i)}^{r_i}$. Note that P_i can compute K even if $P_{\text{aunt}}(i) \notin \mathtt{ActiveList_2}$. On the other hand, P_i in level $\ell_i \geq 2$ picks up a set of indices from $\mathtt{ActiveList_2}$ whose parties form a path from the (reset) parent of P_i to P_1 or P_2, where the set is denoted by $\underline{\text{ancestor}(i)}$: $\underline{\text{ancestor}(i)} = \{\text{parent}(i), \cdots, 1 \text{ or } 2\}$. Actually, the set consists of $\ell_i - 1$ indices of each party from the level $\ell_i - 1$ to 1. If either sister is alive in a binary tree, then $\underline{\text{ancestor}(i)}$ exists. A shared key is given as

$$K = z_{\text{parent}(i)}^{r_i} \cdot \Pi_{j \in \underline{\text{ancestor}(i)}} Y_j = g^{r_1 r_2},$$

where $Y_j = y_{j[\text{parent}(j), j]} y_{j[j, \text{child}(j)]} = g^{r_{\text{parent}(j)} r_j - r_j^2} g^{r_j^2 - r_{\text{child}(j)} r_j} = g^{r_{\text{parent}(j)} r_j - r_{\text{child}(j)} r_j}$ (for $j \neq 1, 2$), $Y_j = y_{j[\text{aunt}(j), j]} y_{j[j, \text{child}(j)]} = g^{r_{\text{aunt}(j)} r_j - r_j^2} g^{r_j^2 - r_{\text{child}(j)} r_j} = g^{r_{\text{aunt}(j)} r_j - r_{\text{child}(j)} r_j}$ (for $j = 1$ or 2), and $\text{parent}(j)$ (resp. $\text{child}(j)$) is the (reset) parent (resp. child) of j in $\underline{\text{ancestor}(i)}$.

Protocol 4 satisfies correctness. Example 1 shows how Party 14 computes a shared key in CH-GKE among 14 parties. See Figure 4.3, where black or white nodes correspond to parties alive or dead in the 2nd round, respectively; big nodes correspond to parties in the path $\underline{\text{ancestor}(14)}$; bold edges correspond to the path $\underline{\text{ancestor}(14)}$ and dotted lines represent a flow where auxiliary elements have not been sent in the 2nd round.

Example 1 Let $n = 14$; and $\mathtt{ActiveList_2} = \{1, 3, 4, 5, 7, 8, 9, 10, 11, 12, 13, 14\}$. Then, Party 14 computes a shared key as follows. In this case, the reset parent of P_{14} is P_5 and the reset parent of P_5 is P_1, which becomes the end of the path, and thus, $\underline{\text{ancestor}(14)} = \{5, 1\}$. So, P_{14} computes $Y_5 = g^{r_1 r_5 - r_5^2} g^{r_5^2 - r_{14} r_5} = g^{r_1 r_5 - r_5 r_{14}}$; $Y_1 = g^{r_2 r_1 - r_1^2} g^{r_1^2 - r_1 r_5} = g^{r_2 r_1 - r_1 r_5}$; and thus, results in $K = z_5^{r_{14}} Y_1 Y_5 = g^{r_1 r_2}$.

Remarks 1. 1. Parties with low computational resources are arranged to nodes in leaves. Then, they can skip the 2nd round. Those parties executes 1 exponentiation in the round 1 and the computation of the shared key.
2. Parties with large computational resources are arranged to inner nodes. They need to execute both the 1st and the 2nd rounds.

4.3   Generalized CH-GKE

We show the generalization of Protocol 4 as Protocol 5. Figure 4.4 shows the relations of parties in the generalized CH-GKE. In Protocol 5, a concept of children and parent of a party in level ℓ_i are generalized to $\{G.child(i)_j\}$ and $\{G.parent(i)_j\}$: a party in level ℓ_i makes auxiliary elements for O_i parties in level $\ell_i + 1$, denoted by $G.child(i)_1, \cdots, G.child(i)_{O_i}$; and I_i parties in level ℓ_i-1, denoted by $G.parent(i)_1, \cdots, G.parent(i)_{I_i}$, make P_i's auxiliary elements, respectively. Remark that the number of I_i is less than that of parties in level $\ell_i - 1$. The generalized CH-GKE tolerates all patterns of party faults if one party in $\{G.parent(i)_j\}_i$ for each party P_i is alive in a binary tree. The detailed protocol is given in the below, which is the same as Protocol 4 except using $\{G.child(i)_j\}$ and $\{G.parent(i)_j\}$.

Protocol 5 (Generalized CH-GKE)

1. *Each party P_i computes $z_i = g^{r_i}$ for a (private) uniformly and randomly chosen $r_i \in \mathbb{Z}_q^*$ and sends it to its neighbors.*
2. *Let* ActiveList$_1$ *be the list of indices of all parties who complete the 1st round. Then, P_i ($i \in \{1, 2\}$) computes 5 auxiliary elements $y_{i[parent(i),\, i]}$, $y_{i[i,\, 1.child(i)]}$, $y_{i[i,\, r.child(i)]}$, $y_{i[i,\, 1.niece(i)]}$, and $y_{i[i,\, r.niece(i)]}$, and multi-casts them to parties in levels ≥ 2. Let P_i be an inner-node party in level $\ell_i \geq 2$. Then, P_i computes $I_i + O_i$ auxiliary elements, $\{y_{i[G.parent(i)_j,\, i]}\}_{j=1,\cdots,I_i}$ and $\{y_{i[i,\, G.child(i)_j]}\}_{j=1,\cdots,O_i}$, and multicasts them to parties in levels $\geq \ell_i + 1$, where*

$$y_{i[G.parent(i)_j,\, i]} = \left(\frac{z_{G.parent(i)_j}}{z_i}\right)^{r_i} = g^{r_{G.parent(i)_j} \cdot r_i - r_i^2}; \quad y_{i[i,\, G.child(i)_j]} = \left(\frac{z_i}{z_{G.child(i)_j}}\right)^{r_i} = g^{r_i^2 - r_{G.child(i)_j} \cdot r_i}.$$

3. *Let* ActiveList$_2$ *be the list of indices of all parties who complete the 2nd round. Then, P_i ($i \in \{1, 2\}$) in level 1 computes $K = z_{aunt(i)}^{r_i}$. P_i in level $\ell_i \geq 2$ picks up a set of indices from* ActiveList$_2$ *whose parties form a path from the (reset) parent of P_i to P_0 or P_1, where the set is denoted by* ancestor(i). *If one of $\{G.parent(i)_j\}_i$ is alive in a binary tree, then* ancestor(i) *exists. A shared key is given as*

$$K = z_{parent(i)}^{r_i} \cdot \Pi_{j \in \underline{ancestor(i)}} Y_j = g^{r_1 r_2},$$

where $Y_j = y_{j[parent(j),\, j]} y_{j[j,\, child(j)]}$ and parent(j) *and* child(j) *are the (reset) parent and child of j in* ancestor(i), *respectively.*

Remarks 2. 1. *For P_i in level $\ell_i \geq 2$, the number of generalized parents I_i (resp. children O_i) needs to be $I_i \leq 2^{\ell_i - 1}$ (resp. $O_i \leq 2^{\ell_i + 1}$) to satisfy correctness.*
2. *The more auxiliary elements $I_i + O_i$ are generated, the more party faults CH-GKE tolerates. However, it needs to locate positions where parties have failed.*

4.4   Security of CH-GKE

Theorem 3 that a passive adversary breaks CH-GKE (Protocol 4) is used to solve the Square-DDH Problem.

Theorem 3. *Assuming the Square-DDH over $\mathbb{G}$ is hard, CH-GKE (basic) (Protocol 4), denoted simply by Π, is a secure group GKE protocol. Namely,*

$$Adv_\Pi^{GKE}(t, q_{ex}) \leq Adv_\mathbb{G}^{Square-DDH}(t'),$$

where $\mathsf{Adv}_{\Pi}^{\mathsf{GKE}}(t, q_{ex})$ *is an adversary to* Π *with* $q_{ex}\mathsf{Execute}$ *queries and in* t *time, and* $\mathsf{Adv}_{\mathsf{G}}^{\mathsf{Square-DDH}}(t')$ *is an adversary to Square-DDH in* $t' = t + q_{ex}(13n - 15)EM$ *time for the number of parties* n.

Proof: Given an algorithm $\mathcal{A}$ against Π running in time t, we show how to construct an adversary $\mathcal{B}$ against the Square-DDH.

A tuple $(g, y, h) \in \mathsf{G} \times \mathsf{G} \times \mathsf{G}$ is given to $\mathcal{B}$, where $y = g^x$ with unknown x to $\mathcal{B}$ and $h = g^{x^2}$ or a random number in G. Then, $\mathcal{B}$ runs $\mathcal{A}$ to decide whether $h = g^{x^2}$ or not. $\mathcal{B}$ sets $z_1 = y$. Next, choose $c_2, \cdots, c_n \in \mathbb{Z}_p$ randomly, and set $z_i = z_{i-1} \cdot g^{-c_i} = g^{x - \sum_{j=2}^{i} c_j}$ for $i \geq 2$. z_i can be computed since z_{i-1} was computed before. Note that $\mathcal{B}$ knows z_i but does not know the logarithm r_i of $z_i = g^{r_i}$, where $r_i = x - \sum_{j=2}^{i} c_j$. From this, $\mathcal{B}$ can compute $y_{i[\text{parent}(i), i]}$, $y_{i[\text{aunt}(i), i]}$, $y_{i[i, 1.\text{child}(i)]}$, $y_{i[i, r.\text{child}(i)]}$, $y_{i[i, 1.\text{niece}(i)]}$, and $y_{i[i, r.\text{niece}(i)]}$ as follows.

In the case of P_1, set

$$y_{i[\text{aunt}(1), 1]} = \left(\frac{z_{\text{aunt}(1)}}{z_1}\right)^{r_1} = g^{r_2 r_1 - r_1^2} = g^{(x-c_2)x - x^2} = y^{-c_2}$$

which is computable since c_2 is known to $\mathcal{B}$. Let us discuss the other 4 auxiliary elements. Set the number of $1.\text{child}(1)$ to $\text{num.lc}(1)$. Then, $\text{num.lc}(1) > 2$ holds and

$$y_{1[1, 1.\text{child}(1)]} = \left(\frac{z_1}{z_{1.\text{child}(1)}}\right)^{r_1} = g^{r_1^2 - r_{1.\text{child}(1)} r_1} = g^{x^2 - (x - \sum_{j=2}^{\text{num.lc}(1)} c_j)x} = g^{(\sum_{j=2}^{\text{num.lc}(1)} c_j)x} = y^{\sum_{j=2}^{\text{num.lc}(1)} c_j},$$

which is computable since $\sum_{j=2}^{\text{num.lc}(1)} c_j$ is known to $\mathcal{B}$. The computation of the other 3 auxiliary elements follows the above.

In the case of P_2, set

$$y_{2[\text{aunt}(2), 2]} = \left(\frac{z_{\text{aunt}(2)}}{z_2}\right)^{r_2} = g^{r_1 r_2 - r_2^2} = g^{x(x - c_2) - (x - c_2)^2} = y^{c_2} g^{-c_2^2},$$

which is computable since c_2 is known to $\mathcal{B}$. The other 4 auxiliary elements can be computed in the same way.

In the case of $P_i (i \geq 3)$, $\mathcal{B}$ can compute 6 auxiliary elements in the same way as above. Let us discuss two auxiliary elements of $y_{i[\text{parent}(i), i]}$ and $y_{i[\text{aunt}(i), i]}$. Set the number of $\text{parent}(i)$ to $\text{num.par}(i)$. Then, $\text{num.par}(i) < i$ holds and

$$y_{i[\text{parent}(i), i]} = \left(\frac{z_{\text{parent}(i)}}{z_i}\right)^{r_i} = g^{r_{\text{parent}(i)} r_i - r_i^2} = g^{(x - \sum_{j=2}^{\text{num.par}(i)} c_j)(x - \sum_{j=2}^{i} c_j) - (x - \sum_{j=2}^{i} c_j)^2}$$

$$= y^{-\sum_{j=\text{num.par}(i)+1}^{i} c_j} g^{(\sum_{j=\text{num.par}(i)+1}^{i} c_j)(\sum_{j=2}^{i} c_j)}$$

which is computable since $\forall c_j$ is known to $\mathcal{B}$. $y_{i[\text{aunt}(i), i]}$ can be computed in the same way as above. Let us discuss the other 4 auxiliary elements. Set the number of $1.\text{child}(i)$

to $\mathrm{num.lc}(i)$. Then, $\mathrm{num.lc}(i) > i$ holds and

$$y_{i[i,\,1.\mathrm{child}(i)]} = \left(\frac{z_i}{z_{1.\mathrm{child}(i)}}\right)^{r_i} = g^{r_i^2 - r_{1.\mathrm{child}(i)} r_i} = g^{(x - \sum_{j=2}^{i} c_j)^2 - (x - \sum_{j=2}^{\mathrm{num.lc}(i)} c_j)(x - \sum_{j=2}^{i} c_j)}$$

$$= y^{-\sum_{j=i+1}^{\mathrm{num.lc}(i)} c_j} g^{(\sum_{j=2}^{i} c_j)(\sum_{j=i+1}^{\mathrm{num.lc}(i)} c_j)}$$

which is computable since $\forall c_j$ is known to $\mathcal{B}$. The computation of the other 3 auxiliary elements follows the above.

As the $c_i (i \geq 2)$ are distributed uniformly at random, the distribution of z_i and $y_{i[i,\,j]}$ is identical to that in Π. The transcript consists of

$$T = \{z_i,\, y_{i[\mathrm{parent}(i),\,i]},\, y_{i[\mathrm{aunt}(i),\,i]},\, y_{i[i,\,1.\mathrm{child}(i)]},\, y_{i[i,\,r.\mathrm{child}(i)]},\, y_{i[i,\,1.\mathrm{niece}(i)]},\, y_{i[i,\,r.\mathrm{niece}(i)]}\},$$

for each party P_i. Let $\mathrm{ActiveList}_2$ be the list of indices of all parties who complete the 2nd round. Upon the Test request, $\mathcal{B}$ issues the shared key K as follows,

$$K = g^{r_1 r_2} = g^{x(x - c_2)} = g^{x^2} y^{-c_2} = h y^{-c_2}.$$

If K is the shared group key, then $h = g^{x^2}$, i.e. (g, y, h) is a valid Square-DDH set. Therefore, $\mathcal{B}$ succeeds with the same advantage as $\mathcal{A}$ by $(13n - 15)EM$ additional computational time to generate T.

We consider the case in which $\mathcal{A}$ makes a single Execute query, since $\mathcal{B}$ can easily generate another set of (g^r, y^r, h^{r^2}) of the same type as (g, y, h) for a random exponent $r \in \mathbb{Z}_p$. Bounding the number n by the total number of parties, the claim follows. ∎

In the same way, we can show that a passive adversary that breaks the generalized CH-GKE (Protocol 5) is used to solve the Square-DDH Problem, whose proof will be shown in the final paper.

Theorem 4. *Assuming the Square-DDH over $\mathbb{G}$ is hard, CH-GKE (general) (Protocol 5), denoted simply by Π, is a secure group GKE protocol. Namely,*

$$\mathrm{Adv}_{\Pi}^{\mathrm{GKE}}(t, q_{ex}) \leq \mathrm{Adv}_{\mathbb{G}}^{\mathrm{Squre\text{-}DDH}}(t'),$$

where $\mathrm{Adv}_{\Pi}^{\mathrm{GKE}}(t, q_{ex})$ is an adversary to Π with q_{ex} Execute queries and in t time; $\mathrm{Adv}_{\mathbb{G}}^{\mathrm{Squre\text{-}DDH}}(t')$ is an adversary to Square-DDH in $t' = t + q_{ex}((2\mathrm{max}_{\Pi} + 1)n + 9 - 4\mathrm{max}_{\Pi})EM$ time for the number of parties n; and max_{Π} is the maximum number of auxiliary elements constructed by a single party.

5   T-Robust GKE with $O(\log n)$ Complexity

CH-GKE, shown in Section 4, achieves robustness with $O(\log n)$ message size. The generalized CH-GKE tolerates if one of paths from any alive party to P_1 or P_2 is alive. However, it needs to locate positions where parties have failed. In fact, either P_1 or P_2 need to be alive. On the other hand, JKT tolerates any T-party faults in any position, however, it needs $O(n)$ computational complexity.

In this section, we present T-robust GKE with $O(\log n)$ communicational and computational complexity. Our idea is to combine both JKT and CH-GKE considering their advantages: JKT is a symmetric structure and tolerates any T-party fault in any position but works in $O(n)$ computational complexity, while CH-GKE is an asymmetric structure and works in $O(\log n)$ communicational and computational complexity. Our strategies to combine both T-robust JKT and CH-GKE are: from the point of any T-party fault, the shared key is computed in the procedure of T-robust JKT. Then, auxiliary elements for parties not in JKT procedure to compute the shared key are generated and multicasted in the procedure of CH-GKE. According to this strategy, parties are arranged to a circle of JKT or trees of CH-GKE; and compute JKT-and-CH-GKE-like auxiliary elements or CH-GKE-like auxiliary elements, respectively. Figure 5.1 shows an arrangement of parties, where parties in a circle execute JKT part, parties in trees execute CH-GKE part, dotted lines in the circle represent a flow that the party 1 computes auxiliary elements in JKT part, and bold lines from the party 1 to nodes in level 1 of trees represent a flow that the party 1 computes auxiliary elements in CH-GKE part. Let us show how we combine JKT and CH-GKE briefly, then present our T-robust GKE (Protocol 6):

1. *JKT part*

   $T + 2$ parties are arranged in the circle of JKT[5]. They compute auxiliary elements to execute T-robust JKT, which are in total for $T + 1$ parties. They also compute ones to execute CH-GKE, which are in total for $2(T + 1)$ parties, because any party in level ℓ_i of CH-GKE needs to get auxiliary elements sent by $T + 1$ different parties in level $\ell_i - 1$, while the number of parties in level ℓ_i is twice as large as one in level $\ell_i - 1$. Those auxiliary elements are called JKT-and-CH-GKE-like auxiliary elements.

2. *CH-GKE part*

   Arrange $T + 2$ trees of CH-GKE under each party in the circle of JKT. $\frac{n-(T+2)}{T+2}$ parties are arranged in each tree of CH-GKE, where the height of tree is $\ell = \lceil \log_2 \frac{n}{T+2} + 1 \rceil - 1$. They execute CH-GKE and compute auxiliary elements for $2(T + 1)$ parties in total in the same reason as above.

Protocol 6 (T-robust GKE) *T-robust GKE among n parties is a combination of T-robust JKT among $T + 2$ parties and T-robust CH-GKE among $n - T - 2$ parties.*
INITIALIZATION: ARRANGE PARTIES TO JKT OR CH-GKE
Set $T + 2$ parties in the circle in JKT, where they are numbered from 1 to $T + 2$. Arrange $T + 2$ trees of CH-GKE by setting each party in the circle of JKT to each root. Finally, set $\frac{n-T-2}{T+2}$ parties in each tree.
GROUP KEY EXCHANGE PROTOCOL

1. *Each party P_i computes $z_i = g^{r_i}$ for a (private) uniformly and randomly chosen $r_i \in \mathbb{Z}_q^*$ and sends it to its neighbors.*
2. *Let* ActiveList$_1^{JKT}$ *(resp.* ActiveList$_1^{CH}$*) be the list of indices of parties in the circle (resp. trees) who complete the 1st round.*

---

[5] This is the smallest number of parties arranged in the circle, which can be set to more than $T + 2$.

PARTIES IN THE CIRCLE OF JKT: *Let P_i be a party in the circle, who has general-ized children $\{\mathtt{G.child}(i)_j\}$ in trees. P_i computes and broadcasts two types of aux-iliary elements: auxiliary elements $x_{[k,\,i]}$ seen in the 2nd round in JKT(Protocol 2) and those seen in children parts of the 2nd round in CH-GKE between P_i itself and generalized $2(T+1)$ children in the tree, $P_{\mathtt{G.child}(i)_j}$ (see Section 4.3 for $\mathtt{G.child}(i)_j$):*

$$x_{[k,\,i]} = \left(\frac{z_i}{z_k}\right)^{r_i} = g^{r_i^2 - r_k r_i} \; (\mathtt{ActiveList}_1^{JKT} \ni k \neq i)$$

$$y_{i[i,\,\mathtt{G.child}(i)_j]} = \left(\frac{z_i}{z_{\mathtt{G.child}(i)_j}}\right)^{r_i} = g^{r_i^2 - r_i r_{\mathtt{G.child}(i)_j}} \; (\mathtt{ActiveList}_1^{CH} \ni \mathtt{G.child}(i)_j).$$

PARTIES IN TREES OF CH-GKE: *Let P_i be an inner-node party in level ℓ_i in a tree, who has generalized parents $\{\mathtt{G.parent}(i)_j\}$ in the circle (resp. trees) if $\ell_i = 1$ (resp. $\ell_i \geq 2$) and generalized children $\{\mathtt{G.child}(i)_j\}$ in trees. P_i computes $(T+1)+2(T+1)$ auxiliary elements and multi-casts them to parties in level $\geq \ell_i + 1$ in the same way as in Protocol 5:*

$$y_{i[\mathtt{G.parent}(i)_j,\,i]} = \left(\frac{z_{\mathtt{G.parent}(i)_j}}{z_i}\right)^{r_i} = g^{r_{\mathtt{G.parent}(i)_j} r_i - r_i^2}; \; y_{i[i,\,\mathtt{G.child}(i)_j]} = \left(\frac{z_i}{z_{\mathtt{G.child}(i)_j}}\right)^{r_i} = g^{r_i^2 - r_{\mathtt{G.child}(i)_j} r_i}.$$

3. *Let $\mathtt{ActiveList}_2^{JKT}$ (resp. $\mathtt{ActiveList}_2^{CH}$) be the list of indices of parties in the circle (resp. trees) who complete the 2nd round. Here we set $\#\mathtt{ActiveList}_2^{JKT} = T'$. Assume that the alive parties in $\mathtt{ActiveList}_2^{JKT}$ are sorted in the same order as before and ordered $\{P_{a_1}, P_{a_2}, \cdots, P_{a_{T'}}\}$.*
PARTIES IN THE CIRCLE OF JKT: *Each P_{a_i} in the circle computes a shared key*

$$K = z_{a_{i-1}}^{r_{a_i} \cdot T'} \cdot X_{a_i}^{T'-1} \cdot X_{a_{i+1}}^{T'-2} \cdots X_{a_{i-2}} = g^{r_{a_1} r_{a_2} + r_{a_2} r_{a_3} + \cdots + r_{a_{T'}} r_{a_1}},$$

where $X_{a_j} = x_{[a_{j-1},\,a_j]} \cdot (x_{[a_{j+1},\,a_j]})^{-1} = g^{r_{a_j} r_{a_{j+1}} - r_{a_{j-1}} r_{a_j}} \; (j \in \{1, \cdots, T'\})$.
PARTIES IN TREES OF CH-GKE: *Each P_i in level 1 picks up a party $P_{a_i} \in \mathtt{ActiveList}_2^{JKT}$ whose corresponding auxiliary element, $y_{a_i[a_i,\,i]}$, has been sent to P_i. A shared key is given as follows:*

$$K = (K')^{T'} \cdot X_{a_i}^{T'-1} \; X_{a_{i+1}}^{T'-2} \cdots X_{u_{i-2}},$$

where $K' = z_{a_i}^{r_i} \cdot y_{a_i[a_i,\,i]} \cdot (x_{[a_{i+1},\,a_i]})^{-1} = g^{r_{a_i} r_{a_{i+1}}}$.
Each P_i in level $\ell_i \geq 2$, first, picks up a set of indices from $\mathtt{ActiveList}_2^{CH}$ P_i whose parties form a path from the (reset) parent of P_i to the (reset) ancestor in level 1. The set is denoted by $\underline{\mathtt{ancestor}(i)}$ in the same way as Protocols 4 and 5. We also denote the index of the party in level 1 in $\underline{\mathtt{ancestor}(i)}$ by $\underline{\mathtt{ancestor}(i)[1]}$. Then, $\underline{\mathtt{ancestor}(i)} = \{\mathtt{parent}(i), \cdots, \underline{\mathtt{ancestor}(i)[1]}\}$. If the number of fault par-ties is less than or equal to T, then $\underline{\mathtt{ancestor}(i)}$ exists. Next, P_i picks up a party $P_{a_i} \in \mathtt{ActiveList}_2^{JKT}$ whose corresponding auxiliary element, $y_{a_i[a_i,\,\underline{\mathtt{ancestor}(i)[1]}]}$, has been sent to $P_{\underline{\mathtt{ancestor}(i)[1]}}$. A shared key is given as follows:

$$K = (K')^{T'} \cdot X_{a_i}^{T'-1} \cdot X_{a_{i+1}}^{T'-2} \cdots X_{a_{i-2}},$$

where $K' = z_{\underline{parent}(i)}^{r_i} \cdot \left(\Pi_{j \in \underline{ancestor}(i)} Y_j\right) \cdot y_{a_i[a_i, \underline{ancestor}(i)[1]]} \cdot (x_{[a_i+1, a_i]})^{-1} = g^{r_{a_i} r_{a_{i+1}}}$
and $Y_j = g^{r_{\underline{parent}(j)} r_j - r_{\underline{child}(j)} r_j}$.

Protocol 6 satisfies correctness. Example 2 shows how Party 28 computes a shared key in 5-robust GKE among 49 parties. See Figure 5.2, where black or white nodes correspond to parties alive or dead in the 2nd round, respectively; big nodes correspond to parties in the path $\underline{ancestor}(28)$; bold edges correspond to the path of $\underline{ancestor}(28)$ and $P_{a_i} \in$ ActiveList$_2^{JKT}$; and bold dotted edges correspond to the path of ActiveList$_2^{JKT}$, where a shared key is computed.

Example 2 *Let S be a set of parties with* $\#S = n = 49$, *and F be a set of fault parties, where* $F = \{1, 2, 5, 7, 9, 10, 11, 24, 31\}$. *Then* ActiveList$_2^{JKT} = \{3, 4, 6\}$, $T' = 3$, *and a shared key K is computed to* $K = g^{r_3 r_4 + r_4 r_6 + r_6 r_3}$. *Party 28 computes the shared key as follows. The reset parent of* P_{28} *is* P_8 *and the reset parent of* P_8 *is* P_3, *which becomes the end of the path, and thus,* $\underline{ancestor}(28) = \{8\}$. P_{28} *picks up a party* $P_3 \in$ ActiveList$_2^{JKT}$, *and computes* $K' = (z_{\underline{parent}(28)})^{r_{28}} \cdot Y_8 \cdot y_{3[3, 8]} \cdot x_{[6, 3]}^{-1} = g^{r_8 r_{28}} \cdot g^{r_3 r_8 - r_8 r_{28}} \cdot g^{r_3^2 - r_3 r_8} \cdot g^{r_6 r_3 - r_3^2} = g^{r_6 r_3}$ *for* $Y_8 = y_{8[3, 8]} \cdot y_{8[8, 28]} = g^{r_3 r_8 - r_8 r_{28}}$, *and thus, results in* $K = (K')^3 \cdot X_3^2 \cdot X_4 = g^{3(r_6 r_3)} \cdot g^{2(r_3 r_4 - r_6 r_3)} \cdot g^{r_4 r_6 - r_3 r_4} = g^{r_3 r_4 + r_4 r_6 + r_6 r_3}$.

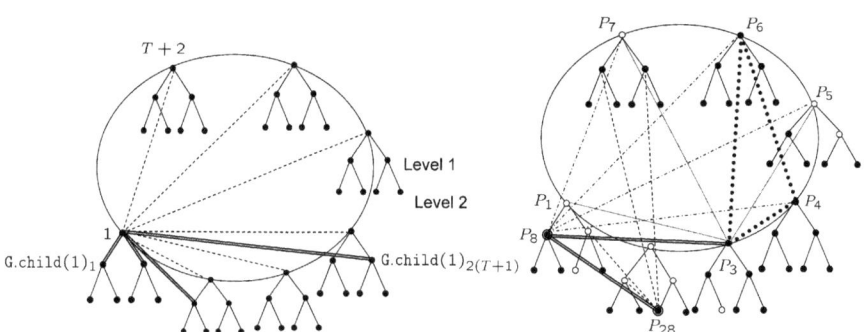

Fig. 5.1. *T*-robust GKE

Fig. 5.2. Example of 5-robust GKE among 49 parties

The security of Protocol 6 is given in Theorem 5, whose proof will be shown in the final paper

Theorem 5. *Assuming the DDH and Square-DDH over* $\mathbb{G}$ *are hard, Protocol 6, denoted simply by* Π, *is a secure group GKE protocol.*

Table 1. Sent/received message complexity of several GKEs among n parties

Party Type	Sent Messages (Large / Low)	Received Message (Large / Low Computational Resources)
BDI	$b + 2p$	$(n-1)b + 2p$
BDII	$2b + 3p$ / p	$\log_2 nb + 3p$ / $\log_2 nb + p$
JKT(cycle)	$2(T+1)b + 2(T+1)p$	$2(n-1)b + 2(T+1)p$
JKT(path)	$2Tb + 2Tp$	$2(n-2)b + 2Tp$
R-TDH1	nb / b	$2(n-1)b$ / b
ours	$3(T+1)b + 3(T+1)p$ / $(T+1)p$	$2(T + \log_2 n + 1)b + 3(T+1)p$ / $2(T + \log_2 n + 1)b + (T+1)p$

Table 2. Computational complexity and robustness(max faults) of GKE among n parties

Party Type	Large Computational Resources			Low Computational Resources			Robustness
	#EM	#I	#M	#EM	#I	#M	
BDI	3	1	$2(n-1)$	3	1	$2(n-1)$	0
BDII	4	2	$\log_2 n$	2	0	$\log_2 n$	0
JKT(cycle)	$2(T+2)$	2	$6n+8T-4$	$2(T+2)$	2	$6n+8T-4$	T
JKT(path)	$2(T+1)$	2	$4(3n+T-6)$	$2(T+1)$	2	$4(3n+T-6)$	T
R-TDH1	$3(n-1)$	0	0	4	0	0	$n-2$
ours	$3T+5$	3	$18T+9$	2	2	$6T+2\log_2 n+5$	T

6   Comparison

This section compares our scheme with previous schemes from the view point of efficiency and robustness. Table 1 (resp. Table 2) summarizes the communicational (resp. computational) complexity per user and robustness of ours (Section 5), JKT, R-TDH1 (basic)[6], BDI, and BDII. Note that, T-robust GKE means GKE tolerates all patterns of party faults up to T. JKT (cycle) or JKT (path) means Hamilton-cycle JKT or Hamilton-path JKT, respectively (See Section 3 for the detailed differences). When we do not have to distinguish JKT (cycle) from JKT (path), JKT is used simply. The notation of p and b is defined in Section 2.2. In an asymmetric party setting seen in BDII and ours, parties can be in different environment and have different computational resources since efficiency is different to each party type. In such GKEs, comparison is done by parties with large or low computational resources. Here after we focus on efficiency of JKT and our GKE since these are the same paradigm, while neither BDI nor BDII is robust and R-TDH1 is fully robust. Our GKE has advantages over JKT in the received message complexity for any party type and computational complexity for parties with low computational resources. In fact, the size of our received message is $O(T+\log n)$, while that of JKT is $O(n)$. As for sent message complexity, ours is slightly worse than that of JKT. On the other hand, the order of our computational complexity for parties with low computational resources is $O(T + \log n)$, while that of JKT is $O(T + n)$, due to our scalable party arrangement.

---

[6] In our comparison, R-TDH1 in [3] is simplified to only key establishment for the estimation of its basic complexity.

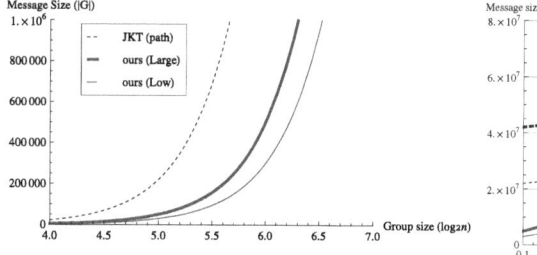

Fig. 6.1. Received message size ($T/n = 0.1$)

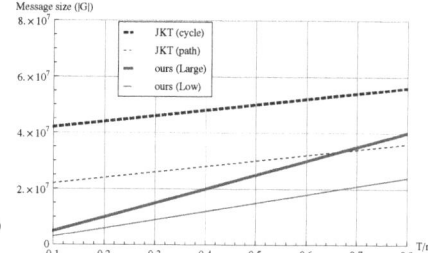

Fig. 6.2. Received message size ($n = 10^7$)

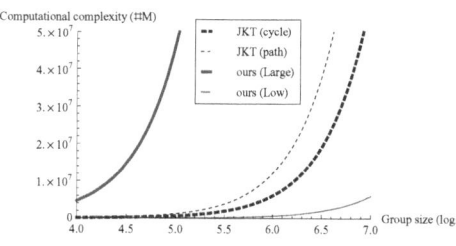

Fig. 6.3. Computational complexity ($T/n = 0.1$)

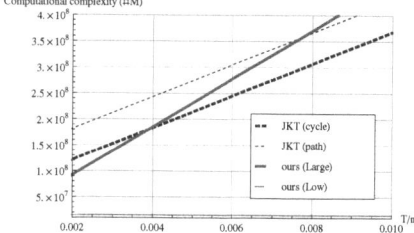

Fig. 6.4. Computational complexity ($n = 10^7$)

Let us compare both our GKE and JKT by using concrete parameter of (n, T). We firstly demonstrate the received message size comparison in Figures 6.1 and 6.2. Figure 6.1 simulates the case that the ratio of fault parties, T/n, is fixed to 0.1 and group size changes from 10^4 to 10^7. In any case, our GKE has better performance than JKT (path)[7]. Figure 6.2 simulates the case that the group size is fixed to $n = 10^7$, and T/n changes from 0.1 to 0.8. When $T/n < 0.67$, received message size of our GKE for parties with large resources is better than that of JKT. As for parties with low resources, our GKE is more efficient than JKT in any T/n.

We next compare them in the computational complexity, shown in Figures 6.3 and 6.4. Note that, the computational complexity is estimated with complexity of a single multiplication on $\mathbb{G}$ (M), where estimation is done by: $|\mathbb{G}| = 1,024$ bits, $EM = 1,536M$, and $I = 30M$[8]. Figure 6.3 simulates the case that T/n is fixed to 0.1 and group size changes from 10^4 to 10^7. In any case computational complexity of our GKE for parties with low resources is smaller than that of JKT. That for parties are larger than that of JKT. Figure 6.4 simulates the case that the group size n is fixed to $n = 10^7$ and T/n changes from 0.002 to 0.01. In the same way as Figure 6.3, the computational complexity for parties with low resources in our GKE is extremely reduced than that of JKT. That for parties with large resources is slightly better than JKT in the range of $T/n < 0.004$.

---

[7] JKT (path) is slightly better than JKT (cycle) in the received message size, and, thus, only JKT (path) is simulated.

[8] The basic binary method is assumed for an exponentiation in $\mathbb{G}$.

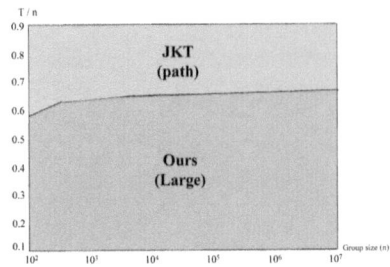
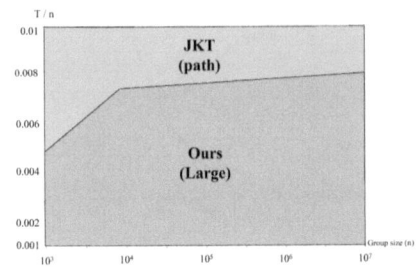

Fig. 6.5. Optimal protocol (received message size)

Fig. 6.6. Optimal protocol (computational complexity)

The above evaluations can be summarized as follows: (i) From the view point of received message complexity, our protocol has an advantage over JKT in received message complexity for parties with low computational resources, while JKT has an advantage over ours in sent message complexity (although both are $O(T)$). (ii) From the view point of computational complexity, our protocol has an advantage over JKT for parties with low computational resources, while JKT has an advantage over our GKE for parties with large resources. (iii) Our GKE has very nice scalability in both computational and communicational complexity and, thus, can be available to parties with relatively low CPU or battery.

Finally, we show the optimal robust GKE for given parameter (n, T), seen in Figures 6.5 and 6.6. Note that, our GKE is more efficient than JKT (cycle) in even received message size in any range of $(n, T/n)$. Thus, our GKE is compared with only JKT (path) from the point of view of received message size and computational complexity. JKT (path) has smaller received message size and computational complexity than our GKE for parties with large resources only if T/n is rather high. Note that, in any range, received message size and computational complexity of our GKE for parties with low resources is smaller than those of JKT. By using our results, we can choose the optimal T-robust GKE for given $(n, T/n)$.

7   Conclusions

JKT was developed to achieve a robust GKE based on BDI, and thus, it suffers communicational and computational complexity $O(n)$ per party for the group size n. Another robust GKE [3] also suffers communicational complexity $O(n^2)$ although it satisfies fully robustness. Note that, up to now, GKE with communicational and computational complexity $O(\log n)$ does not have any robustness.

We have proposed a new robust GKE, CH-GKE, with communicational and computational complexity $O(\log n)$. We have also shown that our robust GKE is secure in the standard model under the Square-DDH assumption. By combining both CH-GKE and JKT, we have proposed T-robust GKE with communicational and computational complexity $O(\log n)$, which tolerates any T-party fault in any position.

References

1. Amir, Y., Kim, Y., Nita-Rotaru, C., Schultz, J., Stanton, J., Tsudik, G.: Exploring robustness in group key agreement. In: ICDCS 2001, pp. 399–409. IEEE CS, Los Alamitos (2001)
2. Abdel-Hafez, A., Miri, A., Orozco-Barbosa, L.: Authenticated group key agreement protocols for ad hoc wireless networks. International Journal of Network Security 4(1), 90–98 (2007)
3. Brecher, T., Bresson, E., Manulis, M.: Fully Robust Tree-Diffie-Hellman Group Key Exchange. In: Garay, J.A., Miyaji, A., Otsuka, A. (eds.) CANS 2009. LNCS, vol. 5888, pp. 478–497. Springer, Heidelberg (2009)
4. Burmester, M., Desmedt, Y.: A secure and efficient conference key distribution system. In: De Santis, A. (ed.) EUROCRYPT 1994. LNCS, vol. 950, pp. 275–286. Springer, Heidelberg (1995)
5. Burmester, M., Desmedt, Y.: Efficient and secure conference key distribution. In: Lomas, M. (ed.) Security Protocols 1996. LNCS, vol. 1189, pp. 119–130. Springer, Heidelberg (1997)
6. Cachin, C., Strobl, R.: Asynchronous group key exchange with failures. In: Proceedings of PODC 2004, pp. 357–366. ACM Press, New York (2004)
7. Desmedt, Y., Lange, T., Burmester, M.: Scalable authenticated tree based group key exchange for ad-hoc groups. In: Dietrich, S., Dhamija, R. (eds.) FC 2007 and USEC 2007. LNCS, vol. 4886, pp. 104–118. Springer, Heidelberg (2007)
8. Desmedt, Y., Miyaji, A.: Redesigning Group Key Exchange Protocol based on Bilinear Pairing Suitable for Various Environments. In: Inscrypt 2010. Springer, Heidelberg (2010) (to appear)
9. Jarecki, S., Kim, J., Tsudik, G.: Robust Group Key Agreement Using Short Broadcast. In: Proceedings of ACM CCS 2007, pp. 411–420. ACM, New York (2007)
10. Kim, Y., Perrig, A., Tsudik, G.: Group Key Agreement Efficient in Communication. IEEE Trans. on Comp. 53(7), 905–921 (2004)
11. Katz, J., Yung, M.: Scalable Protocols for Authenticated Group Key Exchange. In: Boneh, D. (ed.) CRYPTO 2003. LNCS, vol. 2729, pp. 110–125. Springer, Heidelberg (2003)
12. Konstantinou, E.: Cluster-based group key agreement for wireless ad hoc networks. In: Proceedings of ARES 2008, pp. 550–557 (2008)

Application-Binding Protocol in the User Centric Smart Card Ownership Model

Raja Naeem Akram, Konstantinos Markantonakis, and Keith Mayes

Information Security Group Smart card Centre, Royal Holloway,
University of London
Egham, Surrey, United Kingdom
{R.N.Akram,K.Markantonakis,Keith.Mayes}@rhul.ac.uk

Abstract. The control of the application choice is delegated to the smart card users in the User Centric Smart Card Ownership Model (UCOM). There is no centralised authority that controls the card environment, and it is difficult to have implicit trust on applications installed on a smart card. The application sharing mechanism in smart cards facilitates corroborative and interrelated applications to co-exist and augment each other's functionality. The already established application sharing mechanisms (e.g. in Java Card and Multos) do not fully satisfy the security requirements of the UCOM that require a security framework that provides runtime authentication, and verification of an application. Such a framework is the focus of this paper. To support the framework, we propose a protocol that is verified using CasperFDR. In addition, we implemented the protocol and provide a performance comparison with existing protocols.

1 Introduction

On a multi-application smart card, the application sharing mechanism achieves optimised memory usage, data and service sharing between applications [20]. A major concern in such a mechanism is the unauthorised inter-application communication. The framework that ensures that the application sharing is secure and reliable even in adverse conditions (i.e. malicious application, developer's mistake, or design oversight) is referred as a smart card firewall [36]. In this paper, the term firewall or smart card firewall is used interchangeably.

The predominant business model in the smart card based service industry (e.g. banking, transport, and Telecom, etc.) is referred as the Issuer Centric Smart Card Ownership Model (ICOM) [13] and in this model card issuers retain the control of smart cards. Applications installed on an ICOM based smart card requires prior authorisation from the respective card issuer. This establishes an implicit trust relationship between installed applications as they are authorised/trusted by the card issuer. Furthermore, card issuers ensure that their smart card platform, and installed applications are secure, reliable, and trustworthy. Traditionally, card issuers have a business agreement with an application provider before they sanction the application installation. This agreement

U. Parampalli and P. Hawkes (Eds.): ACISP 2011, LNCS 6812, pp. 208–225, 2011.
© Springer-Verlag Berlin Heidelberg 2011

dictates the behaviour of an application and sanctioned actions that it may perform. Such an assumption is difficult to conclude in the User Centric Smart Card Ownership Model (UCOM) [14].

In the UCOM, cardholders have the choice to install, or/and delete any application from their smart cards for which they are entitled [13]. The entitlement to install an application is attained once a customer is registered with a Service Provider (SP). An SP is an organisation that develops a smart card application(s) to assist in their service architecture (i.e. bankcards in banking, and SIM cards in mobile telecom sector, etc.) and issues it to their customers. Customers can then download the application after their smart cards satisfy the Application Lease Policy (ALP) of the SP, and utilise it to access sanctioned services [11]. The ALP stipulates the minimum set of security and operational requirements that a smart card has to satisfy before the SP lease its application(s). Therefore, if an SP leases its application then it has ascertained a certain level of trustworthiness of the host platform. Furthermore, it does not imply that the SP also trusts other applications installed on the same smart card.

Firewall mechanisms are well-defined [22, 17, 8, 1] and studied [29, 18, 37, 15] in the ICOM. However, the ICOM based firewalls do not satisfy the criteria for the UCOM [14]. This paper illustrates the issue of gaining assurance and validation between communicating applications that are not satisfied (dynamically at runtime) by the ICOM based firewall mechanisms. The runtime assurance and validation of communicating applications, and subsequent generation of a cryptographic key between them is termed as application-binding. The open and dynamic nature of the UCOM requires that application sharing should be allowed only after participating applications verify, and validate current-state to be secure, and authenticate each other's identity/credentials. The protocol that achieves these requirements is referred as Application-Binding Protocol (ABP).

In section two, a short introduction to traditional smart card firewall mechanisms is provided. The discussion is then extended to the UCOM framework. Section three emphasises on the threat model, and requirements of the ABP; ending the section with a description of the proposed protocol. Protocol analysis is provided in section four that includes model checking (using CasperFDR), analytical analysis, and practical implementation results. Section five provides the concluding remarks along with future research directions.

2   Application Sharing Mechanism

In this section, we open the discussion with a short description of the existing firewall mechanisms. Next, we describe the UCOM and its firewall mechanism together with the UCOM card architecture.

2.1   Smart Card Firewall Mechanism

The firewall mechanism prohibits any cross application communication; unless sanctioned by the communicating applications. As depicted in figure 1, application A is authorised to access the resources shared by application B, where depicted application C cannot access application B.

In subsequent sections, we will discuss the Java Card, and the Multos firewall mechanisms from a large set of smart card platforms. The reason for this is: 1) these two frameworks are among the most deployed in the smart card industry and 2) they illustrate two contrasting approaches of smart card firewall implementation. We will not discuss the GlobalPlatform speci-

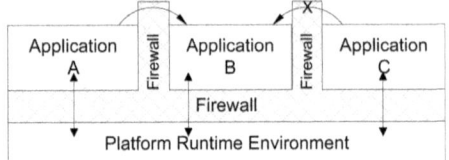

Fig. 1. A Generic Application Sharing Mechanism

fication [7] in this paper as "GlobalPlatform Card Security Requirement Specification" [4, see section 5.9.2] states that the GlobalPlatform relies on the underlying platform's (e.g. Java Card, and Multos) implementation of the firewall mechanism.

Java Card. In Java Card, cross application communication is achieved by the Shareable Interface Objects (SIO), and communication between applications and platform is controlled by the Java Card Runtime Environment (JCRE) Entry Point Objects [19], as shown in figure 2. The Entry Point Objects are instances of Java Card APIs used by applications to access platform services. These objects provide a secure way for applications to execute privileged commands.

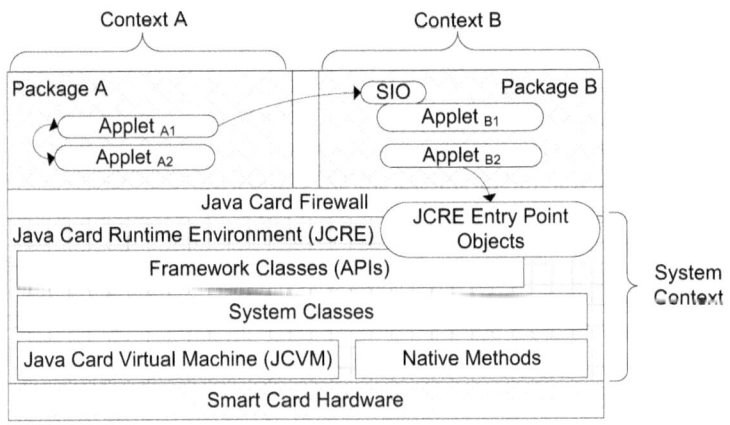

Fig. 2. Java Card Firewall Architecture

The SIO enables an application to share its data and resources (functionality) with other authorised applications. To use the SIO functionality a server application (Package B in figure 2) should implement the shareable interface. The data and functionality defined in the shareable interface then becomes available to the client application (Package A in figure 2). Each application on a smart card has an Application Identifier (AID) [5] that in theory is unique to it across all the smart cards on which it is installed. In Java Card, a client application invokes the `JCSystem.getAppletShareableInterface (Server AID, parameter)` to request sharing with the server application. The client application needs to specify the AID of the respective server application and a parameter (that can be an authorisation token encrypted with a pre-shared key) to the firewall. The firewall then asks the server application whether or not it will allow the sharing request. This framework is fit for the purpose in the ICOM; however, in an open cardx scenario [41], a malicious user can masquerade the AID of a client application and replay the application sharing request message (section 3.2).

Multos. The firewall mechanism on Multos cards is implemented with the assumption that the applications installed on them have prior authorisation from a centralised authority (i.e. the card issuer or/and Multos Certification Authority) [9]. The security of the firewall mechanism in the Multos is actually managed by the off-card agreements between different application providers. The firewall in Multos is based on the concept of Delegation. In the delegation mechanism, the client application is called the delegator and the server application is called the delegate. The delegation process is described as below.

1. Application A creates an Application Protocol Data Unit (APDU) [34] in the public memory that has the AIDs of A and B along with the request.
2. Application A initiates the delegate command, requesting Multos operating system to invoke application B, which then reads and processes the APDU.
3. On completion; application B generates an APDU in the public memory.
4. The Multos operating system switches back to the execution of the application A, which retrieves the results of the request from the public memory.

2.2   User Centric Smart Card Firewall

The firewall mechanism in the UCOM assumes that the smart card may be under the control of a malicious user, or/and there might be malicious applications on it. However, this does not imply that the underlying card platform is either compromised or malicious. We assume that card platform is not malicious; its trustworthy and has necessary functionality to provide tangible and verifiable evidence of the implemented security mechanisms [12]. The UCOM firewall is summarised in figure 3 and in subsequent sections, we will discuss critical components which are relevant to this paper.

Resource Manager. The request for an application's shareable resource is handled by its Application Resource Manager (ARM) and the Runtime Resource Manager (R2M) handles the access to the platform's resources (APIs): figure 3.

When a client application requests shareable resources, the firewall invokes the ARM of the server application. The ARM then verifies, and validates the client application's credentials, and current state as secure for sanctioning the application sharing (as part of the ABP). If successful, the ARM issues the shareable resources to the requesting application.

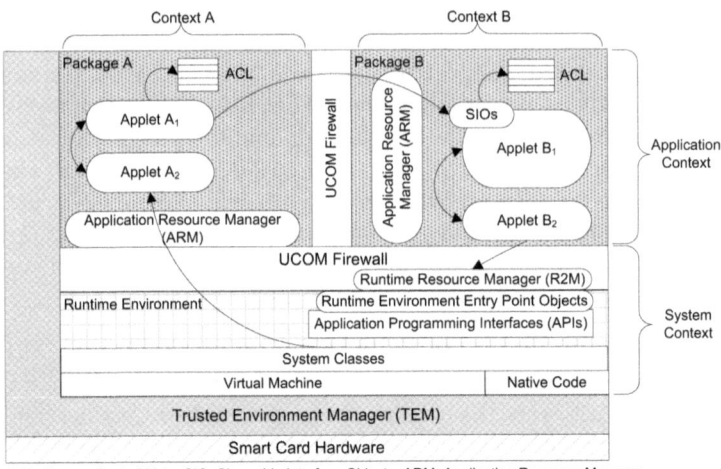

Fig. 3. UCOM Smart Card and Firewall Architecture

Trusted Environment Manager (TEM). The complete architecture and functionality of a TEM [10] for smart cards is still under research but for completeness we discuss those elements of the design that are most relevant to this paper.

The TEM provides assurance and validation of installed applications and the underlying smart card platform. Each application, and platform can have a security evaluation performed by a third party; most notably under the Common Criteria (CC) scheme [6]. In the ICOM, CC certificates are static and offline [42, 12]. However, in the UCOM a CC certificate can be digital that facilitates a dynamic assurance and validation mechanism [12]. The digital certificate will have a digest of the secure-evaluated state of the platform/application. During the ABP, both applications provide their certificates, and the TEM validates the current state as secure as it was at the time of the CC evaluation. In the UCOM, the security evaluation is only mandatory for the smart card platform.

When an application is installed onto a smart card, the TEM calculates the hash of the application and, then establishes a secure relation (shared key) with the installed application. The TEM does not calculate the hash of an application, unless it is authorised by the application itself, a cardholder or the application's SP. When an application authorises the TEM to generate its hash value; it generates a message encrypted with the Application-TEM shared symmetric key $(K_{App-TEM})^1$. The authorisation message generated by an application is referred as an Integrity Verification Authorisation (IVA) message[2].

---

[1] The Application-TEM shared Key $(K_{App-TEM})$ is generated at the time of the application installation. Both the TEM and the application have this key, and it is used to encrypt communication between them.

[2] The IVA structure is $E_{App_A-TEM}(App_A, App_B, RandomNumber_{App_A})$; contents of the messages are the identity of the authorising application and application for whom TEM generates the hash value, and a random number [35].

3   Applications-Binding Protocol (ABP)

In this section, we open the discussion by explaining the concept of application-binding along with the threat model for the ABP. Next we discuss the application enrolment process, and finally describe the proposed protocol.

3.1   Application-Binding

The application-binding is a process in which two applications (on the same smart card) establish trust in each other's identity, credentials, and current state. Following from that they generate a cryptographic key that they will use in all future communications. The generated key binds a client application with the corresponding server application and acts as an authentication credential.

Why is application-binding required in the UCOM, but not needed in the ICOM? The obvious reason for not being included in the ICOM is the centralised control of smart cards. The issuers always know which applications are installed on their smart cards along with having a trust relationship, so there is an offline understanding (or agreement) that applications will not be malicious.

Nevertheless, in the UCOM such assumptions are difficult to sustain. In the absence of a centralised authority an application is unable to verify and authenticate the identity of other applications. Therefore, an application can masquerade as either a server or client application (section 3.2). Avoiding masquerading is possible if: a) the AID allocation is centralised, and b) have an AID enforcement mechanism to forbid the application installation with an unauthorised AID.

Similarly, to the ICOM it might not be necessary to have an ABP in the UCOM; only if an SP knows what other applications are installed on a smart card before leasing its application, and there is a secure AID enforcement mechanism in place. One solution is to scan the smart card to analyse installed applications but this approach violates the privacy requirement of the UCOM [13].

3.2   Threat Model

The following threat model is mapped to the practical attacks demonstrated in [16, 47, 32, 37]; that an UCOM firewall also has to deal with.

Application Masquerading. In this scenario, a malicious application can masquerade as a server or client application. As an example, in Java Card when a client application sends the request for application sharing it generates the request that contains the server application's AID. Now if a malicious application

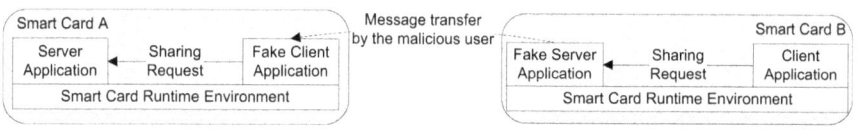

Fig. 4. Application Masquerading and Relay Attack Scenario

is masquerading as a server application; it only has to inform the firewall that it accepts the application sharing request without validating that it has the knowledge of the shared secret. Thus the client application thinks that it is accessing the shared resource of the server application; whereas it is communicating with a malicious application. Now the fake server application can resend the application sharing request message to a genuine server application on another smart card and gain access to shared resources; this scenario is illustrated in figure 4.

Unresolved Binding Instances. When an application is deleted the application sharing with other applications may not be revoked. In this scenario, a malicious application can masquerade as either a client or a server application and try to communicate with other applications on the smart card. Obviously, for such an attack the malicious user requires the knowledge of the object reference to the shareable resources. Nonetheless, in the UCOM it is necessary that bindings are revoked when one of the participating applications is deleted. The deletion of a binding instance is performed by simply deleting the application binding instance in the participating application's ACL (figure 3).

Different User's Applications. Consider a scenario in which we have two users and two applications. One is a malicious user M_u while the other is an authorised user A_u. The two applications are App_A (server application) and App_B (client application) that have a client-server relationship.

Both users are authorised to download application App_A, however M_u is not authorised to download application App_B. Now at some point, the M_u obtains the App_B's credentials for the A_u and manages to download App_B on to his or her smart card. The application sharing between the M_u's App_A and the A_u's App_B can be established. This can lead to some financial benefits for the M_u to which he or she was not entitled.

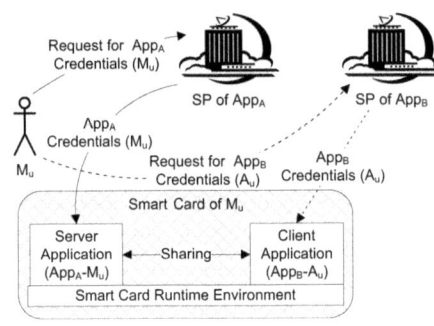

Fig. 5. Application Sharing among Different User's Applications

3.3   Requirements for the Protocol

Based on the threat model described in the previous section and on general UCOM requirements; the ABP should support the following features:

1. Provide mutual entity authentication.
2. Provide protection against application masquerading and relay attacks.
3. A malicious user should not be able to deduce the binding (share secret) by eavesdropping on the communication between client and server application.
4. Binding should be between two applications that belong to the same user.
5. Both applications should mutually provide assurance and validation that their current state is the same as expected (trusted star).

6. Binding should be unique to a specific instance of applications. If one of the applications is deleted the binding should be revoked. When the deleted application is installed again a new binding should be generated.
7. A shared secret key is generated at the successful conclusion of the ABP.
8. Entire protocol should execute on the smart card without using any external entity.

The requirement four may appear counter intuitive as communication buses on smart cards are encrypted [34], and it is difficult to monitor the communication over them if not impossible. However, a configuration similar to the one illustrated in figure 4 can be used to monitor the communication.

3.4   Enrolment Process

During the enrolment process, SPs of a client and server application agree on the business and technical terms for sharing their application resources on an UCOM based smart card.

In this process, an SP of a client application provides assurance and validation from a third party evaluation [6] to an SP of a server application, and vice versa. If third party evaluation is not available then both client and server application's SPs can decide on any other adequate way to establishing trust in each other's application and its functionality. During this process, they decide the details of the ABP, such as: to perform an on-card verification and validation of applications; the SP of the server application issues a certificate to the client application, and vice versa.

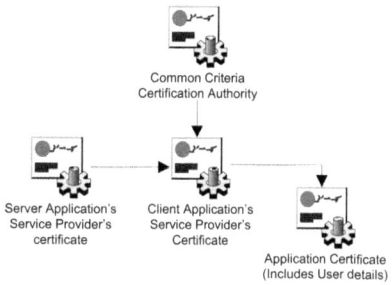

Fig. 6. Hierarchy of a Client Application's Certificate

The certificate hierarchy in the ABP is illustrated in figure 6. In the absence of the CC evaluation; the certificate hierarchy shown in figure 6 will not include "Common Criteria Certification Authority". The client application certificate has the hash value (generated either by the CC or by the SP of the server application) of the application and user's details to which the application is issued. Similar contents will also be included in the server application certificate. Basically, the enrolment process defines the restrictions and mechanisms (i.e. certificates, and cryptographic algorithms, etc.) that a client/server application's SPs agree for the ABP.

3.5   Proposed Application-Binding Protocol

The aim of the ABP is to facilitate both the client and server applications to establish trust in each other's identity, and current state. Enabling them to establish a unique binding for future communications. Figure 7 depicts a generic representation of the proposed ABP and subsequently explanation along with

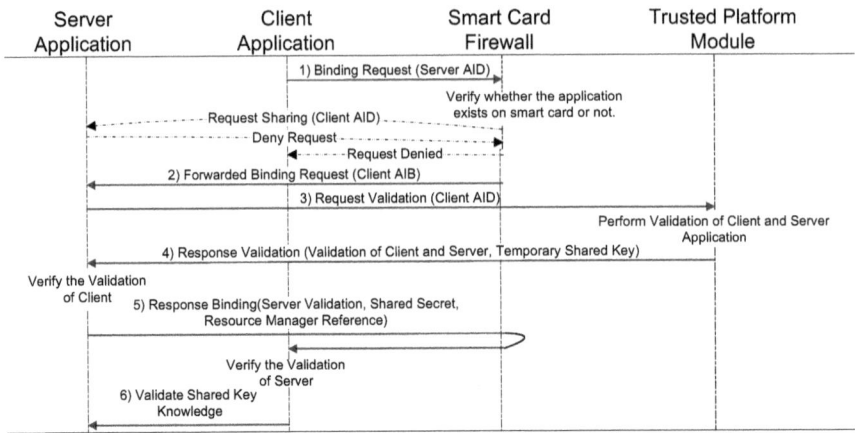

Fig. 7. Generic Representation of the Application-Binding Protocol

Table 1. Protocol Notation

Notation	Description
S	Represents the identity of the server application.
C	Represents the identity of the client application.
TEM	Represents the identity of a TEM on a smart card.
$Firewall$	Represents the identity of the UCOM firewall on a smart card.
K_{A-B}	Long term symmetric key shared between entity A and B.
K_{S-C}^t	Session key generated by TEM.
S_A, V_A	Signature and verification key pair of entity A.
$Cert_A$	Signature key pair certificate of entity A.
$Cert_{A-B}$	Certificate for entity B issued by entity A.
n_x	Random number generated by an entity X.
$n_x + num$	Random number incremented by the value of num is a natural number
$A \rightarrow B$	Message sent by an entity A to an entity B.
$X\|Y$	Represents the concatenation of the data items X, and Y.
$X\|\|Y$	Represents the XOR binary operation on the data items X, Y.
$E_K(Z)$	Result of encrypting data Z with the key K using a symmetric algorithm.
$Sign_K(Z)$	Signature on data Z with the key K using a signature algorithm [23].
$Hash(Z)$	Is the result of generating a hash of data Z.

the description of individual messages is provided. Before we illustrate the ABP, we first describe the notation in table 1 that is used in the protocol description.

Based on the generic ABP framework in figure 7 a number of different protocols can be deployed. The proposed protocol is our attempt to provide a secure, and robust implementation of the ABP. In addition, the proposed protocol provides the trust assurance and validation; enabling it to be a secure and trusted secure channel protocol [24]. The rationale for not opting for protocols like SSL/TLS [21], and Kerberos [38] is twofold; first they have a large set of options and lengthy handshake messages that might slow down the performance on a smart card (see section 4.3), and secondly they do not support trust assurance and validation mechanism for the communicating parties. This is true to some extent for other protocols that are designed for the internet communication [24]. Even those protocols that are designed for the smart card environment; partially run on the smart card and they try to balance the computation load by performing computational intense processes offcard [7, 33, 43, 39].

The messages listed below have a one-to-one relation with the generic protocol illustrated in figure 7.

1. $C \rightarrow Firewall : C|S|Sign_C(C|S|nc|E_{K_{C-TEM}}(C, S, nc))|Cert_C$
The request message contains the identities of the client and server application together with a random number [35] generated by the client application. In addition, the client application also creates an IVA message (i.e. $E_{K_{C-TEM}}(C, S, nc)$) for the TEM (see section 2.2). The client application signs the message and appends its certificate.

2. $Firewall \rightarrow S : C|S|Sign_C(C|S|nc|E_{K_{C-TEM}}(C, S, nc))|Cert_C$
The firewall mechanism receives the application-binding request and it will query the server application. If the server application wants to proceed with the ABP, it forwards the message; otherwise, it registers an exception.

3. $S \rightarrow TEM : C|S|E_{K_{C-TEM}}(C, S, n_c)|E_{K_{S-TEM}}(S, C, n_s)$
The server application verifies the client's signature. If successful, it generates an IVA message for the client application. The server application then sends the message to the TEM that contains the identities and IVA messages from both the client and server applications.

4. $TEM \rightarrow S : E_{K_{C-TEM}}(Hash(S), K_{S-C}^t, n_c+1)|E_{K_{S-TEM}}(Hash(C), K_{S-C}^t, n_s + 1)$
The TEM verifies the IVA messages from both the client and server application. Then it will calculate the hash value of the server application, encrypt it with the client-TEM shared key and sends it to the client application, and vice versa. The encrypted messages also contains a session key generated by the TEM; this key is valid only during the ABP run.

5. $S \rightarrow C : S|C|E_{K_{C-TEM}}(Hash(S), K_{S-C}^t, n_c+1)|E_{K_{S-C}^t}(K_{S-C}, n_c+2, n_s)|Sign_S(S|C|n_s|E_{K_{S-C}}(AccessPermission, ObjectReference, n_c||n_s))|Cert_S$
Following the message 4; the server application verifies the hash value of the client application to be the same as listed either by the server application's SP or by the CC evaluation authority. It then generates an application-binding key; encrypts it with the session key. In addition, the message also contains the object reference to the server application's shared resources and access permission. The client application directly calls the server application's shared resource in all subsequent requests, using the binding key for authentication and authorisation. The access mechanism for shared resource is beyond the scope of this paper.

6. $C \rightarrow S : C|S|E_{K_{S-C}}(AccessPermission|(n_c||n_s) + 1)$

This message gives the assurance to the server application that the client also has the same key thus achieving mutually key verification.

4   Proposed Protocol Analysis

We open the discussion by analytically reviewing the protocol then the formal analysis using CasperFDR tool, and finally providing the implementation results.

4.1   Analytical Analysis

In this section, we consider the proposed protocol and analyse it with respect to the threat model, and the protocol requirements listed in section 3.2 and 3.3.

– Masquerading: A malicious application can be installed with either a server or a client application's AID. However, the ABP does not allow a malicious application to masquerade as a server or client application because to prove the identity of an application; the ABP does not rely on the AIDs. It has a dynamic mechanism with bi-directional exchanges of messages that ascertain the entity and verify its credentials (based on cryptographic certificate and signature generation/verification). Therefore, for a masquerading application it might be difficult to match the cryptographic hash (generated by the TEM) and have the signature key of the genuine application.

– Replay attack: A malicious user can relay the binding request messages, but when these messages are forwarded to the TEM to generate the hash of the client and server application. A malicious application's hash will not match the certified hash of the client and server application. This problem is equivalent to violating the 2nd pre-image property of the hash functions [35]. In addition, IVA messages include random numbers that effectively prevent any replay attacks.

– Mutual Authentication: The server and client applications authenticate one another. The authentication is achieved through signing the messages along with communicating the application's certificate. The authentication gives an assurance to each of the participant applications that the other application is genuine (effectively avoiding masquerading).

– State Verification: Although an application may have genuine credentials but its state might be modified since it was last evaluated by respective SP(s) or the CC evaluation laboratory. To verify whether the state of an application is secure enough to initiate an application sharing. The ABP requires the TEM to generate a hash of both applications and encrypt them with the corresponding keys. The applications have no influence on the outcome of the hash generation; so they cannot fake their current state. If the current state is considered to have deviated from the stated secure state in the application certificate [12]. The recipient can then decide whether to continue the protocol or not.

– Different User Applications: The application certificate contains the user details to whom the application was issued. Therefore, if a client application tries to establish an application sharing with a server application, but their customer credential does not match; the request is denied. This avoids application sharing between two applications from different users.

– Unresolved Binding Instances: The binding is based on the application-binding key that is the outcome of the ABP. If one of the applications (client or server) is to be deleted by the user. The application deletion process will notify the other application that simply deletes the instance of the application-binding key from its ACL.

The ABP provides a framework that facilitates the process; enabling the participant applications to establish a trust relationship on an open, and dynamic environment of the UCOM.

4.2 Protocol Verification by CasperFDR

The CasperFDR approach was adopted to test the soundness of the proposed protocol under the defined security properties. In this approach, the Casper compiler [2] takes a high-level description of the protocol, together with its security requirements. It then translates the description into the process algebra of Communicating Sequential Processes (CSP) [28]. The CSP description of the protocol can be machine-verified using the Failures-Divergence Refinement (FDR) model checker [40]. The intruder's capability modelled in the Casper script (Appendix A) for the proposed protocol is as below:

1. An intruder can masquerade any application's identity in the network.
2. An intruder is not allowed to masquerade the identity of any SP or TEM.
3. An intruder application has a trust relationship with the TEM.
4. It can read the messages transmitted by each entity in the network.
5. An intruder cannot influence the internal process of an agent in the network.

The security specification for which the CasperFDR evaluates the network is as shown below. The listed specifications are defined in the # Specification section of Appendix A:

1. Session and application-binding keys are not revealed to an unauthorised entity.
2. The protocol run is fresh and both applications were alive.
3. The key generated by the server application is known only to the client application.
4. Applications mutually authenticate each other and have mutual key assurance at the conclusion of the protocol.

The protocol description defined in the Casper script (Appendix A) is a simplified representation of the proposed protocol. The off-card agents like SPs of client and server applications are not model in the Casper script as they do not play active role in the protocol run. The CasperFDR tool evaluated the protocol and did not find any feasible attack(s).

4.3 Practical Implementation

The proposed protocol in section 3.5 does not specify actual details of the cryptographic algorithms that are left to the respective SPs. However, in our implementation we used AES 128bit key in Cipher Block Chaining mode with padding [30]. The signature algorithm was chosen to be RSA with 512bit key [35] and SHA-256 [3] for generating hash values .

Our implementation model was based on three applets taking the roles of the TEM, client, and server application on a Java Card (16bit smart card). At the time of writing the paper; we did not have access to a smart card platform that will enable us to implement the TEM at the underlying platform level. We mplement the TEM at the application level and consider that similar or better performance can be attained if the TEM is implemented as part of the platform (which we plan to do in future). As the application level implementation of the TEM cannot have memory access to measure the hash value of the client and server applications. Therefore, we generated the hash values of a fixed array of size 556 bytes to represent an application state. The performance of the hash algorithm is based on the size of the input data and in real deployment of the protocol scenario it definitively depends on the size of the applications.

The proposed protocol's raw implementation running on a 16bit Java Card takes 2484 ms (2.484 seconds approximately) to complete, and we consider that with adequate code optimisation we can achieve better results.

For the sake of comparison, the Kilobyte SSL [26] (KSSL: is a small footprint SSL for hand-held devices) running on a 20Mhz Palm CPU with RSA keys of length 768 and 1024 took 10-13 seconds for only server side authentication [25]. A Kerberos' implementation on a mobile device as performed by Herbitter et al. [27] (performance measures were taken from a mobile device with 100MHz CPU and 16MB of RAM) showed that the best performance was 4.240 seconds; however, based on a trusted proxy architecture the performance was 10.506 seconds. Kambourakis et al. [31] provided performance measures for the SSL based AKA mechanism that took 10 seconds to complete the protocol. For above performance measures; in the SSL implementations the server, and for Kerberos both the Key Distribution Centre (KDC) server and second communication entity, were on desktop computers.

In implementations, where smart cards act as a node in a communication protocol Pascal Urien [44] showed that a high-end SSL Smart Card establishes a SSL session in 4.2 seconds , and for smart cards as a TLS-based network node the performance was in the range of 4.3 seconds (for 32bit smart card) and 26.8 seconds (for 8bit smart card) [45, 46, 48]. The protocol performance mentioned in this section either do not rely on smart cards or partially base their implementation on smart cards. However, if we implement these protocols with all nodes on a smart card along with trust assurance, there performance will degrade.

At the time of writing this paper, the authors were not aware of any performance measures of these protocols implemented in full (all communicating nodes) on a smart card. It can be argued that the above mentioned performance measures cannot be comparable as the complete protocol were not executing on a smart card. This is a valid argument but the reason we mention them here is to augment the rationale based on the computational restrictions that prohibited us from implementing these protocols as part of the ABP.

Nevertheless, the proposed protocol performance is considered adequate. The performance measure is only for the reference of our implementation, as the actual performance will vary depending upon the size of the client and server applications (i.e. hash generation), and performance of public key operation, symmetric encryption, and random number generation.

5   Conclusion and Future Research Direction

In this paper, we discussed the application sharing mechanism from the point of view of two contrasting smart card ownership models. The firewall mechanism in the ICOM is fit for the purpose and is designed with the underlying assumption that the smart card remains under the control of a centralised authority. These firewall mechanisms and associated frameworks to establish application sharing can be considered as the state of the art in the ICOM.

Nevertheless, in the UCOM such assumptions are invalid and this requires a different set of requirements for the application sharing mechanism. We have discussed these requirements along with the threat model and provided a possible approach to resolve them. The proposed protocol meets these requirements under the assumption of the threat model both as a generic model, and then providing a practical protocol based on the generic model. We have verified the security properties for the proposed protocol in the Casper/FDR. Furthermore, an analytical analysis of the protocol is described. Finally, we implemented the protocol to provide the performance measure for the proposed approach.

As part of the future research directions we will concentrate on the architecture and functionality of the TEM for smart cards. The TEM on an UCOM based smart card provides security assurance and validation service to the smart card platform and installed applications. Furthermore, we also like to extend the TEM's capability to not only provide static assurances, but also dynamically ensure the runtime security and reliability of the platform.

6   Acknowledgements

We would like to extend our appreciation to the anonymous reviewers for their valuable time and feedback. Additionally, thanks to Min Chen for patience while proof reading drafts.

References

1. Multos: The Multos Specification
2. Casper: A Compiler for the Analysis of Security Protocols, Journal of Computer Security (June 1998)
3. FIPS 180-2: Secure Hash Standard, SHS (2002)
4. GlobalPlatform Card Security Requirement Specification 1.0 (May 2003)

5. ISO/IEC 7816-5, Information Technology - Identification cards - Integrated Circuit(s) cards with contacts - Part 5: Numbering systems and registration procedure for application identifiers, International Organization for Standardization (2004)
6. Common Criteria for Information Technology Security Evaluation, Part 1: Introduction and general model, Part 2: Security functional requirements, Part 3: Security assurance requirements (August 2006)
7. GlobalPlatform: GlobalPlatform Card Specification, Version 2.2 (March 2006)
8. Java Card Platform Specification; Application Programming Interface, Runtime Environment Specification, Virtual Machine Specification (March 2006)
9. Multos: Guide to Loading and Deleting Applications. Tech. Rep. MAO-DOC-TEC-008 v2.21, MAOSCO (2006)
10. Trusted Module Specification 1.2: Part 1- Design Principles, Part 2- Structures of the TPM, Part 3- Commands (July 2007)
11. Akram, R.N., Markantonakis, K., Mayes, K.: Application Management Framework in User Centric Smart Card Ownership Model. In: Youm, H.Y., Yung, M. (eds.) WISA 2009. LNCS, vol. 5932, pp. 20–35. Springer, Heidelberg (2009)
12. Akram, R.N., Markantonakis, K., Mayes, K.: A Dynamic and Ubiquitous Smart Card Security Assurance and Validation Mechanism. In: Rannenberg, K., Varadharajan, V., Weber, C. (eds.) SEC 2010. IFIP Advances in Information and Communication Technology, vol. 330, pp. 161–172. Springer, Heidelberg (2010)
13. Akram, R.N., Markantonakis, K., Mayes, K.: A Paradigm Shift in Smart Card Ownership Model. In: Apduhan, B.O., Gervasi, O., Iglesias, A., Taniar, D., Gavrilova, M. (eds.) Proceedings of the 2010 International Conference on Computational Science and Its Applications (ICCSA 2010), pp. 191–200. IEEE Computer Society, Fukuoka (2010)
14. Akram, R.N., Markantonakis, K., Mayes, K.: Firewall Mechanism in a User Centric Smart Card Ownership Model. In: Gollmann, D., Lanet, J.L., Iguchi-Cartigny, J. (eds.) CARDIS 2010. LNCS, vol. 6035, pp. 118–132. Springer, Heidelberg (2010)
15. Andronick, J., Chetali, B., Ly, O.: Using COQ to Verify Java Card Applet Isolation Properties. In: Basin, D., Wolff, B. (eds.) TPHOLs 2003. LNCS, vol. 2758, pp. 335–351. Springer, Heidelberg (2003)
16. Barbu, G., Thiebeauld, H., Guerin, V.: Attacks on Java Card 3.0 Combining Fault and Logical Attacks. In: Gollmann, D., Lanet, J.-L., Iguchi-Cartigny, J. (eds.) CARDIS 2010. LNCS, vol. 6035, pp. 148–163. Springer, Heidelberg (2010)
17. Bernardeschi, C., Martini, L.: Enforcement of Applet Boundaries in Java Card Systems. In: IASTED Conf. on Software Engineering and Applications, pp. 96–101 (2004)
18. Caromel, D., Henrio, L., Serpette, B.P.: Context Inference for Static Analysis of Java Card Object Sharing. In: Attali, S., Jensen, T. (eds.) E-SMART 2001. LNCS, vol. 2140, pp. 43–57. Springer, Heidelberg (2001)
19. Chen, Z.: Java Card Technology for Smart Cards: Architecture and Programmer's Guide. Addison-Wesley Longman Publishing Co., Inc., Boston (2000)
20. Deville, D., Galland, A., Grimaud, G., Jean, S.: Smart Card Operating Systems: Past, Present and Future. In: Proceedings of the 5th NORDU/USENIX Conference (2003)
21. Dierks, T., Rescorla, E.: RFC 5246 - The Transport Layer Security (TLS) Protocol Version 1.2. Tech. rep (August 2008)
22. Éluard, M., Jensen, T., Denne, E.: An Operational Semantics of the Java Card Firewall. In: Attali, S., Jensen, T. (eds.) E-SMART 2001. LNCS, vol. 2140, pp. 95–110. Springer, Heidelberg (2001)

23. Furlani, C.: FIPS 186-3 : Digital Signature Standard (DSS) (June 2009)
24. Gasmi, Y., Sadeghi, A.R., Stewin, P., Unger, M., Asokan, N.: Beyond Secure Channels. In: STC 2007: Proceedings of the 2007 ACM workshop on Scalable trusted computing, pp. 30–40. ACM, New York (2007)
25. Gupta, V., Gupta, S.: Securing the Wireless Internet. IEEE Communications 39(12), 68–74 (2001)
26. Gupta, V., Gupta, S.: KSSL: Experiments in Wireless Internet Security. Tech. rep., Mountain View, CA, USA (2001)
27. Harbitter, A., Menascé, D.A.: The Performance of Public Key-Enabled Kerberos Authentication in Mobile Computing Aplications, pp. 78–85 (2001)
28. Hoare, C.A.R.: Communicating Sequential Processes, vol. 21. ACM, New York (1978)
29. Huisman, M., Gurov, D., Sprenger, C., Chugunov, G.: Checking Absence of Illicit Applet Interactions: A Case Study. In: Wermelinger, M., Margaria-Steffen, T. (eds.) FASE 2004. LNCS, vol. 2984, pp. 84–98. Springer, Heidelberg (2004)
30. Daemen, J., Rijmen, V.: The Design of Rijndael: AES - The Advanced Encryption Standard. Springer, Berlin (2002)
31. Kambourakis, G., Rouskas, A., Gritzalis, S.: Experimental Analysis of an SSL-Based AKA Mechanism in 3G-and-Beyond Wireless Networks. Wirel. Pers. Commun. 29, 303–321 (2004)
32. Lanet, J.L., Iguchi-Cartigny, J.: Developing a Trojan applet in a Smart Card. Journal in Computer Virology 6(1) (2009)
33. Markantonakis, K., Mayes, K.: A Secure Channel Protocol for Multi-application Smart Cards based on Public Key Cryptography. In: Chadwick, D., Prennel, B. (eds.) CMS 2004 - Eight IFIP TC-6-11 Conference on Communications and Multimedia Security, pp. 79–96. Springer, Heidelberg (2004)
34. Mayes, K., Markantonakis, K.: Smart Cards, Tokens, Security and Applications. Springer, Heidelberg (2008)
35. Menezes, A.J., van Oorschot, P.C., Vanstone, S.A.: Handbook of Applied Cryptography. CRC, Boca Raton (1996)
36. Montgomery, M., Krishna, K.: Secure Object Sharing in Java Card. In: WOST 1999: Proceedings of the USENIX Workshop on Smartcard Technology. USENIX Association, Berkeley (1999)
37. Mostowski, W., Poll, E.: Malicious Code on Java Card Smartcards: Attacks and Countermeasures. In: Grimaud, G., Standaert, F.-X. (eds.) CARDIS 2008. LNCS, vol. 5189, pp. 1–16. Springer, Heidelberg (2008)
38. Neuman, C., Hartman, S., Raeburn, K.: RFC 4120: The Kerberos Network Authentication Service (V5). Tech. rep (July 2005)
39. Rantos, K., Markantonakis, C.: An Asymmetric Cryptography Secure Channel Protocol for Smart Cards. In: Deswarte, Y., Cuppens, F., Jajodia, S., Wang, L. (eds.) Security and Protection in Information Processing Systems, IFIP 18th WorldComputer Congress, TC11 19th International Information Security Conference, Toulouse, August 22-27, pp. 351–366. Kluwer, Dordrecht (2004)
40. Ryan, P., Schneider, S.: The Modelling and Analysis of Security Protocols: the CSP Approach. Addison-Wesley Professional, Reading (2000)
41. Sauveron, D.: Multiapplication Smart Card: Towards an Open Smart Card? Inf. Secur. Tech. Rep. 14(2), 70–78 (2009)
42. Sauveron, D., Dusart, P.: Which Trust Can Be Expected of the Common Criteria Certification at End-User Level? Future Generation Communication and Networking 2, 423–428 (2007)

43. Sirett, W.G., MacDonald, J.A., Mayes, K., Markantonakis, K.: Design, Installation and Execution of a Security Agent for Mobile Stations. In: Domingo-Ferrer, J., Posegga, J., Schreckling, D. (eds.) CARDIS 2006. LNCS, vol. 3928, pp. 1–15. Springer, Heidelberg (2006)
44. Urien, P.: Collaboration of SSL Smart Cards within the WEB2 Landscape. In: International Symposium on Collaborative Technologies and Systems, pp. 187–194 (2009)
45. Urien, P., Elrharbi, S.: Tandem Smart Cards: Enforcing Trust for TLS-Based Network Services. In: International Workshop on Applications and Services in Wireless Networks, pp. 96–104 (2008)
46. Urien, P., Marie, E., Kiennert, C.: An Innovative Solution for Cloud Computing Authentication: Grids of EAP-TLS Smart Cards. In: International Conference on Digital Telecommunications, pp. 22–27 (2010)
47. Vetillard, E., Ferrari, A.: Combined Attacks and Countermeasures. In: Gollmann, D., Lanet, J.-L., Iguchi-Cartigny, J. (eds.) CARDIS 2010. LNCS, vol. 6035, pp. 133–147. Springer, Heidelberg (2010)
48. Yu, D., Chen, N., Tan, C.: Design and Implementation of Mobile Security Access System (MSAS) Based on SSL VPN. In: International Workshop on Education Technology and Computer Science, vol. 3, pp. 152–155 (2009)

A   Casper/FDR Script

```
# Free variables
S, C, spS, spC : Agent
TEM : Server
nc, ns, nm : Nonce
ksc, abKsc : SessionKey
f : HashFunction
ServerKey : Agent -> ServerKeys
VKey : Agent -> Publickey
SKey : Agent -> SecretKey
realAgent : Server -> Bool
InverseKeys = (ksc, ksc), (abKsc, abKsc), (ServerKey, ServerKey),(VKey, SKey)

#Actual variables
CApp, SApp, MAppl · Agent
TM : Server
Nc, Ns, Nm : Nonce
Ksc, ABKsc : SessionKey
InverseKeys = (Ksc, Ksc), (ABKsc, ABKsc)

#Processes
INITIATOR(C, TEM, S, nc) knows f(S), ServerKey(C), SKey(C), VKey
RESPONDER(S, TEM, C, ns, abKsc) knows f(C), ServerKey(S), SKey(S), VKey
SERVER(TEM, ksc) knows ServerKey

#System
INITIATOR(CApp,TM, SApp, Nc)
RESPONDER(SApp,TM, CApp, Ns, ABKsc)
```

```
SERVER(TM, Ksc)

#Protocol description
0.              -> C : S
1.    C -> S : C, S, {C, S, nc, {C, S, nc}{ServerKey(C)}% mTEM}{SKey(C)}
2.      S -> TEM : S, TEM, C, {S, C, ns}{ServerKey(S)},mTEM % {C,S,nc}{ServerKey(C
[realAgent(TEM)]
3.    TEM -> S : TEM, S, {f(S), ksc, nc}{ServerKey(C)}%TEMC
[realAgent(TEM)]
3a.     TEM -> S : TEM, {f(C), ksc, ns}{ServerKey(S)}
4.    S -> C : S, C, TEMC % {f(S), ksc, nc} {ServerKey(C)}
4a.      S -> C : {abKsc, nc, ns}{ksc},{S, C, nc(+)ns}{abKsc}
5.      C -> S : C, S,{nc(+)ns}{abKsc}

#Specification
StrongSecret(TEM, ksc, [S,C])
StrongSecret(S, abKsc, [C])
Aliveness(S, C)
Aliveness(C, S)
Agreement(S, C, [abKsc])
Agreement(C, S, [abKsc])

#Inline functions
symbolic ServerKey
symbolic VKey, SKey
realAgent(TM)=true
realAgent(_)=false

#Intruder Information
Intruder = MAppl
IntruderKnowledge = {CApp, SApp, MAppl, Nm, ServerKey(MAppl), SKey(MAppl),VKey}
```

Security in Depth through Smart Space Cascades

Benjamin W. Long

Defence Science and Technology Organisation,
Edinburgh, South Australia
benjamin.long@dsto.defence.gov.au

Abstract. Security in depth relies on controlled access across a layering of protective barriers. We introduce *smart space cascades*, a framework in which access control is applied to a hierarchy of smart spaces, as a way of achieving security in depth in the context of highly automated work environments.

1   Introduction

Security in depth (also known as *defence in depth*) is a multi-layered approach to security in which security measures are combined to form a succession of barriers, all of which must be penetrated for resources (or *targets*) to be acquired, reducing the opportunity for unauthorised access [5, 26].

An effective security in depth strategy relies on two factors:

- a suitable layering of barriers, and
- an access control framework capable of supporting it.

A *smart space* [11–13, 19, 23, 28] is a room (or indeed any area) that provides automated control over electronic components within the space, simplifying our technology-enabled environments and empowering users to realise the full potential of the combined technologies.

By applying access control to a smart space, we can restrict access to such electronic components, and use those components to protect our critical assets. For example, access control applied to a 'room space' could prevent particular personnel from opening the door and accessing sensitive materials.

We find then that a hierarchy (or cascade) of smart spaces with access control applied provides a succession of barriers, forming the multi-layered framework required to achieve security in depth. For instance, a room might control access to an electronic safe which controls access to sensitive documents.

However, with the possibility of mobile components, this hierarchy may change at any time. This is significant because the decision to allow access to a space will depend on the resources within it (its *dependent targets*). *Ad hoc smart spaces* [8] allow smart space components to establish connections dynamically and interact with each other to provide functionality specific to the particular combination of components. We adopt this ad hoc nature in our framework and, in our case,

U. Parampalli and P. Hawkes (Eds.): ACISP 2011, LNCS 6812, pp. 226–240, 2011.

allow components to interact with each other to make the best access decision given the hierarchy at the particular time of each request.

In this paper, based on related work, we identify desired characteristics for access control in smart space cascades where dynamic dependencies between targets demand interaction between their access control components. Subsequently, we present a framework based on the ISO/IEC 10181-3 access control standard to support these characteristics, introducing *dependency trees* as a core element to the framework. The components of the framework and the relationships between them are specified using the Z specification language [27] to aid our explanations. Finally, we describe the potential for a real-time analysis of the layers formed by a smart space cascade, to assist in determining whether resources within a cascade are suitably protected.

2   Related Work

Some smart space solutions manage access control centrally [1, 2, 10, 25]. However, research suggests that future development in access control should be towards the composition of independent local access control policies [17, 30]. Furthermore, Altunay et al. [3] argue that targets may have policies that are confidential and should make their own access control decisions too [3]. Decentralisation reduces overhead from round-trip communication to a central server and enables meaningful enforcement of some access control policies even if there is a partial disconnection [6], allowing targets to operate autonomously [21].

However, once access control decisions are decentralised, due to dependencies, targets within a smart space cascade will need to interact with each other in order to make a decision that suits all of them. Some smart space frameworks allow targets in one space to interact with targets in another [19, 29]. Access control in these frameworks is not discussed in depth, although Marsá-Maestre et al. [19] do allow for the possibility of hierarchy of smart spaces.

McCarthy and Thredgold [20] discuss an approach in which access control components negotiate positive outcomes for users. For example, when a user requests to access a room, if a dependent electronic safe is currently open, the room could interact with the safe to close it automatically, and then grant the user access when it would otherwise be denied. Altunay et al.'s [3] approach negotiates positive outcomes for users by searching for targets to replace ones that will not or cannot participate in the collaborative decision making process.

Chandershekarapuram et al. [9] apply 'single sign on' to devices in mobile ad-hoc networks (MANETs) for a more convenient approach to access control involving multiple targets. Incorporating this kind of functionality in smart spaces would allow authorised users to initiate a single request for convenient access to multiple dependent targets across several layers of the security in depth strategy (although it would be more secure to restrict this feature to a single layer).

Ideally, then, a comprehensive framework for access control in the context of smart spaces cascades will:

- support the handling of decentralised requests,
  - allowing targets to exercise access control autonomously and
  - allowing interaction between targets for a comprehensive decision,
- negotiate positive outcomes for users, and
- be amenable to single sign on.

3   Access Control for Smart Space Cascades

In this section we present a decentralised framework, based on the ISO/IEC 10181-3 standard, to support the desired characteristics for access control in smart space cascades, allowing for the required interactions between access control components associated with targets and their dependents. We use the Z specification language [27] to specify the components of the framework and the relationships between them.

3.1   Background: ISO/IEC 10181-3

Part three of the ISO/IEC 10181 standard (ISO/IEC 10181-3) [16] describes an access control framework suggesting the existence of four components for any given request for access: the *initiator*, the *target*, an *access control enforcement function* (AEF) that controls access to the target based on a decision, and an *access control decision function* (ADF) that makes a decision for the AEF, as illustrated by Figure 1.

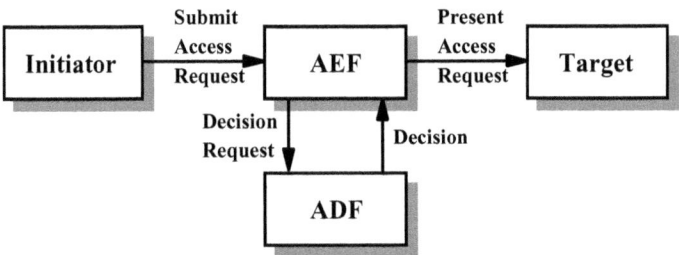

Fig. 1. ISO/IEC 10181 access control

3.2   Targets in Practice

Figure 1 illustrates the intention that every target is to be accessed via an AEF. However, in practice, the primary targets that we want to protect (such as paper documents, electronic documents, and munitions) cannot be controlled directly by an AEF. They must be protected by a target (such as a safe, a file system, or a secure room) that can be controlled by an AEF. Without this distinction, those targets for which we are most concerned are not represented within the framework. Therefore, we introduce:

- *primary targets*—targets that do not provide an interface to which an AEF can communicate; and
- *protective targets*—targets on which an AEF can invoke operations, controlling access to other (primary or protective) targets.

Given the set of all targets, *TARGET*, a smart space cascade will have disjoint sets of primary targets and protective targets, described by the following Z schema.

$$
\begin{array}{l}
\underline{\quad SSC1 \quad\qquad\qquad\qquad\qquad\qquad\qquad\qquad\qquad\qquad} \\
prim\,Targets : \mathbb{P}\ TARGET \\
prot\,Targets : \mathbb{P}\ TARGET \\
\overline{\qquad\qquad\qquad\qquad\qquad} \\
prim\,Targets \cap prot\,Targets = \varnothing
\end{array}
$$

3.3 Dependent Targets

In smart space cascades, with multiple technologies tightly integrated, access to one target exposes the next layer of *dependent targets*. For example, access to a room could expose an electronic display and a safe, potentially also providing access to the sensitive materials they contain (or display).

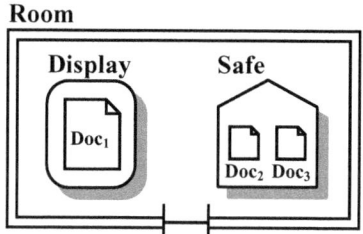

Fig. 2. A room environment

Indeed, ISO/IEC 10181-3 discusses *containment* in which one target resides inside another target T and, therefore, depends on T for protection. In general, a target could depend on multiple targets or circular dependencies could exist; however, for the purpose of this discussion we are only concerned with dependencies resulting from containment and exclude others from our framework.

Therefore, we say that in any given smart space cascade there is a collection of *directed dependency trees*. For example, consider the dependency tree in Figure 3 based on the scenario in Figure 2.

Continuing with the specification (inheriting *SSC1*), the following schema describes a partial function *dependsOn*, allowing each target in a subset of all targets to depend on another.

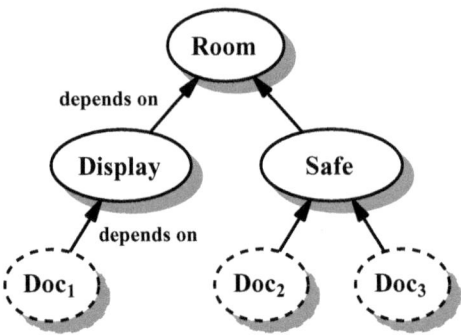

Fig. 3. Cascade dependencies

$$
\begin{array}{|l|}
\hline
\,\underline{SSC2}\\
\quad SSC1\\
\quad dependsOn : TARGET \twoheadrightarrow TARGET\\
\hline
\quad \forall\, t : TARGET \bullet (t, t) \notin dependsOn^{+}\\
\quad \mathrm{dom}\ dependsOn \subseteq primTargets \cup protTargets\\
\quad \mathrm{ran}\ dependsOn \subseteq protTargets\\
\hline
\end{array}
$$

The first predicate below the line states that there is no target that depends on itself within the transitive closure, forming a tree structure among the dependencies. The second predicate restricts the domain of the function to a subset of those primary and protective targets in the system, and the third predicate ensures that targets can only depend on protective targets (i.e., targets cannot depend on primary targets). Note that we consider, for example, a stack of sensitive documents inside a locked cupboard to be a single primary target for the purpose of our discussion.

In practice, the dependents of any given target in a particular scenario could be updated dynamically with registration/deregistration protocols using wireless technologies such as Bluetooth, much like the registration process adopted by EgoSpaces [17]. Primary targets would be entered in to the system manually or through the use of a tracking technology such as RFID.

3.4   Access Control Enforcement Functions

The purpose of the AEF is to act as a protective interface between initiators and targets, enforcing the decision made by the ADF (to grant or deny access) by controlling access to targets accordingly.

ISO/IEC 10181-3 states that an AEF is placed between each initiator-target instance so that the initiator can act on a target only through the AEF. We assume this means each target is associated with a single AEF; this avoids potential inconsistencies in enforcement for a given target. However, we find no reason to prevent a single AEF from enforcing access for multiple targets.

Given the set AEF of all AEFs, the following continuation of the specification, describes a partial function aef associating each target from a subset of all targets to an AEF. The predicate ensures all protective targets are controlled by an AEF, and that only protective targets are controlled by AEFs.

$$\begin{array}{|l}
\hline
\_\_SSC3_____ \\
SSC2 \\
aef : TARGET \rightarrow\!\!\!\!\!\rightarrow AEF \\
\hline
protTargets = \mathrm{dom}\; aef \\
\hline
\end{array}$$

The set of targets controlled by AEFs depends on the level of granularity desired. In a centralised system, it is sufficient for access to all protective targets to be controlled by a single enforcement function ($\#(\mathrm{ran}\; aef) = 1$). However, in order to allow targets to operate autonomously and move dynamically from one space to another, we can decentralise the enforcement (like Altunay et al. [3] and Moloney and Weber [21]) by having a single AEF manage access for each portion of the dependency tree that we want operating autonomously.

3.5   Access Control Decision Functions

An ADF provides a decision to an AEF based on a decision request from the AEF. The ADF bases its decisions on information such as policy, context, history, and additional information provided by the initiator.

Figure 1 suggests each AEF interacts with a single ADF. However, we suggest a single ADF can be used for a particular type of request to access a particular target. Therefore, given the set ADF of all ADFs and request types $REQTYPE$, we continue with the specification and include a function adf providing a single (but not necessarily unique) ADF for every combination of protective target and request type.

$$\begin{array}{|l}
\hline
\_\_SSC4_____ \\
SSC3 \\
adf : (TARGET \times REQTYPE) \rightarrow\!\!\!\!\!\rightarrow ADF \\
\hline
protTargets = \mathrm{dom}(\mathrm{dom}\; adf) \\
\hline
\end{array}$$

4   Smart Space Interactions

Within our framework, a smart space encompasses a portion of targets (and associated access control components) in a dependency tree, and a smart space cascade (see Figure 4) emerges as a hierarchical collection of smart spaces for a particular environment.

It is our intention that targets and associated access control components will change dynamically throughout the life of a smart space cascade. For instance, when an electronic safe is added to a room, the safe will become a dependent of the room, and associated components will form part of the hierarchy appropriately.

In Section 2 we identified that a comprehensive framework for access control in the context of smart spaces will support decentralised requests, negotiate positive outcomes for users, and be amenable to single sign on. The following sections describe how the framework introduced in Section 3 supports these, noting that we leave safe recovery from failure as a topic for future work.

4.1   Decentralised Requests

When an initiator sends an *access request* to the AEF protecting a set of targets, its associated ADF might need to consider ADFs belonging to dependent targets to ensure it does not make a decision that violates any of their policies.

Therefore, the AEF's ADF becomes an initiator and sends its own *cascaded decision request* to the dependents' AEFs. This starts a decision-making procedure that traverses the entire dependency tree. (We believe scaling will not be an issue due to the limited number of levels to traverse given the nature of the environments).

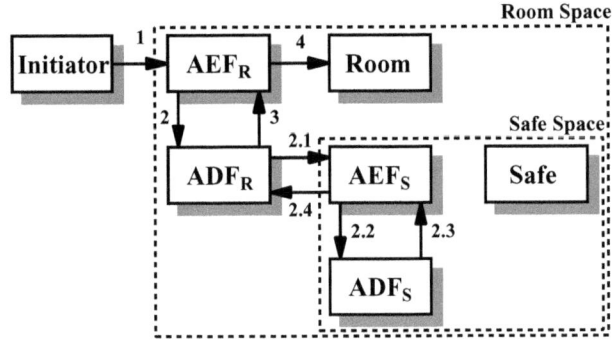

Fig. 4. Requesting access to the room

For example, when the room's AEF, AEF_R, receives a request for entry from an initiator, it will be aware of the dependency of the safe. Subsequently (as illustrated by Figure 4), in step 2.1 the room's ADF, ADF_R, sends a decision request to the safe's AEF, AEF_S. In step 2.2 AEF_S will seek a decision from the safe's ADF, ADF_S. In step 2.3 ADF_S will send a decision to AEF_S. Then, finally, because it is a cascaded decision request and not an access request, instead of following the ISO/IEC 10181-3 standard and providing access to the safe, in step 2.4 AEF_S will forward the decision back to ADF_R—a necessary extension to the standard [7]. From here, ADF_R makes a final *decision* in step 3, based on the results of the cascaded request, and access is granted or denied in step 4.

For example, if none of the dependent ADFs object to the request, the user is granted access (and denied otherwise).

Although the scope of our specification is limited to the relationship between components (but not the interactions between them), we can specify an operation *identifyComponents* that determines for a given protective target *target?*, type of request *req?*, and AEF *aef?*, the access control components (the ADF and the set of AEFs based on dependent targets) with which that AEF will interact.

$$
\begin{array}{|l}
\underline{\;identifyComponents\;}\\[2pt]
\Xi SSC4\\
req? : REQTYPE;\; target? : TARGET;\; aef? : AEF\\
adf! : ADF;\; dependentAEFs! : \mathbb{P}\,AEF\\
\hline
target? \in protTargets\\
(target?, aef?) \in aef\\
((target?, req?), adf!) \in adf\\
dependentAEFs! = \mathrm{ran}((\mathrm{dom}(dependsOn \rhd \{target?\})) \lhd aef)
\end{array}
$$

Although decision information flows up the hierarchy, context information will need to flow down the hierarchy to ensure suitable decisions are made when requests are initiated on targets below the root of the dependency tree.

It is also important to note here that the trustworthiness of components to provide accurate information is out of the scope of this paper.

4.2   Negotiating Positive Outcomes

Using the same scenario, although the user may be allowed to access the room, he/she might not be allowed to access the safe which, in this case, is currently open. In order to negotiate a positive outcome for the user, ADF_S will include preconditions accompanying its decision in step 2.3 (see Figure 4) regarding necessary changes to the safe's state. These preconditions will be sent back through steps 2.4 and 3 to be managed by AEF_R, assuming all ADFs are satisfied. That is (in Figure 5), AEF_R initiates an access request (step 4.1) with AEF_S to close the safe (which must be successful) before the original request for access to the room is granted in step 4.6.

4.3   Smart Space Sign on

Single sign on (SSO) allows users to authenticate only once for a set of software components, although we believe a similar concept will soon become more popular for hardware components. Indeed, Chandershekarapuram et al. [9] propose a framework for 'device single sign on' in mobile ad-hoc networks. Additionally, Nishiki and Tanaka [22] hint at the idea of smart space sign on for context-aware services.

To enable *smart space sign on*, an attempt for access will involve multiple requests; for example, one to gain entry to the room, and one to automatically open the safe. These requests may be set as presets against the user's profile (maintained by the room or the initiator's identity management device).

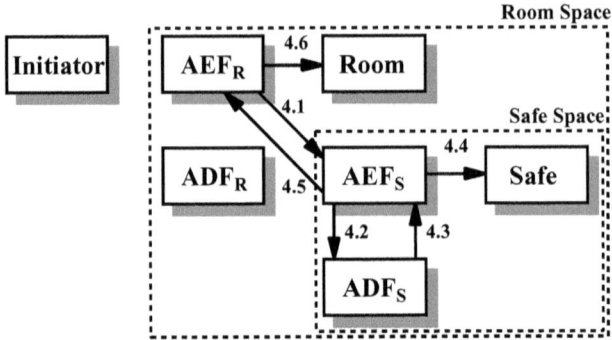

Fig. 5. Subsequent request for actions based on preconditions

Again, using the same scenario, the request to access the safe would be included as part of the request in step 1 and forwarded to AEF_S in step 2.1 (see Figure 4). To avoid potential conflicts between ADFs, the request for access to the safe will need to happen after the final decision has been made. Therefore, ADF_R could include the request for access to the safe as a precondition accompanying its decision in step 3 (see Figure 4) to be acted upon in step 4.1 (see Figure 5).

We note that caution must be taken when allowing initiators to sign on to multiple targets through various layers of the cascade in a single request. For example, if an initiator's credentials are compromised, an unauthorised user may gain instant access to multiple layers within the security in depth strategy. Indeed, Price [24] warns against *cascading failure*. This threat could be minimised by limiting the targets to which the initiator can access via SSO and by insisting on stronger authentication (including multi-factor authentication [18, pp. 235-237]).

5   Protective Layer Analysis

Drawing on elements from the research of Hitchens [14] and Dowell [15], we provide an example approach to demonstrate how we could ensure in real-time that targets within a smart space cascade are protected by a suitable layering of barriers.

The example approach is based on risk analysis, calculated as a function over consequence and likelihood. In this case, consequence and likelihood refer to *asset criticality* and *layer vulnerability*, respectively.

In the context of access control, we require a single value to reflect an overall consequence of unauthorised access to targets. We suggest asset criticality to be an appropriate measure and encourage the use of an asset criticality analysis [4] for this step.

Guidelines for establishing asset criticality is not in the scope of this paper; however, for the purpose of our discussion we will assume asset criticality is a measure between 0 and 1, where 0 represents no (or negligible) consequence and higher values indicate more serious consequences as a result of unauthorised access.

In the context of security in depth, the likelihood of a threat depends on the vulnerability of the layer. Again, although guidelines for establishing layer vulnerability is not in our scope, we assume a value between 0 and 1 indicates the vulnerability of a protective layer in a security in depth strategy, this time where 0 represents an impenetrable barrier and 1 represents the absence of any protection. In general, a layer's vulnerability will be based on factors such as structural materials, security mechanisms and the presence of guards, security cameras and motion sensors.

5.1 Applying the Method

Ultimately, primary targets are the critical targets we need to protect. Therefore, we suggest an asset criticality is determined for each primary target and a layer vulnerability for each protective target. Criticality values for protective targets can be derived based on the criticality of their contents and the level of protection they provide.

Hitchens [14] multiplies layer vulnerabilities to determine an overall vulnerability of the combined independent layers. For example the combined strengths of two layers with a vulnerability of 0.3 each, result in a combined vulnerability of 0.09.

Using a similar approach, the asset criticality value for a given protective target could be calculated by a function over the criticalities of its dependents and its own layer vulnerability. For example, it might be feasible to suggest that a safe with no protection (vulnerability $= 1$) would have a criticality equal to the most critical target it contains. Then as the vulnerability decreases, the criticality also decreases. Therefore, a safe with layer vulnerability $V_S = 0.9$ containing two documents with asset criticalities $C_2 = 0.9$ and $C_3 = 0.8$ could have a derived criticality of $max(C_2, C_3) * V_S = 0.81$. Then the safe's criticality C_S and the display's criticality C_D would be used to calculate the criticality of the room C_R (see Figure 6).

Given the set $LEVEL$ of all real numbers between 0 and 1, such a scheme within our framework is specified as follows.

```
┌─ SSC5 ──────────────────────────────────────────────────────────
│  SSC4
│  vulnerability : TARGET ↠ LEVEL
│  criticality : TARGET → LEVEL
├──────────────────────────────────────────────────────────────────
│  dom vulnerability = protTargets
│  ∀ t : TARGET • t ∈ ran dependsOn ⟹ criticality(t) =
│      max(ran((dom(dependsOn ▷ {t})) ◁ criticality)) * vulnerability(t)
└──────────────────────────────────────────────────────────────────
```

Taking the maximum of the dependent targets' criticalities is consistent with current practice in defence environments. However, another option would be to take the sum of the criticalities, using an unbounded real number for criticality levels.

Although targets in our dependency tree are dependent on each other for protection, the example probabilistic approach above assumes that the mechanisms used by protective targets are independent. Independence can be increased by using different mechanisms at each layer; for example, users could be prompted for an RFID swipe card at one layer and a biometric at the next. Nevertheless, future work involves exploring Bayesian inferencing to cater for such dependencies.

Access by initiators to a smart space can then be restricted based on the criticality of the targets exposed directly by the space. For instance, a user who is cleared to be exposed to targets with criticality levels of 0.85 and below, will be granted access to the room in Figure 6 containing the closed safe ($C_S < 0.85$).

Note that when the safe is open $V_S = 1$ and $C_S = 0.9$, in which case the user would not be allowed access to the room unless the system negotiates a positive outcome for the user by closing the safe first.

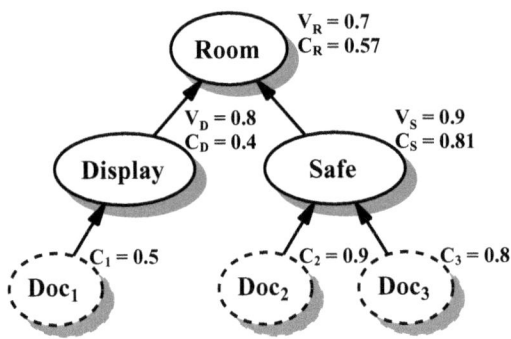

Fig. 6. Derived asset criticalities over a dependency tree

Continuing with our specification, we specify that each user from the set *USER* of all users has a level to which they are cleared, and that users will be exposed to zero or more targets. The predicate below the line ensures that all targets to which a user is exposed have a criticality lower than or equal to that user's clearance.

$$
\begin{array}{|l}
\hline
\_SSC6 _____ \\
SSC5 \\
clearedTo : USER \rightarrow LEVEL \\
exposedTo : USER \leftrightarrow TARGET \\
\hline
\forall u : USER;\ t : TARGET \bullet (u, t) \in exposedTo \Rightarrow \\
\quad clearedTo(u) \geq criticality(t) \\
\hline
\end{array}
$$

In order to prevent the derived criticality for a protective target from increasing beyond a reasonable limit, a criticality threshold could be associated with each protective target. However, a target's threshold could not exceed that which is bound by its parent's threshold and its own vulnerability. That is, the threshold T_S of the safe must be less than or equal to the threshold T_R of the room divided by the vulnerability V_S of the safe ($T_S \leq \frac{T_R}{V_S}$). This is expressed in the following schema.

$$
\begin{array}{|l}
\hline
\_\_SSC7_____ \\
\quad SSC6 \\
\quad threshold : TARGET \nrightarrow LEVEL \\
\hline
\quad \text{dom } threshold = protTargets \\
\quad \forall\, s, r : TARGET \bullet (s, r) \in dependsOn \wedge s \in protTargets \Rightarrow \\
\qquad threshold(s) \leq \frac{threshold(r)}{vulnerability(s)} \\
\quad \forall\, t : TARGET \bullet t \in protTargets \Rightarrow criticality(t) \leq threshold(t) \\
\hline
\end{array}
$$

For example, the room could be placed in an unprotected environment (layer vulnerability $= 1$) with a criticality threshold $T_E = 0.6$, determining the acceptable risk given the potential threats in that particular environment. The room's threshold T_R would have to be less than or equal to $T_E / V_R = 0.6/0.7 = 0.86$. And then the safe's threshold would have to be less than or equal to $T_R / V_S = 0.86/0.9 = 0.96$. So the safe would refuse any request to protect documents with a criticality value greater than 0.96.

Access control mechanisms associated with each target, as discussed in Sections 3 and 4 can enforce this kind of real-time protection layer analysis, ensuring that targets are protected suitably.

6   Conclusion

In this paper we presented a framework for security in depth based on a hierarchy (or cascade) of ad hoc smart spaces with access control applied.

Firstly, we introduced the distinction between primary and protective targets, and dependency trees to describe the protective layering between targets. Then we introduced accompanying access control components based on ISO/IEC 10181-3 and, within a Z specification, stated various design restrictions on the relationships between targets and their accompanying access control components.

We described how the components within smart space cascades can interact with each other to enable: decentralised decisions involving dependent targets; dependent targets to specify preconditions to negotiate positive outcomes for users; and single sign on for user convenience.

Finally, we described an approach for a real-time analysis of the series of layers protecting targets in smart space cascades. Following a risk based approach, we have suggested adopting asset criticality as a measure of consequence and layer vulnerability as a measure of likelihood.

In future work we will further investigate open issues we identified in relation to cascaded interactions, and explore suitable policies within cascaded spaces in defence environments.

Acknowledgements

I would like to thank Paul Montague and Damian Marriott for reviewing multiple drafts of this paper and for their support and valuable contributions. Also Angela Billard, Clare Saddler and David Adie for preliminary discussions, and the anonymous referees for their comments.

References

1. Al-Muhtadi, J., Ranganathan, A., Campbell, R., Mickunas, M.D.: Cerberus: A context-aware security scheme for smart spaces. In: Proceedings of the First IEEE International Conference on Pervasive Computing and Communications, pp. 489–496. IEEE Computer Society, Los Alamitos (2003)
2. Al-Qutayri, M., Barada, H., Al-Mehairi, S., Nuaimi, J.: A framework for an end-to-end secure wireless smart home system. In: Proceedings of Annual IEEE International Systems Conference, pp. 1–7. IEEE Computer Society, Los Alamitos (2003)
3. Altunay, M., Brown, D.E., Byrd, G.T., Dean, R.A.: Collaboration policies: Access control management in decentralized heterogeneous workflows. Journal of Software 1(1), 11–22 (2006)
4. Anderson, D., Keleher, P., Smith, P.: Towards and assessment tool for strategic management of asset criticality. Australian Journal of Mechanical Engineering 5(2), 115–126 (2008)
5. Australian Government: Protective security policy framework (2011)
6. Balasubramanian, M., Bhatnagar, A., Chaturvedi, N., Chowdhury, A.D., Ganesh, A.: A framework for decentralized access control. In: Proceedings of the 2nd ACM symposium on Information, Computer and Communications Security (ASIACCS 2007), pp. 93–104. ACM, New York (2007)
7. Billard, A., Long, B.: Dynamic security architectures: Architecture and case studies. DSTO Technical Report (in review), Defence Science and Technology Organisation (July 2009)
8. Brodt, A., Sathish, S.: Together we are strong—towards ad-hoc smart spaces. In: Proceedings of IEEE International Conference on Pervasive Computing and Communications (PerCom 2009), pp. 1–4. IEEE Computer, Los Alamitos (2009)
9. Chandershekarapuram, A., Vogiatzis, D., Vassilaras, S., Yovanof, G.S.: Architecture framework for device single sign on in personal area networks. In: Meersman, R., Tari, Z., Herrero, P. (eds.) OTM 2006 Workshops. LNCS, vol. 4278, pp. 1367–1379. Springer, Heidelberg (2006)
10. Corradi, A., Montanari, R., Tibaldi, D., Toninelli, A.: A context-centric security middleware for service provisioning in pervasive computing. In: Proceedings of the 2005 Symposium on Applications and the Internet, pp. 421–429. IEEE Computer Society, Los Alamitos (2005)

11. Das, S.K., Cook, D.J.: Designing and modelling smart environments. In: Proceedings of the 2006 International Symposium on a World of Wireless, Mobile and Multimedia Networks (WoWMoM 2006), pp. 490–494. IEEE Computer Society, Los Alamitos (2006)

12. Dimakis, N., Soldatos, J.K., Polymenakos, L., Fleury, P., Cuřín, J., Kleindienst, J.: Integrated development of context-aware applications in smart spaces. Pervasive Computing 7(4), 71–79 (2008)

13. Helal, S., Mann, W., El-Zabadani, H., King, J., Kaddoura, Y., Jansen, E.: The Gator Tech Smart House: A programmable pervasive space. Computer 38(3), 50–60 (2005)

14. Hitchins, D.K.: Secure systems—defence in depth. In: Proceedings of the European Convention on Security and Detection, pp. 34–39. IEEE Computer Society, Los Alamitos (1995)

15. Dowell III, A.M.: Layer of protection analysis for determining safety integrity level. ISA Transactions 37, 155–165 (1998)

16. International Standardization Organization: ISO/IEC 10181-3:1996(E): Information technology — open systems interconnection — security frameworks for open systems: Access control framework (1996)

17. Julien, C., Roman, G.C., Payton, J.: Context-sensitive access control for open mobile agent systems. In: Proceedings of the 3rd International Workshop on Software Engineering for Large-Scale Multi-Agent Systems, co-located with ICSE 2004, pp. 42–48 (2004)

18. Kizza, J.: Computer network security. Springer, Heidelberg (2005)

19. Marsá-Maestre, I., de la Hoz, E., Alarcos, B., Velasco, J.R.: A hierarchical, agent-based approach to security in smart offices. In: Proceedings of the International Conference on Ubiquitous Computing: Applications, Technology and Social Issues, ICUC 2006 (2006)

20. McCarthy, J., Thredgold, J.: Modelling smart security for classified rooms with DOVE. In: Proceedings of the Conference on Application and Theory of Petri Nets, pp. 135–144. Australian Computer Society (2002)

21. Moloney, M., Weber, S.: A context-aware trust-based security system for ad hoc networks. In: Workshop of the 1st International Conference on Security and Privacy for Emerging Areas in Communication Networks, pp. 153–160. IEEE Computer Society, Los Alamitos (2005)

22. Nishiki, K., Tanaka, E.: Authentication and access control agent framework for context-aware services. In: Proceedings of the 2005 Symposium on Applications and the Internet Workshops, pp. 200–203. IEEE Computer Society, Los Alamitos (2005)

23. Phillips, M.: Livespaces technical overview. DSTO Technical Report (draft), Defence Science and Technology Organisation (2008)

24. Price, S.M.: A defense-in-depth security architecture strategy inspired by antiquity. Information Systems Security Association 8(3), 10–16 (2010)

25. Sampemane, G., Naldurg, P., Campbell, R.H.: Access control for active spaces. In: Proceedings of the 18th Annual Computer Security Applications Conference (ACSAC 2002), pp. 343–352. IEEE Computer Society, Los Alamitos (2002)

26. Smith, C.L.: Understanding concepts in the defence in depth strategy. In: Proceedings of the 37th Annual 2003 International Carnahan Conference on Security Technology, pp. 8–16. IEEE Computer Society, Los Alamitos (2003)

27. Spivey, J.M.: The Z Notation: A Reference Manual. Prentice Hall International Series In Computer Science. Prentice Hall, London (1992)

28. Stanford, V., Garofolo, J., Galibert, O., Michel, M., Laprun, C.: The NIST smart space and meeting room projects: Signals, acquisition, annotation, and metrics. In: Proceedings of IEEE International Conference on Acoustics, Speech, and Signal Processing (ICASSP 2003), vol. 4, pp. 736–739. IEEE Computer Society, Los Alamitos (2003)
29. Suo, Y., Shi, Y.: Towards initiative smart space model. In: Proceedings of the Third International Conference on Pervasive Computing and Applications, pp. 747–752. IEEE Computer Society, Los Alamitos (2008)
30. Zhou, W., Meinel, C., Raja, V.H.: A framework for supporting distributed access control policies. In: Proceedings of the 10th IEEE Symposium on Computers and Communications (ISCC 2005), pp. 442–447. IEEE Computer Society, Los Alamitos (2005)

GeoEnc: Geometric Area Based Keys and Policies in Functional Encryption Systems

Mingwu Zhang[1,2] and Tsuyoshi Takagi[1]

[1] Institute of Mathematics for Industry, Kyushu University,
Fukuoka, 819-0395, Japan
[2] College of Informatics, South China Agricultural University, 510642, China
{mwzhang,takagi}@imi.kyushu-u.ac.jp

Abstract. Functional encryption provides more sophisticated and flexible expression between the encryption key ek and decryption key dk by deriving from attribute vectors $\overrightarrow{x}$ and policy vector $\overrightarrow{v}$, respectively. There is a function $f(\overrightarrow{x}, \overrightarrow{v})$ that determines what type of a user with a secret key dk can decrypt the ciphertext encrypted under ek. This allows an encryptor to specify a functional formula as a decryptable policy describing what users can learn from the ciphertext without knowing the decryptor's identities or public keys.

In this paper, we explore two geometric-area-based key generation and functional encryption schemes (GeoEnc), where secret keys are associated with a point on a planar coordinate system and encrypt policies are associated with a line (*GeoEncLine* scheme) or a convex polygon (*GeoEncHull* scheme). If the attribute point lies on the line or inside the convex hull, the decryption key holder can decrypt the ciphertext associated with the geometric policy such as the line or the convex polygon. The proposed schemes have *policy hiding* as well as *payload hiding* characteristics. To the best of our knowledge, they are the first functional encryptions using geometric-area-based keys and policies. We give an evaluation of key distribution in a practical coordinate system and also give a security analysis with a hybrid model. The proposed schemes have many applications as sources for keys generation and policies encryption such as computer graphics security, network topology protection, secure routing and mobile networking, secure multiparty computation, secure GPS/GIS, military area protection, etc.

Keywords: geometric-based key, functional encryption, convex hull.

1   Introduction

Public keys [8], identities [9] and *attributes* [11] are used in public key cryptography. Public key cryptography is a cryptographic approach that involves the use of asymmetric key algorithms to ensure the confidentiality and integrity of a message, i.e., by using a public key to encrypt a message that can only be decrypted using the corresponding secret key. The public key is derived from a trusted public-key infrastructure.

U. Parampalli and P. Hawkes (Eds.): ACISP 2011, LNCS 6812, pp. 241–258, 2011.
© Springer-Verlag Berlin Heidelberg 2011

In identity-based cryptography (IBC) and attribute-based cryptography (ABC), the identities and attributes might include users' names and IP addresses, or biometric data such as fingerprints and iris-scans.

In many situations, however, answers to *"where s/he is"* questions describe users' *"what he can"* abilities. The positional relationship means the user is able to get the information from an encrypted ciphertext. The following examples describe our applications:

1. Someone who must stay in a library may download documents or watch movies that have been stored on the server. However, under our scheme, even though s/he has the secret key, s/he is unable to obtain the documents or movies once s/he goes outside the library.

2. If a combatant wants to attack an enemy's goal site by using a modern weapon such as a guided missile, s/he has the weapon to launch, whereby in our scheme, s/he can launch the weapon only if s/he receives the instruction while in a 'safe' area. His or her location (i.e., longitude, latitude and altitude) can thus be used as an attribute to decrypt the confidential instruction to launch an attack. The result if that s/he can not launch the weapon when s/he is outside the safe area even though s/he has the secret key stored in the weapon.

3. In mobile communication systems, sometimes we can constrain the sorts of communications that can be made in meeting rooms, classrooms, etc. To ensure that someone in a meeting room cannot communicate with others by using his mobile phone. The policy for the meeting room *"whether there is a meeting now"* and for the mobile phone *"whether the phone is inside the meeting room"* should be associated with the geometric attributes to allow the phone to communicate.

1.1   Our Results

In this paper, we explore two functional encryption schemes that use geometric area as an attribute to describe decryption key and encryption policy. Our contribution provides several benefits:

1. We extend functional encryption to a geometric space. To obtain the consistency of finite fields in the encryption scheme, we use a special finite point to describe a geometric graph such as a line or a convex polygon. We also convert the planar coordinate system into a finite field model to describe the point and line calculations.

2. We give the formal structures and security definitions for a geometric-area-based encryption scheme using geometric position or area as the attributes to generate the secret key. The decryption keys are associated with a point on the planar coordinate system and the encrypting policies are associated with a geometric graph, i.e., a line or a polygon.

3. We construct two concrete geometric-area encryption schemes: GeoEncLine and GeoEncHull. In the GeoEncLine scheme, the ciphertext is associated

with a line equation and the secret key is associated with a point coordinate. One can decrypt the ciphertext if and only if one carries a valid decryption key and one's position is on the line, i.e., (a) the point (decryption role) is a solution of the line equation; (b) the decryptor's position is authenticated if it is confirmed to be on the line. We resolve the physical position authenticity problem by using the position-based cryptography described in [7, 26], which is a building block to provide position attribute authenticity. In the GeoEncHull scheme, the decryption policy is the relationship of a point inside a convex polygon.

4. The GeoEnc schemes have the payload hiding property (they guarantee confidentiality of the encrypted message); they also support the attribute hiding (security for the encryption policy) in such a way that the policy of the point/polygon on a planar coordinate is also confidential.

1.2   Related Work

Chandran et al. [7] introduced the identity (or other credentials and inputs) of a party are derived from its *geographic location*, which answers the question *"Can you convince others about where you are?"*. They designed two protocols to support secure positioning and position-based key exchanging. In these protocols, the key exchange between the verifiers and the devices at position P that is enclosed within a tetrahedron formed between four verifiers in three-dimensional space, which is provably secure against any number of (possibly computationally unbounded) adversaries colluding together.

Sobrado and Birget [24] first proposed a graphical human identification protocol that utilizes the properties of a convex hull. The main part of the protocol involves the user mentally forming a convex hull of secret icons in a set of graphical icons and then clicking randomly within this convex hull. A variant of this protocol was later proposed in [28]. Wiedenbeck et al. [27] presented a detailed description of the protocol in [24] and a usability analysis employing human participants. Asghar et al. [1] analyzed the security of this convex hull based protocol, and gave two probabilistic attacks that reveal the user's secret key after observation of only a handful of authentication sessions.

Attribute-based encryption (ABE) was first introduced by Sahai and Waters [19]. Goyal et al. [11] formulated two complimentary forms of the ABE scheme: ciphertext-policy ABE (CP-ABE) and key-policy ABE (KP-ABE). ABE schemes have the desirable functionality, but have one limitation in that the structure of the ciphertext is revealed to users who cannot be allowed to reveal it to others. For example, in a CP-ABE system, a user who cannot decrypt the ciphertext can still learn the formula associated with the ciphertext. This is unacceptable for applications where the access policy must also be kept secret. Also, in many of ABE schemes [11, 15, 20], the attributes are described as identity strings, not geometric areas.

Spatial encryption (SE) was first proposed in [2], which is a new instance of the generalized identity-based encryption (GIBE) to construct IBE systems with different properties. GIBE is close to predicate encryption except that it incorporates the delegation property in HIBE. The GIBE scheme allows a sender to encrypt a message under a certain policy of set P. Users hold secret keys corresponding to roles. Roles are organized in a partial ordered set R, i.e., a set endowed with a *reflexive*, *transitive*, and *antisymmetric* relation $\succeq$ that have certain geometric properties. Given a key $SK_{\rho 1}$ which means that ρ_1's affine space contains ρ_2's space, there is a delegation algorithm that can produce the key $SK_{\rho 2}$, as long as $SK_{\rho_1} \succeq SK_{\rho_2}$. GIBE can be derived from spatial encryptions such as hierarchical IBE [5, 10, 15, 21], broadcast IBE [2], and forward security scheme [6]. However, the proposed SE schemes are only for payload hiding. They have not been considered for use in practical geometric coordinate environments.

Predicate encryption (PE) [12] can overcome the limitation of ABE to keep the secret of the access policy. Predicate encryption provides an ability that is *attribute-hiding* (policy confidentiality) for inner-product predicates that is stronger than the basic security requirement, *payload-hiding* (message confidentiality). Roughly speaking, attribute-hiding requires that a ciphertext conceal the associated attribute as well as the plaintext, while payload-hiding only requires that a ciphertext conceal the plaintext. If attributes are identities, i.e., PE is IBE [3, 9, 25], attribute-hiding PE implies anonymous IBE [5, 10, 16, 21]. Informally, secret keys in a PE scheme correspond to predicates in some class $\mathcal{F}$, and a sender associates a ciphertext with attribute in set $\sum$; a ciphertext associated with an attribute $I \in \sum$ can be decrypted using a secret key SK_f corresponding to predicate $f \in \mathcal{F}$ if and only if $f(I) = 1$.

Because encryption does not require a secret key, an attacker can encrypt any plaintext of his choice and evaluate a policy on the resulting ciphertext to learn whether the plaintext satisfies the predicate associated with the token. To obtain predicate confidentiality and privacy, Shen et al. [22] proposed a predicate privacy encryption system based on [12]. Constructions of such schemes are currently known for relatively few classes of predicates. An important research direction is to construct functional encryption schemes for function classes $\mathcal{F}$ that are as expressive as possible, with the ultimate goal being to handle all polynomial-time predicates [23].

Shi and Waters [23] presented a delegation mechanism for a class of PE, but the admissible predicates are a class of equality tests for HVE (Hidden Vector Encryption), which are more limited than inner-product predicates. The selective security proof is given in [23, 17].

A functional encryption (FE) scheme is for a class of functions F on the message space. Roughly speaking, functional encryption supports restricted secret keys that enable a key holder to learn a specific function of encrypted data, but learn nothing else about the plaintext. IBE, HIBE, PE, SE, and ABE schemes are the instance of a functional encryption. Lewko et al. [15] presented two fully secure functional encryption schemes: a fully secure attribute-based encryption and a fully secure (attribute-hiding) predicate encryption for inner-product predicates. The formal definition of a functional encryption is described by Boneh et al. [4]. In particular, O'Neill [18] introduced various security notions for functional encryption and studied the relationships among them.

1.3 Organization

We present the preliminaries and blocks in Section 2 and construct a line policy functional encryption scheme and give the scheme's security model and security analysis in Section 3. In Section 4, we give a coordinate evaluation of the practical geometric functional encryption system. In Section 5, we describe a polygon-based policy encryption scheme wherein a secret key associated with a point inside the convex polygon may extract the message. We draw our conclusions in Section 6.

2 Preliminaries and Blocks

Throughout this paper, we shall use the following notation. Let U be a set; $x \xleftarrow{R} U$ denotes that x is chosen uniformly at random from U. We denote a finite field of order p by $\mathbb{F}_p$, and $\mathbb{F}_p \setminus \{0\}$ by $\mathbb{F}_p^{\times}$. Consider a vector by $\hat{X}$ the vector. We shall denote the cardinality of $\hat{X}$ by $|\hat{X}|$, and $\hat{X}[i]$ the i-th component of $\hat{X} = (x_1, x_2, \ldots, x_{|\hat{X}|}) \in \mathbb{F}_N^{|\hat{X}|}$. Furthermore, we shall denote inner product $\sum_{i=1}^{n} x_i y_i$ of two vectors $\hat{X} = (x_1, x_2, \ldots, x_n)$ by $\langle \hat{X}, \hat{Y} \rangle = \hat{X}^T \cdot \hat{Y}$ and $\hat{Y} = (y_1, y_2, \ldots, y_n)$ by $\langle \hat{X}, \hat{Y} \rangle = \hat{X}^T \cdot \hat{Y}$. The zero vector is $\hat{0} = \{0, 0, \ldots, 0\}$.

Let $f : \sum_e \times \sum_d \to \{0, 1\}$ be a boolean function where $\sum_e$ and $\sum_d$ denote a key attribute space and ciphertext policy space. Let $P \subseteq \sum_e$ and $W \subseteq \sum_d$; we say that $P \models W$ iff $f(P, W) = 1$, otherwise $P \nvDash W$ iff $f(P, W) = 0$.

We define $P(u, v)$ as a point in a two-dimensional planar coordinate system where u, v is the x-coordinate and y-coordinate, respectively. Let the value of the determinant be denoted as $D(P_0, P_1, P_2) = \begin{vmatrix} u & v & 1 \\ x_1 & y_1 & 1 \\ x_2 & y_2 & 1 \end{vmatrix}$ where P_0, P_1 and P_2 are the pairs of $(u, v), (x_1, y_1)$ and (x_2, y_2), respectively. Let $P_0(u, v), P_1(x_1, y_1)$, and $P_2(x_2, y_2)$ be three points in a planar coordinate system, $D(P_0, P_1, P_2) = 0$ means that these points form a line. Additionally, we identify $\overrightarrow{P_1 P_2}$ as a directed line segment from P_1 to P_2, and $L = \overline{P_1 P_2}$ as the line L generated by two points P_1 and P_2.

2.1 Geometric Polygon

Definition 1. *Geometric polygon. A geometric polygon is a piece-wise linear, closed curve in a coordinate plane. The straight line segments forming the closed curve are called the sides of the polygon. A point joining two consecutive sides is called a vertex.*

We say that a geometric polygon is a simple polygon if it does not cross itself, and a simple polygon is convex if all points on the line segment joining any two points in its boundary or interior lie in the polygon.

Definition 2. *Interior, exterior and boundary of geometric polygon. The set of points in the planar coordinate system that lies outside a simple polygon is called the exterior; the set of points lying on the polygon forms its boundary; the set of points inside the boundary of the polygon is called the interior. If a point P lies on the boundary or in the interior of a polygon, we say that the polygon contains P or P is contained in the geometric polygon.*

Theorem 1. *The intersection of two convex polygons is a convex.*

Proof. Let S_1, S_2 be two convex sets, and denote the intersection of S_1 and S_2 by S. Let P_1 and P_2 be two points in S. Obviously, P_1 and P_2 are also points in S_1 and S_2 since $S = S_1 \cap S_2$. For S_1 and S_2 are convex polygons, then the entire line segment $\overline{P_1 P_2}$ is inside the S. Hence, the intersection two convex polygons is also a convex polygon.

2.2   Bilinear Maps in Composite Order Group

Definition 3. *Bilinear Maps in Composite Order* Let $\mathbb{G} = \langle g \rangle$ and $\mathbb{G}_T$ be two cyclic multiplicative groups of composite order $N = pqr$, i.e., $|\mathbb{G}| = |\mathbb{G}_T| = N$. Let $\hat{e}$ be an admissible bilinear map from $\mathbb{G}^2$ to $\mathbb{G}_T$; i.e., for all $u, v \in \mathbb{G}$ and $a, b \in \mathbb{F}_N$, it holds that $\hat{e}(u^a, v^b) = \hat{e}(u^b, v^a) = \hat{e}(u, v^b)^a = \hat{e}(u, v)^{ab}$ and $\hat{e}$ is non-trivial, i.e., $\hat{e}(g, g) \neq 1_{\mathbb{G}_T}$.

We use the notation $\mathbb{G}_p, \mathbb{G}_q$ and $\mathbb{G}_r$ to denote the subgroups of order p, q, r of $\mathbb{G}$, respectively, and we use the notation $\mathbb{G}_{T,p}, \mathbb{G}_{\mathbb{G},q}$ and $\mathbb{G}_{T,r}$ to denote as the subgroups of $\mathbb{G}_T$. Then $\mathbb{G} = \mathbb{G}_p \times \mathbb{G}_q \times \mathbb{G}_r$, and $\mathbb{G}_T = \mathbb{G}_{T,p} \times \mathbb{G}_{T,q} \times \mathbb{G}_{T,r}$, respectively.

2.3   Complexity Assumptions

ASSUMPTION 1 For a given composite order group generating $\mathcal{G}$, let $P(\lambda)$ to be

$$(p, q, r, \mathbb{G}_p, \mathbb{G}_q, \mathbb{G}_r, \mathbb{G}_{T,p}, \mathbb{G}_{T,q}, \mathbb{G}_{T,r}, \hat{e}) \xleftarrow{R} \mathcal{G}(\lambda)$$
$$\mathbb{G} = \mathbb{G}_p \times \mathbb{G}_q \times \mathbb{G}_r, \mathbb{G}_T = \mathbb{G}_{T,p} \times \mathbb{G}_{T,q} \times \mathbb{G}_{T,r}, N = pqr$$
$$g_p \xleftarrow{R} \mathbb{G}_p, g_q, Q_1, Q_2, Q_3 \xleftarrow{R} \mathbb{G}_q, g_r, R_1, R_2, R_3 \xleftarrow{R} \mathbb{G}_r, a, b, s \xleftarrow{R} \mathbb{F}_p$$
$$\Theta \leftarrow N, \mathbb{G}, \mathbb{G}_T, \hat{e}, g_p, g_r, g_q R_1, g_p^s, g_p^b, g_p^{b^2}, g_p^{ab} R_1, g_p^{bs} R_2$$
$$\eta \leftarrow \{0, 1\}, T_\eta \xleftarrow{R} \mathbb{G}_p^{b^2 s} Q_3^\eta R_3$$

We call (Θ, T_η) the challenge pair of Assumption 1. After giving the challenge pair to an attacker $\mathcal{A}$, $\mathcal{A}$ guesses $\eta \in \{0, 1\}$ in Assumption 1.

ASSUMPTION 2 For a given composite order group generating $\mathcal{G}$, let $P(\lambda)$ to be

$$(p, q, r, \mathbb{G}_p, \mathbb{G}_q, \mathbb{G}_r, \mathbb{G}_{T,p}, \mathbb{G}_{T,q}, \mathbb{G}_{T,r}, \hat{e}) \xleftarrow{R} \mathcal{G}(\lambda)$$

$$\mathbb{G} = \mathbb{G}_p \times \mathbb{G}_q \times \mathbb{G}_r, \mathbb{G}_T = \mathbb{G}_{T,p} \times \mathbb{G}_{T,q} \times \mathbb{G}_{T,r}, N = pqr$$

$$g_p, \omega, h \xleftarrow{R} \mathbb{G}_p, Q_1, Q_2 \xleftarrow{R} \mathbb{G}_q, g_r \xleftarrow{R} \mathbb{G}_r, s, r \xleftarrow{R} \mathbb{F}_N$$

$$\Theta \leftarrow N, g_p, g_q, g_r, h, \mathbb{G}, \mathbb{G}_T, \hat{e}, g_p^s, h^s Q_1, g_p^r Q_2, \hat{e}(g_p, h)^r$$

$$T_0 \leftarrow \hat{e}(g_p, \omega)^{rs}, T_1 \xleftarrow{R} \mathbb{G}_T$$

We call (Θ, T_0, T_1) the challenge pair of the Assumption 2, where $(\Theta, T_0, T_1) \xleftarrow{R} P(\lambda)$. After giving the challenge pair to attacker $\mathcal{A}$, $\mathcal{A}$ distinguishes between T_1 and T_2 in Assumption 2.

2.4 Framework of GeoEnc Scheme

Here, we shall describe two frameworks of geometric-area-based functional encryption schemes, called GeoEncLine and GeoEncHull, respectively. Let $P(x, y) \in \mathbb{F}_N^2$ be a point in the planar coordinate system. As $N = pqr$ can not be factored, we will discuss the relationship between the coordinate x, y and integer N in section 5.

GeoEncLine Scheme. In the GeoEncLine scheme, we encrypt a message with a line equation where the line is described as two different points on the line (because we use the slope-direct to describe the line equation, we assume that two points have different x-coordinate values, i.e, the line is not vertical). The secret key is associated with a point in the planar coordinate system.

One can extract a valid message from a ciphertext encrypted by the GeoEncLine algorithm if and only if one's current position is on the line and one carries a valid decryption key. Informally, one can require point secret key queries in a two-dimensional planar coordinate system by using the PointKeyGen algorithm. The sender may encrypt a message associated with a line equation. If the point (hidden in the secret key) is a solution of the line equation, then the receiver can decrypt the ciphertext. For instance, in Fig.1(a), points P and P' require secret keys, but only P can decrypt the message encrypted by line L ($L=\overline{P_1 P_2}$).

Position Authenticity. The decryption attribute point P can be located by using physical equipment such as sensor networks, global positioning systems (GPS), or radar systems [26, 7]. In order to keep the physical position authenticable, we shall use a secure anti-collusion attack positioning protocol [7] that only needs four verifiers in a three-dimensional space. We shall apply it to a two-dimensional planar coordinate system with three verifiers. The authentication protocol is as follows.

1. Verifiers V_1, V_2 and V_3 pick keys K_1, K_2 and $K_3 \xleftarrow{R} \{0, 1\}^m$ and send them to the other parties.
2. V_1 broadcasts key K_1 at time $T - t_1$, while V_2 broadcasts X_1 at time $T - t_2$ simultaneously also broadcasts $K_2' = PRG(X_1, K_1) \oplus K_2$, where $PRG : \{0, 1\}^m \times \{0, 1\}^m \rightarrow \{0, 1\}^m$ is an ϵ-secure pseudorandom generator. Similarly, V_3 broadcasts $(X_2, K_3' = PRG(X_2, K_2) \oplus K_3)$ at time $T - t_3$.

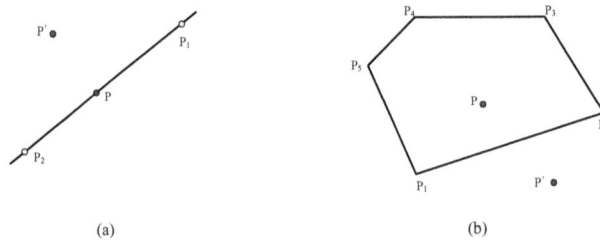

Fig. 1. Geometric area encryption   (a)Line policy; (b)Polygon policy

3. At time T, the prover at position P computes messages $K_{i+1} = PRG(X_i, K_i) \oplus K'_{i+1}$ for $i = 1, 2$ and returns K_3 to all verifiers.
4. The three verifiers check that the string K_3 is received at time $(T + t_i)$ and that it equals the K_3 of their pre-chosen. If these verifications succeed, the position claim of the prover is accepted and the prover is assumed to be indeed at position P. Otherwise, the position claim is rejected.

Because we use this physical position authenticity protocol as a building block to prove that the point location is convincing, we assume that this block is executed before the secret key is requested and the ciphertext is decrypted.

GeoEncHull Scheme. We construct another geometric-area-based functional encryption scheme called GeoEncHull. In the GeoEncHull scheme, the ciphertext is associated with a convex polygon and the secret key is associated with a point. If the point that produces the secret key is inside the polygon and the position is authenticated, it can decrypt the ciphertext. In Fig.1(b), points P and P' require secret keys, but only P can extract the message from the ciphertext by using its secret key. To determine whether a point is inside the convex polygon, we introduce a new operator in a finite group Ψ (see Section 5).

3   GeoEncLine Scheme

We first construct a line-policy-based functional encryption scheme whereby the ciphertext CT is generated from a line and the key SK is derived from a point on the planar coordinate system. The GeoEncLine scheme is comprised of the $SysGen, LineEnc, PointKeyGen,$ and $PointDec$ algorithms.

3.1   Syntax of GeoEncLine

- **SysGen**$(\lambda) \rightarrow (params, msk)$ The SysGen algorithm generates system public parameters, denoted as $params$, and a corresponding master secret key, denoted as msk.

- **LineEnc**($params, M, L=\overline{P_1(x_1, y_1)P_2(x_2, y_2)}$) $\rightarrow$ CT_L The LineEnc algorithm takes the public parameters, a message M, and a line L defined by two points $P_1(x_1, y_1)$, $P_2(x_2, y_2)$ $(x_i, y_i \in \mathbb{F}_N, i = 1, 2$ and $x_1 \neq x_2)$ as input, and it produces a ciphertext CT_L such that only users whose secret key associated with this attribute point on the line are able to extract the message M.
- **PointKeyGen**($msk, params, P(u, v)$) $\rightarrow SK_{u,v}$ The PointKeyGen algorithm takes the master secret key msk, public parameters $params$ and a planar coordinate point $P(u, v)$ $(u, v \in \mathbb{F}_N)$ as input, and it outputs a secret key $SK_{P(u,v)}$.
- **PointDec**($CT_L, params, SK_{P(u,v)}$) $\rightarrow M|\bot$. After being given the input of a ciphertext CT_L, public parameters $params$, and a geometric point secret key $SK_{P(u,v)}$, this algorithm outputs the message M if the point position is convincing and $P(u, v)$ satisfies the line equation $P \models L$; i.e., (u, v) is a root of the equation: $y = \frac{y_1 - y_2}{x_1 - x_2}(x - x_1) + y_1 \mod N$. Otherwise, it returns the distinguished symbol $\bot$.

The GeoEncLine scheme should have the following consistency and correctness properties: for all correctly produced $params$ and $SK_{P(u,v)}$, generate $CT_L \leftarrow LineEnc(params, M, \overline{P_1 P_2})$ and $M' = PointDec(params, SK_{P(u,v)}, CT_L)$. If P satisfies the line equation $P \models L$, then $M = M'$. Otherwise, $M \neq M'$, except for a negligible probability, i.e.,

$$Pr \begin{bmatrix} (params, msk) \leftarrow \mathcal{G}(\lambda), \ P \models \overline{P_1 P_2} \\ SK_P = PointKey(msk, params, P(u, v)) \\ CT_L = LineEnc(params, M, \overline{P_1 P_2}) \\ M' = PointDec(CT_L, params, SK_P) \\ M' = M \end{bmatrix} > 1 - \epsilon(\lambda)$$

where $\epsilon(\lambda)$ is a negligible function such that there exists an integer k that for every $\lambda > k$ satisfies $f(\lambda) < 1/\epsilon(k)$.

3.2 Security Model of GeoEncLine

We define GeoEncLine security by using a game that captures the strong privacy property including payload hiding and policy hiding. The GeoEncLine scheme is semantically secure, as the following game between a challenger $\mathcal{C}$ and an attacker $\mathcal{A}$.

Payload Hiding

1. Init Attacker $\mathcal{A}$ commits a line $\hat{L}$ specified by two distinguished points $\hat{P}_1(\tilde{x}_1, \tilde{y}_1)$, $\hat{P}_2(\tilde{x}_2, \tilde{y}_2)$ to the attack.
2. SysGen The challenger $\mathcal{C}$ runs the SysGen algorithm to produce the public parameters $params$ and master secret key, and sends the $params$ to the attacker.
3. Adaptive phase-I Attacker $\mathcal{A}$ makes a bounded number of queries to $\mathcal{C}$ for secret keys corresponding to the point $P_1(x_1, y_1), P_2(x_2, y_2), \ldots, P_t(x_t, y_t)$, with

only the restriction being that none of these queried points on line $\hat{L}$ which satisfies the equation in (1). i.e., x_i, y_i ($1 \leq i \leq t$) does not satisfy the following equation

$$\hat{L} = y_i = \tilde{y}_1 + \frac{\tilde{y}_1 - \tilde{y}_2}{\tilde{x}_1 - \tilde{x}_2}(x_i - \tilde{x}_1) \mod N \tag{1}$$

4. **Challenge** Once attacker $\mathcal{A}$ decides that phase-I is over, it outputs two equal length messages M_0, M_1. The challenger $\mathcal{C}$ flips a random coin $b \in \{0, 1\}$, and encrypts M_b under the line with points $\hat{P}_1, \hat{P}_2$. It then sends the ciphertext $CT_{\hat{L}}$ to $\mathcal{A}$.
5. **Adaptive phase-II** $\mathcal{A}$ can query the challenger for secret keys corresponding to point set $P_{t+1}(x_{t+1}, y_{t+1}), \ldots, P_k(x_k, y_k)$ like in query-I.
6. **Guess** The attacker $\mathcal{A}$ outputs a guess b' for b.

The advantage of an attacker $\mathcal{A}$ in this game is defined as $|Pr[b' = b] - \frac{1}{2}|$ where the probability is taken over the random bits used by the challenger and the attacker.

Definition 4. *The geometric line policy encryption scheme (GeoEncLine) is semantically secure for payload hiding if all polynomial time attackers have at most a negligible advantage in above security game of payload hiding.*

We denote the queried point set $P_1, \ldots, P_k$ as $\sum_e$, and the challenged point set $\hat{P}_1$ and $\hat{P}_2$ as $\sum_d$. We also define the encryption policy function (formula 1). Attacker $\mathcal{A}$'s queries satisfy $P \nvDash \hat{L} := \overline{\hat{P}_1 \hat{P}_2}$, which means the attacker cannot perform queries that satisfy the challenged encryption policy function.

Geometric Policy Hiding. The Geometric policy hiding experiment is payload hiding one, except that the challenge phase is modified as follows.

- *Challenge.* The attacker $\mathcal{A}$ outputs a message M and a line denoted as two points $\ddot{L} := (\overline{P_1^i, P_2^i})$ by its choice with the restriction that $\ddot{L}$ has not been queried in phase-I. Then $\mathcal{C}$ flips a random coin η and a random ciphertext C from the ciphertext space. If $\eta = 1$ then $\mathcal{C}$ encrypts M with $CT_{\hat{L}} \leftarrow$ LineEnc ($params, M, \ddot{L}$), else if $\eta = 0$ he sets $CT_{\hat{L}} \leftarrow C$ as the challenge to attacker $\mathcal{A}$.

Definition 5. *A geometric line policy encryption scheme is secure for policy hiding if all polynomial time attackers have at most a negligible advantage in geometric policy hiding game.*

3.3   Construction of GeoEncLine

We assume that $P_1(x_1, y_1), P_2(x_2, y_2)$ are two points on the line $\hat{L}$, and point $P(u, v)$ is a attribute point to generate secret key. We take a vector $\hat{\boldsymbol{X}} = (u, v, 1)$ as the encryption point vector, and a vector $\hat{\boldsymbol{Y}} = (y_1 - y_2, x_2 - x_1, x_1 y_2 - x_2 y_1)$ as the line policy vector. Obviously, $P_0(u, v)$ is on the line iff $\langle \hat{\boldsymbol{X}}, \hat{\boldsymbol{Y}} \rangle = 0$, i.e.,

$$D(P_0(u,v), P_1(x_1, y_2), P_2(x_2, y_2))$$
$$= x_1y_2 + x_2v + y_1u - x_2y_1 - x_1v - y_2u = \langle \hat{X}, \hat{Y} \rangle$$

If the sign of $D(P_0, P_1, P_2)$ is positive then (P_0, P_1, P_2) forms a counterclockwise cycle, i.e., P_0 lies to the left of directed line segment $\overrightarrow{P_1P_2}$, and negative if and only if (P_0, P_1, P_2) forms a clockwise cycle, i.e., P_0 lies to the right of $\overrightarrow{P_1P_2}$. If $D(P_0, P_1, P_2) = 0$, P_0 lies in on the line $\overline{P_1P_2}$.

SysGen(1^λ): First, given a security parameter λ, this algorithm uses the group generator $\mathcal{G}$ to produce $(N = pqr, p, q, r, \mathbb{G}, \mathbb{G}_T, \hat{e}) \leftarrow \mathcal{G}(1^\lambda)$ with $\mathbb{G} = \mathbb{G}_p \times \mathbb{G}_q \times \mathbb{G}_r$. It randomly picks generators $g_p, g_q, g_r \in \mathbb{G}_p, \mathbb{G}_q, \mathbb{G}_r$, randomly chooses $\omega, g_i \in \mathbb{G}_p$, $h, h_i \in \mathbb{G}_r$, and computes $\Omega = g_q h \in \mathbb{G}_{qr}$, $z_i = g_i h_i \in \mathbb{G}_{pr}$ for $i = 1, \ldots, 6$. Then it publishes the parameters $params = (N, g_p, g_r, \Omega, z_i, \hat{e}(g_p, \omega))$ and keeps the master secret key $(g_q, \omega, h_i)(i = 1, \ldots, 6)$.

LineEnc($params, M, L = \overline{P_1P_2}$): Let $P_1(x_1, y_1), P_2(x_2, y_2)$ be two distinguished points on line L, and $M \in \mathbb{G}_T$ be a message. This algorithm first computes $l = y_1 - y_2 \pmod{N}, m = x_2 - x_1 \pmod{N}, n = x_1y_2 - x_2y_1 \pmod{N}$. Then, it randomly picks $s, \xi, \delta \in \mathbb{F}_N$ and $R_i \in \mathbb{G}_r (i = 1, \ldots, 6)$, and produces the ciphertext $CT_L = (C', C_0, C_{i(i=1,\ldots,6)})$ as follows:

$$
\begin{array}{lll}
C' = M\hat{e}(g_p, \omega)^s & C_0 = g_p^s & \\
C_1 = z_1^s \Omega^{\xi l} R_1 & C_2 = z_2^s \Omega^{\delta l} R_2 & C_3 = z_3^s \Omega^{\xi m} R_3 \quad (2)\\
C_4 = z_4^s \Omega^{\delta m} R_4 & C_5 = z_5^s \Omega^{\xi n} R_5 & C_6 = z_6^s \Omega^{\delta n} R_6
\end{array}
$$

Obviously, the ciphertext CT_L has the same random distribution even though the CT_L is produced from any two different points in L.

PointKeyGen($msk, params, P(u,v)$): Let $P(u,v)$ be a point on the planar coordinate system where $u, v \in \mathbb{F}_N$. This algorithm randomly picks $f_1, f_2, t, r_i \in \mathbb{F}_N$ $(i = 1, \ldots, 6)$, and generates the secret key $SK_{P(u,v)} = (K_0, K_{i(i=1,\ldots,6)})$ as:

$$
\begin{array}{llll}
K_0 = \omega g_q^t \prod_{i=1}^{6} z_i^{-r_i} & K_1 = g_p^{r_1} g_q^{f_1 u} & K_2 = g_p^{r_2} g_q^{f_2 v} & K_3 = g_p^{r_3} g_q^{f_1} \\
K_4 = g_p^{r_4} g_q^{f_2 u} & K_5 = g_p^{r_5} g_q^{f_1 v} & K_6 = g_p^{r_6} g_q^{f_2} &
\end{array} \quad (3)
$$

PointDec($CT_L, params, SK_{P(u,v)}$): Upon input of the line policy ciphertext $CT_L = (C', C_0, C_i|_{i=1}^{6})$ and the point secret key $SK_{P(u,v)} = (K_0, K_i|_{i=1}^{6})$, this algorithm first verifies the validate for the position $P(u,v)$. If the verification succeeds, it extracts the message by using

$$M = \frac{C'}{\prod_{i=0}^{6} \hat{e}\{C_i, K_i\}} \quad (4)$$

3.4   Correctness and Consistency

Assume a ciphertext is well-formed for the decryption key, that is, ciphertext $CT_{L=\overline{P_1(x_1,y_1)P_2(x_2,y_2)}}$ and secret key $SK_{P(u,v)}$ produced as above guarantee that point $P(u,v)$ is on the line $L=\overline{P_1P_2}$ such that $D(P, P_1, P_2) = 0$. The encryption policy and the decryption role satisfies $\langle \hat{\boldsymbol{X}}, \hat{\boldsymbol{Y}} \rangle = 0$ where $\hat{\boldsymbol{X}} = (u, v, 1)$ and $\hat{\boldsymbol{Y}} = (l, m, n) = (y_1 - y_2, x_2 - x_1, x_1 y_2 - x_2 y_1)$.

To prove correctness and consistency, first we check the following equation.

$$
\hat{e}(C_1, K_1) = \hat{e}(z_1^s \Omega^{\xi l} R_1, g_p^{r_1} g_q^{f_1 u}) = \hat{e}(z_1^s \Omega^{\xi l}, g_p^{r_1} g_q^{f_1 u})
$$
$$
= \hat{e}(g_1^s h_1^s g_q^{\xi l} h^{\xi l}, g_p^{r_1} g_q^{f_1 u}) = \hat{e}(g_1^{r_1}, g_p^s)\hat{e}(g_q^{\xi l}, g_q^{f_1 u})
$$

Furthermore, we have the following calculation.

$$
\prod_{i=1}^{6} e(C_i, K_i) = (\prod_{i=1}^{6} \hat{e}(g_i^{r_i}, g_p^s)) \cdot \hat{e}(g_q, g_q)^{\xi f_1 ul + \delta f_2 vl + \xi f_1 m + \delta f_2 u + \xi f_1 vn + \delta f_2 n}
$$
$$
= \hat{e}(\prod_{i=1}^{6} g_i^{r_i}, g_p^s)\hat{e}(g_q, g_q)^{(\xi f_1 + \delta f_2)\langle (u,v,1),(l,m,n)\rangle}
$$
$$
= \hat{e}(\prod_{i=1}^{6} g_i^{r_i}, g_p^s)\hat{e}(g_q, g_q)^{(\xi f_1 + \delta f_2)\langle \hat{\boldsymbol{X}}, \hat{\boldsymbol{Y}}\rangle} \tag{5}
$$

Therefore, if P is on the line L, i.e., $\hat{e}(g_q, g_q)^{(\xi f_1 + \delta f_2)\langle \hat{\boldsymbol{X}}, \hat{\boldsymbol{Y}}\rangle} = 1$ for $\langle \hat{\boldsymbol{X}}, \hat{\boldsymbol{Y}}\rangle = 0$.

$$
\frac{C'}{\prod_{i=0}^{6} \hat{e}(C_i, K_i)} = C' \frac{1}{\hat{e}(C_0, K_0)\prod_{i=1}^{6}\hat{e}(C_i, K_i)}
$$
$$
= C' \frac{1}{\hat{e}(g_p^s, \omega g_q^t)\prod_{i=1}^{6} z_i^{-r_i})\hat{e}(\prod_{i=1}^{6} g_i^{r_i}, g_p^s)}
$$
$$
= \frac{M\hat{e}(g_p, \omega)^s}{\hat{e}(g_p^s, \omega)\hat{e}(g_p^s, \prod_{i=1}^{6}(g_i h_i)^{-r_i})\hat{e}(\prod_{i=1}^{6} g_i^{r_i}, g_p^s))}
$$
$$
= \frac{M\hat{e}(g_p, \omega)^s}{\hat{e}(g_p^s, \omega))\hat{e}(g_p^s, \prod_{i=1}^{6}(g_i)^{-r_i})\hat{e}(\prod_{i=1}^{6} g_i^{r_i}, g_p^s))}
$$
$$
= \frac{M\hat{e}(g_p, \omega)^s}{\hat{e}(g_p^s, \omega)} = M \tag{6}
$$

Otherwise, if $\langle \hat{\boldsymbol{X}}, \hat{\boldsymbol{Y}} \rangle \neq 0$, the decrypt equation in the PointDec algorithm will contain a factor $\hat{e}(g_q, g_q)^\varsigma$ where $\varsigma = (\xi f_1 + \sigma f_2)\prod_i \hat{\boldsymbol{X}}_i \hat{\boldsymbol{Y}}_i)$, which is a uniform distributed value in $\mathbb{G}_{T_q}$. Then $C'/\prod_{i=0}^{6} \hat{e}(C_i, K_i)$ is a uniform distributed element in $\mathbb{G}_T$, that is indistinguishable from $\mathbb{G}_{T_p}$ under Assumption 2.

3.5   Security Analysis

To understand our construction, it would be useful to examine the role of each of the subgroups $\mathbb{G}_p, \mathbb{G}_q, \mathbb{G}_r$. The $\mathbb{G}_p$ subgroup is used to prevent an attacker from manipulating components of either a ciphertext CT_L or a key SK_P and then evaluating a query on the improperly formed inputs. The $\mathbb{G}_p$ subgroup also encodes a message in the bilinear subgroup.

The $\mathbb{G}_q$ subgroup is used to encode a cleartext point coordinate vector $\hat{\boldsymbol{Y}}$ in $C_1, \ldots, C_6$ terms of the attribute policy in the ciphertext CT_L, and the extended point coordinate vector $\hat{\boldsymbol{X}}$ in $K_1, \ldots, K_6$ terms of the point vector in the secret key. When a point for $\hat{\boldsymbol{X}}$ is sure that an encryption of $\hat{\boldsymbol{Y}}$ is good, the inner product $\langle \hat{\boldsymbol{X}}, \hat{\boldsymbol{Y}} \rangle$ is evaluated as an identical element in the $\mathbb{G}_{T_q}$ subgroup.

The $\mathbb{G}_r$ subgroup is used to hide factors from other subgroups and ensure the ciphertext's confidentiality and privacy.

Our construction consists of two parallel sub-systems. Note that C_1, C_3, C_5 and C_2, C_4, C_6 in CT_L (similarly in keys SK_P) play identical roles. Our proof of security will rely on having these two parallel subsystems.

To prove that the case when the challenge ciphertext is associated with $\hat{\boldsymbol{X}}$ (which for $\hat{\boldsymbol{X}} \times \hat{\boldsymbol{X}}$) is indistinguishable, and the case when the challenge ciphertext is associated with $\hat{\boldsymbol{Y}}$ (which corresponds to $\hat{\boldsymbol{Y}} \times \hat{\boldsymbol{Y}}$, we will use a sequence of intermediate hybrid games $(\hat{\boldsymbol{X}}, \hat{\boldsymbol{0}}), (\hat{\boldsymbol{X}}, \hat{\boldsymbol{Y}}), (\hat{\boldsymbol{0}}, \hat{\boldsymbol{Y}})$ and prove indistinguishability in each intermediate case. That is,

$$(\hat{\boldsymbol{X}}, \hat{\boldsymbol{X}}) \leftrightarrow (\hat{\boldsymbol{X}}, \hat{\boldsymbol{0}}) \leftrightarrow (\hat{\boldsymbol{X}}, \hat{\boldsymbol{Y}}) \leftrightarrow (\hat{\boldsymbol{0}}, \hat{\boldsymbol{Y}}) \leftrightarrow (\hat{\boldsymbol{Y}}, \hat{\boldsymbol{Y}})$$

Theorem 2. *Under Assumptions 1 and 2, the GeoEncLine encryption scheme is attribute hiding for geometric policy security.*

proof. The proof is straightforward. Our proof of security is structured as a hybrid experiment over a sequence of games defined as follows:

- *Game-real*: The challenge ciphertext is $\Gamma_0 : CT_L = (C', C_0, C_1, C_2, C_3, C_4, C_5, C_6)$, which is the actual GeoEncLine game;
- *Game-1*: The challenge ciphertext is generated as a proper encryption using $(\hat{\boldsymbol{X}}, \hat{\boldsymbol{0}})$ such that $\Gamma_1 : CT_L = (C', C_0, C_1, C_2 = z_2^s R_2, C_3, C_4 = z_4^s R_2, C_5, C_6 = z_6^s R_6)$;
- *Game-2*: Generate C_2, C_4, C_6 components as if encryption were to be done using $(\hat{\boldsymbol{X}}, \hat{\boldsymbol{Y}})$ such that $\Gamma_2 : CT_L = (C', C_0, C_1, C_2 = z_2^s R_2, C_3, C_4 = z_4^s R_2, C_5, C_6 = z_6^s R_6)$;
- *Game-3*: Generate C_2, C_4, C_6 components using the vector $(\hat{\boldsymbol{0}}, \hat{\boldsymbol{Y}})$;
- *Game-4*: This game Γ_4 is defined symmetrically with respect to Game-3. In Game-4, the C_1, C_3, C_5 components are generated using $\hat{\boldsymbol{0}}$.
- *Game-5*: This game Γ_5 is defined symmetrically with respect to Game-4. In Game-5, the C_1, C_3, C_5 components are generated using $\hat{\boldsymbol{Y}}$.

In Game-5, the challenge ciphertext is a proper encryption with respect to the vector $\hat{\boldsymbol{Y}}$. Thus, the proof of the theorem 2 is conclude by proofs the security with a hybrid model. We show that the attacker cannot distinguish between Game-i and Game-$(i+1)$ for $i = 1, \ldots, 4$.

Intermediate games Game-2 and Game-4 are used to simplify the proof which helps when part of the ciphertext corresponds to an encryption using $\hat{\boldsymbol{0}}$ since this vector is orthogonal. The main difficulty in proof is to answer queries for decryption keys while ensuring the the indistinguishability of Game-1/Game-2 and indistinguishability of Game-4/Game-5.

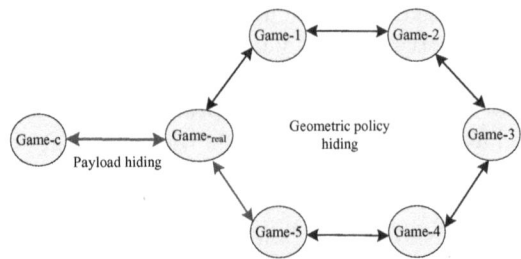

Fig. 2. Hybrid security model for game indistinguishability

We will allow the attacker to have the ability to construct all decryption keys such that it can distinguish an encryption relative to $\hat{X}$ from an encryption relative to $\hat{Y}$. We show that even such keys cannot be used to distinguish a well-formed encryption of $\hat{X}$ or $\hat{Y}$ from a badly formed one.

We give the proof of the indistinguishability between Game_i and Game_{i+1} (i=1,2,3,4) in the full version of this paper.

Theorem 3. *The GeoEncLine scheme is semantically secure under Assumption 2.*

proof. We define the Game_c to be a playload hiding game. The challenge ciphertext is $\Gamma_1 : CT_L = (R, C_0, C_1, C_2, C_3, C_4, C_5, C_6)$, which is the difference between C' and a random $R \in \mathbb{G}_{T_p}$. This can be considered as a bilinear subgroup decisional assumption in Assumption 2. According to the Theorem 2, we show that the components in $\Gamma_0 : CT_L = (C', C_0, C_1, C_2, C_3, C_4, C_5, C_6)$ is indistinguishable from random components in $\Gamma_c : CT_L = (R', R'_0, R'_1, R'_2, R'_3, R'_4, R'_5, R'_6)$, where R' is a random element in $\mathbb{G}_t$ and R'_i ($i = 1, \ldots, 6$) is a random element in $\mathbb{G}$.

4   Practical Coordination Evaluation

Note that we use a composite order $N = pqr$ with primes p, q, r in 2.2 and assume that it is hard to factor the p, q, r of N of an order of group generator $\mathcal{G}$ with 128-bit AES security [14]. Then $|N|=|p|+|q|+|r|$ is at least 1024-bit[1]. We set the numbers p, q and r to be 342-bit, 341-bit and 341-bit, respectively. In a planar coordinate, the line segment $\overline{P_1 P_2}$ vector $\hat{Y}$ is defined as $(y_1, y_2, y_3) = (y_1 - y_2, x_2 - x_1, x_1 y_2 - x_2 y_1)$ (see equation 2). Since we are considering a point on the line in the GeoEncLine scheme, $< \hat{X}, \hat{Y} >= 0$ holds if the decryption role

---

[1] Kleinjung et al. [13] factored a 768-bit RSA modulus in 2010 and they declared that it would be prudent to phase out usage of 1024-bit RSA within the next 3-4 years. Thus, we should also have to expand the size of N in a practical deployment. In this section, we discuss the coordination evaluation between the practical coordination and the theoretical security requirement. Actually, the scope of coordination system will increase if we deploy a more secure order N, i.e. $N = 2048$.

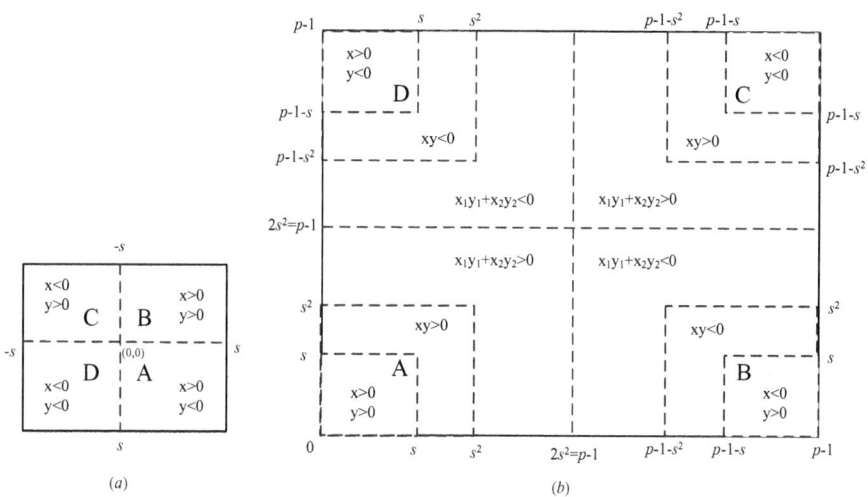

Fig. 3. Map geometric coordinate system to finite field space

(point P_0) satisfies the encryption policy $(\overline{P_1 P_2})$. Thus, one need not to consider the sign of $< \hat{X}, \hat{Y} >$ in the GeoEncLine scheme.

Next, we estimate the coordinate scope in the GeoEncLine scheme under a finite field $\mathbb{F}_p$. Assume the point in the GeoEncLine scheme is $P(x, y) \in \mathbb{F}_\top^2$ (i.e., in a two-dimensional Euclidean space with the maximum x and y coordinate value $\top$). We use $\mathbb{F}_p$ as the planar coordinate points and line calculation scope[2]. In Figure 3(a), the maximum coordinate values of $P(x, y)$ are such that $-\top < x < \top$, $-\top < y < \top$. We extend these coordinate planar (with negative values) to a finite field $\mathbb{F}_p^2$ ($\top < p$, and $\mathbb{F}_p = 1, \ldots, p-1$). In Figure 3(b), we takes $[1, \frac{p}{2} - 1]$ as being positive and $[\frac{p}{2} + 1, p - 1]$ as being negative. Considering a vector $\hat{Y} = (y_1 - y_2, x_2 - x_1, x_1 y_2 - x_2 y_1)$, the maximum value of the third component is $2\top^2$ since the point is in $\mathbb{F}_\top^2$ (i.e., $x_1 = y_2 = \top$, $x_2 = y_1 = -\top$). We set $2\top^2 = p/2$, then $\top = \sqrt{2^{340}} \approx 1.496 \times 10^{51}$, which is large enough for the practical application to a two-dimensional coordinate system.

5   Extension to GeoEncHull Scheme

We first define a map f for an input $x \in [0, N - 1]$ over $\mathbb{F}_N$[3]

$$f(x) = \begin{cases} \boxplus & if\ 0 < x < p/2 \ (mod\ p) \\ \boxdot & if\ x = 0 \ (mod\ p) \\ \boxminus & if\ p/2 \leq x < p - 1 \ (mod\ p) \end{cases} \quad (7)$$

---

[2] $(x, y) \in \mathbb{F}_\top^2$, and two point equation $L_{\overline{P_1 P_2}}$ will overflow the $\mathbb{F}_\top$ because it will introduce the addition and multiplication operators to this field, for instance, $xy \in \mathbb{F}_{\top^2}$.

[3] Note that $N = pqr$ where p, q, r are prime numbers such that $\gcd(p, q) = 1$, $\gcd(p, r) = 1$, and $\gcd(q, r) = 1$.

We define two operators $\oplus, \otimes$ over group $\Psi = (\boxplus, \boxdot, \boxminus)$ such that

$$\boxplus \oplus \boxplus = \boxplus, \boxplus \oplus \boxdot = \boxplus, \boxplus \oplus \boxminus = \boxdot,$$
$$\boxdot \oplus \boxdot = \boxdot, \boxdot \oplus \boxminus = \boxminus, \boxminus \oplus \boxminus = \boxminus$$
$$\boxplus \otimes \boxplus = \boxplus, \boxplus \otimes \boxdot = \boxdot, \boxplus \otimes \boxminus = \boxminus,$$
$$\boxdot \otimes \boxdot = \boxdot, \boxdot \otimes \boxminus = \boxdot, \boxminus \otimes \boxminus = \boxplus$$

Intuitively, the operators $\oplus, \otimes$ in group Ψ satisfy the condition of sign operators $+, 0, -$ in real or integer fields.

Lemma 1. *Suppose that Δ is a triangle on the plane with three points P_0, P_1, P_2. Then the signed area of Δ is half of the determinant $D(P_0, P_1, P_2)$, where the sign of D is positive if (P_0, P_1, P_2) forms a counterclockwise cycle and negative if (P_0, P_1, P_2) forms a clockwise cycle. We say that the path from P_0 through the line segment $\overrightarrow{P_0 P_1}$ to P_1 then through the line segment $\overrightarrow{P_1 P_2}$ to point P_2 is a left turn if $D(P_0, P_1, P_2) > 0$. Otherwise, we say the path makes a right turn.*

Lemma 2. *Point inside polygon.* *If a polygon is convex and a point is always in the same hemispace with regard to the directed edges, the point is inside the polygon.*

Let $\Sigma = (P_1, P_2, \ldots, P_\pi)$ be a convex polygon where $P_i(x_i, y_i)$ $(i = 1, \ldots, \pi)$ are the vertexes and $\overrightarrow{P_i P_{i+1}}$ are the edges of Σ with a counterclockwise cycle. The convex hull vector is produced as follows. Let $P_i = (x_i, y_i)$ and $P_{\pi+1} = P_1$. For $i = 1, \ldots, \pi$, compute

$$\begin{cases} v_{i1} = y_i - y_{i+1} & (\mathrm{mod}\ N) \\ v_{i2} = x_{i+1} - x_i & (\mathrm{mod}\ N) \\ v_{i3} = x_i y_{i+1} - x_{i+1} y_i & (\mathrm{mod}\ N) \end{cases}$$

For $i = 1, \ldots, \pi$, $j = 1, 2, 3$, randomly pick $s, \xi_i, \delta_i \in Z_N$ and $R_{i1j}, R_{i2j} \in G_r$, and compute

$$\begin{cases} C' = M\hat{e}(g_p, \omega)^s \\ C_0 = g_p^s \\ C_{i1j} = z_{1i}^s \Omega^{\xi_i v_{ij}} R_{i1j} \\ C_{i2j} = z_{2i}^s \Omega^{\delta_i v_{ij}} R_{i2j} \end{cases} \qquad (8)$$

Then output the convex-polygon policy based ciphertext $CT_\Sigma = (C', C_0, C_{i1j}, C_{i2j})$ $(1 \leq i \leq \pi, 1 \leq j \leq 3)$.

6 Concluding Remarks

We proposed two geometric-area-based key generation and policy encryption schemes. We modeled the geometric graphic functional encryption schemes and gave a security analysis under composite order groups. Geometric area cryptography can be applied to computer graphics, network topology, secure routing and mobile networking and the military etc. There remain several open problems for geometric based cryptosystems: i.e., finding a solution for the geometric graph in a real number field, encryption using a flexible geometric function, high-dimension encryption, etc.

Acknowledgment. The authors grateful thank the anonymous reviewers for their helpful comments. This work is supported by NSFC (60973134), NSF of Guangdong (1015106420100 0028, 10351806001000000), the Foundation for Distinguished Young Talents in Higher Education of Guangdong (wym09066), and supported by Grant-in-Aid for JSPS Fellows of Japan (22·P10045).

References

1. Asghar, H.J., Li, S., Pieprzyk, J., Wang, H.: Cryptanalysis of the convex hull click human identification protocol. In: Burmester, M., Tsudik, G., Magliveras, S., Ilić, I. (eds.) ISC 2010. LNCS, vol. 6531, pp. 24–30. Springer, Heidelberg (2011)
2. Boneh, D., Hamburg, M.: Generalized identity based and broadcast encryption schemes. In: Pieprzyk, J. (ed.) ASIACRYPT 2008. LNCS, vol. 5350, pp. 455–470. Springer, Heidelberg (2008)
3. Boneh, D., Goh, E.-J., Nissim, K.: Evaluating 2-DNF formulas on ciphertexts. In: Kilian, J. (ed.) TCC 2005. LNCS, vol. 3378, pp. 325–341. Springer, Heidelberg (2005)
4. Boneh, D., Sahai, A., Waters, B.: Functional encryption: Definitions and challenges. In: Ishai, Y. (ed.) TCC 2011. LNCS, vol. 6597, pp. 253–273. Springer, Heidelberg (2011)
5. Boyen, X., Waters, B.: Anonymous hierarchical identity-based encryption (Without random oracles). In: Dwork, C. (ed.) CRYPTO 2006. LNCS, vol. 4117, pp. 290–307. Springer, Heidelberg (2006)
6. Brakerski, Z., Goldwasser, S.: Circular and leakage resilient public-key encryption under subgroup indistinguishability. In: Rabin, T. (ed.) CRYPTO 2010. LNCS, vol. 6223, pp. 1–20. Springer, Heidelberg (2010)
7. Chandran, N., Goyal, V., Moriarty, R., Ostrovsky, R.: Position based cryptography. In: Halevi, S. (ed.) CRYPTO 2009. LNCS, vol. 5677, pp. 391–407. Springer, Heidelberg (2009)
8. Diffie, W., Hellman, M.E.: New directions in cryptography. IEEE Transactions on Information Theory IT-22, 644–654 (1976)
9. Gentry, C.: Practical identity-based encryption without random oracles. In: Vaudenay, S. (ed.) EUROCRYPT 2006. LNCS, vol. 4004, pp. 445–464. Springer, Heidelberg (2006)
10. Gentry, C., Halevi, S.: Hierarchical identity based encryption with polynomially many levels. In: Reingold, O. (ed.) TCC 2009. LNCS, vol. 5444, pp. 437–456. Springer, Heidelberg (2009)
11. Goyal, V., Pandey, O., Sahai, A., Waters, B.: Attribute-based encryption for fine-grained access control of encrypted data. ACM CCS 2006, 89–98 (2006)
12. Katz, J., Sahai, A., Waters, B.: Predicate encryption supporting disjunctions, polynomial equations, and inner products. In: Smart, N.P. (ed.) EUROCRYPT 2008. LNCS, vol. 4965, pp. 146–162. Springer, Heidelberg (2008)

13. Kleinjung, T., Aoki, K., Franke, J., Lenstra, A.K., Thomé, E., Bos, J.W., Gaudry, P., Kruppa, A., Montgomery, P.L., Osvik, D.A., te Riele, H., Timofeev, A., Zimmermann, P.: Factorization of a 768-bit RSA modulus. In: Rabin, T. (ed.) CRYPTO 2010. LNCS, vol. 6223, pp. 333–350. Springer, Heidelberg (2010)

14. Lenstra Jr., H.W.: Factoring integers with elliptic curves. Annals of Mathematics, 649–673 (1987)

15. Lewko, A., Okamoto, T., Sahai, A., Takashima, K., Waters, B.: Fully secure functional encryption: Attribute-based encryption and (Hierarchical) inner product encryption. In: Gilbert, H. (ed.) EUROCRYPT 2010. LNCS, vol. 6110, pp. 62–91. Springer, Heidelberg (2010)

16. Lewko, A., Waters, B.: New techniques for dual system encryption and fully secure HIBE with short ciphertexts. In: Micciancio, D. (ed.) TCC 2010. LNCS, vol. 5978, pp. 455–479. Springer, Heidelberg (2010)

17. Park, J.H.: Inner-product encryption under standard assumption. Des. Codes Cryptogr. 58(3), 235–257 (2011)

18. O'Neill, A.: Definitional issues in functional encryption. Cryptology ePrint Archive, Report 2010/556

19. Sahai, A., Waters, B.: Fuzzy identity-based encryption. In: Cramer, R. (ed.) EUROCRYPT 2005. LNCS, vol. 3494, pp. 457–473. Springer, Heidelberg (2005)

20. Sahai, A., Waters, B.: Fuzzy identities and attributed-based encryption. In: Tuyls, P., Škoric, B., Kevenaar, T. (eds.) Security with noisy data, Springer, Heidelberg (2007)

21. Seo, J.H., Kobayashi, T., Ohkubo, M., Suzuki, K.: Anonymous hierarchical identity-based encryption with constant size ciphertext. IEICE Trans. on Fundamentals E94-A(1), 45–56 (2011)

22. Shen, E., Shi, E., Waters, B.: Predicate privacy in encryption systems. In: Reingold, O. (ed.) TCC 2009. LNCS, vol. 5444, pp. 457–473. Springer, Heidelberg (2009)

23. Shi, E., Waters, B.: Delegating capabilities in predicate encryption systems. In: Aceto, L., Damgård, I., Goldberg, L.A., Halldórsson, M.M., Ingólfsdóttir, A., Walukiewicz, I. (eds.) ICALP 2008, Part II. LNCS, vol. 5126, pp. 560–578. Springer, Heidelberg (2008)

24. Sobrado, L., Birget, J.C.: Graphical passwords. The Rutgers Scholar, 4 (2002), http://rutgersscholar.rutgers.edu/volume04/sobrbirg

25. Waters, B.: Dual system encryption: Realizing fully secure IBE and HIBE under simple assumptions. In: Halevi, S. (ed.) CRYPTO 2009. LNCS, vol. 5677, pp. 619–636. Springer, Heidelberg (2009)

26. Waters, B., Felten, E.W.: Secure, privacy proof of locations. TR-667-03

27. Wiedenbeck, S., Waters, J., Sobrado, L., Birget, J.C.: Design and evaluation of a shoulder-surfing resistant graphical password scheme. In: AVI 2006, pp. 177–184. ACM, New York (2006)

28. Zhao, H., Li, X.: S3PAS: A scalable shoulder-surfing resistant textual-graphical password authentication scheme. In: AINAW 2007, pp. 467–472. IEEE Computer Society, Los Alamitos (2007)

An Efficient Rational Secret Sharing Scheme Based on the Chinese Remainder Theorem

Yun Zhang[1,2], Christophe Tartary[3], and Huaxiong Wang[1]

[1] Division of Mathematical Sciences, School of Physical and Mathematical Sciences,
Nanyang Technological University, Singapore
ZHAN0233@e.ntu.edu.sg
[2] School of Mathematical Science, Yangzhou University, Yangzhou, 225002,
People's Republic of China
ctartary@mail.tsinghua.edu.cn
[3] Institute for Interdisciplinary Information Sciences, Institute for Theoretical
Computer Science, Tsinghua University, Beijing, 100084,
People's Republic of China
hxwang@ntu.edu.sg

Abstract. The design of rational cryptographic protocols is a recently created research area at the intersection of cryptography and game theory. At TCC'10, Fuchsbauer *et al.* introduced two equilibrium notions (computational version of strict Nash equilibrium and stability with respect to trembles) offering a computational relaxation of traditional game theory equilibria. Using trapdoor permutations, they constructed a rational t-out-of n sharing technique satisfying these new security models. Their construction only requires standard communication networks but the share bitsize is $2n|s|+O(k)$ for security against a single deviation and raises to $(n-t+1) \cdot (2n|s|+O(k))$ to achieve $(t-1)$-resilience where k is a security parameter. In this paper, we propose a new protocol for rational t-out-of n secret sharing scheme based on the Chinese reminder theorem. Under some computational assumptions related to the discrete logarithm problem and RSA, this construction leads to a $(t-1)$-resilient computational strict Nash equilibrium that is stable with respect to trembles with share bitsize $O(k)$. Our protocol does not rely on simultaneous channel. Instead, it only requires synchronous broadcast channel and synchronous pairwise private channels.

Keywords: rational cryptography, computational strict Nash equilibrium, stability with respect to trembles, Asmuth-Bloom sharing scheme.

1 Introduction

1.1 Preliminaries

In 1979, Shamir [16] and Blakley [4] independently introduced the concept of *secret sharing scheme* (SSS) in order to facilitate the distributed storage of private data in an unreliable environment. Since then, secret sharing has become a major building block for cryptographic primitives in particular in the

U. Parampalli and P. Hawkes (Eds.): ACISP 2011, LNCS 6812, pp. 259–275, 2011.

area of *multiparty computation* (MPC). The goal of a (perfect) SSS is to distribute a secret value s amongst a finite set of participants $\mathcal{P} = \{P_1, \ldots, P_n\}$ in such a way that only specific subsets of $\mathcal{P}$ can reconstruct s while the others have no information about this secret element whatsoever.

Traditional cryptographic models assume that some parties are honest (i.e. they faithfully follow a given protocol) while others are malicious participants against whom the honest players must be protected. However, in many real-world applications, a participant will choose to be dishonest if deviating from the protocol will provide him with some advantage. Game theory can be used to model such a situation where players are *self-interested* (i.e. *rational*). In this representation, each participant P_i has a utility function U_i and the execution of the cryptographic protocol is regarded as a game over $\mathcal{P}$ where the n players' strategies $\sigma_1, \ldots, \sigma_n$ are dictated by their respective utilities $U_1, \ldots, U_n$.

Halpern and Teague introduced the first general approach for rational secret sharing in 2004 [9]. This opened new research directions and many results appeared subsequently [6,1,7,12,13,10,14,2]. In game theory, a *Nash equilibrium* (NE) captures the idea of stable solution for a given game. Indeed, in a NE, no single player P_i can individually improve his welfare by deviating from the strategy σ_i specified by the equilibrium $(\sigma_1, \ldots, \sigma_n)$ if all remaining participants stick to theirs. Most of the rational protocols quoted above focus on achieving a NE surviving iterated deletion of weakly dominated strategies. However, as pointed out in [12], some bad strategies still survive this deletion process. As a remedy, Kol and Naor proposed to use the notion of *strict NE* requiring that each player's strategy is his unique best response to the other players' strategies. This notion is more appealing than a NE in that, in a NE, there is no incentive to deviate while, in a strict NE, there is an incentive *not* to deviate. However, it is difficult to achieve a strict NE in many cases since this notion rules out many cryptographic techniques. In order to balance this tradeoff, Fuchsbauer *et al.* [7] proposed a *computational version of strict NE* (which enables the use of cryptography) and the notion of NE *stable with respect to trembles*. They also provided an efficient construction for standard communication networks achieving such an equilibrium as long as all the players are computationally bounded. However, the bitlength of their shares is $2n|s| + O(k)$ which gets very large especially when n (number of players) or k (security parameter) is large. While not a serious issue in its own right, this may be problematic when their rational SSS is used as a subroutine for rational MPC.

1.2   Our Results

In this paper, we present a protocol for rational t-out-of-n SSS. We only need a synchronous (but non-simultaneous) broadcast channel along with pairwise point-to-point channels. We do not assume any on-line dealer nor do we apply any generic MPC protocol to redistribute the shares of the secret. Instead, we borrow the idea from *Joint Random Secret Sharing* to allow every player to form his "one-time" share at the beginning of each iteration by interactions among the group of $m(m \geq t)$ participants. The main idea is described as follows.

In the share distribution phase, the dealer use the modified version of Asmuth-Bloom SSS proposed by Kaya and Selçuk [11] to generate n shares for the secret s. Suppose there are m players active in the reconstruction phase, say $P_1, \ldots, P_m$. This phase proceeds with several rounds. At the beginning of each iteration, the "one-time" shares for $(s+d) \bmod m_0$ are generated (jointly by the active players) using the technique from *Joint Random Secret Sharing*, where $d = d^{(1)} + \cdots + d^{(m)}$ and each $d^{(i)}$ is chosen independently and uniformly at random from the domain of the secret by P_i. If $d \equiv 0 \bmod m_0$, then all the "one-time" shares are valid for recovering s and, in this sense, the current iteration is called the valid iteration. Otherwise, the current iteration is invalid, which is designed only for catching possible cheaters. Each communicated message carries a commitment with perfect binding and computational hiding (assuming the hardness of computing discrete logarithm). Thus, at every point of our protocol, there is a unique legal message that each player can send (except with negligible probability). This prevents a player from outwardly appearing to follow the protocol while subliminally communicating with other participants.

Then, all the active players are required to open their "one-time" shares. After each player P_i has received the "one-time" shares from all the other active players, he is required to open $d^{(i)}$, which provides a unique way for the participants to jointly identify the valid iteration. If $d \bmod m_0 \neq 0$, then the current iteration is invalid and all the players are asked to restart a new iteration; otherwise, it is valid, the secret s is recovered and the protocol terminates immediately after this iteration. In this way, no player can identify the valid iteration before he opens his "one-time" share. Furthermore, each player can identify the valid iteration only after it has occurred, that is, once a player learns that the current iteration is valid, each player has already got the real secret. Due to this, we do not need simultaneous channels. Our protocol is efficient in that the round complexity and computation complexity are both polynomial (in the security parameter k). It induces a $(t-1)$-resilient computational strict Nash equilibrium that is stable with respect to trembles. However, our protocol relies on the assumption that no player knows auxiliary information about the secret s, which has been proved to be inherent in the non-simultaneous channels model [2].

1.3   Comparison to Fuchsbauer *et al.*'s Scheme

The protocol from [7] provides good point of comparison to ours since both techniques have similar features:

- Both of them induce a $(t-1)$-resilient computational strict NE that is stable with respect to trembles.
- Neither of them relies on simultaneous channels.
- Both of them assume that no player knows any auxiliary information about the secret s. This property has been proved to be inherent to the non-simultaneous channels model [2].
- Both protocols run in time polynomial in k (security parameter) and they have almost the same round complexity.

However, our protocol has smaller share size even when $(t-1)$-resilience to coalitions is required. Our shares are $O(k)$ bits long while those from [7] need $(n-t+1)\,(2n|s|+O(k))$ bits. The latter share length leads to practical efficiency issues when $n-t+1$ is large or when Fuchsbauer *et al.*'s technique is used as a building block within more general rational MPC protocols.

2   Definitions and Background

2.1   Secret Sharing

A *t-out-of-n SSS* with secret domain S is a two-phase protocol (share distribution and secret reconstruction) executed by the dealer and a subgroup of the n players $P_1, \ldots, P_n$ respectively. During the share distribution phase, the dealer chooses a secret $s \in S$ and generates n shares $s_1, \ldots, s_n$ based on a security parameter k. Each s_i is given to P_i secretly. In the secret reconstruction phase, some collection of at least t players jointly reconstruct s from their shares without any interaction with the dealer. We require the following two properties to hold:

- **Correctness.** Any collection of t or more players can uniquely determine the secret by putting their shares together honestly.
- **Privacy.** Any collection of fewer than t players can not recover the secret s.

In this paper, the security will be guaranteed under some computational assumptions related to the discrete logarithm problem and RSA which will be specified in Sect. 3.3. Thus, the security of our rational SSS will be computational.

2.2   Notions of Game-Theoretic Equilibria

As said in Sect. 1.1, in the rational model, each player is self-interested: he does what is in his interest. To formalize rationality, each player P_i is associated to a real-valued utility function U_i modeling the gain that P_i obtains when following his many strategies. For more details, we refer the reader to [1].

We now present the game theoretic concepts our cryptographic construction relies on. We are to design a rational SSS with the expectation that, when rationally played, the secret is revealed to all the players participating in the reconstruction. In the share distribution phase, all n players are silent and the dealer is assumed to be honest. The reconstruction process is to be viewed as a game amongst $m \geq t$ players. We denote $\sigma = (\sigma_1, \ldots, \sigma_m)$ the strategy profile of these players where σ_i is P_i's strategy for $1 \leq i \leq m$. As usual, let σ_{-C} denote the strategy profile of all m players except the players in C and σ_C denote the strategy profile constricted to the coalition $C \subseteq \{1, \ldots, m\}$. Given a strategy profile σ, it induces the utility value $U_i(\sigma)$ for each player P_i expressing his payoff when σ is played by the m players.

In the following, we denote the security parameter by k and it is assumed that the n utility functions are polynomials in k. The definitions appearing in this subsection originate from [7].

Definition 1. *Let $\epsilon : \mathbb{N} \to [0, \infty)$ be a function. We say ϵ is **negligible** if for every positive polynomial $p(\cdot)$ there exists an integer $N_{p(\cdot)} > 0$ such that for all $k > N_{p(\cdot)}$, it holds that $\epsilon(k) < \frac{1}{p(k)}$. We say that ϵ is **noticeable** if there exists a positive polynomial $p(\cdot)$ and an integer $M_{p(\cdot)}$ such that $\epsilon(k) > \frac{1}{p(k)}$ for any $k > M_{p(\cdot)}$.*

Definition 2. *A strategy σ induces an r-**resilient computational NE** if for any coalition C of at most r players and for any probabilistic polynomial time strategy profile σ', it holds:*

$$U_i(k, \sigma'_C, \sigma_{-C}) \le U_i(k, \sigma_C, \sigma_{-C}) + \epsilon(k) \quad \text{for any } i \in C,$$

where ϵ is a negligible function.

Remark 1. When $r = 1$, the definition of r-resilient computational NE coincides with that of the computational NE.

We need to define what it means for two strategies to be equivalent. Although we could refer the reader to [7] for the details, for completeness of our paper, we recall the corresponding notions below. As said before, every player is to be considered as a polynomial-time probabilistic Turning (PPT) machine (as function of the security parameter k). We assume that m players participate in the reconstruction phase. As often in MPC, security will be demonstrated by simulating the views of the different participants [8].

Definition 3. *Denote $P_C := \{P_i | i \in C\}$, $P_{-C} := \{P_i | i \notin C\}$ and the strategy vector of the m players by σ. Define the random variable $\text{View}^{\sigma}_{-C}$ as follows:*

> *Let Trans denote the messages sent by P_C not including any message sent by P_C after they write to their output tapes. $\text{View}^{\sigma}_{-C}$ includes the information given by the dealer to P_{-C}, the random coins of P_{-C} and the (partial) transcript Trans.*

Fix a strategy ρ_C and an algorithm T. Define the random variable $\text{View}^{T, \rho_C}_{-C}$ as follows:

> *When the m players interact, P_C follows ρ_C and P_{-C} follows σ_{-C}. Let Trans denote the messages sent by P_C. Algorithm T, given the entire view of P_C, outputs an arbitrary truncation Trans' of Trans (defining a cut-off point and deleting any messages sent after that point). $\text{View}^{T, \rho_C}_{-C}$ includes the information given by the dealer to P_{-C}, the random coins of P_{-C}, and the (partial) transcript Trans'.*

Strategy ρ_C yields equivalent play with respect to σ, denoted $\rho_C \approx \sigma$, if there exists a PPT algorithm T such that for all PPT distinguishers D:

$$\left| \text{Prob}[D(1^k, \text{View}^{T, \rho_C}_{-C}) = 1] - \text{Prob}[D(1^k, \text{View}^{\sigma}_{-C}) = 1] \right| \le \epsilon(k)$$

where $\epsilon(\cdot)$ is a negligible function.

Definition 4. *A strategy σ is said to be an r-**resilient computational strict NE**, if:*

1. *σ induces an r-resilient computational NE;*
2. *For any coalition C of at most r players and for any probabilistic polynomial time strategy σ'_C with $\sigma'_C \not\approx \sigma$, there is a positive polynomial $p(\cdot)$ such that for any $i \in C$, it holds that $U_i(k, \sigma_C, \sigma_{-C}) \geq U_i(k, \sigma'_C, \sigma_{-C}) + \frac{1}{p(k)}$ for infinitely many values of k, namely, $U_i(k, \sigma_C, \sigma_{-C}) - U_i(k, \sigma'_C, \sigma_{-C})$ is non-negligible.*

Definition 5. *For any coalition C, strategy ρ_C is δ-**close** to strategy σ_C if ρ_C is as follows:*

ρ_C: *With probability $1 - \delta$, players in C play according to σ_C.*
   *With probability δ, players in C follow an arbitrary (possibly correlated) PPT strategy σ'_C (called the **residual strategy** of ρ_C).*

Definition 6. *σ induces an r-**resilient computational NE that is stable with respect to trembles** if:*

1. *σ induces an r-resilient computational NE;*
2. *There is a noticeable function δ such that for any coalition C with $|C| \leq r$, and any vector of PPT strategies ρ_{-C} that is δ-close to σ_{-C}, any PPT strategy ρ_C, there exists a PPT strategy $\sigma'_C \approx \sigma$ such that $U_i(k, \rho_C, \rho_{-C}) \leq U_i(k, \sigma'_C, \rho_{-C}) + \epsilon(k)$, where $\epsilon(\cdot)$ is negligible.*

Remark 2. Intuitively, the strategy vector (σ_C, σ_{-C}) is stable with respect to trembles if σ_C remains a best response even if P_{-C} plays any PPT strategies other than σ_{-C} with some small but noticeable probability δ.

2.3   Assumptions on the Utility Functions

Following most previous works on this topic, we assume the following properties of the utility functions:

- each player P_i first prefers outcomes in which he outputs the real secret;
- each player P_i secondly prefers outcomes in which the fewest of the other players output the real secret.

As in [7], the expected utility is also assumed to be a polynomial of the security parameter k. We distinguish four cases as follows. For each $i \in \{1, \ldots, n\}$, let $U_i(k)$ (respectively, $U_i^+(k)$) be the minimal (respectively, maximal) payoff of P_i when he outputs the correct secret and let $U_i^-(k)$ be his maximal payoff when P_i does not output s. As usually assumed, we consider: $U_i^+(k) > U_i(k) > U_i^-(k)$ for all $i \in \{1, \ldots, n\}$. As in [7], define

$$U_i^r(k) := \frac{1}{|S|} \cdot U_i^+(k) + (1 - \frac{1}{|S|}) \cdot U_i^-(k)$$

which is the expected utility of a player outputting a random guess for the secret (assuming that the other players abort without any outputs, or with wrong outputs). It is reasonable to assume that $U_i(k) > U_i^r(k)$, since otherwise, players hardly have any incentive to execute the secret reconstruction phase at all. Furthermore, it is still reasonable to assume that the difference between $U_i(k)$ and $U_i^r(k)$ is non-negligible for any $1 \leq i \leq n$, that is, there exists a polynomial $p(\cdot)$ such that for infinitely many k's it holds that:

$$U_i(k) \geq U_i^r(k) + \frac{1}{p(k)}.$$

Note that, this assumption is not restrictive in that without it, it is hard to guarantee the players have enough motivation to execute the share reconstruction phase rather than guess the secret locally, especially in the computational setting, where no player cares about negligible difference in utilities. In this paper, we consider coalitions of at most $t - 1$ players. We assume for simplicity that during the whole process of share reconstruction phase, there is at most one coalition which contains a subset of active players and all the players in this coalition share all information they jointly have. Thus, all the players in some coalition are assumed to share a single output.

3   Our Protocol for t-out-of-n Rational Secret Sharing

Our protocol contains two phases: share distribution and secret reconstruction. The first phase is executed by the dealer only while the second phase is designed for all the active players who want to jointly recover the secret without the dealer. Our share distribution phase is similar to the revisited version of the Asmuth-Bloom's non-interactive verifiable SSS [11,3] except with minor but necessary modifications for our needs. The dealer is available only in the initial share distribution phase during which he is assumed to be honest. We assume the existence of synchronous broadcast channels (but non-simultaneous) for all participating players and the presence of private channels between any pair of these players and the dealer.

As said in the previous section, all n players are assumed to be computationally bounded. In the following, let k be a security parameter.

3.1   Initial Share Phase

This is the only phase where the dealer is active. His goal is to distribute s over $\mathcal{P} := \{P_1, \ldots, P_n\}$ using the Asmuth-Bloom SSS with threshold t. As mentioned above, we adopt the modified version of Asmuth-Bloom SSS proposed by Kaya and Selçuk [11] and make further modifications (mainly on the parameters settings) to meet our needs. This initial share phase has two stages.

Remark 3. The value g is the unique integer in $\mathbb{Z}_Q$ satisfying $g_i \equiv g \bmod p_i$, for all $1 \leq i \leq n$. Besides, the order of g in $\mathbb{Z}_{QN}^*$ is at least $\prod_{j=1}^n m_j$ and for each $1 \leq i \leq n$, we have:

$$E(y) \bmod p_i = (g^y \bmod QN) \bmod p_i = g^y \bmod p_i = g_i^{y_i} \bmod p_i$$

Hence, during the whole protocol, we use $(E(y) \mod p_i)$ as a commitment to y_i, which is perfect binding but is computational hiding. That is, the committer cannot commit himself to two values y_i and y'_i by the same commitment value and, under the assumption that computing discrete logarithm is intractable in $\mathbb{Z}_{p_i}$, no PPT player learns y_i from $E(y) \mod p_i$ except with negligible probability in k. This allows players to check the consistency of the received data. Since the dealer is assumed to be honest, $E(y)$ is only used to detect the players' possible malicious behavior during the reconstruction process described in the next section.

Initial Share Phase

1. Parameters Setup

To share a secret s, the dealer chooses $m_0(> s)$ and publishes it. This value m_0 should also be lower bounded by a value depending on players' utilities and discussed later in this paper.

1. The dealer chooses and publishes a set of pairwise coprime integers $m_1, \ldots, m_n$ of bitlength k such that the following requirements are satisfied:
   (a) $m_0 < m_1 < \ldots < m_n$;
   (b) $\prod_{i=1}^{t} m_i > (n+1)m_0^2 \prod_{i=1}^{t-1} m_{n-i+1}$;
   (c) $p_j = 2m_j + 1$ is prime for any $1 \le j \le n$.
2. For any $1 \le i \le n$, let G_i be a subgroup of $\mathbb{Z}_{p_i}^*$ of order m_i and denote g_i a generator of G_i. Let $Q = \prod_{i=1}^{n} p_i$ and $g = (\sum_{i=1}^{n} g_i \cdot Q'_i \cdot \frac{Q}{p_i}) \mod Q$ and, where Q'_i is the inverse of $\frac{Q}{p_i}$ in $\mathbb{Z}_{p_i}^*$, for $1 \le i \le n$. The dealer publishes g.
3. The dealer chooses and publishes an RSA modulus N of length at least k whose factorization is unknown to any of the n players [15].

2. Share Distribution

To share a secret $s \in \mathbb{Z}_{m_0}$ among a group of n players $\{P_1, \ldots, P_n\}$, the dealer executes the following steps.

1. He sets $M := \left\lfloor \frac{\prod_{i=1}^{t} m_i}{n+1} \right\rfloor$. He computes $y = s + A_0 \cdot m_0$ for some positive integer A_0 generated randomly subject to the condition that $0 < y < M$, calculates $y_i = y \mod m_i$ and finally sends the share y_i to player P_i secretly, for $1 \le i \le n$.
2. He computes $E(y) := g^y \mod QN$ and broadcasts $E(y)$.

3.2 Secret Reconstruction Phase

We assume that $m(\ge t)$ players participate in the secret reconstruction phase. For ease of description, we can assume without loss of generality that those players are $P_1, \ldots, P_m$. The reconstruction phase proceeds in a series of iterations, each of which consists of multiple communication rounds among those players. First, we propose two subprotocols to be called upon within the reconstruction phase.

3.2.1 Share Update Phase. This is done by the players participating in the secret reconstruction process, namely, by $P_1, \ldots, P_m$. In this phase, each participating P_i (sorted in index increasing order) plays a similar role to the dealer's (initial share phase) to share a random element $d^{(i)} \in \mathbb{Z}_{m_0}$ and to finally get his "one-time" share for $(s + d^{(1)} + \cdots + d^{(m)}) \bmod m_0$.

In [11], in order to prevent the dealer from distributing inconsistent shares, the range-proof technique proposed from [5] is used to allow the dealer to convince each player that some committed integer lies in a particular interval. This range proof is statistically zero-knowledge in the random-oracle model. Besides, provided that computing discrete logarithm problems is intractable, a cheating dealer can only succeed with negligible probability (in k). We refer to [11,5] for further details.

Here, in order to prevent a player P_i from distributing inconsistent shares for his random chosen $d^{(i)}$, we need to apply this range-proof technique. Throughout this section, we will use $\mathsf{RngPrf}(E(y), M)$ to denote the Cao-Liu's non-interactive range proof that a secret integer y committed with $E(y)$ is in the interval $[0, M)$ [5]. In the following share update phase, we will use $\mathsf{RngPrf}(E(y), M)$ as a black box and we refer to [5] for additional information.

Share Update Phase

1. Each P_i selects a random element $d^{(i)} \in \mathbb{Z}_{m_0}$ uniformly and independently. He computes $y^{(i)} = A_i \cdot m_0 + d^{(i)}$, where A_i is a positive integer chosen randomly conditioned on $0 < y^{(i)} < M$. Then, he computes $y_j^{(i)} = y^{(i)} \bmod m_j$ along with $E(y^{(i)}) := g^{y^{(i)}} \bmod QN$, and he finally sends $y_j^{(i)}$ to player P_j secretly through a secure channel for each $j \neq i$. In addition, P_i broadcasts $E(y^{(i)})$ and $\mathsf{RngPrf}(E(y^{(i)}), M)$.

2. If player P_i only receives partial messages (hereinafter, partial messages including the case of no message at all), then he outputs a random guess of the secret and terminates the protocol. Otherwise, he checks whether $g_i^{y_i^{(j)}} \equiv E(y^{(j)}) \bmod p_i$ and he checks the correctness of $\mathsf{RngPrf}(E(y^j), M)$ for $1 \leq j \neq i \leq m$. If all the checks are successful, then P_i computes $d_i = \sum_{l=1}^{m} y_i^{(l)} \bmod m_i$. Otherwise, he outputs a random guess of the secret and stops the protocol.
   Let $d := d^{(1)} + \cdots + d^{(m)}$. Note that $\{d_1, \ldots, d_m\}$ are the shares for $d \bmod m_0$.

3. Each P_i computes $\widetilde{y}_i := (y_i + d_i) \bmod m_i$ as his "one-time" share for the current iteration. The commitment for $\widetilde{y}_i$ is $E(\widetilde{y}_i) := E(y) \prod_{l=1}^{m} E(y^{(l)}) \bmod p_i$, which can be locally computed by each player.

Proposition 1. *Let $Y := y + y^{(1)} + \ldots + y^{(m)} = (s + d) + (A_0 + \cdots + A_m) \cdot m_0$. Then after the share update phase, $\{\widetilde{y}_1, \ldots, \widetilde{y}_m\}$ are valid shares for $(s + d) \bmod m_0$ as long as all the players follow the protocol honestly. In addition, all the commitments are correctly checked.*

This proposition means that every subset of at least t players uniquely determines $(s + d) \bmod m_0$ (Correctness), while for any subset of $t - 1$ players, every candidate for s or for each $d^{(i)}$ is (approximately) equally likely, and so each candidate for each $(s + d) \bmod m_0$ is (approximately) equally likely (Privacy).

Proposition 2 ([11]). *During the share update phase, any player P_i can not distribute inconsistent shares for $d^{(i)}$ without being detected except with probability negligible in k. In other words, if all checks are successful, then all the shares*

$y_1^{(i)}, \ldots, y_m^{(i)}$ are residues of some integer less than M except with negligible probability which is introduced by the error probability of RngPrf.

Remark 4. Let $T := y^{(1)} + \cdots + y^{(m)}$. Since the "one-time" shares $\{\widetilde{y_1}, \ldots, \widetilde{y_m}\}$ are the shares for $(s + d) \bmod m_0$, they are the shares for s if and only if $d \equiv 0 \bmod m_0$. This is equivalent to $T \equiv 0 \bmod m_0$. In this sense, the iteration in which $T \equiv 0 \bmod m_0$ is called a *valid* iteration. It is called an *invalid* iteration otherwise.

Remark 5. The goals of the Share Update Phase are twofold. On one hand, it makes our protocol proceed with several iterations: all except the last one are invalid iterations, which are designed to catch possible cheaters. During the valid iteration, all active players get the real secret. In addition, no one will know in advance whether the current iteration is going to be the last iteration. On the other hand, since during each iteration all the "one-time" shares are revealed, if the current round is invalid, the players should proceed to the next round with totally new shares, which are provided by the share updating phase. Hence, "one-time" shares are shares used only once (i.e. in the current iteration) and they become meaningless in later iterations.

3.2.2 Combiner Phase. In this phase, each player P_i uses the reconstruction algorithm from the Asmuth-Bloom SSS to recover $(s + d) \bmod m_0$.

Combiner Phase

---

1. Let U be a collection of t shares that player P_i chooses in the reconstruction phase and let V be the corresponding collection of the indices of the players to whom those t shares belong. Let M_V denote $\prod_{j \in V} m_j$.
2. Let $M_{V-\{j\}}$ denote $\prod_{\ell \in V, \ell \neq j} m_\ell$ and let $M'_{V,j}$ be the multiplicative inverse of $M_{V-\{j\}}$ in $\mathbb{Z}^*_{m_j}$. Player P_i computes $Y^{(i)} := \sum_{j \in V} \widetilde{y_j} \cdot M'_{V,j} \cdot M_{V-\{j\}} \bmod M_V$. Finally, let $S^{(i)} := Y^{(i)} \bmod m_0$.

3.2.3 Overview of the Reconstruction Phase. In order for the reader to get an easier understanding of the reconstruction phase, we first give its general view. The full description is in Sect. 3.2.4.

The reconstruction phase proceeds with a sequence of invalid/valid iterations such that the last iteration is valid and each iteration has two stages. During the first stage, players first interact to get their "one-time" shares for $(s+d) \bmod m_0$, where $d = d^{(1)} + \cdots + d^{(m)}$ and each $d^{(i)}$ is chosen randomly by P_i. During the second stage, each player P_i is required to open the value $y^{(i)}$ he chose in the first stage. Thus, since $d^{(i)} = y^{(i)} \bmod m_0$, the players can jointly identify the status of the current iteration: if $d \bmod m_0 \neq 0$, then the current iteration is invalid and all the players are asked to restart a new iteration; otherwise, it is valid, the secret s is recovered and the protocol terminates immediately after this iteration.

The iterations have the following properties:

- <u>invalid iteration</u>: no information about s is revealed since all the revealed shares are the shares for $(s+d) \bmod m_0$. At the beginning of the subsequent iteration, all the shares are updated which guarantees that the "one-time" shares revealed in the current iteration are useless for the next iteration.
- <u>valid iteration</u>: every player recovers s on the assumption that every participant follows the protocol (which will be demonstrated to be the case since they are rational).

The key in this process is the fact that nobody knows before the opening of the "one-time" shares whether the current iteration will be valid. Furthermore, when a given player realizes that the valid iteration occurs, each other player can compute the secret as well. That is why we do not need simultaneous channels.

3.2.4   Secret Reconstruction Phase. We assume that $m(\geq t)$ players participate in the secret reconstruction. As before, we can assume that they are $P_1, \ldots, P_m$. Our reconstruction protocol proceeds with multiple iterations, each of which contains two stages for each of these m players. It is assumed without lost of generality that in each step, each P_i executes his strategy in index increasing order. For each of these m participants P_i, his strategy σ_i is as follows.

Secret Reconstruction Phase

Stage 1

1. Player P_i executes the share update phase to get his "one-time" share $\widetilde{y}_i$ for the value $(s + d) \bmod m_0$, where $d = d^{(1)} + \cdots + d^{(m)}$ and each $d^{(i)}$ is chosen independently and uniformly at random by P_i .
2. Player P_i broadcasts his "one-time" share $\widetilde{y}_i$ obtained at the previous step. If P_i does not receive m shares (including his own), or if he detects that $g_j^{\widetilde{y}_j} \bmod p_j \neq E(\widetilde{y}_j) \bmod p_j$ for some j, he outputs a random guess of the secret and aborts the protocol abruptly.
3. Otherwise, P_i chooses randomly t data from $\{\widetilde{y}_1, \ldots, \widetilde{y}_m\}$ and executes the Combiner Phase.

The second stage is used to recover T so that player P_i can identify the status (valid/invalid) of the current round since $\{\widetilde{y}_1, \cdots, \widetilde{y}_m\}$ are the shares for s if and only if $T \equiv 0 \bmod m_0$.

Stage 2

1. Player P_i broadcasts $y^{(i)}$. If he does not receive m messages (including his own), or if he detects that $g^{y^{(j)}} \bmod QN \neq E(y^{(j)})$ for some j, P_i outputs $S^{(i)}$ he obtains in the Combiner Phase and aborts the whole protocol.
2. Otherwise, P_i computes $T = y^{(1)} + \ldots + y^{(m)}$. If $T \equiv 0 \bmod m_0$, then he outputs $S^{(i)}$ and stops the whole protocol; Otherwise, P_i goes back to Stage 1 and starts another iteration.

3.3   Security of our Rational SSS

The reconstruction phase is a game amongst the m active players. The strategy profile is denoted $\sigma = (\sigma_1, \ldots, \sigma_m)$ where σ_i is P_i's strategy described in the previous section. Let $U_i^*(k) := \frac{1}{m_0} \cdot U_i^+(k) + (1 - \frac{1}{m_0}) \cdot U_i^r(k)$, $1 \le i \le n$. Based on the security requirements of [5], we make the following assumption:

$\mathcal{A}$: The discrete logarithm problem over finite fields is intractable.
   The RSA modulus N is hard to factor; the resulting RSA encryption scheme and Schnorr signature is secure.

Theorem 1. *Assuming that $\mathcal{A}$ holds. σ induces a $(t-1)$-resilient computational NE as long as $U_i(k) - U_i^*(k)$ is non-negligible, for $1 \le i \le n$.*

Theorem 2. *Assuming that $\mathcal{A}$ holds. σ induces a $(t-1)$-resilient computational strict NE provided that $U_i(k) - U_i^*(k)$ is non-negligible, for $1 \le i \le n$.*

Theorem 3. *Assuming that $\mathcal{A}$ holds. σ induces a computational NE that is stable with respect to trembles provided that $U_i(k) - U_i^*(k)$ is non-negligible, for $1 \le i \le n$.*

Remark 6. The expected number of iterations of our protocol is m_0. Note that the requirements for m_0 are that $m_0 > \frac{2[U_i^+(k) - U_i^r(k)]}{U_i(k) - U_i^r(k)}$, for $1 \le i \le n$. Since all the utility functions are polynomial in k and $U_i(k) - U_i^r(k)$ is assumed to be non-negligible, m_0 can be chosen to be a prime less than some polynomial in k. Since all the computations are based on modular arithmetic, they can be executed in polynomial time. Besides, RngPrf can also be verified in polynomial time. All these considerations imply that our protocol is efficient.

4   Conclusion

In this paper, we presented a new protocol for t-out-of-n rational secret sharing based on the CRT in non-simultaneous channels. Our technique leads to a $(t-1)$-resilient computational strict NE that is stable with respect to trembles while having much smaller share size than the protocol proposed by Fuchsbauer *et al.* [7].

Acknowlegments

The authors would like to thanks the reviewers for their suggestions to improve the quality of this paper. Christophe Tartary's work was funded by the National Natural Science Foundation of China grants 61033001, 61061130540, 61073174 and 61050110147 (International Young Scientists program) as well as the National Basic Research Program of China grants 2007CB807900 and 2007CB807901. Christophe Tartary also acknowledges support from the Danish National Research Foundation and the National Natural Science Foundation of China (under the grant 61061130540) for the Sino-Danish Center for the Theory of Interactive Computation (CTIC) within which part of this work was performed. Huaxiong Wang's work was supported by the Singapore National Research Foundation under Research Grant NRF-CRP2-2007-03.

References

1. Abraham, I., Dolev, D., Gonen, R., Halpern, J.: Distributed computing meets game theory: Robust mechanisms for rational secret sharing and multiparty computation. In: 25th Annual ACM Symposium on Principles of Distributed Computing (PODC 2006), pp. 53–62. ACM Press, New York (2006)

2. Asharov, G., Lindell, Y.: Utility dependence in correct and fair rational secret sharing. In: Halevi, S. (ed.) CRYPTO 2009. LNCS, vol. 5677, pp. 559–576. Springer, Heidelberg (2009)

3. Asmuth, C., Bloom, J.: A modular approach to key safeguarding. IEEE Transactions on Information Theory IT-29(2), 208–210 (1983)

4. Blakley, G.R.: Safeguarding cryptographic keys. In: AFIPS 1979 National Computer Conference, pp. 313–317. AFIPS Press (June 1979)

5. Cao, Z., Liu, L.: Boudot's range-bounded commitment scheme revisited. In: Qing, S., Imai, H., Wang, G. (eds.) ICICS 2007. LNCS, vol. 4861, pp. 230–238. Springer, Heidelberg (2007)

6. Dov Gordon, S., Katz, J.: Rational secret sharing, revisited. In: De Prisco, R., Yung, M. (eds.) SCN 2006. LNCS, vol. 4116, pp. 229–241. Springer, Heidelberg (2006)

7. Fuchsbauer, G., Katz, J., Naccache, D.: Efficient rational secret sharing in standard communication networks. In: Micciancio, D. (ed.) TCC 2010. LNCS, vol. 5978, pp. 419–436. Springer, Heidelberg (2010)

8. Goldwasser, S., Micali, S., Rackoff, C.: The knowledge complexity of interactive proof-systems. In: 17th Annual ACM Symposium on Theory of Computing (STOC 1985), pp. 291–304. ACM, New York (1985)

9. Halpern, J., Teague, V.: Rational secret sharing and multiparty computation: Extended abstract. In: 36th Annual ACM Symposium on Theory of Computing (STOC 2004), pp. 623–632. ACM Press, New York (2004)

10. Izmalkov, S., Micali, S., Lepinski, M.: Rational secure computation and ideal mechanism design. In: 46th Annual Symposium on the Foundations of Computer Science (FOCS 2005), pp. 585–594. IEEE Computer Society, Los Alamitos (2005)

11. Kaya, K., Selçuk, A.A.: Secret sharing extensions based on the Chinese reminder theorem. Cryptology ePrint Archive, Report 2010/096 (2010), http://eprint.iacr.org/2010/096

12. Kol, G., Naor, M.: Games for exchanging information. In: 40th Annual ACM Symposium on Theory of Computing (STOC 2008), pp. 423–432. ACM Press, New York (2008)

13. Micali, S., shelat, a.: Purely rational secret sharing (Extended abstract). In: Reingold, O. (ed.) TCC 2009. LNCS, vol. 5444, pp. 54–71. Springer, Heidelberg (2009)

14. Ong, S.J., Parkes, D.C., Rosen, A., Vadhan, S.: Fairness with an honest minority and a rational majority. In: Reingold, O. (ed.) TCC 2009. LNCS, vol. 5444, pp. 36–53. Springer, Heidelberg (2009)

15. Rivest, R.L., Shamir, A., Adleman, L.M.: A method for obtaining digital signatures and public key cryptosystems. Communications of the ACM 21(2), 120–126 (1978)

16. Shamir, A.: How to share a secret. Communications of the ACM 22(11), 612–613 (1979)

A   Proof of Theorem 1

By Proposition 1, our protocol is a valid secret sharing scheme. All the active players will be expected to recover the real secret in $\frac{1}{\Pr[d\equiv 0 \bmod m_0]} = m_0$ iterations, as long as they stick to σ.

Now, we prove that σ induces a $(t-1)$-resilient computational NE. Let C be any coalition of size at most $t-1$. Assume that all the players not in C stick to their prescribed strategies. We focus on PPT deviations from some players in C. There are several possible cases: (1) some player P_i in C deviates during the share update phase; (2) some player P_i in C lies about his "one-time" share or only sends partial messages in Stage 1 - Step 2 ; (3) some player P_i in C either opens a fake $y^{(i)}$ or broadcasts nothing in Stage 2.

Suppose (1) happens. There are two possible deviations. **Case 1.** P_i only sends (or broadcasts) partial messages in Stage 2 of share update phase. However, this will be detected and cause the protocol to terminate. In this case, the only profitable thing he can do is to output a random guess of the secret, which will earn him at most $U_i^r(k)$. Obviously, it is a worse outcome to P_i, since $U_i^r(k) < U_i(k)$. Hence, P_i will send all data, fake or real, as required. **Case 2.** P_i distributes inconsistent shares for his randomly chosen $d^{(i)}$ to some player P_j not in C. Under assumption $\mathcal{A}$, no cheating P_i can convince any other P_j to accept $\mathsf{RngPrf}(E(y^{(i)}, M))$ except with negligible probability $\epsilon'(k)$. Once $\mathsf{RngPrf}(E(y^{(i)}, M))$ is rejected, which happens with probability $1 - \epsilon'(k)$, the protocol terminates immediately and the best that player P_i can do is to output a random guess of the secret. Thus, the expected utility P_i can get by distributing inconsistent shares is at most $\epsilon'(k) \cdot U_i^+(k) + (1 - \epsilon'(k)) \cdot U_i^r(k) = \epsilon'(k) \cdot (U_i^+(k) - U_i^r(k)) + U_i^r(k) < \epsilon(k) + U_i(k)$, where $\epsilon(k) = \epsilon'(k)(U_i^+(k) - U_i^r(k))$ is a negligible function in k, since we assumed that $U_1, \ldots, U_n$ were polynomials in k. That is, using this type of deviation, P_i can only increase his payoff by a negligible amount (if at all). Thus, given our computational setting, no rational player P_i is to deviate by distributing inconsistent shares.

Now, we consider the possible deviations in Step 2 of Stage 1. There are two possible cases. **Case 1.** P_i does not broadcast anything at all. **Case 2.** P_i cheats about his "one-time" share. However, either of these deviations will be detected and cause the protocol to terminate. Hence, we do not distinguish between these two cases. If $(d \bmod m_0) = 0$ (i.e., the current iteration is valid which happens with probability $\frac{1}{m_0}$), then all the players in C will output the real secret and hence P_i will get at most $U_i^+(k)$. If $(d \bmod m_0) \neq 0$ (i.e., the current iteration is invalid which happens with probability $1 - \frac{1}{m_0}$), then the best thing P_i can do is to output a random guess of the secret earning at most $U_i^r(k)$. Thus, the expected payoff of P_i with this type of deviation is at most $\frac{1}{m_0} \cdot U_i^+(k) + (1 - \frac{1}{m_0}) \cdot U_i^r(k) = U_i^*(k)$. It is less than $U_i(k)$ by our assumption. Hence, as a rational player, P_i will not deviate in Step 2 of Stage 1.

Finally, we study what happens if some player in C does not broadcast anything at all or broadcast a fake value in Stage 2. Either deviation will be detected and cause the protocol to terminate abruptly. Since we assume players execute every step of the protocol in ascending order, we can assume without loss of generality that $C = \{P_{m-t+2}, \ldots, P_m\}$. Since all the players in C share their information, for any $m - t + 2 \leq i \leq m$, after receives the message from the players not in C P_i can first compute $T := y^{(1)} + \cdots + y^{(m)}$ to identify whether the current round is valid or not, then determines what to do in this stage. Note that we have proved that, in the computational and rational setting, any player will execute the reconstruction phase honestly up to the end of Stage 1. Therefore, if the current iteration is valid, each $S^{(j)}$ obtained by P_j in the Combiner Phase is indeed the real secret. In this case, regardless of what P_i will do, each player will output the real secret, which will earn $U_i(k)$ to P_i. On the other hand, if the current round is invalid, no one has recovered the real secret yet and either type of deviations will cause the protocol to terminate abruptly resulting in a payoff at most $U_i^r(k)$ to P_i. Hence, P_i is never better off by this deviations.

B Proof of Theorem 2

Suppose C is any subset of $\{1, \ldots, m\}$ of size at most $t - 1$. Let $P_C := \{P_i | i \in C\}$ and $P_{-C} := \{P_i | i \in \{1, \ldots, m\} - C\}$. Since all the players in P_C acts in unison, we can regard P_C as a whole. By Theorem 1, it is sufficient to prove that for any PPT strategy $\rho_C \not\approx \sigma$, there is a positive polynomial $p(\cdot)$ such that for any $i \in C$, $U_i(k, \sigma) \geq U_i(k, \rho_C, \sigma_{-C}) + \frac{1}{p(k)}$ for infinitely many values of k, that is, $U_i(k, \sigma) - U_i(k, \rho_C, \sigma_{-C})$ is positive and non-negligible.

Let Deviate be the event that P_C deviates from σ_C before he can compute his output, that is, before entering the Stage 2 of the valid iteration. Since $\rho_C \not\approx \sigma$, Prob[Deviate] is non-negligible by definition. Now, consider the interaction of ρ_C with σ_{-C}. Let Valid be the event that P_C deviates from σ_C before entering Stage 2 during the valid iteration and let Invalid be the event that P_C deviates from σ_C during an invalid iteration. Let Caught be the event that P_C is caught cheating. Then, for each $i \in C$, we have:

$$U_i(k, \rho_C, \sigma_{-C})$$
$$\leq U_i^+(k) \cdot \text{Prob}[\text{Valid}] + U_i^+(k) \cdot \text{Prob}[\text{Invalid} \wedge \overline{\text{Caught}}]$$
$$\quad + U_i^r(k) \cdot \text{Prob}[\text{Invalid} \wedge \text{Caught}] + U_i(k) \cdot \text{Prob}[\overline{\text{Deviate}}]$$
$$= U_i^+(k) \cdot (\text{Prob}[\text{Valid}|\text{Deviate}] + \text{Prob}[\overline{\text{Caught}}|\text{Invalid}] \cdot \text{Prob}[\text{Invalid}|\text{Deviate}]) \cdot \text{Prob}[\text{Deviate}]$$
$$\quad + U_i^r(k) \cdot \text{Prob}[\text{Caught}|\text{Invalid}] \cdot \text{Prob}[\text{Invalid}|\text{Deviate}] \cdot \text{Prob}[\text{Deviate}] + (1 - \text{Prob}[\text{Deviate}])U_i(k)$$
$$= U_i^+(k) \cdot \left[\frac{1}{m_0} + \epsilon(k)(1 - \frac{1}{m_0}) \right] \cdot \text{Prob}[\text{Deviate}]$$
$$\quad + U_i^r(k) \cdot (1 - \epsilon(k)) \cdot (1 - \frac{1}{m_0}) \cdot \text{Prob}[\text{Deviate}] + U_i(k) - U_i(k) \cdot \text{Prob}[\text{Deviate}]$$
$$= U_i(k) + (U_i^*(k) - U_i(k)) \cdot \text{Prob}[\text{Deviate}] + \eta(k)$$

where $\eta(k) = \epsilon(k) \cdot (1 - \frac{1}{m_0}) \cdot (U_i^+(k) - U_i^r(k)) \cdot \text{Prob}[\text{Deviate}]$ is negligible. It follows

$$U_i(k, \sigma) = U_i(k) \geq U_i(k, \rho_C, \sigma_{-C}) + (U_i(k) - U_i^*(k)) \cdot \text{Prob}[\text{Deviate}] - \eta(k).$$

Since both $U_i(k) - U_i^*(k)$ and Prob[Deviate] are positive and non-negligible, $U_i(k,\sigma) - U_i(k,\rho_C,\sigma_{-C})$ is positive and non-negligible, which completes this proof.

Remark 7. In this proof, we actually show that, for any PPT strategy ρ_C, we have:

$$U_i(k,\sigma) = U_i(k) \geq U_i(k,\rho_C,\sigma_{-C}) + (U_i(k) - U_i^*(k)) \cdot \text{Prob[Deviate]} - \eta(k)$$

where $\eta(\cdot)$ is a negligible function.

C   Proof of Theorem 3

This proof is based on [7]. Let δ be a parameter which we will specify at the end of the proof. Note that δ may depend on k. Since we assumed players execute every step of the protocol in an index increasing order, we can assume without loss of generality that $C = \{m-t+2, \ldots, m\}$. It is sufficient to show that for any $i \in C$, any vector of PPT strategies ρ_{-C} that is δ-close to σ_{-C}, and any PPT strategy ρ_C, there exists a PPT strategy $\sigma_C' \approx \sigma$ such that $U_i(k,\rho_C,\rho_{-C}) \leq U_i(k,\sigma_C',\rho_{-C}) + \epsilon(k)$, where $\epsilon(\cdot)$ is negligible. Let $P_C = \{P_i | i \in C\}$ and $P_{-C} = \{P_i | i \in (\{1, \ldots, m\} - C)\}$. First, we construct a strategy σ_C' for the players in P_C as follows.

---

1. Set Detect:=0.
2. In each iteration:
   (a) Receive the messages from P_{-C} in each possible step. If P_C detects that some player P_j in P_{-C} has deviated from σ_j, set Detect:= 1.
   (b) If Detect= 1, execute the remaining steps according to ρ_C; otherwise σ_C.
3. If Detect= 0, determine the output according to σ_C, otherwise, output whatever ρ_C outputs.

Observe that when σ_C' interacts with σ_{-C}, Detect is never set to be 1. Hence $\sigma_C' \approx \sigma$ and $U_i(k,\sigma_C',\sigma_{-C}) = U_i(k,\sigma_C,\sigma_{-C}) = U_i(k)$ for any $i \in C$. Now, we want to show that $U_i(k,\rho_C,\rho_{-C}) \leq U_i(k,\sigma_C',\rho_{-C}) + \eta(k)$ for any $i \in C$, where $\eta(\cdot)$ is negligible. Let $\widetilde{\rho_{-C}}$ denote the residual strategy of ρ_{-C}. In an interaction where P_C follows strategy ρ_C, let Detected be the event that P_C is detected deviating from σ_C before entering stage 2 of the valid iteration while no player in P_{-C} is detected cheating so far. Also, let $\text{Prob}_{\text{Detected}}(\alpha)$ be the probability of Detected when P_{-C} follows strategy α. Since no player in P_{-C} will be detected cheating when P_{-C} execute σ_{-C}, $\text{Prob}_{\text{Detected}}(\sigma_{-C})$ equals the probability of P_C being detected deviating from σ_C before entering Stage 2 of the valid iteration.

Claim 1. Prob[Deviate]=$\text{Prob}_{\text{Detected}}(\sigma_{-C}) + \epsilon(k) \cdot$ Prob[Deviate], for some negligible function ϵ.

Claim 2. For any $i \in C$,

$$U_i(k, \rho_C, \widetilde{\rho_{-C}}) - U_i(k, \sigma'_C, \widetilde{\rho_{-C}}) \leq \text{Prob}_{\text{Detected}}(\widetilde{\rho_{-C}}) \cdot (U_i^+(k) - U_i^r(k)) + \epsilon(k),$$

where $\epsilon(\cdot)$ is negligible.

Claim 3. $\text{Prob}_{\text{Detected}}(\widetilde{\rho_{-C}}) \leq \text{Prob}_{\text{Detected}}(\sigma_{-C}) + \epsilon(k)$ for some $\epsilon(\cdot)$ negligible.

By Remark 7, we know that for any PPT strategy ρ_C,

$$U_i(k, \sigma) = U_i(k) \geq U_i(k, \rho_C, \sigma_{-C}) + (U_i(k) - U_i^*(k)) \cdot \text{Prob}[\text{Deviate}] - \eta(k).$$

where $\eta(\cdot)$ is a negligible function. Now, we get:

$$
\begin{aligned}
U_i(k, \rho_C, \rho_{-C}) &= (1 - \delta) \cdot U_i(k, \rho_C, \sigma_{-C}) + \delta \cdot U_i(k, \rho_C, \widetilde{\rho_{-C}}) \\
&\leq (1 - \delta) \cdot [U_i(k) + (U_i^*(k) - U_i(k)) \cdot \text{Prob}[\text{Deviate}] + \eta(k)] \\
&\quad + \delta \cdot U_i(k, \rho_C, \widetilde{\rho_{-C}})
\end{aligned}
$$

Also

$$
\begin{aligned}
U_i(k, \sigma'_C, \rho_{-C}) &= (1 - \delta) \cdot U_i(k, \sigma'_C, \sigma_{-C}) + \delta \cdot U_i(k, \sigma'_C, \widetilde{\rho_{-C}}) \\
&= (1 - \delta) \cdot U_i(k) + \delta \cdot U_i(k, \sigma'_C, \widetilde{\rho_{-C}})
\end{aligned}
$$

It follows:

$$U_i(k, \rho_C, \rho_{-C}) - U_i(k, \sigma'_C, \rho_{-C})$$

$\leq \quad (1 - \delta) \cdot (U_i^*(k) - U_i(k)) \cdot \text{Prob}[\text{Deviate}] + \delta \cdot [U_i(k, \rho_i, \widetilde{\rho_{-C}}) - U_i(k, \sigma'_C, \widetilde{\rho_{-C}}] + \eta(k)$

by Claim 2

$\leq \quad (1 - \delta) \cdot (U_i^*(k) - U_i(k)) \cdot \text{Prob}[\text{Deviate}]$

$\quad + \delta \cdot \text{Prob}_{\text{Detected}}(\widetilde{\rho_{-C}}) \cdot (U_i^+(k) - U_i^r(k)) + \delta \cdot \epsilon(k) + \eta(k)$

by Claim 1

$= \quad (1 - \delta) \cdot (U_i^*(k) - U_i(k)) \cdot (\text{Prob}_{\text{Detected}}(\sigma_{-C}) + \epsilon'(k) \cdot \text{Prob}[\text{Deviate}])$

$\quad + \delta \cdot (U_i^+(k) - U_i^r(k)) \cdot \text{Prob}_{\text{Detected}}(\widetilde{\rho_{-C}}) + \delta \cdot \epsilon(k) + \eta(k)$

by Claim 3

$\leq \quad (1 - \delta) \cdot (U_i^*(k) - U_i(k)) \cdot \text{Prob}_{\text{Detected}}(\widetilde{\rho_{-C}})$

$\quad + \delta \cdot (U_i^+(k) - U_i^r(k)) \cdot \text{Prob}_{\text{Detected}}(\widetilde{\rho_{-C}}) + \eta'(k)$

where $\eta'(\cdot)$ is some negligible function. Hence, there exists $\delta > 0$ (may depend on k) such that the above expression is negligible in k for each $i \in C$.

DMIPS - Defensive Mechanism against IP Spoofing

Shashank Lagishetty, Pruthvi Sabbu, and Kannan Srinathan

International Institute of Information and Technology, Hyderabad, India
{shashankl,pruthvireddy.sabbu}@research.iiit.ac.in,
srinathan@iiit.ac.in

Abstract. The usage of internet has increased in all fields of the globe and its size is increasing at a high rate. The network providers are not able to afford enough resources like computation power and bandwidth which are needed to maintain their quality of service. This inability is exploited by the attackers in the form of Denial of Service attacks (DoS) and Distributed Denial of Service attacks (DDoS). The systems trying to mitigate DoS attacks should focus on the technique called IP spoofing. IP Spoofing refers to the creation of IP packets with forged source address. IP spoofing aids the DoS attackers in maintaining their anonymity. IP spoofing is beneficial when the systems use source address for authentication of the packets. Previously, an anti-spoofing method called HCF (Hop Count Filtering) was proposed which could effectively filter the spoofed packets. The HCF works on the basis that the attacker cannot falsify the Hop count (HC), the number of hops an IP packet takes to reach the destination. This HC value can be inferred from the TTL (Time To Live) field in the IP packet. However, the working of HCF has the following problems: 1) Multiple path possibility is ignored. 2) The method of building the HC tables must be more secure. 3) Lack of good renew procedure which can detect network changes. In this paper, we propose a 2 level filtering scheme called DMIPS, based on HCF. DMIPS is secure, resolves the multiple path problem and can filter the spoofed packets effectively. The present scheme can detect the changes in the network and can update the HC values. DMIPS improve the quality of service of the network by minimizing the number of false positives. The network under discussion is of the type server and clients and the server is the point of attack.

Keywords: anonymity, hop count, ip spoofing, network security.

1   Introduction

The internet has become an integral part of everyday life. Many types of public services, social networking and bank transactions are running online. For most of the times, the information transfer in the internet has server and client architecture. Thus, the presence and efficient usage of the servers is of utmost importance. Recently, social networking websites like Twitter and Facebook were brought down by attackers using DDoS attacks [19,20]. The attackers are using

U. Parampalli and P. Hawkes (Eds.): ACISP 2011, LNCS 6812, pp. 276–291, 2011.
© Springer-Verlag Berlin Heidelberg 2011

DoS attacks, more than any other tool to bring down the systems. The uniqueness of DoS or DDoS attack is that it does not involve any information theft or data modification. So, the administrators perceive it as a less of a threat. However it can cause huge losses in terms of reliability and confidentiality from their customers.

The DoS attack is made possible by the usage of techniques like IP spoofing which enables hiding the attacker's source location and impersonating other legitimate hosts. When a packet reaches a server, the server checks whether the source address present in the packet is legitimate for its services. If yes, it processes the packet otherwise drops it. Due to the advent of IP spoofing, attackers are able to impersonate other users and are getting their things done. The attackers are also successful in flooding the resources without being caught. We could counter attack them by efficient filtering techniques. We propose one such solution in this paper. We utilize the information contained in the IP header for filtering the spoofed packets. Our scheme contains 2 levels of filtering, one at the Border Gateway Routers (BGRs) inside the autonomous system (AS) and the other at the server. Although, an attacker can forge any field in the IP header, he cannot falsify the number of hops an IP packet takes to reach its destination. It is solely determined by the Internet routing infrastructure. The hop-count information is indirectly reflected in the TTL field of the IP header, since each intermediate router decrements the TTL value by one before forwarding it to the next hop. The Hop-count heuristic is used for filtering inside the AS. To maintain secure communication outside the AS, we ensure the truthfulness of the advertisements broadcast by various AS during set-up and then filter the packets based on the tags added into the packets by the BGR of source AS.

In the following sections, we describe our scheme discussing various issues followed inside the AS and outside the AS separately. Section 2 deals with the related work, 3.1 and 3.2 deals with set-up and working phase in inside the AS,outside the AS respectively. Section 4 explores the ways in which the scheme can be compromised and how good is our scheme. We evaluate our scheme in terms of CPU usage and accuracy in section 5 and finally end with the conclusion in section 6.

2   Related Work

IP spoofing has annoyed the internet for many years and is still prevalent today [6]. It is an important problem to solve. In [6], the authors discussed the current state of IP spoofing and surveyed various defense schemes. Several schemes have been proposed to mitigate this IP spoofing problem. They can be categorized into one of the following types. 1) Traceback schemes. 2) Schemes which filter in the route. 3) Schemes which filter at the destination.

Traceback schemes are used to trace the real source of the malicious packets when the attack happens. This is useful in identifying the attacker and stopping the attack. Traceback can be done in multiple ways. In schemes such as [1,2,12] the routers add information about the path into the packets. In [21], the routers send ICMP messages to the destinations of the packets. The other possibility of tracing the attackers is to deposit the digests of the packets at the routers, [11,13] are of this type. The recent DoS or DDoS attacks have originated from the hosts which are compromised by the attackers. So, the traceback strategy would lead to bots instead of the real attackers. The traceback schemes do not work well against spoofing.

Filtering in the route, the second category of anti-spoofing methods is more effective. Ingress filtering [3,7] and Unicast Reverse Path Forwarding (uRPF) [18] fall into this category. Ingress filtering works efficiently only when it is deployed on a large scale. Hence stepwise deployment is difficult. The deployment of ingress/egress filtering inflicts significant costs on the ISP implementing it, without being assured significant benefit or protection. Thus the incentive for an ISP to deploy these mechanisms is relatively low. uRPF is based on the idea that a packet destined to a given AS would always traverse a specific route. It does not work when asymmetry routes are present. uRPF also does not support stepwise deployment.

Probabilistic Packet Marking (PPM) [10], Deterministic Packet Marking (DPM) [4], Pi [16], Hop Count Filtering (HCF) [8] and Spoofing Prevention Method (SPM) [5] are the important schemes which do filter at the destination. In PPM, every router marks the packet with partial information. Each router marks their IP address onto the packet with the probability of P along the way the packet is traversed. PPM has high computation overhead and is not effective. DPM unlike PPM marks all the incoming packets at the ingress router interface. The router sends the Incoming interface address in two packets, each containing 16 bits of that address. When marked packets arrive at end host, end host will extract the both halves and check if the match of ingress address and the source address is same. This method has 50% false positive rate. Path Identifier (Pi) proposed a packet marking algorithm to mark each packet with the value of the packet's TTL modulo 16/n before the packet is forwarded (n can be any positive integer less than 16). Pi works well under the network where all routers deploy the Pi marking scheme. Unfortunately, it is rather impossible to have all routers in AS's from different ISP to deploy Pi. StackPi improved Pi's performance by proposing two new marking schemes - Stack-based marking and Write-ahead marking [17]. This Write-ahead marking increases the performance of StackPi against legacy routers. In SPM, a packet is tagged with a key at the border of the source AS, and the key is verified and removed at the border of the destination AS. The problem with SPM is that the interactions between the AS can easily become the target to DoS/DDoS attacks.

The authors of [6] recommended Hop count filtering defense over all other schemes because of its effectiveness against different kinds of spoofing attacks. HCF uses TTL value as a key for each (source, destination) pair. FLF4DoS [9] and probabilistic HCF [14], are based on HCF but uses fuzzy logic and probabilistic approaches respectively. HCF is deployed at end host, hence it is easier to deploy. But, HCF did not consider the case of multiple paths and the IP2HC table creation is also not secure. Our scheme DMIPS is also based on HCF.

We include the multiple path possibility as packets take alternate paths during congestion in the network caused by the DoS attacks. By implementing our improved Hop Count Filtering technique (IHCF) as proposed in section 3.1 inside the AS, we filter most of the spoofed packets inside the AS itself. DMIPS reduces the server overhead in filtering the packets, so that it can provide services to legitimate traffic more effectively. Our IHCF method is secure and saves bandwidth.

3   DMIPS Scheme

The present day internet consists of large ISPs which are Autonomous systems (AS) owned by different administrative authorities. The autonomous systems are free to choose an internal routing architecture and protocols. The routers inside the AS communicate with each other directly and with the outside world, they transmit the packets through border gateway routers (BGRs). DMIPS is a two level filtering scheme and we follow different implementations for inside the AS and outside the AS parts of the network.

3.1   Inside the AS

The improved Hop Count Filtering method (IHCF) is the filtering scheme for inside of the AS and is implemented at the BGRs. We choose to implement the scheme on gateway routers because the packets of an AS can reach other domains only through gateway routers. The border gateway routers/ISPs of the AS initiate IHCF scheme, which considers the case of multiple hop counts. The previous implementations of variants of Hop-Count filtering assumed a single path between server and client. But, new communication approaches like Networks on Chips (NoC) are focused on implementation of multiple path routing. Provision of multiple paths is useful for load balancing and also when the possibility of link failures is high.

In this context, we do not discuss on how the routers assign a particular path to a packet. The multipath routing scheme improves path diversity, thereby minimizing network congestion and traffic bottlenecks. When there is an attack, traffic congestion is bound to happen. Since at each time the packets take the shortest path, the links along this path are continuously used. Sometimes there is no room for new packets. Hence, we consider the possibility of multiple paths and multiple Hop counts. This will decrease the number of false positives in hop count based schemes. False positives are those legitimate packets that are incorrectly identified as spoofed. False negatives are spoofed packets that go undetected by the HCF. Both should be minimized in order to achieve good filtering accuracy.

There are two phases in the implementation of the networks, set-up phase and working phase. The set up phase is for exchanging necessary information at the time of implementation so as to decrease the overhead while running. Working phase refers to the period where the scheme is implemented and is running.

Offline Set-Up phase

Calculating Hop counts: The HC for a packet is calculated from the TTL value in its header. The HC is the difference between initial and final TTL values. TTL is 8-bit field in the packet used to denote the time, the packet can stay in the network.

$$\text{Hop Count} = \text{Initial TTL value - Final TTL value}$$

The receiver after extracting the final TTL value from the packet estimates the initial TTL as there are only some selected initial TTL values. If the final TTL value is 50 then the receiver can assume that the initial TTL value must be 64 since the difference between initial and final TTL values will not be more than 30 which is maximum number of hops between any two nodes in the network. The initial TTL values for most used operating systems are shown in Table 1. We assume that there are only finite operating systems and hence limited initial

Table 1. List of possible initial TTL values

Operating System	Initial TTL values
Windows 95	32
Windows 98	128
Mac OS X	64
Windows XP, Vista	128
NetBSD	255
RedHat 9	64
HP - UX	30

TTL values and their number does not increase significantly in the future. Most of the new operating systems are upgrades of already existing and they preserve the initial TTL value. In our paper, we assume that the hosts do not change their initial TTL value as they have no incentive to modify the default values. But, an attacker can modify its initial TTL values according to his strategies. Attackers intruding into the host system and changing its initial TTL value is out of the scope of this paper.

Construction of IP2HCS tables: IP2HCS table is the mapping between IP addresses and the possible hop counts of the host to the BGR. Whenever the BGR needs the possible HC values to a certain host it initiates RREQ (Route request) packets with the destination address set to it. The BGRs send these RREQ packets to its neighbouring routers in the same AS and these routers will in turn forward it to their neighbours. This process repeats until the packets reaches destination.

The host on receiving the request sends RREP (Route reply) packet with route and hop count information. If the route discovery is successful, the BGR gets the route reply packet listing the sequence of network hops through which the packet can reach the target. The structure of RREQ and RREP packets are shown in Fig 1 and Fig 2 respectively. The RREP packet traverses the same path as RREQ packet but in reverse direction. The originator upon receiving the RREP packet can calculate one of the multiple hop counts possible to that node. As said already, calculating the accurate, secure hop counts is one of our major concerns. Using these routing packets has some possible security concerns.

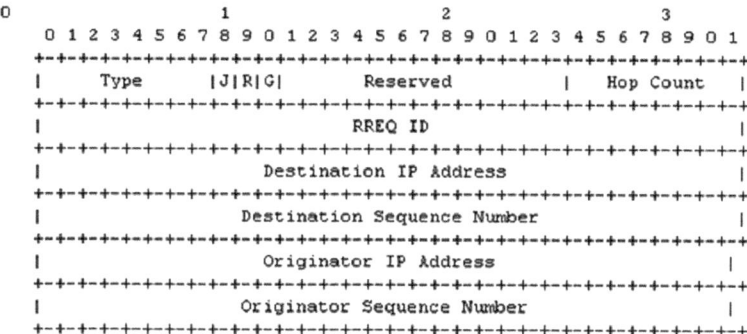

```
0                   1                   2                   3
0 1 2 3 4 5 6 7 8 9 0 1 2 3 4 5 6 7 8 9 0 1 2 3 4 5 6 7 8 9 0 1
+-+-+-+-+-+-+-+-+-+-+-+-+-+-+-+-+-+-+-+-+-+-+-+-+-+-+-+-+-+-+-+
|     Type      |J|R|G|        Reserved        |   Hop Count    |
+-+-+-+-+-+-+-+-+-+-+-+-+-+-+-+-+-+-+-+-+-+-+-+-+-+-+-+-+-+-+-+
|                            RREQ ID                           |
+-+-+-+-+-+-+-+-+-+-+-+-+-+-+-+-+-+-+-+-+-+-+-+-+-+-+-+-+-+-+-+
|                       Destination IP Address                 |
+-+-+-+-+-+-+-+-+-+-+-+-+-+-+-+-+-+-+-+-+-+-+-+-+-+-+-+-+-+-+-+
|                    Destination Sequence Number               |
+-+-+-+-+-+-+-+-+-+-+-+-+-+-+-+-+-+-+-+-+-+-+-+-+-+-+-+-+-+-+-+
|                        Originator IP Address                 |
+-+-+-+-+-+-+-+-+-+-+-+-+-+-+-+-+-+-+-+-+-+-+-+-+-+-+-+-+-+-+-+
|                     Originator Sequence Number               |
+-+-+-+-+-+-+-+-+-+-+-+-+-+-+-+-+-+-+-+-+-+-+-+-+-+-+-+-+-+-+-+
```

Fig. 1. Route Request packet

```
0                   1                   2                   3
0 1 2 3 4 5 6 7 8 9 0 1 2 3 4 5 6 7 8 9 0 1 2 3 4 5 6 7 8 9 0 1
+-+-+-+-+-+-+-+-+-+-+-+-+-+-+-+-+-+-+-+-+-+-+-+-+-+-+-+-+-+-+-+
|     Type      |J|R|G|        Reserved        |   Hop Count    |
+-+-+-+-+-+-+-+-+-+-+-+-+-+-+-+-+-+-+-+-+-+-+-+-+-+-+-+-+-+-+-+
|                            RREQ ID                           |
+-+-+-+-+-+-+-+-+-+-+-+-+-+-+-+-+-+-+-+-+-+-+-+-+-+-+-+-+-+-+-+
|                       Destination IP Address                 |
+-+-+-+-+-+-+-+-+-+-+-+-+-+-+-+-+-+-+-+-+-+-+-+-+-+-+-+-+-+-+-+
|                    Destination Sequence Number               |
+-+-+-+-+-+-+-+-+-+-+-+-+-+-+-+-+-+-+-+-+-+-+-+-+-+-+-+-+-+-+-+
|                        Originator IP Address                 |
+-+-+-+-+-+-+-+-+-+-+-+-+-+-+-+-+-+-+-+-+-+-+-+-+-+-+-+-+-+-+-+
|                     Originator Sequence Number               |
+-+-+-+-+-+-+-+-+-+-+-+-+-+-+-+-+-+-+-+-+-+-+-+-+-+-+-+-+-+-+-+
```

Fig. 2. Route Reply packet

Challenge 1: Some malicious node can impersonate BGR and initiate RREQ message with the aim of polluting IP2HCS table.

Solution: We use Route request and Router reply packets to populate the IP2HCS table. So, when the attacker initiates forged RREQ message, the reply packets reach the attacker and not the BGR. This is because the route reply packets retrace the same path as route request packets and in this way the attacker cannot pollute the IP2HCS table in the BGR. Hence, DMIPS is secure against IP2HCS table modification.

Challenge2: Some routers can misbehave in some cases in which the hop counts in the packet before and after processing, by the router does not differ by 1.

Solution: Though compromising the router is difficult, it is possible to alter its behavior. After receiving the route reply message, the BGR takes the HC value, keeps it as the initial TTL value and sends a query to the host in the reverse path. If the reply comes, then store this hop count in the table otherwise invoke Trace() function as shown in Fig 3.

```
Trace ()
    Get the BGR to host path from RREP packet, x₁ → x₂ → x₃...
    Extract the hop count (l)
    For j in range 1 to l
        Send a query to x_j in the reverse RREP packet path with
initial TTL value set to j-1
        If reply is not received
            Return IP address of j
    Return 1
```

Fig. 3. Trace function

Using this above function, we can trace the adversary or spoiled router trying to pollute the hop count table. The function returns the address of the misbehaving node. If the function returns 1 then the received hop count is not altered by the intermediate routers and is truthful. The other nodes cannot reply to the BGR because we are querying in the same path. A very light weight query is enough for this authentication.

The BGR can now populate its IP2HCS table with the received secure Hop counts in the region indexed by its source address. The BGRs can store all the possible HCs or the best K hop counts. Here, K is the security parameter and each ISP can select its own value. The ISPs can select the best hop counts as the hop counts of those paths which are reliable or having minimum number of hops.

Working

1. Filtering· The filtering process is explained in Fig 4. After IP2HCS table is created, the following algorithm is used to filter the incoming packets. In addition to filtering, our scheme needs to check for link failures or route changes also. To monitor these changes in the network, we use another table called IP2WC.
2. IP2WC: The IP2WC table contains 3 fields, the source IP, wrong HC value and the frequency. If we assume the absence of multiple attackers then we can remove the separate wrong HC field. Whenever we receive a packet which is assumed to be spoofed, the frequency indexed by its address and hop count values is increased. Here the frequency refers to the number of times, a packet with wrong hop count has been received from an IP address.
3. Renew procedure: We maintain IP2WC table to keep track of packets from the hosts with wrong hop counts. It is a mapping from IP address to the

```
For each packet received
  Extract final TTL value (T) and IP address of the source (S)
  Estimate the initial TTL T₀ based on T
  Compute the hop count using HC = T₀- T
  If (HC present in IP2HCS[S])
        Packet is legitimate
  Else
    Spoofed, increment frequency of IP2WC[S][HC]
```

Fig. 4. Algorithm for filtering the spoofed packets

number of packets appearing to be coming from it with wrong HC. If any frequency field in this table becomes sufficiently high, then we assume that these packets are sent by the actual host. The change in the HC may be due to some change in the AS network. So, the BGR re-initiates RREQ packet for the node whose hop counts are conflicting. The basis of this assumption is that the attacker is smart enough to use his resources effectively. He will not waste his resources by continuously spoofing a single source even after knowing that his packets will be filtered. Moreover, by spoofing a single host, the efficiency of the DoS attacks is lost. In DoS/DDoS attacks, the spoofing is done randomly. The administrators can easily monitor non-random spoofed IP addresses.

To make our technique secure against these kinds of attacks (spoofing a single host) on our defense mechanism, we add the following conditions. The ISP administrators can define a threshold count (TC) and a threshold time interval (TTI) so that update is done if and only if the following conditions are satisfied. If the attacker sends the packets with same spoofed address every time, our scheme initiate the Renew procedure but only after TTI time. This decreases the efficiency of the attack. The value of TC depends on the number of possible malicious users in the subnet, the administrators have to keep this in mind while defining the threshold. Most of the updates in the network are made when the network is idle.

– Any frequency field in IP2WC is more than TC and
– Time since last update is more than TTI.

3.2   Outside the AS

The IHCF method implemented at border routers filters most of the spoofed packets and prevents them from going beyond their AS. But, some of the autonomous systems may not implement our scheme. Though we provide services for the hosts inside them, we must make sure that DoS or DDoS attacks does not originate from these autonomous systems.

We provide services, with high priority to the autonomous systems which have installed our scheme. The requests from non DMIPS installed ASes are processed only after processing the high priority packets. Hence, the attackers would try to spoof the hosts which are inside a DMIPS installed AS and the server needs to filter these types of packets. For this we check the path taken by the packet from the source host to the server. If the path taken by the packet is same as what it should be then it is assumed to be legitimate. Hence, we need to authenticate the source AS from where the packet is coming.

The Inter-AS routes in the present day internet are based on BGP. Each BGP route is a chain of autonomous systems, leading to a particular set of destinations. The autonomous system advertises chosen routes to its peer ASes. Upon receiving its peers' advertisements, an AS applies its internal policy to select the best route to every destination and update its route database. To achieve confidentiality on the network, the advertisements of the routers should be truthful and the packets should maintain the path as advertised. The advertisements are verified at set-up phase.

Offline Set-Up phase

Route selection policies are meaningful only if the advertisements on which they are based are truthful. Autonomous systems may have different incentives to misroute data traffic. There may be a substantial financial benefit for an AS to advertise a fast route, which is more likely to be selected by its peers and provide more revenue, but forward data packets using a different, cheaper path. Because route selection depends on the assumption that the advertised route is the one that will actually be used, such inconsistencies will affect the performance of the Internet.

To verify the advertisements, we use a lightweight protocol that enables a router, acting as the verifier, to verify that its data traffic follows a certain route through the Internet. The verification protocol should be designed in such a way that it (i) completely avoids cryptographic operations in the data path, and (ii) does not require the prover to maintain any long-term flow specific state. The internet routers potentially process billions of packets per second. Hence, the method should deliver high performance verification.

The AS which sends the advertisements is called the prover and the other AS which checks its authenticity is called a verifier. The set-up phase is executed using a TLS (transport layer security) protected website which enables the prover and the verifier to share a set of $(2l+1)$ bit random secret tuples (s1,s2,b), where l is the security parameter, s1,s2 are l bit secrets, and b is a random bit. We emphasize that the prover and the verifier do not need to maintain a secure channel outside of this setup. The TLS-protected website shares another 32 bit key in addition to providing secrets so as to authenticate with the server. This 32 bit key is used for tagging during the working phase. Since, the key distribution is done using TLS website a malicious AS cannot masquerade as another AS and set up a secret. We use the techniques proposed in [15] to verify the truthfulness of the advertisements. The authentication is done based on the shared secret tuples. Since, the verifier is checking the authenticity of the advertisements sent by the prover, both the parties need to be involved in the process.

The autonomous systems which install our DMIPS scheme get better services than others in the attack scenarios. Hence, AS/ISPs would like to install this scheme and so have to prove their authenticity even if it incurs overhead on them. The secret tuples corresponding to other autonomous systems are stored at the BGRs of the AS. This information can be added into the forwarding table by inserting another column to enable faster access.

Working

When the network is running, we need to differentiate the spoofed packet from legitimate based on the source AS from which it reaches the destination. We tag each packet with a key associated with source autonomous system and the destination (server). This key is stored in the packet header and is compared with the actual key at the destination to check the authenticity of the packet. The key handling procedures should be light weight as the server receives packets in high volumes. Distributing key information about the routes as required in our method is not new. Due to various Border Gateway Protocol (BGP) security issues such as prefix hijacking there are suggestions to secure BGP by adding certificate keys to BGP announcements, in order to validate them. We follow a similar procedure in our scheme.

In order to tag a packet, a lookup on the destination address is required at source BGR and in order to authenticate a packet, a lookup on the source address is required at the server. The process of extracting the key can be combined with the regular IP-lookup in the table. The information of the other autonomous systems and their corresponding keys are stored as additional fields in the forwarding information table of a BGR. Notice that the cost of tagging a packet is minimal as it is piggybacked on the IP lookup process. Upon arrival at the destination network the key in the packet is extracted and verified. The source autonomous AS is decided based on the global IP address of the source in the packet. The packets coming from non-DMIPS autonomous networks would have no tag or some random tag. Thus the method can verify the authenticity of packets carrying the address s which belongs to AS S. Our scheme works well even if some of the parts of the network do not install DMIPS. These autonomous systems get lower priority in services at the server and all other AS are not affected. Hence, the scheme is scalable and incremental.

4   Attack Scenario

4.1   Inside the AS

Our scheme relies on the fact that the spoofed IP packets often have mismatching IP addresses and hop-counts. The attackers will try to generate spoofed packets with matching IP addresses and HC. We will see how difficult it is for the attackers to evade our scheme. To spoof a packet with the correct HC, the attacker must know the hops between spoofed IP and BGR. Then he must set appropriate initial TTL value. Fig 5 shows how TTL spoofing needs to be done in order to overcome the filtering.

Let H_a, H_l be the Hop counts of attacker and legitimate host respectively. Assuming attacker and legitimate host maintain same initial TTL value (I), the attacker can get H_a value, the hop count from BGR to itself very easily using traceroute. But, due to the random selection of spoofed host it is very difficult for the attacker to know its H_l (the HC between randomized IP address and the BGR). To figure out H_l in real time, the attacker has to build a table before the attack itself. He has to know the HC values for all the IP addresses to spoof effectively. The attacker cannot build such a table easily because he cannot see the final TTL values at the BGR. The attacker can use traceroute and can know H_l only when the host is in the path between itself and the BGR. But, the traceroute is not effective when the attacker is trying to spoof randomly. The alternate approach to build IP2HCS table by the attacker is to figure out

```
Final TTL of a packet from attacker = I - Ha
Final TTL of a packet from legitimate host = I - H₁
If the attacker wants to spoof the legitimate host, he must change
his initial TTL value to I - (H₁ - Ha)
The final TTL value at the BGR would be (I - (H₁ - Ha)) - Ha)
    = I - H₁ which is same as that of legitimate host
```

Fig. 5. Spoofing the victim

the topological positions of IP addresses. The attacker can get this information using snapshots of the network at various times. The inter domain routing in the internet is policy based and the routing policies are not public. So, even after getting the correct topological mapping, the attacker cannot decide the correct HC. The HC is based on many factors like policies, algorithms and not network connectivity alone. So, it is very difficult to construct IP2HCS table for the attackers.

The attacker instead of building the tables may choose to spoof the IP addresses from a small set of compromised machines whose HC can be known. But, this weakens the DDoS attacks. Then the list of spoofed source address will be very less which makes detecting and blocking the attack packets much simpler. If the attacker happens to be the same number of hops from the target as the spoofed source, this method would result in a false negative. The advantage of using HC metric is that the attacker never knows whether the spoofing attempt is successful or not because the reply from server goes to the actual host and not the attacker. He can listen to the reply if and only if the victim is also in the same location. But this can narrow down the search and the attacker will be in danger of getting caught.

4.2   Outside the AS

The authenticity of the packets outside the autonomous systems is based on the tagged keys. The keys are marked at the source BGR and checked at the server. The server removes the keys after verifying the authentication. So, the attackers who have access only to edge devices, cannot read or observe the tagging keys.

But, if the attacker somehow manages to sniff the backbone network then he can be dangerous and can initiate more serious attacks than DDoS spoofing attacks. So, we assume that the backbone the network is secure. To maintain the freshness of the key, it should be changed periodically using TLS website as explained in the offline set up phase of outside the AS in section 3.2.

5 Evaluation

In our scheme, the routers need no extra knowledge about the state of the network, than provided by the standard BGP protocols. Our scheme is incrementally deployable and requires no significant changes to TCP/IP protocols. We simulated our network and autonomous systems using NetSim tool and the network structure is based on the mapping produced by the Internet Mapping project [22].

5.1 Inside the AS

We divide the entire network into ISPs or autonomous systems which contain thousands of hosts inside it. We analyze the efficiency of the improved hop count method by extracting the packets received by the border gateway routers.

We took a random ISP and used a packet generator process to simulate normal internet traffic inside it. This packet generator process sends packets from a randomly selected user. It is observed that the hop-count values for the hosts spans over a range of values. The Gaussian distribution is a good approximation for them. The mean value of the distribution observed is in the range of 13 to 16 with standard deviation of 3 to 5 depending upon the network topology. Hence, the distribution is diverse enough making the hop count filtering schemes effective. But during a DDoS attack, a large percentage of packets received by the BGR will be coming from the attackers. The attackers generate many packets with the aim that at least some of the many randomly chosen hosts would match with its HC. Because of the usage of RREQ and RREP packets to populate the IP2HC table, we are secure against phishing attacks and we get the correct values of hop count.

We selected a random number of attackers of around 1% of total number of hosts in that ISP. These attackers send spoofed packets to the BGRs. It is observed that our improved hop count scheme filters up-to 82% of the spoofed packets. This is the case when we store the 2 best hop counts for each host at the BGR. When we increase it to 3, the accuracy decreased to 75%. Though our scheme's accuracy is less than that of actual HCF [8] which is 90% accurate, we show that the number of false positives is less for our scheme in the following paragraph.

When we increase the number of attackers to 10%, there is heavy congestion in the AS. This resulted in packets taking alternate paths. If the packets reach the server with different hop counts, the HCF scheme will drop it. Storing alternate path's HC in our scheme have increased the Normal Packet Survival Ratio

Table 2. Comparison of HC based scheme

	Hop count table generation	Accuracy(Number of Spoofed packets detected)	False positive rates(values depends on the traffic)	Procedure to detect network changes
HCF	Table population is not secure and maintains single HC per host	90%	20%	No
FLF4DoS	Table population is not secure and maintains single HC per host.	90%	25%	No
Probabilistic HCF	Table population is not secure and maintains single HC per host.	80 - 85%	25%	No
Improved HCF in DMIPS	Stores multiple hop counts for each host and is more secure.	82% (for K=2)	5% (for K=2)	Yes

(NPSR) of the network. The number of false positives in this scenario for HCF and other hop count based schemes which ignored multiple paths is around 20%. In our DMIPS scheme, it is 5% when 2 hop counts are stored and negligible percentage (less than 1%) when 3 hop counts are stored at BGR. We blocked some links to simulate link failures and our Renew technique identified them when the selected best paths are affected by this change. In the context of this paper, we compare our scheme with other hop count based schemes (shown in Table 2).

The storage at the BGR is bounded by the size of IP2HCS table. Each row in the table consists of host identifier (16 bits) and K hop counts (5 bits for each hop count). Since, the BGR already contains the host ID in the forwarding table, we just need extra storage for HCs. Thus, an ISP of 1000 hosts with 3 hop counts selected for each host needs around 1000 * 15 bits = 2 KB approximately.

5.2   Outside the AS

BGRs at the source autonomous systems tag the packet with the shared key. The key look up and tagging is piggybacked with the IP look up process. Hence, there is not much overhead at the BGRs. The server on receiving the packets extracts the key and verifies it with the stored key corresponding to the source AS. The verification of the key requires a look up operation per packet on its source AS. We calculate the number of CPU cycles saved by our scheme as follows.

CPU Overhead: Assuming a total of "N" packets are generated in the network from various autonomous systems in unit time. Out of these, "p" is the fraction of spoofed packets and "q" is the fraction of legitimate packets. During a DDoS attack, the number of spoofed packets generated in the network will be very much greater than the legitimate packets. If X is the efficiency of filtering at BGRs, then (1-X) fraction of these spoofed packets reaches the server and this number would still be high. Hence, $(1 - X)p >> q$.

If t_S and t_L are the processing times of spoofed requests and legitimate requests respectively and T is the processing time per packet of DMIPS at the server. The number of CPU cycles saved is derived in Fig 6.

```
If the server does not install any filter then the number of CPU
cycles consumed = pNts + qNtL.
The number of CPU cycles consumed by DMIPS = qNtL + (1-X)*pNT +
qNT.
The number of CPU cycles saved = Number of CPU cycles without any
filter - Number of CPU cycles with DMIPS installed
= pNts + qNtL - [qNtL + (1-X)pNT + qNT]
= pNts - (1-X)pNT - qNT.
```

Fig. 6. Number of saved CPU cycles

The spoofed packets are designed so as to consume maximum amount of server's CPU usage. Hence, the request processing time (t_S) of a spoofed packet will be significantly larger than the processing time (T) comprising of a look up and a key comparison operation. Hence, $t_S >> T$.

Using the above conditions, the number of CPU cycles saved can be approximated as pNt_S - qNT = pNt_S Hence, the savings is proportional to the attack magnitude or the fraction of spoofed packets received at the server. The additional storage incurred by our scheme is the mapping between ASes and their corresponding keys. There are about 2^{16} autonomous systems in the internet and each AS id is encoded with 2 bytes. Each AS is associated with a 32 bit key (4 bytes). Hence, additional storage required by our scheme is bounded by 2^{16} * (6 bytes) = 384 KB. The server can afford this extra storage for the CPU cycles it is saving.

6   Conclusion

Choosing Hop Count or TTL value as the parameter for filtering the spoofed packets has several advantages. The HC based filtering techniques does not require much change in the existing network architecture. Most of the earlier hop count based schemes are based on the assumptions that

1. When a packet is sent between two hosts, as long as the same route is taken, the number of hops will be the same.
2. Packets will take the same route to the destination always.
3. Routes change infrequently.

If these assumptions do not hold, the described methods may result in false positives, that is, valid packets may appear to be spoofed. For traffic engineering in the ISPs, provision of multiple paths becomes a critical issue for Quality of Service (QoS). Hence, our scheme DMIPS is developed by taking multiple paths into account. This reduces contention and retransmission of packets.

The percentage of erroneous packets to be dropped in our scheme DMIPS (82%) is less than that of HCF, which is 90% accurate, but the false positive rate is negligible(less than 1%) in our scheme when compared to 20% in HCF. The attacker can spoof only the hosts in its AS because of the key based filtering at the server. Hence, the number of IP addresses an attacker can successfully spoof is decreased. We have a good Renew procedure through which we can easily identify any link failures or network changes. The normal packet survival ratio (NPSR) and network performance of DMIPS is more than that of other hop count based schemes. In our scheme, most of the spoofed packets are filtered in the AS itself. It has two advantages. Firstly, it avoids congestion at several points in the network during the attack. Secondly, it minimizes the server overhead in filtering.

Our scheme enjoys step-wise deployment also. The autonomous systems which install DMIPS can get full benefits from the server even if some parts of the network do not install the scheme. The server provides relatively more benefits to the DMIPS installed autonomous systems by providing more quality services and higher availability in attack scenarios. This would attract the administrators of AS/ISP to install DMIPS.

References

1. Adler, M.: Tradeoffs in probabilistic packet marking for IP traceback. In: Proceedings of Thirty-Fourth Annual ACM Symposium on Theory of Computing, pp. 407–418. ACM, New York (2002)
2. Amin, S.O., Kang, M.S., Hong, C.S.: A lightweight IP traceback mechanism on iPv6. In: Zhou, X., Sokolsky, O., Yan, L., Jung, E.-S., Shao, Z., Mu, Y., Lee, D.C., Kim, D.Y., Jeong, Y.-S., Xu, C.-Z. (eds.) EUC Workshops 2006. LNCS, vol. 4097, pp. 671–680. Springer, Heidelberg (2006)
3. Baker, F., Savola, P.: Ingress Filtering for Multihomed Networks. RFC 3704 (2004)
4. Belenky, A., Ansari, N.: IP traceback with deterministic packet marking. Proceedings of IEEE Communication Letters 7(4), 162–164 (2003)
5. Bremler-Barr, A., Levy, H.: Spoofing Prevention Method. In: Proceedings of IEEE Infocom (2005)
6. Ehrenkranz, T., Li, J.: On the State of IP Spoofing Defense. Proceedings of ACM Transactions on Internet Technology 9(2) (2009)
7. Ferguson, P., Senie, D.: Network Ingress Filtering: Defeating Denial of Service attacks which employ IP source address spoofing. RFC 2827 (2000)
8. Jin, C., Wang, H., Shin, K.G.: Hop-count filtering: An effective defense against spoofed DDoS traffic. In: Proceedings of the 10th ACM conference on Computer and Communications Security, ACM CCS, New York, pp. 30–41 (2003)
9. Rodriguez, J.C., Briones, A.P., Nolazco, J.A.: FLF4DoS. Dynamic DDoS Mitigation based on TTL field using fuzzy logic. In: Proceedings of 17th International Conference on Electronics. IEEE computer Society, Washington, DC (2007)

10. Savage, S., Wetherall, D., Karlin, A., Anderson, T.: Practical network support for IP traceback. Computer Communication Review 30, 295–306 (2000)
11. Snoeren, A.C., Craig, P., Luis, A.S., Christine, E.J., Fabrice, T., Beverly, S., Stephen, K., Strayer, W.: Single-packet IP traceback. In: Proceedings of ACM/IEEE Transactions on Networking (2002)
12. Song, D.X., Perrig, A.: Advanced and authenticated marking schemes for IP traceback. In: Proceedings of IEEE Infocom (2001)
13. Strayer, T.W., Christine, E.J., Fabrice, T., Regina, R.H.: SPIE-IPv6: Single IPv6 Packet Traceback. In: Proceedings of 29th Annual IEEE Conference on Local Computer Networks, Washington, pp. 118–125 (2004)
14. Swain, B.R., Sahoo, B.: Mitigating DDoS attack and Saving Computational Time using a Probabilistic approach and HCF method. In: Proceedings of IEEE International Advance Computing Conference (2009)
15. Wong, E.L., Balasubramanian, P., Alvisi, L., Gouda, M.G., Shmatikov, V.: Truth in Advertising: Lightweight Verification of Route Integrity. In: Proceedings of 26th Annual ACM symposium on Principles of Distributed Computing, PODC, New York, pp. 147–156 (2007)
16. Yaar, A., Perrig, A., Song, D.: Pi: A path identification mechanism to defend against DDoS attacks. In: Proceedings of IEEE Computer Society Symposium on Research in Security and Privacy, pp. 93–107 (2003)
17. Yaar, A., Perrig, A., Song, D.: StackPi: New packet marking and filtering mechanisms for DDoS and IP spoofing defense. Proceedings of IEEE Journal on Selected Areas in Communications 24, 1853–1863 (2006)
18. Unicast reverse path forwarding, Cisco IOS (1999),
    http://www.cisco.com/web/about/security/intelligence/unicast-rpf.html
19. Denial-of-Service attack knocks Twitter Offline (updated) (2009),
    http://www.wired.com/epicenter/2009/08/twitter-apparently-down/
20. Facebook Confirms Denial-of-Service Attack (updated) (2009),
    http://www.wired.com/epicenter/2009/08/
    facebook-apparently-attacked-in-addition-to-twitter/
21. Icmp traceback messages (2003),
    http://tools.ietf.org/html/draft-ietf-itrace-04
22. Internet Mapping Project, http://www.lumeta.com/research/

Provably Secure Key Assignment Schemes from Factoring

Eduarda S.V. Freire* and Kenneth G. Paterson**

Information Security Group,
Royal Holloway, University of London, U.K.

Abstract. We provide constructions for key assignment schemes that are provably secure under the factoring assumption in the standard model. Our first construction is for simple "chain" hierarchies, and achieves security against key recovery attacks with a tight reduction from the problem of factoring integers of a special form. Our second construction applies for general hierarchies, achieves the stronger notion of key indistinguishability, and has security based on the hardness of factoring Blum integers. We compare our constructions to previous schemes, in terms of security and efficiency.

Keywords: Key assignment scheme, general poset, provably secure, factoring, access control.

1 Introduction

A key assignment scheme is a method for implementing access control policies by assigning encryption keys and private information to each security class in a hierarchy, with the hierarchy being represented by a partially ordered set (poset). The encryption key could be used to protect or restrict access to some information, whereas, the private information will be used to derive the keys of any descendant class in the hierarchy. The scheme is administered by a trusted authority, who is responsible for generating and distributing keys and private information, generating public data, and managing changes to the access control policies.

Such key assignment schemes can be used to implement access control policies in many applications where some users have more access rights than others. These schemes can be useful, for example, for content distribution, management of databases containing sensitive information, government communications and broadcast services (such as cable TV).

The use of cryptographic techniques to solve the problem of key management in hierarchical access control was first suggested by Akl and Taylor [1] in 1983. Due to its simplicity, the scheme has been widely proposed for use to implement access control in different areas. Since then, a number of schemes such as

---

* This author supported by CAPES Foundation/Brazil on grant 0560/09-0 and Royal Holloway, University of London.
** This author supported by EPSRC Leadership Fellowship EP/H005455/1.

U. Parampalli and P. Hawkes (Eds.): ACISP 2011, LNCS 6812, pp. 292–309, 2011.

those of MacKinnon *et al.* [2], Harn-Lin [3] and many others (see [4]) have been proposed to improve existing key assignment schemes. The Akl-Taylor scheme has also been used as a starting point to enforce the problem of access control for more general policies, including schemes with time-dependent constraints [5,6,7]. Despite the large number of publications on this topic, many schemes lack formal security analysis or are of limited use (hierarchies are limited to trees, for example) [8,9,10,11]. A recent survey by Crampton *et al.* [12] defines five generic key assignment schemes and classifies existing schemes as instances of these generic schemes. These schemes offer different trade-offs in terms of the amount of storage required and the complexity of key derivation.

More recently, Crampton *et al.* [13] proposed a new and intriguing approach to constructing key assignment schemes for arbitrary posets using *chain partitions*. In this approach, the poset is partitioned into a collection of chains, and the scheme is built by combining, in a particular way, separate schemes for each of the chains. This idea was instantiated using two different cryptographic bases: collision-resistant hash functions and the RSA primitive. The latter instantiation, called the Multi-key RSA scheme in [13], makes use of multiple RSA moduli, one modulus for each chain in the partition, leading to the scheme potentially requiring a substantial amount of public data. Interestingly, these new schemes do not fit into the taxonomy of generic key assignment schemes proposed in [12]. Unfortunately, none of the schemes in [13] comes with a formal security analysis.

Indeed, despite the long history of research on key assignment schemes, work to formalise security of such schemes began only recently. Two different notions of security for key assignment schemes were proposed by Atallah *et al.* [4]: security against key recovery attacks (KR-security) and security with respect to key indistinguishability (KI-security). Informally, KR-security captures the notion that an adversary is not able to compute a key to which it should not have access. In the stronger notion of KI-security, the adversary is not able to learn *any* information about a key to which it should not have access, that is, the adversary is not able to distinguish between the real key and a random string of the same length.

Atallah *et al.* [4] gave a construction for KR-secure key assignment schemes for arbitrary posets based on any pseudorandom function. They also gave a construction for KI-secure key assignment schemes for arbitrary posets based on any symmetric encryption scheme secure against chosen ciphertext attacks. Ateniese *et al.* [5] proposed a key assignment scheme that is KI-secure under the Bilinear Decisional Diffie-Hellman Assumption and another whose security relies on the OW-CPA (One-Wayness against Chosen-Plaintext Attacks) security of a symmetric encryption scheme. D'Arco *et al.* [14] proved the Akl and Taylor [1], MacKinnon *et al.* [2], and Harn-Lin [3] key assignment schemes to be secure against key recovery attacks under the RSA assumption. They also gave a construction yielding KI-secure key assignment schemes using as components KR-secure schemes and the Goldreich-Levin hard-core bit (GL bit) [15]. This construction can be used to construct KI-secure schemes for quite general posets (assuming suitable KR-secure schemes can be obtained). For example,

D'Arco *et al.* used it to obtain KI-secure schemes based on the RSA assumption by using their construction with the Akl-Taylor scheme. As another example, a pseudorandom function with security based on the hardness of factoring can be obtained (e.g. by converting the BBS pseudorandom generator to a pseudorandom function using the results of [16]). Thus a key assignment scheme for general posets having KI-security under the factoring assumption can be obtained by combining the construction of [14] and the construction for KR-secure schemes based on any pseudorandom function from [4]. However, the construction of [14] involves a "blow-up" of the poset for the desired KI-secure scheme by a factor directly related to the security parameter. Moreover, the KR-secure scheme of [4] involves a large amount of public information (one string per edge in the graph associated with the hierarchy).

1.1   Our Contributions

We provide two constructions for key assignment schemes with security provably based on the hardness of factoring integers in the standard model (*i.e.* without using random oracles). The first construction yields KR-secure schemes for posets that are chains (totally ordered hierarchies), and enjoys a tight security reduction for static adversaries from the factoring problem, albeit for integers of special form. This construction can be extended to produce KR-secure schemes for arbitrary posets and via the construction of [14], KI-secure schemes for arbitrary posets. We omit the details in order to focus on our second construction, which we consider to be our main contribution.

Our second construction *directly* yields KI-secure schemes for arbitrary posets, with security based on the hardness of factoring Blum integers. In addition, this construction is relatively efficient, avoiding the blow-up associated with the construction of [14] and not relying on the Goldreich-Levin method for the construction of hard-core bits. As we shall see, our construction is obtained by extending the ideas of [13], using chain partitions in combination with the Blum-Blum-Shub (BBS) generator. We are then able to apply well-known results relating the security of the BBS generator to the factoring problem to complete our security analysis. In contrast to the Multi-key RSA scheme of [13], our analysis shows that using one RSA modulus suffices, enabling us to shrink the size of the public data relative to the Multi-key RSA scheme of [13]. The amount of private information stored for each class is still related to the number of chains in the partition. By an application of Dilworth's Theorem [17], this can be as small as the *width w* of the poset, that is, the cardinality of the largest antichain in the poset. Typical posets arising in applications have sufficient structure that chain partitions containing w chains can be computed efficiently.

Thus our paper provides a formal security analysis of schemes closely related to those presented in [13], as well as a natural and relatively efficient construction for key assignment schemes whose KI-security is provably based on the hardness of factoring, in the standard model.

1.2   Organization

In the next section, Section 2, we define the Factoring Assumption and give a definition of a key assignment scheme (KAS) and its security with respect to key indistinguishability and key recovery attacks. In Section 3 we introduce our first scheme, the Basic Scheme, and prove its security against key recovery attacks for static adversaries under a tight reduction from factoring. Section 4 presents a description of our second scheme (for the cases of one chain and more general posets), the FP scheme, along with its security proof with respect to key indistinguishability. Finally, in Section 5 we make some concluding remarks.

2   Preliminaries

The security of most key assignment schemes depends on the difficulty of solving a computational problem. In this section, we first recall the cryptographic assumption on which the security of our schemes relies, the factoring assumption. We then define hierarchical key assignment schemes, following [14], considering two security notions: security against key recovery and security with respect to key indistinguishability. We also distinguish static from dynamic adversaries.

In this paper we take the asymptotic approach, which views the running time of the adversary as well as its success probability as functions of a security parameter ρ. Then, the notion of *efficient algorithms* is equivalent to probabilistic algorithms running in time polynomial in ρ, PPT algorithms. The notion of *small probability of success* is substituted by negligible probability of success, meaning that the probability is smaller than any inverse polynomial in ρ, i.e., for every constant c, the adversary's success probability is smaller than ρ^{-c} for large enough values of ρ. However, all the results in the paper can be made concrete in a straightforward manner.

2.1   Factoring Assumption

Let Gen_F be some polynomial-time algorithm that, on input 1^ρ, outputs (N, p, q) where $N = pq$, and p and q are ρ-bit primes, possibly with additional constraints. The factoring assumption relative to Gen_F states that given N, it is computationally infeasible to obtain the prime factors p and q, except with negligible probability in ρ. More formally, following Hofheinz and Kiltz [18], we have:

Definition 1 (Factoring Assumption). *For an algorithm A_F, we define its factoring advantage to be*

$$Adv_{Gen_F,A_F}^{fac}(\rho) = \Pr[(N, p, q) \leftarrow Gen_F(1^\rho) : A_F(N) = \{p, q\}].$$

The factoring assumption *(with respect to Gen_F) states that $Adv_{Gen_F,A_F}^{fac}(\rho)$ is negligible in ρ for every PPT A_F.*

In what follows, we will consider two instances of Gen_F: Gen_{Blum}, which outputs N with random ρ-bit factors p, q such that $p = q = 3 \bmod 4$, and Gen_S, which outputs N with random ρ-bit factors $p = 1 \bmod 2^n$ and $q = 3 \bmod 4$ for some parameter n. Note that the generation of random ρ-bit primes with the additional constraint $p = 1 \bmod 2^n$ can be achieved by generating values $p = T \cdot 2^n + 1$, where T is some random integer having bitlength $\rho - n$, and then testing for primality.

2.2   Key Assignment Schemes

A *partially ordered hierarchy* is a pair $(V, \leq)$ where V is a set of disjoint classes, called *security classes*, and $\leq$ is a binary relation on V. A security class can represent a person, a department or a user group in an organization. The binary relation $\leq$ is defined in accordance with authority for each class in V. For any two classes u and v we write $v \leq u$ or $u \geq v$ to indicate that users in class u can access the data of users in class v. We say u covers v, denoted $v \lessdot u$ or $u \gtrdot v$, if $v < u$ and there does not exist $c \in V$ such that $v < c < u$. V is a *total order* (or chain) if for all $u, v \in V$, either $v \leq u$ or $u \leq v$. We say $A \subseteq V$ is an *antichain* if for all $u, v \subset A$, $v \not\leq u$ and $v \not\geq u$. The partially ordered hierarchy $(V, \leq)$ can be represented by the directed acyclic graph $G = (V, E)$, where each class corresponds to a vertex in the graph and there is an edge from class u to class v if and only if $v \leq u$.

The problem of key management for such hierarchies consists of assigning symmetric encryption keys and private information to each class in the hierarchy in such a way that the encryption keys will be used to protect or access data, whereas the private information will be used to efficiently derive the keys for any descendant class in the hierarchy.

A *partition* of a set V is a collection of sets $\{V_1, \ldots, V_s\}$ such that (i) $V_i \subseteq V$ (ii) $V_1 \cup \cdots \cup V_s = V$, and (iii) $V_i \cap V_j \neq \emptyset$ if and only if $i = j$.

A hierarchical key assignment scheme for a family of graphs Γ corresponding to partially ordered hierarchies is defined as follows.

Definition 2 (Key Assignment Scheme). *A hierarchical key assignment scheme (KAS) for Γ is a pair of algorithms* (Gen, Derive) *statisfying the following conditions:*

1. Gen *is a probabilistic polynomial-time algorithm that takes as inputs the security parameter 1^ρ and an access graph $G = (V, E) \in \Gamma$, and outputs*

    (a) *private information S_u and key k_u, for any class $u \in V$;*
    (b) *public information pub.*

    We denote by (S, k, pub) the output of the algorithm Gen *on inputs 1^ρ and G, where S and k denote the sets of private information and keys, respectively. We assume that the size of each key k_u can be represented using $\ell(\rho)$ bits.*

2. Derive *is a deterministic polynomial-time algorithm that takes as inputs the security parameter 1^ρ, an access graph $G = (V, E) \in \Gamma$, two classes u and v such that $v \leq u$, the private information S_u assigned to class u, and the public information pub, and outputs the key k_v assigned to class v.*

For correctness, for all classes $u \in V$, $v \leq u$, it must hold that

$$\Pr[k_v = Derive(G, pub, u, v, S_u)] = 1. \tag{1}$$

In order to evaluate the security of key assignment schemes, we consider static adversaries and use the result in [5] that security against dynamic adversaries is polynomially equivalent to security against static adversaries to ensure that our schemes are automatically secure against the corresponding dynamic adversaries (albeit with a less tight overall security reduction)[1]. We note that this result is also implicit in [4].

A static adversary, A_{stat}, first chooses a class $u \in V$ to attack, and then is allowed to access the private information assigned to all classes not allowed to compute the key k_u associated with class u, as well as the public information pub. Let $Corrupt_u$ be an algorithm which, on input the private information S generated by the algorithm Gen, outputs the secret values S_v associated with *all* classes $v \in V$ such that $u \not\leq v$. We denote by corr the output of $Corrupt_u$.

A dynamic (also called adaptive) adversary, A_{dyn}, first gets access to all public information generated by Gen, and then chooses, in an adaptive manner, a number of classes to be corrupted. It is assumed that a challenger provides the adversary with the private information S_v held by each of the corrupted classes v. The adversary then chooses a class $u \in V$ to attack, such that $u \not\leq v$, for all classes v in the corrupted set. After this, the adversary is still allowed to corrupt classes of its choice, subject to the constraint that $u \not\leq v$.

We consider two different security goals: security against key recovery and security with respect to key indistinguishability.

In the key recovery case, an adversary on input a security parameter ρ, a directed access graph G, the public information pub generated by $Gen(1^\rho, G)$ and the private information corr, held by corrupted users, outputs a string k'_u and succeeds if $k'_u = k_u$. Here u is the target class specified by the adversary. For security, we require that the adversary will succeed with probability only negligible in the security parameter ρ. More formally, we have the following definition:

Definition 3 (KR-ST). *Let Γ be a family of graphs corresponding to partially ordered hierarchies, let $G = (V, E) \in \Gamma$ be a graph, and let (Gen, Derive) be a hierarchical key assignment scheme for Γ. Let A be a static adversary which attacks a class $u \in V$. Consider the following experiment:*

---

[1] This result is trivial to prove: in the reduction, the static adversary simply guesses which class will be the subject of the dynamic adversary's query, and aborts if it guesses incorrectly; this reduction succeeds with probability $1/|V|$.

$$\text{Experiment } \text{Exp}_A^{KR-ST}(1^\rho, G):$$
$$u \leftarrow A(1^\rho, G)$$
$$(S, k, pub) \leftarrow Gen(1^\rho, G)$$
$$corr \leftarrow Corrupt_u(S)$$
$$k'_u \leftarrow A(1^\rho, G, pub, corr)$$
$$return \ k'_u$$

The advantage of A is defined as $Adv_A^{KR-ST}(1^\rho, G) = \Pr[k'_u = k_u]$. The scheme is said to be secure in the sense of key recovery with respect to static adversaries (KR-ST-secure) if, for every graph $G = (V, E) \in \Gamma$ and each class $u \in V$, the function $Adv_A^{KR-ST}(1^\rho, G)$ is negligible for every adversary whose time complexity is polynomial in ρ.

Note here that we require the KR-ST adversary to output the actual key for class u in its attack, not merely a key that is *consistent* with the information received in *corr*.

For security in the sense of key indistinguishability, two experiments are considered. In the first, the adversary is given as a challenge the key k_u, where u is the target class, whereas in the second, it is given a random string r, having the same length as k_u. It is the adversary's job to determine whether the value given corresponds to the real key k_u or not.

Definition 4 (KI-ST). *Let Γ be a family of graphs corresponding to partially ordered hierarchies, let $G = (V, E) \in \Gamma$ be a graph, and let (Gen, Derive) be a hierarchical key assignment scheme for Γ. Let A be a static adversary which attacks a class $u \in V$. Consider the following two experiments:*

$$\text{Experiment } \text{Exp}_A^{KI-ST-1}(1^\rho, G):$$
$$u \leftarrow A(1^\rho, G)$$
$$(S, k, pub) \leftarrow Gen(1^\rho, G)$$
$$corr \leftarrow Corrupt_u(S)$$

$$d \leftarrow A(1^\rho, G, pub, corr, k_u)$$
$$return \ d$$

$$\text{Experiment } \text{Exp}_A^{KI-ST-0}(1^\rho, G):$$
$$u \leftarrow A(1^\rho, G)$$
$$(S, k, pub) \leftarrow Gen(1^\rho, G)$$
$$corr \leftarrow Corrupt_u(S)$$
$$r \leftarrow \{0, 1\}^{\ell(\rho)}$$
$$d \leftarrow A(1^\rho, G, pub, corr, r)$$
$$return \ d$$

The advantage of A is defined as

$$Adv_A^{KI-ST}(1^\rho, G) = \left| \Pr[\text{Exp}_A^{KI-ST-1}(1^\rho, G) = 1] - \Pr[\text{Exp}_A^{KI-ST-0}(1^\rho, G) = 1] \right|.$$

The scheme is said to be secure in the sense of key indistinguishability with respect to static adversaries (KI-ST-secure) if, for each graph $G = (V, E) \in \Gamma$ and each class $u \in V$, the function $Adv_A^{KI-ST}(1^\rho, G)$ is negligible for every adversary A whose time complexity is polynomial in ρ.

It is easy to see that KI-ST-secure scheme is also KR-ST-secure, so KI-ST security is the stronger notion. Informally, a scheme satisfying this stronger notion can be securely composed with schemes making use of the encryption keys k_v, which is not the case for schemes possessing only KR-ST security.

3   A Basic Scheme

In this section, we describe a basic scheme and prove it KR-secure under the factoring assumption. This scheme is limited to hierarchies that are total orders (chains) on n classes. For this scheme we do not distinguish private information from encryption keys. That is, $S_u = k_u$ for all u. The public information, N, consists of a product of two large primes p and q, where $p = 1 \bmod 2^n$ and $q = 3 \bmod 4$. Computation of a key lower in the hierarchy is done using repeated squaring modulo N. A variant of this scheme, which is provably secure in the sense of key indistinguishability (KI) under the factoring assumption, is given in the next section.

Let $\varGamma$ be a family of graphs corresponding to totally ordered hierarchies, let $G = (V, E) \in \varGamma$ be a graph, where $V = \{u_0, \cdots, u_{n-1}\}$ and $u_{i+1} < u_i$ for all i, and consider a security parameter ρ. Our scheme works as follows.

Algorithm $Gen(1^\rho, G)$:

1. Run $Gen_S(1^\rho)$ to obtain two ρ-bit primes p and q, where $p = 1 \bmod 2^n$ and $q = 3 \bmod 4$, and compute $N = pq$;
2. Let $pub = N$ be the public information;
3. Randomly choose a secret value γ from Z_N^*;
4. For each class $u_i \in V$ set $k_{u_i} = S_{u_i} = \gamma^{2^i} \bmod N$;
5. Let S and k be the sets of private information and keys, respectively, computed in the previous step;
6. Output (S, k, pub).

Notice that, except possibly for the first class, the keys associated with the classes in the chain are quadratic residues modulo N.

Algorithm $Derive(G, pub, u_i, u_j, k_{u_i})$:

1. For $j > i$, compute $k_{u_j} = (k_{u_i})^{2^{j-i}} \bmod N$;
2. Output k_{u_j}.

As we can see, the key derivation is done by exponentiation modulo N. Basically, each key in the hierarchy is the square of the key immediately above. Due to the nature of the key derivation, a user who has access to class u_i can derive keys associated with all the descendant classes in the hierarchy, but cannot obtain access to any of its ancestor classes unless he can obtain a particular square root modulo N of γ^{2^i}, namely $\gamma^{2^{i-1}}$. We will show that as a consequence of this, the security of the scheme relies on the difficulty of factoring N. Before proving this, we have the following:

Lemma 1. *Let $N = pq$, where $p = 1 \bmod 2^n$ and $q = 3 \bmod 4$ are two primes, γ a random value in Z_N^* and i an integer with $0 \le i < n-1$. Then, every element ζ of the form $\zeta = \gamma^{2^{i+1}} \bmod N$, has exactly two square roots that are themselves 2^i-th powers modulo N.*

Proof. Let $\gamma \in Z_N^*$ with $N = pq$, where p and q are odd prime numbers. We write $\gamma \leftrightarrow (\gamma_p, \gamma_q)$ for $\gamma_p = [\gamma \bmod p]$ and $\gamma_q = [\gamma \bmod q]$ via the Chinese Remainder Theorem. Now let $\zeta \leftrightarrow (\zeta_p, \zeta_q)$ be an element of the form $\zeta = \gamma^{2^{i+1}} \bmod N$. Notice that $\gamma^{2^i} \bmod N$ is a trivial square root of $\zeta = \gamma^{2^{i+1}} \bmod N$, which is itself a 2^i-th power modulo N. Let $\gamma^{2^i} \bmod N \leftrightarrow (\alpha_p, \alpha_q)$. Then the four square roots of ζ can be written as:

$$(\alpha_p, \alpha_q), (-\alpha_p, \alpha_q), (\alpha_p, -\alpha_q), (-\alpha_p, -\alpha_q).$$

We claim that exactly two of these are 2^i-th powers modulo N. To see this, first notice that -1 is not a quadratic residue modulo q and thus is not a 2^i-th power modulo q. Since an element in Z_N^* is a square modulo N if and only if it is a square modulo p and modulo q, it follows that $(\alpha_p, -\alpha_q)$ and $(-\alpha_p, -\alpha_q)$ cannot be 2^i-th powers modulo N. Conversely, for $p = 1 \bmod 2^n$, -1 is a 2^i-th power modulo p. To see this, let β be a primitive element in Z_p^*. Then, $\beta^{p-1} = 1 \bmod p$ and $\beta^{\frac{p-1}{2^n}}$ is a 2^n-th root of 1 modulo p, not equal to 1. Hence, $(\beta^{\frac{p-1}{2^n}})^{2^{n-1}} = -1 \bmod p$. It follows that -1 is a 2^{n-1}-th power of $\beta^{\frac{p-1}{2^n}}$ and hence it is a 2^i-th power modulo p for all $i \le n-1$. It is easy to see now that the two square roots of $\zeta = \gamma^{2^{i+1}} \bmod N$, which are themselves 2^i-th powers modulo N, are (α_p, α_q) and $(-\alpha_p, \alpha_q)$. $\square$

We now prove the security of our scheme against key recovery attacks. We show here that breaking the KR security of our basic scheme is computationally equivalent to factoring N. We use the well known result that if factoring is hard so is the problem of computing square roots modulo N. More formally, we use the following lemma.

Lemma 2. *Let $N = pq$ with p, q distinct, odd primes. Given a, $\hat{a}$, such that $a^2 = b = \hat{a}^2 \bmod N$ but $u \ne \pm \hat{a} \bmod N$, it is possible to factor N in time polynomial in $\log_2 N$. Indeed, both $\gcd(N, a + \hat{a})$ and $\gcd(N, a - \hat{a})$ are equal to one of the prime factors of N.*

Theorem 1 (KR security of the basic scheme). *Assume the factoring assumption relative to Gen_S holds (Definition 1). Then our basic scheme is KR-ST secure.*

Proof. Let Γ be a family of graphs corresponding to totally ordered hierarchies and let $G = (V, E)$, where $V = \{u_0, \cdots, u_{n-1}\}$, be any graph in Γ. Assume there exists a static adversary A against our basic scheme attacking class $u_i \in V$. Assume A is able to compute, with non-negligible advantage $Adv_A^{\text{KR-ST}}$, the key k_{u_i} associated with the class u_i. We construct a polynomial time adversary A_F that uses the adversary A to factor the modulus N with some non-negligible advantage. Algorithm A_F simulates the environment of A in a way that A cannot tell if it is dealing with its own challenge for the attacked class in the basic scheme or not. The first step in A_F's simulation consists of setting up the access hierarchy for graph G in such a way that A_F will be able to handle all A's corrupt queries. Eventually, A will output its guess k'_{u_i} for the encryption key corresponding to class u_i and A_F will use this output to factor N with non-negligible advantage. More formally, the algorithm A_F on input the modulus N works as follows.

Algorithm $A_F(N)$:

1. Run A with input $(1^\rho, G)$ to get $u_i \in V$, A's choice of target class;
2. Choose random $\gamma \leftarrow Z_N^*$ and compute $z = \gamma^{2^i} \bmod N$;
3. Run A with inputs $(1^\rho, G, N, corr)$, where $corr = \{\gamma^{2^{i+1}} \bmod N,$ $\gamma^{2^{i+2}} \bmod N, \ldots, \gamma^{2^{n-1}} \bmod N\}$, to obtain an output k'_{u_i};
4. If $(k'_{u_i})^2 = \gamma^{2^{i+1}} \bmod N$ and $k'_{u_i} \neq \pm\gamma^{2^i} \bmod N$, then factor N using Lemma 2.

Conditioned on the fact that A has no information about which value γ was chosen by A_F initially, other than the data in $corr$, the value output by A is equally likely to be each of the two square roots of $\gamma^{2^{i+1}} \bmod N$ which are also 2^i-th powers modulo N (see Lemma 1). However, only one of these outputs, namely $\gamma^{2^i} \bmod N$, allows A to win the KR-ST security game, while the other value allows A_F to factor the modulus N. Thus, we see that A's advantage $Adv_A^{\text{KR-ST}}$ is equal to the advantage of A_F in factoring N. Since we assumed that $Adv_A^{\text{KR-ST}}$ is non-negligible, then $Adv_{\text{Gen}_S, A_F}^{\text{fac}}$ is also non-negligible, and the theorem follows. $\qquad\square$

The basic scheme above uses a modulus N which is the product of two primes $p = 1 \bmod 2^n$ and $q = 3 \bmod 4$. This unusual form enabled us to get a tight reduction to factoring in the KR-ST security model. An alternative would be to weaken the requirements on p at the cost of introducing a stronger hardness assumption, namely a higher residuosity assumption modulo N (this being a generalisation of the quadratic residuosity assumption). Suppose, for example, $p = 1 \bmod 4$ but $p \neq 1 \bmod 8$. Then $\gamma^{2^{i+1}} \bmod N$ still has 2 square roots which are quadratic residues modulo N. One of these is the key $k_{u_i} = \gamma^{2^i} \bmod N$ associated with the class being attacked, but the other may not be a 2^i-th power, and so in this case would not be output by a correct adversary A that can distinguish 2^i-th powers from elements that are not 2^i-th powers. Suppose, however, that an appropriate higher residuosity assumption holds, with the implication that no efficient adversary can distinguish the 2 distinct square roots of $\gamma^{2^{i+1}} \bmod N$ that are themselves squares with non-negligible advantage. Then we can argue that, given A's view, either square root is (almost) equally likely to be output by A, and we can go on to construct a factoring algorithm as before. Hence, if factoring the modulus N of the given form is hard, and the appropriate higher residuosity assumption holds, then our scheme is KR-secure under these assumptions. Because the higher residuosity assumption is stronger than the factoring assumption, we obtain a proof of security under the higher residuosity assumption. Following the same line of argument, it can be shown that if $p = 3 \bmod 4$, we can obtain a reduction from the standard quadratic residuosity assumption because then exactly one square root of $\gamma^{2^{i+1}} \bmod N$ will be a square.

The basic scheme presented in this section fails to achieve key indistinguishability: an adversary will be able to test whether the challenge key it is given is the real key associated with the challenge class u_i or a random value in Z_N^* with overwhelming probability, simply by squaring it and comparing to the key for the class u_{i+1}.

4   The FP Scheme

We now show how to construct a key assignment scheme for arbitrary posets with KI-ST-security based on the hardness of factoring. We call this scheme the *FP scheme* for ease of reference. In our scheme N is an arbitrary Blum integer, that is, $N = pq$, where $p = q = 3 \bmod 4$. For expositional reasons, we first present a construction for totally ordered hierarchies and then we build on ideas from [13] to obtain a scheme for arbitrary posets. We begin with some preliminaries.

Definition 5 (BBS pseudorandom number generator). *Let N be a Blum integer (that is, $N = pq$ where p, q are distinct primes both congruent to $3 \bmod 4$). Let x be a quadratic residue mod N. We establish the following notation: $LSB_N(x) = x \bmod 2$ (the least significant bit of x). The BBS pseudorandom number generator applied to x and modulus N is defined to have output:*

$$BBS_N(x) = (LSB_N(x), LSB_N(x^2), \ldots, LSB_N(x^{2^{\ell-1}})) \in \{0,1\}^\ell,$$

where the output consists of ℓ bits.

We now recall the definition of the advantage of a distinguisher D in breaking the BBS pseudorandom number generator.

Definition 6 (Security of BBS generator). *Let $\mathrm{Gen}(1^\rho)$ be a probabilistic polynomial-time algorithm that on input a security parameter ρ, runs $\mathrm{Gen}_{\mathrm{Blum}}$ with input 1^ρ to generate an integer N of the form $N = pq$, where $p = q = 3 \bmod 4$, and which selects an integer $x \in QR_N$ (the set of quadratic residues modulo N). Let D be a distinguisher. Consider the following two experiments:*

Experiment $\mathrm{Exp}_D^{BBS-1}(1^\rho)$:	Experiment $\mathrm{Exp}_D^{BBS-0}(1^\rho)$:
$x, N \leftarrow Gen(1^\rho)$	$x, N \leftarrow Gen(1^\rho)$
	$r \leftarrow \{0,1\}^{\ell(\rho)}$
$d \leftarrow D(N, z = x^{2^\ell} \bmod N, BBS_N(x))$	$d \leftarrow D(N, z = x^{2^\ell} \bmod N, r)$
return d	return d

The advantage of D is defined as

$$Adv_D^{BBS}(\rho) = \left| \Pr[\mathrm{Exp}_D^{BBS-1}(1^\rho) = 1] - \Pr[\mathrm{Exp}_D^{BBS-0}(1^\rho) = 1] \right|.$$

We say that the BBS pseudorandom number generator is secure if the advantage of any polynomial time distinguisher D is negligible, that is, if $Adv_D^{BBS}(\rho)$ is negligible for any algorithm D running in polynomial time in ρ.

The result below, proved in [18] and based on results from [19,20] states that any BBS-distinguisher can be used to factor Blum integers.

Theorem 2 (BBS-distinguisher $\Rightarrow$ factoring algorithm). *For every PPT algorithm D that succeeds in breaking the BBS generator with advantage Adv_D^{BBS} running in time t_{BBS}, there exists a PPT algorithm A_F that factors Blum integers with advantage $Adv_{Gen_{Blum},A_F}^{fac}(\rho)/\ell$, where ℓ is the size of the BBS output. Algorithm A_F runs in time $t_{fac} \approx \rho^4 t_{BBS}/(Adv_D^{BBS})^2$.*

4.1 The FP Scheme for a Single Chain

We now describe our scheme for a single chain. It contains the essential ideas needed to understand the general version of the FP scheme that follows.

Let Γ be a family of graphs corresponding to totally ordered hierarchies, let $G = (V, E) \in \Gamma$ be a graph, where $V = \{u_0, \cdots, u_{n-1}\}$ and $u_{i+1} < u_i$ for all i, and consider a security parameter ρ. The *Gen* and *Derive* algorithms work as follows.

Algorithm $Gen(1^\rho, G)$:

1. Run Gen_{Blum} on input 1^ρ to obtain two random ρ-bit primes p and q, with $p = q = 3 \bmod 4$, and compute $N = pq$. Set $pub = N$;
2. Choose γ at random from QR_N; (Note that to choose a random quadratic residue, it is sufficient to choose γ' uniformly at random from Z_N^* and square it modulo N.)
3. For each class $u_i \in V$, $0 \le i \le n-1$, set $S_{u_i} = \gamma^{2^{i\ell}} \bmod N$ and $k_{u_i} = BBS_N(S_{u_i})$;
4. Let S and k be the sets of private information and encryption keys, respectively;
5. Output (S, k, pub).

Algorithm $Derive(1^\rho, G, u_i, u_j, S_{u_i}, N)$:

1. For $j > i$, compute $S_{u_j} = S_{u_i}^{2^{(j-i)\ell}} \bmod N$;
2. Output $k_{u_j} = BBS_N(S_{u_j})$.

The security of this scheme follows as a special case of our security analysis for the more general scheme in the next section.

4.2 The FP Scheme for General Posets

We now build on ideas from [13], specifically, the idea of using chain partitions, to obtain the FP scheme, a key assignment scheme for arbitrary posets with

KI-security based on the hardness of factoring Blum integers. We begin with an informal description.

Given a partially ordered hierarchy $(V, \leq)$, represented by the directed acyclic graph $P = (V, E)$, we select a partition of V into chains $\{C_0, \ldots, C_{w-1}\}$. Dilworth's Theorem [17] asserts that every partially ordered set $(V, \leq)$ can be partitioned into w chains, where w is the *width* of V, that is, the cardinality of the largest antichain in V. The length of C_i, $0 \leq i \leq w-1$, is denoted by l_i. The maximum class of C_i is regarded as the first class in C_i and the minimum class as the last class. Since $\{C_0, \ldots, C_{w-1}\}$ is a partition of V, each $u \in V$ belongs to precisely one chain.

Let $C = u_0 > \ldots > u_m$ be any chain in V. Then any chain of the form $u_j > \ldots > u_m$, $0 < j \leq m$ is said to be a *suffix* of C. Now, for any $u \in V$, the set $\downarrow u := \{v \in V : v \leq u\}$ has non-empty intersection with one or more chains $C_0, \ldots, C_{w-1}$. It is proved in [13] that the intersection of $\downarrow u$ and the chain C_i is a suffix of C_i or the empty set. This enables us to define the private information that should be given to a user with label u. We explain this next.

Since $\{C_0, \ldots, C_{w-1}\}$ is a partition of V into chains, $\{\downarrow u \cap C_0, \ldots, \downarrow u \cap C_{w-1}\}$ is a disjoint collection of chain suffixes. Additionally, the private information for each class in V should be chosen so that the private information for the j-th class of a chain can be used to compute keys for all lower classes in that chain. Hence, we can see that a user with label u should be given the private information for the maximal classes in the non-empty suffixes $\downarrow u \cap C_0, \ldots, \downarrow u \cap C_{w-1}$. Given $u \in V$, let $\hat{u}_0, \ldots, \hat{u}_{w-1}$ denote these maximal classes, with the convention that $\hat{u}_i = \perp$ if $\downarrow u \cap C_i = \emptyset$. Let u_j^i denote the j-th class in the chain C_i, where $0 \leq j \leq l_i - 1$.

We now provide a formal description of the FP scheme.

The FP Scheme:

Let Γ be a family of graphs corresponding to partially ordered hierarchies, let $P = (V, E) \in \Gamma$ be a graph, and consider a security parameter ρ. The *Gen* and *Derive* algorithms of the FP scheme are as follows.

Algorithm $Gen(1^\rho, P)$:

1. Run Gen_{Blum} to randomly obtain two distinct ρ-bit primes p and q, with $p = q = 3 \bmod 4$, and compute $N = pq$. Set $pub = N$;
2. Select a chain partition of V into w chains $C_0, \ldots, C_{w-1}$, so that C_i contains classes $u_0^i, u_1^i, \ldots, u_{l_i-1}^i$;
3. Select w values $\gamma_0, \ldots, \gamma_{w-1}$ at random from QR_N, the set of quadratic residues modulo N;
4. For each $u_j^i \in V$, $0 \leq j < l_i$, compute $T_{u_j^i} = \gamma_i^{2^{j\ell}} \bmod N$;
5. For each $u \in V$, define the private information S_u to be $\{T_{\hat{u}_i} : \hat{u}_i \neq \perp, 0 \leq i \leq w-1\}$ and the encryption key k_u to be $BBS_N(T_u)$.

Algorithm $Derive(1^\rho, P, u_j^i, u_h^g, S_{u_j^i}, N)$:

1. For $u_j^i \geq u_h^g$, find $\hat{u}_g$, the maximal class in $\downarrow u_j^i \cap C_g$. This class is in chain C_g. We denote it by u_r^g, where $0 \leq r < l_g$. Note that, by construction, $u_r^g \leq u_j^i$ and $T_{u_r^g} \in S_{u_j^i}$;

2. Compute $T_{u_h^g} = (T_{u_r^g})^{2^{(h-r)\ell}} \mod N$ and output $k_{u_h^g} = BBS_N(T_{u_h^g})$.

Theorem 3 (KI-ST security of the FP scheme for general posets). *Assume the factoring assumption relative to Gen_{Blum} holds. Then the FP scheme for general posets is KI-ST secure.*

We split the proof of this theorem into two parts. First we recall that the BBS generator is pseudorandom if factoring Blum integers is hard – see Theorem 2. In the next theorem, we show that any successful adversary A against the KI-ST security of the FP scheme implies a successful BBS-distinguisher D. Combining both parts yields Theorem 3.

Theorem 4 (KI-ST adversary $\Rightarrow$ BBS-distinguisher). *For every KI-ST adversary A that breaks the FP scheme with advantage $\mathrm{Adv}_A^{\text{KI-ST}}$, there exists a PPT algorithm D that breaks the BBS generator with the same advantage. Moreover, the running times of A and D are essentially the same.*

Proof. Assume we have a KI-ST adversary A against the FP scheme for general posets that attacks a class u_j^i. Assume the adversary is able to distinguish between the real key $k_{u_j^i}$ associated with class u_j^i and a random string having the same length. We describe below how to construct an algorithm D that, using A as a black box, is able to distinguish between $BBS_N(x)$ and a random ℓ-bit string r. Algorithm D plays the BBS game described in Definition 6, and is thus given access to an ℓ-bit string R that is either random or the output of the BBS generator. Algorithm D is also given inputs N and $z = x^{2^\ell} \mod N$. In order to use algorithm A, D simulates the environment of A in a way that A's view is indistinguishable from its view when playing the indistinguishability game described in Definition 4. More formally, algorithm D on inputs $(N, z = x^{2^\ell} \mod N, R)$ works as follows.

Algorithm D:
1. Run A with input $(1^\rho, G)$ to get A's choice of target class u_j^i (recall that u_j^i is the j-th class in chain C_i);
2. Pick a random value $\gamma_t \in QR_N$ for each chain $C_t \neq C_i$ and set $T_{u_0^t} = \gamma_t$, where u_0^t is the maximal class in C_t;
3. Run A with inputs $(1^\rho, G, N, corr, R)$, where $corr = \{S_{u_h^g} : u_h^g \not\geq u_j^i\}$, to obtain a bit b. (We will explain below how D can compute the set $corr$ in this simulation.) Here b is A's guess as to whether it was given the real key associated with class u_j^i or a random string having the same length;
4. Output b.

Here, algorithm D setups up values T_u and associates them with classes $u \in V$ in such a way that the key at class u_j^i should be $BBS_N(x)$. Note that D does not know the value $T_{u_j^i}$, which corresponds to x. However, it does know $z = x^{2^\ell} \bmod N$, which corresponds to $T_{u_{j+1}^i}$ and thus D can compute all the values $T_{u_h^i}$, $j < h < l_i$. Further, A cannnot corrupt any class $u_h^i \in C_i$ for $h \leq j$. Moreover, by our setup of variables D knows all the other values $T_{u_h^t}$, with $t \neq i$ and $0 \leq h < l_t$. Combining these two observations, it is easy to see that D can compute all the private information that A can request through its corruption query, denoted in step 3 by $corr$.

If A outputs $b = 1$, guessing that it was given the real key $k_{u_j^i}$ associated with class u_j^i, then D will also output $b = 1$, guessing that it was given the $BBS_N(x)$. Conversely, if A outputs $b = 0$, guessing that it was given a random ℓ-bit string, D will also output $b = 0$, guessing that it was given a random value. It is easy to see that the advantage of D in breaking the security of the BBS pseudorandom generator is the same as that of A in winning the KI-ST security game against the FP scheme. Thus

$$Adv_D^{BBS} = Adv_A^{KI\text{-}ST}.$$

The theorem now follows.   □

We remind the reader that using the results proven in [5], we can immediately conclude that the FP scheme for general posets is also secure against dynamic adversaries in the sense of key indistinguishability, under the factoring assumption for Blum integers.

4.3   A Scheme with Faster Key Derivation

Our description of the FP scheme involves the extraction of a single bit in each iteration of the BBS generator; consequently, because of the structure of the scheme, this means that $\ell(h - r)$ modular squarings are needed to derive the private value $T_{u_h^g}$ from the private value $T_{u_r^g}$ during key derivation. However, it is known that it is possible to extract more than one bit at a time from each iteration of the BBS generator. For example, the generator which extracts the $O(\log \log N)$ least significant bits of each value is known to be secure under the factoring assumption [21]. This can be used to speed up key derivation in the FP scheme. For N having 1024 bits (and assuming the implicit constant can be set to 1), the speed-up is roughly a factor of 10. However, this comes at the cost of a looser reduction to KI-ST security of the scheme from the factoring problem.

4.4   Efficiency Considerations

Our main construction provides schemes having a trade-off between storage of private information and efficiency of key derivation. Users associated with a certain class will have to store a number of private values, with that number depending on how the poset $(V, \leq)$ is partitioned into chains. The maximum number of values to be stored is equal to the width w of the poset. The overall efficiency of key derivation in the FP scheme is bounded by the length of the longest chain in the partition: as many as ℓ times this length of squaring modulo N operations are needed to derive a key. Thus, we need to find a good choice of chain partition to balance the efficiency of key derivation and the private information storage requirements. Aside from the cost of key derivation, the FP scheme preserves all of the benefits of the Multi-key RSA scheme of [13], but has reduced public data and benefits from a formal security analysis establishing its KI security against static and dynamic adversaries. For further discussion and comparison of the schemes of [13] to other schemes in the literature, we refer to [13].

D'Arco *et al.* [14] gave a general construction for KI-secure schemes, using the Goldreich-Levin hard-core bit and an underlying KR-secure scheme. They evaluated an instantiation of their construction using the Akl-Taylor scheme with the MacKinnon *et al.* assignment to obtain KI-secure schemes under the RSA assumption. The FP scheme compares favorably to this instantiation. In particular, the FP scheme is proven to be KI-secure under a weaker assumption, the factoring assumption, compared to the RSA assumption required in [14]. The construction used to obtain the FP scheme is arguably also more natural, easier to understand and enjoys a simpler security analysis. Additionally, the FP scheme has much smaller public information: The instantiation in [14] requires $|V|(1 + \ell) + 2$ public values, while the FP scheme requires a single value, N. We also avoid the intrinsic blow-up that is involved in the construction given in [14]: that construction needs a KR-secure scheme for a poset having ℓ times as many classes than are in the final KI-secure scheme, where ℓ is the length of keys. On the other hand, the construction given in [14] is fully generic, relying only on the existence of KR-secure schemes for certain posets. For reasons of space, we omit a full comparison with all schemes in the literature known to be KI-secure.

5   Concluding Remarks

In this paper we have described key assignment schemes with provable security based on the factoring assumption. Our main construction extends ideas from [13] and achieves schemes for arbitrary posets with shorter public data and having KI security based on the factoring assumption. These schemes could be useful in a wide range of applications where access hierarchies arise, including, for example, management of databases and secure broadcast services. Our main construction appears to yield the first reasonably efficient schemes with security provably based on the factoring problem, in the standard model.

Acknowledgements

We thank Jason Crampton for many illuminating discussions and valuable feedback on an earlier version of this paper. We also thank the anonymous referees for their helpful comments.

References

1. Akl, S.G., Taylor, P.D.: Cryptographic solution to a problem of access control in a hierarchy. ACM Transactions on Computer Systems 1, 239–248 (1983)
2. MacKinnon, S.J., Taylor, P.D., Meijer, H., Akl, S.G.: An optimal algorithm for assigning cryptographic keys to control access in a hierarchy. IEEE Transactions on Computers 34, 797–802 (1985)
3. Harn, L., Lin, H.Y.: A cryptographic key generation scheme for multilevel data security. Computers & Security 9, 539–546 (1990)
4. Atallah, M.J., Blanton, M., Fazio, N., Frikken, K.B.: Dynamic and efficient key management for access hierarchies. In: ACM Conference on Computer and Communications Security, pp. 190–202 (2006)
5. Ateniese, G., Santis, A.D., Ferrara, A.L., Masucci, B.: Provably-secure time-bound hierarchical key assignment schemes. In: ACM Conference on Computer and Communications Security, pp. 288–297 (2006)
6. Tzeng, W.G.: A secure system for data access based on anonymous authentication and time-dependent hierarchical keys. In: ACM Symposium on Information, Computer and Communications Security, pp. 223–230 (2006)
7. Wang, S.Y., Laih, C.S.: An efficient solution for a time-bound hierarchical key assignment scheme. IEEE Transactions on Dependable and Secure Computing 3, 91–100 (2006)
8. Chen, T.S., Chung, Y.F.: Hierarchical access control based on chinese remainder theorem and symmetric algorithm. Computers & Security 21, 565–570 (2002)
9. Shen, V.R.L., Chen, T.S.: A novel key management scheme based on discrete logarithms and polynomial interpolations. Computers & Security 21, 164–171 (2002)
10. Wu, T.C., Chang, C.C.: Cryptographic key assignment scheme for hierarchical access control. International Journal of Computer Systems Science and Engineering 16, 25–28 (2001)
11. Yeh, J.-H., Chow, R., Newman, R.: A key assignment for enforcing access control policy exceptions. In: International Symposium on Internet Technology, pp. 54–59 (1998)
12. Crampton, J., Martin, K.M., Wild, P.R.: On key assignment for hierarchical access control. In: Computer Security Foundations Workshop, pp. 98–111 (2006)
13. Crampton, J., Daud, R., Martin, K.M.: Constructing key assignment schemes from chain partitions. In: Foresti, S., Jajodia, S. (eds.) DBSec 2010. LNCS, vol. 6166, pp. 130–145. Springer, Heidelberg (2010)
14. D'Arco, P., Santis, A.D., Ferrara, A.L., Masucci, B.: Variations on a theme by Akl and Taylor: Security and tradeoffs. Theoretical Computer Science 411, 213–227 (2010)
15. Goldreich, O., Levin, L.A.: A hard-core predicate for all one-way functions. In: ACM STOC, pp. 25–32 (1989)
16. Goldreich, O., Goldwasser, S., Micali, S.: How to construct random functions. J. ACM 33, 792–807 (1986)
17. Dilworth, R.P.: A decomposition theorem for partially ordered sets. Annals of Mathematics 51, 161–166 (1950)
18. Hofheinz, D., Kiltz, E.: Practical chosen ciphertext secure encryption from factoring. In: Joux, A. (ed.) EUROCRYPT 2009. LNCS, vol. 5479, pp. 313–332. Springer, Heidelberg (2009)

19. Blum, L., Blum, M., Shub, M.: A simple unpredictable pseudo-random number generator. SIAM Journal on Computing 15(2), 364–383 (1986)
20. Alexi, W., Chor, B., Goldreich, O., Schnorr, C.P.: RSA and Rabin functions: Certain parts are as hard as the whole. SIAM Journal on Computing 17, 194–209 (1988)
21. Vazirani, U.V., Vazirani, V.V.: Efficient and secure pseudo-random number generation. In: Blakely, G.R., Chaum, D. (eds.) CRYPTO 1984. LNCS, vol. 196, pp. 193–202. Springer, Heidelberg (1985)

Efficient CCA-Secure CDH Based KEM Balanced between Ciphertext and Key

Yamin Liu, Bao Li, Xianhui Lu, and Dingding Jia

State Key Laboratory of Information Security,
Graduate University of Chinese Academy of Sciences,
No.19A Yuquan Road, 100049 Beijing, China
{ymliu,lb,xhlu,ddjia}@is.ac.cn

Abstract. In this paper we construct an efficient CCA-secure key encapsulation scheme in the standard model. The new scheme is based
on the computational Diffie-Hellman assumption and the twinning technique, which has been widely discussed in recent years. Compared with
previous schemes of the same kind, the new scheme is more generic, and
offers a simple approach for reconciling ciphertext length and key size by
altering a parameter. Choosing a reasonable value for the parameter, a
balance between the ciphertext length and key size could be achieved.

Keywords: computational Diffie-Hellman, twin Diffie-Hellman, key encapsulation mechanism, standard model.

1 Introduction

Indistinguishability against adaptive chosen-ciphertext attack (IND-CCA)
[12,13] is the standard rule in public-key encryption, and constructing practical
CCA-secure encryption schemes is one of the most important tasks, especially
based on reasonable intractability assumptions such as the computational Diffie-
Hellman (CDH) assumption in the standard model and without resorting to the
convenient but controversial random oracle model [3,4].

However, recognizing correct solutions to the CDH problem is not an easy
task, thus checking ciphertext validity in the security reduction becomes a hard
nut. As a result, some strong assumptions are used, such as the Strong Diffie-
Hellman (SDH) assumption [1], which allows a decisional oracle for checking the
validity of solutions to the CDH problem.

The twinning technique, which is invented by Cash, Kiltz and Shoup [5] in
2008, provides a convenient approach for basing security of encryption schemes
on the intractability of the standard CDH problem. The twin Diffie-Hellman
(2DH) assumption [5] is proved to be equivalent to CDH, while it still allows
a decisional oracle for recognizing correct answers to the CDH problem, thus
consistency of ciphertexts could be easily checked in the security reduction.

With the twinning technique, Cash, Kiltz and Shoup constructed a CCA-
secure encryption scheme based on the CDH assumption in the standard model
[5]. However, efficiency of the scheme in [5] is not very satisfactory since both

U. Parampalli and P. Hawkes (Eds.): ACISP 2011, LNCS 6812, pp. 310–318, 2011.
© Springer-Verlag Berlin Heidelberg 2011

the ciphertext and the key require polynomial group elements. A CCA-secure key encapsulation mechanism (KEM) under the same assumption proposed by Haralambiev, Jager, Kiltz and Shoup [9] in 2010, improves greatly in the length of the ciphertext, which only comprises three group element, while the key size still remains polynomial. Later, Wee introduced the extractable hash proof system [14], which encompasses the CDH based CCA-secure schemes of [5,9], and gave another CDH based CCA-secure KEM. The KEM in [14] costs only four group elements in the public key and the secret key respectively, while still requires a linear number of group elements in the ciphertext. However, as Wee pointed out in [14], such a scheme might be preferable in the circumstance of encrypting long messages via the hybrid encryption.

Independent of the twinning framework, Hanaoka and Kurosawa constructed a CDH-based IND-CCA secure encryption scheme in [10] by employing a broadcast encryption (BE) scheme. Akin to the scheme in [9], the scheme in [10] also costs only three group elements in ciphertext, but the key size still remains polynomial.

1.1   Our Contributions

In this paper we construct an efficient CCA-secure encryption schemes in the standard model, based on the CDH assumption and the twinning technique. The new scheme can be viewed as a hybrid of the schemes in [9] and [14], since the scheme in [9] is preferable for short ciphertext, and the scheme in [14] is superior in short keys.

The new scheme is similar to the schemes in [5,9,14], and it also complies with the extractable hash proof framework in [14]. The difference is that a parameter for controlling key size and ciphertext length is introduced. By altering the parameter, the new scheme flexibly switches between ciphertext length and key size, and by choosing the parameter reasonably, a balance between the ciphertext length and key size could be achieved. This method especially improves the ciphertext length and computation efficiency of the scheme in [14].

Organization. The paper is organized as follows. Section 2 provides some notations and definitions. In Section 3 the new scheme and its security proof and efficiency analysis are elaborated. Finally, section 4 is the conclusion.

2   Preliminaries

For a positive integer n, $[n]$ denotes the set $\{1, ..., n\}$. $x \xleftarrow{\$} S$ means that x is randomly chosen from the set S. For a randomize algorithm $\mathcal{A}$, $x \xleftarrow{\$} \mathcal{A}(\cdot)$ means that x is assigned the output of $\mathcal{A}$. An algorithm is efficient if it runs in polynomial time in its input length. A function $f(\kappa)$ is negligible if it decreases faster than any polynomial, and is denoted as $f(\kappa) \leq \epsilon(\kappa)$. PPT is the short form of probabilistic polynomial time. $\bot$ is the error symbol.

2.1   Key Encapsulation Mechanisms

A public-key encapsulation mechanism $KEM = (KGen, Enc, Dec)$ [6,7] consists of three polynomial-time algorithms, wherein $KGen$ generates a public key and secret key pair (pk, sk), Enc produces a ciphertext C encapsulating a corresponding session key K; with sk, Dec recovers K from C. K is the key of a symmetric encryption scheme called the data encapsulation mechanism (DEM) [6,7].

The IND-CCA security of KEM is described by the following game.

$$\mathrm{Exp}_{KEM, \mathcal{A}}^{\text{IND-CCA}}(\kappa)$$

$$(pk, sk) \xleftarrow{\$} Kg(1^\kappa); K_1^* \xleftarrow{\$} KeySp(\kappa); (K_0^*, C^*) \xleftarrow{\$} Enc(pk)$$

$$\sigma \xleftarrow{\$} \{0, 1\}; \sigma' \xleftarrow{\$} \mathcal{A}^{DecO(sk,\cdot)}(pk, K_\sigma^*, C^*)$$

$\mathcal{A}$ wins the game if $\sigma = \sigma'$. Its advantage is defined as

$$Adv_{KEM, \mathcal{A}}^{\text{IND-CCA}}(1^\kappa) = |Pr[\sigma = \sigma'] - \frac{1}{2}|$$

Definition 1. *(IND-CCA Security) A key encapsulation mechanism $KEM = (KGen, Enc, Dec)$ is said to be IND-CCA secure if for all PPT adversary $\mathcal{A}$, $Adv_{KEM, \mathcal{A}}^{IND-CCA}(1^\kappa)$ is negligible.*

2.2   Diffie-Hellman Assumptions

Consider a cyclic group $\mathbb{G}$ with a generator g. The computational Diffie-Hellman assumption states that given $A = g^a$, $B = g^b$, computing $C = g^{ab}$ for random $A, B \in \mathbb{G}$ is intractable for all efficient algorithms. We use notations from previous literatures such as [5,9], and define $C = dh(A, B)$.

The twin Diffie-Hellman problem is given random $A_1, A_2, B \in \mathbb{G}$ and a decisional oracle $2dhp(A_1, A_2, \cdot, \cdot, \cdot)$, computing $(dh(A_1, B), dh(A_2, B))$. The oracle $2dhp(A_1, A_2, \cdot, \cdot, \cdot)$, on input (B', C_1', C_2'), judges whether there is $C_1' = dh(A_1, B')$ and $C_2' = dh(A_2, B')$, where $B' \neq B$. The 2DH assumption asserts the intractability of the 2DH problem for all efficient algorithms, and it is proved to be equivalent to the CDH assumption in [5].

2.3   Goldreich-Levin Hardcore Function

The Goldreich-Levin hardcore function [8] is a hardcore function for all one-way functions. Let $f_{gl} : \mathbb{G} \times \{0, 1\}^u \mapsto \{0, 1\}^v$ denote a Goldreich-Levin hardcore function for CDH problem over group $\mathbb{G}$. The following lemma from [5,9] is needed for security analysis in this paper.

Lemma 1. *Let $\mathbb{G}$ be a prime-order group generated by g. Let $A_1, A_2, B \xleftarrow{\$} \mathbb{G}$ be random group elements, $R \xleftarrow{\$} \{0, 1\}^u$, and let $K = f_{gl}(dh(A_1, B), R)$. Let*

$U_v \xleftarrow{\$} \{0,1\}^v$ *be uniformly random. Suppose that there exists a PPT algorithm* $\mathcal{B}$ *having access to an oracle computing* $2dhp(A_1, A_2, \cdot, \cdot, \cdot)$ *and distinguishing the distributions*

$$\Delta_{dh} = (g, A_1, A_2, B, K, R) \text{ and } \Delta_{rand} = (g, A_1, A_2, B, U_v, R)$$

with non-negligible advantage. Then there exists a probabilistic polynomial-time algorithm computing $dh(A, B)$ *on input* (A, B) *with non-negligible success probability.*

3   The New Scheme

In this section a new method for constructing efficient CCA-secure encryption schemes based on CDH assumption is proposed.

Let κ be the security parameter, and let $\mathbb{G}$ be a group of prime order q. Let $T_s : \mathbb{G} \mapsto \mathbb{Z}_q$ be a target collision resistant hash function indexed by s. The Goldreich-Levin hard-core function is $f_{gl} : \mathbb{G} \times \{0,1\}^u \mapsto \{0,1\}^v$. Let $N = nv$ be the length of the encapsulated session key, where $n = n(\kappa)$ is a polynomial-sized integer, and let c be an integer ranging from 1 to n and $c|n$. Set $w = n/c$. The new scheme offers a variable ciphertext length and key size by altering c, and is defined as follows.

$KGen(1^\kappa)$ Randomly choose a generator $g \xleftarrow{\$} \mathbb{G}$, a string $R \xleftarrow{\$} \{0,1\}^u$, and a function index s. Randomly choose integers $x, x', y, y' \xleftarrow{\$} \mathbb{Z}_q$, and $z_i \xleftarrow{\$} \mathbb{Z}_q$ for $i \in [c]$, and compute $X = g^x, X' = g^{x'}, Y = g^y, Y' = g^{y'}$ and $Z_i = g^{z_i}$ for $i \in [c]$. Set

$$pk = (X, X', Y, Y', Z_1, ..., Z_c, R, s),$$
$$sk = (x, x', y, y', z_1, ..., z_c),$$

and return (pk, sk).

$Enc(pk)$ To encapsulate a key of length N, sample $r_1, ..., r_w \xleftarrow{\$} \mathbb{Z}_q$. Compute

$$C_0 = (g^{r_1}, ..., g^{r_w}), t = T_s(C_0),$$
$$C_1 = ((X^t X')^{r_1}, ..., (X^t X')^{r_w}),$$
$$C_2 = ((Y^t Y')^{r_1}, ..., (Y^t Y')^{r_w}),$$
$$k_i = (f_{gl}(Z_i^{r_1}, R), ..., f_{gl}(Z_i^{r_w}, R)), i \in [c],$$
$$K = k_1 || ... || k_c,$$

and return $((C_0, C_1, C_2), K)$.

$Dec(sk, (C_0, C_1, C_2))$ Compute $t = T_s(C_0)$. Parse (C_0, C_1, C_2) as w triples $(C_{0,i}, C_{1,i}, C_{2,i})$, and check whether $C_{1,i} = C_{0,i}^{xt+x'}$ and $C_{2,i} = C_{0,i}^{yt+y'}$ for $i \in [w]$. If not then return $\perp$. Otherwise compute

$$k_i = (f_{gl}(C_{0,1}^{z_i}, R), ..., f_{gl}(C_{0,w}^{z_i}, R)), i \in [c],$$
$$K = k_1 || ... || k_c$$

and return K.

3.1   Security Proof

Theorem 1. *Let T_s be a target collision-resistant hash function and assume the intractability of the computational Diffie-Hellman assumption over group $\mathbb{G}$. Then the above described KEM is IND-CCA secure.*

The security proof employs the game sequence technique, all-but-one simulation, and the twinning technique. These techniques are used in previous literatures such as [2,11,5,9] and have been concluded by Wee as the extractable hash proofs [14].

Proof. Let (C_0^*, C_1^*, C_2^*) be the challenge ciphertext encapsulating the key K_0^*, let K_1^* be the random key chosen by the IND-CCA challenger, and let $t^* = T_s(C_0^*)$.

Let S_i be the event that the IND-CCA adversary $\mathcal{A}$ wins in Game i. Here is the sequence of games.

Game 0. This is the original IND-CCA game, and there is

$$Pr[S_0] = \frac{1}{2} + Adv_{KEM, \mathcal{A}}^{IND\text{-}CCA}(1^\kappa)$$

by definition.

Game 1. Define **Game 1** as identical with **Game 0**, except that the decapsulation mechanism of **Game 1** outputs $\perp$ on query $C = (C_0, C_1, C_2)$ where $C_0 = C_0^*$. The probability that $\mathcal{A}$ queries such a ciphertext before receiving the challenge ciphertext is bounded by Q_d/q, where Q_d is the total number of decapsulation queries of $\mathcal{A}$. Thus Q_d/q is negligible. If C is queried after the challenge with $C_0 = C_0^*$ but $C \neq C^*$, then C is an invalid ciphertext and should be rejected. Hence

$$|Pr[S_1] - Pr[S_0]| \leq \epsilon(\kappa).$$

Game 2. Proceed as in **Game 1**, except that the decapsulation mechanism outputs $\perp$ on query $C = (C_0, C_1, C_2)$ where $C_0 \neq C_0^*$ while $T_s(C_0) = t^*$. Since T_s is target collision resistant, there is

$$|Pr[S_2] - Pr[S_1]| < \epsilon(\kappa).$$

Game 3. Let **Game 3** be identical with **Game 2** except that $K_0^* \xleftarrow{\$} \{0,1\}^N$ is uniformly chosen at random. Obviously

$$Pr[S_3] = \frac{1}{2},$$

since now both K_0^* and K_1^* are uniformly random and irrelevant to C^*.

Assuming the intractability of the computational Diffie-Hellman problem, we claim that

$$|Pr[S_3] - Pr[S_2]| \leq \epsilon(\kappa),$$

and prove it employing a hybrid argument.

Define a sequence of hybrid games $H_0, ..., H_n$, where H_0 equals **Game 2** and H_n equals **Game 3**. In H_i the first iv bits of K_0^* are sampled as independent random bits, and the rest bits of K_0^* are computed as in H_{i-1}.

Denote the event that the adversary $\mathcal{A}$ outputs 1 in H_i as E_i. If

$$|Pr[S_3] - Pr[S_2]| = 1/p(\kappa)$$

for a polynomial $p(\cdot)$, i.e.,

$$|Pr[E_0] - Pr[E_n]| = 1/p(\kappa),$$

then there must exist an index j such that

$$|Pr[E_{j-1}] - Pr[E_j]| = 1/p'(k)$$

for a polynomial $p'(\cdot)$ according to the Pigeonhole Principle. Then we could construct an algorithm $\mathcal{B}$, giving access to the adversary $\mathcal{A}$ and a 2DH oracle, distinguishing the distributions Δ_{dh} and Δ_{rand}.

Given a challenge $\delta = (g, A_1, A_2, B, L, R)$ and a twin Diffie-Hellman oracle $2\text{dhp}(A_1, A_2, \cdot, \cdot, \cdot)$, $\mathcal{B}$ guesses the index $j \in [n]$ with a probability at least $1/n$, and simulates a hybrid game as follows.

Key Generation and the Challenge. Set $a = \lceil j/w \rceil$ and $b = j \mod w$, if $b = 0$ then reset $b = w$. $\mathcal{B}$ generates the public key and the challenge ciphertext as follows:

1. Randomly pick $d, e, f \xleftarrow{\$} \mathbb{Z}_q$, $r_i^* \xleftarrow{\$} \mathbb{Z}_q$ for $i \in [w]\backslash\{b\}$, and $z_i \xleftarrow{\$} \mathbb{Z}_q$ for $i \in [c]\backslash\{a\}$. The hash function T_s is chosen as in **Game 0**.
2. Compute $C_{0,i}^* = g^{r_i^*}$ for $i \in [w]\backslash\{b\}$, set $C_{0,b}^* = B$ and $t^* = T_s(C_0^*)$.
3. Set $X = A_1^e$, $X' = A_1^{-et^*} g^d$, $Y = A_2, Y' = A_2^{-t^*} g^f$. Compute $Z_i = g^{z_i}$ for $i \in [c]\backslash\{a\}$, and set $Z_a = A_1$. Then $(X, X', Y, Y', Z_1, ..., Z_c, s, R)$ is sent to $\mathcal{A}$ as the public key.
4. Compute $C_{1,i}^* = (X^{t^*} X')^{r_i^*}, C_{2,i}^* = (Y^{t^*} Y')^{r_i^*}$ for $i \in [w]\backslash\{b\}$, and set $C_{1,b}^* = B^d$, $C_{2,b}^* = B^f$. (C_0^*, C_1^*, C_2^*) is the challenge ciphertext.
5. Compute $k_i^* = (f_{gl}(C_{0,1}^{*z_i}, R), ..., f_{gl}(C_{0,w}^{*z_i}, R))$ for $i \in [c]\backslash\{a\}$. For $k_a^* = (k_{a,1}^*, ..., k_{a,w}^*)$, compute $k_{a,l}^* = f_{gl}(A_1^{r_l^*}), l \in [w]\backslash\{b\}$ and set $k_{a,b}^* = L$.
6. Set $K_0^* = k_1^* || ... || k_c^*$.

According to the key generation step, $X, X', Y, Y', Z_1, ..., Z_c$ are independent and distribute uniformly random in $\mathbb{G}$.

Note that

$$(X^{t^*} X')^{\log_g B} = ((A_1^e)^{t^*} A_1^{-et^*} g^d)^{\log_g B} = B^d,$$

and similarly $(Y^{t^*} Y')^{\log_g B} = B^f$, the triple $(C_{0,b}^*, C_{1,b}^*, C_{2,b}^*)$ is consistent, thus $C^* = (C_0^*, C_1^*, C_2^*)$ is a valid ciphertext.

If $L = f_{gl}(dh(C_{0,b}^*, Z_a), R) = f_{gl}(dh(A_1, B), R)$, then $\mathcal{B}$ simulates H_{j-1} for $\mathcal{A}$, otherwise $\mathcal{B}$ simulates H_j. Thus $\mathcal{B}$ can distinguish $\delta \in \Delta_{dh}$ from $\delta \in \Delta_{rand}$ if $\mathcal{A}$ distinguishes H_{j-1} and H_j.

Decapsulation Simulation. Let $C = (C_0, C_1, C_2)$ be an arbitrary decapsulation query of $\mathcal{A}$. $\mathcal{B}$ computes $t = T_s(C_0)$. For every triple $(C_{0,i}, C_{1,i}, C_{2,i})_{i \in [w]}$, $\mathcal{B}$ computes

$$\tilde{X}_i = (C_{1,i}/C_{0,i}^d)^{1/(et-et^*)}, \text{ and } \tilde{Y}_i = (C_{2,i}/C_{0,i}^f)^{1/(t-t^*)}.$$

If $t \neq t^*$ and C is a valid ciphertext, then

$$\begin{aligned}
\tilde{X}_i &= ((X^t X')^{r_i}/(g^{r_i})^d)^{1/(et-et^*)} \\
&= (A_1^{r_i(et-et^*)} g^{r_i d}/g^{r_i d})^{1/(et-et^*)} \\
&= A_1^{r_i} = dh(A_1, C_{0,i}),
\end{aligned}$$

similarly there is $\tilde{Y}_i = A_2^{r_i}$. Then $\mathcal{B}$ checks the consistency of $(C_{0,i}, C_{1,i}, C_{2,i})$ by querying $2dhp(A_1, A_2, C_{0,i}, \tilde{X}_i, \tilde{Y}_i)$.

If all these tests are passed, for $k_i = (k_{i,1}, ..., k_{i,w})$, $i \in [c]\backslash\{a\}$, $\mathcal{B}$ computes $k_{i,l} = f_{gl}(C_{0,l}^{z_i}, R)$ for $l \in [w]$; $\mathcal{B}$ then computes $k_{a,l} = f_{gl}(\tilde{X}_l, R)$ for $l \in [w]$, and returns $K = k_1 ||...|| k_c$.

Obviously $\mathcal{B}$ can answer all decapsulation queries correctly if $t \neq t^*$, and the probability of $t = t^*$ have been shown to be negligible in **Game 2**.

Thus, if $\mathcal{A}$ has a non-negligible advantage in distinguishing H_{j-1} and H_j, then $\mathcal{B}$ has a non-negligible advantage in distinguishing Δ_{dh} and Δ_{rand}, and this contradicts the CDH assumption according to Lemma 1.   □

3.2   Efficiency Analysis

Here is a comparison among the new scheme and some previous CDH-based schemes constructed with the twinning technique. Some figures are from [9]. Let $N = nv$ be the expected length of the session key. In the comparison of key size and ciphertext length, we omit R and s and just list the number of group elements. Similarly, in the comparison of computation efficiency, we just count the numbers of exponentiations, since the exponentiations are the main time-consumer.

Table 1. Efficiency comparison of related schemes

	key size (pk,sk)	ciphertext length	efficiency #exp (enc,dec)
[5] CKS08	$2n + 2, 2n + 2$	$n + 2$	$3n + 1, 2n + 1$
[9] HJKS10	$n + 4, n + 4$	3	$n + 5, n + 2$
[14] Wee10	$4, 4$	$3n$	$6n, 3n$
Our scheme	$c + 4, c + 4$	$3n/c$	$n + 5n/c, n + 2n/c$

Note that when $c = n$, the new method yields exactly the CDH-based encryption scheme in [9]; when $c = 1$, the resulting scheme is essentially the same as the CDH-based encryption scheme in [14]. By choosing a reasonable value for c, the protocol yields an encryption scheme balanced between key size and ciphertext length. The value of c can be chosen according to the requirement of different application circumstances.

Especially, our method significantly improves the efficiency of the scheme in [14]. For example, when $c = 3$, the ciphertext length shrinks 2/3 compared to the scheme in [14], while only 3 group elements are added to the public key and the secrete key respectively. And as to the computation efficiency, when $c = 3$, the number of exponentiations in the encryption and decryption also decreases 5/9 and 4/9 respectively, compared to [14].

4   Conclusion

We propose a new efficient CCA-secure key encapsulation mechanism based on the CDH assumption and the twinning technique in the standard model. The new scheme complies with the extractable hash proof framework of [14] and offers a choice between ciphertext length and key size.

Acknowledgment. We are grateful to anonymous reviewers for their invaluable comments. Besides, we thank Kunpeng Wang, Haixia Xu and Qixiang Mei for helpful discussions. This work is supported by the National Natural Science Foundation of China (No. 61070171), the National High-Tech Research and Development Plan of China (863 project, No.2006AA01Z427) and the National Basic Research Program of China (973 project, No.2007CB311201).

References

1. Abdalla, M., Bellare, M., Rogaway, P.: The Oracle Diffie-Hellman Assumptions and an Analysis of DHIES. In: Naccache, D. (ed.) CT-RSA 2001. LNCS, vol. 2020, pp. 143–158. Springer, Heidelberg (2001)
2. Boneh, D., Boyen, X.: Efficient selective-ID Secure Identity Based Encryption without Random Oracles. In: Cachin, C., Camenisch, J.L. (eds.) EUROCRYPT 2004. LNCS, vol. 3027, pp. 223–238. Springer, Heidelberg (2004)
3. Bellare, M., Rogaway, P.: Random Oracles are Practical: A Paradigm for Designing Efficient Protocols. In: Proceedings of the 1st ACM Conference on Computer and Communications Security, pp. 62–73. ACM Press, New York (1993)
4. Bellare, M., Rogaway, P.: Optimal Asymmetric Encryption – How to Encrypt with RSA. In: De Santis, A. (ed.) EUROCRYPT 1994. LNCS, vol. 950, pp. 92–111. Springer, Heidelberg (1995)
5. Cash, D., Kiltz, E., Shoup, V.: The Twin Diffie-Hellman Problem and Applications. In: Smart, N.P. (ed.) EUROCRYPT 2008. LNCS, vol. 4965, pp. 127–145. Springer, Heidelberg (2008)
6. Cramer, R., Shoup, V.: A Practical Public Key Cryptosystem Provably Secure against Adaptive Chosen Ciphertext Attack. In: Krawczyk, H. (ed.) CRYPTO 1998. LNCS, vol. 1462, pp. 13–25. Springer, Heidelberg (1998)
7. Cramer, R., Shoup, V.: Design and Analysis of Practical Public-Key Encryption Schemes Secure against Adaptive Chosen Ciphertext Attack. SIAM Journal on Computing 33(1), 167–226 (2004)

8. Goldreich, O., Levin, L.: A Hard-core Predicate for All One-way Functions. In: 21st STOC. ACM, New York (1989)
9. Haralambiev, K., Jager, T., Kiltz, E., Shoup, V.: Simple and Efficient Public-Key Encryption from Computational Diffie-Hellman in the Standard Model. In: Nguyen, P.Q., Pointcheval, D. (eds.) PKC 2010. LNCS, vol. 6056, pp. 1–18. Springer, Heidelberg (2010)
10. Hanaoka, G., Kurosawa, K.: Efficient Chosen Ciphertext Secure Public Key Encryption under the Computational Diffie-Hellman Assumption. In: Pieprzyk, J. (ed.) ASIACRYPT 2008. LNCS, vol. 5350, pp. 308–325. Springer, Heidelberg (2008)
11. Kiltz, E.: Chosen-Ciphertext Secure Key-Encapsulation Based on Gap Hashed Diffie-Hellman. In: Okamoto, T., Wang, X. (eds.) PKC 2007. LNCS, vol. 4450, pp. 282–297. Springer, Heidelberg (2007)
12. Naor, M., Yung, M.: Public-Key Cryptosystems Provably Secure against Chosen Ciphertext Attacks. In: Proceedings of the 22nd STOC, pp. 427–437. ACM, New York (1990)
13. Rackoff, C., Simon, D.R.: Non-Interactive Zero-Knowledge Proof of Knowledge and Chosen Ciphertext Attack. In: Feigenbaum, J. (ed.) CRYPTO 1991. LNCS, vol. 576, pp. 433–444. Springer, Heidelberg (1992)
14. Wee, H.: Chosen-Ciphertext Security via Extractable Hash Proofs. In: Rabin, T. (ed.) CRYPTO 2010. LNCS, vol. 6223, pp. 314–332. Springer, Heidelberg (2010)

Generic Construction of Strongly Secure Timed-Release Public-Key Encryption

Atsushi Fujioka[1], Yoshiaki Okamoto[2], and Taiichi Saito[2]

[1] NTT Information Sharing Platform Laboratories
fujioka.atsushi@lab.ntt.co.jp
[2] Tokyo Denki University
{okamoto@crypt.,taiichi@}c.dendai.ac.jp

Abstract. This paper provides a sufficient condition to construct timed-release public-key encryption (TRPKE), where the constructed TRPKE scheme guarantees strong security against malicious time servers, proposed by Chow et al., and strong security against malicious receivers, defined by Cathalo et al., in the random oracle model if the component IBE scheme is IND-ID-CPA secure, the component PKE scheme is IND-CPA secure, and the PKE scheme satisfies negligible γ-uniformity for every public key. Chow et al. proposed a strongly secure TRPKE scheme, which is concrete in the standard model. To the best of our knowledge, the proposed construction is the first generic one for TRPKE that guarantees strong security even in the random oracle model.

Keywords: timed-release public-key encryption, public-key encryption, identity-based encryption, random oracle model.

1   Introduction

Timed-Release Public-Key Encryption. Timed-release public-key encryption (TRPKE) [8] provides a public-key encryption mechanism through which one cannot decrypt a ciphertext even with one's own secret key before a specific time. It has many applications in constructing secure protocols in which information is revealed to several users after a specific time, e.g., releasing a new movie, distributing an examination paper, sealed-bit auction, and e-voting.

A TRPKE system consists of three entities: *time server*, *sender*, and *receiver*. The sender encrypts a plaintext with the receiver's and the time server's public keys, designating a time T after which the receiver is allowed to decrypt the ciphertext (throughout this paper, we call this time information *time period*, according to [6]). The time server periodically broadcasts a *time signal* s_T to all users including the receiver. At each T, the time server generates an s_T corresponding to T and broadcasts it. Until the designated T comes and the time server broadcasts the corresponding s_T. The receiver cannot decrypt the ciphertext even with his/her own secret key as he/she does not obtain s_T. When the receiver obtains s_T at T, the receiver can decrypt the ciphertext with his/her own secret key and s_T. It is well known that TRPKE can be constructed from *identity-based encryption* (IBE) and *public-key encryption* (PKE) [9,10].

U. Parampalli and P. Hawkes (Eds.): ACISP 2011, LNCS 6812, pp. 319–336, 2011.
© Springer-Verlag Berlin Heidelberg 2011

Related to TRPKE, a similar primitive is known as *certificateless encryption* (CLE) [2,13], which provides a public-key encryption mechanism that overcomes the drawbacks of both IBE and PKE. The key generation center in IBE has the power to decrypt every ciphertext, and every public key in PKE needs to be guaranteed by a certificate authority. CLE also can be constructed by combining IBE and PKE (see a detailed survey of CLE [13]). In fact, Chow, Roth, and Rieffel [11] proposed a generic construction that converts any *general certificateless encryption* scheme to a TRPKE scheme.[1]

1.1   Background

Security of Timed-Release Public-Key Encryption. The security of timed-release public-key encryption is considered from two aspects: *security against malicious receivers* and *security against malicious time servers*. The security against malicious receivers means that no receiver can derive any information of the plaintext from a ciphertext without the corresponding time signal. A formal definition, called *indistinguishability against chosen time period and ciphertext attacks* (IND-CTCA security), was presented by Cathalo, Libert, and Quisquater [6]. IND-CTCA security is defined in an indistinguishability game between a challenger and an adversary in which the adversary can send a release query consisting of a time period to the challenger to obtain the corresponding time signal and can send a decryption query consisting of a ciphertext, a time period, and the user public key used in encryption to the challenger to obtain the plaintext encrypted in the ciphertext. Note that the adversary can ask a decryption query even if it does not know any user secret key corresponding to the user public key in the decryption query. Another definition called IND-RTR-CCA2 security was independently given by Cheon, Hopper, Kim, and Osipkov [9], but is weaker than IND-CTCA security.

The security against malicious time servers means that a time server cannot derive any information of plaintext from a ciphertext without the receiver's secret key. A formal definition of security against malicious time servers was also given by Cheon et al. [9], and is called IND-CCA2 security. Another definition of security against time servers was presented by Cathalo et al. [6] and is called IND-CCA security, which is essentially the same as IND-CCA2 security. We call IND-CCA security defined by Cathalo et al. *indistinguishability against chosen ciphertext attacks for time server* (IND-CCA-TS security).

Recently, Chow et al. defined two security models for general CLE: *security against Type-I attackers* and *security against Type-II attackers*. Based on these models, they also discussed security models for TRPKE. Type-I attackers are related to malicious receivers and Type-II attackers are related to malicious time servers in the context of TRPKE. We should note that both Type-I and Type-II

---

[1] While each identity in CLE is linked with the user's public key, every public key in general CLE can be independent from the identity. Therefore, general CLE can be converted to TRPKE since a time period can be assigned to an identity and need not be related to any user public key.

attackers can access a strong decryption oracle, which must answer to a decryption query encrypted with a public key chosen by the attacker. While security against Type-I attackers is essentially equivalent to IND-CTCA security, security against Type-II attackers is stronger than IND-CCA-TS security because the adversary in the IND-CCA-TS game is not allowed to access the strong decryption oracle.Later, Chow and Yiu redefined these security notions in the context of TRPKE [7].

In this paper, we call security against malicious time servers with the strong decryption oracle *indistinguishability against strong chosen ciphertext attacks for time server* (IND-SCCA-TS security), and explicitly describe it in the context of TRPKE. Note that IND-SCCA-TS security is essentially the same as security against TYPE-II attackers.

Table 1. Security Notions

	notion	against	query
Cheon et al. [9]	IND-RTR-CCA2	receiver	(usk, T, c)
	IND-CCA2	time server	(T, c)
Cathalo et al. [6]	IND-CTCA	receiver	(upk, T, c)
	IND-CCA	time server	(T, s_T, c)
Chow et al. [11]	IND-CTCA	receiver	(upk, T, c)
	IND-SCCA-TS	time server	(upk, T, c)

c is a ciphertext, usk is a user's secret key, upk is a user's public key, T is a time period, and s_T is the time signal at T.

Strong Security Models. IND-CTCA and IND-SCCA-TS security models allow adversary to send to challenger decryption queries (upk, T, c) consisting of a ciphertext c, a user's public key upk and a time period T used in generating c, and the challenger to return a plaintext encrypted in c.

IND-CTCA and IND-SCCA-TS security might look too strong since it allows decryption queries (upk, T, c) to vary in a user's public key upk and adversaries to ask (upk, T, c) even without knowing a secret key corresponding to the upk. However, since TRPKE has many applications for constructing secure protocols that involve multiple receivers, it seems natural to allow upks of not only a fixed single receiver but also of distinct multiple receivers to appear in (upk, T, c).

IND-CTCA and IND-SCCA-TS security may not represent a realistic attack scenario. Though, when we consider a real attacker's abilities, we need to confirm that they sufficiently reflect a security model. Considering strong models is not only important in practical sense that it can ensure security beyond realistic attacks but also interesting in theoretical sense whether there exists a scheme that archives the strong security. A similar discussion was done in the context of CLE (see Sec 2.3.1 of [13]).

Note also that in the IND-CTCA and IND-SCCA-TS security models, the behavior of the challenger for decryption queries (upk, T, c) in the case of invalid upk is not defined. Since, in the IND-CTCA secure scheme presented by Cathalo et al. [6], the upk is a group element in a prime order group and is uniformly distributed over the group, all elements are possible and upks are valid.

Thus, this scheme does not require a validity check of upk in (upk, T, c). However, when IND-CTCA or IND-SCCA-TS security is applied to other schemes, the behavior of the challenger for (upk, T, c) including invalid upk, should be defined, or the behavior of the adversary should be restricted. We assume that adversaries are not allowed to ask (upk, T, c) including invalid upks. Although it is possible to avoid the assumption and adopt the underlying PKE scheme that allows to check the validity of public keys, this restriction may lose generality of the construction.

Previous TRPKE Schemes. Chan and Blake [8] proposed the first TRPKE scheme but did not present a formal security definition for TRPKE. Later, Cathalo et al. [6], Chalkias et al. [7], and Hristu-Varsakelis et al. [20] proposed TRPKE schemes and proved IND-CTCA security of their schemes in the random oracle model [4]. It should be noted that they all are concrete TRPKE schemes based on specific number theoretic assumptions and are not generated by a generic construction of TRPKE. Independently, Cheon et al. [10] proposed a generic construction of TRPKE and showed that it produces a TRPKE scheme that is IND-RTR-CCA2 and IND-CCA2 secure in the standard model from an IND-ID-CCA secure IBE, IND-CCA secure PKE, and SUF-CMA secure one-time signature schemes. Furthermore, Chow et al. [11] proposed a concrete TRPKE scheme that satisfies both IND-CTCA and IND-SCCA-TS security in the standard model.[2] Recently, Paterson and Quaglia extended TRPKE to *public-key time-specific encryption*, and presented a generic construction secure in the standard model [24]. However, as well as in IND-CCA-TS security, the security model for time servers does not allow adversaries to access the strong decryption oracle and the security model for receivers is considered only in the chosen-plaintext scenarios (and the security is called IND-CPA-CR security).

To the best of our knowledge, no generic construction of TRPKE has been proposed satisfying the known strongest security against receivers, IND-CTCA security, and the known strongest security against time servers, IND-SCCA-TS security. In other words, a sufficient condition of component primitives for constructing a TRPKE scheme that achieves IND-CTCA and IND-SCCA-TS security has not been found.

1.2   Our Contributions

We propose a generic construction of TRPKE based on an IBE scheme and a PKE scheme and call it *IBE-then-PKE construction* because the encryption algorithm in the resulting TRPKE scheme is constructed as follows:

$$\dot{c} = \mathsf{IBE.Enc}(params, T, m||r; H_{h_1}(pk||T||m||r)),$$
$$c = \mathsf{PKE.Enc}(pk, \dot{c}; H_{h_2}(pk||T||\dot{c})),$$

where IBE.Enc is the encryption algorithm of the IBE scheme, PKE.Enc is the encryption algorithm of the PKE scheme, H_{h_1}, H_{h_2} are hash functions, m is

---

[2] Chow et al. presented a generic construction of TRPKE from general CLE, which is different from CLE. They proposed a *concrete* scheme of general CLE.

a plaintext, r is a random string, *params* is a public parameter, and $\dot{c}$ is an intermediate ciphertext. The final output c is a ciphertext of m in the resulting TRPKE scheme. A designated time period T is input to the encryption algorithm of the IBE scheme, instead of identity.

In our IBE-then-PKE construction, the second encryption (PKE.Enc) is processed in a deterministic manner; namely, the computation in PKE.Enc depends only on the output $\dot{c}$ of IBE.Enc and does not use internal randomness. Fujioka, Okamoto, and Saito used a similar technique, called the *bound randomness construction*, for multiple encryptions [16], which is based on the Fujisaki-Okamoto conversion [17]. However, only the bounded randomness construction might not ensure security in a model allowing the strong decryption oracle since security of multiple encryption is considered only with public keys generated by a challenger. Moreover, to prove IND-CTCA security, we input to the hash function H_{h_i} not only $\dot{c}$ but also *upk* and T. At this point, our IBE-then-PKE construction is different from the bound randomness construction.[3] On the other hand, the first encryption (IBE.Enc) has the form similar to the Fujisaki-Okamoto conversion, and each intermediate ciphertext $\dot{c}$ is generated with *upk* in addition to T. This enables us to simulate the strong decryption oracle in the IND-SCCA-TS game. We prove that the constructed TRPKE scheme satisfies both IND-CTCA and IND-SCCA-TS security in the random oracle model if the underlying IBE scheme is IND-ID-CPA secure and the underlying PKE scheme is IND-CPA secure with a property that is easily satisfied, which is described later.

Our IBE-then-PKE construction is the first generic construction of TRPKE that achieves both the known strongest security against receivers, IND-CTCA security, and the known strongest security against time servers, IND-SCCA-TS security. A sufficient condition for the constructed TRPKE scheme to achieve security is that the underlying IBE scheme is IND-ID-CPA secure and the underlying PKE scheme is IND-CPA secure with a property related to γ-uniformity [17,18] and defined as negligible γ-uniformity satisfied for every *upk*.

Security against the Decrypt-then-Encrypt-again Attack. It may be natural to combine IBE and PKE schemes in the following way to construct a TRPKE scheme:

$$c = \mathsf{PKE.Enc}(upk, \mathsf{IBE.Enc}(params, T, m; r_1); r_2),$$

where a ciphertext of a plaintext m is c, r_1 and r_2 are random strings, and a designated time period is T, which is inputted to IBE.Enc as identity. However, under *adaptively chosen ciphertext attack* environments and against adversaries who have the secret key of PKE, this TRPKE scheme would no longer be secure, even if the underlying IBE and PKE schemes are IND-ID-CCA and IND-CCA secure, respectively [26,15]. Given the challenge ciphertext c^* where $c^* = \mathsf{PKE.Enc}(upk, \mathsf{IBE.Enc}(params, T^*, m_b; r_1^*); r_2^*)$, an adversary can know $\dot{c} = \mathsf{IBE.Enc}(params, T^*, m_b; r_1^*)$ by decrypting c^* with the secret key, and encrypt

---

[3] Similar techniques have appeared in the context of certificateless encryption and deterministic encryption [21,3].

$\dot{c}$ again with another randomness r'_2 to obtain a new ciphertext c' such that $c' = \mathsf{PKE.Enc}(upk, \dot{c}; r'_2)$. Then, the adversary can ask c' to the decryption oracle and obtain a plaintext m_b. This means that the TRPKE scheme based on the above simple construction is insecure in the IND-CTCA model. We call this specific attack *decrypt-then-encrypt-again attack*. This scenario describes a situation where secret key exposure occurs in the chosen-ciphertext attack environment. It is noted [26,15] that the decrypt-then-encrypt-again attack can break naive multiple encryption scheme like the above TRPKE scheme in the scenario.

We outline why our construction produces a TRPKE scheme secure against the decryption-then-encrypt-again attack. Suppose that an adversary has the secret key corresponding to upk^* and is given a challenge ciphertext c^* that has the form $\dot{c}^* = \mathsf{IBE.Enc}(params, T^*, m_b||r^*; H_{h_1}(upk^*||T^*||m_b||r^*))$, $c^* = \mathsf{PKE.Enc}(upk^*, \dot{c}^*; H_{h_2}(upk^*||T^*||\dot{c}^*))$. Note that the adversary in the IND-CTCA game can partially decrypt c^* and obtain $\dot{c}^*$ as it has the secret key.

If the adversary encrypts $\dot{c}^*$ again in a legitimate manner, the generated ciphertext becomes equal to the challenge ciphertext c^* because the second encryption is deterministic and does not depend on randomness. Then the generated ciphertext is not allowed to be inputted into the decryption oracle.

If the adversary encrypts $\dot{c}^*$ with a different upk, the generated ciphertext becomes different from the challenge ciphertext c^*, and may be allowed to ask the decryption oracle. If the final decrypted value is equal to $(m_b||r^*)$, it fails to pass the re-encryption check (Step 7 in TR.Dec) since using a different upk implies using a different r.

On the other hand, if the adversary does not encrypt $\dot{c}^*$ in a legitimate manner (e.g., using some randomness), it obtains another ciphertext c' different from the challenge ciphertext c^* and sends it to the decryption oracle. Then the decryption oracle first decrypts c' to obtain $\dot{c}^*$ and checks whether $c' = \mathsf{PKE.Enc}(upk^*, \dot{c}^*; H_{h_2}(upk^*||T^*||\dot{c}^*))$ holds. However, since $c' \neq c^*$, the check equation does not hold and the decryption oracle returns $\perp$.

Consequently, the decrypt-then-encrypt-again attack cannot be applied to the TRPKE scheme produced from our construction.

γ-pk-Uniformity and Decryption Query. In IND-CTCA and IND-SCCA-TS security, adversaries are allowed to ask (upk, T, c) including any upk and then ask special decryption queries including a special upk for which the component PKE scheme does not has negligible γ-uniformity. In the proof of IND-CTCA and IND-SCCA-TS security, if a decryption query including such upk is asked, the simulation for decryption query fails. For avoiding such a problem, we require that the underlying PKE scheme is negligible γ-uniform for any upk, i.e., it has negligible γ-pk-uniformity.

Comparison. We compare the existing concrete TRPKE schemes [6,7,20,11] and TRPKE schemes based on generic constructions [10,23,22] with the TRPKE scheme that our generic construction produces.

Table 2. Scheme Comparison

	generic or concrete	security notion	security model
CLQ [6]	concrete	IND-CTCA & IND-CCA-TS	ROM
CHS [7]	concrete	IND-CTCA & IND-CCA-TS	ROM
HCS [20]	concrete	IND-CTCA & IND-CCA-TS	ROM
CHKO [10]	generic	IND-RTR-CCA2 & IND-CCA-TS	SM
NMKM [23]	generic	IND-TRPC-CPA & IND-CCA-TS	SM
MNM1 [22]	generic	IND-TRPC-CPA & IND-CCA-TS	SM
MNM2 [22]	generic	IND-TRPC-CPA & IND-CCA-TS	ROM
CRR [11]	concrete	IND-CTCA & IND-SCCA-TS	SM
PQ [24]	generic	IND-CPA-CR & IND-CCA-TS	SM
ours	generic	IND-CTCA & IND-SCCA-TS	ROM

"concrete" indicates that the paper proposes concrete TRPKE scheme(s), and "generic" indicates that the paper proposes generic construction of TRPKE from IBE and PKE schemes. "ROM" indicates that security is proved in the random oracle model, and "SM" indicates that security is proved in the standard model.

We note that Nakai et al. [23] and Matsuda et al. [22] also proposed generic constructions, which produce TRPKE schemes with *pre-open capability*, however these schemes archive weak security (IND-TRPC-CPA security), which treats a case against chosen plaintext attacks, defined by [14]. The pre-open capability enables the sender to release trapdoor information, called *pre-open key*, that allows the intended receiver to decrypt the ciphertext even before the time period.

The resulting scheme from our IBE-then-PKE construction does not have pre-open capability. However, the PKE-then-IBE construction, a variant of our construction, would be expected to have pre-open capability, where the intermediate ciphertext is given as the pre-open key.

Organization. Section 2 defines TRPKE and related security notions. In **Section 3**, we describe a generic construction of TRPKE and then prove the security of the constructed scheme. We discuss security in **Section 4** and conclude in **Section 5**.

2   Definitions

2.1   Components

Public-Key Encryption. Let k be a security parameter. A *public-key encryption scheme* $\mathcal{PKE} = (\mathsf{PKE.KGen}, \mathsf{PKE.Enc}, \mathsf{PKE.Dec})$ consists of three probabilistic polynomial-time algorithms. The key generation algorithm PKE.KGen takes 1^k as input, and outputs a public key pk and a secret key sk. The encryption algorithm PKE.Enc takes a pk, a message $m \in \{0,1\}^*$ and a random string $r \in \{0,1\}^{pke.rlen(k)}$ as inputs, and outputs a ciphertext $c \in \{0,1\}^*$ where $pke.rlen(k)$ is a polynomial in k. The decryption algorithm PKE.Dec takes a sk

and a ciphertext $c' \in \{0,1\}^*$ as inputs, and outputs the plaintext $m' \in \{0,1\}^*$ or $\bot$. These algorithms are assumed to satisfy that if $(pk, sk) = $ PKE.KGen then PKE.Dec$(sk,$ PKE.Enc$(pk, m; r)) = m$ for any m and r.

Identity-Based Encryption. Let k be a security parameter. An *identity-based encryption scheme* $\mathcal{IBE} = ($IBE.Setup, IBE.Extract, IBE.Enc, IBE.Dec$)$ consists of four probabilistic polynomial-time algorithms. The setup algorithm IBE.Setup takes 1^k as input, and outputs a public parameter *params* and a master secret key *msk*. The extract algorithm IBE.Extract takes a *params*, *msk*, and an arbitrary string (identity) ID $\in \{0,1\}^*$ as inputs, and outputs a decryption key d_{ID}. The encryption algorithm IBE.Enc takes a *params*, ID a message $m \in \{0,1\}^*$ and a random string $r \in \{0,1\}^{ibe.rlen(k)}$ as inputs, and outputs a ciphertext $c \in \{0,1\}^*$ where $ibe.rlen(k)$ is a polynomial in k. The decryption algorithm IBE.Dec takes as inputs *params*, a ciphertext $c' \in \{0,1\}^*$ and a decryption key d_{ID} and outputs the plaintext $m' \in \{0,1\}^*$ or $\bot$. These algorithms are assumed to satisfy that if $(params, msk) = $ IBE.Setup(1^k) and $d_{ID} = $ IBE.Extract$(params, msk,$ ID$)$, then IBE.Dec$(params, d_{ID},$ IBE.Enc$(params,$ ID$, m; r)) = m$ for any m and r.

IND-CPA Security. We describe the IND-CPA security [19] for PKE scheme $\mathcal{PKE}$ based on the following IND-CPA game between a challenger and adversary A: At beginning of the game, the challenger takes a security parameter k, runs the key generation algorithm $(pk, sk) = $ PKE.KGen(1^k), and gives adversary A the public key pk. Adversary A gives the challenger two messages x_0, x_1, such that $|x_0| = |x_1|$. The challenger randomly chooses $r \in \{0,1\}^{pke.rlen(k)}$ and $b \in \{0,1\}$ and gives adversary A a challenge ciphertext $c^* = $ PKE.Enc$(pk, x_b; r)$. Adversary A finally outputs a guess $b' \in \{0,1\}$ and wins the game if $b = b'$.

We define the advantage of A in the IND-CPA game as $Adv_{\mathcal{PKE},A}^{ind\text{-}cpa}(1^k) = |2\Pr[b = b'] - 1|$, where the probability is taken over the random coins used by the challenger and A. We say that the PKE scheme $\mathcal{PKE}$ is IND-CPA *secure* if, for any probabilistic polynomial-time adversary A, the function $Adv_{\mathcal{PKE},A}^{ind\text{-}cpa}(1^k)$ is negligible in k.

IND-ID-CPA Security. We describe the IND-ID-CPA security [5] for IBE scheme $\mathcal{IBE}$ based on the following IND-ID-CPA game between a challenger and adversary A: At beginning of the game, the challenger takes a security parameter k, runs the setup algorithm $(params, msk) = $ IBE.Setup(1^k), and gives adversary A the public parameter *params*. The adversary gives the challenger two messages x_0, x_1 such that $|x_0| = |x_1|$, an identity to be challenged ID*. Then the challenger randomly chooses $r \in \{0,1\}^{ibe.rlen(k)}$ and $b \in \{0,1\}$ and gives adversary A a challenge ciphertext $c^* = $ IBE.Enc$(params,$ ID$^*, x_b; r)$. Adversary A finally outputs a guess $b' \in \{0,1\}$ and wins the game if $b = b'$.

During the game, adversary A can issue extraction queries ID$_i$ to the challenger to obtain the decryption key IBE.Extract$(params, msk,$ ID$_i)$. The extraction queries ID$_i$ must differ from the challenged identity ID*.

We define the advantage of A in the IND-ID-CPA game as $Adv_{\mathcal{IBE},A}^{ind\text{-}id\text{-}cpa}(1^k) = |2\Pr[b = b'] - 1|$, where the probability is taken over the random coins used by the challenger and A. We say that the IBE scheme $\mathcal{IBE}$ is IND-ID-CPA *secure if*, for any probabilistic polynomial-time adversary A, the function $Adv_{\mathcal{IBE},A}^{ind\text{-}id\text{-}cpa}(1^k)$ is negligible in k.

γ-Uniformity. Fujisaki and Okamoto proposed γ-uniformity for evaluating the probability that a fixed ciphertext is generated in probabilistic encryption of a fixed message [17,18].

Let k be a security parameter and $\mathcal{PKE} = (\mathsf{PKE.KGen}, \mathsf{PKE.Enc}, \mathsf{PKE.Dec})$ be a PKE scheme. For given $x \in \{0,1\}^*$, $y \in \{0,1\}^*$ and pk generated by $\mathsf{PKE.KGen}(1^k)$, let $g(x,y)$ denote $\Pr[r \leftarrow \{0,1\}^{pke.rlen(k)} : y = \mathsf{PKE.Enc}(pk, x; r)]$. We say $\mathcal{PKE}$ is γ-*uniform* if $g(x,y) \leq \gamma$ for any $x \in \{0,1\}^*$, $y \in \{0,1\}^*$ and pk generated by $\mathsf{PKE.KGen}(1^k)$.

We should note that the probability in this γ-uniformity is also taken over the coin flips in the key generation, and we define the following stronger property, γ-*pk-uniformity*. The definition of γ-pk-uniformity is that the γ-uniformity holds for any upk. Although an IND-CPA secure encryption does not generally have γ-pk-uniformity where γ is negligible in k, any IND-CPA secure encryption can be converted to one that has negligible γ-pk-uniformity [18].

Yang et al. extended γ-uniformity to IBE as follows [25]. Let k be the security parameter and $\mathcal{IBE} = (\mathsf{IBE.Setup}, \mathsf{IBE.Extract}, \mathsf{IBE.Enc}, \mathsf{IBE.Dec})$ be an IBE scheme. For given $x \in \{0,1\}^*$, $y \in \{0,1\}^*$, $\mathsf{ID} \in \{0,1\}^*$ and $params$ generated by $\mathsf{IBE.Setup}(1^k)$, let $g'(x,y,\mathsf{ID})$ denote $\Pr[r \leftarrow \{0,1\}^{ibe.rlen(k)} : y = \mathsf{IBE.Enc}(params, \mathsf{ID}, x; r)]$. We say $\mathcal{IBE}$ is γ-*uniform* if $g'(x,y,\mathsf{ID}) \leq \gamma$ for any $x \in \{0,1\}^*$, $y \in \{0,1\}^*$, $\mathsf{ID} \in \{0,1\}^*$ and $params$ generated by $\mathsf{IBE.Setup}(1^k)$. Note that any IND-ID-CPA secure encryption has negligible γ-uniformity in k.

Hash Function. A hash function H_h is a polynomial-time algorithm that is parameterized by an index $h \in \{0,1\}^{ilen(k)}$, and on arbitrary long input m outputs $hlen(k)$ bit string $H_h(m)$, where $ilen(k)$, $hlen(k)$ are polynomials in k.

2.2 Timed-Release Public-Key Encryption

The TRPKE is formally defined as follows: Let k be a security parameter. A *timed-release public-key encryption scheme* $\mathcal{TRPKE} = (\mathsf{TR.Setup}, \mathsf{TR.KGen}, \mathsf{TR.Release}, \mathsf{TR.Enc}, \mathsf{TR.Dec})$ consists of five probabilistic polynomial-time algorithms. The setup algorithm $\mathsf{TR.Setup}$ takes 1^k as input, and outputs a time server's public key tpk and the corresponding secret key tsk. The user key generation algorithm $\mathsf{TR.KGen}$ takes a tpk as input, and outputs a user's public key upk and the corresponding secret key usk. The release algorithm $\mathsf{TR.Release}$ takes a tpk, tsk, and a time period $T \in \{0,1\}^*$ as inputs, and outputs a time signal s_T. The encryption algorithm $\mathsf{TR.Enc}$ takes a tpk, T, a receiver's upk, a message $m \in \{0,1\}^*$ and a random string $r \in \{0,1\}^{tr.rlen(k)}$ as inputs, and outputs a ciphertext $c \in \{0,1\}^*$, where $tr.rlen(k)$ is a polynomial in k that gives the upper bound of the length of an r used in $\mathsf{TR.Enc}$. The decryption algorithm $\mathsf{TR.Dec}$ takes a tpk, s_T, a receiver's secret key usk and a ciphertext $c' \in \{0,1\}^*$ as inputs, and outputs the plaintext $m' \in \{0,1\}^*$ or $\perp$. These algorithms are assumed to satisfy $\mathsf{TR.Dec}(tpk, s_T, usk, \mathsf{TR.Enc}(tpk, T, upk, m; r)) = m$ for any m and r if $(tpk, tsk) = \mathsf{TR.Setup}(1^k)$, $(upk, usk) = \mathsf{TR.KGen}(tpk)$, and $s_T = \mathsf{TR.Release}(tpk, tsk, T)$ hold.

In a TRPKE system, the time server does not need to bilaterally interact with users. At the start of the system, the time server generates a tpk with $\mathsf{TR.Setup}$ and publishes it. At each time after the start, all the time server has to do is to periodically generate a s_T with $\mathsf{TR.Release}$ and broadcast it. After the start, any sender need not interact with the server in encryption, and any receiver need only receive the s_T but not have any other interaction with the time server in decryption.

2.3 IND-CTCA Security

We review the known strongest security against malicious receivers, called *indistinguishability against chosen time period and ciphertext attacks* (IND-CTCA security) [6].

We formally describe the IND-CTCA security for TRPKE scheme $\mathcal{TRPKE}$ based on the following IND-CTCA game between a challenger and adversary A: The challenger takes a security parameter k, runs the setup algorithm $(tpk, tsk) = \mathsf{TR.Setup}(1^k)$, and gives adversary A the server's public key tpk. Adversary A gives the challenger two messages x_0, x_1 such that $|x_0| = |x_1|$, a time period T^*, a user's public key upk^*. The challenger randomly chooses $r \in \{0, 1\}^{tr.rlen(k)}$ and $b \in \{0, 1\}$, computes a challenge ciphertext $c^* = \mathsf{TR.Enc}(tpk, T^*, upk^*, x_b; r)$, and gives it to adversary A. Adversary A finally outputs a guess $b' \in \{0, 1\}$ and wins the game if $b = b'$. During the game, adversary A can issues release queries T_i to obtain the time signal $\mathsf{TR.Release}(tpk, tsk, T_i)$, and also can issue decryption queries (upk_j, T_j, c_j) to obtain the decrypted message $\mathsf{TR.Dec}(tpk, s_{T_j}, usk_j, c_j)$ where $s_{T_j} = \mathsf{TR.Release}(tpk, tsk, T_j)$ and usk_j is the secret key corresponding to upk_j. The release queries T_i must differ from the challenged time period T^*, and the decryption queries (upk_j, T_j, c_j) must differ from the tuple (upk^*, T^*, c^*) consisting of the challenged user's public key upk^*, the challenged time period T^* and the challenge ciphertext c^*. We define the advantage of A in the IND-CTCA game as $Adv_{\mathcal{TRPKE},A}^{ind\text{-}ctca}(1^k) = |2\Pr[b = b'] - 1|$, where the probability is taken over the random coins used by the challenger and A. We say that the TRPKE scheme $\mathcal{TRPKE}$ is IND-CTCA *secure* if, for any probabilistic polynomial-time adversary A, the function $Adv_{\mathcal{TRPKE},A}^{ind\text{-}ctca}(1^k)$ is negligible in k.

2.4 IND-SCCA-TS Security

We define the security against malicious time servers, *indistinguishability against strong chosen ciphertext attacks for time server* (IND-SCCA-TS security).

We formally describe the IND-SCCA-TS security for TRPKE scheme $\mathcal{TRPKE}$ based on the following IND-SCCA-TS game between a challenger and adversary A: The challenger takes a security parameter k, runs the setup algorithm $(tpk, tsk) = \mathsf{TR.Setup}(1^k)$ and the user key generation algorithm $(upk^*, usk^*) = \mathsf{TR.KGen}(tpk)$. It gives the user's public key upk^* and the server's public and secret keys (tpk, tsk) to adversary A. Adversary A finally gives the challenger two messages x_0, x_1 such that $|x_0| = |x_1|$, and a time period T^*. The challenger randomly chooses $r \in \{0,1\}^{tr.rlen(k)}$ and $b \in \{0,1\}$, computes a challenge cipher-text $c^* = \mathsf{TR.Enc}(tpk, T^*, upk^*, x_b; r)$, and gives it to adversary A. Adversary A finally outputs a guess $b' \in \{0,1\}$ and wins the game if $b = b'$. During the game, adversary A can issue decryption queries (upk_j, T_j, c_j) to obtain the decrypted message $\mathsf{TR.Dec}(tpk, s_{T_j}, usk_j, c_j)$ where $s_{T_j} = \mathsf{TR.Release}(tpk, tsk, T_j)$ and usk_j is the secret key corresponding to upk_j. The decryption queries (upk_j, T_j, c_j) must differ from (upk^*, T^*, c^*) consisting of the user's public key upk^* given at the beginning of game, the challenged time period T^* and the challenge ciphertext c^*. We define the advantage of A in the IND-SCCA-TS game as $Adv_{TRPKE,A}^{ind\text{-}scca\text{-}ts}(1^k) = |2\Pr[b = b'] - 1|$, where the probability is taken over the random coins used by the challenger and A. We say that the TRPKE scheme $\mathcal{TRPKE}$ is IND-SCCA-TS *secure* if, for any probabilistic polynomial-time adversary A, the function $Adv_{TRPKE,A}^{ind\text{-}cca\text{-}ts}(1^k)$ is negligible in k.

3   Construction of Timed-Release Public-Key Encryption

3.1   IBE-then-PKE Construction

Let $\mathcal{PKE} = (\mathsf{PKE.KGen}, \mathsf{PKE.Enc}, \mathsf{PKE.Dec})$ be a public-key encryption scheme, and $\mathcal{IBE} = (\mathsf{IBE.Setup}, \mathsf{IBE.Extract}, \mathsf{IBE.Enc}, \mathsf{IBE.Dec})$ be an identity-based encryption scheme. Let $hlen_1(k)$, $hlen_2(k)$, $ibe.rlen(k)$, $pke.rlen(k)$ and $tr.rlen(k)$ are polynomials in k. Let H_{h_1} be a hash function from $\{0,1\}^*$ to $\{0,1\}^{ibe.rlen(k)}$, and H_{h_2} be also a hash function from $\{0,1\}^*$ to $\{0,1\}^{pke.rlen(k)}$, such that $hlen_1(k) = ibe.rlen(k)$ and $hlen_2(k) = pke.rlen(k)$.

We assume that the PKE scheme, the IBE scheme and the hash function can handle inputs of arbitrary length. The assumptions for the PKE and IBE schemes are reasonable because most practical schemes can be transformed into the ones allowing arbitrary length input, by using the Fujisaki-Okamoto integration [18,25]. We also assume that k can be derived from pk or $params$ as well as in most IBE and PKE schemes.

The proposed timed-release public-key encryption scheme $\mathcal{TRPKE} = (\mathsf{TR.Setup}, \mathsf{TR.KGen}, \mathsf{TR.Release}, \mathsf{TR.Enc}, \mathsf{TR.Dec})$ is constructed as follows.

Setup TR.Setup: The input is 1^k (k is a security parameter).
  Step 1: Run IBE.Setup on input 1^k to generate $(params, msk)$.
  Step 2: Randomly choose indices $h_1 \in \{0,1\}^{ilen_1(k)}$ and $h_2 \in \{0,1\}^{ilen_2(k)}$.
  Step 3: Set $tpk = (params, h_1, h_2)$ and $tsk = msk$. Step 4: Return (tpk, tsk).
The output tpk is a time server's public key of $\mathcal{TRPKE}$, and the output tsk is the corresponding time server's secret key.

User Key Generation TR.KGen: The input is tpk.
  Step 1: Parse tpk as $(params.h_1, h_2)$ and derive the security parameter k from $params$.
  Step 2: Run PKE.KGen on input 1^k to generate (pk, sk).
  Step 3: Set $upk = pk$ and $usk = (pk, sk)$.
  Step 4: Return (upk, usk).

The output upk is a user's public key of $\mathcal{TRPKE}$, and the output usk is the corresponding user's secret key.

Release TR.Release: The inputs are time server's public and secret keys (tpk, tsk) and a time period T.
  Step 1: Parse tpk as $(params, h_1, h_2)$.
  Step 2: Run IBE.Extract$(params, tsk, T)$ to obtain d_T.
  Step 3: Set $s_T = d_T$ and return s_T.
The output s_T is a time signal at the time period T.

Encryption TR.Enc: The inputs are a time server's public key tpk, a time period T, a user's public key $upk(=pk)$, a message m, and a random string $r \in \{0,1\}^{tr.rlen(k)}$.
  Step 1: Parse tpk as $(params, h_1, h_2)$.
  Step 2: Compute $\dot{c} = $ IBE.Enc$(params, T, m||r; H_{h_1}(pk||T||m||r))$.
  Step 3: Compute $\ddot{c} = $ PKE.Enc$(pk, \dot{c}; H_{h_2}(pk||T||\dot{c}))$.
  Step 4: Set $c = \ddot{c}$ and return c.
The output c is a ciphertext of m.

Decryption TR.Dec: The inputs are a time server's public key tpk, a time signal s_T, a user's secret key usk and a ciphertext c'.
  Step 1: Parse tpk as $(params, h_1, h_2)$.
  Step 2: Parse usk as (pk, sk).
  Step 3: Set $\ddot{c}' = c'$.
  Step 4: Compute $\dot{c}' = $ PKE.Dec$(sk, \ddot{c}')$.
  Step 5: Check whether $\ddot{c}' = $ PKE.Enc$(pk, \dot{c}'; H_{h_2}(pk||T||\dot{c}'))$ holds,
  and if it does not hold, return $\perp$ and stop.
  Step 6: Compute $m' = $ IBE.Dec$(params, s_T, \dot{c}')$.
  Step 7: Check whether $\dot{c}' = $ IBE.Enc$(params, T, m'; H_{h_1}(pk||T||m'))$ holds,
  and if it does not hold, return $\perp$ and stop.
  Step 8: Parse m' as $m''||r''$.
  Step 9: Return m''.
The output m'' is a decrypted message.

This completes the description of the proposed construction for TRPKE.

3.2   IND-CTCA Security

The following theorem holds for the security against malicious receivers of the constructed TRPKE scheme $\mathcal{TRPKE}$.

Theorem 1. *Suppose that* $\mathcal{IBE}$ *is an* IND-ID-CPA *secure IBE scheme and that* $\mathcal{PKE}$ *is an* IND-CPA *secure PKE scheme with* γ_{pke}-*pk-uniformity where* γ_{pke} *is negligible in the security parameter* k. *Then, the constructed* $\mathcal{TRPKE}$ *is* IND-CTCA *secure in the random oracle model.*

Let $\mathcal{IBE}$ have γ_{ibe}-uniformity. We note that γ_{ibe} is negligible in k since $\mathcal{IBE}$ is an IND-ID-CPA secure IBE scheme.

To prove that the $\mathcal{TRPKE}$ is IND-CTCA secure, we need to show that the advantage $Adv_{\mathcal{TRPKE},A}^{ind\text{-}ctca}(1^k)$ is negligible. We assume for contradiction that there is a polynomial-time IND-CTCA adversary, A, for $\mathcal{TRPKE}$ with non-negligible advantage. Then we show that with adversary A, a polynomial-time IND-ID-CPA adversary, B, for $\mathcal{IBE}$ with non-negligible advantage can be constructed. The existence of the constructed adversary for $\mathcal{IBE}$ contradicts the assumption, and thus **Theorem 1** is proved. We will give the detailed proof in the full paper.

Sketch of Proof.
We construct an IND-ID-CPA adversary, B, which uses A as a blackbox and also simulates the challenger for adversary A in the IND-CTCA game and the random oracles H_{h_i} as follows.

Setup: The challenger in the IND-ID-CPA game for $\mathcal{IBE}$ runs IBE.Setup(1^k), obtains ($params, msk$), and inputs $params$ to B. Then B randomly chooses two indices of hash functions h_1, h_2, and gives a time server's public key tpk ($=$ ($params, h_1, h_2$)) to adversary A in the IND-CTCA game for $\mathcal{TRPKE}$.

Hash queries: B responds to hash queries while maintaining two query-answer lists $\mathcal{T}_{h_1}$ and $\mathcal{T}_{h_2}$ in the usual manner, except for the case that an H_{h_1} hash query is $(upk^*||T^*||x_b||r_b)$. If B receives such an H_{h_1} query $(upk^*||T^*||x_b||r_b)$ for some $b \in \{0,1\}$, B stops A and, without going to **Guess**, outputs the corresponding b as an answer in the IND-ID-CPA game.

Release queries: For a release query, B uses the outer extraction oracle in the IND-ID-CPA game.

Decryption queries: For a decryption query (upk_j, T_j, c_j), B searches the list $\mathcal{T}_{h_2}$ to find an entry $(upk_j||T_j||\sigma, \tau) \in \mathcal{T}_{h_2}$ satisfying $c_j = $ PKE.Enc($upk_j, \sigma; \tau$). If such an entry is found, B searches the list $\mathcal{T}_{h_1}$ to find an entry $(upk_j||T_j||m||r, \mu) \in \mathcal{T}_{h_1}$ satisfying $\sigma = $ IBE.Enc($params, T_j, m||r; \mu$). If such an entry is found, B answers m to A. Otherwise B returns $\perp$.

Challenge: For two messages (x_0, x_1), a time period T^* and a user's public key upk^* to be challenged, B randomly chooses r_0, r_1 and passes two messages $(x_0||r_0, x_1||r_1)$ and the identity T^* to the challenger for the IND-ID-CPA game. The challenger randomly chooses $b \in \{0,1\}$ and $r^* \in \{0,1\}^{ibe.rlen(k)}$, creates a challenge ciphertext $\dot{c}^* = $ IBE.Enc($params, T^*, x_b||r_b; r^*$) and returns $\dot{c}^*$ to B. B creates $\ddot{c}^* = $ PKE.Enc($upk^*, \dot{c}^*; H_{h_2}(upk^*||T^*||\dot{c}^*)$) and returns c^* ($= \ddot{c}^*$) to A as a challenge ciphertext of $\mathcal{TRPKE}$.

Guess: A outputs b' as an answer to the challenge in the IND-CTCA game. Then, B receives b' and outputs it as B's answer to the challenge in the IND-ID-CPA game.

We note that, while B does not know only the correct hash value r^* for $(upk^*||T^*||x_b||r_b)$ (i.e., $r^* = H_{h_1}(upk^*||T^*||x_b||r_b)$) in the generation of challenge ciphertext, B can independently choose hash values for all hash queries except for the H_{h_1} hash query $(upk^*||T^*||x_b||r_b)$.

We observe the simulation for the hash queries in the IND-CTCA game. If A makes an H_{h_1} query $(upk^*||T^*||x_b||r_b)$ coincident with $(upk^*||T^*||x_0||r_0)$ or $(upk^*||T^*||x_1||r_1)$, B stops the simulation and judges the corresponding b to be equal the random coin flipped by the challenger. Intuitively speaking, this judgment is correct with high probability because, if the challenger chooses b, A does not see $x'_{\bar{b}} = $ IBE.Enc$(params, T^*, x_{\bar{b}}||r_{\bar{b}}; H_{h_1}(upk^*||T^*||x_{\bar{b}}||r_{\bar{b}}))$, then A's view is independent of the random $r_{\bar{b}}$ for generating $x'_{\bar{b}}$. Thus the probability that A guesses $r_{\bar{b}} \in \{0,1\}^{tr.rlen(k)}$ and asks $(upk^*||T^*||x_{\bar{b}}||r_{\bar{b}})$ as an H_{h_1} query is not over $2^{-tr.rlen(k)}$.

We observe the simulation for the decryption queries in the IND-CTCA game. We recall that in the IND-CTCA game, the decryption query (upk_j, T_j, c_j) must differ from (upk^*, T^*, c^*).

In the case that there exist entries $(upk_j||T_j||\sigma, \tau) \in \mathcal{T}_{h_2}$, $(upk_j||T_j||m||r, \mu) \in \mathcal{T}_{h_1}$ such that $c_j = $ PKE.Enc$(upk_j, \sigma; \tau)$ and $\sigma = $ IBE.Enc$(params, T_j, m||r; \mu)$ for a decryption query (upk_j, T_j, c_j), the check equations in Step 5 and Step 7 of TR.Dec hold. Thus B's answer m is correct and the simulation is perfect.

In the case that there exists no entry $(upk_j||T_j||\sigma, \tau) \in \mathcal{T}_{h_2}$ such that $c_j = $ PKE.Enc$(upk_j, \sigma; \tau)$ for a decryption query (upk_j, T_j, c_j), B is designed to answer $\bot$. Since $\mathcal{PKE}$ has γ_{pke}-pk-uniformity and is γ_{pke}-uniform for upk_j, the probability that this simulation is incorrect is not over γ_{pke}.

In the case that there exist an entry $(upk_j||T_j||\sigma, \tau) \in \mathcal{T}_{h_2}$ such that $c_j = $ PKE.Enc$(upk_j, \sigma; \tau)$, but no entry $(upk_j||T_j||m||r, \mu) \in \mathcal{T}_{h_1}$ such that $\sigma = $ IBE.Enc$(params, T_j, m||r; \mu)$ for a decryption query (upk_j, T_j, c_j), B is designed to answer $\bot$. The probability that this simulation is incorrect is not over γ_{ibe} where the IBE scheme has γ_{ibe}-uniformity.

Consequently, since γ_{pke} and γ_{ibe} are negligible, the whole simulation succeeds with overwhelming probability.

Note that the proof uses the assumption that the PKE scheme has γ_{pke}-pk-uniformity where γ_{pke} is negligible but not the IND-CPA security of the PKE scheme.

3.3   IND-SCCA-TS Security

The following theorem holds for the security against malicious time servers of the constructed TRPKE scheme $\mathcal{TRPKE}$.

Theorem 2. *Suppose that $\mathcal{IBE}$ is an* IND-ID-CPA *secure IBE scheme and that $\mathcal{PKE}$ is an* IND-CPA *secure PKE scheme with γ_{pke}-pk-uniformity where γ_{pke} is negligible in the security parameter k. Then, the constructed $\mathcal{TRPKE}$ is* IND-SCCA-TS *secure in the random oracle model.*

Let $\mathcal{IBE}$ have γ_{ibe}-uniformity. We note that γ_{ibe} is negligible in k since $\mathcal{IBE}$ is an IND-ID-CPA secure IBE scheme.

To prove that this $\mathcal{TRPKE}$ is IND-SCCA-TS secure, we need to show that the advantage $Adv_{\mathcal{TRPKE}, A}^{ind\text{-}scca\text{-}ts}(1^k)$ is negligible. We assume for contradiction that there is a polynomial-time IND-SCCA-TS adversary, A, for $\mathcal{TRPKE}$ with non-negligible advantage. Then we show that with adversary A, a polynomial-time

IND-CPA adversary, C, for $\mathcal{PKE}$ with non-negligible advantage can be constructed. The existence of the constructed adversary for $\mathcal{PKE}$ contradicts the assumption, an thus **Theorem 2** is proved. We will give the detailed proof in the full paper.

Sketch of Proof.
We construct an IND-CPA adversary, C, which uses A as a blackbox and also simulates the challenger for adversary A in the IND-SCCA-TS game and the random oracles H_{h_i} as follows.

Setup: The challenger in the IND-CPA game for $\mathcal{PKE}$ runs PKE.KGen(1^k) to obtain (pk, sk), and inputs pk to C. Then C runs IBE.Setup(1^k) to obtain $(params, msk)$, randomly chooses two indices of hash functions h_1, h_2, and gives a time server's public key tpk ($= (params, h_1, h_2)$), a time server's secret key $tsk(= msk)$ and a user's public key $upk^*(= pk)$ to adversary A in the IND-SCCA-TS game for $\mathcal{TRPKE}$.

Hash Queries: C responds to hash queries while maintaining two query-answer lists $\mathcal{T}_{h_1}$ and $\mathcal{T}_{h_2}$ in the usual manner, except for the case that an H_{h_2} hash query is $(upk^*||T^*||x'_b)$. If C receives such an H_{h_2} query $(upk^*||T^*||x'_b)$ for some $b \in \{0,1\}$, C stops A and, without going to **Guess**, outputs the corresponding b as an answer in the IND-CPA game.

Decryption Queries: For a decryption query (upk_j, T_j, c_j), C searches the list $\mathcal{T}_{h_2}$ to find an entry $(upk_j||T_j||\sigma, \tau) \in \mathcal{T}_{h_2}$ satisfying $c_j = $ PKE.Enc($upk_j, \sigma; \tau$). If such an entry is found, C searches the list $\mathcal{T}_{h_1}$ to find an entry $(upk_j||T_j||m||r, \mu) \in \mathcal{T}_{h_1}$ satisfying $\sigma = $ IBE.Enc($params, T_j, m||r; \mu$). If such an entry is found, C answers m to A. Otherwise C returns $\perp$.

Challenge: For two messages (x_0, x_1) and a time period T^* to be challenged in the IND-SCCA-TS game, C randomly chooses r_0 and r_1 to compute $x'_b = $ IBE.Enc($params, T^*, x_b||r_b; H_{h_1}(upk^*||T^*||x_b||r_b)$) ($b \in \{0,1\}$), and passes the two messages (x'_0, x'_1) to the challenger for the IND-CPA game for $\mathcal{PKE}$. The challenger randomly chooses $b \in \{0,1\}$ and $r^* \in \{0,1\}^{pke.rlen(k)}$, creates a challenge ciphertext $\ddot{c}^* = $ PKE.Enc($upk^*, x'_b; r^*$) and returns $\ddot{c}^*$ to C. C returns c^* ($= \ddot{c}^*$) to A as a challenge ciphertext of $\mathcal{TRPKE}$.

Guess: A outputs b' as an answer to the challenge in the IND-SCCA-TS game. Then, C receives b' and outputs it as C's answer to the challenge in the IND-CPA game.

We note that, while C does not know only the correct hash value r^* for $(upk^*||T^*||x'_b)$ (i.e., $r^* = H_{h_2}(upk^*||T^*||x'_b)$) in the generation of challenge ciphertext, C can independently choose hash values for all hash queries except for the H_{h_2} hash query $(upk^*||T^*||x'_b)$.

We observe the simulation for the hash queries in the IND-SCCA-TS game. If A makes an H_{h_2} query $(upk^*||T^*||x'_b)$ coincident with either $(upk^*||T^*||x'_0)$ or $(upk^*||T^*||x'_1)$, C stops the simulation and judges the corresponding b to be equal the random coin flipped by the challenger. Intuitively speaking, this judgment is correct with high probability because, if the challenger chooses b, A does not see $x'_{\bar{b}} = \mathsf{IBE.Enc}(params, T^*, x_{\bar{b}}||r_{\bar{b}}; H_{h_1}(upk^*||T^*||x_{\bar{b}}||r_{\bar{b}}))$, then A's view is independent of the random $r_{\bar{b}}$ for generating $x'_{\bar{b}}$. Thus, when we let $S = \{y | y = \mathsf{IBE.Enc}(params, T^*, x_{\bar{b}}||r; H_{h_1}(upk^*||T^*||x_{\bar{b}}||r)), r \in \{0,1\}^{tr.rlen(k)}\}$, the probability that A guesses $x'_{\bar{b}} \in S$ and asks $(upk^*||T^*||x'_{\bar{b}})$ as an H_{h_2} query is not over $2^{-tr.rlen(k)}$.

We observe the simulation for the decryption queries in the IND-SCCA-TS game. We recall that the decryption query (upk_j, T_j, c_j) must differ from (upk^*, T^*, c^*) in the IND-SCCA-TS game.

In the case that there exist entries $(upk_j||T_j||\sigma, \tau) \in \mathcal{T}_{h_2}$, $(upk_j||T_j||m||r, \mu) \in \mathcal{T}_{h_1}$ such that $c_j = \mathsf{PKE.Enc}(upk_j, \sigma; \tau)$ and $\sigma = \mathsf{IBE.Enc}(params, T_j, m||r; \mu)$ for a decryption query (upk_j, T_j, c_j), the check equations in Step 5 and Step 7 of TR.Dec hold. Thus C's answer m is correct and the simulation is perfect.

In the case that there exists no entry $(upk_j||T_j||\sigma, \tau) \in \mathcal{T}_{h_2}$ such that $c_j = \mathsf{PKE.Enc}(upk_j, \sigma; \tau)$ for a decryption query (upk_j, T_j, c_j), C is designed to answer $\perp$. Since $\mathcal{PKE}$ has γ_{pke}-pk-uniformity and is γ_{pke}-uniform for upk_j, the probability that this simulation is incorrect is not over γ_{pke}.

In the case that there exist an entry $(upk_j||T_j||\sigma, \tau) \in \mathcal{T}_{h_2}$ such that $c_j = \mathsf{PKE.Enc}(upk_j, \sigma; \tau)$, but no entry $(upk_j||T_j||m||r, \mu) \in \mathcal{T}_{h_1}$ such that $\sigma = \mathsf{IBE.Enc}(params, T_j, m||r; \mu)$ for a decryption query (upk_j, T_j, c_j), C is designed to answer $\perp$. The probability that this simulation is incorrect is not over γ_{ibe} where the IBE scheme has γ_{ibe}-uniformity.

Consequently, since γ_{pke} and γ_{ibe} are negligible, the whole simulation succeeds with overwhelming probability.

Note that the proof uses the assumption that the IBE scheme has γ_{ibe}-uniformity where γ_{ibe} is negligible and this comes from the IND-ID-CPA security of the IBE scheme.

4   Conclusion

We provided a generic construction of TRPKE, the IBE-then-PKE construction, based on IBE and PKE schemes. The resulting TRPKE scheme satisfies both IND-CTCA and IND-SCCA-TS security in the random oracle model if the component IBE scheme is IND-ID-CPA secure and the component PKE scheme is IND-CPA secure with negligible γ-pk-uniformity. The proposed construction is the first generic one for TRPKE that guarantees strong security, i.e., both IND-CTCA and IND-SCCA-TS security. We conclude by listing several open problems for further research:

- Can we have a provably secure construction in the standard model?
- Is it possible to remove the restriction on invalid public keys?[4]

---

[4] *Robustness encryption* [1] may solve this problem.

Acknowledgments. We would like to thank Fumitaka Hoshino for his comment on the re-encryption attack with a replaced public key and Ryo Kikuchi for his comment on invalid public keys in preliminary version of this paper. We also would like to thank anonymous referees for their comments that helped us to improve this paper.

References

1. Abdalla, M., Bellare, M., Neven, G.: Robust Encryption. In: Micciancio, D. (ed.) TCC 2010. LNCS, vol. 5978, pp. 480–497. Springer, Heidelberg (2010)
2. Al-Riyami, S.S., Paterson, K.G.: Certificateless Public Key Cryptography. In: Laih, C.-S. (ed.) ASIACRYPT 2003. LNCS, vol. 2894, pp. 452–473. Springer, Heidelberg (2003)
3. Bellare, M., Boldyreva, A., O'Neill, A.: Deterministic and Efficiently Searchable Encryption. In: Menezes, A. (ed.) CRYPTO 2007. LNCS, vol. 4622, pp. 535–552. Springer, Heidelberg (2007)
4. Bellare, M., Rogaway, P.: Random Oracles are Practical: A Paradigm for Designing Efficient Protocols. In: 1st ACM Conference on Computer and Communications Security, pp. 62–73. ACM, New York (1993)
5. Boneh, D., Franklin, M.: Identity-Based Encryption from the Weil Pairing. In: Bellare, M. (ed.) CRYPTO 2001. LNCS, vol. 2139, pp. 213–229. Springer, Heidelberg (2001)
6. Cathalo, J., Libert, B., Quisquater, J.-J.: Efficient and Non-Interactive Timed-Release Encryption. In: Qing, S., Mao, W., López, J., Wang, G. (eds.) ICICS 2005. LNCS, vol. 3783, pp. 291–303. Springer, Heidelberg (2005)
7. Chalkias, K., Hristu-Varsakelis, D., Stephanides, G.: Improved Anonymous Timed-Release Encryption. In: Biskup, J., López, J. (eds.) ESORICS 2007. LNCS, vol. 4734, pp. 311–326. Springer, Heidelberg (2007)
8. Chan, A.C.-F., Blake, I.F.: Scalable, Server-Passive, User-Anonymous Timed Release Public Key Encryption from Bilinear Pairing. In: 25th International Conference on Distributed Computing Systems, pp. 504–513. IEEE, Los Alamitos (2005), full version of this paper, http://eprint.iacr.org/2004/211
9. Cheon, J.H., Hopper, N., Kim, Y., Osipkov, I.: Timed-Release and Key-Insulated Public Key Encryption. In: Di Crescenzo, G., Rubin, A. (eds.) FC 2006. LNCS, vol. 4107, pp. 191–205. Springer, Heidelberg (2006), full version of this paper http://eprint.iacr.org/2004/231
10. Cheon, J.H., Hopper, N., Kim, Y., Osipkov, I.: Provably Secure Timed-Release Public Key Encryption. ACM Trans. Inf. Syst. Secur. 11(2), 1–44 (2008)
11. Chow, S.S.M., Roth, V., Rieffel, E.G.: General Certificateless Encryption and Timed-Release Encryption. In: Ostrovsky, R., De Prisco, R., Visconti, I. (eds.) SCN 2008. LNCS, vol. 5229, pp. 126–143. Springer, Heidelberg (2008)
12. Chow, S.S.M., Yiu, S.-M.: Timed-Release Encryption Revisited. In: Baek, J., Bao, F., Chen, K., Lai, X. (eds.) ProvSec 2008. LNCS, vol. 5324, pp. 38–51. Springer, Heidelberg (2008)
13. Dent, A.W.: A Survey of Certificateless Encryption Schemes and Security Models. Int. J. Inf. Sec. 7(5), 349–377 (2008)
14. Dent, A.W., Tang, Q.: Revisiting the Security Model for Timed-Release Encryption with Pre-open Capability. In: Garay, J.A., Lenstra, A.K., Mambo, M., Peralta, R. (eds.) ISC 2007. LNCS, vol. 4779, pp. 158–174. Springer, Heidelberg (2007)

15. Dodis, Y., Katz, J.: Chosen-Ciphertext Security of Multiple Encryption. In: Kilian, J. (ed.) TCC 2005. LNCS, vol. 3378, pp. 188–209. Springer, Heidelberg (2005)
16. Fujioka, A., Okamoto, Y., Saito, T.: Security of Sequential Multiple Encryption. In: Abdalla, M., Barreto, P.S.L.M. (eds.) LATINCRYPT 2010. LNCS, vol. 6212, pp. 20–39. Springer, Heidelberg (2010)
17. Fujisaki, E., Okamoto, T.: How to Enhance the Security of Public-Key Encryption at Minimum Cost. In: Imai, H., Zheng, Y. (eds.) PKC 1999. LNCS, vol. 1560, pp. 53–68. Springer, Heidelberg (1999)
18. Fujisaki, E., Okamoto, T.: Secure Integration of Asymmetric and Symmetric Encryption Schemes. In: Wiener, M.J. (ed.) CRYPTO 1999. LNCS, vol. 1666, pp. 537–554. Springer, Heidelberg (1999)
19. Goldwasser, S., Micali, S.: Probabilistic Encryption. J. Comput. Syst. Sci. 28(2), 270–299 (1984)
20. Hristu-Varsakelis, D., Chalkias, K., Stephanides, G.: Low-Cost Anonymous Timed-Release Encryption. In: 3rd International Symposium on Information Assurance and Security, pp. 77–82. IEEE, Los Alamitos (2007); An extended version of this paper appears in Journal of Information Assurance and Security 3(1), 80–88 (2008)
21. Libert, B., Quisquater, J.-J.: On Constructing Certificateless Cryptosystems from Identity Based Encryption. In: Yung, M., Dodis, Y., Kiayias, A., Malkin, T. (eds.) PKC 2006. LNCS, vol. 3958, pp. 474–490. Springer, Heidelberg (2006)
22. Matsuda, T., Nakai, Y., Matsuura, K.: Efficient Generic Constructions of Timed-Release Encryption with Pre-open Capability. In: Joye, M., Miyaji, A., Otsuka, A. (eds.) Pairing 2010. LNCS, vol. 6487, pp. 225–245. Springer, Heidelberg (2010)
23. Nakai, Y., Matsuda, T., Kitada, W., Matsuura, K.: A Generic Construction of Timed-Release Encryption with Pre-open Capability. In: Takagi, T., Mambo, M. (eds.) IWSEC 2009. LNCS, vol. 5824, pp. 53–70. Springer, Heidelberg (2009)
24. Paterson, K.G., Quaglia, E.A.: Time-Specific Encryption. In: Garay, J.A., De Prisco, R. (eds.) SCN 2010. LNCS, vol. 6280, pp. 1–16. Springer, Heidelberg (2010)
25. Yang, P., Kitagawa, T., Hanaoka, G., Zhang, R., Matsuura, K., Imai, H.: Applying Fujisaki-Okamoto to Identity-Based Encryption. In: Fossorier, M., Imai, H., Lin, S., Poli, A. (eds.) AAECC 2006. LNCS, vol. 3857, pp. 183–192. Springer, Heidelberg (2006)
26. Zhang, R., Hanaoka, G., Shikata, J., Imai, H.: On the Security of Multiple Encryption or CCA-security+CCA-security=CCA-security? In: Bao, F., Deng, R.H., Zhou, J. (eds.) PKC 2004. LNCS, vol. 2947, pp. 360–374. Springer, Heidelberg (2004)

Identity-Based Server-Aided Decryption*

Joseph K. Liu, Cheng Kang Chu, and Jianying Zhou

Institute for Infocomm Research
Singapore
{ksliu,ckchu,jyzhou}@i2r.a-star.edu.sg

Abstract. Identity-Based Cryptosystem plays an important role in the modern cryptography world, due to the elimination of the costly certificate. However, all practical identity-based encryption schemes require pairing operation in the decryption stage. Pairing is a heavy mathematical algorithm, especially for resource-constrained devices such as smart cards or wireless sensors. In other words, decryption can hardly be done in these devices if identity-based cryptosystem is employed. We solve this problem by proposing a new notion called *Identity-Based Server-Aided Decryption*. It is similar to normal identity-based encryption scheme, but it further enables the receiver to decrypt the ciphertext without needing to compute pairing with the assistance of an external server. Secure mechanisms are provided to detect whether the server has computed correctly and prevent the server from getting any information about the plaintext or the user secret key. We give two concrete instantiations of this notion.

1 Introduction

THE MOTIVATION. Since the introduction of Identity-Based Cryptosystem by Shamir [12] in 1984, people tried to find a practical way to implement Identity-Based Encryption (IBE) scheme. It was not invented until Boneh and Franklin [1] proposed a pairing based practical IBE scheme in 2001. Since then, further IBEs were proposed. They have great improvement over efficiency and security. However, all existing practical IBEs require pairing, at least in the decryption stage. Non-pairing based IBEs such as [7,2] are inefficient to be used. They can only encrypt a single bit at one time. Thus they are only of theoretical interest.

Although pairing can facilitate the implementation of practical IBE, the main drawback is the heavy computation requirement. Resource-constrained devices such as wireless sensors or smart cards are inefficient, or even unable to execute pairing algorithm. That makes them impossible to decrypt a ciphertext if the underlying infrastructure is Identity-Based Cryptosystem.

THE CONCEPT. In this paper, we introduce a new notion called *Identity-Based Server-Aided Decryption* (IBSAD). There is no difference between an IBE and

---

* The work is supported by A*STAR project SEDS-0721330047.

U. Parampalli and P. Hawkes (Eds.): ACISP 2011, LNCS 6812, pp. 337–352, 2011.
© Springer-Verlag Berlin Heidelberg 2011

IBSAD in the encryption process. On the other side, there is an additional server which helps the receiver to decrypt the ciphertext. The receiver delegates a special key to the server and sends (part of) the ciphertext to the server. It generates a token after executing all necessary pairing computations and sends the token back to the receiver. Finally the receiver computes the plaintext from his own secret key and this token, without needing to compute any pairing.

One may worry that the server will have too much power. The worry may fall into two sides. First, does the server know any information about the plaintext or the user secret key? Second, if the server wrongly generates a token (instead of computing it according to the algorithm), does the receiver be aware of this cheating behaviour? Our scheme can protect the receiver in both aspects. Although the server gets the delegated secret key, it gives no additional information about the plaintext or the user secret key. Besides, the receiver can also check whether the server has computed the token correctly.

APPLICATIONS. This can be very useful in many applications. For example, in the case of smart card, the card itself has only very limited resource. It may not be able to execute heavy computation such as pairing. It may need to rely on some third parties, such as smart card readers, to assist for any expensive algorithm. If Identity-Based Cryptosystem is used for the infrastructure, that means the smart card cannot carry out any decryption of ciphertext by itself. It will widely limit the applications of smart card. Our IBSAD provides an excellent solution for this deadlock, by delegating the card reader for some heavy computation part, yet the smart card itself does not leak any secret information.

The outcome is more conspicuous in the case of a mobile phone SIM card. Nowadays many people are using smart phones, such as iPhone or Android phones. These phones are no longer just a mobile phone, but a small computer with 1GHz CPU and 16G internal storage. On the other side, the SIM card inside is still a SIM card. Most of the computations are still executed by the phone instead of the SIM card itself. Previously if it needs to decrypt a ciphertext, the easiest way is to decrypt the ciphertext within the SIM card. However, the SIM card has only very limited resource. On the other side, it is impossible to give the secret key to the mobile phone as the phone is regarded as an untrusted party. In order to carry out faster decryption, or decryption for large ciphertext in a secure way, our IBSAD can be a suitable solution.

Another application scenario is wireless sensor network. A wireless sensor has only very limited resource. A signature generation or encryption process is just enough to be executed by a sensor. An asymmetric decryption process is too heavy respectively. It is generally believed that a sensor may not carry out asymmetric decryption as it is insecure to give its secret key for the surrounding party for computation. Although sensors can carry out symmetric decryption, it takes additional key exchange process for both parties to establish a session key. Moreover, sensors may receive ciphertext from different parties or base stations. In this way, the symmetric encryption is not suitable, as it may take a number of key exchange processes. Asymmetric encryption is the only possible solutions in this scenario. By using our IBSAD, it can facilitate the secure asymmetric decryption process by delegating a special key to any third party without leaking any secret information. This greatly increases the security level and efficiency of the entire wireless sensor network.

1.1 Related Works

Although there exist some IBEs that can be splitted into online and offline stage for encryption [9,11,6,5], where the online stage does not require any heavy computation, they do not differ with normal IBEs in the decryption stage. Pairing is still required. So far there is no practical IBE that does not require pairing for the decryption stage. However, there exists some *pairing delegation protocols* [10,13,4] that allows a third party to compute the pairing part of an algorithm. We can apply these protocols in any IBE for the decryption part, so that pairing can be eliminated for the receiver. In terms of functionality and security, they can achieve the same level as our IBSAD. Nevertheless, these protocols are for generic purpose. Namely, they can be used in signature verification, IBE decryption, authenticated key exchange etc. Thus the efficiency is far behind our IBSAD. In order to save 1 pairing, they require to have at least 4 additional exponentiations. Although exponentiation is not as expensive as pairing in terms of computation complexity, it is also desirable to be reduced as much as possible. In contrast, our scheme just requires 1 exponentiation to replace 1 pairing.

1.2 Contribution

In this paper, we propose a new notion called *Identity-Based Server-Aided Decryption* (IBSAD). It is similar to normal IBE, but it further enables the receiver to decrypt the ciphertext without needing to compute pairing with the assistance of an external server. We provide secure mechanism to detect whether server has computed correctly. Furthermore, the server knows nothing about the plaintext or the user secret key.

We provide two efficient implementations. The first one is based on Gentry's IBE [8] which is proven secure in the standard model. The second implementation is based on Boneh-Franklin's IBE [1]. It is very efficient in the decryption stage, while the security can be proven in the random oracle model.

2 Definitions

2.1 Pairings and Related Intractability Assumption

Let $\mathbb{G}$ and $\mathbb{G}_T$ be two multiplicative cyclic groups of prime order q. Let g be a generator of $\mathbb{G}$. We define $e : \mathbb{G} \times \mathbb{G} \to \mathbb{G}_T$ to be a bilinear pairing if it has the following properties:

1. *Bilinearity*: For all $u, v \in \mathbb{G}$, and $a, b \in \mathbb{Z}$, $e(u^a, v^b) = e(u, v)^{ab}$.
2. *Non-degeneracy*: $e(g, g) \neq 1$.
3. *Computability*: It is efficient to compute $e(u, v)$ for all $u, v \in \mathbb{G}$.

We define the truncated decision q-ABDHE problem [8] as follows:

Definition 1 (Truncated Decision q-Augmented Bilinear Diffie-Hellman Exponent Assumption (q-ABDHE)). *Given a vector of $q + 3$ elements:*

$$\left(\tilde{g}, \tilde{g}^{(\alpha)^{q+2}}, g, g^{\alpha}, g^{(\alpha)^2}, \ldots, g^{(\alpha)^q}\right) \in \mathbb{G}^{q+3}$$

and an element $Z \in \mathbb{G}_T$ as input, output 1 if $Z = e(g^{(\alpha)^{q+1}}, \tilde{g})$ and output 0 otherwise. We say that the decision (t, ϵ, q)-ABDHE assumption holds in $(\mathbb{G}, \mathbb{G}_T)$ if no t-time algorithm has advantage at least ϵ over random guessing in solving the decision q-ABDHE problem in $(\mathbb{G}, \mathbb{G}_T)$.

2.2 Building Blocks

We introduce two building blocks used in our schemes:

1. AN IND-CCA SECURE SYMMETRIC ENCRYPTION SCHEME.
   The IND-CCA security of a symmetric encryption scheme $\mathsf{SE}_\kappa = (\mathsf{SEnc}, \mathsf{SDec})$ with key length κ is captured by defining the advantage of an adversary $\mathcal{A}$ as

   $$\mathsf{Adv}_{\mathcal{A},\mathsf{SE}}^{\mathrm{CCA}}(\lambda) = 2 \cdot \Pr[\; \beta' = \beta : K \in_R \{0,1\}^\kappa; \beta \in_R \{0,1\};$$
   $$\beta' \leftarrow \mathcal{A}^{\mathsf{SEnc}_K, \mathsf{SDec}_K, \mathsf{Chal}_{K,\beta}}(1^\lambda) \;] - 1.$$

   In the above, $\mathsf{Chal}_{K,\beta}(m_0, m_1)$ returns $\mathsf{SEnc}_K(m_\beta)$. Moreover, $\mathcal{A}$ is allowed to issue only one query to the Chal oracle, and is not allowed to query SDec_K on the ciphertext returned by it. The symmetric encryption scheme is IND-CCA secure if $\mathsf{Adv}_{\mathsf{SE}}^{\mathrm{CCA}}(\lambda) = max_{\mathcal{A}}\{\mathsf{Adv}_{\mathcal{A},\mathsf{SE}}^{\mathrm{CCA}}(\lambda)\}$ is negligible for any adversary $\mathcal{A}$.

2. A SECURE KEY DERIVATION FUNCTION.
   A key derivation function $\mathsf{D} : \mathbb{G}_T \rightarrow \{0,1\}^\kappa$ on a random input outputs a κ-bit string which is computationally indistinguishable from a random string. We define the advantage of an adversary $\mathcal{A}$ in distinguishing two distributions as

   $$\mathsf{Adv}_{\mathcal{A},\mathsf{D}}^{KDF}(\lambda) = \Pr[\mathcal{A}(1^\lambda, \mathsf{D}(x)) = 1] - \Pr[\mathcal{A}(1^\lambda, r) = 1]$$

   where $x \in \mathbb{G}_T, r \in_R \{0,1\}^\kappa$ and λ is the security parameter determining κ. So the key derivation function is *KDF-secure* if $\mathsf{Adv}_{\mathsf{D}}^{KDF}(\lambda) = max_{\mathcal{A}}\{\mathsf{Adv}_{\mathcal{A},\mathsf{D}}^{KDF}(\lambda)\}$ is negligible for any adversary $\mathcal{A}$.

2.3 Framework of ID-Based Server-Aided Decryption

An ID-based Server-Aided Decryption (IBSAD) scheme is the same as an ordinary IBE, except with the additional of an (untrusted) server for the processing of part of the decryption. It consists of the following six probabilistic polynomial time (PPT) algorithms:

- (param, msk) ← Setup(1^λ) takes a security parameter $\lambda \in \mathbb{N}$ and generates param the global public parameters and msk the master secret key of the KGC.
- sk_{ID} ← Extract(1^λ, param, msk, ID) takes a security parameter λ, a global parameters param, a master secret key msk and an identity ID to generate a user secret key sk_{ID} corresponding to this identity.
- $\mathfrak{C}$ ← Encrypt(1^λ, param, m, ID) takes a security parameter λ, a global parameters param, a message m, an identity ID to generate a ciphertext $\mathfrak{C}$.
- (dsk_{ID}, dpk_{ID}) ← DelegatedKeyGen(1^λ, param, sk_{ID}) takes a security parameter λ, a global parameters param and a user secret key sk_{ID} to generate a delegated secret key dsk_{ID} and a delegated public key dpk_{ID}. The delegated secret key should be kept secret together with the user secret key, while the delegated public key should be given to the server for further processing.
- $\mathfrak{T}$ ← DelegatedCompute(1^λ, param, dpk_{ID}, $\mathfrak{C}'$) takes a security parameter λ, a global parameters param, a delegated public key dpk_{ID}, and a subset of ciphertext $\mathfrak{C}' \subseteq \mathfrak{C}$ to generate a delegated token $\mathfrak{T}$.
- ($m/ \perp$) ← LightDecrypt(1^λ, param, $\mathfrak{C}$, $\mathfrak{T}$, sk_{ID}, dsk_{ID}) takes a security parameter λ, a global parameters param, a ciphertext $\mathfrak{C}$, a delegated token $\mathfrak{T}$, a secret key of the receiver sk_{ID} and a delegated secret key dsk_{ID} to generate a message m or $\perp$ which indicates the failure of decryption. Note that the failure of decryption may due to the malformation of the original plaintext, or the misbehaviour of the server.[1]

For simplicity, we omit the notation of 1^λ and param from the input arguments of the above algorithms in the rest of this paper.

2.4   Security of ID-Based Server-Aided Decryption

Next we define the security of IBSAD. It has two kinds of security:

1. The first one is very similar to normal CCA or CPA of normal IBE, except that the adversary is allowed to have the delegated public key of the challenged identity. It is to model the untrusted server which tries to distinguish the plaintext given the delegated public key.

   Definition 2 (Chosen Ciphertext Security (CCA)). *An IBSAD scheme is semantically secure against chosen ciphertext insider attack (IND − ID − CCA) if no PPT adversary has a non-negligible advantage in the following game:*
   (a) The challenger $\mathcal{C}$ runs Setup and gives the resulting param to adversary $\mathcal{A}$. It keeps msk secret.
   (b) In the first stage, $\mathcal{A}$ makes a number of queries to the following oracles simulated by $\mathcal{C}$. Note that without loss of generality, we assume $\mathcal{C}$ performs DelegatedKeyGen for each ID once only and maintains a list of the form (ID, dsk_{ID}, dpk_{ID}) for consistency.

---

[1] We do not require the receiver to distinguish between these two cases. Yet the receiver is able to do so, if he computes the token and decrypt the original ciphertext by himself.

i. OExt(·): $\mathcal{A}$ submits an identity ID to the extraction oracle for $(sk_{\text{ID}}, dsk_{\text{ID}}, dpk_{\text{ID}})$ where $sk_{\text{ID}} \leftarrow$ Extract(msk, ID) and $(dsk_{\text{ID}}, dpk_{\text{ID}}) \leftarrow$ DelegatedKeyGen(sk_{ID}).

ii. OEdpk(·): $\mathcal{A}$ submits an identity ID to the dpk extraction oracle for dpk_{ID} from the result of DelegatedKeyGen(sk_{ID}) where $sk_{\text{ID}} \leftarrow$ Extract(msk, ID).

iii. ODec(·, ·, ·): $\mathcal{A}$ submits a ciphertext $\mathfrak{C}$, a receiver identity ID and a token $\mathfrak{T}$ to the decryption oracle for the result of LightDecrypt($\mathfrak{C}, \mathfrak{T}, sk_{\text{ID}}, dsk_{\text{ID}}$) where $(dsk_{\text{ID}}, dpk_{\text{ID}}) \leftarrow$ DelegatedKeyGen(sk_{ID}) and $sk_{\text{ID}} \leftarrow$ Extract(msk, ID). The result is made of a message if the decryption is successful. Otherwise, a symbol $\perp$ is returned for rejection.

These queries can be asked adaptively. That is, each query may depend on the answers of previous ones.

(c) $\mathcal{A}$ produces two messages m_0, m_1 and an identity ID^* where ID^* is not queried to the extraction oracle before. $\mathcal{C}$ chooses a random bit $b \in \{0, 1\}$ and sends an encrypted ciphertext $\mathfrak{C}^* = $ Encrypt(m_b, ID^*). $\mathfrak{C}^*$ to $\mathcal{A}$.

(d) $\mathcal{A}$ makes a number of new queries as in the first stage with the restriction that it cannot query the extraction oracle with ID^* and the decryption oracle with $(\mathfrak{C}^*, \text{ID}^*, \cdot)$.

(e) At the end of the game, $\mathcal{A}$ outputs a bit b' and wins if $b' = b$.

An IBSAD scheme is $(t, \epsilon, q_{\text{ID}}, q_C)$ IND − ID − CCA secure if all t-time IND −ID − CCA adversaries making at most q_{ID} extraction oracle queries and at most q_C decryption queries have advantage at most ϵ in winning the above game.

There is another weaker version of security, the Chosen Plaintext Security (CPA). It is the same as CCA, except there is no decryption oracle in the CPA game.

2. The second is *Detectability*. It is to make sure that the user can detect if the server computes wrongly. The definition is given below.

Definition 3. An IBSAD scheme is detectable if no PPT adversary has a non-negligible advantage in the following game:

(a) The challenger $\mathcal{C}$ runs Setup, Extract, Encrypt, DelegatedKeyGen according to the algorithm to obtain sk_{ID}, dsk_{ID}, dpk_{ID}, $\mathfrak{C}$.

(b) $\mathcal{C}$ gives $\mathfrak{C}', dpk_{\text{ID}}$ to the adversary $\mathcal{A}$ where $\mathfrak{C}' \subseteq \mathfrak{C}$. $\mathcal{C}$ also runs DelegatedCompute($dpk_{\text{ID}}, \mathfrak{C}'$) to obtain $\mathfrak{T}$.

(c) $\mathcal{A}$ produces a token $\mathfrak{T}'$. $\mathcal{A}$ wins if

i. $\mathfrak{T} \neq \mathfrak{T}'$, and

ii. $\perp \neq$ LightDecrypt($\mathfrak{C}, \mathfrak{T}', sk_{\text{ID}}, dsk_{\text{ID}}$).

$\mathcal{A}$'s advantage is defined as $\mathbf{Adv}^{\text{Detect}}(\mathcal{A}) = \Pr[\mathcal{A} \ wins\ ]$.

3   CCA-Secure ID-Based Server-Aided Decryption scheme from Gentry's IBE

Our scheme presented in this section is based on Gentry's IBE [8]. Note that the Setup and Extract algorithms are the same as Gentry's.

3.1   Construction

Let $\mathbb{G}$ and $\mathbb{G}_T$ be groups of order p, and let $e : \mathbb{G} \times \mathbb{G} \to \mathbb{G}_T$ be the bilinear map. The scheme works as follows.

<u>Setup:</u> The KGC picks random generators $g, h_1, h_2, h_3 \in \mathbb{G}$ and a random $\alpha \in \mathbb{Z}_p$. It sets $g_1 = g^\alpha \in \mathbb{G}$. It chooses a hash function H from a family of universal one-way hash functions, where $H : \{0,1\}^* \to \mathbb{Z}_p$, an IND-CCA secure symmetric encryption scheme $\mathsf{SE}_\kappa = (\mathsf{SEnc}, \mathsf{SDec})$ with key length κ and a KDF-secure key derivation function $\mathsf{D} : \mathbb{G}_T \to \{0,1\}^\kappa$. The public param and master secret key msk are given by

$$\mathsf{param} = (g, g_1, h_1, h_2, h_3, H, \mathsf{SE}, \mathsf{D}) \qquad \mathsf{msk} = \alpha$$

<u>Extract:</u> To generate a secret key for a user with identity $\mathsf{ID} \in \mathbb{Z}_p$, the KGC generates random $r_{\mathsf{ID},i} \in \mathbb{Z}_p$ for $i \in \{1,2,3\}$, and outputs the user secret key

$$sk_{\mathsf{ID}} = \{(r_{\mathsf{ID},i}, h_{\mathsf{ID},i}) : i \in \{1,2,3\}\}, \qquad \text{where} \qquad h_{\mathsf{ID},i} = \left(h_i g^{-r_{\mathsf{ID},i}}\right)^{\frac{1}{\alpha - \mathsf{ID}}}$$

If $\mathsf{ID} = \alpha$, the KGC aborts. We require that the KGC always use the same random value $\{r_{\mathsf{ID},i}\}$ for ID.

<u>Encrypt:</u> To encrypt a message $m \in \{0,1\}^{\ell_m}$ for some security length ℓ_m using identity $\mathsf{ID} \in \mathbb{Z}_p$, the sender generates random $s \in \mathbb{Z}_p$, $k \in \mathbb{G}_T$ and sends the ciphertext

$$\mathfrak{C} = \left(g_1^s g^{-s \cdot \mathsf{ID}} , \ e(g,g)^s , \ k \cdot e(g,h_1)^{-s} , \ e(g,h_2)^s e(g,h_3)^{s\beta} , \ \mathsf{SEnc}_{\mathsf{D}(k)}(m)\right)$$

Above, for $\mathfrak{C} = (u, v, w, y, z)$, we set $\beta = H(u, v, w)$. Encryption does not require any pairing computations once $e(g,g)$ and $\{e(g,h_i)\}$ have been pre-computed or alternatively included in param.

<u>DelegatedKeyGen:</u> To generate a delegated key pair from the secret key sk_{ID}, the user randomly generates $\hat{x}, \hat{x}' \in \mathbb{Z}_p$ and outputs

$$dsk_{\mathsf{ID}} = \{\hat{x}, \hat{x}'\} \qquad \text{and} \qquad dpk_{\mathsf{ID}} = \{x_{\mathsf{ID},i} : i \in \{1,2,3\}\}$$

where

$$x_{\mathsf{ID},1} = (h_{\mathsf{ID},1})^{\hat{x}} , \ x_{\mathsf{ID},2} = (h_{\mathsf{ID},2})^{\hat{x}'} , \ x_{\mathsf{ID},3} = (h_{\mathsf{ID},3})^{\hat{x}'}$$

dpk_{ID} is sent to the server while dsk_{ID} is kept secret. We require that the user always uses the same random value $\{\hat{x}, \hat{x}'\}$ for dsk_{ID}.

<u>DelegatedCompute:</u> On upon the received ciphertext $\mathfrak{C}' = (u, v, w) \subset \mathfrak{C}$, the receiver computes $\beta = H(u, v, w)$ and sends (u, β) to the server for delegated computation. The server computes

$$T_1 = e(u, x_{\mathsf{ID},2}(x_{\mathsf{ID},3})^\beta), \qquad T_2 = e(u, x_{\mathsf{ID},1})$$

and outputs $\mathfrak{T} = (T_1, T_2)$ to the user.

LightDecrypt: To decrypt a ciphertext $\mathfrak{C} = (u, v, w, y, z)$ using the secret key sk_{ID}, delegated secret key dsk_{ID} and the delegated token $\mathfrak{T}$, the user computes the following steps:

1. Sets $\beta = H(u, v, w)$.
2. Tests whether $y = T_1^{1/\hat{x}'} v^{r_{\mathsf{ID},2} + r_{\mathsf{ID},3}\beta}$. If the check fails, outputs $\bot$.
3. Computes $k = w \cdot T_2^{1/\hat{x}} v^{r_{\mathsf{ID},1}}$.
4. Outputs $\mathsf{SDec}_{\mathsf{D}(k)}(z)$.

Note that $\mathsf{SDec}_{\mathsf{D}(k)}(z)$ will output $\bot$ if a different key $K' \neq \mathsf{D}(k)$ is used for the symmetric encryption of the plaintext.

3.2   Security Analysis

We now prove that our scheme is $\mathsf{IND-ID-CCA}$ secure under the truncated decision q-ABDHE assumption. We follow the approach from [8].

Theorem 1. *Assume the truncated decision (t, ϵ, q)-ABDHE assumption holds for $(\mathbb{G}, \mathbb{G}_T, e)$. Then the above IBSAD scheme is $(t', \epsilon', q_{\mathsf{ID}}, q_C)$ $\mathsf{IND-ID-CCA}$ secure for $t' = t - \mathcal{O}(t_{exp} \cdot q^2)$, $\epsilon' = \epsilon + 4q_C/p$ and $q = q_{\mathsf{ID}} + 2$, where t_{exp} is the time required to compute exponentiation in $\mathbb{G}$.*

Proof. Let $\mathcal{A}$ be an adversary that $(t', \epsilon', q_{\mathsf{ID}}, q_C)$-breaks the $\mathsf{IND-ID-CCA}$ security. We construct an algorithm $\mathcal{B}$, that solves the truncated decision q-ABDHE problem. $\mathcal{B}$ takes as input a random truncated decision q-ABDHE challenge $(g', g'_{q+2}, g, g_1, \ldots, g_q, Z)$, in which Z is either $e(g_{q+1}, g')$ or a random element of $\mathbb{G}_T$, where we use g_i and g'_i to denote $g^{(\alpha^i)}$ and $g'^{(\alpha^i)}$. Algorithm $\mathcal{B}$ proceeds as follows.

<u>Setup</u>: $\mathcal{B}$ generates random polynomials $f_i(x) \in \mathbb{Z}_p[x]$ of degree q for $i \in \{1, 2, 3\}$. It sets $h_i = g^{f_i(x)}$. It also chooses a hash function H, an IND-CCA secure symmetric encryption scheme SE and a KDF-secure key derivation function D. It sends the public param $(g, g_1, h_1, h_2, h_3, H, \mathsf{SE}, \mathsf{D})$ to $\mathcal{A}$. Since g, α and $f_i(x)$ for $i \in \{1, 2, 3\}$ are chosen uniformly at random, h_1, h_2, h_3 are uniformly random and the public key has a distribution identical to that in the actual construction.

<u>Oracle Simulation</u>:

1. *Extraction Oracle* $\mathsf{OExt}(\cdot)$: $\mathcal{B}$ responds to a query on $\mathsf{ID} \in \mathbb{Z}_p$ as follows. If $\mathsf{ID} = \alpha$, $\mathcal{B}$ uses α to solve the truncated decision q-ABDHE immediately. Otherwise, to generate a pair $(r_{\mathsf{ID},i}, h_{\mathsf{ID},i})$ for $i \in \{1, 2, 3\}$ such that $h_{\mathsf{ID},i} = (h_i g^{-r_{\mathsf{ID},i}})^{1/(\alpha+\mathsf{ID})}$, $\mathcal{B}$ sets $r_{\mathsf{ID},i} = f_i(\mathsf{ID})$ and computes $h_{\mathsf{ID},i} = g^{F_{\mathsf{ID},i}(\alpha)}$, where $F_{\mathsf{ID},i}(x)$ denote the $(q-1)$-degree polynomial

$$\frac{f_i(x) - f_i(\mathsf{ID})}{x - \mathsf{ID}}$$

This is a valid secret key for ID, since

$$g^{F_{\text{ID},i}(\alpha)} = g^{(f_i(\alpha) - f_i(\text{ID}))/(\alpha - \text{ID})} = (h_i g^{-r_{\text{ID},i}})^{1/(\alpha + \text{ID})}$$

$\mathcal{B}$ also generates $(dsk_{\text{ID}}, dpk_{\text{ID}})$ according to the algorithm. It maintains a list for the record of $(\text{ID}, sk_{\text{ID}}, dsk_{\text{ID}}, dpk_{\text{ID}})$. It outputs $(sk_{\text{ID}}, dsk_{\text{ID}}, dpk_{\text{ID}})$ as the response to the query.

2. *dpk Extraction Oracle* $\text{OEdpk}(\cdot)$: To respond to a *dpk* extraction oracle query on ID, $\mathcal{B}$ checks the list whether ID has been queried before. If yes, it returns the value dpk_{ID}. Otherwise, it executes the extract oracle and outputs dpk_{ID}.
3. *Decryption Oracle* $\text{ODec}(\cdot, \cdot, \cdot)$: To respond to a decryption oracle query on $(\text{ID}, \mathfrak{C}, \mathfrak{T})$, $\mathcal{B}$ generates the secret key and delegated secret key as above. It then decrypts $\mathfrak{C}$ according to the algorithm.

Output: $\mathcal{A}$ outputs identities ID^* and messages m_0, m_1. Let $f'(x) = x^{q+2}$ and let

$$F'_{\text{ID}^*}(x) = \frac{f'(x) - f'(\text{ID}^*)}{x - \text{ID}^*}$$

which is a polynomial of degree $q + 1$. $\mathcal{B}$ computes a secret key $\{(r_{\text{ID}^*,i}, h_{\text{ID}^*,i}) : i \in \{1, 2, 3\}\}$ for ID^*. It randomly generates $b \in \{0, 1\}$, $k \in \mathbb{G}_T$ and sets

$$u = g'^{f'(\alpha) - f'(\text{ID}^*)}$$

$$v = Z \cdot e(g', \prod_{j=0}^{q} g^{(F'_{\text{ID}^*,(j)}) \cdot (\alpha^j)})$$

$$w = \frac{k}{e(u, h_{\text{ID}^*,1}) v^{r_{\text{ID}^*,1}}}$$

$$y = e(u, h_{\text{ID}^*,2} h_{\text{ID}^*,3}^{\beta}) v^{r_{\text{ID}^*,2} + r_{\text{ID}^*,3}\beta}$$

$$z = \text{SEnc}_{D(k)}(m_b)$$

where $F'_{\text{ID}^*,(j)}$ is the coefficient of x^j in $F'_{\text{ID}^*}(x)$ and $\beta = H(u, v, w)$. It sends (u, v, w, y, z) to $\mathcal{A}$ as the challenge ciphertext.

Let $s = (\log_g g') F'_{\text{ID}^*}(\alpha)$. If $Z = e(g_{q+1}, g')$, then

$$u = g^{s(\alpha - \text{ID}^*)}$$

$$v = e(g, g)^s$$

$$k/w = e(u, h_{\text{ID}^*,1}) v^{r_{\text{ID}^*,1}} = e(g, h)^s$$

Thus (u, v, w, y, z) is a valid ciphertext for m_b under randomness s. Since $\log_g g'$ is uniformly random, s is uniformly random, and so (u, v, w, y, z) is a valid, appropriately-distributed challenge to $\mathcal{A}$.

Probability Analysis: If $Z = e(g_{q+1}, g')$, the simulation is perfect, and $\mathcal{A}$ will guest the bits b correctly with probability $1/2 + \epsilon'$. Else, Z is uniformly random, and thus (u, v) is uniformly random and independent element of $\mathbb{G} \times \mathbb{G}_T$. In this case, the inequality $v \neq e(u,g)^{1/(\alpha - \mathsf{ID}^*)}$ holds with probability $1 - 1/p$. When this inequality holds, the value of

$$e(u, h_{\mathsf{ID}^*,1}{}^{\hat{x}})^{1/\hat{x}} v^{r_{\mathsf{ID}^*,1}}$$
$$= e(u, (hg^{-r_{\mathsf{ID}^*,1}})^{1/(\alpha - \mathsf{ID}^*)}) v^{r_{\mathsf{ID}^*,1}}$$
$$= e(u, h)^{\alpha - \mathsf{ID}^*} \left(\frac{v}{e(u,g)^{1/(\alpha - \mathsf{ID}^*)}} \right)^{r_{\mathsf{ID}^*,1}}$$

is uniformly random and independent from $\mathcal{A}$'s view (except for the value w), since $r_{\mathsf{ID}^*,1}$ is uniformly random and independent from $\mathcal{A}$'s view (except for the value w). Thus w is uniformly random and independent, and (u,v,w,y,z) can impart no information regarding the bit b.

Next, we need to show that the decryption oracle, in the simulation and in the actual construction, rejects all invalid ciphertexts under identities not queried by $\mathcal{A}$, except with probability q_c/p. This is exactly the proof of Lemma 1 and Lemma 2 of [8]. We skip here and readers may refer to the proof there.

Time Complexity: In the simulation, $\mathcal{B}$'s overhead is dominated by computing $g^{F_{\mathsf{ID},i}(\alpha)}$ in response to $\mathcal{A}$'s key generation query on ID, where $F_{\mathsf{ID},i}(x)$ are polynomials of degree $q - 1$. Each such computation requires $\mathcal{O}(q)$ exponentiations in $\mathbb{G}$. Since $\mathcal{A}$ makes at most $q - 1$ such queries, $t = t' + \mathcal{O}(t_{exp} \cdot q^2)$. $\square$

Theorem 2. *The above IBSAD scheme is detectable.*

Proof. Assume the adversary $\mathcal{A}$ outputs a token $\mathfrak{T}' = (T_1', T_2')$ which is not equal to $\mathfrak{T} = (T_1, T_2)$, the one generated according to the algorithm. There are two cases, namely $T_1' \neq T_1$ or $T_2' \neq T_2$.

For case 1, we have

$$T_1' \neq T_1$$
$$T_1' \neq e(u, x_{\mathsf{ID},2}(x_{\mathsf{ID},3})^\beta)$$
$$T_1' \neq e(u, (h_{\mathsf{ID},2})^{\hat{x}'}(h_{\mathsf{ID},3})^{\hat{x}'\beta})$$

$$T_1'^{1/\hat{x}'} v^{r_{\mathsf{ID},2}+r_{\mathsf{ID},3}\beta} \neq e(u, h_{\mathsf{ID},2}(h_{\mathsf{ID},3})^\beta) \cdot e(g,g)^{s(r_{\mathsf{ID},2}+r_{\mathsf{ID},3}\beta)}$$

$$T_1'^{1/\hat{x}'} v^{r_{\mathsf{ID},2}+r_{\mathsf{ID},3}\beta} \neq e\left(g_1^s g^{-s \cdot \mathsf{ID}}, (h_2 g^{-r_{\mathsf{ID},2}})^{\frac{1}{\alpha - \mathsf{ID}}} (h_3 g^{-r_{\mathsf{ID},3}})^{\frac{1}{\alpha - \mathsf{ID}} \cdot \beta}\right) \cdot \cdot e(g,g)^{s(r_{\mathsf{ID},2}+r_{\mathsf{ID},3}\beta)}$$

$$T_1'^{1/\hat{x}'} v^{r_{\mathsf{ID},2}+r_{\mathsf{ID},3}\beta} \neq e\left(g^{s(\alpha - \mathsf{ID})}, (h_2 g^{-r_{\mathsf{ID},2}})(h_3 g^{-r_{\mathsf{ID},3}})^\beta\right)^{\frac{1}{\alpha - \mathsf{ID}}} \cdot e(g,g)^{s(r_{\mathsf{ID},2}+r_{\mathsf{ID},3}\beta)}$$

$$T_1'^{1/\hat{x}'} v^{r_{\mathsf{ID},2}+r_{\mathsf{ID},3}\beta} \neq e\left(g^s, (h_2 g^{-r_{\mathsf{ID},2}})(h_3 g^{-r_{\mathsf{ID},3}})^\beta\right) \cdot e(g,g)^{s(r_{\mathsf{ID},2}+r_{\mathsf{ID},3}\beta)}$$

$$T_1'^{1/\hat{x}'} v^{r_{\mathsf{ID},2}+r_{\mathsf{ID},3}\beta} \neq e(g^s, h_2 h_3^\beta) \cdot e(g^s, g^{-r_{\mathsf{ID},2}-r_{\mathsf{ID},3}\beta}) \cdot e(g,g)^{s(r_{\mathsf{ID},2}+r_{\mathsf{ID},3}\beta)}$$

$$T_1'^{1/\hat{x}'} v^{r_{\mathsf{ID},2}+r_{\mathsf{ID},3}\beta} \neq e(g, h_2)^s \cdot e(g, h_3)^{s\beta}$$

$$T_1'^{1/\hat{x}'} v^{r_{\mathsf{ID},2}+r_{\mathsf{ID},3}\beta} \neq y$$

Thus LightDecrypt will output $\perp$ as it fails the checking in step 2.

For case 2, we have

$$T_2' \neq T_2$$
$$T_2' \neq e(u, x_{\mathsf{ID},1})$$
$$T_2' \neq e(u, h_{\mathsf{ID},1}^{\hat{x}})$$
$$w \cdot T_2'^{1/\hat{x}} v^{r_{\mathsf{ID},1}} \neq k \cdot e(g, h_1)^{-s} \cdot e\left(g_1^s g^{-s \cdot \mathsf{ID}}, (h_1 g^{-r_{\mathsf{ID},1}})^{\frac{1}{\alpha - \mathsf{ID}}}\right) \cdot e(g, g)^{s r_{\mathsf{ID},1}}$$
$$w \cdot T_2'^{1/\hat{x}} v^{r_{\mathsf{ID},1}} \neq k \cdot e(g, h_1)^{-s} \cdot e\left(g^{s(\alpha - \mathsf{ID})}, h_1 g^{-r_{\mathsf{ID},1}}\right)^{\frac{1}{\alpha - \mathsf{ID}}} \cdot e(g, g)^{s r_{\mathsf{ID},1}}$$
$$w \cdot T_2'^{1/\hat{x}} v^{r_{\mathsf{ID},1}} \neq k \cdot e(g, h_1)^{-s} \cdot e(g^s, h_1 g^{-r_{\mathsf{ID},1}}) \cdot e(g, g)^{s r_{\mathsf{ID},1}}$$
$$w \cdot T_2'^{1/\hat{x}} v^{r_{\mathsf{ID},1}} \neq k \cdot e(g, h_1)^{-s} \cdot e(g^s, h_1) \cdot e(g^s, g^{-r_{\mathsf{ID},1}}) \cdot e(g, g)^{s r_{\mathsf{ID},1}}$$
$$w \cdot T_2'^{1/\hat{x}} v^{r_{\mathsf{ID},1}} \neq k$$
$$\mathsf{D}(w \cdot T_2'^{1/\hat{x}} v^{r_{\mathsf{ID},1}}) \neq \mathsf{D}(k)$$

Thus SDec will output $\bot$, as $\mathsf{D}(w \cdot T_2'^{1/\hat{x}} v^{r_{\mathsf{ID},1}})$ is not the key for the encryption.

$\square$

4 CCA-Secure ID-Based Server-Aided Decryption scheme from Boneh-Franklin's IBE

Our scheme presented in this section is based on Boneh-Franklin's IBE [1]. Note that the Setup, Extract and Encrypt algorithms are the same as Boneh-Franklin's.

4.1 Construction

Let $\mathbb{G}$ and $\mathbb{G}_T$ be groups of order p, and let $e : \mathbb{G} \times \mathbb{G} \to \mathbb{G}_T$ be the bilinear map. The scheme works as follows.

<u>Setup:</u> The KGC picks random generators $g \in \mathbb{G}$ and a random $s \in \mathbb{Z}_p$. It sets $g_1 = g^s \in \mathbb{G}$. Let $\mathcal{M} = \{0, 1\}^n$ be the message space. It chooses 4 hash functions:

$$H_1 : \{0, 1\}^* \to \mathbb{G}$$
$$H_2 : \mathbb{G}_T \to \{0, 1\}^n$$
$$H_3 : \{0, 1\}^n \times \{0, 1\}^n \to \mathbb{Z}_p$$
$$H_4 : \{0, 1\}^n \to \{0, 1\}^n$$

The public param and master secret key msk are given by

$$\mathsf{param} = (g, g_1, H_1, H_2, H_3, H_4) \qquad \mathsf{msk} = s$$

<u>Extract:</u> To generate a secret key for a user with identity $\mathsf{ID} \in \{0, 1\}^*$, the KGC generates the user secret key

$$sk_{\mathsf{ID}} = H_1(\mathsf{ID})^s$$

Encrypt: To encrypt a message $m \in \{0,1\}^n$ using identity $\mathsf{ID} \in \mathbb{Z}_p$, the sender generates random $\sigma \in \{0,1\}^n$, sets $r = H_3(\sigma, m)$ and sends the ciphertext

$$\mathfrak{C} = \left(g^r \,,\, \sigma \oplus H_2\Big(e\big(H_1(\mathsf{ID}), g_1\big)^r\Big) \,,\, m \oplus H_4(\sigma) \right) = (u, v, w)$$

to the receiver.

DelegatedKeyGen: To generate a delegated key pair from the secret key sk_{ID}, the user randomly generates $\hat{x} \in \mathbb{Z}_p$ and outputs

$$dsk_{\mathsf{ID}} = \hat{x} \qquad \text{and} \qquad dpk_{\mathsf{ID}} = (sk_{\mathsf{ID}})^{\hat{x}}$$

dpk_{ID} is sent to the server while dsk_{ID} is kept secret. We require that the user always uses the same random value $\{\hat{x}\}$ for dsk_{ID}.

DelegatedCompute: On upon the recevied ciphertext $\mathfrak{C}' = u \subset \mathfrak{C}$, the receiver sends u to the server for delegated computation. The server computes and outputs

$$\mathfrak{T} = e(dpk_{\mathsf{ID}}, u)$$

to the user.

LightDecrypt: To decrypt a ciphertext $\mathfrak{C} = (u, v, w)$ using the secret key sk_{ID}, delegated secret key dsk_{ID} and the delegated token $\mathfrak{T}$, the user computes the following steps:

1. Computes $\sigma = v \oplus H_2(\mathfrak{T}^{1/\hat{x}})$
2. Computes $m = w \oplus H_4(\sigma)$
3. Computes $r = H_3(\sigma, m)$
4. If $u = g^r$, outputs m. Otherwise outputs $\bot$.

4.2   Security Analysis

Theorem 3. *Assume the Boneh-Franklin IBE is* $\mathsf{IND - ID - CCA}$ *secure, our scheme is also* $\mathsf{IND - ID - CCA}$ *secure.*

Proof. Let $\mathcal{A}$ be an adversary that breaks the $\mathsf{IND - ID - CCA}$ security of our scheme. We construct an algorithm $\mathcal{B}$, that breaks the $\mathsf{IND - ID - CCA}$ security of Boneh-Franklin scheme. Algorithm $\mathcal{B}$ proceeds as follows.

Setup: $\mathcal{B}$ receives the setup environment and parameters from the simulator of the Boneh-Franklin security game, denoted by $\mathcal{S}_{BF}$. $\mathcal{B}$ forwards all these parameters to $\mathcal{A}$.

Oracle Simulation:

1. *Extraction Oracle* $\mathsf{OExt}(\cdot)$: $\mathcal{B}$ responds to a query on $\mathsf{ID} \in \mathbb{Z}_p$ as follows. $\mathcal{B}$ asks the extraction oracle from $\mathcal{S}_{BF}$ for ID to obtain sk_{ID}. $\mathcal{B}$ also generates $(dsk_{\mathsf{ID}}, dpk_{\mathsf{ID}})$ according to the algorithm. It maintains a list for the record of $(\mathsf{ID}, sk_{\mathsf{ID}}, dsk_{\mathsf{ID}}, dpk_{\mathsf{ID}})$. It outputs $(sk_{\mathsf{ID}}, dsk_{\mathsf{ID}}, dpk_{\mathsf{ID}})$ as the response to the query.

2. *dpk Extraction Oracle* OEdpk($\cdot$)*:* To respond to a *dpk* extraction oracle query on ID, $\mathcal{B}$ checks the list whether ID has been queried before. If yes, it returns the value dpk_{ID}. Otherwise, it executes the extract oracle and outputs dpk_{ID}. If ID is the challanged identity, $\mathcal{B}$ randomly generates $\hat{X} \in \mathbb{G}$ and returns it as dpk_{ID}. It stores $(\text{ID}, \bot, \bot, \hat{X})$ to the list.

3. *Decryption Oracle* ODec($\cdot, \cdot, \cdot$)*:* To respond to a decryption oracle query on $(\text{ID}, \mathfrak{C}, \mathfrak{T})$, $\mathcal{B}$ first queries OEdpk to get dpk_{ID}, then it checks whether $\mathfrak{T} = e(dpk_{\text{ID}}, u)$. If not, it outputs $\bot$. Otherwise, $\mathcal{B}$ asks the decryption oracle from $\mathcal{S}_{BF}$ for $(\text{ID}, \mathfrak{C})$ to get m or $\bot$, and outputs the answer directly.

Output: $\mathcal{A}$ outputs identities ID^* and messages m_0, m_1. $\mathcal{B}$ forwards (ID^*, m_0, m_1) to $\mathcal{S}_{BF}$ to get a challenged ciphertext $\mathfrak{C}^*$. $\mathcal{B}$ forwards $\mathfrak{C}^*$ to $\mathcal{A}$.
  $\mathcal{A}$ outputs a bit b and $\mathcal{B}$ forwards b to $\mathcal{S}_{BF}$.

Probability Analysis and Time Complexity: The successful probability and time complexity of $\mathcal{B}$ should be the same as $\mathcal{A}$.   □

Theorem 4. *The above IBSAD scheme is detectable.*

Proof. Assume the adversary $\mathcal{A}$ outputs a token $\mathfrak{T}'$ which is not equal to $\mathfrak{T}$, the one generated according to the algorithm. We have:

$$\sigma' = v \oplus H_2(\mathfrak{T}'^{1/\hat{x}})$$
$$m' = w \oplus H_4(\sigma')$$
$$r' = H_3(\sigma', m')$$
$$\because \mathfrak{T}' \neq \mathfrak{T}$$
$$\therefore \sigma' \neq \sigma, \qquad m' \neq m, \qquad r' \neq r$$

where v, w, σ, m, r are generated according to the algorithm

$$\because r' \neq r \qquad \therefore \text{Output} \perp$$

                                                                    □

5   Comparison

In this section we compare our scheme with some generic pairing delegation protocols. In order to save 1 pairing, the following computations should be added:

– Computation at decryptor side: Scalar Multiplication (SM), Point Addition (PA), $\mathbb{G}_T$ exponentiation (GE) and $\mathbb{G}_T$ multiplication (GM).
– Computation at server side: Pairing

According to Boyen [3], 1 pairing computation is approximately equal to 10 SM and GE operations, and more than 100 PA and GM operations. Other computations such as hashing or symmetric decryption are negligible when compared to pairing. Thus we only compare SM, PA, GE and GM operations. The comparisons are shown in the following tables.

Table 1 shows the overall computation done by the decryptor side and the server side for our scheme 1 (modified from Gentry's IBE) and different generic protocols. Table 2 shows the overall computation done by the decryptor side and the server side for our scheme 2 (modified from Boneh-Franklin's IBE) and different generic protocols.

Table 3 shows the extra computation required to compute 1 pairing (on average).

Table 1. Overall SM, PA, GE, GM operations required for the entire decryption process for scheme 1

	Computation at decryptor				Server (♯ of pairing)
	SM	PA	GE	GM	
Kang *et al.* [10]	7	2	10	5	8
Chevallier *et al.* [4]	7	2	8	5	4
Tsang *et al.* [13]	7	2	8	5	4
Our Scheme 1	3	0	4	3	2

Table 2. Overall SM, PA, GE, GM operations required for the entire decryption process for scheme 2

	Computation at decryptor				Server (♯ of pairing)
	SM	PA	GE	GM	
Kang *et al.* [10]	4	1	4	1	4
Chevallier *et al.* [4]	4	1	3	1	2
Tsang *et al.* [13]	4	1	3	1	2
Our Scheme 2	1	0	1	0	1

Table 3. Additional SM, PA, GE, GM operations in order to save 1 pairing at the decryptor side

	Computation at decryptor				Server (♯ of pairing)
	SM	PA	GE	GM	
Kang *et al.* [10]	3	1	4	1	4
Chevallier *et al.* [4]	3	1	3	1	2
Tsang *et al.* [13]	3	1	3	1	2
Our Scheme 1	1	0	1	0	1
Our Scheme 2	0	0	1	0	1

The computations in Table 3 are measured at the equivalence of having 1 pairing at the decryptor side. Note that for our scheme, the entire decryption process takes more computations (as shown in Table 1 and 2) than the data shown in Table 3. The reason is that: If we use the original decryption algorithm by Gentry's IBE, the decryptor needs to compute 2 pairings, 1 SM, 2 GE and 3 GM. The overall computation for our scheme requires 0 pairing, 3 SM, 4 GE and 3 GM at the decryptor side. We calculate the difference based on saving 1 pairing on average. The result is: $(3-1)/2 = 1$ SM, $(4-2)/2 = 1$ GE and $(3-3)/2 = 0$ GM. The same calculation is done for Boneh-Franklin's IBE.

6   Conclusion

In this paper, we have proposed a new notion called *Identity-Based Server-Aided Decryption*. It is similar to normal IBE, but it further allows the decryptor to delegate a third party for computing the heavy pairing operation. As all practical IBE schemes require pairing for the decryption stage, resource-constrained devices can hardly handle it. By using our proposed scheme, decryption in the identity-based cryptosystem for these devices can be realized. Besides efficiency advantages, it also achieves the highest level of security. Although the pairing operation is delegated to a third party, it cannot get any information about the plaintext or the user secret key. This can protect the user intensively. We provided two concrete implementations of our new notion. One is a modification from Gentry's IBE. We proved the security in the standard model. Another one is a modification from Boneh-Franklin's IBE. It is very efficient in the decryption stage and we proved the security in the random oracle model.

We also compared our scheme with some existing pairing delegation protocols, that can achieve the same function as ours. However, the efficiency is far behind us. We can achieve at least 3 to 4 times more efficient than any one of them.

References

1. Boneh, D., Franklin, M.: Identity-Based Encryption from the Weil Pairing. In: Kilian, J. (ed.) CRYPTO 2001. LNCS, vol. 2139, pp. 213–229. Springer, Heidelberg (2001)
2. Boneh, D., Gentry, C., Hamburg, M.: Space-Efficient Identity Based Encryption Without Pairings. In: FOCS 2007, pp. 647–657. IEEE Computer Soceity, Los Alamitos (2007)
3. Boyen, X.: A tapestry of identity-based encryption: practical frameworks compared. IJACT 1(1), 3–21 (2008)
4. Chevallier-Mames, B., Coron, J.-S., McCullagh, N., Naccache, D., Scott, M.: Secure delegation of elliptic-curve pairing. In: Gollmann, D., Lanet, J.-L., Iguchi-Cartigny, J. (eds.) CARDIS 2010. LNCS, vol. 6035, pp. 24–35. Springer, Heidelberg (2010)
5. Chow, S.S.M., Liu, J.K., Zhou, J.: Identity-based online/offline key encapsulation and encryption. In: ASIACCS 2011 (to Appear, 2011)
6. Chu, C.-K., Liu, J.K., Zhou, J., Bao, F., Deng, R.H.: Practical ID-based Encryption for Wireless Sensor Network. In: ASIACCS 2010, pp. 337–340. ACM, New York (2010)
7. Cocks, C.: An identity based encryption scheme based on quadratic residues. In: Honary, B. (ed.) Cryptography and Coding 2001. LNCS, vol. 2260, pp. 360–363. Springer, Heidelberg (2001)

8. Gentry, C.: Practical identity-based encryption without random oracles. In: Vaudenay, S. (ed.) EUROCRYPT 2006. LNCS, vol. 4004, pp. 445–464. Springer, Heidelberg (2006)
9. Guo, F., Mu, Y., Chen, Z.: Identity-based online/Offline encryption. In: Tsudik, G. (ed.) FC 2008. LNCS, vol. 5143, pp. 247–261. Springer, Heidelberg (2008)
10. Kang, B.G., Lee, M.S., Park, J.H.: Efficient delegation of pairing computation. Cryptology ePrint Archive, Report 2005/259 (2005), http://eprint.iacr.org/
11. Liu, J.K., Zhou, J.: An efficient identity-based online/Offline encryption scheme. In: Abdalla, M., Pointcheval, D., Fouque, P.-A., Vergnaud, D. (eds.) ACNS 2009. LNCS, vol. 5536, pp. 156–167. Springer, Heidelberg (2009)
12. Shamir, A.: Identity-Based Cryptosystems and Signature Schemes. In: Blakely, G.R., Chaum, D. (eds.) CRYPTO 1984. LNCS, vol. 196, pp. 47–53. Springer, Heidelberg (1985)
13. Tsang, P.P., Chow, S.S.M., Smith, S.W.: Batch pairing delegation. In: Miyaji, A., Kikuchi, H., Rannenberg, K. (eds.) IWSEC 2007. LNCS, vol. 4752, pp. 74–90. Springer, Heidelberg (2007)

A Generic Variant of NIST's KAS2 Key Agreement Protocol

Sanjit Chatterjee[1], Alfred Menezes[2], and Berkant Ustaoglu[3]

[1] Indian Institute of Science, India
sanjit@csa.iisc.ernet.in
[2] University of Waterloo, Canada
ajmeneze@uwaterloo.ca
[3] Sabanci University, Turkey
bustaoglu@cryptolounge.net

Abstract. We propose a generic three-pass key agreement protocol that is based on a certain kind of trapdoor one-way function family. When specialized to the RSA setting, the generic protocol yields the so-called KAS2 scheme that has recently been standardized by NIST. On the other hand, when specialized to the discrete log setting, we obtain a new protocol which we call DH2. An interesting feature of DH2 is that parties can use different groups (e.g., different elliptic curves). The generic protocol also has a hybrid implementation, where one party has an RSA key pair and the other party has a discrete log key pair. The security of KAS2 and DH2 is analyzed in an appropriate modification of the extended Canetti-Krawczyk security model.

1 Introduction

In 2009, the U.S. government's National Institute of Standards and Technology (NIST) published SP 800-56B [17], a standard that specifies several RSA-based key establishment schemes. SP 800-56B mirrors the earlier SP 800-56A standard [16] which described discrete log-based key establishment mechanisms.

SP 800-56B refines the schemes described in ANSI X9.44 [1] and introduces some new ones. Two key agreements protocols, KAS1 and KAS2, are presented in [17], as well as two key transport protocols. The KAS2 scheme, called 'KAS2-bilateral-confirmation' in [17], is a three-pass protocol that offers key confirmation. Three variants of KAS2 are described: a two-pass protocol called 'KAS2-basic' which does not offer key confirmation, 'KAS2-responder-confirmation' which provides unilateral key confirmation of the responder to the initiator, and 'KAS2-initiator-confirmation' which provides unilateral key confirmation of the initiator to the responder. SP 800-56B also specifies a two-pass protocol KAS1 (called 'KAS1-responder-confirmation' in [17]) that provides unilateral authentication and key confirmation of the responder to the initiator, and a variant of KAS1 (called 'KAS1-basic' in [17]) without responder key confirmation.

We have chosen to present and analyze the KAS2-bilateral-confirmation protocol because it offers the most security attributes of the four KAS2 variants and is most likely to be deployed in applications that wish to be compliant with

U. Parampalli and P. Hawkes (Eds.): ACISP 2011, LNCS 6812, pp. 353–370, 2011.
© Springer-Verlag Berlin Heidelberg 2011

SP 800-56B. We begin in §2 by introducing a generic three-pass key agreement protocol based on a certain kind of trapdoor one-way function family. We present in §3 a variant of the extended Canetti-Krawczyk security model for key agreement [5,12] that we believe captures all the essential security properties of the generic protocol. The security of the generic protocol can be argued under appropriate assumptions on the trapdoor one-way function family. For the sake of concreteness, we omit the reductionist security proof of the generic protocol and focus instead on two specific instantiations.

When specialized to the RSA setting, the generic protocol yields the KAS2 scheme which is presented and analyzed in §4. When specialized to the discrete log setting, we obtain a new protocol which we call DH2 and analyze in §5. DH2 is similar to the KEA+ protocol studied in [13]. An interesting feature of DH2 is that parties can use different groups (e.g., different elliptic curves) provided, of course, that each party is capable of performing operations in the other party's group. The generic protocol also has a hybrid implementation, where one party has an RSA key pair and the other party has a discrete log key pair. The hybrid protocol, the KAS1 protocol, and some concerns with reusing static key pairs in more than one protocol are briefly discussed in §6.

2   A Generic Protocol

The generic protocol utilizes a family of trapdoor one-way functions which we informally define next. Each function $f : Z \to Z$ from the family is bijective and has the following properties: (i) there is an efficient algorithm that outputs $(X, f(X))$ with $X \in_R Z$;[1] (ii) given $f(X)$ for $X \in_R Z$, it is infeasible to determine X; (iii) there exists some trapdoor data T_f, knowledge of which allows one to efficiently compute X given $f(X)$ for $X \in_R Z$.

An example of such a trapdoor one-way function is $f_{N,e} : \mathbb{Z}_N \to \mathbb{Z}_N$ defined by $f_{N,e}(m) = m^e \bmod N$, where (N, e) is an RSA public key. The trapdoor data is the corresponding RSA private key d.

Another example comes from discrete log cryptography. Let $\mathbb{G} = \langle g \rangle$ be a cyclic group of prime order q, let $a \in_R [1, q-1]$, and let $A = g^a$. Then $f_A : \mathbb{G} \to \mathbb{G}$ defined by $f(g^x) = A^x$ is such a trapdoor one-way function with trapdoor data a. Inversion of f without knowledge of a is infeasible provided that the following *Diffie-Hellman division (DHD) problem* is intractable: given $g, A^x, A \in \mathbb{G}$, determine g^x [2].

In the generic protocol, depicted in Figure 1, party $\hat{A}$'s static public key is a trapdoor function $f_A : Z_A \to Z_A$, and the corresponding trapdoor data T_A is her static private key. Similarly, party $\hat{B}$'s static public key is the trapdoor function $f_B : Z_B \to Z_B$ and the corresponding trapdoor data T_B is his static public key. We let MAC denote a secure message authentication code algorithm such as HMAC, and denote by $\mathcal{I}$ and $\mathcal{R}$ the constant strings "KC_2_U" and "KC_2_V" [17].

---

[1] Requirement (i) is different than the usual notion of one-wayness, which is the existence of an efficient algorithm for computing $f(X)$ *given* $X \in_R Z$.

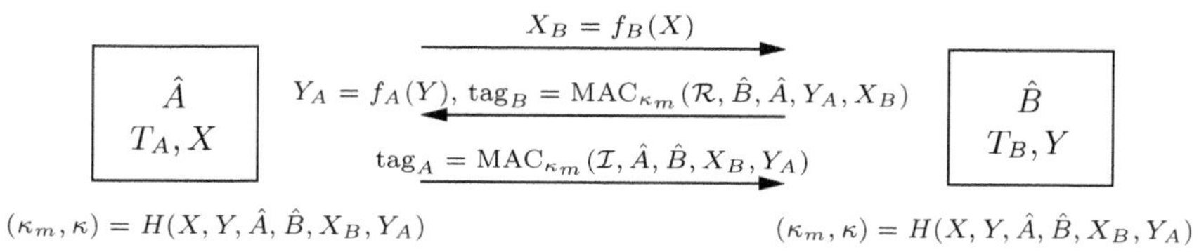

$$X_B = f_B(X)$$

$$\boxed{\begin{array}{c} \hat{A} \\ T_A, X \end{array}} \quad \begin{array}{c} Y_A = f_A(Y),\ \mathrm{tag}_B = \mathrm{MAC}_{\kappa_m}(\mathcal{R}, \hat{B}, \hat{A}, Y_A, X_B) \\[2mm] \mathrm{tag}_A = \mathrm{MAC}_{\kappa_m}(\mathcal{I}, \hat{A}, \hat{B}, X_B, Y_A) \end{array} \quad \boxed{\begin{array}{c} \hat{B} \\ T_B, Y \end{array}}$$

$(\kappa_m, \kappa) = H(X, Y, \hat{A}, \hat{B}, X_B, Y_A)$ $\qquad\qquad\qquad\qquad\qquad$ $(\kappa_m, \kappa) = H(X, Y, \hat{A}, \hat{B}, X_B, Y_A)$

Fig. 1. A generic three-pass protocol

Definition 1 (generic protocol). The generic protocol proceeds as follows:

1. Upon receiving $(\hat{A}, \hat{B})$, party $\hat{A}$ (the initiator) does the following:
   (a) Select $(X, X_B = f_B(X))$ with $X \in_R Z_B$; X is $\hat{A}$'s ephemeral private key and $f_B(X)$ is the corresponding ephemeral public key.
   (b) Initialize the session identifier to $(\hat{A}, \hat{B}, \mathcal{I}, X_B)$.
   (c) Send $(\hat{B}, \hat{A}, \mathcal{R}, X_B)$ to $\hat{B}$.
2. Upon receiving $(\hat{B}, \hat{A}, \mathcal{R}, X_B)$, party $\hat{B}$ (the responder) does the following:
   (a) Verify that $X_B \in Z_B$.
   (b) Select $(Y, Y_A = f_A(Y))$ with $Y \in_R Z_A$; Y is $\hat{B}$'s ephemeral private key and $f_A(Y)$ is the corresponding ephemeral public key.
   (c) Compute $X = f_B^{-1}(X_B)$ using trapdoor data T_B.
   (d) Compute $(\kappa_m, \kappa) = H(X, Y, \hat{A}, \hat{B}, X_B, Y_A)$.
   (e) Compute $\mathrm{tag}_B = \mathrm{MAC}_{\kappa_m}(\mathcal{R}, \hat{B}, \hat{A}, Y_A, X_B)$.
   (f) Compute $\mathrm{tag}_A = \mathrm{MAC}_{\kappa_m}(\mathcal{I}, \hat{A}, \hat{B}, X_B, Y_A)$ and store it.
   (g) Destroy X, Y and κ_m.
   (h) Send $(\hat{A}, \hat{B}, \mathcal{I}, X_B, Y_A, \mathrm{tag}_B)$ to $\hat{A}$.
   (i) Set the session identifier to $(\hat{B}, \hat{A}, \mathcal{R}, Y_A, X_B)$.
3. Upon receiving $(\hat{A}, \hat{B}, \mathcal{I}, X_B, Y_A, \mathrm{tag}_B)$, party $\hat{A}$ does the following:
   (a) Verify that an active session $(\hat{A}, \hat{B}, \mathcal{I}, X_B)$ exists and $Y_A \in Z_A$.
   (b) Compute $Y = f_A^{-1}(Y_A)$ using trapdoor data T_A.
   (c) Compute $(\kappa_m, \kappa) = H(X, Y, \hat{A}, \hat{B}, X_B, Y_A)$.
   (d) Verify that $\mathrm{tag}_B = \mathrm{MAC}_{\kappa_m}(\mathcal{R}, \hat{B}, \hat{A}, Y_A, X_B)$.
   (e) Compute $\mathrm{tag}_A = \mathrm{MAC}_{\kappa_m}(\mathcal{I}, \hat{A}, \hat{B}, X_B, Y_A)$.
   (f) Destroy X, Y and κ_m.
   (g) Send $(\hat{B}, \hat{A}, \mathcal{R}, Y_A, X_B, \mathrm{tag}_A)$ to $\hat{B}$.
   (h) Update the session identifier to $(\hat{A}, \hat{B}, \mathcal{I}, X_B, Y_A)$ and complete the session by accepting κ as the session key.
4. Upon receiving $(\hat{B}, \hat{A}, \mathcal{R}, Y_A, X_B, \mathrm{tag}_A)$, party $\hat{B}$ does the following:
   (a) Verify that an active session $(\hat{B}, \hat{A}, \mathcal{R}, Y_A, X_B)$ exists.
   (b) Verify that the received tag_A is equal to the one stored.
   (c) Complete session $(\hat{B}, \hat{A}, \mathcal{R}, Y_A, X_B)$ by accepting κ as the session key.

3   Security Model

This section describes a security model and associated security definition that aims to capture the essential security assurances provided by the generic key agreement protocol presented in §2. A characteristic feature of this protocol is that the session key is computed by hashing individual ephemeral private keys and some public information. In particular, the session key does not depend on the static keys of the participating parties. (Static private keys are used as trapdoors to extract the other party's ephemeral private key from its ephemeral public key.) We follow the eCK model [5,12], but the definition of a fresh session deviates from the standard definition and is specifically crafted keeping the above characteristic in mind (cf. Remarks 1, 2 and 3).

Session creation. A party $\hat{A}$ can be activated via an incoming message to create a session. The incoming message has one of the following forms: (i) $(\hat{A}, \hat{B})$ or (ii) $(\hat{A}, \hat{B}, \mathcal{R}, In)$. If $\hat{A}$ was activated with $(\hat{A}, \hat{B})$ then $\hat{A}$ is the session *initiator*; otherwise $\hat{A}$ is the session *responder*.

Session initiator. If $\hat{A}$ is the session initiator then $\hat{A}$ creates a separate session state where session-specific short-lived data is stored, and prepares a reply $Out = (f_B(X), OtherInfo)$, where f_B is $\hat{B}$'s static public key, X is $\hat{A}$'s ephemeral private key, and $OtherInfo$ is additional data that the protocol may specify. The session is labeled *active* and identified via a (temporary and incomplete) session identifier $s = (\hat{A}, \hat{B}, \mathcal{I}, f_B(X))$. The outgoing message is $(\hat{B}, \hat{A}, \mathcal{R}, Out)$.

Session responder. If $\hat{A}$ is the session responder then $\hat{A}$ creates a separate session state and prepares a reply Out that includes $f_B(X)$ where f_B is $\hat{B}$'s static public key and X is $\hat{A}$'s ephemeral private key. The session is labeled active and identified via a session identifier $s = (\hat{A}, \hat{B}, \mathcal{R}, f_B(X), f_A(Y))$, where $f_A(Y)$ is the ephemeral public key in the incoming message In. The outgoing message is $(\hat{B}, \hat{A}, \mathcal{I}, f_A(Y), Out)$.

Session update. A party $\hat{A}$ can be activated to update a session via an incoming message of the form $(\hat{A}, \hat{B}, role, f_B(X), f_A(Y), In)$, where $role \in \{\mathcal{I}, \mathcal{R}\}$. Upon receipt of this message, $\hat{A}$ checks that she owns an active session with identifier $s = (\hat{A}, \hat{B}, role, f_B(X), f_A(Y))$. Since ephemeral keys are chosen uniformly at random from the appropriate domain, except with negligible probability $\hat{A}$ can own at most one such session. If no such session exists then the message is rejected; otherwise $\hat{A}$ follows the protocol specifications. Initiator $\hat{A}$ can also be activated to update a session with incomplete session identifier $(\hat{A}, \hat{B}, \mathcal{I}, f_B(X))$ with an incoming message of the form $(\hat{A}, \hat{B}, \mathcal{I}, f_B(X), f_A(Y), In)$ where In is any message specified by the protocol. In this case $\hat{A}$ performs the required validations before updating the session identifier to $(\hat{A}, \hat{B}, \mathcal{I}, f_B(X), f_A(Y))$.

Completed sessions. If the protocol stipulates that no further messages are to be received then the session owner accepts a session key and marks the session as completed.

Aborted sessions. A protocol may require parties to perform some checks on incoming messages. For example, a party may be required to perform some form of public key validation or verify a message authentication tag. If a party is activated to create a session with an incoming message that does not meet the protocol specifications, then that message is rejected and no session is created. If a party is activated to update an active session with an incoming message that does not meet the protocol specifications, then the party deletes all information specific to that session (including the session state and the session key if it has been computed) and *aborts* the session. Abortion occurs before the session identifier is updated.

Matching sessions. Since ephemeral keys are selected at random on a per-session basis, session identifiers are unique except with negligible probability. Party $\hat{A}$ is said to be the *owner* of a session $(\hat{A}, \hat{B}, role, *, *)$, where $role \in \{\mathcal{I}, \mathcal{R}\}$. For a session $(\hat{A}, \hat{B}, role, *, *)$ we call $\hat{B}$ the session *peer*; together $\hat{A}$ and $\hat{B}$ are referred to as the *communicating parties*. Let s be a session with complete session identifier $(\hat{A}, \hat{B}, role_A, f_B(X), f_A(Y))$ where $role_A \in \{\mathcal{I}, \mathcal{R}\}$. A session s^* with session identifier $(\hat{C}, \hat{D}, role_C, f_D(U), f_C(V))$, where $role_C \in \{\mathcal{I}, \mathcal{R}\}$, is said to be *matching* to s if $\hat{A} = \hat{D}$, $\hat{B} = \hat{C}$, $role_A \neq role_C$, $f_B(X) = f_C(V)$ and $f_A(Y) = f_D(U)$. A session s with incomplete session identifier $(\hat{A}, \hat{B}, \mathcal{I}, f_B(X))$ is matching to any session $s^* = (\hat{C}, \hat{D}, \mathcal{R}, f_D(U), f_C(V))$ with $\hat{A} = \hat{D}$, $\hat{B} = \hat{C}$ and $f_B(X) = f_C(V)$; s^* is also matching to s. Since ephemeral keys are selected at random on a per-session basis, only sessions with incomplete session identifiers can have more than one matching session.

Adversary. The adversary $\mathcal{M}$ is modeled as a probabilistic Turing machine and controls *all* communications. Parties submit outgoing messages to $\mathcal{M}$, who makes decisions about their delivery. The adversary presents parties with incoming messages via *Send*(message), thereby controlling the activation of parties. The adversary does not have immediate access to a party's private information, however in order to capture possible leakage of private information $\mathcal{M}$ is allowed to make the following queries:

- *StaticKeyReveal*($\hat{A}$): $\mathcal{M}$ obtains $\hat{A}$'s static private key.
- *Expire*(s): The owner of s deletes the session key associated with s if one exists, and labels the session expired. We henceforth assume that $\mathcal{M}$ issues this query only to completed sessions. At any point in time a session is in exactly one of the following states: active, completed, aborted, expired.
- *EphemeralKeyReveal*(s): $\mathcal{M}$ obtains the ephemeral private key held by session s. We will henceforth assume that $\mathcal{M}$ issues this query only to sessions that hold an ephemeral private key.
- *SessionKeyReveal*(s): If s has completed and has not been expired, $\mathcal{M}$ obtains the session key held by s. We will henceforth assume that $\mathcal{M}$ issues this query only to sessions that have completed and have not been expired.
- *EstablishParty*($\hat{A}, A$): This query allows $\mathcal{M}$ to register an identifier $\hat{A}$ and a static public key A on behalf of a party. The adversary totally controls that party, thus permitting the modeling of attacks by malicious insiders. Parties that were established by $\mathcal{M}$ using *EstablishParty* are called *corrupted* or *adversary controlled*. If a party is not corrupted it is said to be *honest*.

Adversary's goal. To capture indistinguishability $\mathcal{M}$ is allowed to make a special query *Test*(s) to a 'fresh' session s. In response, $\mathcal{M}$ is given with equal probability either the session key held by s or a random key. If $\mathcal{M}$ guesses correctly whether the key is random or not, then the adversary is said to be successful and meets its goal. Note that $\mathcal{M}$ can continue interacting with the parties after issuing the *Test* query, but must ensure that the test session remains fresh throughout $\mathcal{M}$'s experiment.

Definition 2 (fresh session). Let s be the identifier of a completed session, owned by an honest party $\hat{A}$ with peer $\hat{B}$, who is also honest. Let s^* be the identifier of the matching session of s, if the matching session exists. Define s to be *fresh* if none of the following conditions hold:

1. $\mathcal{M}$ issued *SessionKeyReveal(s)* or *SessionKeyReveal(s*)* (if s^* exists).
2. s^* exists and $\mathcal{M}$ issued one of the following:
   (a) Both *StaticKeyReveal($\hat{A}$)* and *EphemeralKeyReveal(s)*.
   (b) Both *StaticKeyReveal($\hat{B}$)* and *EphemeralKeyReveal(s*)*.
   (c) Both *StaticKeyReveal($\hat{A}$)* and *StaticKeyReveal($\hat{B}$)*.
   (d) Both *EphemeralKeyReveal(s)* and *EphemeralKeyReveal(s*)*.
3. s^* does not exist and $\mathcal{M}$ issued one of the following:
   (a) *EphemeralKeyReveal(s)*.
   (b) *StaticKeyReveal($\hat{B}$)* before *Expire(s)*.

Definition 3 (secure key agreement protocol). A key agreement protocol is said to be *secure* in the above model if the following conditions hold:

1. If two honest parties complete matching sessions then, except with negligible probability, they both compute the same session key.
2. No polynomially bounded adversary $\mathcal{M}$ can distinguish the session key of a fresh session from a randomly chosen session key with probability greater than $\frac{1}{2}$ plus a negligible fraction.

Remark 1. (comparing Definition 2 with the notion of freshness in [12]) Our definition of fresh session is more restrictive than the corresponding definition of fresh session in the eCK model [12]. We have added two more sub-conditions, namely 2(c) and 2(d) and also made condition 3(a) more restrictive. Conditions 2(c) and 2(d) are needed because of the nature of the generic protocol wherein the only secret inputs to the key derivation function are the ephemeral private keys, and the static keys are only used to extract the ephemeral private keys from the ephemeral public keys that are exchanged. Condition 3(a) is defined this way because an active adversary who learns the ephemeral private key of a party for a particular session can impersonate others to the party in that session.

Remark 2. (comparing Definition 2 with the notion of freshness in [5]) Our model is stronger than the CK model [5] in that it incorporates resistance to key-compromise impersonation (KCI) attacks [10]; that is, an adversary who learns a party's static private key is unable to impersonate other entities to that party. The model also covers half-forward secrecy, wherein the security of a session key is preserved even if an adversary subsequently learns the static private keys of one (but not both) of the communicating parties.

Remark 3. (EphemeralKeyReveal vs. SessionStateReveal) Unlike the CK model, our model is not equipped with a *SessionStateReveal* query with which the adversary can learn all the secret information contained in an active session. This deficiency is partly mitigated by inclusion of the *EphemeralKeyReveal* query and by considering the session state to consist of the ephemeral private key of the session's owner. Observe also that if our model were to incorporate a *SessionStateReveal* query, then the protocol must specify that the session state cannot include the peer's ephemeral private key. Otherwise, the adversary could compute the session key of the *Test* session (thereby breaking the protocol) by replaying the ephemeral public keys to the relevant parties, and subsequently learning the ephemeral private keys with *SessionStateReveal* queries; the adversary would then have all elements needed to compute the session key of the Test session.

4 The RSA Setting

Let λ be a security parameter. On input 1^λ, a party selects an RSA static public key (N, e) by randomly selecting two primes p and q of the same bitlength (determined by λ) and choosing an arbitrary integer $e \in [3, N-2]$ relatively prime to $\phi(N)$; the party's corresponding static private key is $d = e^{-1} \bmod \phi(N)$. Party $\hat{A}$'s static key pair is denoted by (N_A, e_A) and d_A. Similarly, party $\hat{B}$'s static key pair is denoted by (N_B, e_B) and d_B. A certifying authority issues certificates that binds a party's identifier to its static public key. The protocol description will omit the exchange of certificates.

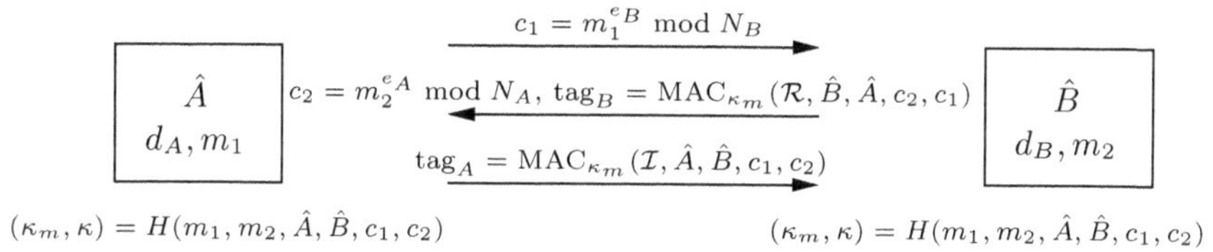

$$c_1 = m_1^{e_B} \bmod N_B$$

$\hat{A}$ d_A, m_1 $c_2 = m_2^{e_A} \bmod N_A,\ \mathrm{tag}_B = \mathrm{MAC}_{\kappa_m}(\mathcal{R}, \hat{B}, \hat{A}, c_2, c_1)$ $\hat{B}$ d_B, m_2

$$\mathrm{tag}_A = \mathrm{MAC}_{\kappa_m}(\mathcal{I}, \hat{A}, \hat{B}, c_1, c_2)$$

$(\kappa_m, \kappa) = H(m_1, m_2, \hat{A}, \hat{B}, c_1, c_2)$ $(\kappa_m, \kappa) = H(m_1, m_2, \hat{A}, \hat{B}, c_1, c_2)$

Fig. 2. The KAS2 protocol

Definition 4 (KAS2 protocol [17]). The KAS2 protocol proceeds as follows:

1. Upon receiving $(\hat{A}, \hat{B})$, party $\hat{A}$ (the initiator) does the following:
   (a) Select an ephemeral private key $m_1 \in_R [2, N_B - 2]$ and compute the ephemeral public key $c_1 = m_1^{e_B} \bmod N_B$.
   (b) Initialize the session identifier to $(\hat{A}, \hat{B}, \mathcal{I}, c_1)$.
   (c) Send $(\hat{B}, \hat{A}, \mathcal{R}, c_1)$ to $\hat{B}$.
2. Upon receiving $(\hat{B}, \hat{A}, \mathcal{R}, c_1)$, party $\hat{B}$ (the responder) does the following:
   (a) Verify that $c_1 \in [2, N_B - 2]$ and compute $m_1 = c_1^{d_B} \bmod N_B$.
   (b) Select an ephemeral private key $m_2 \in_R [2, N_A - 2]$ and compute the ephemeral public key $c_2 = m_2^{e_A} \bmod N_A$.
   (c) Compute $(\kappa_m, \kappa) = H(m_1, m_2, \hat{A}, \hat{B}, c_1, c_2)$.

    (d) Compute $\text{tag}_B = \text{MAC}_{\kappa_m}(\mathcal{R}, \hat{B}, \hat{A}, c_2, c_1)$.

    (e) Compute $\text{tag}_A = \text{MAC}_{\kappa_m}(\mathcal{I}, \hat{A}, \hat{B}, c_1, c_2)$ and store it.

    (f) Destroy m_1, m_2 and κ_m.

    (g) Send $(\hat{A}, \hat{B}, \mathcal{I}, c_1, c_2, \text{tag}_B)$ to $\hat{A}$.

    (h) Set the session identifier to $(\hat{B}, \hat{A}, \mathcal{R}, c_2, c_1)$.

3. Upon receiving $(\hat{A}, \hat{B}, \mathcal{I}, c_1, c_2, \text{tag}_B)$, party $\hat{A}$ does the following:

    (a) Verify that an active session $(\hat{A}, \hat{B}, \mathcal{I}, c_1)$ exists.

    (b) Verify that $c_2 \in [2, N_A - 2]$ and compute $m_2 = c_2^{d_A} \bmod N_A$.

    (c) Compute $(\kappa_m, \kappa) = H(m_1, m_2, \hat{A}, \hat{B}, c_1, c_2)$.

    (d) Verify that $\text{tag}_B = \text{MAC}_{\kappa_m}(\mathcal{R}, \hat{B}, \hat{A}, c_2, c_1)$.

    (e) Compute $\text{tag}_A = \text{MAC}_{\kappa_m}(\mathcal{I}, \hat{A}, \hat{B}, c_1, c_2)$.

    (f) Destroy m_1, m_2 and κ_m.

    (g) Send $(\hat{B}, \hat{A}, \mathcal{R}, c_2, c_1, \text{tag}_A)$ to $\hat{B}$.

    (h) Update the session identifier to $(\hat{A}, \hat{B}, \mathcal{I}, c_1, c_2)$ and complete the session by accepting κ as the session key.

4. Upon receiving $(\hat{B}, \hat{A}, \mathcal{R}, c_2, c_1, \text{tag}_A)$, party $\hat{B}$ does the following:

    (a) Verify that an active session $(\hat{B}, \hat{A}, \mathcal{R}, c_2, c_1)$ exists.

    (b) Verify that the received tag_A is equal to the one stored.

    (c) Complete session $(\hat{B}, \hat{A}, \mathcal{R}, c_2, c_1)$ by accepting κ as the session key.

4.1 Comparisons

We note that the key derivation function H in [17] also includes an integer keydatalen that indicates the bitlength of the secret keying material to be generated, a bit string AlgorithmID that indicates how the derived keying material will be parsed and for which algorithm it will be used, and two optional strings SuppPubInfo and SuppPrivInfo. We have chosen to include (c_1, c_2) in the SuppPubInfo field as it simplifies the security reduction. The strings keydatalen, AlgorithmID and SuppPrivInfo are omitted because they are not relevant to our security analysis.

    An important difference between KAS2 and KAS2-basic (the two-pass variant of KAS2 without the key confirmation messages tag_A and tag_B) is that KAS2-basic provides a weaker notion of half-forward secrecy than KAS2. Namely, if the adversary learns the static private key of one of the communicating parties of a KAS2-basic session after the session key has been established, then security of the session key is only guaranteed if the session was 'clean', i.e., was free from active adversarial intrusion.

4.2 Security Argument

The *RSA problem* is to determine the integer $m \in [2, N-2]$ such that $c \equiv m^e \pmod{N}$ given an RSA public key (N, e) and an integer $c \in_R [2, N-2]$. The *RSA assumption* is that no polynomially-bounded algorithm exists that solves the RSA problem with non-negligible probability of success.

Theorem 1. *Suppose that (i) the* RSA *assumption holds; (ii) the* MAC *scheme is secure; and (iii) H is a random oracle. Then the KAS2 key agreement protocol is secure.*

Proof. It is easy to see that matching sessions produce the same session key. We will verify that for a security parameter λ, no polynomially-bounded adversary $\mathcal{M}$ can distinguish the session key of a fresh session from a randomly chosen session key with probability $\frac{1}{2} + p(\lambda)$ for some non-negligible function $p(\lambda)$.

Let M denote the event that $\mathcal{M}$ succeeds in the distinguishing game, and suppose that $\Pr(M) = \frac{1}{2} + p(\lambda)$ where $p(\lambda)$ is non-negligible. We assume that $\mathcal{M}$ operates in an environment with n parties, and where each party is activated at most t times to create a new session. We will show how $\mathcal{M}$ can be used to construct a polynomial-time algorithm $\mathcal{S}$ that, with non-negligible probability of success, either solves an instance of the RSA problem or produces a MAC forgery.

Since H is modeled as a random function, $\mathcal{M}$ has only two strategies for winning the distinguishing game with probability significantly greater than $\frac{1}{2}$:

(i) induce two non-matching sessions to establish the same session key, set one as the test session, and thereafter issue a *SessionKeyReveal* query to the other; or
(ii) query oracle H with $(c_1^{d_B} \bmod N_B, c_2^{d_A} \bmod N_A, \hat{A}, \hat{B}, c_1, c_2)$ where the test session is either $(\hat{A}, \hat{B}, \mathcal{I}, c_1, c_2)$ or $(\hat{B}, \hat{A}, \mathcal{R}, c_2, c_1)$.

Since the input to the key derivation function includes the identities of the communicating parties and the exchanged ephemeral public keys, non-matching completed sessions produce different session keys except with negligible probability of H collisions. This rules out strategy (i).

Now, let H^* denote the event that $\mathcal{M}$ queries H with $(c_1^{d_B} \bmod N_B, c_2^{d_A} \bmod N_A, \hat{A}, \hat{B}, c_1, c_2)$ where the test session is either $(\hat{A}, \hat{B}, \mathcal{I}, c_1, c_2)$ or $(\hat{B}, \hat{A}, \mathcal{R}, c_2, c_1)$. Since H is a random function, we have $\Pr(M|\overline{H^*}) = \frac{1}{2}$ where negligible terms are ignored. Hence

$$\Pr(M) = \Pr(M \wedge H^*) + \Pr(M|\overline{H^*})\Pr(\overline{H^*}) \leq \Pr(M \wedge H^*) + \frac{1}{2},$$

so $\Pr(M \wedge H^*) \geq p(\lambda)$. The event $M \wedge H^*$ will henceforth be denoted by M^*.

Let s^t denote the test session selected by $\mathcal{M}$, and let s^m denote its matching session (if it exists). Consider the following complementary events:

1. Event E_1: s^m exists and $\mathcal{M}$ issues neither *StaticKeyReveal*$(\hat{A})$ nor *EphemeralKeyReveal*(s^m).
2. Event E_2: either s^m does not exist, or s^m exists and $\mathcal{M}$ issues *StaticKeyReveal*$(\hat{A})$ or *EphemeralKeyReveal*(s^m).

We have $M^* = (M^* \wedge E_1) \vee (M^* \wedge E_2)$. Since $\Pr(M^*)$ is non-negligible, it must be the case that either $p_1 = \Pr(M^* \wedge E_1)$ or $p_2 = \Pr(M^* \wedge E_2)$ is non-negligible. The events E_1 and E_2 are analyzed separately.

We will show how to construct a solver $\mathcal{S}$ that takes as input an RSA challenge (N_V, e_V, c_V), has access to a MAC oracle with unknown key $\tilde{\kappa}_m$ and to an adversary $\mathcal{M}$, and produces a solution to the RSA challenge or a MAC forgery.

Setup. Algorithm $\mathcal{S}$ begins by establishing n parties. One of these parties, denoted $\hat{V}$, is selected at random and assigned the static public key (N_V, e_V).

The remaining parties are assigned static key pairs as specified by the protocol. Furthermore, $\mathcal{S}$ selects an integer $u \in_R [1, nt]$. The u'th session created will be called s^u. For this session, $\mathcal{S}$ deviates from the protocol description as follows: if the peer of s^u is $\hat{V}$ then c_V is chosen as the outgoing ephemeral public key; otherwise, $\mathcal{S}$ aborts with failure. For all other sessions, $\mathcal{S}$ selects ephemeral key pairs as specified by the protocol.

χ-function. During the simulation, $\mathcal{S}$ constructs a secret function $\chi : [2, N_V - 2] \to [2, N_V - 2]$. At the beginning of the simulation, $\chi(c)$ is undefined for all $c \in [2, N_V - 2]$. At any stage of the simulation, if $\mathcal{S}$ selects $m \in [2, N_V - 2]$ and computes $c = m^{ev} \bmod N_V$ as an outgoing ephemeral public key with intended recipient $\hat{V}$, then $\chi(c)$ is defined to be m. If χ is ever invoked by $\mathcal{S}$ for its value at an input c, and $\chi(c)$ is undefined, then $\chi(c)$ is set equal to a randomly selected integer in $[2, N_V - 2]$; in this case $\chi(c)$ is said to 'represent' $c^{1/ev} \bmod N_V$. Except with negligible probability, $\mathcal{M}$ will not detect that χ is being used.

Event E_1. The simulation of $\mathcal{M}$'s environment proceeds as follows:

1. $Send(\hat{A}, \hat{B})$. $\mathcal{S}$ answers the query faithfully with the following exception. If the session activated is s^u then $\mathcal{S}$ proceeds as stipulated in the Setup (and aborts if $\hat{B} \neq \hat{V}$).

2. $Send(\hat{B}, \hat{A}, \mathcal{R}, c_1)$. $\mathcal{S}$ answers the query faithfully with the following exceptions. (a) If $\hat{B} = \hat{V}$ then $\mathcal{S}$ sets $m_1 = \chi(c_1)$. (b) If the session activated is s^u then $\mathcal{S}$ proceeds as stipulated in the Setup and sets $m_2 = \chi(c_V)$ (and aborts if $\hat{A} \neq \hat{V}$).

3. $Send(\hat{A}, \hat{B}, \mathcal{I}, c_1, c_2, \text{tag}_B)$. $\mathcal{S}$ answers the query faithfully with the following exceptions. (a) If $\hat{A} - V$ then $\mathcal{S}$ sets $m_2 = \chi(c_2)$. (b) If the session activated is s^u then $\mathcal{S}$ sets $m_1 = \chi(c_V)$.

4. $Send(\hat{B}, \hat{A}, \mathcal{R}, c_2, c_1, \text{tag}_A)$. $\mathcal{S}$ answers the query faithfully.

5. $H(m_1, m_2, \hat{A}, \hat{B}, c_1, c_2)$.
   (a) If (i) $\hat{A} = \hat{V}$, $c_2 = c_V$, and $m_2^{ev} \equiv c_V \pmod{N_V}$, or (ii) $\hat{B} = \hat{V}$, $c_1 = c_V$, and $m_1^{ev} \equiv c_V \pmod{N_V}$, then $\mathcal{S}$ terminates $\mathcal{M}$ and successfully completes by outputting m_2 or m_1, respectively.
   (b) If $\hat{A} = \hat{V}$, $m_2 \neq \chi(c_2)$, and $m_2^{ev} \equiv c_2 \pmod{N_V}$, then $\mathcal{S}$ responds with $H(m_1, \chi(c_2), \hat{A}, \hat{B}, c_1, c_2)$; otherwise, $\mathcal{S}$ simulates a random oracle in the usual way[2].
   (c) If $\hat{B} = \hat{V}$, $m_1 \neq \chi(c_1)$, and $m_1^{ev} \equiv c_1 \pmod{N_V}$, then $\mathcal{S}$ responds with $H(\chi(c_1), m_2, \hat{A}, \hat{B}, c_1, c_2)$; otherwise, $\mathcal{S}$ simulates a random oracle in the usual way.
   (d) $\mathcal{S}$ simulates a random oracle in the usual way.

6. $SessionKeyReveal(s)$. $\mathcal{S}$ answers the query faithfully.

7. $StaticKeyReveal(\hat{A})$. If $\hat{A} = \hat{V}$ then $\mathcal{S}$ aborts with failure; otherwise, $\mathcal{S}$ answers the query faithfully.

8. $EphemeralKeyReveal(s)$. If $s = s^u$ then $\mathcal{S}$ aborts with failure; otherwise, $\mathcal{S}$ answers the query faithfully.

---

[2] i.e., $\mathcal{S}$ returns random values for new queries and replays answers if the queries were previously made.

9. *Expire(s)*. $\mathcal{S}$ answers the query faithfully.
10. *EstablishParty*. $\mathcal{S}$ answers the query faithfully.
11. *Test(s^t)*. If s^t is not owned by $\hat{V}$ or if s^t is not matching to s^u, then $\mathcal{S}$ aborts with failure; otherwise, $\mathcal{S}$ answers the query faithfully.
12. $\mathcal{M}$ outputs a guess γ. $\mathcal{S}$ aborts with failure.

Event E_1 analysis. The simulation of $\mathcal{M}$'s environment can be seen to be perfect except with negligible probability. The probability that $\mathcal{M}$ selects a test session owned by $\hat{V}$ with matching session s^u is at least $1/n^2 t$. Suppose that this is indeed the case and suppose that event $M^* \wedge E_1$ occurs. Since the test session owner is $\hat{V}$ and the matching session is s^u, $\mathcal{M}$ does not abort as in Steps 1, 2 and 11. Furthermore, under event E_1 the adversary $\mathcal{M}$ does not query for the static private key of the test session owner or for the ephemeral private key of the matching session. Therefore, abortions as in Steps 7 and 8 do not occur. Under event M^*, the adversary $\mathcal{M}$ queries H with $c_V^{1/e_V} \bmod N_V$ before outputting a guess γ and hence $\mathcal{S}$ is successful in Step 5a before a failure in Step 12 occurs. The probability that $\mathcal{S}$ successfully outputs a solution to the RSA challenge is thus bounded by

$$\Pr(S) \geq \frac{p_1}{n^2 t}. \tag{1}$$

Event E_2. The Setup and the definition of the χ-function are the same as for Event E_1. During the simulation, $\mathcal{S}$ also accesses a MAC oracle with key $\tilde{\kappa}_m$ that is unknown to $\mathcal{S}$. The simulation of $\mathcal{M}$'s environment proceeds as follows:

1. *Send($\hat{A}, \hat{B}$)*. $\mathcal{S}$ answers the query faithfully with the following exception. If the session activated is s^u then $\mathcal{S}$ proceeds as stipulated in the Setup (and aborts if $\hat{B} \neq \hat{V}$).
2. *Send($\hat{B}, \hat{A}, \mathcal{R}, c_1$)*. $\mathcal{S}$ answers the query faithfully with the following exceptions. (a) If $\hat{B} = \hat{V}$ then $\mathcal{S}$ sets $m_1 = \chi(c_1)$. (b) If the session activated is s^u then $\mathcal{S}$ proceeds as stipulated in the Setup (and aborts if $\hat{A} \neq \hat{V}$); furthermore, instead of querying the key derivation function H to generate a MAC key and a session key, $\mathcal{S}$ selects a random session key κ, sets the MAC key κ_m equal to the (unknown) key $\tilde{\kappa}_m$ of the MAC oracle, queries the MAC oracle with $(\mathcal{R}, \hat{B}, \hat{A}, c_2, c_1)$, and sets tag_B equal to the oracle's response; tag_A is not computed.
3. *Send($\hat{A}, \hat{B}, \mathcal{I}, c_1, c_2, \text{tag}_B$)*. $\mathcal{S}$ answers the query faithfully with the following exceptions. (a) If $\hat{A} = \hat{V}$ then $\mathcal{S}$ sets $m_2 = \chi(c_2)$. (b) If the session activated is s^u then $\mathcal{S}$ selects a random session key κ, sets the MAC key κ_m equal to the (unknown) key $\tilde{\kappa}_m$ of the MAC oracle, queries the MAC oracle with $(\mathcal{I}, \hat{A}, \hat{B}, c_1, c_2)$, and sets tag_A equal to the oracle's response, and completes without verifying tag_B.
4. *Send($\hat{B}, \hat{A}, \mathcal{R}, c_2, c_1, \text{tag}_A$)*. $\mathcal{S}$ answers the query faithfully. However, if the session activated is s^u then $\mathcal{S}$ completes without verifying tag_A.
5. $H(m_1, m_2, \hat{A}, \hat{B}, c_1, c_2)$.
   (a) If (i) $\hat{A} = \hat{V}$, $c_2 = c_V$, and $m_2^{e_V} \equiv c_V \pmod{N_V}$, or (ii) $\hat{B} = \hat{V}$, $c_1 = c_V$, and $m_1^{e_V} \equiv c_V \pmod{N_V}$, then $\mathcal{S}$ terminates $\mathcal{M}$ and successfully completes by outputting m_2 or m_1, respectively.

(b) If $\hat{A} = \hat{V}$, $m_2 \neq \chi(c_2)$, and $m_2^{ev} \equiv c_2 \pmod{N_V}$, then $\mathcal{S}$ responds with $H(m_1, \chi(c_2), \hat{A}, \hat{B}, c_1, c_2)$; otherwise, $\mathcal{S}$ simulates a random oracle in the usual way.

(c) If $\hat{B} = \hat{V}$, $m_1 \neq \chi(c_1)$, and $m_1^{ev} \equiv c_1 \pmod{N_V}$, then $\mathcal{S}$ responds with $H(\chi(c_1), m_2, \hat{A}, \hat{B}, c_1, c_2)$; otherwise, $\mathcal{S}$ simulates a random oracle in the usual way.

(d) $\mathcal{S}$ simulates a random oracle in the usual way.

6. *SessionKeyReveal(s)*. $\mathcal{S}$ answers the query faithfully.
7. *StaticKeyReveal($\hat{A}$)*. If $\hat{A} = \hat{V}$ then $\mathcal{S}$ aborts with failure; otherwise, $\mathcal{S}$ answers the query faithfully.
8. *EphemeralKeyReveal(s)*. If $s = s^u$ then $\mathcal{S}$ aborts with failure; otherwise, $\mathcal{S}$ answers the query faithfully.
9. *Expire(s)*. $\mathcal{S}$ answers the query faithfully. However, if $s = s^u$ and s^u has no matching session, then $\mathcal{S}$ aborts with success and outputs as its MAC forgery the key confirmation tag received by s^u and the associated message.
10. *EstablishParty*. $\mathcal{S}$ answers the query faithfully.
11. *Test(s^t)*. If s^t is not s^u with peer $\hat{V}$, then $\mathcal{S}$ aborts with failure; otherwise, $\mathcal{S}$ answers the query faithfully.
12. $\mathcal{M}$ outputs a guess γ. $\mathcal{S}$ aborts with failure.

Event E_2 analysis. The simulation of $\mathcal{M}$'s environment can be seen to be perfect except with negligible probability. The probability that $\mathcal{M}$ selects s^u as the test session and s^u has peer $\hat{V}$ is at least $1/n^2 t$. Suppose that this is indeed the case and suppose that event $M^* \wedge E_2$ occurs. Since the test session is s^u and the session peer is $\hat{V}$, $\mathcal{M}$ does not abort as in Steps 1, 2 and 11. Now, by definition of a fresh session and of event E_2, the adversary $\mathcal{M}$ does not query for the ephemeral private key of the test session and so abortion as in Step 8 does not occur. Furthermore, $\mathcal{M}$ is allowed to query for the static private key of the test session peer only after expiring the test session; therefore, before an abortion can occur in Step 7, $\mathcal{S}$ will be successful in Step 9. Under event M^*, the adversary $\mathcal{M}$ queries H with $c_V^{1/ev} \bmod N_V$ before outputting a guess γ and hence $\mathcal{S}$ is successful in Step 5a before a failure in Step 12 occurs. The probability that $\mathcal{S}$ successfully outputs a solution to the RSA challenge or a valid MAC forgery is thus bounded by

$$\Pr(S) \geq \frac{p_2}{n^2 t}. \tag{2}$$

Overall analysis. By combining (1) and (2), we see that the success probability of $\mathcal{S}$ is bounded by

$$\Pr(S) \geq \frac{\max(p_1, p_2)}{n^2 t}. \tag{3}$$

During the simulation, $\mathcal{S}$ performs modular exponentiations and simulates a random oracle. All operations take polynomial time and hence $\mathcal{S}$'s running time is bounded by

$$\mathcal{T}_S \leq (4\mathcal{T}_{\bmod N} + 3\mathcal{T}_H)\mathcal{T}_{\mathcal{M}}, \tag{4}$$

where $\mathcal{T}_{\text{mod } N}$, $\mathcal{T}_H$, $\mathcal{T}_M$, respectively, denote the time to perform a modular exponentiation, the time to respond to an H query, and the running time of $\mathcal{M}$. Together (3) and (4) show that $\mathcal{S}$ is a polynomially-bounded algorithm that succeeds with non-negligible probability in either solving the RSA instance or in forging a MAC tag. This contradicts the assumptions of the theorem, thereby completing the argument. ☐

5   The Discrete Log Setting

Let λ be a security parameter. We let $\mathbb{G}$ be a cyclic group with security parameter λ; that is, $\mathbb{G}$ has prime order q with $2^{2\lambda} \leq q < 2^{2\lambda+1}$, and the fastest algorithm known for solving the discrete logarithm problem in $\mathbb{G}$ has running time approximately 2^λ. Examples of such groups include the group of points on carefully-chosen elliptic curves. Let $\mathcal{G}_\lambda = \{\mathbb{G}\}_k$ be a set of cyclic groups with security parameter λ and indexed by $k \in S_\lambda \subset \mathbb{N}$. We assume that DH2 users select a group $\mathbb{G} \in \mathcal{G}_\lambda$ uniformly at random, and subsequently select a generator $g \in_R \mathbb{G}$. For example, $\mathcal{G}_\lambda$ could consist of all cryptographically strong prime-order elliptic curves defined over prime fields $\mathbb{F}_p$ where p has bitlength 2λ; this corresponds to the case where users randomly generate their own elliptic curve parameters.

On input 1^λ, party $\hat{A}$ selects a cyclic group $\mathbb{G}_1 = \langle g_1 \rangle \in_R \mathcal{G}_\lambda$ of order q_1. Her static private key is $a \in_R [1, q_1 - 1]$ and her static public key is $A = g_1^a$. Similarly, party $\hat{B}$ selects a cyclic group $\mathbb{G}_2 = \langle g_2 \rangle \in_R \mathcal{G}_\lambda$ of order q_2. His static private key is $b \in_R [1, q_2 - 1]$ and his static public key is $B = g_2^b$. A certifying authority issues certificates that binds a party's identifier to its static public key (and also the group parameters if these are not clear from context). The protocol description will omit the exchange of certificates.

Definition 5 (DH2 protocol). The DH2 protocol proceeds as follows:

1. Upon receiving $(\hat{A}, \hat{B})$, party $\hat{A}$ (the initiator) does the following:
   (a) Select $x \in_R [1, q_2 - 1]$ and compute the ephemeral private key $X = g_2^x$ and the ephemeral public key $X_B = B^x$.
   (b) Destroy x.
   (c) Initialize the session identifier to $(\hat{A}, \hat{B}, \mathcal{I}, X_B)$.
   (d) Send $(\hat{B}, \hat{A}, \mathcal{R}, X_B)$ to $\hat{B}$.
2. Upon receiving $(\hat{B}, \hat{A}, \mathcal{R}, X_B)$, party $\hat{B}$ (the responder) does the following:
   (a) Verify that $X_B \in \mathbb{G}_2^*$.
   (b) Select $y \in_R [1, q_1 - 1]$ and compute the ephemeral private key $Y = g_1^y$ and the ephemeral public key $Y_A = A^y$.
   (c) Compute $X = (X_B)^{(1/b)}$.
   (d) Compute $(\kappa_m, \kappa) = H(X, Y, \hat{A}, \hat{B}, X_B, Y_A)$.
   (e) Compute $\text{tag}_B = \text{MAC}_{\kappa_m}(\mathcal{R}, \hat{B}, \hat{A}, Y_A, X_B)$.
   (f) Compute $\text{tag}_A = \text{MAC}_{\kappa_m}(\mathcal{I}, \hat{A}, \hat{B}, X_B, Y_A)$ and store it.
   (g) Destroy y, X, Y and κ_m.
   (h) Send $(\hat{A}, \hat{B}, \mathcal{I}, X_B, Y_A, \text{tag}_B)$ to $\hat{A}$.
   (i) Set the session identifier to $(\hat{B}, \hat{A}, \mathcal{R}, Y_A, X_B)$.

3. Upon receiving $(\hat{A}, \hat{B}, \mathcal{I}, X_B, Y_A, \mathrm{tag}_B)$, party $\hat{A}$ does the following:
   (a) Verify that an active session $(\hat{A}, \hat{B}, \mathcal{I}, X_B)$ exists and $Y_A \in \mathbb{G}_1^*$.
   (b) Compute $Y = (Y_A)^{1/a}$.
   (c) Compute $(\kappa_m, \kappa) = H(X, Y, \hat{A}, \hat{B}, X_B, Y_A)$.
   (d) Verify that $\mathrm{tag}_B = \mathrm{MAC}_{\kappa_m}(\mathcal{R}, \hat{B}, \hat{A}, Y_A, X_B)$.
   (e) Compute $\mathrm{tag}_A = \mathrm{MAC}_{\kappa_m}(\mathcal{I}, \hat{A}, \hat{B}, X_B, Y_A)$.
   (f) Destroy X, Y and κ_m.
   (g) Send $(\hat{B}, \hat{A}, \mathcal{R}, Y_A, X_B, \mathrm{tag}_A)$ to $\hat{B}$.
   (h) Update the session identifier to $(\hat{A}, \hat{B}, \mathcal{I}, X_B, Y_A)$ and complete the session by accepting κ as the session key.
4. Upon receiving $(\hat{B}, \hat{A}, \mathcal{R}, Y_A, X_B, \mathrm{tag}_A)$, party $\hat{B}$ does the following:
   (a) Verify that an active session $(\hat{B}, \hat{A}, \mathcal{R}, Y_A, X_B)$ exists.
   (b) Verify that the received tag_A is equal to the one stored.
   (c) Complete session $(\hat{B}, \hat{A}, \mathcal{R}, Y_A, X_B)$ by accepting κ as the session key.

5.1   Comparisons

The DH2 protocol, or more precisely its two-pass variant (called DH2-basic) without the key confirmation tags, is similar to the MTI/C0 protocol [14,11]. In both protocols, the messages exchanged are A^y and B^x. However, in MTI/C0 the shared secret is g^{xy}, whereas in DH2-basic it is (g^x, g^y). The MTI/C0 protocol is more efficient – each party performs two exponentiations compared to three exponentiations in DH2-basic. However, one notable advantage of DH2-basic over MTI/C0 is that the communicating parties can use different groups. Moreover, DH2-basic can be used in a hybrid fashion with the RSA-based KAS2 protocol (cf. §6.1).

DH2-basic is also similar to the two-pass KEA+ protocol [13] where the messages exchanged are X and Y and the session key is $H(g^{ay}, g^{bx}, \hat{A}, \hat{B})$. Unlike the case of DH2-basic, in KEA+ the initiator does not need to know the responder's static public key when it initiates a session of the protocol. Thus, in the situation where all parties use the same group and where certificates have not already been exchanged, DH2-basic is in fact a three-pass protocol (an extra round is needed in which the initiator obtains the responder's certificate) whereas KEA+ is a two-pass protocol. Our analysis of DH2 complements the analysis of KEA+ in [13] by considering the case where parties can select their own groups. Additionally, in contrast with [13], our security model allows the adversary to learn the ephemeral private key of either the Test session or its matching session.

5.2   Security Argument

Recall that the DHD problem in a cyclic group $\mathbb{G}$ of prime order q is the problem of determining $g^{u/v}$, given $g, g^u, g^v \in_R \mathbb{G}$. Our reductionist security proof for DH2 relies on the *gap DHD (GDHD) assumption* which asserts that the DHD problem is intractable even when the solver is given a Decision DHD (DDHD) oracle which, on input a quadruple (h, h^a, h^b, h^c), determines whether $c \equiv a/b$ (mod q). The following lemma establishes that the GDHD assumption is equivalent to the more familiar gap Diffie-Hellman (GDH) assumption [15] which asserts that computing g^{uv} from $g, g^u, g^v \in_R \mathbb{G}$ is intractable even when the solver is given a Decision DH (DDH) oracle which, on input a quadruple (h, h^a, h^b, h^c), determines whether $c \equiv ab \pmod{q}$.

Lemma 1. *The GDHD and GDH assumptions are equivalent.*

Proof. We show that the GDH problem reduces to the GDHD problem, i.e., given a GDHD solver $\mathcal{A}$ we construct an algorithm $\mathcal{B}$ that solves GDH. The reduction from GDHD to GDH is similar.

Given a CDH instance (g, g^u, g^v), we construct the DHD instance (g^u, g^v, g) and give it to the GDHD solver $\mathcal{A}$. When $\mathcal{A}$ queries its DDHD oracle with (h, h^a, h^b, h^c), we construct a DDH instance (h^b, h, h^a, h^c) and give it to the DDH oracle that is provided with the CDH instance. We return to $\mathcal{A}$ whatever the DDH oracle returns. Finally, when $\mathcal{A}$ outputs its solution, $\mathcal{B}$ outputs the same as the solution of the given CDH instance.

Let $k = g^u$. Then $(g^u, g^v, g) = (k, k^{v/u}, k^{1/u})$, so $\mathcal{A}$ returns $k^{(v/u)u} = g^{uv}$ as required. Similarly, letting $\ell = h^b$, we can write the DDHD oracle query (h, h^a, h^b, h^c) made by $\mathcal{A}$ as $(\ell^{1/b}, \ell^{a/b}, \ell, \ell^{c/b})$, and the corresponding DDH query made by $\mathcal{B}$ as $(h^b, h, h^a, h^c) = (\ell, \ell^{1/b}, \ell^{a/b}, \ell^{c/b})$. One can check that the latter is a valid DH quadruple if and only if $c \equiv a/b \pmod{q}$. Hence, $\mathcal{B}$'s simulation of $\mathcal{A}$'s DDHD oracle is perfect. $\square$

The GDHD problem for $\mathcal{G}_\lambda$ is to determine $g^{v/u}$, given $g, g^u, g^v \in_R \mathbb{G}$ and a DDHD oracle for $\mathbb{G}$, where $\mathbb{G} \in_R \mathcal{G}_\lambda$. The *GDHD assumption for $\mathcal{G}_\lambda$* is that no polynomially-bound algorithm exists that solves the GDHD problem for $\mathcal{G}_\lambda$ with non-negligible probability of success.

Theorem 2. *Suppose that (i) the GDHD assumption for $\mathcal{G}_\lambda$ holds; (ii) the* MAC *scheme is secure; and (iii) H is a random oracle. Then the DH2 key agreement protocol is secure.*

The proof of Theorem 2 is deferred to the full version of this paper [8]. While the proof is similar to that of Theorem 1, a significant difference is that a DH oracle is needed in order to provide consistent answers to H-oracle queries. Thus, unlike the case of KAS2, the reductionist security proof for DH2 relies on a gap assumption.

Remark 4. (fixed generators vs. random generators) In the description of DH2, each party selects its own group and generator. Another scenario worthy of consideration is where each group $\mathbb{G}$ has a fixed generator g. For example, $\mathcal{G}_\lambda$ could consist of a single elliptic curve (and corresponding generator) from the list specified by NIST [9], which corresponds to the case where all parties use the same elliptic curve. In the remainder of the paper, we will consider the case where generators of each group $\mathbb{G}$ are selected uniformly at random. We note that Theorem 2 and its proof can be easily modified to the case of fixed generators. However, it is worth pointing out that the GDHD assumption with fixed generators is not known to be equivalent to the GDH assumption with fixed generators.

6   Miscellaneous Notes

6.1   Hybrid Protocol

The KAS2-DH2 hybrid protocol is depicted in Figure 3. Party $\hat{A}$ has a DH2 key pair $(A = g_1^a, a)$ where $\mathbb{G}_1 = \langle g_1 \rangle$, whereas party $\hat{B}$ has a KAS2 key pair $((N_B, e_B), d_B)$. The protocol can be useful is scenarios where one communicating

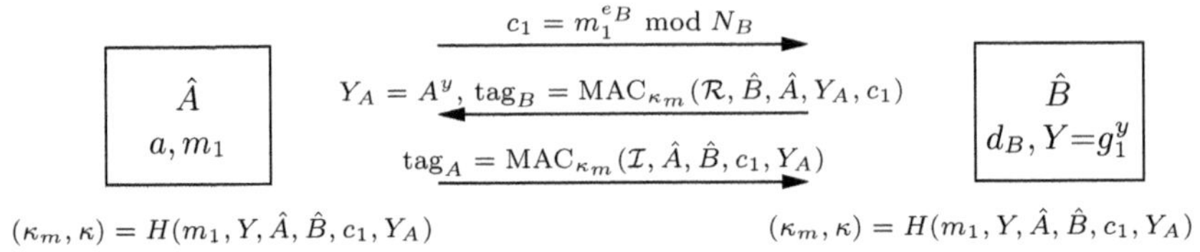

$$c_1 = m_1^{e_B} \bmod N_B$$

$$Y_A = A^y, \text{tag}_B = \text{MAC}_{\kappa_m}(\mathcal{R}, \hat{B}, \hat{A}, Y_A, c_1)$$

$$\text{tag}_A = \text{MAC}_{\kappa_m}(\mathcal{I}, \hat{A}, \hat{B}, c_1, Y_A)$$

$\hat{A}$ \
a, m_1

$\hat{B}$ \
$d_B, Y = g_1^y$

$(\kappa_m, \kappa) = H(m_1, Y, \hat{A}, \hat{B}, c_1, Y_A)$ $(\kappa_m, \kappa) = H(m_1, Y, \hat{A}, \hat{B}, c_1, Y_A)$

Fig. 3. The KAS2-DH2 hybrid protocol

party only has an RSA certificate, whereas the other communicating party only has a discrete log certificate. The security of KAS2-DH2 can be established by combining the proofs of Theorems 1 and 2.

We note that Boyd et al. [3,4] designed a generic two-pass protocol using key encapsulation mechanisms (KEMs). The protocol allows users to employ different primitives to implement the KEM, and even permits identity-based primitives. In contrast to our protocol, the analysis of their protocol is in the standard model.

6.2   KAS1

The KAS1 protocol [17] is depicted in Figure 4. In this protocol, the initiator $\hat{A}$ contributes only an ephemeral key pair whereas the responder $\hat{B}$ contributes only a static key pair and a nonce. KAS1 provides unilateral authentication and key confirmation of $\hat{B}$ to $\hat{A}$. In KAS1, the constant string $\mathcal{R}$ is "KC_1_V", which is different from the string "KC_2_V" used in KAS2. The KAS1 protocol is suitable in applications such as SSL/TLS where the initiator typically does not have a static key pair.

As stated in [17], only the KAS1 initiator has assurances that third parties cannot recover the session key. The KAS1 responder obtains no cryptographic

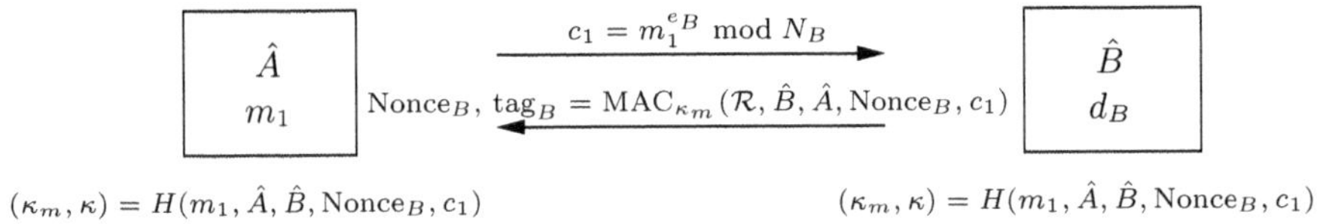

$$c_1 = m_1^{e_B} \bmod N_B$$

$$\text{Nonce}_B, \text{tag}_B = \text{MAC}_{\kappa_m}(\mathcal{R}, \hat{B}, \hat{A}, \text{Nonce}_B, c_1)$$

$\hat{A}$ \
m_1

$\hat{B}$ \
d_B

$(\kappa_m, \kappa) = H(m_1, \hat{A}, \hat{B}, \text{Nonce}_B, c_1)$ $(\kappa_m, \kappa) = H(m_1, \hat{A}, \hat{B}, \text{Nonce}_B, c_1)$

Fig. 4. The KAS1 protocol

assurances about the true identity of its peer. If the responder's static private key is compromised, then previously-established session keys are easily recoverable. Similarly, if the initiator's ephemeral private key is exposed, then the secrecy of the session key is compromised. Inclusion of the nonce assures the responder that the session key is fresh.

Since the session initiator does not contribute a static key pair to the key establishment, and since the responder obtains no assurances about the identity of its peer, some security attributes such as key-compromise impersonation re-silience are not applicable to KAS1. Consequently, the model proposed in §3 is not suitable for analyzing KAS1.

6.3   Key Reusage

Contrary to conventional wisdom, SP 800-56B explicitly permits a party to use its static key pair in more than one of the key establishment schemes specified in [17]. This is a little surprising since the KAS1 and KAS2 protocols have no-ticeably different security attributes and, as observed in [6], interference attacks on the runs of two protocols can render one of the protocols insecure. Following [7], it would be worthwhile to specify a shared security model that incorporates the individual security attributes of KAS1 and KAS2, and formally verify that the protocols are secure even when static key pairs are reused.

Acknowledgements. We thank the anonymous referees for their valuable comments.

References

1. ANSI X9.44, Public Key Cryptography for the Financial Services Industry: Key Establishment Using Integer Factorization Cryptography, American National Standards Institute (2007)
2. Bao, F., Deng, R., Zhu, H.: Variations of Diffie-Hellman problem. In: Qing, S., Gollmann, D., Zhou, J. (eds.) ICICS 2003. LNCS, vol. 2836, pp. 301–312. Springer, Heidelberg (2003)
3. Boyd, C., Cliff, Y., Nieto, J., Paterson, K.: Efficient one-round key exchange in the standard model. In: Mu, Y., Susilo, W., Seberry, J. (eds.) ACISP 2008. LNCS, vol. 5107, pp. 69–83. Springer, Heidelberg (2008)
4. Boyd, C., Cliff, Y., Nieto, J., Paterson, K.: One-round key exchange in the standard model. International Journal of Applied Cryptography 1, 181–199 (2009)
5. Canetti, R., Krawczyk, H.: Analysis of key-exchange protocols and their use for building secure channels. In: Pfitzmann, B. (ed.) EUROCRYPT 2001. LNCS, vol. 2045, pp. 453–474. Springer, Heidelberg (2001), http://eprint.iacr.org/2001/040
6. Chatterjee, S., Menezes, A., Ustaoglu, B.: Reusing static keys in key agreement protocols. In: Roy, B., Sendrier, N. (eds.) INDOCRYPT 2009. LNCS, vol. 5922, pp. 39–56. Springer, Heidelberg (2009), http://www.cacr.math.uwaterloo.ca/techreports/2009/cacr2009-36.pdf
7. Chatterjee, S., Menezes, A., Ustaoglu, B.: Combined security analysis of the one- and three-pass unified model key agreement protocols. In: Gong, G., Gupta, K.C. (eds.) INDOCRYPT 2010. LNCS, vol. 6498, pp. 49–68. Springer, Heidelberg (2010)

8. Chatterjee, S., Menezes, A., Ustaoglu, B.: A generic variant of NIST's KAS2 key agreement protocol, full version, Technical Report CACR 2011-09, http://www.cacr.math.uwaterloo.ca/techreports/2011/cacr2011-09.pdf
9. FIPS 186-3, Digital Signature Standard (DSS), Federal Information Processing Standards Publication 186-3, National Institute of Standards and Technology (2009)
10. Just, M., Vaudenay, S.: Authenticated multi-party key agreement. In: Kim, K.-c., Matsumoto, T. (eds.) ASIACRYPT 1996. LNCS, vol. 1163, pp. 36–49. Springer, Heidelberg (1996)
11. Kunz-Jacques, S., Pointcheval, D.: About the security of MTI/C0 and MQV. In: De Prisco, R., Yung, M. (eds.) SCN 2006. LNCS, vol. 4116, pp. 156–172. Springer, Heidelberg (2006)
12. LaMacchia, B., Lauter, K., Mityagin, A.: Stronger security of authenticated key exchange. In: Susilo, W., Liu, J.K., Mu, Y. (eds.) ProvSec 2007. LNCS, vol. 4784, pp. 1–16. Springer, Heidelberg (2007)
13. Lauter, K., Mityagin, A.: Security analysis of KEA authenticated key exchange. In: Yung, M., Dodis, Y., Kiayias, A., Malkin, T. (eds.) PKC 2006. LNCS, vol. 3958, pp. 378–394. Springer, Heidelberg (2006)
14. Matsumoto, T., Takashima, Y., Imai, H.: On seeking smart public-key distribution systems. The Transactions of the IECE of Japan E69, 99–106 (1986)
15. Okamoto, T., Pointcheval, D.: The gap-problem: a new class of problems for the security of cryptographic schemes. In: Kim, K.-c. (ed.) PKC 2001. LNCS, vol. 1992, pp. 104–118. Springer, Heidelberg (2001)
16. SP 800-56A, Special Publication 800-56A, Recommendation for Pair-Wise Key Establishment Schemes Using Discrete Logarithm Cryptography (Revised), National Institute of Standards and Technology (March 2007)
17. SP 800-56B, Special Publication 800-56B, Recommendation for Pair-Wise Key Establishment Schemes Using Integer Factorization Cryptography, National Institute of Standards and Technology (August 2009)

A Single Key Pair is Adequate for the Zheng Signcryption

Jia Fan[1,2], Yuliang Zheng[2], and Xiaohu Tang[1]

[1] Southwest Jiaotong University, 610031, P.R.China
[2] University of North Carolina at Charlotte, NC 28223, USA
`fanjia@mars.swjtu.edu.cn, yzheng@uncc.edu, xhutang@ieee.org`

Abstract. We prove that the original Zheng signcryption scheme published at Crypto'97, with a couple of minor tweaks, requires only a single public/private key pair for each user. That is the user can employ the same public/private key pair for both signcryption and unsigncryption in a provably secure manner. We also prove that the Zheng signcryption scheme allows a user to securely signcrypt a message to himself. Our first result confirms a long-held belief that signcryption reduces the overhead associated with public keys, while our second result foretells potential applications in cloud storage where one with a relatively less resourceful storage device may wish to off-load data to an untrusted remote storage network in a secure and unforgeable way.

Keywords: Public key, Security proof, Signcryption, Single key pair.

1 Introduction

The concept and first instantiation of signcryption were proposed by Zheng in 1997 [9]. As a cryptographic primitive, signcryption combines both the functions of public key encryption and those of digital signature, in such a way that its overhead is far less than that required by performing encryption and signature separately. At PKC'02, Baek, Steinfeld and Zheng [2] successfully established a security model for signcryption, and proved that with a couple of minor tweaks, the original Zheng signcryption scheme was indeed provably secure under commonly accepted computational assumptions. In the journal version [3] of the same paper, their security model was further enhanced and security proofs were made more rigorous. Their papers, however, still leave two interesting questions unanswered.

The first question has to do with the number of public/private key pairs a user has to keep in order to apply the Zheng signcryption in a provably secure manner. The security models and proofs presented in [2,3] all assume that a user holds two separate public/private key pairs. One of the two key pairs serves as a *sender signcryption key pair* for signcrypting messages originated from that user to other users, while the other key pair serves as a *receiver unsigncryption*

U. Parampalli and P. Hawkes (Eds.): ACISP 2011, LNCS 6812, pp. 371–388, 2011.
© Springer-Verlag Berlin Heidelberg 2011

key pair for unsigncrypting ciphertexts received by that user from other users. A natural question is whether the requirement of two separate public/private key pairs can be relaxed to a single key pair. An obvious benefit of the use of a single key pair is that it will minimize the cost associated with the creation and maintenance of public/private key pairs, especially the cost of public key verification prior to the execution of signcryption and unsigncryption.

The second question is whether the signcryption scheme can be employed by a user to securely signcrypt a message to the user himself. An ability to do so would have applications in emerging computing and communicating platforms such as cloud storage. Cloud storage is a model of networked data storage where data is stored on multiple virtual servers, generally hosted by third parties, rather than being hosted on dedicated servers. In practice, users with limited storage may wish to store data on a not always trusted cloud in a secure and unforgeable manner. In such a scenario, the user could signcrypt the data to himself first, then store the signcryptext to the cloud. When this user downloads the signcryptext from the cloud, it may check whether the signcryptext is valid, and decrypt the signcryptext if it is.

It turns out that the security models in [2,3] in their original forms do not appear to be capable to address the two open questions. This calls for new ideas in security proofs, especially new security models that capture a real-world scenario where a single public/private key pair is used by each user as well as scenario where one wishes to signcrypt messages to oneself.

Our main contributions are to give affirmative answers to both questions outlined above. To this end we define a strengthened security model for signcryption allowing a user to have only a single public/private key pair and also allowing a user to signcrypt a message to himself. We then prove that the Zheng signcryption scheme, with a minor tweak, is indeed secure in that model, under commonly accepted assumptions including the Gap Diffie-Hellman, the Gap Discrete Logarithm and the random oracle assumptions.

2   Overview of the Zheng Signcryption Scheme

We focus our discussions on the SDSS-1 signcryption scheme proposed by Zheng [9]. Our security proofs apply to other schemes in the same family, including SDSS-2 and counterparts of SDSS-1 and SDSS-2 in other groups such as groups of points on an elliptic curve over a finite field [10].

We follow [3] in describing the Zheng scheme. A minor technical difference between our version and the version in [3] is that we add both sender and receiver's public keys as input to the G hash function. This minor tweak is useful during proof reductions which will become clear later in our description of proofs.

The signcryption scheme with the tweak is described in Tables 1 and 2. We use k to indicate a security parameter that determines other parameters such as the size of a key, the output length of a hash function and ultimately, the level of security of a concrete instantiation of a signcryption scheme in practice.

Table 1. Setup & KeyGen Algorithms

$Setup(1^k)$ by Trusted Authority TA:
1. Choose a random prime q of l_q bits.
2. Choose a random prime p of l_p bits such that $q|(p-1)$.
3. Choose an element $g \in Z_p^*$ such that $Ord_{Z_p^*}(g) = q$.
4. Choose a one-way hash function $G : \{0,1\}^* \rightarrow \{0,1\}^{l_G}$.
5. Choose a one-way hash function $H : \{0,1\}^* \rightarrow Z_q$.
6. Choose a symmetric key encryption scheme $\mathcal{SKE} = (E, D)$.
7. Let $cp = (k, p, q, g, G, H, \mathcal{SKE})$ be the common parameter.

$KeyGen(cp)$ by User U:
1. Choose $x_U \in Z_q^*$ uniformly at random.
2. Compute $y_U \leftarrow g^{x_U} \mod p$.
3. Let the public key pk_U be y_U and the private key sk_U be (x_U, y_U).

In this table, $l_p : N \rightarrow N$, $l_q : N \rightarrow N$ and $l_G : N \rightarrow N$ are functions of k determining the lengths in bits of p, q and an output of G respectively. $Ord_{Z_p^*}(g) = q$ means that the order of g in the multiplicative group of Z_p^* is q, and $\mathcal{SKE} = (E, D)$ is a one-time symmetric key encryption scheme with message, key and ciphertext spaces being $\mathcal{SP}_m$, $\{0,1\}^{l_G}$ and $\mathcal{SP}_c$ respectively.

Table 2. Signcryption & Unsigncryption Algorithms

$Signcryption(cp, m, sk_S, pk_R)$	$Unsigncryption(cp, \sigma, pk_S, sk_R)$
by Sender S:	by Receiver R:
1. Parse sk_S as (x_S, y_S), pk_R as y_R.	1. Parse sk_R as (x_R, y_R), pk_S as y_S.
2. Choose $x \in Z_q^*$ uniformly at random.	2. Parse σ as (c, r, s).
3. Compute $K \leftarrow y_R^x \mod p$.	3. Compute $w \leftarrow (y_S \cdot g^r)^s \mod p$.
4. Compute $\tau \leftarrow G(y_S, y_R, K)$.	4. Compute $K \leftarrow w^{x_R} \mod p$.
5. Compute $c \leftarrow E_\tau(m)$.	5. Compute $\tau \leftarrow G(y_S, y_R, K)$.
6. Compute $r \leftarrow H(m, y_S, y_R, K)$.	6. Compute $m \leftarrow D_\tau(c)$.
7. If $r + x_S = 0 \mod q$, return to Step 2.	7. If $H(m, y_S, y_R, K) = r$, return m;
8. Compute $s \leftarrow x/(r + x_S) \mod q$.	otherwise return $Reject$.
9. Output $\sigma \leftarrow (c, r, s)$ as the signcryptext.	

We assume $m \in \mathcal{SP}_m$ in the signcryption algorithm, and $\sigma \in \mathcal{SP}_c \times Z_q \times Z_q^*$ in the unsigncryption algorithm. In practice appropriate tests are carried out first to ensure that these conditions are met. With the unsigncryption algorithm, $Reject$ is interpreted as a special symbol indicating that the signcryptext is invalid.

3   Security Model

We now introduce a stronger security model that is extended from the model proposed by Baek *et al.* [3]. Major differences between the two security models are outlined below.

First, our new model allows the use of a single public/private key pair by a user. This is achieved by permitting a target user (with a single public/private key pair) in an attack game to be both a sender and a receiver. This modification makes it possible for an adversary to make signcryption queries with any target user as a sender and unsigncryption queries with any target user as a receiver. We note that in the original security model by Baek *et al*, a target user always has a fixed role, being either a sender or a receiver.

Second, our model adds a security consideration for the case where one signcrypts a message to oneself. In an attack game for confidentiality, an adversary is given two target users, A and B. We allow the adversary to attack on $(S^*, R^*) \in \{(A, B), (A, A), (B, A), (B, B)\}$, where S^* is the sender and R^* is the receiver. By contrast, the model by Baek *et al*. allows only $(S^*, R^*) = (A, B)$. And in an attack game for unforgeability, an adversary is given one target user A. We allow the adversary to attack on (S^*, R^*) where $S^* = A$ and R^* can be an arbitrary user including $R^* = A$, while the model by Baek *et al*. does not allow $R^* = A$.

According to the adversary's capability, An *et al*. [1] divide the security model into two classes, called the *insider* setting and the *outsider* setting respectively. In the outsider setting, an adversary has access to neither sk_{S^*} nor sk_{R^*}. In comparison, the only restriction on an adversary in the insider setting is that it is not allowed to have access to sk_{R^*}. Since our main goal in this paper is to prove the Zheng signcryption scheme is secure in the "outsider" setting for confidentiality, we will define unforgeability in the insider setting, and confidentiality in the outsider setting.

Throughout this paper we will use the term of a *negligible* function to indicate any function in an appropriate security parameter k that vanishes faster than the inverse of any integer-valued polynomial in the same parameter k when k is sufficiently large.

3.1   Syntax of Signcryption

A generic signcryption system $\mathcal{SC}$ consists of four algorithms as follows:

- $Setup(1^k)$: It takes as input a security parameter 1^k and generates a common parameter cp for an entire system under consideration. It is run by a trusted authority.
- $KeyGen(cp)$: It takes as input a system-wide common parameter cp, outputs a pair of public/private keys (pk_U, sk_U) for a user U. This algorithm is run by users within the system, independently of one another.
- $Signcryption(cp, m, sk_S, pk_R)$: When a sender S plans to communicate a message $m \in \mathcal{SP}_m$ to a receiver R, where $\mathcal{SP}_m$ is the message space, he runs this algorithm to generate a signcryptext σ from m, a common parameter cp, his private key sk_S and the receiver R's public key pk_R.
- $Unsigncryption(cp, \sigma, pk_S, sk_R)$: When a receiver R receives a signcryptext σ from a sender S, he runs this algorithm with σ, the public parameter cp, the sender S's public key pk_S, and his private key sk_R as input. The algorithm outputs a message m if σ is valid, or a special symbol *Reject* otherwise.

For a signcryption scheme to be useful in practice, we further require that for any plaintext m, any sender S and any receiver R, we have

$$m = Unsigncryption(cp, Signcryption(cp, m, sk_S, pk_R), pk_S, sk_R).$$

3.2 Definition of Confidentiality

We follow an established definition, called indistinguishability under chosen ciphertext attack (IND-CCA) to define confidentiality for signcryption as indistinguishability under chosen signcryptext and plaintext attack (IND-CSPA). This is done by defining an attack game, called an IND-CSPA game.

Let k be the security parameter of the scheme, A and B be two target users. The IND-CSPA game is played between an IND-CSPA adversary and its environment Σ which contains an IND-CSPA challenger and two oracles, namely a signcryption oracle and an unsigncryption oracle. Specifically, the IND-CSPA game proceeds as follows:

- Stage 1: The challenger computes $cp \leftarrow Setup(1^k)$; $(pk_A, sk_A) \leftarrow KeyGen(cp)$; $(pk_B, sk_B) \leftarrow KeyGen(cp)$. It then equips the signcryption and unsigncryption oracles with (sk_A, sk_B) and gives (cp, pk_A, pk_B) to the adversary.
- Stage 2: The adversary makes a sequence of adaptive queries. Each query is one of two types:
  1. Signcryption query: the adversary submits (m, pk_S, pk_R) to the challenger, where $m \in \mathcal{SP}_m$, $pk_S \in \{pk_A, pk_B\}$ and pk_R can be an arbitrary public key in the system including $pk_R \in \{pk_A, pk_B\}$. The challenger forwards
     (m, pk_S, pk_R) to the signcryption oracle which then returns to the challenger with an outcome of $Signcryption(cp, m, sk_S, pk_R)$. Finally, the challenger passes this answer to the adversary.
  2. Unsigncryption query: the adversary submits (σ, pk_S, pk_R) to the challenger, where σ is a signcryptext, $pk_R \in \{pk_A, pk_B\}$, and pk_S can be an arbitrary public key in the system including $pk_S \in \{pk_A, pk_B\}$. The challenger forwards (σ, pk_S, pk_R) to the unsigncryption oracle which then returns to the challenger with an outcome of $Unsigncryption$ (cp, σ, pk_S, sk_R). Finally, the challenger passes this answer to the adversary.
- Stage 3: The adversary submits $(m_0, m_1, pk_{S^*}, pk_{R^*})$ to the challenger where $m_0, m_1 \in \mathcal{SP}_m$ are of equal length, and $pk_{S^*}, pk_{R^*} \in \{pk_A, pk_B\}$. The challenger chooses a random bit $\beta \in \{0, 1\}$. Then it forwards $(m_\beta, pk_{S^*}, pk_{R^*})$ to the signcryption oracle which then returns to the challenger with a signcryptext σ^* which is an outcome of $Signcryption(cp, m_\beta, sk_{S^*}, pk_{R^*})$. The challenger then passes σ^* to the adversary as a challenge signcryptext.
- Stage 4: This is identical to Stage 2, except that the adversary can not query an unsigncryption with $(\sigma^*, pk_{S^*}, pk_{R^*})$.
- Stage 5: The adversary outputs a bit β' as his guess for β and pass it over to the challenger. The challenger then checks whether $\beta = \beta'$. If it is, the adversary wins the challenge.

For an IND-CSPA adversary $\mathcal{A}$ running in time t, making at most j_s sign-cryption queries and j_u unsigncryption queries, we define the advantage of $\mathcal{A}$ in winning the challenge as $Adv_{SC,\mathcal{A}}^{ind-cspa}(t, j_s, j_u) = |Pr[\beta = \beta'] - 1/2|$. And we define $\epsilon_{t,j_s,j_u}^{ind-cspa}$ to be the maximum value of $Adv_{SC,\mathcal{A}}^{ind-cspa}(t, j_s, j_u)$ over all IND-CSPA adversaries with the same *resources* parameter (t, j_s, j_u).

Definition 1. *We say that a signcryption scheme SC is IND-CSPA secure if for any IND-CSPA adversary that runs in time t, makes at most j_s signcryption queries and j_u unsigncryption queries, the maximum advantage $\epsilon_{t,j_s,j_u}^{ind-cspa}$ is negligible in k, where t, j_s and j_u are all polynomials in k.*

3.3 Definition of Unforgeability

Unforgeability is defined as existential unforgeability against chosen signcryptext and plaintext attack (EUF-CSPA), which follows the established definition of existential unforgeability against chosen message attack (EUF-CMA). This is done by defining an attack game, called an EUF-CSPA game.

Let k be the security parameter of the scheme, A be a target user. The EUF-CSPA game is played between an EUF-CSPA adversary and its environment Σ which contains an EUF-CSPA challenger and two oracles, one being a sign-cryption oracle and the other an unsigncryption oracle. The EUF-CSPA game proceeds as follows:

- Stage 1: The challenger computes $cp \leftarrow Setup(1^k)$; $(pk_A, sk_A) \leftarrow KeyGen(cp)$. It then equips the signcryption and unsigncryption oracles with sk_A and gives (cp, pk_A) to the adversary.
- Stage 2: It is mostly the same as Stage 2 in the IND-CSPA game described above, except that in this case there is no pk_B.
- Stage 3: The adversary passes $(\sigma^*, pk_{S^*}, pk_{R^*}, sk_{R^*})$ to the challenger, where σ^* is a signcryptext, $pk_{S^*} = y_A$, pk_{R^*} can be an arbitrary public key in the system including $pk_{R^*} = pk_A$, and sk_{R^*} is the corresponding private key of pk_{R^*}. The challenger then checks whether the outcome of $Unsigncryption(cp, \sigma^*, pk_{S^*}, sk_{R^*})$ is a special symbol *Reject* or a message $m^* \in SP_m$. If the outcome is m^* and the adversary has never made a signcryption query on $(m^*, pk_{S^*}, pk_{R^*})$, then the adversary wins the challenge.

For an adversary $\mathcal{A}$ running in time t, making at most j_s signcryption queries and j_u unsigncryption queries, we define the advantage of $\mathcal{A}$ in winning the challenge as $Adv_{SC,\mathcal{A}}^{euf-cspa}(t, j_s, j_u) = Pr[\mathcal{A}\ wins]$, where "$\mathcal{A}\ wins$" denotes an event that adversary $\mathcal{A}$ wins the challenge in the above attack game. And we define $\epsilon_{t,j_s,j_u}^{euf-cspa}$ to be the maximum of $Adv_{SC,\mathcal{A}}^{euf-cspa}(t, j_s, j_u)$ over all EUF-CSPA adversaries with the same resource parameter (t, j_s, j_u).

Definition 2. *We say that a signcryption scheme SC is EUF-CSPA secure if for any EUF-CSPA adversary running in time t, and making at most j_s signcryption queries and at most j_u unsigncryption queries, the maximum advantage $\epsilon_{t,j_s,j_u}^{euf-cspa}$ is negligible in k, where t, j_s and j_u are all polynomials in k.*

4 Assumptions and Primitives

4.1 Problems and Assumptions

Let $\mathcal{G}$ be a finite multiplicative group with g being a generator of the group. The Discrete Logarithm (DL) problem is one where an attacker is given $(g, y) \in \mathcal{G}^2$, asked to find an x such that $y = g^x$ in $\mathcal{G}$. The well-known Diffie-Hellman (DH) problem has two different flavors: a computational one and a decisional one. With the Computational Diffie-Hellman (CDH) problem, an attacker is given three elements $(g, g^a, g^b) \in \mathcal{G}^3$ for unknown a and b, and asked to compute g^{ab}. In contrast, with the Decisional Diffie-Hellman (DDH) problem an attacker is given four elements $(g, g^a, g^b, z) \in \mathcal{G}^4$, for unknown a and b, and asked to tell whether $z = g^{ab}$.

The CDH problem has a gap based version in which an attacker is granted access to a powerful oracle, named DDH oracle, that solves the DDH problem [7]. This new problem is called the Gap Diffie-Hellman (GDH) problem. A gap based version of the DL problem can be obtained in a similar way. In this paper, we follow [3] to define the Gap Discrete Logarithm (GDL) problem as one in which an attacker has access to a *restricted* oracle for the DDH problem. Similar to the DDH oracle, a restricted DDH oracle also answers whether a given quadruple is a DH quadruple or not. However, the restricted DDH oracle only accepts queries on $(g, y, ., .) \in \mathcal{G}^4$ where (g, y) is the input of the adversary.

The CDH problem has a number of interesting variants. In one variant, an attacker is given $(g, g^a) \in \mathcal{G}^2$ with an unknown a and asked to compute $y = g^{a^2}$. It turns out that this variant is equivalent to the CDH problem [4].

In our proofs we will employ a new variant of the CDH problem in which an attacker is given $(g, g^a, g^b) \in \mathcal{G}^3$ for unknown a and b, and attempts to output one of $(g^{a^2}, g^{b^2}, g^{ab})$. The attacker is considered successful as long as its output is one of the three possible values. We call this new problem the *extended Computational Diffie-Hellman* problem or the eCDH problem for short. Clearly the eCDH problem is computationally equivalent to the CDH problem. A gap based version of the eCDH problem is defined by allowing an attacker to have access to a DDH oracle. Let us call it the *extended GDH* problem or the eGDH problem for short. Naturally, the equivalence of the CDH and eCDH problems is carried over to their gap based versions. That is, the following lemma is true.

Lemma 1. *The GDH problem and the eGDH problem are equivalent.*

Assumptions related to the above mentioned problems are defined by stating that no attacker that runs in polynomial time in the size of the group can successfully solve the respective problem with a non-negligibly success probability. In particular, according to Lemma 1, we claim that the eGDH assumption are equivalent to the GDH assumption.

4.2 One-Time Symmetric Key Encryption

A one-time symmetric key encryption system $\mathcal{SKE}$ [6] consists of two bijective and deterministic algorithms E and D.

- $E_\tau(m)$: On input a key τ, a plaintext m, it outputs a ciphertext $c \leftarrow E_\tau(m)$.
- $D_\tau(c)$: On input a key τ, a ciphertext c, it outputs a plaintext $m \leftarrow D_\tau(c)$.

In the above, $\tau \in \mathcal{SP}_\tau$, $m \in \mathcal{SP}_m$ and $c \in \mathcal{SP}_c$, where the sizes of spaces $\mathcal{SP}_\tau$, $\mathcal{SP}_m$ and $\mathcal{SP}_c$ are all determined by a security parameter k. And it is required that for all $m \in \mathcal{SP}_m$ and $\tau \in \mathcal{SP}_\tau$, $m = D_\tau(E_\tau(m))$.

We will use a one-time symmetric key encryption with the security of passive indistinguishability of $\mathcal{SKE}$ (PI-SKE). In a PI-SKE attack game, a passive attacker is given $(k, \mathcal{SKE})$, and then submits two equal length messages (m_0, m_1) to get a ciphertext c where $c \leftarrow E_\tau(m_\beta)$, β is a random bit. PI-SKE security states that, any passive attacker running in polynomial time cannot determine which of the two messages was chosen.

4.3 One-Way Hash Functions

Our proofs rely on the random oracle methodology [5]. In other words we assume that each one-way hash function used in the Zheng signcryption scheme behaves like a random oracle, a mathematical function mapping every possible query to a random response from its output domain.

5 Security Proofs

Our proofs for confidentiality and unforgeability apply the game based technique. For each proof, we describe a sequence of $n+1$ games, from Game 0 to Game n (n is a constant). Game 0 is the normal attack game in the security definition. We use a sequence of simulators (from Game 1 to Game n) to replace the challenger. Game $i+1$ and Game i ($0 \leq i \leq n-1$) are mostly the same, except that the simulator's behavior in Game $i+1$ is a little bit different from the simulator's (or the challenger's when $i = 0$) behavior in Game i.

Define S_i to be an event that the adversary wins the challenge in Game i. To analyze the relation between $Pr[S_i]$ and $Pr[S_{i+1}]$, we make use of two techniques introduced by Shoup [8], namely bridging step and transition based on a failure event.

1. Bridging Step: The change from Game i to Game $i+1$ is a bridging step means that the change is only conceptual. From the adversary's point of view, these two games proceed identically. Therefore, in this case we have $Pr[S_i] = Pr[S_{i+1}]$.
2. Transition Based on a Failure Event: The change from Game i to Game $i+1$ is a transition based on a failure event means that from the adversary's point of view, these two games proceed identically unless a certain "failure event" occurs. We can then apply a so-called Difference Lemma [8]:

   Lemma 2. *(Difference Lemma): Let S_1, S_2 and F be events defined on some probability spaces. Suppose that the event $S_1 \wedge \neg F$ occurs if and only if $S_2 \wedge \neg F$ occurs. Then $\mid Pr[S_1] - Pr[S_2] \mid \leq Pr[F]$.*

We have in this case $|Pr[S_{i+1}] - Pr[S_i]| \leq Pr[Failure\ Event\ Occurs]$.

In each proof, we make sure that for all i $(0 \leq i \leq n - 1)$, the change from Game i to Game $i+1$ is either a bridging step or a transition based on a failure event which occurs with at most a negligible probability in k. In Game n, we show that the adversary's advantage in winning the challenge is negligible in k. Finally, from the results in all the games, we can arrive at our desired conclusion that the adversary's advantage in winning Game 0 (the normal attack game) is also negligible in k.

We define $\langle g \rangle$ be a group generated by g. The security proofs of unforgeability and confidentiality for the Zheng signcryption are as follows.

5.1 Proof of Unforgeability

Theorem 1. *Let H and G be two hash functions modeled as random oracles. Then under the GDL assumption in $\langle g \rangle$, the Zheng signcryption scheme is EUF-CSPA secure. Specifically, let k be a security parameter of signcryption, $\mathcal{A}$ be an EUF-CSPA adversary that runs in time t, and makes at most j_s signcryption queries, j_u unsigncryption queries, j_g hash queries to G and j_h hash queries to H, where t, j_s, j_u, j_g and j_h are all polynomials in k. Then the maximum advantage $\epsilon_{t,j_s,j_u}^{euf-cspa}$ of the adversary satisfies the following condition:*

$$\epsilon_{t,j_s,j_u}^{euf-cspa} \leq \frac{j_s(j_g + j_h + 3j_u + 2j_s) + 2j_h + j_u + 1}{q} + 2 \cdot \sqrt{j_h \cdot \epsilon_{t_{gdl},j_{gdl}}^{gdl}}$$

where $\epsilon_{t_{gdl},j_{gdl}}^{gdl}$ is negligible in k for all sufficiently large k.

Before diving into details of the proof of Theorem 1, we review in Table 3 an assumption introduced in [3], which is renamed as the Random Beacon GDL assumption (or rbGDL assumption for short). The following Lemma 3 is also from [3] which shows an equivalence relationship between the rbGDL assumption and the GDL assumption.

Lemma 3. *Any algorithm $\mathcal{A}_{rbgdl}$ attacking the rbGDL assumption with run-time t_{rbgdl}, j_{rbgdl} restricted DDH queries, j_r Random Beacon queries, and success probability $Adv_{\mathcal{RBGDL},\mathcal{A}_{rbgdl}}^{rbgdl}(t_{rbgdl}, j_{rbgdl}, j_r) \geq 2j_r/q$ can be converted into an algorithm $\mathcal{A}_{gdl}$ attacking the GDL assumption with run-time $t_{gdl} = 2t_{rbgdl} + O(q^2)$, $j_{gdl} = 2j_{rbgdl}$ restricted DDH queries, and success probability*

$$Adv_{\mathcal{GDL},\mathcal{A}_{gdl}}^{gdl}(t_{gdl}, j_{gdl}) \geq \frac{1}{j_r}\left(\frac{Adv_{\mathcal{RBGDL},\mathcal{A}_{rbgdl}}^{rbgdl}(t_{rbgdl}, j_{rbgdl}, j_r)}{2} - \frac{j_r}{q}\right)^2.$$

Proof of Theorem 1. We describe our proof in a sequence of seven games, from Game 0 to Game 6 as follows. We define $\mathcal{S}_i$ $(1 \leq i \leq 6)$ to be the simulator in Game i.

Table 3. The rbGDL Assumption

Random Beacon Gap Discrete Logarithm (rbGDL) Assumption [3]
Given a pair of elements (g, g^a) in $\mathcal{G}$, g is a generator of $\mathcal{G}$, $Ord_{\mathcal{G}}(g) = q$, $a \in \{0, ..., q-1\}$. With the help of a Restricted DDH Oracle and a Random Beacon,
A Random Beacon takes as input a pair of elements $(y[i], K[i]) \in \mathcal{G}^2$ $(y[i] \neq 1)$, outputs uniformly a random independent number $r[i] \in \{0, ..., q-1\}$. $i \in \{1, ..., j_r\}$ where j_r is total number of Random Beacon queries been made.
it is computationally intractable to compute the value of $(r[i^*], s^*, i^*)$, satisfying $K[i^*] = y[i^*]^{s^*(r[i^*]+a)}$.

The difference between a random beacon and a random oracle is that a random beacon returns a random and independent response even for the same input.

Game 0 (EUF-CSPA Game in the Random Oracle Model): This game is the EUF-CSPA game defined in Section 3.3 in the random oracle model. Therefore, we have

$$Adv_{SC,\mathcal{A}}^{euf-cspa}(t, j_s, j_u) = Pr[S_0]. \tag{1}$$

Game 1 (Apply G_{sim} to Simulate the G Random Oracle): In this game, S_1 behaves mostly the same as C, except that S_1 additionally runs an algorithm G_{sim} to simulate the G random oracle. In order to simulate the G random oracle, S_1 holds two lists, called $Glist_1$ and $Glist_2$ respectively, which are both initially empty. Records on $Glist_1$ are generated by G_{sim}, while records on $Glist_2$ are generated by $Signcryption_{sim}$ and $Unsigncryption_{sim}$ which will be applied in later games. The ν-th record on $Glist_1$ is in form of $(y_{S_\nu}, y_{R_\nu}, K_\nu, \tau_\nu)$, and the μ-th record on $Glist_2$ is in form of $(r_\mu, s_\mu, y_{S_\mu}, y_{R_\mu}, \tau_\mu)$. l_{Glist_1} and l_{Glist_2} denote the total number of records on $Glist_1$ and $Glist_2$ respectively.

– When the i-th hash query is made on (y_S, y_R, K) to the G random oracle, G_{sim} runs the following steps:
  1. If $Glist_1$ is not empty, then from $\nu = 1$ to $\nu = l_{Glist_1}$ do
     (a) take the value of $(y_{S_\nu}, y_{R_\nu}, K_\nu, \tau_\nu)$ which is the ν-th record on $Glist_1$;
     (b) if $(y_{S_\nu}, y_{R_\nu}, K_\nu) = (y_S, y_R, K)$, return τ_ν;
     (c) $\nu = \nu + 1$.
  2. If $Glist_2$ is not empty, then from $\mu = 1$ to $\mu = l_{Glist_2}$ do
     (a) take the value of $(r_\mu, s_\mu, y_{S_\mu}, y_{R_\mu}, \tau_\mu)$ which is the μ-th record on $Glist_2$;
     (b) if $y_S = y_A$ and $(y_S, y_R) = (y_{S_\mu}, y_{R_\mu})$, check whether a quadruple $(g, y_S, y_R{}^{s_\mu}, \frac{K}{y_R{}^{s_\mu \cdot r_\mu}})$ is a DH quadruple in $\langle g \rangle$; if it is, return τ_μ;
     (c) if $y_R = y_A$ and $(y_S, y_R) = (y_{S_\mu}, y_{R_\mu})$, check whether a quadruple $(g, y_R, (y_S g^{r_\mu})^{s_\mu}, K)$ is a DH quadruple in $\langle g \rangle$; if it is, return τ_μ;
     (d) $\mu = \mu + 1$.
  3. Choose $\tau \in \{0, 1\}^{l_G}$ uniformly at random, add (y_S, y_R, K, τ) to the end of $Glist_1$ and return τ.

It is easy to check that the change from Game 0 to Game 1 is a bridging step, therefore,

$$Pr[S_1] = Pr[S_0]. \tag{2}$$

Game 2 (Apply H_{sim} to Simulate the H Random Oracle): In this game, S_2 behaves mostly the same as S_1, except that S_2 additionally runs an algorithm H_{sim} to simulate the H random oracle. In order to simulate the H random oracle, S_2 holds another two lists, called $Hlist_1$ and $Hlist_2$ respectively, which are both initially empty. Records on $Hlist_1$ are generated by H_{sim}, while records on $Hlist_2$ are generated by $Signcryption_{sim}$ which will be applied in later games. l_{Hlist_1} and l_{Hlist_2} denote the total number of records on $Hlist_1$ and $Hlist_2$ respectively. The ν-th record on $Hlist_1$ is in form of $(m_\nu, y_{S_\nu}, y_{R_\nu}, K_\nu, r_\nu)$, and the μ-th record on $Glist_2$ is in form of $(r_\mu, s_\mu, m_\mu, y_{S_\mu}, y_{R_\mu}, r'_\mu)$.

- When the i-th hash query is made on (m, y_S, y_R, K) to the H random oracle, H_{sim} runs the following steps:
  1. If $Hlist_1$ is not empty, then from $\nu = 1$ to $\nu = l_{Hlist_1}$ do
     (a) take the value of $(m_\nu, y_{S_\nu}, y_{R_\nu}, K_\nu, r_\nu)$ which is the ν-th record on $Hlist_1$;
     (b) if $(m_\nu, y_{S_\nu}, y_{R_\nu}, K_\nu) = (m, y_S, y_R, K)$, then return r_ν;
     (c) $\nu = \nu + 1$.
  2. If $Hlist_2$ is not empty, then from $\mu = 1$ to $\mu = l_{Hlist_2}$ do
     (a) take the value of $(r_\mu, s_\mu, m_\mu, y_{S_\mu}, y_{R_\mu}, r'_\mu)$ which is the μ-th record on $Hlist_2$;
     (b) if $y_S = y_A$ and $(m, y_S, y_R) = (m_\mu, y_{S_\mu}, y_{R_\mu})$, check whether a quadruple $(g, y_S, y_R^{s_\mu}, \frac{K}{y_R^{s_\mu \cdot r_\mu}})$ is a DH quadruple in $\langle g \rangle$; if it is, return r'_μ;
     (c) if $y_R = y_A$ and $(m, y_S, y_R) = (m_\mu, y_{S_\mu}, y_{R_\mu})$, check whether a quadruple $(g, y_R, (y_S g^{r_\mu})^{s_\mu}, K)$ is a DH quadruple in $\langle g \rangle$; if it is, return r'_μ;
     (d) $\mu = \mu + 1$.
  3. If $y_S = y_A$, it computes $r \leftarrow R'(y_R, K)$, otherwise it chooses $r \in Z_q$ uniformly at random; add (m, y_S, y_R, K, r) to the end of $Hlist_1$ and return r.
     Here, R' is an algorithm that has the same output distribution as a random beacon R. For any input (even with the same input as before), R' chooses $r \in Z_q$ uniformly at random, and outputs r.

It is also easy to check that the change from Game 1 to Game 2 is a bridging step, therefore,

$$Pr[S_2] = Pr[S_1]. \tag{3}$$

Game 3 (Apply $Signcryption_{sim}$ to Simulate the Signcryption Oracle): In this game S_3 behaves mostly the same as S_2, except that S_3 additionally runs an algorithm $Signcryption_{sim}$ to simulate the signcryption oracle.

- When the i-th signcryption query is made on (m, pk_S, pk_R) to the signcryption oracle, $Signcryption_{sim}$ runs as follows:
  1. Parse pk_S as y_S, pk_R as y_R.
  2. Choose $r \in Z_q$, $s \in Z_q^*$, $\tau \in \{0,1\}^{l_G}$ uniformly at random.
  3. If $g^r y_S = 1 \mod p$, jump to Step 2.
  4. If $Glist_2$ is not empty, then from $\mu = 1$ to $\mu = l_{Glist_2}$ do
     (a) take the value of $(r_\mu, s_\mu, y_{S_\mu}, y_{R_\mu}, \tau_\mu)$ which is the μ-th record on $Glist_2$;
     (b) if $(y_{S_\mu}, y_{R_\mu}) = (y_S, y_R)$, $(y_S g^r)^s = (y_S g^{r_\mu})^{s_\mu}$ and $\tau_\mu \neq \tau$, then return $Reject$;
     (c) $\mu \leftarrow \mu + 1$.
  5. If $Hlist_2$ is not empty, then from $\mu = 1$ to $\mu = l_{Hlist_2}$ do
     (a) take the value of $(r_\mu, s_\mu, m_\mu, y_{S_\mu}, y_{R_\mu}, r'_\mu)$ which is the μ-th record on $Hlist_2$;
     (b) if $(m_\mu, y_{S_\mu}, y_{R_\mu}) = (m, y_S, y_R)$, $(y_S g^r)^s = (y_S g^{r_\mu})^{s_\mu}$ and $r'_\mu \neq r$, then return $Reject$;
     (c) $\mu = \mu + 1$.
  6. Add (r, s, y_S, y_R, τ) to the end of $Glist_2$, (r, s, m, y_S, y_R, r) to the end of $Hlist_2$;
  7. Compute $c \leftarrow E_\tau(m)$;
  8. Return $\sigma = (c, r, s)$.

If the following four conditions are all satisfied, then it is easy to check that $Signcryption_{sim}$ has the same output distribution as the signcryption oracle:

- $Signcryption_{sim}$ does not change the output distribution of G_{sim} and H_{sim}.
- $Signcryption_{sim}$ does not return $Reject$.
- $\tau = G_{sim}(y_S, y_R, K)$ with $K = g^{(x_S+r) \cdot s \cdot x_R}$ when $Signcryption_{sim}$ does not return $Reject$.
- $r = H_{sim}(m, y_S, y_R, K)$ with $K = g^{(x_S+r) \cdot s \cdot x_R}$ when $Signcryption_{sim}$ does not return $Reject$.

In the following, we analyze all the above conditions one by one.

1. Adding (r, s, y_S, y_R, τ) to $Glist_2$, (r, s, m, y_S, y_R, r) to $Hlist_2$ does not change the output distribution of G_{sim} and H_{sim}, since $\tau \in \{0,1\}^{l_G}$ which may be used as an output for G_{sim} and $r \in Z_q$ which may be used as an output for H_{sim} are all chosen uniformly at random. Therefore, $Signcryption_{sim}$ does not change the output distribution of G_{sim} and H_{sim}.
2. For the i-th signcryption query, the probability that it returns $Reject$ at Step 4 is at most $\frac{j_s + j_u}{q}$, since $(y_S g^r)^s$ is uniformly and randomly distributed in $\langle g \rangle$ and $l_{Glist_2} \leq j_s + j_u$. Similarly, we have for the i-th signcryption query, the probability that it returns $Reject$ at Step 5 is also at most $\frac{j_s}{q}$ since $l_{Hlist_2} \leq j_s$. Therefore, for the i-th signcryption query, the probability $Signcryption_{sim}$ does not return $Reject$ is at most $\frac{2j_s + j_u}{q}$. Then, the probability that the second condition is not satisfied during some signcryption query is at most $\frac{j_s(2j_s + j_u)}{q}$.

3. For the i-th signcryption query in which $Signcryption_{sim}$ does not return $Reject$, $\tau \neq G_{sim}(y_S, y_R, K)$ if and only if G_{sim} has been run on (y_S, y_R, K) before the i-th signcryption query and the corresponding output does not equal to τ. The probability that G_{sim} has been run on (y_S, y_R, K) before the i-th signcryption query is at most $\frac{j_u + j_g}{q}$, since K is randomly and uniformly distributed in $\langle g \rangle$ and G_{sim} must have been run for at most $j_u + j_g$ times (j_u times called by the unsigncryption oracle, j_g times called directly by the challenger) before the i-th signcryption query. Therefore, the probability that the third condition is not satisfied during some signcryption query is at most $\frac{j_s(j_u + j_g)}{q}$.

4. Following a very similar analysis as for the third condition, we have the probability that the fourth condition is not satisfied during some signcryption query is at most $\frac{j_s(j_u + j_h)}{q}$.

We define a certain event F_1 to be that at least one of the above conditions is not satisfied. From the above analysis, we have

$$Pr[F_1] \leq \frac{j_s(j_g + j_h + 3j_u + 2j_s)}{q}. \tag{4}$$

Now it is clear that the signcryption oracle and $Signcryption_{sim}$ has the same output distribution unless F_1 occurs. Moreover, $Signcryption_{sim}$ does not change the output distribution of G_{sim} and H_{sim}. Thus, the change from Game 2 and Game 3 is a transition based on a failure event F_1. We have

$$|Pr[S_3] - Pr[S_2]| \leq Pr[F_1], \tag{5}$$

Game 4 (Apply $Unsigncryption_{sim}$ to Simulate the Unsigncryption Oracle): In this game S_4 behaves mostly the same as S_3, except that S_4 additionally runs an algorithm $Unsigncryption_{sim}$ to simulate the unsigncryption oracle as follows:

- When the i-th unsigncryption query is made on (σ, pk_S, pk_R) to the unsigncryption oracle, $Unsigncryption_{sim}$ runs as follows:
  1. Parse pk_S as y_S, pk_R as y_R.
  2. Parse σ as (c, r, s).
  3. Compute $w \leftarrow (y_S g^r)^s \mod p$.
  4. If $Glist_1$ is not empty, then from $\nu = 1$ to $\nu = l_{Glist_1}$ do
     (a) take the value of $(y_{S_\nu}, y_{R_\nu}, K_\nu, \tau_\nu)$ which is the ν-th record on $Glist_1$;
     (b) if $(y_{S_\nu}, y_{R_\nu}) = (y_S, y_R)$ and (g, y_R, w, K_ν) is a DH tuple in $\langle g \rangle$, compute $\hat{\tau} \leftarrow \tau_\nu$ and jump to Step 7;
     (c) $\nu \leftarrow \nu + 1$.
  5. If $Glist_2$ is not empty, then from $\mu = 1$ to $\mu = l_{Glist_2}$ do
     (a) take the value of $(r_\mu, s_\mu, y_{S_\mu}, y_{R_\mu}, \tau_\mu)$ which is the μ-th record on $Glist_2$;
     (b) if $(y_{S_\mu}, y_{R_\mu}) = (y_S, y_R)$ and $(y_S g^{r_\mu})^{s_\mu} = (y_S g^r)^s$, then compute $\hat{\tau} \leftarrow \tau_\mu$ and jump to Step 7;

(c) $\mu \leftarrow \mu + 1$.
6. Choose $\tau \in \{0,1\}^{l_G}$ uniformly at random, add (r, s, y_S, y_R, τ) to the end of $Glist_2$, and compute $\hat{\tau} \leftarrow \tau$.
7. Compute $m \leftarrow D_{\hat{\tau}}(c)$.
8. If $Hlist_1$ is not empty, then from $\nu = 1$ to $\nu = l_{Hlist_1}$ do
   (a) take the value of $(m_\nu, y_{S_\nu}, y_{R_\nu}, K_\nu, r_\nu)$ which is the ν-th record on $Hlist_1$;
   (b) if $(m_\nu, y_{S_\nu}, y_{R_\nu}) = (m, y_S, y_R)$ and (g, y_R, w, K_ν) is a DH tuple in $\langle g \rangle$, then compute $\hat{r} \leftarrow r_\nu$ and jump to Step 11;
   (c) $\nu \leftarrow \nu + 1$.
9. If $Hlist_2$ is not empty, then from $\mu = 1$ to $\mu = l_{Hlist_2}$ do
   (a) take the value of $(r_\mu, s_\mu, m_\mu, y_{S_\mu}, y_{R_\mu}, r'_\mu)$ which is the μ-th record on $Hlist_2$;
   (b) if $(m_\mu, y_{S_\mu}, y_{R_\mu}) = (m, y_S, y_R)$ and $(y_S g^{r_\mu})^{s_\mu} = (y_S g^r)^s$, then compute $\hat{r} \leftarrow r'_\mu$ and jump to Step 11;
   (c) $\mu \leftarrow \mu + 1$;
10. Return $Reject$;
11. Check whether $r = \hat{r}$; if it is, return m, otherwise return $Reject$.

We define a certain event F_2 to be that for some unsigncryption query, $Unigncryption_{sim}(\sigma, pk_S, pk_R) = Reject$ and $H_{sim}(m, y_S, y_R, K) = r$ where $\sigma = (c, r, s)$, $m = D_\tau(c)$, $\tau = G_{sim}(y_S, y_R, K)$, and $K = w^{x_R}$.

It is easy verify that the unsigncryption oracle and $Unsigncryption_{sim}$ has the output distribution unless F_2 occurs. Therefore, the change from Game 3 to Game 4 is a transition based on a failure event F_2. We have

$$|Pr[S_4] - Pr[S_3]| \leq Pr[F_2]. \tag{6}$$

In this proof, F_2 occurs if and only if $Unsigncryption_{sim}$ returns $Reject$ at Step 10, while $H_{sim}(m, y_S, y_R, K) = r$. According to the description of $Unsigncryption_{sim}$, in this case H_{sim} has never been run on (m, y_S, y_R, K) before this unsigncryption query and there is no record on $Hlist_2$ satisfies the output condition. According to the H_{sim} algorithm, H_{sim} will generate and return a random value at Step 3. For each unsigncryption query, the probability that r equals to that random value is $\frac{1}{q}$. Considering all j_u unsigncryption queries, the probability that $Unsigncryption_{sim}$ returns $Reject$ at Step 10, while $H_{sim}(m, y_S, y_R, K) = r$ in that case is $\frac{j_u}{q}$. Therefore, we have

$$Pr[F_2] = \frac{j_u}{q}. \tag{7}$$

Game 5 (Replace H_{sim} with H' at Stage 3): In this game S_5 behaves mostly the same as S_4, except that S_5 replaces H_{sim} with another algorithm H' at Stage 3. On input $(m^*, y_{S^*}, y_{R^*}, K^*)$, H' chooses $\bar{r}^* \in Z_q$ uniformly at random and outputs $\bar{r}^*$.

Since $\bar{r}^*$ is chosen uniformly at random from Z_q, the probability that $r^* = \bar{r}^*$ is $\frac{1}{q}$. Therefore, we have

$$Pr[S_5] = \frac{1}{q}. \tag{8}$$

We define a certain event F_3 to be that at Stage 2, H_{sim} is run on input $(m^*, y_{S^*}, y_{R^*}, K^*)$ where $K^* = y_{R^*}{}^{s^*(r^*+x_{S^*})}$.

If F_3 does not occur, then from $\mathcal{A}$'s point of view, Game 5 and Game 4 proceeds identically. Therefore, the change from Game 4 to Game 5 is a transition based on a failure event F_3. Then, we have

$$|Pr[S_5] - Pr[S_4]| \leq Pr[F_3] \qquad (9)$$

Game 6 (Change the Way to Generate an Input to $\mathcal{A}$): In this game $\mathcal{S}_6$ behaves mostly the same as $\mathcal{S}_5$, except that $\mathcal{S}_6$ runs in a different way at Stage 1, and at Stage 2 it calls for a restricted DDH oracle to check whether a quadruple is a DH quadruple and makes use of a random beacon R to replace the R' algorithm, where both the restricted DDH oracle and the random beacon R are provided by the rbGDL problem. In this game $\mathcal{S}_6$ (which can also be regarded as an adversary $\mathcal{A}_{rbgdl}$) prepares to take up the challenge of attacking the rbGDL problem in a group $\langle g \rangle$ with an input (g, g^a). Particularly, at Stage 1, $\mathcal{S}_6$ runs as follows:

1. Set (p, q, g) as the same as in the rbGDL problem.
2. Set $G, H, \mathcal{SKE}$ according to the Setup algorithm.
3. Set $y_A \leftarrow g^a$.
4. Give (cp, pk_A) to $\mathcal{A}$, where $cp = (p, q, g, G, H, \mathcal{SKE})$, $pk_A = y_A$.

It is obvious that the changes are only conceptual. In other words, from $\mathcal{A}$'s point of view, Game 6 and Game 5 proceeds identically. Therefore, F_3 in Game 6 and Game 5 occurs with the same probability.

Now we analyze the probability that F_3 occurs. In this proof, records on $Hlist_2$ are only be generated by $Signcryption_{sim}$ and according to the rule of the game, $\mathcal{A}$ is not allowed to make a signcryption query on $(m^*, pk_{S^*}, pk_{R^*})$ which implies there will be no such a record $(m_\mu, r_\mu, s_\mu, y_{S_\mu}, y_{R_\mu}, r'_\mu)$ on $Hlist_2$ satisfying $(m_\mu, y_{S_\mu}, y_{R_\mu}) = (m^*, y_{S^*}, y_{R^*})$. Therefore, when H_{sim} is queried on $(m^*, y_{S^*}, y_{R^*}, K^*)$, the output value will never be returned at Step 2. That is, the output value of $H_{sim}(m^*, y_{S^*}, y_{R^*}, K^*)$ is generated at Step 3 when it is first queried. In this case, according to the H_{sim} algorithm, $r^* = R(y_{R^*}, K^*)$. Therefore, $\mathcal{S}_6$ can solve the rbGDL problem by outputting (r^*, s^*, i^*) where i^* denotes R runs on input (y_{R^*}, K^*) at the i-th time. From the above analysis, we have

$$Pr[F_3] \leq Adv_{\mathcal{A}_{rbgdl}}^{rbgdl}(t_{rbgdl}, j_{rbgdl}, j_r) \qquad (10)$$

where $Adv_{\mathcal{EGDH}, \mathcal{A}_{egdh}}^{egdh}(t_{egdh}, j_{egdh})$ is the advantage of $\mathcal{A}_{rbgdl}$ running in time t_{rbgdl} and making at most j_{rbgdl} restricted DDH queries, and at most j_r random beacon queries. According to the execution of $\mathcal{S}_6$ in Game 6, we can compute that $t_{rbgdl} = t + t'_c$ where $t'_c = O((j_s + j_u)^2 + j_h^2 + j_g^2)$ is the simulation time of $\mathcal{S}_6$, $j_{rbgdl} = O((j_g + j_h)(j_s + j_u))$ and $j_r \leq j_h$. Therefore, t_{rbgdl}, j_{rbgdl}, and j_r are all polynomials in k.

By Lemma 3, we can construct an algorithm $\mathcal{A}_{gdl}$ to attack the GDL assumption that runs in time $t_{gdl} = 2t_{rbgdl} + O(q^2)$ and makes $j_{gdl} = 2j_{rbgdl}$ restricted DDH queries, with a success probability

$$Adv_{\mathcal{GDL},\mathcal{A}_{gdl}}^{gdl}(t_{gdl}, j_{gdl}) \geq \frac{1}{j_r}\left(\frac{Adv_{\mathcal{RBGDL},\mathcal{A}_{rbgdl}}^{rbgdl}(t_{rbgdl}, j_{rbgdl}, j_r)}{2} - \frac{j_r}{q}\right)^2. \quad (11)$$

Here t_{gdl} and j_{gdl} are also polynomials in k, since t_{rbgdl} and j_{rbgdl} are polynomials in k. Recall that in Lemma 3, we have

$$Adv_{\mathcal{RBGDL},\mathcal{A}_{rbgdl}}^{rbgdl}(t_{rbgdl}, j_{rbgdl}, j_r) \geq \frac{2j_r}{q},$$

as a result, (11) can be expressed as

$$Adv_{\mathcal{RBGDL},\mathcal{A}_{rbgdl}}^{rbgdl}(t_{rbgdl}, j_{rbgdl}, j_r) \leq 2(\sqrt{j_r \cdot Adv_{\mathcal{GDL},\mathcal{A}_{gdl}}^{gdl}(t_{gdl}, j_{gdl})} + \frac{j_r}{q}).(12)$$

Combining (10) and (11), with $j_r \leq j_h$, the probability for F_3 to occur is

$$Pr[F_3] \leq \frac{2j_h}{q} + 2 \cdot \sqrt{j_h \cdot Adv_{\mathcal{GDL},\mathcal{A}_{gdl}}^{gdl}(t_{gdl}, j_{gdl})}. \quad (13)$$

Arrive at our conclusion: Combining the formulas from (1) to (9), and formula (13), we have

$$Adv_{\mathcal{SC},\mathcal{A}}^{euf-cspa}(t, j_s, j_u) \quad (14)$$

$$\leq \frac{j_s(j_g + j_h + 3j_u + 2j_s) + 2j_h + j_u + 1}{q} + 2\sqrt{j_h \cdot Adv_{\mathcal{GDL}\mathcal{A}_{gdl}}^{gdl}(t_{gdl}, j_{gdl})}. \quad (15)$$

Let $\epsilon_{t_{gdl},j_{gdl}}^{gdl}$ be the maximum of $Adv_{\mathcal{GDL},\mathcal{A}_{gdl}}^{gdl}(t_{gdl}, j_{gdl})$ over all algorithms attacking the GDL problem that runs in time t_{gdl} and makes at most j_{gdl} restricted DDH queries to a DDH oracle. From the analysis of Game 6, we get that t_{gdl} and j_{gdl} are polynomials in k. Therefore, under the GDL assumption, $\epsilon_{t_{gdl},j_{gdl}}^{gdl}$ must be negligible in k.
Taking a maximum over all EUF-CSPA adversaries with appropriate resource parameters, we get our conclusion that

$$\epsilon_{t,j_s,j_u}^{euf-cspa} \leq \frac{j_s(j_g + j_h + 3j_u + 2j_s) + 2j_h + j_u + 1}{q} + 2 \cdot \sqrt{j_h \cdot \epsilon_{t_{gdl},j_{gdl}}^{gdl}}. \quad (16)$$

Finally, we remark that the minor tweak we made is useful in Step 4(b) of $Signcryption_{sim}$, and Step 5(b) of $Unsigncryption_{sim}$. This tweak takes y_S and y_R as part of the input to the G hash function. Therefore, in these two cases we are sure that $(y_{S_\mu}, y_{R_\mu}) = (y_S, y_R)$.

5.2 Proof of Confidentiality

Theorem 2. *Let H and G be two hash functions modeled as random oracles. Then under the GDH assumption in $\langle g \rangle$ which is a subgroup of Z_p^* generated by g, and the assumption that the $\mathcal{SKE}$ is PI-SKE secure, the Zheng signcryption scheme is IND-CSPA secure.*

Specifically, let k be a security parameter of the Zheng signcryption, $\mathcal{A}$ be an IND-CSPA adversary that runs in time t, and makes at most j_s signcryption queries, j_u unsigncryption queries, j_g hash queries to G and j_h hash queries to H, where t, j_s, j_u, j_g, j_h are all polynomials in k. Then the maximum advantage $\epsilon_{t,j_s,j_u}^{ind-cspa}$ of the adversary satisfies the following condition:

$$\epsilon_{t,j_s,j_u}^{ind-cspa} \leq \epsilon_{t_{egdh},j_{egdh}}^{egdh} + \epsilon_{t_{ske}}^{pi-ske} + \frac{j_s(j_g + j_h + 6j_u + 3j_s + 2)}{q}$$

where $\epsilon_{t_{egdh},j_{egdh}}^{egdh}$, $\epsilon_{t_{ske}}^{pi-ske}$ are negligible in k for all sufficiently large k.

The proof for this theorem follows a similar path to that for Theorem 1. We leave details of the proof to a full version of this paper.

6 Relationships with Proofs by Baek, Steinfeld and Zheng

The proof of confidentiality by Baek, Steinfeld and Zheng can be naturally extended to the single key pair setting, with the exception that in the new model, more cases need to be considered. As a result, all the games in the proof should be properly described and probabilities for all the events need to be carefully analyzed by taking into account all the added cases throughout the whole proof.

The proof of unforgeability in the new model can not be naturally derived from the proof by Baek, Steinfeld and Zheng. For example, in Game 3 of the proof by Baek, Steinfeld and Zheng, when $(m^*, y_S{}^*, y_R{}^*, K^*)$ is presented to $HSim$, it is the same as that R (the random beacon) has been run on $(y_R{}^*, K^*)$ which implies the rbGDL problem has been resolved. When it is extended to the single key pair setting, there should be unsigncryption queries which (according to their proof for confidentiality) may add records to $Hlist_2$. In this case $(m^*, y_S{}^*, y_R{}^*, K^*)$ is presented to $HSim$ which can be different from R being run on $(y_R{}^*, K^*)$, since the result of $HSim(m^*, y_S{}^*, y_R{}^*, K^*)$ may come from $Hlist_2$. To ensure that unforgeability can be reduced to the GDL assumption under the new model, we had to resolve a number of technical issues, including the use of a random beacon, the way to add records to $Hlist_2$, and the way to simulate the unsigncryption oracle among many other minor technical issues.

Acknowledgment. We thank Joonsang Baek and Ron Steinfeld for thoroughly reading the early version of this paper and providing helpful comments. We also thank the anonymous reviewers of ACISP 2011 for their valuable comments.

References

1. An, J.H., Dodis, Y., Rabin, T.: On the security of joint signature and encryption. In: Knudsen, L.R. (ed.) EUROCRYPT 2002. LNCS, vol. 2332, pp. 83–107. Springer, Heidelberg (2002)
2. Baek, J., Steinfeld, R., Zheng, Y.: Formal proofs for the security of signcryption. In: Naccache, D., Paillier, P. (eds.) PKC 2002. LNCS, vol. 2274, pp. 80–98. Springer, Heidelberg (2002)
3. Baek, J., Steinfeld, R., Zheng, Y.: Formal proofs for the security of signcryption. J. Cryptology 20(2), 203–235 (2007)
4. Bao, F., Deng, R.H., Zhu, H.: Variations of Diffie-Hellman problem. In: Qing, S., Gollmann, D., Zhou, J. (eds.) ICICS 2003. LNCS, vol. 2836, pp. 301–312. Springer, Heidelberg (2003)
5. Bellare, M., Rogaway, P.: Random oracles are practical: A paradigm for designing efficient protocols. In: Proceedings of the First ACM Conference on Computer and Communications Security, New York, pp. 62–73. The Association for Computing Machinery (November 1993)
6. Cramer, R., Shoup, V.: Design and analysis of practical public-key encryption schemes secure against adaptive chosen ciphertext attack. SIAM Journal on Computing 33, 167–226 (2003)
7. Okamoto, T., Pointcheval, D.: The gap-problems: A new class of problems for the security of cryptographic schemes. In: Kim, K.-c. (ed.) PKC 2001. LNCS, vol. 1992, pp. 104–118. Springer, Heidelberg (2001)
8. Shoup, V.: Sequences of games: A tool for taming complexity in security proofs (2004), http://eprint.iacr.org/2004/332
9. Zheng, Y.: Digital signcryption or how to achieve cost (Signature & encryption) << cost(Signature) + cost(Encryption). In: Kaliski Jr., B.S. (ed.) CRYPTO 1997. LNCS, vol. 1294, pp. 165–179. Springer, Heidelberg (1997)
10. Zheng, Y., Imai, H.: Efficient signcryption schemes on elliptic curves. In: IFIP/SEC 1998: Proceedings of the IFIP 14th International Information Security Conference, New York, pp. 75–84. Chapman and Hall, Boca Raton (1998)

Towards Public Key Encryption Scheme Supporting Equality Test with Fine-Grained Authorization

Qiang Tang

DIES, Faculty of EEMCS
University of Twente, the Netherlands
`q.tang@utwente.nl`

Abstract. In this paper we investigate a new category of public key encryption schemes which supports equality test between ciphertexts. With this primitive, two users, who possess their own public/private key pairs, can issue token(s) to a proxy to authorize it to perform equality test between their ciphertexts. We provide a formulation and a corresponding construction for this primitive, and our formulation provides fine-grained authorization policy enforcements for users. With the increasing popularity of outsourcing data and computations to third-party service providers, this primitive will be an important building block in designing privacy protection solutions supporting operations on encrypted data.

1 Introduction

Today, more and more IT applications outsource the storage and business transactions of corporate/personal database to third-party service providers. For such applications, it is a big challenge to design mechanisms, which simultaneously achieve the intended business objectives and provide a maximal level of privacy guarantee on the sensitive data. Within the information security community, a lot of research efforts have been dedicated to cryptographic techniques supporting operations on encrypted data. In this paper, we are interested in Public Key Encryption schemes which support Equality Test between ciphertexts. This primitive is formally referred to as PKEET, and an informal functional description is as follows.

Given a public key encryption scheme (KeyGen, Enc, Dec), *suppose that two users possess their public/private key pairs* (PK, SK) *and* (PK', SK') *respectively. If this public key encryption scheme belongs to the category of PKEET, then the two users can authorize a third-party proxy to perform the following test: Given* Enc(M, PK) *and* Enc(M', PK') *for any* M *and* M', *test whether* $M = M'$ *without knowing* M *or* M'.

As mentioned in [20], PKEET is a useful building block in construct privacy-preserving applications, such as outsourced databases. Besides, we can foresee more applications in the emerging computing scenarios. For example, in an

U. Parampalli and P. Hawkes (Eds.): ACISP 2011, LNCS 6812, pp. 389–406, 2011.

Internet-based PHR application [17], a PKEET cryptosystem can achieve the following: (1) patients can encrypt their attributes (2) a semi-trusted proxy can match patients' encrypted attributes and recommend them to each other, without knowing the plaintext attributes.

1.1   Related Work

The concept of PKEET cryptosystem was proposed by Yang *et al.* [20]. However, their formulation lacks an authorization mechanism for users to specify who can perform equality test between their ciphertexts, and in fact any entity can perform the equality test. The consequence is that standard semantic security or IND-CPA security cannot be achieved against any entity, when considering the fact that ciphertexts are public information. In addition, if the message space is polynomial size or the min-entropy of the message distribution is much lower than the security parameter, then any entity can potentially mount an offline message recovery attack. This attack is similar to the offline keyword guessing attack in the case of PEKS (or searchable encryption) [11,18].

The concepts of PKEET has a close nature to that of Public key encryption with keyword search (PEKS) [8] and public key encryption with registered keyword search (PERKS) [18]. With a PEKS or PERKS scheme, a user can enable a server to perform equality test between the keywords embedding in a tag and a ciphertext, and the user enforces her authorization by issuing a token to the server. The difference is that, instead of keywords, PKEET is concerned with the equality test of plaintexts which are encrypted under different public keys. Another related concept is order preserving encryption (OPE) scheme, which is a primitive firstly proposed by Agrawal *et al.* [1] and then further investigated by Boldyreva *et al.* [6]. With an OPE scheme, the order of ciphertexts always remains the same as that of the corresponding plaintexts. Therefore, given a set of ciphertexts, any entity can directly compare the plaintexts. The order-preserving property of an OPE scheme holds only for the ciphertexts generated under the same public key, which differs from the purpose of PKEET.

1.2   Our Contribution

To mitigate the potential vulnerabilities of PKEET, we integrate a fine-grained authorization policy enforcement mechanism into PKEET and propose an enhanced primitive, namely FG-PKEET. With an FG-PKEET cryptosystem, two users, say Alice and Bob, need to run the authorization algorithm together to issue a token to a semi-trusted proxy, which will then be authorized to perform equality test between their ciphertexts. Without the token, the equality test cannot be performed. With this primitive, users gain more control over the operations on their encrypted data.

- A user has tight control over who can perform equality test on her ciphertexts, by choosing the semi-trusted proxies.
- A user has tight control over with whose ciphertexts that her ciphertexts can be tested with, by choosing with which user to run the authorization algorithm.

For FG-PKEET, we consider two types of adversaries: Type-I adversary which represents the semi-trusted proxies, and Type-II adversary which represents all malicious entities. With respect to a Type-I adversary, we provide OW-CCA (i.e. one-way CCA) and OW-CPA (i.e. one-way CPA) security definitions; while with respect to a Type-II adversary, we provide standard IND-CCA and IND-CPA security definitions. Furthermore, a fine-grained authorization property is defined for FG-PKEET. Informally, this property means that a proxy cannot perform equality test between two users' ciphertexts unless it receives a token assigned by these two users together. For example, a proxy cannot compare the ciphertexts of Bob and Charlie, even if it has received a token to compare the ciphertexts of Alice and Bob together with another token to compare the ciphertexts of Alice and Charlie. We propose an FG-PKEET cryptosystem, which achieves all the security properties defined in our security model.

In the extreme situation, when the message space is polynomial size or the min-entropy of the message distribution is much lower than the security parameter, for FG-PKEET, only a Type-I adversary is capable of mounting an offline message recovery attack which is unavoidable due to the desired equality test functionality. However, compared with the formulation in [20], where any adversary can mount the attack, our formulation achieves a significant security improvement. Furthermore, based on computational client puzzles [14], we propose an enhancement to mitigate this type of attack.

1.3   Organization

The rest of the paper is organized as follows. In Section 2, we formulate the concept of FG-PKEET. In Section 3, we propose an FG-PKEET cryptosystem. In Section 4, we analyse the proposed cryptosystem and provide an enhancement. In Section 5, we conclude the paper.

2   Formulation of FG-PKEET

In this section, we first provide a formal description for FG-PKEET, and then present the security model. Throughout the paper, we use "$\|$" to denote the concatenation operator and use $x \in_R X$ to denote that x is chosen from X uniformly at random.

2.1   Description of FG-PKEET

An FG-PKEET cryptosystem consists of algorithms (KeyGen, Enc, Dec, Aut, Com), where (KeyGen, Enc, Dec) define a standard public key encryption scheme while (Aut, Com) define the equality test functionality.

- KeyGen(ℓ): This algorithm takes a security parameter ℓ as input, and outputs a public/private key pair (PK, SK). Let $\mathcal{M}$ denote the message space.
- Enc(M, PK): This algorithm takes a message $M \in \mathcal{M}$ and the public key PK as input, and outputs a ciphertext C.

- Dec(C, SK): This algorithm takes a ciphertext C and the private key SK as input, and outputs the plaintext M or an error message $\perp$.

Let all the potential users be denoted as U_i ($1 \le i \le N$), where N is an integer, and they adopt the above public key encryption scheme. For any i, suppose that U_i's key pair is denoted as (PK_i, SK_i). Suppose that U_i and U_j want to enable a proxy to perform equality test between their ciphertexts, the Aut and Com algorithms are defined as follows.

- Aut($SK_i; SK_j; \cdot$): This algorithm is interactively run among U_i, U_j and the proxy, and the two users use their private keys as their secret inputs. At the end of the algorithm execution, the proxy receives a token $T_{i,j}$ as the output, while U_i and U_j receive no explicit output. We assume that the communication channels among the participants are confidential.
- Com($C_i, C_j, T_{i,j}$): This algorithm takes two ciphertexts C_i, C_j and the token $T_{i,j}$ as input, and outputs 1 if $M_i = M_j$ or 0 otherwise. Note that C_i, C_j are two ciphertexts encrypted under PK_i and PK_j respectively.

In the algorithm definitions, besides the explicitly specified parameters, other public parameters could also be specified and be implicitly part of the input. We omit those parameters for the simplicity of description. Note that, under our definition of Aut, $T_{i,j}$ and $T_{j,i}$ are exactly the same thing.

It is worth noting that the Aut algorithm is supposed to run interactively among two users and the proxy. The interactive nature of this algorithm may seem to be a drawback, but it in fact reflects the process that the two users together authorize the semi-trusted proxy to perform equality test between their ciphertexts. Moreover, this algorithm only needs to be run once for any selected proxy, which will then be able to compare all ciphertexts of the two users. Therefore, the interactive nature of the the Aut algorithm will not be a performance bottleneck in practice.

Similar to other cryptographic primitives, the basic requirement for FG-PKEET is soundness. Informally, this property means that the algorithms Dec and Com work properly with valid inputs. Formally, it is defined as follows.

Definition 1. *An FG-PKEET cryptosystem achieves (unconditional) soundness if the following two equalities hold for any $i, j \ge 1$ and $M, M' \in \mathcal{M}$. Let $(PK_i, SK_i) = $ KeyGen(ℓ) and $(PK_j, SK_j) = $ KeyGen(ℓ).*

1. Dec(Enc(M, PK_i), SK_i) $= M$ *and* Dec(Enc(M', PK_j), SK_j) $= M'$.
2. Com(Enc(M, PK_i), Enc(M', PK_j), Aut($SK_i; SK_j; \cdot$)) *is equal to 1 if* $M = M'$, *and 0 otherwise.*

As a remark, in the definitions of Aut and Com, we implicitly assume that $i \ne j$ because we are only interested in testing the equality of the ciphertexts of two different users.

2.2 The Security Model

To facilitate our formal discussions, we make the following assumptions.

1. First of all, all users honestly generate their public/private key pairs and the execution of the Aut algorithm will be carried out through secure channels between the involved entities.
2. Secondly, the proxies are semi-trusted (or, honest-but-curious) to the users who have chosen them. They will faithfully follow the protocol specifications, but will try to deduce some information from the acquired data. In addition, one proxy can serve multiple pairs of users to perform equality test.
3. Thirdly, there is no overlap between the user set and the proxy set, namely no user will be allowed to act as a proxy for another two users. This will greatly simplify our discussion. Yet, we leave it as a future work to investigate FG-PKEET in the case where this assumption is not true.

With respect to an FG-PKEET cryptosystem, for an honest user U_t, where $1 \leq t \leq N$, we consider two categories of adversaries, namely Type-I and Type-II adversaries as illustrated in Fig. 1.

1. Type-I adversary represents the semi-trusted proxies with which U_t has run the algorithm Aut with. Referring to Fig. 1, Proxy I and Proxy L are Type-I adversary.
2. Type-II adversary represents all possibly malicious entities in the system from the perspective of U_t, namely U_i $(1 \leq i \leq N, i \neq t)$. In fact, all proxies with which U_t has not run the algorithm Aut should also be regarded as a malicious adversary, because U_t do not even semi-trust them. For example, Proxy T in Fig. 1 is such an entity. However, taking them into account will not give the Type-II adversary extra power, so that we simply ignore them.

As to a Type-I adversary, it is involved in the executions of the Aut algorithm as the proxy with U_t, and obtains the tokens, and it may also obtain some

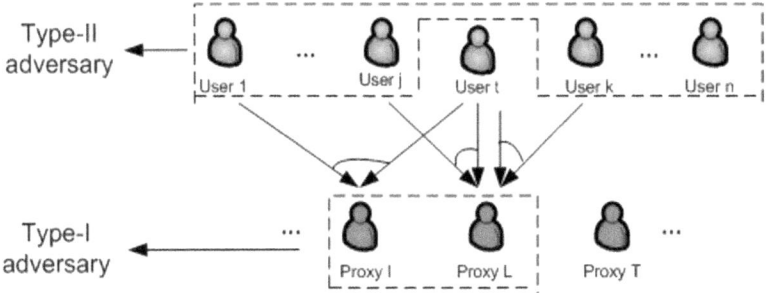

Fig. 1. An Illustration of Adversaries for FG-PKEET

information about U_t's plaintexts through accessing U_t's decryption oracle. Clearly, in the presence of a Type-I adversary, standard indistinguishability notions, such as IND-CCA and IND-CPA, cannot be achieved. Against a Type-I adversary, we consider the following two security properties.

1. OW-CCA (i.e. one-wayness under a chosen ciphertext attack), which implies that an adversary cannot recover the plaintext from a ciphertext $C_t^* = \mathsf{Enc}(M_t, PK_t)$ even if it is allowed to query the decryption oracle with any ciphertext except for C_t^*. This is the best achievable security guarantee considering the desired equality test functionality.
2. Fine-grained authorization property, which means that if two users have not authorized a proxy to perform equality test between their ciphertexts then the proxy should not be able to do so. Referring to Fig. 1, U_t and U_n have not authorized Proxy L to perform equality test between their ciphertexts, so that it should not be able to do so even if U_t has authorized it to perform equality test between her ciphertexts and those of U_j and U_k. It is worth noting this is an analog to the collusion resistance property in the attribute-based encryption schemes [15].

As to the power of a Type-II adversary, it is involved in the executions of the Aut algorithm as the other user with U_t, so that it may learn some information about U_t's private key. Moreover, it may also obtain some information about U_t's plaintexts through accessing U_t's decryption oracle. In the presence of a Type-II adversary, we define the standard IND-CCA security.

2.3   OW-CCA Security against a Type-I Adversary

Definition 2. *An FG-PKEET cryptosystem achieves OW-CCA security against a Type-I adversary, if, for any $1 \leq t \leq N$, any polynomial-time adversary has only a negligible advantage in the attack game shown in Fig. 2, where the advantage is defined to be $\Pr[M_t' = M_t]$.*

It is worth noting that, strictly speaking, the notion of OW-CCA is neither weaker nor stronger than IND-CPA [3]. One one hand, an IND-CPA secure scheme may not be OW-CCA. For instance, many homomorphic encryption schemes, such as Elgamal [12] and Paillier scheme [13], are IND-CPA but they are clearly not OW-CCA. On the other hand, an OW-CCA secure scheme may not be IND-CPA. For instance, the scheme proposed in Section 3 is OW-CCA but it is not IND-CPA.

2.4   Fine-Grained Authorization Property

Definition 3. *An FG-PKEET cryptosystem achieves the fine-grained authorization property against a Type-I adversary, if, for any $1 \leq t \leq N$, any polynomial-time adversary has only a negligible advantage in the attack game shown in Fig. 3, where the advantage is defined to be $|\Pr[b' = b] - \frac{1}{2}|$.*

1. The challenger runs KeyGen to generate public/private key pairs (PK_i, SK_i) for all $1 \leq i \leq N$.
2. Phase 1: The adversary is allowed to issue the following types of oracle queries.
   (a) Dec query with data C as input for the index i: the challenger returns $\text{Dec}(C, SK_i)$.
   (b) Aut query with two integer indexes i, j as input: the challenger runs the Aut algorithm with the adversary which plays the role of the proxy.
   At some point, the adversary asks the challenger for a challenge for an index t.
3. Challenge phase: The challenger chooses a message $M_t \in_R \mathcal{M}$ and sends $C_t^* = \text{Enc}(M_t, PK_t)$ to the adversary.
4. Phase 2: The adversary is allowed to issue the same types of oracle queries as in Phase 1. In this phase, the adversary's activities should adhere to the following restriction: *The Dec oracle should not have been queried with the data C_t^* for the index t.* At some point, the adversary terminates by outputting a guess M_t'.

Fig. 2. The Game for OW-CCA

1. The challenger runs KeyGen to generate public/private key pairs (PK_i, SK_i) for all $1 \leq t \leq N$.
2. Phase 1: The adversary is allowed to issue the following types of oracle queries.
   (a) Dec query with data C as input for the index i: the challenger returns $\text{Dec}(C, SK_i)$.
   (b) Aut query with two integer indexes i, j as input: the challenger runs the Aut algorithm with the adversary which plays the role of the proxy.
   At some point, the adversary sends two integer indexes t, w to the challenger for a challenge. In this phase, the Aut oracle should not have been queried with two integer indexes t, w.
3. Challenge phase: The challenger randomly chooses two different messages M_0, M_1 from $\mathcal{M}$ and a random bit b. If $b = 0$, send $C_t^* = \text{Enc}(M_0, PK_t)$ and $C_w^* = \text{Enc}(M_0, PK_w)$ to the adversary, otherwise send $C_t^* = \text{Enc}(M_0, PK_t)$ and $C_w^* = \text{Enc}(M_1, PK_w)$.
4. Phase 2: The adversary is allowed to issue the same types of oracle queries as in Phase 1. In this phase, the adversary's activities should adhere to the restriction described in Phase 1, together with the following one: *The Dec oracle should not have been queried with the data C_t^* and index t or with the data C_w^* and index w.* At some point, the adversary terminates by outputting a guess b'.

Fig. 3. The Game for the Fine-grained Authorization Property

In the attack game, it is clear that $b = 0$ ($b = 1$) implies the challenge ciphertexts do (not) contain the same plaintext. As a result, the adversary's ability of determining b is equivalent to determining the equality of ciphertexts of U_t and U_w. The adversary is not allowed to access $T_{t,w}$ because we assume the adversary is not authorized by U_t and U_w to perform the equality test.

Note the fact that a FG-PKEET cryptosystem can only achieve OW-CCA but not IND-CPA or IND-CCA. If the adversary is allowed to choose M_0, M_1 in the game, then it can trivially win the game. Therefore, different from a typical

IND (indistinguishability) security definition, where the adversary is allowed to choose M_0, M_1, in this game the challenger chooses both messages.

2.5   IND-CCA Security against a Type-II Adversary

Definition 4. *An FG-PKEET cryptosystem achieves IND-CCA security against a Type-II adversary, if, for any $1 \leq t \leq N$, any polynomial-time adversary has only a negligible advantage in the attack game shown in Fig. 4, where the advantage is defined to be $|\Pr[b' = b] - \frac{1}{2}|$.*

---

1. The challenger runs KeyGen to generate public/private key pairs (PK_i, SK_i) for all $1 \leq t \leq N$.

2. Phase 1: The adversary is allowed to issue the following types of oracle queries.
    (a) KeyRetrieve query with an integer index i as input: the challenger returns SK_i to the adversary.
    (b) Dec query with data C as input for the index i: the challenger returns $\mathsf{Dec}(C, SK_i)$.
    (c) Aut query, defined as below.
    At some point, the adversary sends an integer index t and two messages M_0, M_1 from $\mathcal{M}$ to the challenger for a challenge. In this phase, the adversary's activities should adhere to the following criteria.
    (a) The KeyRetrieve oracle should not have been queried with the index t.
    (b) For any $i \neq t$, the adversary is allowed to issue Aut oracle queries with indexes i, t as input, for any $i \neq t$, where the adversary plays the role of U_i.

3. Challenge phase: The challenger selects $b \in_R \{0, 1\}$ and sends $C_t^* = \mathsf{Enc}(M_b, PK_t)$ to the adversary.

4. Phase 2: The adversary is allowed to issue the same types of oracle queries as in Phase 1. In this phase, the adversary's activities are subject to the restrictions described in Phase 1, together with the following one: *The Dec oracle should not have been queried with the data C_t^* and index t.* At some point, the adversary terminates by outputting a guess b'.

Fig. 4. The Game for IND-CCA

In this game, the challenger generates all key pairs while the adversary is allowed to adaptively retrieve all private keys except SK_t. This formulation faithfully describe the power of a Type-II adversary in our security model, as defined in Section 2.2. In particular, the adversary is allowed to issue Aut oracle queries, which reflects the fact that U_t may interactively run the Aut algorithm with a Type-II adversary. A PKEET is IND-CCA secure against a Type-II adversary implies that, for U_t, the execution of the Aut algorithm leaks no information to other users.

3   A New FG-PKEET Cryptosystem

The proposed cryptosystem has $(\ell, \mathbb{G}, g, p, \mathsf{H}_1, \hat{e}, \mathbb{G}_1, \mathbb{G}_2, g_1, g_2, \mathbb{G}_T, q, \mathsf{H}_2, \mathsf{H}_3)$ as the global parameters which are defined as follows.

1. ℓ is the security parameter, $\mathbb{G}$ is a multiplicative group of prime order p, g is a generator of $\mathbb{G}$, and $H_1 : \{0,1\}^* \to \{0,1\}^\ell$ is a cryptographic hash function.
2. $\hat{e} : \mathbb{G}_1 \times \mathbb{G}_2 \to \mathbb{G}_T$ is a bilinear map, where $\mathbb{G}_1$ and $\mathbb{G}_2$ are multiplicative groups of prime order q, and they have g_1 and g_2 as their generators respectively. $H_2 : \{0,1\}^* \to \{0,1\}^{m+d_1}$, $H_3 : \{0,1\}^* \to \mathbb{G}_1$ are two cryptographic hash functions, where m is a polynomial in ℓ, $\{0,1\}^m$ is the message space and d_1 is the bit-length of p.

Note the fact that, in a PKEET cryptosystem, a ciphertext allows the receiver to decrypt and also allows a proxy to perform equality test. Hence, the intuition behind our construction is to integrate some extra components into a standard public key encryption scheme, so that these components will facilitate the equality test functionality. Specifically, in the encryption algorithm of the proposed scheme described in next subsection, the extra components are $C^{(2)}$ and $C^{(4)}$.

3.1 The Public Key Encryption Scheme

With the above global parameters defined, we first define the public key encryption algorithms (KeyGen, Enc, Dec).

- KeyGen(ℓ): This algorithm outputs a private key $SK = (x, y)$, where $x \in_R \mathbb{Z}_p$ and $y \in_R \mathbb{Z}_q$, and the corresponding public key is $PK = (g^x, g_1^y)$. Note that the message space is $\mathcal{M} = \{0,1\}^m$.
- Enc(M, PK): This algorithm outputs $C = (C^{(1)}, C^{(2)}, C^{(3)}, C^{(4)}, C^{(5)})$, where

$$u \in_R \mathbb{Z}_p, \ C^{(1)} = g^u, \ C^{(3)} = H_2(g^{ux}) \oplus M||u, \ v \in_R \mathbb{Z}_q,$$

$$C^{(2)} = g_1^v, \ C^{(4)} = g_1^{vy} \cdot H_3(M), \ C^{(5)} = H_1(C^{(1)}||C^{(2)}||C^{(3)}||C^{(4)}||M||u).$$

- Dec(C, SK): This algorithm first computes $M'||u' = C^{(3)} \oplus H_2((C^{(1)})^x)$, and then check that $g^{u'} = C^{(1)}$ and $H_1(C^{(1)}||C^{(2)}||C^{(3)}||C^{(4)}||M'||u') = C^{(5)}$ hold. If all checks pass, output M', otherwise output an error message $\perp$.

Suppose that every user U_i, for $1 \le t \le N$, adopts the above public key encryption scheme. To facilitate our description, we use the index i for all the variables in defining U_i's data. For example, U_i's key pair is denoted as (PK_i, SK_i), where $SK = (x_i, y_i)$ and $PK = (g^{x_i}, g_1^{y_i})$, and U_i's ciphertext $C_i = (C_i^{(1)}, C_i^{(2)}, C_i^{(3)}, C_i^{(4)}, C_i^{(5)})$ is written in the following form.

$$u_i \in_R \mathbb{Z}_p, \ C_i^{(1)} = g^{u_i}, \ C_i^{(3)} = H_2(g^{u_i x_i}) \oplus M_i||u_i, \ v_i \in_R \mathbb{Z}_q,$$

$$C_i^{(2)} = g_1^{v_i}, \ C_i^{(4)} = g_1^{v_i y_i} \cdot H_3(M_i), \ C_i^{(5)} = H_1(C_i^{(1)}||C_i^{(2)}||C_i^{(3)}||C_i^{(4)}||M_i||u_i).$$

3.2   The Token Generation Algorithm

Suppose that U_i and U_j want a proxy to perform equality test between their ciphertexts, then they run the following Aut algorithm to generate the token $T_{i,j}$ for the proxy.

- Aut($SK_i, SK_j, \cdot$): This algorithm results in a token $T_{i,j} = (g_2^{r_{i,j}}, g_2^{y_i r_{i,j}}, g_2^{y_j r_{i,j}})$ for the proxy. In more details, the token is interactively generated as follows.
  1. U_i and U_j generate $r_{i,j} \in_R \mathbb{Z}_q$ together.
  2. U_i sends $g_2^{r_{i,j}}, g_2^{y_i r_{i,j}}$ to the proxy, and U_j sends $g_2^{y_j r_{i,j}}$ to the proxy.

Note that, there can be many different ways for U_i and U_j to generate $r_{i,j}$ in implementing this algorithm. For instance, they can use a interactive coin flipping protocol, such as that of Blum [5]. Or, simply they can exchanges two nonces and set $r_{i,j}$ to be the hash value of them. In addition, the security properties will not be affected if U_j is required to send $g_2^{r_{i,j}}$ to the proxy.

3.3   The Equality Test Algorithm

Suppose a proxy has received the token $T_{i,j}$, then it can run the following Com algorithm to perform equality test between the ciphetexts C_i and C_j, which are encrypted under PK_i and PK_j respectively.

- Com($C_i, C_j, T_{i,j}$): This algorithm outputs 1 if $x_i = x_j$ or 0 otherwise, where

$$x_i = \frac{\hat{e}(C_i^{(4)}, g_2^{r_{i,j}})}{\hat{e}(C_i^{(2)}, g_2^{y_i r_{i,j}})} \qquad\qquad x_j = \frac{\hat{e}(C_j^{(4)}, g_2^{r_{i,j}})}{\hat{e}(C_j^{(2)}, g_2^{y_j r_{i,j}})}$$

$$= \frac{\hat{e}(g_1^{v_i y_i} \cdot \mathsf{H}_3(M_i), g_2^{r_{i,j}})}{\hat{e}(g_1^{v_i}, g_2^{y_i r_{i,j}})} \qquad = \frac{\hat{e}(g_1^{v_j y_j} \cdot \mathsf{H}_3(M_j), g_2^{r_{i,j}})}{\hat{e}(g_1^{v_j}, g_2^{y_j r_{i,j}})}$$

$$= \hat{e}(\mathsf{H}_3(M_i), g_2)^{r_{i,j}} \qquad\qquad = \hat{e}(\mathsf{H}_3(M_j), g_2)^{r_{i,j}}$$

In this construction, the group $\mathbb{G}$ can be any multiplicative group which holds the CDH assumption. In face, it can be set to be $\mathbb{G}_1$ or $\mathbb{G}_2$, in which case $p = q$. We keep it the present way for a general construction.

4   Comprehensive Security Analysis

In this section, we first prove that the proposed cryptosystem in Section 3 is secure in our security model. Then, we show how to improve its security against a Type-I adversary.

4.1   Preliminary

Following the work by Bellare and Rogaway [4], we use random oracle to model hash functions in our security analysis. A function $P(k) : \mathbb{Z} \to \mathbb{R}$ is said to be negligible with respect to k if, for every polynomial $f(k)$, there exists an integer N_f such that $P(k) < \frac{1}{f(k)}$ for all $k \geq N_f$. We say that the CDH (computational

Diffie-Hellman) assumption holds in $\mathbb{G}$ of prime order p if, given g^a, g^b where g is a group generator and $a, b \in_R \mathbb{Z}_p$, an adversary has only a negligible advantage in computing g^{ab}. We say that the DDH (decisional Diffie-Hellman) assumption holds in $\mathbb{G}_1$ of prime order q, if an adversary has only a negligible advantage in distinguishing (g_1^a, g_1^b, g_1^{ab}) from (g_1^a, g_1^b, g_1^c) where g_1 is a group generator and $a_1, b_1, c_1 \in_R \mathbb{Z}_q$. In the pairing setting, namely there is an efficient and non-degenerate bilinear map $\hat{e} : \mathbb{G}_1 \times \mathbb{G}_2 \to \mathbb{G}_T$, the DDH assumption in $\mathbb{G}_1$ is also referred to as the XDH (external Diffie-Hellman) assumption [7].

In order to prove the fine-grained authorization property, we need a new assumption, referred to as extended DBDH (decisional bilinear Diffie-Hellman) assumption. Let a pairing setting be $\hat{e} : \mathbb{G}_1 \times \mathbb{G}_2 \to \mathbb{G}_T$, where the order of groups is a prime q. The extended DBDH problem is formulated as follows.

1. The challenger selects $g_1, g_4, g_5 \in_R \mathbb{G}_1$, and $g_2, g_3 \in_R \mathbb{G}_2$, and $x_1, y_1, \in_R \mathbb{Z}_p$, and $\alpha, \beta \in_R \mathbb{G}_1$. The challenger flips a coin $b \in_R \{0, 1\}$ and sends X_b to the adversary, where

$$X_0 = (g_1^{x_1}, g_2^{x_1}, g_4^{x_1} \cdot \alpha, g_1^{y_1}, g_3^{y_1}, g_5^{y_1} \cdot \alpha), X_1 = (g_1^{x_1}, g_2^{x_1}, g_4^{x_1} \cdot \alpha, g_1^{y_1}, g_3^{y_1}, g_5^{y_1} \cdot \beta).$$

2. The adversary's outputs a guess b'. The adversary's advantage is $|\Pr[b = b'] - \frac{1}{2}|$.

The extended DBDH problem is at most as hard as the XDH problem in a Type-3 pairing setting [10]. In other words, if there is an algorithm to solve the XDH problem then there must be an algorithm to solve the extended DBDH problem, but it is not clear whether the vise-versa is true. Nonetheless, similar to the proof of the implicit XDH assumption in [2], we can show the extended DBDH assumption is hard in the generic group model. We leave the details to the full paper.

4.2 Proof Results

It is straightforward to verify that the soundness property is achieved, namely the Dec and Com work properly. We skip the details here. The following security proofs are done through a sequence of games [16].

Theorem 1. *The proposed FG-PKEET cryptosystem is OW-CCA secure against a Type-I adversary in the random oracle model based on the CDH assumption in* $\mathbb{G}$.

Proof sketch. Suppose an adversary has the advantage ϵ in the attack game shown in Fig. 2.

Game$_0$: In this game, the challenger faithfully simulates the protocol execution and answers the oracle queries from the adversary, and all hash functions are treated as random oracles. Let $\epsilon_0 = \Pr[M_t' = M_t]$. Clearly, $\epsilon_0 = \epsilon$ holds.

Game$_1$: In this game, the challenger performs identically to that in Game$_0$ except that the following. For any index i, if the adversary queries the decryption oracle Dec with C_i, the challenger computes $M_i \| u_i = H_2(g^{u_i x_i}) \oplus C_i^{(3)}$ and verifies $g^{u_i} = C_i^{(1)}$. If the verification fails, return $\bot$. Then, the challenger checks whether there exists an input query $C_i^{(1)}||C_i^{(2)}||C_i^{(3)}||C_i^{(4)}||M_i||u_i)$ to H_1, which outputs $C_i^{(5)}$. If such an input query exists, return M_i; otherwise return $\bot$. Let the event Ent_1 be that, for some C_i, a fresh input $C_i^{(1)}||C_i^{(2)}||C_i^{(3)}||C_i^{(4)}||M_i||u_i$ to H_1 results in $C_i^{(5)}$. Clearly, This game is identical to Game_0 unless the event Ent_1 occurs. It is straightforward that $\Pr[Ent_1]$ is negligible if H_1 is modeled as a random oracle. Let $\epsilon_1 = \Pr[M_t' = M_t]$ in this game. From the Difference Lemma in [16], we have $|\epsilon_1 - \epsilon_0| \leq \Pr[Ent_1]$.

Game_2: In this game, the challenger performs identically to that in Game_1 except that, for any index i, if the adversary queries the decryption oracle Dec with C_i, the challenger does the following. Try to obtain the query to the oracle H_1 with the input $C_i^{(1)}||C_i^{(2)}||C_i^{(3)}||C_i^{(4)}||M_i||u_i$ satisfying

$$M_i||u_i = \mathsf{H}_2(g^{u_i x_i}) \oplus C_i^{(3)}, \ g^{u_i} = C_i^{(1)}, \ \mathsf{H}_1(C_i^{(1)}||C_i^{(2)}||C_i^{(3)}||C_i^{(4)}||M_i||u_i) = C_i^{(5)}.$$

If such a query cannot be found, return $\bot$. Otherwise, return M_i. This game is indeed identical to Game_1. Let $\epsilon_2 = \Pr[M_t' = M_t]$, then we have $\epsilon_2 = \epsilon_1$.

Game_3: In this game, the challenger performs identically to that in Game_2 except that the challenge C_t^* is generated as follows.

$$C_t^{(1)} = g^{u_t}, \ C_t^{(2)} = g_1^{v_t}, \delta \in_R \{0,1\}^{m+d_1}, \ C_t^{(3)} = \delta,$$

$$C_t^{(4)} = g_1^{v_t y_t} \cdot \mathsf{H}_3(M_t), \ C_t^{(5)} = \mathsf{H}_1(C_t^{(1)}||C_t^{(2)}||C_t^{(3)}||C_t^{(4)}||M_t||u_t).$$

This game is identical to Game_2 unless the event Ent_2 occurs, namely $g^{u_t x_t}$ is queried to the random oracle H_2. Note that the private key x_t is never used to answer the adversary's queries. Therefore, $\Pr[Ent_2]$ is negligible based on the CDH assumption in $\mathbb{G}$. Let $\epsilon_3 = \Pr[M_t' = M_t]$ in this game. From the Difference Lemma in [16], we have $|\epsilon_3 - \epsilon_2| \leq \Pr[Ent_2]$.

Since H_1 and H_3 are modeled as random oracles, it is clear that ϵ_3 is negligible. From the above analysis, we have that $\epsilon \leq \Pr[Ent_1] + \Pr[Ent_2] + \epsilon_3$, which is negligible in the random oracle model based on the CDH assumption in $\mathbb{G}$. The theorem now follows. $\square$

Theorem 2. *The proposed FG-PKEET cryptosystem achieves fine-grained authorization property against a Type-I adversary in the random oracle model based on the CDH assumption in $\mathbb{G}$ and the extended DBDH assumption.*

Proof sketch. Suppose an adversary has the advantage ϵ in the attack game shown in Fig. 3.

Game_0: In this game, the challenger faithfully simulates the protocol execution and answers the oracle queries from the adversary, and all hash functions are treated as random oracles. Let $\epsilon_0 = \Pr[b' = b]$. Clearly, $\epsilon_0 = \epsilon$ holds.

Game_1: In this game, the challenger performs identically to that in Game_0 except that the following. For any index i, if the adversary queries the decryption oracle Dec with C_i, the challenger computes $M_i||u_i = \mathsf{H}_2(g^{u_i x_i}) \oplus C_i^{(3)}$ and verifies $g^{u_i} = C_i^{(1)}$. If the verification fails, return $\bot$. Then, the challenger checks whether there exists an input query $C_i^{(1)}||C_i^{(2)}||C_i^{(3)}||C_i^{(4)}||M_i||u_i$ to H_1, which outputs $C_i^{(5)}$. If such an input query exists, return M_i; otherwise return $\bot$. Let the event Ent_1 be that, for some C_i, a fresh input $C_i^{(1)}||C_i^{(2)}||C_i^{(3)}||C_i^{(4)}||M_i||u_i$ to H_1 results in $C_i^{(5)}$. Clearly, This game is identical to Game_0 unless the event Ent_1 occurs. it is straightforward that $\Pr[Ent_1]$ is negligible if H_1 is modeled as a random oracle. Let $\epsilon_1 = \Pr[b' = b]$ in this game. From the Difference Lemma in [16], we have $|\epsilon_1 - \epsilon_0| \leq \Pr[Ent_1]$.

Game_2: In this game, the challenger performs identically to that in Game_1 except that, for any index i, if the adversary queries the decryption oracle Dec with C_i, the challenger does the following. Try to obtain the query to the oracle H_1 with the input $C_i^{(1)}||C_i^{(2)}||C_i^{(3)}||C_i^{(4)}||M_i||u_i$ satisfying

$$M_i||u_i = H_2(g^{u_i x_i}) \oplus C_i^{(3)}, \ g^{u_i} = C_i^{(1)}, \ H_1(C_i^{(1)}||C_i^{(2)}||C_i^{(3)}||C_i^{(4)}||M_i||u_i) = C_i^{(5)}.$$

If such a query cannot be found, return $\bot$. Otherwise, return M_i. This game is indeed identical to Game_1. Let $\epsilon_2 = \Pr[b' = b]$, then we have $\epsilon_2 = \epsilon_1$.

Game_3: In this game, the challenger performs identically to that as in Game_2 except the following. The challenge C_t^* is generated as follows.

$$C_t^{(1)} = g^{u_t}, \ C_t^{(2)} = g_1^{v_t}, \delta_t \in_R \{0,1\}^{m+d_1}, \ C_t^{(3)} = \delta_t,$$

$$C_t^{(4)} = g_1^{v_t y_t} \cdot H_3(M_0), \ C_t^{(5)} = H_1(C_t^{(1)}||C_t^{(2)}||C_t^{(3)}||C_t^{(4)}||M_t||u_t).$$

The challenge C_w^* is generated as follows.

$$C_w^{(1)} = g^{u_w}, \ C_w^{(2)} = g_1^{v_w}, \delta_w \in_R \{0,1\}^{m+d_1}, \ C_w^{(3)} = \delta_w,$$

$$C_w^{(4)} = g_1^{v_w y_w} \cdot H_3(M_b), \ C_w^{(5)} = H_1(C_w^{(1)}||C_w^{(2)}||C_w^{(3)}||C_w^{(4)}||M_b||u_w).$$

This game is identical to Game_2 unless the event Ent_2 occurs, namely $g^{u_t x_t}$ or $g^{u_w x_w}$ is queried to the random oracle H_2. Note that the private keys x_t, x_w are never used to answer the adversary's queries. Therefore, $\Pr[Ent_2]$ is negligible based on the CDH assumption in $\mathbb{G}$. Let $\epsilon_3 = \Pr[b' = b]$ in this game. From the Difference Lemma in [16], we have $|\epsilon_3 - \epsilon_2| \leq \Pr[Ent_2]$.

Game_4: In this game, the challenger performs identically to that as in Game_2 except for answering the Aut queries. For U_t and U_w, the challenger chooses $h_i, h_w \in_R \mathbb{Z}_q$ at the beginning of the game. On receiving an Aut query with the inputs i, t, the challenger returns $(g_2^{h_i r}, g_2^{h_i y_i r}, g_2^{h_i y_j r})$, where $r \in_R \mathbb{Z}_q$, and does something similar to answering the query with the input i, w. Let $\epsilon_4 = \Pr[b' = b]$ in this game. It is clear that this game is identical to Game_3, therefore $\epsilon_4 = \epsilon_3$ holds.

Game_5: In this game, the challenger performs identically to that in Game_4 except the following. The challenge C_t^* is generated as follows.

$$C_t^{(1)} = g^{u_t}, \ C_t^{(2)} = g_1^{v_t}, \delta_t \in_R \{0,1\}^{m+d_1}, \ C_t^{(3)} = \delta_t,$$

$$k_t \in_R \mathbb{Z}_q, \ C_t^{(4)} = g_1^{v_t y_t k_t}, \ C_t^{(5)} = \mathsf{H}_1(C_t^{(1)}||C_t^{(2)}||C_t^{(3)}||C_t^{(4)}||M_t||u_t).$$

The challenge C_w^* is generated as follows.

$$C_w^{(1)} = g^{u_w}, \ C_w^{(2)} = g_1^{v_w}, \delta_w \in_R \{0,1\}^{m+d_1}, \ C_w^{(3)} = \delta_w,$$

$$C_w^{(4)} = g_1^{v_w y_w X}, \ C_w^{(5)} = \mathsf{H}_1(C_w^{(1)}||C_w^{(2)}||C_w^{(3)}||C_w^{(4)}||M_b||u_w).$$

The value of X is set to be k_t if $b = 0$, and otherwise k_w is randomly chosen from $\mathbb{Z}_q$. Let $\epsilon_5 = \Pr[b' = b]$ in this game. It is clear that this game is identical to Game_4, therefore $\epsilon_5 = \epsilon_4$ holds. Let $C_0 = (C_t^*, C_w^*)$ when $b = 0$, and $C_1 = (C_t^*, C_w^*)$ when $b = 1$. Distinguishing C_0 and C_1 is equivalent to distinguishing the following tuples.

$$(g_1^{y_t}, g_1^{v_t}, g_1^{y_t v_t k_t}, g_2^{h_t}, g_2^{h_t y_t}, g_1^{y_w}, g_1^{v_w}, g_1^{y_w v_w k_t}, g_2^{h_w}, g_2^{h_w y_w})$$

$$(g_1^{y_t}, g_1^{v_t}, g_1^{y_t v_t k_t}, g_2^{h_t}, g_2^{h_t y_t}, g_1^{y_w}, g_1^{v_w}, g_1^{y_w v_w k_w}, g_2^{h_w}, g_2^{h_w y_w})$$

It is straightforward to prove that to distinguish the above tuples is equivalent to distinguishing the extended DBDH tuples. Therefore, similar to proving semantic security of ElGamal scheme [16], it is straightforward to verify that $\epsilon_5 - \frac{1}{2}$ is negligible based on the extended DBDH assumption.

From the above analysis, we have that $|\epsilon_0 - \epsilon_5| \le \Pr[Ent_1] + \Pr[Ent_2]$, which is negligible in the random oracle model based on the CDH assumption in $\mathbb{G}$ and the extended DBDH assumption. Note that $\epsilon = |\epsilon_0 - \frac{1}{2}|$ and $|\epsilon_5 - \frac{1}{2}|$ is negligible, then ϵ is negligible. The theorem now follows. □

Theorem 3. *The proposed FG-PKEET cryptosystem is IND-CCA secure against a Type-II adversary in the random oracle model based on the CDH assumption in $\mathbb{G}$ and the DDH assumption in $\mathbb{G}_1$.*

Proof sketch Suppose that an adversary has the advantage ϵ in the attack game shown in Fig. 4.

Game_0 and Game_1: They are the same as in the proof of Theorem 2.

Game_2: In this game, the challenger performs identically to that in Game_1 except that, for any index i, if the adversary queries the decryption oracle Dec with U_i, the challenger does the following. Try to obtain the query to the oracle H_1 with the input $C_i^{(1)}||C_i^{(2)}||C_i^{(3)}||C_i^{(4)}||M_i||u_i$ satisfying

$$M_i||u_i = \mathsf{H}_2(g^{u_i x_i}) \oplus C_i^{(3)}, \ g^{u_i} = C_i^{(1)}, \ \mathsf{H}_1(C_i^{(1)}||C_i^{(2)}||C_i^{(3)}||C_i^{(4)}||M_i||u_i) = C_i^{(5)}.$$

If such a query cannot be found, return $\perp$. Otherwise, return M_i. This game is indeed identical to Game_1. Let $\epsilon_2 = \Pr[b' = b]$, then we have $\epsilon_2 = \epsilon_1$.

Game_3: In this game, the challenger performs identically to that in Game_2 except that the challenge C_t^* is generated as follows.

$$C_t^{(1)} = g^{u_t}, \ C_t^{(2)} = g_1^{v_t}, \delta \in_R \{0,1\}^{m+d_1}, \ C_t^{(3)} = \delta,$$

$$C_t^{(4)} = g_1^{v_t y_t} \cdot \mathsf{H}_3(M_b), \ C_t^{(5)} = \mathsf{H}_1(C_t^{(1)}||C_t^{(2)}||C_t^{(3)}||C_t^{(4)}||M_b||u_t).$$

This game is identical to Game_2 unless the event Ent_2 occurs, namely $g^{u_t x_t}$ is queried to the random oracle H_2. Note that the private key x_t is never used to answer the adversary's queries. Therefore, $\Pr[Ent_2]$ is negligible based on the CDH assumption in $\mathbb{G}$. Let $\epsilon_3 = \Pr[b' = b]$ in this game. From the Difference Lemma in [16], we have $|\epsilon_3 - \epsilon_2| \leq \Pr[Ent_2]$.

Game_4: In this game, the challenger performs identically to that in Game_3 except that the challenge C_t^* is generated as follows.

$$C_t^{(1)} = g^{u_t}, \ C_t^{(2)} = g_1^{v_t}, \delta \in_R \{0,1\}^{m+d_1}, \ C_t^{(3)} = \delta,$$

$$C_t^{(4)} = g_1^{v_t y_t} \cdot \mathsf{H}_3(M_b), \ \gamma \in_R \{0,1\}^\ell, \ C_t^{(5)} = \gamma.$$

This game is identical to Game_3 unless $C_t^{(1)}||C_t^{(2)}||C_t^{(3)}||C_t^{(4)}||M_b||u_t$ is queried to the random oracle H_1, referred to as the event Ent_3. Let $\epsilon_4 = \Pr[b' = b]$ in this game. Based on the CDH in $\mathbb{G}$, we have $|\epsilon_4 - \epsilon_3| \leq \Pr[Ent_3]$ is negligible.

Just the same as in proving the semantic security of ElGamal scheme [16], it is straightforward to verify that $\epsilon_4 - \frac{1}{2}$ is negligible based on the DDH assumption in $\mathbb{G}_1$. From the above analysis, we have that $|\epsilon_0 - \epsilon_4| \leq \Pr[Ent_1] + \Pr[Ent_2] + \Pr[Ent_3]$, which is negligible in the random oracle model based on the CDH assumption in $\mathbb{G}$ and the DDH assumption in $\mathbb{G}_1$. Note that $\epsilon = |\epsilon_0 - \frac{1}{2}|$ and $|\epsilon_4 - \frac{1}{2}|$ is negligible, then ϵ is negligible. The theorem now follows. $\square$

4.3 Potential Vulnerability and Enhancement

Note that since a Type-I adversary has access to a token $T_{i,t}$, then given a ciphertext $\mathsf{Enc}(M, PK_t)$ it can test whether $M' = M$ holds for any M' by checking the following equality $\mathsf{Com}(\mathsf{Enc}(M', PK_i), \mathsf{Enc}(M, PK_t), T_{i,t}) = 1$. Therefore, in the extreme situation when the actual message space $\mathcal{M}$ is polynomial size or the min-entropy of the message distribution is much lower than the security parameter, for FG-PKEET, a Type-I adversary (or, semi-trusted proxies) is capable of mounting an offline message recovery attack by checking every $M' \in \mathcal{M}$.

This type of attack is unavoidable due to the desired plaintext equality test functionality, similar to the offline keyword guessing attack in the case of PEKS (or searchable encryption) [11,18]. However, compared with the formulation in [20], where any adversary can mount the attack, our formulation achieves a significant security improvement because a Type-II adversary is unable to mount the attack. Although an offline message recovery attack is theoretically unavoidable in the presence of a Type-I adversary, but, depending on the specific cryptosystem, certain countermeasure can be employed to mitigate such an attack. One possible countermeasure is shown as below.

As in the original cryptosystem proposed in Section 3, the enhanced cryptosystem requires the same global parameters, namely

$$(\ell, \mathbb{G}, g, p, \mathsf{H}_1, \hat{e}, \mathbb{G}_1, \mathbb{G}_2, g_1, g_2, \mathbb{G}_T, q, \mathsf{H}_2, \mathsf{H}_3).$$

In addition, $Q \cdot T$, a puzzle hardness parameter L (detailed below), and a hash function $\mathsf{UH} : \{0,1\}^* \to \mathbb{Z}_{Q \cdot T}^*$ are also published, where Q, T are two large primes. These additional parameters are required by the computational client puzzle scheme [14], which is employed because it is deterministic and immune to parallel attacks [19]. Note that the generation of $Q \cdot T$ could be bootstrapped by a party trusted by all users in the system, and threshold techniques (e.g. [9]) can be used to improve the security. Nevertheless, this trust assumption is not required for achieving the existing security properties.

The algorithm KeyGen is identical to that in the original scheme, while the algorithms Enc and Dec are redefined as follows.

- Enc(M, PK): This algorithm outputs a ciphertext $C = (C^{(1)}, C^{(2)}, C^{(3)}, C^{(4)}, C^{(5)})$, where

$$u \in_R \mathbb{Z}_p, \ C^{(1)} = g^u, \ C^{(3)} = \mathsf{H}_2(g^{ux}) \oplus M||u, \ v \in_R \mathbb{Z}_q, \ C^{(2)} = g_1^v,$$

$$C^{(4)} = g_1^{vy} \cdot \mathsf{H}_3((\mathsf{UH}(M))^{2^L} \bmod Q \cdot T)), C^{(5)} = \mathsf{H}_1(C^{(1)}||C^{(2)}||C^{(3)}||C^{(4)}||M||u).$$

- Dec(C, SK): This algorithm first computes $M'||u' = C^{(3)} \oplus \mathsf{H}_2((C^{(1)})^x)$, and then check that $g^{u'} = C^{(1)}$ and $\mathsf{H}_1(C^{(1)}||C^{(2)}||C^{(3)}||C^{(4)}||M'||u') = C^{(5)}$ hold. If all checks pass, output M', otherwise output an error message $\perp$.

Compared with the original encryption and decryption algorithms, the main difference is in computing $C^{(4)}$, where the encryptor needs to perform L multiplications in $\mathbb{Z}_{Q \cdot T}^*$ in order to compute $(\mathsf{UH}(M))^{2^L} \bmod Q \cdot T$ to form $C^{(4)}$. Let every user U_i, for $i \geq 1$, adopt the above public key encryption scheme, and U_i's key pair be denoted as (PK_i, SK_i). The algorithms Aut is identical to that in the original cryptosystem, but the Com algorithm is defined as follows.

- Com$(C_i, C_j, T_{i,j})$: This algorithm outputs 1 if $x_i = x_j$ or 0 otherwise, where

$$x_i = \frac{\hat{e}(C_i^{(4)}, g_2^{r_{i,j}})}{\hat{e}(C_i^{(2)}, g_2^{y_i r_{i,j}})}$$

$$= \frac{\hat{e}(g_1^{v_i y_i} \cdot \mathsf{H}_3((\mathsf{UH}(M_i))^{2^L} \bmod Q \cdot T)), g_2^{r_{i,j}})}{\hat{e}(g_1^{v_i}, g_2^{y_i r_{i,j}})}$$

$$= \hat{e}(\mathsf{H}_3((\mathsf{UH}(M_i))^{2^L} \bmod Q \cdot T)), g_2)^{r_{i,j}}$$

$$x_j = \frac{\hat{e}(C_j^{(4)}, g_2^{r_{i,j}})}{\hat{e}(C_j^{(2)}, g_2^{y_j r_{i,j}})}$$

$$= \frac{\hat{e}(g_1^{v_j y_j} \cdot \mathsf{H}_3((\mathsf{UH}(M_j))^{2^L} \bmod Q \cdot T)), g_2^{r_{i,j}})}{\hat{e}(g_1^{v_j}, g_2^{y_j r_{i,j}})}$$

$$= \hat{e}(\mathsf{H}_3((\mathsf{UH}(M_j))^{2^L} \bmod Q \cdot T)), g_2)^{r_{i,j}}$$

As to this enhanced cryptosystem, the existing properties still hold, and their security proofs remain exactly the same. If a proxy is given U_t's ciphertext Enc(M, PK_t) and token $T_{i,t}$, then it can obtain $\mathsf{H}_3((\mathsf{UH}(M))^{2^L} \bmod Q \cdot T)$. To test any M', the most efficient approach for the proxy is to compute $(\mathsf{UH}(M'))^{2^L} \bmod Q \cdot T$ and perform a comparison based on its hash value. Since every test will cost L multiplications, then by setting an appropriate L the offline message recovery attack will be made computationally very expensive.

It is worth noting that, in this enhanced cryptosystem, the encryptor needs to perform L multiplications to mask the message in the encryption. This may be a computational bottleneck for some application scenarios. How to overcome this drawback while still mitigating the attack is an interesting future work.

5 Conclusion

In this paper, we have proposed a new formulation for PKEET, namely FG-PKEET. Compared with the formulation in [20], we have introduced a fine-grained authorization mechanism for users to specify who can perform equality test between their ciphertexts and successfully mitigate the possible drawbacks. We believe that the new formulation suits theoretical and practical security requirements better, and will be an important building block in designing privacy protection solutions supporting operations on encrypted data. Beyond this work, there are many interesting future research directions. One is to investigate the security implications when the user set and the proxy set overlap in the case of FG-PKEET. Our feeling is that in that case OW-CCA is the strongest security we can achieve. Another line of research is to investigate the practical countermeasures against offline message recovery attacks in the extreme situation, when the message space is polynomial size or the min-entropy of the message distribution is much lower than the security parameter.

Acknowledgement. The author would like to thank Dr. Liqun Chen for her help on clarifying the pairing assumptions, and thank the anonymous reviewers for their valuable comments.

References

1. Agrawal, R., Kiernan, J., Srikant, R., Xu, Y.: Order preserving encryption for numeric data. In: SIGMOD 2004: Proceedings of the 2004 ACM SIGMOD International Conference on Management of Data, pp. 563–574. ACM, New York (2004)
2. Ballard, L., Green, M., de Medeiros, B., Monrose, F.: Correlation-resistant storage via keyword-searchable encryption. Technical Report Report 2005/417, IACR (2005), http://eprint.iacr.org/2005/417
3. Bellare, M., Desai, A., Pointcheval, D., Rogaway, P.: Relations among notions of security for public-key encryption schemes. In: Krawczyk, H. (ed.) CRYPTO 1998. LNCS, vol. 1462, pp. 26–45. Springer, Heidelberg (1998)
4. Bellare, M., Rogaway, P.: Random oracles are practical: a paradigm for designing efficient protocols. In: Proceedings of the 1st ACM Conference on Computer and Communications Security, pp. 62–73. ACM Press, New York (1993)
5. Blum, M.: Coin flipping by telephone a protocol for solving impossible problems. SIGACT News 15(1), 23–27 (1983)
6. Boldyreva, A., Chenette, N., Lee, Y., O'Neill, A.: Order-Preserving Symmetric Encryption. In: Joux, A. (ed.) EUROCRYPT 2009. LNCS, vol. 5479, pp. 224–241. Springer, Heidelberg (2009)
7. Boneh, D., Boyen, X., Shacham, H.: Short group signatures. In: Franklin, M.K. (ed.) CRYPTO 2004. LNCS, vol. 3152, pp. 41–55. Springer, Heidelberg (2004)
8. Boneh, D., Di Crescenzo, G., Ostrovsky, R., Persiano, G.: Public Key Encryption with Keyword Search. In: Cachin, C., Camenisch, J.L. (eds.) EUROCRYPT 2004. LNCS, vol. 3027, pp. 506–522. Springer, Heidelberg (2004)
9. Boneh, D., Franklin, M.K.: Efficient generation of shared rsa keys (extended abstract). In: Kaliski Jr., B.S. (ed.) CRYPTO 1997. LNCS, vol. 1294, pp. 425–439. Springer, Heidelberg (1997)

10. Boyen, X.: The uber-assumption family. In: Galbraith, S.D., Paterson, K.G. (eds.) Pairing 2008. LNCS, vol. 5209, pp. 39–56. Springer, Heidelberg (2008)
11. Byun, J.W., Rhee, H.S., Park, H., Lee, D.H.: Off-Line Keyword Guessing Attacks on Recent Keyword Search Schemes over Encrypted Data. In: Jonker, W., Petković, M. (eds.) SDM 2006. LNCS, vol. 4165, pp. 75–83. Springer, Heidelberg (2006)
12. El Gamal, T.: A public key cryptosystem and a signature scheme based on discrete logarithms. In: Blakely, G.R., Chaum, D. (eds.) CRYPTO 1984. LNCS, vol. 196, pp. 10–18. Springer, Heidelberg (1985)
13. Paillier, P.: Public-key cryptosystems based on composite degree residuosity classes. In: Stern, J. (ed.) EUROCRYPT 1999. LNCS, vol. 1592, pp. 223–238. Springer, Heidelberg (1999)
14. Rivest, R.L., Shamir, A., Wagner, D.A.: Time-lock puzzles and timed-release crypto. Technical Report MIT/LCS/TR-684, Massachusetts Institute of Technology (1996)
15. Sahai, A., Waters, B.: Fuzzy identity-based encryption. In: Cramer, R. (ed.) EUROCRYPT 2005. LNCS, vol. 3494, pp. 457–473. Springer, Heidelberg (2005)
16. Shoup, V.: Sequences of games: a tool for taming complexity in security proofs (2006), http://shoup.net/papers/
17. Sittig, D.F.: Personal health records on the internet: a snapshot of the pioneers at the end of the 20th century. I. J. Medical Informatics 65(1), 1–6 (2002)
18. Tang, Q., Chen, L.: Public-key encryption with registered keyword search. In: Martinelli, F., Preneel, B. (eds.) EuroPKI 2009. LNCS, vol. 6391, pp. 163–178. Springer, Heidelberg (2010)
19. Tang, Q., Jeckmans, A.: On non-parallelizable deterministic client puzzle scheme with batch verification modes. Technical Report TR-CTIT-10-02, CTIT, University of Twente (2010), http://eprints.eemcs.utwente.nl/17107/
20. Yang, G., Tan, C., Huang, Q., Wong, D.S.: Probabilistic public key encryption with equality test. In: Pieprzyk, J. (cd.) CT-RSA 2010. LNCS, vol. 5985, pp. 119–131. Springer, Heidelberg (2010)

Lattice-Based Completely Non-malleable PKE in the Standard Model (Poster)

Reza Sepahi, Ron Steinfeld, and Josef Pieprzyk

Department of Computing, Macquarie University,
Sydney, Australia

Abstract. This paper presents ongoing work toward constructing efficient completely non-malleable public-key encryption scheme based on lattices in the standard (common reference string) model. An encryption scheme is completely non-malleable if it requires attackers to have negligible advantage, even if they are allowed to transform the public key under which the related message is encrypted. Ventre and Visconti proposed two *inefficient* constructions of completely non-malleable schemes, one in the common reference string model using non-interactive zero-knowledge proofs, and another using interactive encryption schemes. Recently, two efficient public-key encryption schemes have been proposed, both of them are based on pairing identity-based encryption.

1 Introduction

The notion of complete non-malleability in the context of public-key encryption [Fis05] ensures that an adversary who knows both the public key pk and the ciphertext c is unable to find another ciphertext $c' = \mathsf{Encrypt}_{pk'}(m')$ in such a way that there is a polynomially computable relation R between m, m', pk and pk'. In the work [VV08], Ventre and Visconti introduced two (quite inefficient) constructions of complete non-malleable encryptions. Recently, Libert and Yung [LY10] and Barbosa and Farshim [BF10] proposed two efficient completely non-malleable encryptions in the standard model based on pairing.

The initial motivation for complete non-malleability relied on constructing higher level protocols using asymmetric encryption schemes as building blocks (e.g., non-malleable commitment on top of public-key encryption). Also, impossibility results of Fischlin [Fis05] and inefficient constructions of Ventre and Visconti [VV08] is another motivation for constructing efficient completely non-malleable encryption schemes in standard model with possibly some relaxations (i.e., not in plain model).

Libert and Young [LY10] mentioned briefly (without a security proof) a scheme based on lossy trapdoor functions, which may be instantiated based on lattice problems as shown by Peikert et al [PW08].

1.1 Contributions of the Paper

In this paper, we propose the first concrete scheme that efficiently achieves complete non-malleability based on lattices. The security proof for our scheme is

U. Parampalli and P. Hawkes (Eds.): ACISP 2011, LNCS 6812, pp. 407–411, 2011.
© Springer-Verlag Berlin Heidelberg 2011

in standard model (without random oracles). What distinguishes this paper from previous ones is that it is the first lattice-based scheme with a full security proof and based on a novel combination of techniques used in lattice-based identity-based encryption (IBE) schemes, and not on lossy trapdoor functions. So, as there are currently no known quantum algorithms for solving lattice problems that perform significantly better than the best known classical (i.e., non-quantum) algorithms, it will remain secure even for post-quantum world.

2   Background and Definitions

Throughout the paper, when S is a set, $x \leftarrow S$ denotes the action of choosing x uniformly at random in S. By $a \in \text{poly}(\lambda)$, we mean that a is a polynomial in λ while $b \in \text{negl}(\lambda)$ says that b is a negligible function of λ (i.e., a function that decreases faster than the inverse of any $a \in \text{poly}(\lambda)$). When A is a possibly probabilistic algorithm, $b \leftarrow A(x)$ denotes the event that A outputs the value b when fed with the input x. Column vectors are named by lower-case bold letters (e.g., $\boldsymbol{x}$) and matrices by upper-case letters (e.g., X). We identify a matrix X with the ordered set $\{\boldsymbol{x}_j\}$ of its column vectors, and let $X|X'$ denote the (ordered) concatenation of the sets X and X'. For a set X of real vectors, we define $\|X\| = \max_j \|\boldsymbol{x}_j\|$, where $\| \cdot \|$ denotes the Euclidean norm.

NOTE: we refer the reader to [ABB10] for more background and notational materials. Functions TrapGen, ExtendBasis, and SamplePre have been described in [ABB10] as well.

3   A Construction Based on LWE Problem

Our scheme can be obtained using a somewhat(see next paragraph for differences) analogous methodology of Libert-Yung's [LY10] in the lattice setting. More specifically, it starts from the CCA2-secure cryptosystems constructed by the CHK methodology [CHK04] from selective-id IBE schemes proposed by Agrawal, Boneh and Boyen [ABB10] (we call it ABB-CHK scheme from now on). To turn the output of ABB-CHK method into a completely non-malleable scheme, we assume that all parties have access to a common reference string CRS comprising the description of three random matrices A_0, A_1, B, and a strongly unforgeable one-time signature. Now, we must also sign the receiver's public key along with other ciphertext components when generating the one-time signature. The one-time signature acts as a "checksum" binding all the pieces of the ciphertext together.

CONSTRUCTION. The system uses parameters q, n, m, δ, α specified below. Also, the function H refers to the FRD map $H : \mathbb{Z}_q^n \rightarrow Z_q^{n \times n}$ defined in [ABB10]. The system also makes use of a strongly unforgeable one-time signature scheme $\text{Sig} = (\text{Gen}, \text{Sign}, \text{Verify})$.

CRSGen(λ): given a security parameter $\lambda \in N$, do the following steps:

1. Set the parameters m, q, δ, α as follows:

$$m = O\left(n \log n\right); \quad \delta = O\left(\sqrt{n \log n}\right) \cdot \sqrt{m} \cdot w\left(\log m\right);$$

$$q = O\left(n^{c_q}\right); \quad \alpha = \left(m^2 \cdot O\left(\sqrt{n \log n}\right) \cdot w\left(\log^{1.5} m\right)\right)^{-1}$$

2. Select A_0, A_1, B, three uniformly random matrices in $\mathbb{Z}_q^{n \times m}$.

The common reference string is $\Sigma = \{\lambda, q, n, m, \delta, \alpha, A_0, A_1, B\}$.

KeyGen(Σ): given the common reference string Σ, do the following steps:

1. Select a uniformly random n-vector $\boldsymbol{u} \xleftarrow{\$} \mathbb{Z}_q^n$
2. Set $(A_2, T_{A_2}) \leftarrow$ TrapGen(q, n).
3. Secret key is T_{A_2} and public key is $(\boldsymbol{u}, A_2)$.

Encrypt(pk, Σ, b): given the public key $pk = (\boldsymbol{u}, A_2)$ and the common reference string Σ, to encrypt a single bit $b \in \{0, 1\}$, do the following steps:

1. Set the one-time signature key pair $(SK, VK) \leftarrow$ Sig.Gen(λ).
2. Set $A_{VK} \leftarrow A_1 + H(VK) \cdot B$
3. Set $F_{VK} \leftarrow [A_0 \mid A_{VK} \mid A_2] \in \mathbb{Z}_q^{n \times 3m}$.
4. Choose a uniformly random vector $\boldsymbol{s} \xleftarrow{\$} \mathbb{Z}_q^n$.
5. Choose a uniformly random matrix $R \xleftarrow{\$} \{-1, 1\}^{m \times m}$.
6. Choose noise vectors $\boldsymbol{x} \xleftarrow{\overline{\Psi}_\alpha} \mathbb{Z}_q$, $\boldsymbol{y} \xleftarrow{\overline{\Psi}_\alpha^m} \mathbb{Z}_q^m$ and $\boldsymbol{w} \xleftarrow{\overline{\Psi}_\alpha^m} \mathbb{Z}_q^m$ and set $\boldsymbol{z} \leftarrow R^T \boldsymbol{y} \in \mathbb{Z}_q^m$. If the magnitude of $\boldsymbol{x}$ or any coordinate of $\boldsymbol{y}$ or $\boldsymbol{w}$ exceeds $\alpha q \cdot w(\sqrt{\log m})$, or if $\|\boldsymbol{z}\|$ exceeds $\alpha q m \cdot w(\log m)$, then restart this step.
7. Set the ciphertext components as follow:

$$c_0 \leftarrow \boldsymbol{u}^T \boldsymbol{s} + x + b \left\lfloor \frac{q}{2} \right\rfloor \tag{1}$$

$$\boldsymbol{c}_1 \leftarrow F_{VK}^T \cdot \boldsymbol{s} + \begin{bmatrix} \boldsymbol{y} \\ \boldsymbol{z} \\ \boldsymbol{w} \end{bmatrix} \tag{2}$$

8. Sign the ciphertext components $c_0, \boldsymbol{c}_1$ along with the corresponding public key pk as $\sigma =$ Sig.Sign$(SK, \langle c_0, \boldsymbol{c}_1, pk \rangle)$.
9. Output the ciphertext $c := (VK, c_0, \boldsymbol{c}_1, \sigma)$.

Decrypt(sk, Σ, c): given the secret key $sk = T_{A_2}$ and the common reference string Σ, to decrypt the ciphertext $c = (VK, c_0, \boldsymbol{c}_1, \sigma)$, do the following steps:

1. if Sig.Verify$_{VK}(\sigma, \langle c_0, \boldsymbol{c}_1, pk \rangle) \neq 1$, return $\bot$.
2. otherwise, do the following steps:
3. Set $A_{VK} \leftarrow A_1 + H(VK) \cdot B$
4. Set $F_{VK} \leftarrow [A_0 \mid A_{VK} \mid A_2] \in \mathbb{Z}_q^{n \times 3m}$.
5. Run ExtendBasis(T_{A_2}, F_{VK}) to obtain trapdoor basis $T_{F_{VK}}$ for $\Lambda_q^{\perp}(F_{VK})$.

6. Set $e_{VK} \leftarrow \mathsf{SamplePre}(F_{VK}, T_{F_{VK}}, u, \delta)$.
7. Set $\Delta_c \leftarrow c_0 - e_{VK}^{\mathsf{T}} c_1$.
8. output the plaintext

$$b = \begin{cases} 1 & \text{if } |\Delta_c - \lfloor \frac{q}{2} \rfloor| < \lfloor \frac{q}{4} \rfloor \text{ in } \mathbb{Z}, \\ 0 & \text{otherwise.} \end{cases}$$

4   Correctness and Security

Given a valid ciphertext output by Encrypt, Step 7 of decryption procedure computes Δ_c as follows:

$$\Delta_c = b \left\lfloor \frac{q}{2} \right\rfloor + \underbrace{x - e_{VK}^{\mathsf{T}} \begin{bmatrix} y \\ z \\ w \end{bmatrix}}_{\text{error term}}.$$

Hence, decryption succeeds (i.e. correctness holds) if the magnitude of the error term above is less than $q/5$.

Lemma 1 (Correctness). *If* $\delta \geq \sqrt{n \log q} \cdot \omega(\sqrt{\log m})$ *and* $\alpha \delta m^{1.5} \cdot \omega(\sqrt{\log m})$ *< 1, then correctness holds, except with negligible probability.*

The security of our scheme is based on hardness of LWE problem and the strong unforgeability of the one-time signature:

Theorem 1 (Security). *Assume that* $\alpha < (\sqrt{n \log q} \cdot m^{1.5} \cdot \omega(\log^{1.5} m))^{-1}$, $\delta \geq \sqrt{n \log q} \cdot \omega(\sqrt{\log m})$ *and* $m \geq 6n \log q$. *The public-key scheme* PKE *derived in Section 3 is NM*-CCA2 assuming the hardness of the LWE problem and the strong unforgeability of the one-time signature. More precisely, the advantage of any NM*-CCA2 adversary* $\mathcal{A}$ *is bounded by*

$$\mathbf{Adv}_{\mathcal{A}, PKE}^{\mathrm{nm}^*\text{-}\mathrm{cca2}}(\lambda) \leq 2 \cdot \left(\mathbf{Adv}^{\mathrm{OTS}}(\lambda) + \mathbf{Adv}^{\mathrm{LWE}}(\lambda) \right).$$

The following corollary is the direct result of applying [Reg05, Theorem 1.1] and Theorem 1:

Corollary 1. *Let* $\alpha < (\sqrt{n \log q} \cdot m^{1.5} \cdot \omega(\log^{1.5} m))^{-1}$, $\delta \geq \sqrt{n \log q} \cdot \omega(\sqrt{\log m})$, $m \geq 6n \log q$ *and* $q > 2\sqrt{n}/\alpha$. *Assuming the strong unforgeability of the one-time signature and worst-case hardness of the* $\widetilde{O}(n/\alpha)$-SIVP *problem, the public-key scheme* PKE *derived in Section 3 is NM*-CCA2.*

5   Conclusion and Open Problems

We constructed a completely non-malleable public-key encryption scheme. The construction is based on lattices and its security proof is in the standard model from the LWE assumption.

One interesting open problem is to generalize both our and LY's construction to find general conditions on an IBE scheme to allow it to be modified into NM*-CCA (i.e. to formulate an NM* version of the general CHK transform). Another open problem is to improve the efficiency of our scheme by constructing a scheme based on the worst-case hardness of shortest vector problem (SVP) in ideal lattices.

Acknowledgement. Reza Sepahi was supported by a Macquarie University MQRES scholarship and Josef Pieprzyk and Ron Steinfeld were supported by the Australian Research Council grant DP0987734. Ron Steinfeld was also supported by ARC grant DP110100628

References

[ABB10] Agrawal, S., Boneh, D., Boyen, X.: Efficient lattice (H)IBE in the standard model. In: Gilbert, H. (ed.) EUROCRYPT 2010. LNCS, vol. 6110, pp. 553–572. Springer, Heidelberg (2010)

[BF10] Barbosa, M., Farshim, P.: Relations among notions of complete non-malleability: Indistinguishability characterisation and efficient construction without random oracles. In: Steinfeld, R., Hawkes, P. (eds.) ACISP 2010. LNCS, vol. 6168, pp. 145–163. Springer, Heidelberg (2010)

[CHK04] Canetti, R., Halevi, S., Katz, J.: Chosen-ciphertext security from identity-based encryption. In: Cachin, C., Camenisch, J. (eds.) EUROCRYPT 2004. LNCS, vol. 3027, pp. 207–222. Springer, Heidelberg (2004)

[Fis05] Fischlin, M.: Completely non-malleable schemes. In: Caires, L., Italiano, G.F., Monteiro, L., Palamidessi, C., Yung, M. (eds.) ICALP 2005. LNCS, vol. 3580, pp. 779–790. Springer, Heidelberg (2005)

[LY10] Libert, B., Yung, M.: Efficient completely non-malleable public key encryption. In: Abramsky, S., Gavoille, C., Kirchner, C., Meyer auf der Heide, F., Spirakis, P.G. (eds.) ICALP 2010. LNCS, vol. 6198, pp. 127–139. Springer, Heidelberg (2010)

[PW08] Peikert, C., Waters, B.: Lossy trapdoor functions and their applications. In: Proceedings of the Thirty-Seventh Annual ACM Symposium on Theory of Computing, STOC 2008, pp. 187–196. ACM, New York (2008)

[Reg05] Regev, O.: On lattices, learning with errors, random linear codes, and cryptography. In: Proceedings of the Thirty-Seventh Annual ACM Symposium on Theory of Computing, STOC 2005, pp. 84–93. ACM, New York (2005)

[VV08] Ventre, C., Visconti, I.: Completely non-malleable encryption revisited. In: Cramer, R. (ed.) PKC 2008. LNCS, vol. 4939, pp. 65–84. Springer, Heidelberg (2008)

Compliance or Security, What Cost? (Poster)

Craig Wright

Springer-Verlag, Computer Science Editorial, Tiergartenstr. 17,
69121 Heidelberg, Germany
{cwrigh20}@postoffice.csu.edu.au

Abstract. This paper presents ongoing work toward measuring the effectiveness of audit and assessment as an information security control. The trend towards the application of security control measures which are employed to demonstrate compliance with legislation or regulations, rather than to actually detect or prevent breaches occurring is demonstrated to result in a misallocation of funds. Information security is a risk function. Paying for too much security can be more damaging in economic terms than not buying enough. This research reveals several major misconceptions among businesses about what security really means and that compliance is pursued to the detriment of security. In this paper, we look at some of the causes of compliance based audit failures and why these occur. It is easier to measure compliance than it is to measure security and spending money to demonstrate compliance does not in itself provide security. When the money spent on achieving compliance reduces the funding available for control measures that may actually improve security problems may arise.

Keywords: Audit, Economics, Incentives, Risk, Security, Compliance.

1 Introduction

Information security is a risk function [1]. Paying for too much security can be more damaging in economic terms than not buying enough. This paper presents ongoing work toward measuring the effectiveness of audit and assessment as an information security control. In this paper, we demonstrate that the trend towards the application of security control measures which are employed to express compliance with legislation or regulations, rather than to actually detect or prevent breaches occurring results in a misallocation of funds. Information security is a risk function. Paying for too much security can be more damaging in economic terms than not buying enough. This research reveals several major misconceptions among businesses about what security really means and that compliance is pursued to the detriment of security. In this paper, we look at some of the causes of compliance based audit failures and why these occur. The major point of the paper is that it is easier to measure compliance than it is to measure security, and that spending money to demonstrate compliance does not in itself provide security.

This extends to include a look at the misalignment of audit to security. This misalignment is demonstrated to result from the drawing of funds from security in

U. Parampalli and P. Hawkes (Eds.): ACISP 2011, LNCS 6812, pp. 412–416, 2011.

order to provide compliance with little true economic gain. Funds are moved to alternate uses with no further funds allocated.

This paper presents the early research into an empirical study of data collected by the authors from 2,361 information systems audits in the period 1998 to 2010. These audit reports were collected from 894 Australian and US organizations in the Finance, Gaming, Media, FMCG, and Mining sectors as well as both Federal and State Government departments. Reports from Chartered audit firms, security companies and internal audit contractors are included. The composition of Australian organizations varies greatly. All US organizations consist of medium or larger listed companies with requirements under the Sarbanes Oxley Act (Sect 3.2 & 404). The audit reports from Australian organizations include PCI-DSS, APRA, BASELII, AML-CTF and those required for listed company financial reporting. This research incorporated the financial data for 451 of the organizations. An analysis of 210 incidents that resulted in a compromise will be analyzed in a forthcoming examination.

2   Misaligned Incentives: Audit and the Failure to Determine Risk

The existing audit industry provides compliance services under the guise of security. These services provide little if any increase in security and yet consumers purchase them. In addition, it is demonstrable that these services are extremely inelastic for large organizations[1]. There are several reasons for this. First, government[2] or commercial groups (e.g. PCI-DSS) mandate many compliance regimes. Next, negligence rules and the governance functions of companies require that boards and senior management take action to protect the value of the company. Unfortunately, this also means using reports that demonstrate compliance from audit companies in place of a real effort to ensure that data protection occurs.

In a review of 1,878 audit and risk reports collected by the authors on Australian firms by the top 8 international audit and accounting firms, 29.8% of tests evaluated the effectiveness of the control process. The security of systems were validated to any level in only 6.5% of reports. Of these, the process rarely tested for effectiveness, but instead tested that the controls met the documented process. Audit practice in US and UK based audit firms does not differ significantly.

Installation guidelines provided by the Centre for Internet Security (CIS)[3] openly provide system benchmarks and scoring tools that contain the "*consensus minimum due care security configuration recommendations*" for the most widely deployed operating systems and applications in use. The baseline templates will not themselves stop a determined attacker, but can to demonstrate minimum due care and diligence. Only 32 of 542 organizations analyzed in this paper deploy this form of implementation standards.

---

[1] Although these services may remain highly elastic for many smaller organizations who may choose not to control risk when budgets are tight.
[2] This includes SOX, APRA, FISMA, and many other compliance regimes.
[3] CIS benchmark and scoring tools are available from http://www.cisecurity.org/

The information systems employees within an organization also have a misaligned set of incentives. A large component of any audit involves discussions with the employees and management at the examined organization. The term auditor essentially derives from the act of listening as *'one who listens'*. Listening to the assertions of employees remains a large component of any information systems audit. Those interviewed in this process include the employees who are responsible for the maintenance of the system audited. These same employees commonly have incentives that align with the audit results. For instance, in 1,325 of the audits review that directly include firewalls, 798 (60.2%) of these audits involved direct interviews with firewall administrators who either had bonuses tied to the outcome of the audit or whose employment was in some manner conditional on the outcome of the audit.

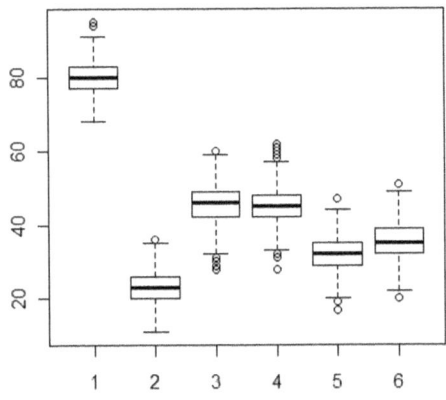

1. The employee has no knowledge of extra testing by the auditor and no extra tests are to be conducted
2. The client has knowledge of testing (from prior experience) conducted by the auditor.
3. Prior tests have occurred, but new auditors have done the testing.
Audit with no employee incentives
4. The employee has no knowledge of extra testing by the auditor and no extra tests are to be conducted
5. The client has knowledge of testing (from prior experience) conducted by the auditor.
6. Prior tests have occurred, but new auditors have done the testing.

Fig. 1. Misaligned incentives and a lack of accuracy delivered to the auditor (%) in an audit with employee incentives

The consequence of these misaligned incentives is obvious, misinformation. Fig. 1 displays the results of the audit when the employee has incentives and knowledge or neither. Further analysis associated to the assignment of a new auditor followed. The differences between the audits of a known tested system known and of a system excluded from testing were statistically significant at the $\alpha = 5\%$ level. At this level, we have a confidence interval of (77, 83) with a corresponding confidence interval of (20, 26) when the employee has incentives and knows that the statements they offer will be tested. The distinction from an employee with incentives who has been audited and had their assertions validated (42, 48) when a new auditor is assigned do not differ significantly from the employee with no incentives (42, 49).

2.1 Patching and Validation

Patching is a common test for compliance. Auditors assert that this compliance test aligns to good security practice [5]. A correctly patched system is less likely to experience issues and be more secure [2]. This is agreed. The question is what is "*correctly patched*" and "*have the patches been applied correctly*"? Audits generally test for the application of patches. The problem is that this is generally limited to testing the existence of operating system that has all required patches applied. Application patches are another matter.

Tests of the patching processes for Windows Servers, clients, applications, routers, switches and firewalls are reported in Table 1. The 95% confidence intervals for patching times for each of these systems have been recorded. The patch date is determined as the difference in time between when the software vendor has released the patch to the installation of the patch on the system. In a few instances, this result is statistically censored due to the lack of patching. This can take place where the system is installed and left running without the application of updates. In this case, the difference between the installation date of the device and the date of the patch or update that should be applied is used to determine the interval. This situation was found to be most common in network equipment (with several routers and switches never having been patched or updated) as well as with selected examples of user application software.

Table 1. Patching Analysis of Audited Systems

	No. Analyzed	95% Confidence Interval of days between patching (Mean)	Average Policy Patch time (CI)	% Prior Reports noting patching
Windows Server	1571	41.1, 122.4 (86.2)	55.5, 87.9	98.4%
Windows Client	13,951	22.8, 69.3 (48.1)	29.6, 49.4	96.6%
Other Windows Applications	30,290	58.1, 181.8 (125.2)	68.1% NA	18.15%
Internet Facing Routers	515	58.2, 164.1 (114.2)	58.1% NA	8.7%
Internal Routers	1,323	129.3, 384.6 (267.8)	73.2 NA	3.99%
Internal Switches	452	139.9, 483.9 (341.2)	87.5 NA	1.2%
Firewalls	1,562	21.5, 65.7 (45.4)	24.5, 108.2	70.7%

A further analysis of prior audit reports was conducted to note how many of these had included patch levels for each of the various hosts and systems deployed at the audited client. Nearly all audit reports note the inclusion or exclusion of operating system patches (98.4% and 96.6% for server and client systems respectively). The majority of these reports included no testing of the network devices and little tests of the application software in use by a client. Network switches were the least analyzed device. The mean time between patching on these devices was recorded at 341.2 days. It was uncommon for organizations to have a policy requiring the patching of network devices. The majority of organizations have policies in place for the patching of Servers (with a range of 55.5 to 87.9 days) and Client operating systems (with a range of 29.6 to 49.4 days). All results and Confidence intervals are reported at a 95% CI.

Operating system patches for client systems and firewalls are generally applied and tested within 60 days. The patching rates for network equipment vary significantly. Again, it is clear that the incentives to ensure compliance result in insecure systems. The audit process checked policy statements against a sample of systems, but did nothing to validate those systems not included in the policy. The result is an overwhelming focus on selected systems that are incorporated within a checklist at the expense of excluding many essential systems.

The patching of client applications was problematic with a mean of 125.2 days between patching of these applications and a 95% confidence interval of (58.1, 181.8) days. This varied widely not only across hosts and organizations, but also within the same host. Only 2.18% of hosts have patched at least 95% of applications within 120 days. The development systems analyzed exhibited the worst results. Compilers and IDE (integrated development software) were patched at a rate of between (82.0, 217.3) days. These systems were also generally not included within the audit report. The consequence being that there is little incentive for the organization to ensure that they are maintained sufficiently.

4 Conclusion

Compliance is easier than security. It would seem costs of normal compliance auditing do not benefit the bottom line financial posture of organizations seeking to be both secure and compliant. An appropriate view would be to seek to be secure in place of appearing secure. This leads to an endless cycle of continual audit satisfying the needs of compliance and the bottom lines of financial firms, but with few other true paybacks. So we are led to ask, at what cost?

The practice of implementing monitoring controls that do not report on breaches, but which do satisfy the compliance needs of an organization can cost far more in the long term [1,3]. Businesses need to demand more thorough audits and results that are more than simply meeting a compliance checklist. These must include not only patching for all levels of software (both system and applications) as well as the hardware these run on. This failure of audits to "*think outside the box*" and only act as a watchdog could ultimately be perceived as negligence for all parties.

Compliance at the expense of security in the global economy is a practice that is difficult to overcome, but a challenge that we have to meet. It may be easier to measure compliance than it is to measure security, but spending money to demonstrate compliance does not in itself provide security. When the money spent on achieving compliance reduces the funding available for control measures that may actually improve security problems may arise.

References

1. Anderson, R.: Why information security is hard – an economic perspective. In: 17th Annual Computer Security Applications Conference, pp. 358–365 (2001)
2. Halderman, J.: To Strengthen Security, Change Developers' Incentives. IEEE Security and Privacy 8(2), 79–82 (2010)
3. Katz, M.L., Shapiro, C.: Network externalities, competition, and compatibility. The American Economic Review 75, 424 (1985)
4. Roese, N.J., Olson, J.M.: Better, stronger, faster: Self-serving judgment, affect regulation, and the optimal vigilance hypothesis. Perspectives on Psychological Science 2, 124–141 (2007)
5. Turcato, L.M.: Use of COBIT as a Risk Management & Audit Framework for Access Compliance, San Francisco ISACA Fall Conference (2004)

Preimage Attacks on Full-ARIRANG (Poster)

Chiaki Ohtahara[1], Keita Okada[1], Yu Sasaki[2], and Takeshi Shimoyama[3]

[1] Chuo-University
{cohtahara,kokada}@chao.ise.chuo-u.ac.jp
[2] NTT Corporation
sasaki.yu@lab.ntt.co.jp
[3] Fujitsu Laboratories LTD.
shimo@labs.fujitsu.com

Abstract. This paper presents ongoing work toward the first preimage attacks on hash function ARIRANG, which is one of the first round candidates in the SHA-3 competition. ARIRANG has an unique design where the feed-forward operation is computed not only after the last step but also in a middle step. In fact, this design prevents previous preimage attacks. We apply a meet-in-the-middle preimage attacks to ARIRANG. Specifically, we propose a new initial-structure optimized for ARIRANG and overcome the middle feed-forward.

Keywords: ARIRANG, SHA-3, preimage, meet-in-the-middle.

1 Introduction

ARIRANG [1] is a hash function submitted to the SHA-3 competition. Although ARIRANG could not go into the second round, its security is not broken yet.

The designers claim that the middle feed-forward, which is an unique design for ARIRANG, raises the resistance against preimage attacks [1, Sect. 6.6].

Hong *et al.* proposed preimage attacks on reduced steps [2,3] by applying the meet-in-the-middle preimage attack [4]. In this attack, how to separate the computation into two independent parts (called *chunks*) is the most important. In both of [2,3], there is one strong limitation; the first step and the step where the middle feed-forward is computed must be included in the same chunk. Due to this limitation, previous work principally cannot attack full steps.

In this paper, we propose preimage attacks on full ARIRANG-256 and -512. We introduce an improved matching technique which checks the match of linear relations among several variables. The attack results are shown in Table 1.

2 Description of ARIRANG

ARIRANG is a family of hash functions. In this paper, we deal with ARIRANG-256 and ARIRANG-512. ARIRANG uses a narrow-pipe Merkle-Damgård structure and its compression function uses a modified Davies-Meyer construction.

U. Parampalli and P. Hawkes (Eds.): ACISP 2011, LNCS 6812, pp. 417–422, 2011.
© Springer-Verlag Berlin Heidelberg 2011

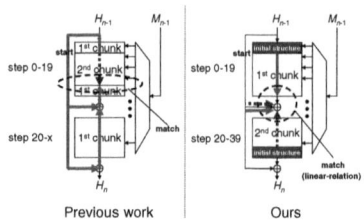

Fig. 1. Compression function and strategies to attack middle feed-forward

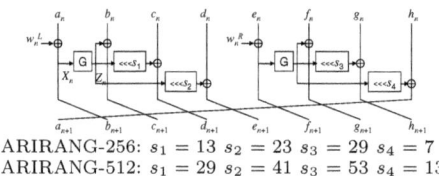

ARIRANG-256: $s_1 = 13$ $s_2 = 23$ $s_3 = 29$ $s_4 = 7$
ARIRANG-512: $s_1 = 29$ $s_2 = 41$ $s_3 = 53$ $s_4 = 13$

Fig. 2. Step function of ARIRANG

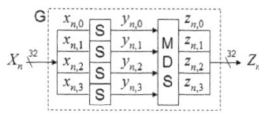

Fig. 3. Function G of ARIRANG-256

Table 1. Summary of attack results

Target	Attack	Steps	Comp. for each output size (Time, Mem.)				Ref.
			224-bit	256-bit	384-bit	512-bit	
CF	Free-start coll	40 (full)	$(2^{24}, Neg)$		$(1, Neg)$		[5]
CF	Free-start near-coll	40 (full)		$(1, Neg)$		$(1, Neg)$	[5]
CF	Collision	26		$(1, Neg)$		$(1, Neg)$	[5]
Hash	Preimage	33		$(2^{241}, 2^{32})$		$(2^{481}, 2^{64})$	[2]
Hash	Preimage	35		$(2^{240.94}, 2^{32})$		$(2^{480.94}, 2^{64})$	[3]
Hash	Preimage	40 (full)		$(2^{254}, 2^6)$		$(2^{505}, 2^{16})$	Ours

The step function updates eight working variables $a_n, b_n, c_n, d_n, e_n, f_n, g_n$ and h_n on n-th step as shown in Fig. 2. The function G is composed of the S-box and a linear mapping MDS for AES. Fig. 3 shows the function G.

The message schedule of ARIRANG generates 16 extra words $w_i (16 \leq i \leq 31)$ from the 16 input message words $w_i (0 \leq i \leq 15)$. Refer to [1], for the whole description.

3   Preimage Attacks on Full ARIRANG-256/-512

3.1   Chunk Separation

We first separate the target into two independent chunks. We coded the neutral-word-search algorithm, and determined to use (w_0, w_4) and (w_5, w_{11}) with a condition $w_0 = w_4$ and $w_5 = w_{11}$. The chunk separation is shown in Fig. 4.

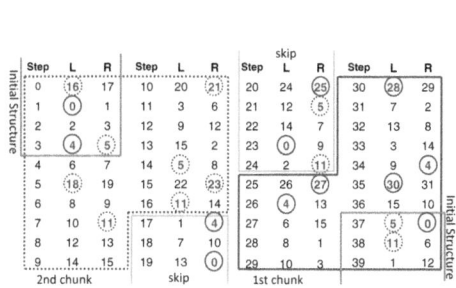

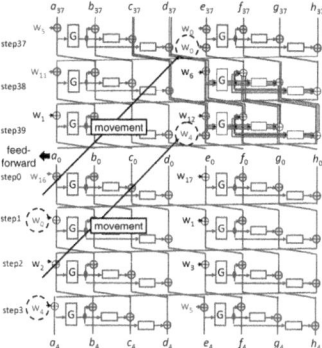

Fig. 4. Chunk separation for full ARIRANG   **Fig. 5.** Overview of initial structure

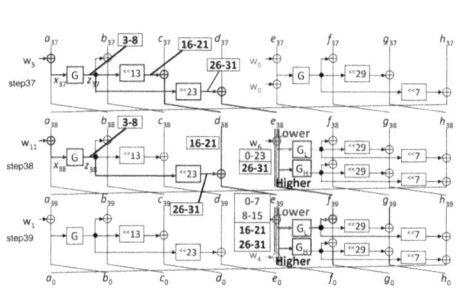

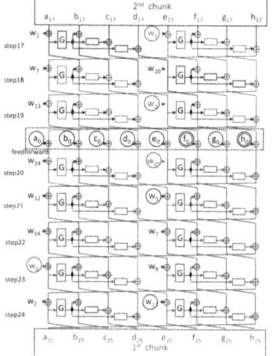

Fig. 6. Details of initial-structure for steps 37–39   **Fig. 7.** Skipping 8 steps

3.2 Preimage Attacks on ARIRANG-256

Initial Structure. The overview of the initial structure is shown in Fig. 5. We only activate a part of bits so that the impact from two chunks never reach the same bit positions. Assume that the first chunk impacts to the lower 2 bytes of the input variable to G in step 39 (represented by $x_{39,0}$ and $x_{39,1}$) and the second chunk impacts to the higher 2 bytes ($x_{39,2}$ and $x_{39,3}$). Then, the impacts from the first and second chunks are written as $MDS(S(x_{39,0})\|S(x_{39,1})\|0\|0)$ and $MDS(0\|0\|S(x_{39,2})\|S(x_{39,3}))$, respectively, which can be computed independently.

First chunk (Backward). We choose the values of neutral words w_0 and w_4 so that only the lower 2 bytes and the lower 3 bytes of the input variable to G in steps 39 and 38 are influenced, respectively. Therefore, we only activate the lower 2 bytes of w_4. Furthermore, we need to ensure that the impacts never goes to the highest byte through the G function in step 39. The impacts

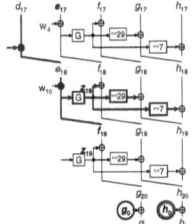

Fig. 8. Indirect partial-matching

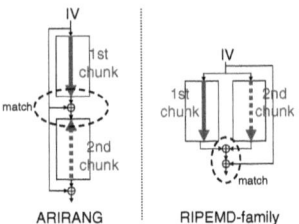

Fig. 9. Another look of middle feed-forward

to $z_{39,3}$ from $MDS(y_{39,0}\|y_{39,1}\|0\|0)$ is computed as $z_{39,3} = (03 \cdot y_{39,0}) \oplus y_{39,1}$. Therefore, every time we choose $y_{39,0}$, we set $y_{39,1}$ to $(03 \cdot y_{39,0})$. We stress that the freedom degree for the first chunk is 8 bits.

Second Chunk (Forward). We fix a_{37} and a_{38} to be identical. We activate bit positions 3 to 8 of z_{37}. This will impact to bits 26 to 31 of the input of G in step 38 and bits 16 to 21 of the input of G in step 39. We then compute x_{37} and compute w_5 by $x_{37} \oplus a_{37}$. In step 38, because $w_{11} = w_5$ and $a_{37} = a_{38}$, $x_{38} = x_{37}$ is always satisfied and thus $z_{38} = z_{37}$. Namely, only bits 3–8 of z_{38} are influenced. This will impact to bits 26 to 31 of the input of G in step 38. We stress that the freedom degree for the second chunk is 6 bits.

Partial-match. Steps 17–24 cannot be fully computed. Hence, we can only partially match the results. Fig. 7 shows the partial computations. We denote variables just before the middle feed-forward by $a_{20}, b_{20}, \ldots, h_{20}$ and immediately after the middle feed-forward by $a_*, b_*, \ldots, h_*$. In the backward computation from step 24, after we compute the inverse of the step function in step 20, we know the values of 4 right most variables (e_*, f_*, g_*, and h_*). Then, initial value p_0 is added by the middle feed-forward operation. From Fig. 5, two variables f_0 and h_0 are influenced by the first chunk and all variables are influenced by the second chunk. Hence, more independent computations are impossible.

We similarly consider the forward computation as shown in Fig 7. We then consider the match of a linear-relation among several variables. The indirect partial-matching technique [6] can be applied to perform this match.

We denote the impact on h_0 from the first and second chunks by h_0^{1st} and h_0^{2nd}, respectively. Then, the equations to compute the values of g_* and h_* can be written as Eq. (1), and then Eq. (2). Note that notations $\underline{V}_1$ and $\underline{V}_2$ represent that a variable V is computed in the first and second chunks, respectively.

$$\underline{g_{*}}_1 = \underline{f_{19}}_2 \oplus z_{19} \oplus \underline{g_{0}}_2, \quad \underline{h_{*}}_1 = \underline{z_{18}}_2 \oplus \underline{e_{17}}_2 \oplus \underline{w_4}_1 \oplus (z_{19})^{\lll 29} \oplus \underline{h_0^{2nd}}_2 \oplus \underline{h_0^{1st}}_1 \tag{1}$$

$$\underline{f_{19}}_2^{\lll 29} \oplus \underline{g_0}_2^{\lll 29} \oplus \underline{z_{18}}_2 \oplus \underline{e_{17}}_2 \oplus \underline{h_0^{2nd}}_2 = \underline{g_*}_1^{\lll 29} \oplus \underline{w_4}_1 \oplus \underline{h_0^{1st}}_1 \oplus \underline{h_{*}}_1. \tag{2}$$

By computing each side of Eq. 2 in each chunk independently, we can efficiently match the 32-bit linear relations of the results.

The Attack Procedure

1. In the second chunk, for all 2^6 choices of bit positions 3 to 8 of z_{37}, compute the initial structure and then compute the step function from step 0 to 16.
2. For the match, calculate the left side of Eq. 2 and store them in a table T.
3. In the first chunk, for all 2^8 choices of $y_{39,0}$ compute the initial structure, and then compute the step function from step 39 to 25.
4. Calculate the right side of Eq. 2 and check if that value is stored in T.
5. If the match is found, check the match of other bits.

Finally, pseudo-preimages are converted to preimages with $2^{\frac{256+250}{2}+1} = 2^{254}$ operations with a generic conversion.

3.3 Preimage Attacks on ARIRANG-512

The strategy for attacking ARIRANG-512 is almost the same as that of ARIRANG-256. Due to the page limitation, we omit the details. Pseudo-preimages of ARIRANG-512 can be obtained with 2^{496} complexity and 2^{16} memory, and preimages can be obtained with 2^{505} complexity and 2^{16} memory.

4 Concluding Remarks

Our strategy can be regarded to be the same as the one for the double-branch structure (e.g. RIPEMD-family), which is illustrated in Fig. 9. Specifically, it is the same as [7, Approach 1] and the attacks on RIPEMD-128/-160 [8]. Assume that a hash function using the modified DM-mode with the middle feed-forward like ARIRANG consists of r steps. Then, the number of steps from the first step to the middle and the last step to the middle are both $\frac{r}{2}$. Finally, we can say that the security of the r-step DM-mode with the middle feed-forward is almost the same as the $\frac{r}{2}$-step double-branch structure.

References

1. Chang, D., Hong, S., Kang, C., Kang, J., Kim, J., Lee, C., Lee, J., Lee, J., Lee, S., Lee, Y., Lim, J., Sung, J.: Arirang: Sha-3 proposal (2008)
2. Hong, D., Kim, W.H., Koo, B.: Preimage attack on ARIRANG. Cryptology ePrint Archive, Report 2009/147 (2009)
3. Hong, D., Koo, B., Kim, W.H., Kwon, D.: Preimage attacks on reduced steps of ARIRANG and PKC98-hash. In: Lee, D., Hong, S. (eds.) ICISC 2009. LNCS, vol. 5984, pp. 315–331. Springer, Heidelberg (2010)

4. Aoki, K., Sasaki, Y.: Preimage attacks on one-block MD4, 63-step MD5 and more. In: Avanzi, R.M., Keliher, L., Sica, F. (eds.) SAC 2008. LNCS, vol. 5381, pp. 103–119. Springer, Heidelberg (2009)

5. Guo, J., Matusiewicz, K., Knudsen, L.R., Ling, S., Wang, H.: Practical pseudo-collisions for hash functions ARIRANG-224/384. In: Jacobson Jr., M.J., Rijmen, V., Safavi-Naini, R. (eds.) SAC 2009. LNCS, vol. 5867, pp. 141–156. Springer, Heidelberg (2009)

6. Aoki, K., Guo, J., Matusiewicz, K., Sasaki, Y., Wang, L.: Preimages for step-reduced SHA-2. In: Matsui, M. (ed.) ASIACRYPT 2009. LNCS, vol. 5912, pp. 578–597. Springer, Heidelberg (2009)

7. Sasaki, Y., Aoki, K.: Meet-in-the-middle preimage attacks on double-branch hash functions: Application to RIPEMD and others. In: Boyd, C., González Nieto, J. (eds.) ACISP 2009. LNCS, vol. 5594, pp. 214–231. Springer, Heidelberg (2009)

8. Ohtahara, C., Sasaki, Y., Shimoyama, T.: Preimage attacks on step-reduced RIPEMD-128 and RIPEMD-160. In: Preproceedings of INSCRYPT 2010, pp. 191–208 (2010)

Finding Collisions for Reduced *Luffa*-256 v2 (Poster)

Bart Preneel[2,3], Hirotaka Yoshida[1,2,3], and Dai Watanabe[1]

[1] Yokohama Research Laboratory, Hitachi, Ltd,
292 Yoshida-cho, Totsuka-ku, Yokohama-shi, Kanagawa-ken, 244-0817 Japan
[2] Katholieke Universiteit Leuven, Dept. ESAT/SCD-COSIC,
Kasteelpark Arenberg 10, B–3001 Heverlee, Belgium
[3] Interdisciplinary Institute for BroadBand Technology (IBBT), Belgium

Abstract. This paper presents ongoing work toward analysis of a second round SHA-3 candidate *Luffa*. This article analyses the collision resistance of reduced-round versions of *Luffa*-256 v2 which is the 256-bit hash function in the *Luffa* family. This paper focuses on the hash function security. To the best of our knowledge, this is the first collision analysis for fixed initial vector of *Luffa*. We show that collisions for 4 out of 8 steps of *Luffa*-256 v2 can be found with complexity 2^{90} using sophisticated message modification techniques.

Keywords: Hash functions, collision attack, message modification.

1 Introduction

Recent cryptanalytic results focus on the collision resistance of hash functions. In response to the collision attack [9] on SHA-1 [6], NIST launched the SHA-3 competition [7] to find an alternative to the SHA-2 family. NIST received more than 60 candidate hash functions and it currently focuses on the 5 final round candidates.

Luffa is a family of cryptographic hash functions that has been selected as one of the 14 second round SHA-3 candidate. The hash function *Luffa* adopts the structure of a sponge function and a wide-pipe strategy. In the previous results on *Luffa*, its building blocks have been extensively analyzed: the designers found a differential path for the internal permutation of *Luffa*. Aumasson and Meier [1] constructed an algebraic zero-sum distinguisher for the same component. Watanabe *et. al* [8] constructed a higher order distinguisher for 7-steps of the compression function of *Luffa* v1. Khovratovich *et. al* [5] found a semi-free start collision for 7 steps of the compression function of *Luffa*-256 v2.

This article analyses the collision resistance of reduced-round versions of *Luffa* which is the 256-bit hash function in the *Luffa* family. We show how collision attacks, using sophisticated message modification techniques, can be mounted on reduced variants of *Luffa*-256 v2. We present an attack on *Luffa*-256 v2 reduced from 8 to 4 steps with a complexity of 2^{90}.

The outline of this paper is as follows. In Sect. 2, we give a short description of *Luffa*-256 v2. In Sect. 3, the results of the collision attacks on 4-step variant of *Luffa*-256 v2 are presented. Section 4 concludes the paper.

U. Parampalli and P. Hawkes (Eds.): ACISP 2011, LNCS 6812, pp. 423–427, 2011.
© Springer-Verlag Berlin Heidelberg 2011

2   Specification of *Luffa*-256 v2

In this section, we introduce a part of the specification of *Luffa* to describe the
attack. The reader is referred to [4] for the details of the specification.

2.1   Chaining and Round Function

The chaining of *Luffa* is a variant of a sponge function [2], that processes 256
message bits in each iteration. The message is padded with 10...0 to ensure
that the padded message has a length divisible by 256.

The round function is a composition of a message injection function MI and
three permutations Q_j of 256 bits input. Let the input of the i-th round be
$(H_0^{(i-1)}, H_1^{(i-1)}, H_2^{(i-1)})$, then the output of the i-th round is given by

$$H_j^{(i)} = Q_j(X_j), \quad 0 \le j < 3,$$
$$X_0||X_1||X_2 = MI(H_0^{(i-1)}, H_1^{(i-1)}, H_2^{(i-1)}, M^{(i)}),$$

where $H_j^{(0)} = V_j$. The MI function is linear over $GF(2^8)$ and can be represented
by a matrix over the ring $GF(2^8)^{32}$. The map from an 8-word value $(a_0, \dots, a_7)$
to an element of the ring is defined by $(\sum_{0 \le k < 8} a_{k,l} x^k)_{0 \le l < 32}$.

2.2   Non-linear Permutation

At the beginning of the step function process in the permutation Q_j, 256 bits
of data are stored in 8 32-bit registers denoted by $a_k^{(r)}$ for $0 \le k < 8$. The
permutation Q_j is defined as the composition of an input tweak and iterations
of a step function Step. which consists of the following three functions: SubCrumb,
MixWord, AddConstant. The number of iterations of a step function is 8.

In permutation Q_j, the input tweak rotates the least significant four words to
the left by j bits. SubCrumb substitutes the bits of a_0, a_1, a_2, a_3 (or a_4, a_5, a_6, a_7)
by a 4-bit S-box S. Let the output of SubCrumb be x_0, x_1, x_2, x_3 (or x_4, x_5, x_6, x_7).
Then SubCrumb is given by $x_{3,l}||x_{2,l}||x_{1,l}||x_{0,l} = S[a_{3,l}||a_{2,l}||a_{1,l}||a_{0,l}]$ and
$x_{4,l}||x_{7,l}||x_{6,l}||x_{5,l} = S[a_{4,l}||a_{7,l}||a_{6,l}||a_{5,l}], (0 \le l < 32)$.

MixWord is a linear permutation of two words. Let the output words be y_h
and y_{k+4} where $0 \le k < 4$. Then MixWord is given by the following equations:

$$y_{k+4} = x_{k+4} \oplus x_k, y_k = x_k \lll 2, y_k = y_k \oplus y_{k+4}, y_{k+4} = y_{k+4} \lll 14,$$
$$y_{k+4} = y_{k+4} \oplus y_k, y_k = y_k \lll 10, y_k = y_k \oplus y_{k+4}, y_{k+4} = y_{k+4} \lll 1.$$

3   The Collision Attack on 4-Step *Luffa*-256 v2

We here present a collision attack on 4-step *Luffa*-256 v2. We give a general
idea of how the attack works: there are three round function calls, meaning that
the attack uses three message blocks which are used in the following manner:
the attack uses the first message block $M^{(1)}$ with no difference for finding a good

value for the second round function input $(H_0^{(1)}, H_1^{(1)}, H_2^{(1)})$, the second message block pair $(M^{(2)}, M^{(2)} \oplus \Delta)$ introduces the differences conforming the differential path for each permutation and those differences are erased with the third message block pairs $(M^{(3)}, M^{(3)} \oplus \Delta')$. The attack first constructs a differential path producing a collision and then applies the message modification [9] to reduce the complexity. We adopt the attacking principle in [5] in the following sense:

1. We apply the modification technique on S-box level.
2. We store the degrees of freedom as the information on the set of message inputs which give the right input for the active S-boxes.

On the other hand, the techniques used in the rebound attack such as match-in-the middle and multi-path are difficult to apply in our attack because we consider the hash function security where IV is fixed. We apply the basic modification using single message bundle for each active S-box and advanced modification [9] using multiple message bundles for each active S-box respectively. In order to reduce the attack complexity, we attempt to take the following ideas:

1. Maximize the number of applications of basic modification.
2. Minimize the number of involved message bundles for advanced modification.

For this optimization purpose, we took some heuristic approach where message bundles which have been used before have higher priorities to be used in the following step than the others.

As for preliminary, we view the 256-bit message block as 32 8-bit bundles and consider their positions t $(0 \leq t < 32)$, to which we will refer as *message bundle* and *message bundle position* respectively. In other words, We handle the message on 8-bit level. Each of these bundles is obtained in a bit-slice manner as adopted in *Luffa*-256 v2: one bit of a bundle is taken from one 32-bit word of in the message block. For the same reason, we will view the 256-bit chaining variable input of the permutation Q_j as 64 4-bit bundles, each of which is taken as input to S-box, and consider their positions u $(0 \leq u < 64)$, to which we will refer as *S-box position*.

3.1   The Differential Path

Our attack first constructs a good differential path for the second round. The overview of how we derive our differential path is that first, we find a good truncated differential path on the permutation Q_j by considering the linear function `MixWord` and the Tweak function and then, we determine the best input output differences of the active Sboxes when the constraint due to the message injection function MI is taken into account. We performed experiments to find a good truncated differential path for the permutation Q_j. The best one we found has 49 active S-boxes shown in Table 1. From this truncated path, we derive our differential path where differential probabilities are 2^{-7}, 2^{-7}, and 2^{-6} for the first step, the second step, and the third step, respectively.

Table 1. The truncated differential path for Q_j

Step	Weight	0	1	2	3	4	5	6
		0123456789	0123456789	0123456789	0123456789	0123456789	0123456789	0123
0	07	0000000100	0100000001	0101010000	0100000000	0000000000	0000000000	0000
1	08	0100000100	0000010001	0100000101	0000000000	0000000000	0000000000	0010
2	19	0001010100	0100011100	1001000100	1100000000	0000000110	1010000000	1110
3	15	0000010000	0001010100	0101000100	0010110010	0010000000	0000000010	1010
4	(42)	0011011000	1110110001	0110011010	1111101110	1111110111	1110011110	0111

3.2   Message Modification

As for the first step, after applying the basic message modification to 7 active S-boxes, the remaining degrees of freedom in the second message block is 221 bits out of 256 bits. We face more difficult situations at the second and the third steps due to the effect of the MixWord which ensures that the input to an S-box depends on multiple message bundles and that one message bundle may affect multiple active S-boxes. It follows that, even if a condition on the input of an active S-box is fulfilled by means of a modification with some message bundle at some step, this fulfillment can be afterwards destroyed by means of a modification with the same message bundle at the following step. Hence the important problem we have to solve is:

For each active S-box, from which message bundles we assign their degrees of freedom to it in order to optimize the attack complexity?

Our strategy to find an optimal (or nearly optimal) solution to this problem is to search for correspondences between active S-boxes and message bundles where each modification per active S-box performed uses message bundles in such a way that the total number of used message bundles including the ones which previous modifications have already used is as small as possible, instead of exhaustively searching for correspondences. Our search considers not only the degrees of freedom left step by step but also the orders in which active S-boxes are dealt with: the earlier modifications deal with the active S-boxes which are more restricted than the others in terms of the total degrees of freedom of the corresponding message bundles. Our strategy allows us to perform some message modifications independently of the others. This helps to optimize the complexity of the message modification performed in total. After applying the message modification to 8 active S-boxes and 19 active S-boxes, the degrees of freedom in the second message block is 165 bits (out of 256 bits) remaining after the second step and 51 bits remaining after the third step. Table 2 indicates the position correspondence between the active S-boxes and the message bundles.

As for the fourth step, there are 15 active S-boxes and the product of differential probabilities for S-boxes in the same (S-box) position over Q_j is 2^{-6}. Hence, in this step, we would need to control 90 bits in the message block. Since the degrees of freedom in the second message block is 51 bits in this step, our random attempt uses the first message block with a complexity of 2^{90-51}. As a result, we expect to find a collision for 4-step *Luffa*-256 v2 with a total

Table 2. Position correspondence between active S-boxes and message bundles

Second	S-box pos.	27	7	62	21	29	1	15	19		
step	Mes. bundle pos.	7,19,21	1,19,31	15	1,15	9,23	25,27	9,27	11, 13		
Third	S-box pos.	27	62	3	30	5	17	52	20	7	11
step	Mes. bundle pos.	2	2,13	9,24	3,24,26	26	29	29,30	3,30	10	10,12
	S-box pos.	47	31	15	16	23	61	48	50	60	
	Mes. bundle pos.	5,12	3,5	28	0,28	0,4	4	6	8	14	

complexity of $2^{90} \approx 2^{39}(2^{14} + 2^{16} + 2^{41} + 2^{51})$ where 2^{14}, 2^{16}, 2^{41}, and 2^{51} are the complexities for the message modifications at the first, second, third, and fourth steps respectively.

4 Conclusion

By taking a simple and effective approach of applying message modification, we show that collisions for 4 steps of *Luffa*-256 v2 can be found with complexity 2^{90}. This is the first analysis of *Luffa* regarding the hash function security.

References

1. Aumasson, J.P., Meier, W.: Zero-sum distinguishers for reduced Keccak-f and for the core functions of Luffa and Hamsi (2009)
2. Bertoni, G., Daemen, J., Peeters, M., Van Assche, G.: On the Indifferentiability of the Sponge Construction. In: Smart, N.P. (ed.) EUROCRYPT 2008. LNCS, vol. 4965, pp. 181–197. Springer, Heidelberg (2008)
3. De Cannière, C., Sato, H., Watanabe, D.: Hash Function Luffa: Supporting Document. Submission to NIST SHA-3 Competition (2008)
4. De Cannière, C., Sato, H., Watanabe, D.: Hash Function Luffa: Specification. Submission to NIST SHA-3 Competition (2008)
5. Khovratovich, D., Plasencia, M.N., Roeck, A., Schlaeffer, M.: Cryptanalysis of Luffa v2 components. In: Biryukov, A., Gong, G., Stinson, D.R. (eds.) SAC 2010. LNCS, vol. 6544, pp. 388–409. Springer, Heidelberg (2011)
6. National Institute of Standards and Technology, Secure hash standard, Federal Information Processing Standards Publication 180-2 (August 2002), http://csrc.nist.gov/publications/fips/fips180-2/fips180-2.pdf
7. National Institute of Standards and Technology, Announcing request for candidate algorithm nominations for a new cryptographic hash algorithm (SHA-3) family (November 2007), http://csrc.nist.gov/groups/ST/hash/documents/
8. Watanabe, D., Hatano, Y., Yamada, T., Kaneko, T.: Higher Order Differential Attack on Step-Reduced Variants of Luffa v1. In: Hong, S., Iwata, T. (eds.) FSE 2010. LNCS, vol. 6147, pp. 270–285. Springer, Heidelberg (2010)
9. Wang, X., Yin, Y.L., Yu, H.: Finding collisions in the full SHA-1. In: Shoup, V. (ed.) CRYPTO 2005. LNCS, vol. 3621, pp. 17–36. Springer, Heidelberg (2005)

Improved Security Analysis of Fugue-256 (Poster)⋆

Praveen Gauravaram[1],[⋆⋆], Lars R. Knudsen[1],
Nasour Bagheri[2], and Lei Wei[3],[⋆⋆⋆]

[1] Department of Mathematics, Technical University of Denmark, Denmark
[2] Shahid Rajaee Teacher Training University, Iran
[3] Nanyang Technological University (NTU), Singapore

Abstract. We present some improved analytical results as part of the ongoing work on the analysis of Fugue-256 hash function, a second round candidate in the NIST's SHA3 competition. First we improve Aumasson and Phans' integral distinguisher on the 5.5 rounds of the *final transformation* of Fugue-256 to 16.5 rounds. Next we improve the designers' meet-in-the-middle preimage attack on Fugue-256 from 2^{480} time and memory to 2^{416}. Finally, we comment on possible methods to obtain free-start distinguishers and free-start collisions for Fugue-256.

1 Fugue-256 Hash Function

Fugue-256, denoted **F**-256, parses the 256-bit initial value (IV) as eight 4-byte words $IV_0, \ldots, IV_7$. It initializes a state S of 30 4-byte words S_i for $i = 0, \ldots, 29$, as a 4×30 matrix by assigning $S_j = 0$ for $j \in [0, 21]$ and $S_j = IV_{j-22}$ for $j \in [22, 29]$. This state is called *initial state*. Hereafter, we denote by $S_{i \sim j}$ the consecutive words of a state S from the index i to j. Streams of 4-byte message words are processed from this state using *round transformation* **R**. If the input message is not a multiple of 32 then it is padded with sufficient 0 bits and it is appended with two 4-byte words that represent the binary encoding of the length of the unpadded message in big-endian notation. Then the *final transformation* **G** is applied to the internal state to obtain an *output state* of 30 words, of which the eight words $S_{1 \sim 4}, S_{15 \sim 18}$ are used as the digest. The transforms **R** and **G** are discussed below where the operation $+$ is the same as 32-bit exclusive-or.

Round transformation **R**: It takes a state S and a 4-byte message word m as inputs and outputs a new thirty column state. **R** calls a sequence of functions: **TIX**(m), **ROR3**, **CMIX**, **SMIX**, **ROR3**, **CMIX**, **SMIX**.

- **TIX**(m): $S_{10}+ = S_0$; $S_0 = m$; $S_8+ = S_0$; $S_1+ = S_{24}$

⋆ This work has been supported in part by the European Commission through the ICT programme under contract ICT-2007-216676 ECRYPT II.
⋆⋆ Author has been sponsored by the Danish Council for Independent Research: Technology and Production Sciences (FTP) grant 09-066486/FTP.
⋆⋆⋆ Author is supported under the Singapore National Research Foundation under Research Grant NRF-CRP2-2007-03.

U. Parampalli and P. Hawkes (Eds.): ACISP 2011, LNCS 6812, pp. 428–432, 2011.
© Springer-Verlag Berlin Heidelberg 2011

- **ROR3** is defined by $S_i = S_{i-3 \bmod 30}$.
- **CMIX**: $S_0+ = S_4$; $S_1+ = S_5$; $S_2+ = S_6$; $S_{15}+ = S_4$; $S_{16}+ = S_5$; $S_{17}+ = S_6$
- **SMIX** transformation combines the non-linear **SBox** with a Super-Mix linear transformation **SMIX-T**. **SMIX** operates only on the first four columns $S_{0\sim3}$ of the state S that are viewed as a 4×4 matrix of 16 words. Each byte of these columns first undergoes an **SBox** transform as in AES and the resulting matrix undergoes an **SMIX-T** transform denoted by a 16×16 matrix **N** of 256 bytes. That is, $S'_{0\sim3} = \mathbf{N}.(S_{0\sim3})$ where **N** is multiplied (.) with a 16-byte 4×1 column matrix output of **SBox**. Similarly, $(S_{0\sim3}) = \overline{\mathbf{N}}.(S'_{0\sim3})$ where $(S'_{0\sim3})$ is a 16-byte 4×1 column matrix. The inverse operations of **SBox** and **SMIX** are denoted by $\overline{\mathbf{SBox}}$ and $\overline{\mathbf{SMIX}}$ respectively.

Final transformation **G:** It takes the output S of the **R** transform and produces a *final state* of 30 words. The function **G** consists of 5 rounds of **G1**, 13 rounds of **G2** and a binary addition of two state words.

- **G1**: ROR3, CMIX, SMIX, ROR3, CMIX, SMIX.
- **G2**: $S_4+ = S_0$; $S_{15}+ = S_0$; ROR15; SMIX; $S_4+ = S_0$; $S_{16}+ = S_0$; ROR14; SMIX
- $S_4+ = S_0$; $S_{15}+ = S_0$

The resultant state is called *final state* from which $S_{1\sim4}$ and $S_{15\sim18}$ are used as digest. In any round i of **R**, the internal state is denoted by *State-i* and its words are denoted by $S_0^i, S_1^i, \ldots, S_{29}^i$, i.e, $S_{1\sim29}^i$. The internal state words after the first **SMIX** in a round i are denoted by $S_0^{i.5}, \ldots, S_{29}^{i.5}$, i.e, $S_{0\sim29}^{i.5}$. In any round i of **R**, the internal state words after the first **ROR3**, **CMIX** and **SBox** transformations are denoted by $x_0'^i, \ldots, x_{29}'^i$, $x_0^i, \ldots, x_{29}^i$ and $\hat{x}_0^i, \ldots, \hat{x}_{29}^i$ respectively and those after the second **ROR3**, **CMIX** and **SBox** transformations are denoted by $y_0'^i, \ldots, y_{29}'^i$, $y_0^i, \ldots, y_{29}^i$ and $\hat{y}_0^i, \ldots, \hat{y}_{29}^i$ respectively. A message word inserted in the ith round of **R** is denoted by m^i.

2   Integral Distinguisher for 16.5 Rounds of G

Our integral attack is a first order integral attack. We follow the notation of [4] for the bytes included in the integral as follows: The symbol $\mathcal{C}$ (for Constant) in the ith byte means that the values of all i^{th} bytes in the attack are equal. The symbol $\mathcal{A}$ (for All) means that all bytes in the attack are different, and the symbol $\mathcal{S}$ (for Sum) means that the sum of all i^{th} bytes is predictable and we write ? when the sum of the bytes is not predictable. We count the rounds of the **G** transform from 0 to 17 and a state in any round i where $i = 0, 0.5, 1, \ldots, 16, 16.5, 17$ is denoted by S^i and the words of S^i by $S_{0\sim29}^i$.

Aumasson and Phan [1] presented an integral distinguisher for 5.5 rounds of the **G** function. Their distinguisher fixes all the bytes of the state S^0 except for the first byte of S_2^0 at the start of the **G** transform. All possible values are assigned to the first byte of S_2^0. They have shown that for $S_2^0 = \mathcal{A}\|\mathcal{C}\|\mathcal{C}\|\mathcal{C}$ one would receive $S_0^{5.5} =?\|?\|?\|?$, $S_1^{5.5} = \mathcal{A}\|?\|?\|?$, $S_2^{5.5} = \mathcal{S}\|?\|?\|?$, $S_3^{5.5} = \mathcal{S}\|?\|?\|?$ which presumably shows a non-randomness property in the first 5.5 rounds of the **G** function. We improve their attack to 16.5 rounds out of 18 rounds of **G**.

Improved attack. A closer analysis of integrals reveals that the values of the integral before the **ROR** functions of **G2** play a crucial role on the success of the distinguisher. It turns out that this word remains unchanged through many rounds of **G2** before being affected by other words. However, for the given integral, all bytes of $S_0^{5.5}$ are unknown ('?') and out of control of the adversary. Hence, the integral of Aumasson and Phan does not seem to extend to more rounds of the **G** transform. Our analysis revealed an integral that runs for more rounds. Note that values with notation $\mathcal{A}$ and $\mathcal{C}$ in our integral are unchanged through **SBox**, but values with notation $\mathcal{S}$ are unknown (?) after **SBox**.

In our integral, we fix whole state bytes of S^0, except for the second byte of S_4^0 where we consider all possible values. The word S_4^0 propagates to S_{28}^5 with probability 1. Hence, the **ROR3** transform in the 5th round of **G1** (i.e 4th round of **G**) shifts this word as one of the inputs to the **SMIX**. Hence, we obtain $S_0^{5.5} = ?\|?\|\mathcal{S}\|?$. It means that we know the sum of the values ($\mathcal{S}$) for this word. On the other hand, this word is propagated to S_4^{16} with probability 1.

In the next step, we have $S_4^{16}+ = S_0^{16}$ which destroys our integral. However, after the **ROR15** function in the 16th round of **G**, S_0^{16} and S_4^{16} are propagated to S_{15}^{16} and S_{19}^{16} respectively. Now if we assume that the adversary has also access to S_{19}^{16} then he can combine S_{15}^{16} and S_{19}^{16} and retrieve the integral values as $S_4^{16.5} = S_{15}^{16.5} \oplus S_{19}^{16.5}$.

Hence, we have an integral which applies to 16.5 out of 18 rounds of the **G** transform. The complexity of attack is 256 evaluations of 16.5 rounds of the **G** transform and memory is 256 bytes. The probability of receiving an $\mathcal{S}$ byte at $S_4^{16.5}$ for **G** is 1 whereas this probability for a random permutation is 2^{-8}. Hence the success probability to distinguish 16.5 rounds of the **G** function from a random permutation is $1 - 2^{-8}$. Our findings illustrate the weak diffusion of the **G** transform.

3   Improved Meet-in-the-Middle Preimage Attack on F-256

The designers of Fugue noted the application of a generic meet-in-the-middle (MIM) preimage attack on any t-bit instance of Fugue [3, p.77] with n-bit internal state in $2^{n/2}$ time and memory complexity [3, p.77].

Improved generic MIM preimage attack on Fugue. Let *State-i'* be the internal state in any round i of **R** after the step $S_{10}^i = S_0^i \oplus S_{10}^i$. This is a $(n-32)/32$-word internal state without the word S_0^i and except for the word S_{10}^i in the *State-i'* all other words are the same as in the $n/32$-word *State-i*. Instead of looking for a *collision match* at *State-i* as in the generic MIM attack, if we look for it at *State-i'*, we can improve the generic attack complexity by a factor of 2^{16}. Briefly, in the *Backward process* of the attack we first establish the required length-padding words in the last two rounds of **R** and then generate required number of messages with that length-padding and corresponding internal states. We use freedom in the words S_0 and S_{10} to accomplish this task. These states are matched with the states generated in the *Forward process* of the attack till *State-i'* which is called *middle state*. We omit details of the attack and refer to the full version of the paper [2]. As an illustration, the improved attack finds a preimage of size at least 29 (resp. 35) message words with a complexity of 2^{464} (resp. 2^{560}) time and memory for **F-256** (resp. **F-512**).

Improved MIM preimage attack on F-256. We further improve the preimage attack on **F-256** by exerting control over 3 words of the 29-word *middle state*. This technique allows us to use a birthday attack to match only 26 words of the *middle state*, thereby reducing the complexity of the attack to 2^{416} from 2^{464}. Let 0 be the round of the **R** transform at which we aim for a *collision match*. Let $-1, -2, -3, \ldots$ and $1, 2, 3, \ldots$ be the respective rounds of the **R** transforms from the 0th round in the *Forward process* and *Backward process* of the attack. The attack is outlined below:

We show that the words S_{17}^0, S_{23}^0 and S_{27}^0 in the *middle state* (i.e *State-0'*) can be controlled such that the internal states evolving from the *initial state* and the *final state* of **F-256** can be matched in these words deterministically with a probability of 1 by solving a simple system of equations. We do this by first assigning fixed values (could be distinct) to the words S_{17}^0, S_{23}^0 and S_{27}^0 that are controlled by using the **R** transforms $-3, -2$ and -1 in the *Forward process* and the **R** transforms 3, 2 and 1 in the *Backward process*. In the *Forward process*, the desired value for the words S_{27}^0, S_{23}^0 and S_{17}^0 is obtained consecutively by using the freedom available in the message words m^{-3}, m^{-2} and m^{-1} in the **R** transforms of the rounds $-3, -2$ and -1 respectively. In the *Backward process*, the desired values for the words S_{17}^0, S_{23}^0 and S_{27}^0 are obtained consecutively by using the freedom available in the words S_0^3, S_0^2 and S_0^1 in the **R** transforms of the rounds 3, 2 and 1 respectively. Below we explain how the word S_{17}^0 can be controlled and a similar explanation follows for controlling the words S_{23}^0 and S_{27}^0.

Controlling the word S_{17}^0: Below we will show how we can obtain the desired word S_{17}^0 of the *middle state* from the *final state* and *initial state* of **F-256** through the *Backward process* and *Forward process* respectively.

Backward process: The word S_{17}^0 in the *middle state* of the 0th round **R** transform will be the word S_{29}^2 in the *State-2'* of the 2nd round **R** transform. Now $x_2'^2 = S_{17^0}$, $x_2^2 = x_2'^2 \oplus x_6'^2$. Now $\hat{x}_2^2 = \mathbf{SBox}(x_2^2)$. Note that $(S_1^{2.5}, S_2^{2.5}, S_3^{2.5}) = (S_4^3, S_5^3, S_6^3)$. For the *final state* of **F-256** inverted till the **R** transform in round 3, the *State-3'* of the 3rd round **R** transform is fixed. Therefore, we can only use $S_0^{2.5}$ input to $\overline{\mathbf{N}}$ to obtain the desired $\hat{x}_2^2$ and therefore, we can obtain the desired S_{17}^0. The matrix **N** has a property that by controlling one of the input words, we can obtain one desired output word by solving a system of 4 equations in 4 unknowns for a negligible complexity. This property is also applicable for $\overline{\mathbf{N}}$. Hence, we can find a $S_0^{2.5}$ such that $\overline{\mathbf{N}}.(S_0^{2.5}, S_1^{2.5}, S_2^{2.5}, S_3^{2.5})$ produces the desired word $\hat{x}_2^2$. This process also determines the message word m^2 which is $\overline{\mathbf{SBox}}(\hat{x}_2^2) = x_3^2 = x_3'^2$. Note that $S_0^{2.5} = y_3'^2 = y_3^2$, $\hat{y}_3^2 = \mathbf{SBox}(y_3^2)$ and the state words $S_{1 \sim 29}^3$ of the state *State-3'* are determined by the *final state* of **F-256**. Now we vary S_0^3 such that $\overline{\mathbf{N}}.(S_0^3, S_1^3, S_2^3, S_3^3)$ produces the desired $\hat{y}_3^2$. Once we have found the candidate S_0^3, we can determine $S_{10}^3 = S_0^3 \oplus x_{13}'^3$ of the state *State-3*.

Forward process: For an *initial state* of **F**-256 processed till the end of the -2nd **R** transform by using *Forward process*, the *State-(-1)* of the -1th round **R** transform is fixed. This implies that $y_{17}'^{-1}$ has already been fixed. To obtain the desired value of S_{17}^0, we need to control $y_6'^{-1}$ which is $y_{17}'^{-1} \oplus S_{17}^0$. The word $y_6'^{-1}$ is the same as $S_3^{-1.5}$. Note that the words $S_{1\sim29}^{-1}$ had already been fixed. Hence, we can determine the words $(\hat{x}_0^{-1}, \hat{x}_1^{-1}, \hat{x}_2^{-1})$, the first three word input to **N**, as follows: $\hat{x}_0^{-1} = \mathbf{SBox}(S_{27}^{-1} \oplus S_1^{-1} \oplus S_{24}^{-1})$; $\hat{x}_1^{-1} = \mathbf{SBox}(S_{28}^{-1} \oplus S_2^{-1})$; $\hat{x}_2^{-1} = \mathbf{SBox}(S_{29}^{-1} \oplus S_3^{-1})$. Having determined the words $\hat{x}_0^{-1}$, $\hat{x}_1^{-1}$ and $\hat{x}_2^{-1}$, we can use the freedom available in the message word m^{-1} to determine the candidate $\hat{x}_3^{-1}$ such that we obtain the desired $S_3^{-1.5} = y_6^{-1}$ and therefore, we obtain the desired word $S_{17}^0 = y_6^{-1} \oplus y_{17}'^{-1}$ in the *middle state*.

Similarly, we can use the freedom available in the words S_0^2 (resp. S_0^1) and m^{-2} (resp. m^{-3}) to obtain the desired word S_{23}^0 (resp. S_{27}^0). This attack produces a preimage of size at least 32 words excluding length-padding words.

4 Concluding Remarks

In this paper we have developed a further understanding of the design of **F**-256. In addition to the above results on **F**-256, we also considered the differential characteristic of the **G** transform proposed by the designers in the PRF analysis of **F**-256 [3, §12.4.2] and improved it by exploiting the differential properties of the inverse **SMIX** such that it produces sparse input state differences for **G** transform. Our improved differential path can be used to mount an *inside-out* distinguisher for **G** similar to of [1]. This distinguisher produces a free-start distinguisher for **F**-256 which appears to produce a sparser state differences when it is extended to a few rounds of **R** compared to the similar extension of the distinguisher for **G** of [1]. Both these results also produce free-start collisions for **F**-256. We remark that a closer analysis of the design revealed that free-start collisions for the length-padded **F**-256 are possible without even inverting the *final transformation* **G**. For details we refer to [2].

Acknowledgments. We thank Charanjit Jutla, Shai Halevi, Søren Thomsen, JP Aumasson, Raphael Phan and Christian Rechberger for comments.

References

1. Aumasson, J.-P., Phan, R.C.-W.: On the Cryptanalysis of the Hash Function Fugue: Partitioning and Inside-Out Distinguishers. To appear in IPL Journal (2011)
2. Gauravaram, P., Knudsen, L.R., Bagheri, N., Wei, L.: Improved security analysis of Fugue-256 (Extended version) (2011), To be available as a MAT, DTU Technical report, http://www2.mat.dtu.dk/pg-projects/
3. Halevi, S., Hall, W.E., Jutla, C.S.: The Hash Function Fugue. Submission to NIST (2009) (updated)
4. Knudsen, L., Wagner, D.: Integral cryptanalysis. In: Daemen, J., Rijmen, V. (eds.) FSE 2002. LNCS, vol. 2365, pp. 112–127. Springer, Heidelberg (2002)

Improved Meet-in-the-Middle Cryptanalysis of KTANTAN (Poster)*

Lei Wei[1], Christian Rechberger[2], Jian Guo[3], Hongjun Wu[1],
Huaxiong Wang[1], and San Ling[1]

[1] Nanyang Technological University, Singapore
{weil0005,wuhj,hxwang,lingsan}@ntu.edu.sg
[2] Katholieke Universiteit Leuven, ESAT/COSIC and IBBT, Belgium
christian.rechberger@groestl.info
[3] Institute for Infocomm Research, A*STAR, Singapore.
ntu.guo@gmail.com

Abstract. This paper presents ongoing work towards extensions of meet-in-the-middle (MITM) attacks on block ciphers. Exploring developments in MITM attacks in hash analysis such as: (i) the *splice-and-cut* technique; (ii) the *indirect-partial-matching* technique. Our first contribution is that we show corrections to previous cryptanalysis and point out that the key schedule is more vulnerable to MITM attacks than previously reported. Secondly we further improve the time complexities of previous attacks with (i) and (ii), now the 80-bit secret key of the full rounds KTANTAN-{32, 48, 64} can be recovered at time complexity of $2^{72.9}$, $2^{73.8}$ and $2^{74.4}$ respectively, each requiring 4 chosen-plaintexts.

1 Introduction

We study the KTANTAN family [3] on resistance to MITM attacks. In the attack due to Bogdanov and Rechberger [2], the key schedule was reported to be weak and the key of full KTANTAN-32 can be recovered slightly faster than brute force. In this paper, we first point out that the previous analysis was not correct (as it was found based on a wrong key schedule), the actual key schedule is even weaker than reported. Based on the corrections, we further examine how developements on MITM preimage attacks in hash analysis can improve the attack. Indeed, with *splice-and-cut* and *indirect-partial-matching*, we find chosen-plaintext key recovery attacks with improved time complexities, faster than brute force by $2^{7.1}$, $2^{6.2}$ and $2^{5.6}$ for block size 32, 48 and 64.

The paper is organized as follows: techniques in MITM attacks are discussed in Section 2, the previous attack versus our experiment results are discussed in Section 3, our improve attacks to KTANTAN family with hash analysis techniques are shown in Section 4 and we conclude in Section 5.

2 Developments in MITM Attacks

MITM attacks were originally developed from cryptanalysis of block ciphers. In 2008, Sasaki and Aoki noticed that the MITM attacks could be applied to hash

---

* Full version at http://www1.spms.ntu.edu.sg/~weil0005/mitm2.pdf

U. Parampalli and P. Hawkes (Eds.): ACISP 2011, LNCS 6812, pp. 433–438, 2011.
© Springer-Verlag Berlin Heidelberg 2011

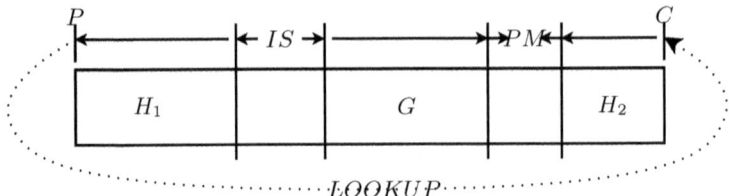

Fig. 1. A general setup for MITM attacks

functions, to find (second) preimages faster than brute-force [4]. The attacks and further developments have successfully broken the one-wayness of many designs. We briefly introduce the techniques relevant to our attacks with Fig. 1.

In a Davies-Meyer construction from a block cipher E keyed by message m, the feedforward $\oplus h$ is used to compute h' as $E_m(h) \oplus h$. To *splice-and-cut*, $E(h)$ is written as three sub-ciphers as $E(h) = H_2 \circ G \circ H_1(h)$ (without IS and PM). Let $h_{inter} = H_1(h)$, we have $H_2^{-1}(H_1^{-1}(h_{inter}) \oplus h') = G(h_{inter})$. The output of E is computed as $h \oplus h'$, this is the *splice* that connects the input and output of E. The attack starts at position h_{inter} and *cut* E into $G(\cdot)$ and $H_2^{-1}(H_1^{-1}(\cdot) \oplus h')$. For a block cipher attack, a lookup table $LOOKUP$ is used as a *virtual* feedforward. For *initial-structure* (IS), neighbouring key bits around h_{inter} may be swapped for more neutral key bits. Let $K_c := K_1 \cap K_2$, we compute $M := G_{K_1}(h_{inter})$ and $M' := H_{K_2}^{-1}(h_{inter})$. A *partial-matching* (PM) of m bits of M and M' is sufficient for filtering wrong keys at a ratio of 2^{-m}. After this the PM portion is repeated to filter more wrong keys, and it does not dominate the overall complexity if $(|K_1| + |K_2| - 2|K_c| - m) + \log_2(\alpha) < \max(|K_1| - |K_c|, |K_2| - |K_c|)$, where α is the percentage of PM steps. *Indirect-partial-matching* (IPM) extends PM for more steps. PM usually starts when key bits in $K_2 \setminus K_c$ appear after the end of G (otherwise, one can extend G for more steps). Similarly for the other side, *i.e.*, key bits in $K_1 \setminus K_c$ appear just before H_2. IPM aims to find, from the PM steps, state bits that can be computed by both $G_{K_1} + \phi_{K_2}$ and $H_{K_2} + \mu_{K_1}$ from h_{inter}, such that the matching can be checked between $G_{K_1} - \mu_{K_1}$ and $H_{K_2} - \phi_{K_2}$ instead. ϕ and μ are linear in their key materials.

3   Meet-in-the-Middle Cryptanalysis of KTANTAN

The KTANTAN family of block ciphers, with block sizes of 32, 48 and 64, was proposed at CHES'09 [3]. They accept 80-bit key and share a key schedule for 254 rounds.

3.1   The Previous Meet-in-the-Middle Attack

The attacks reported in [2] work with a few known plaintexts at around the *unicity distance*. The attack to full KTANTAN32 works in 2^{79} encryptions. Although arguably marginal, it is the first key-recovery attack to the full 254 rounds faster than brute force. For block size b of 48 and 64, the attack manages to break 251 and 248 rounds respectively [1].

The attack cuts the R-round KTANTAN-b cipher (b for block size) into three parts for some $\alpha, \beta < R$. Let $K := k_{79}k_{78} \ldots k_1 k_0$ be the key and $A := \{k_0, k_1, \ldots, k_{78}, k_{79}\}$. Let x_i be the state after round i, for $0 \le i \le 254$. Let $\varphi_{i,j}$ be the transformation from round i to round j (inclusive), key bits in $A_1 := \{k_{15}, k_{79}\}$ are neutral to $H := \varphi_{254-\beta+1,254}$ (the backward phase) and $A_2 := \{k_5, k_{37}, k_{69}\}$ neutral to $G := \varphi_{1,\alpha}$ (the forward phase). Let $A_0 := A \setminus (A_1 \bigcup A_2)$. The attack to KTANTAN32 proceeds with a text pair (P, C), for each guess of key bits in A_0, compute 3-bit of x_{128} for independent guesses of A_1 and A_2, from respectively $M := G(P) = \varphi_{1,105}(P)$ and $M' := H^{-1}(C) = \varphi_{137,254}^{-1}(C)$. For a key guess that passes this 3-bit filter, try current and additional pairs of (P, C) one by one. For $\lceil 80/b \rceil$ pairs tested, the correct key can be recovered with probability close to 1. The claimed complexity is at around $2^{|A_0|}(2^{|A_1|}+2^{|A_2|})+2^{80-3} = 2^{79}$ encryptions. A more accurate calculation showes that it works at a time complexity of $2^{|A_0|}(2^{|A_1|} \cdot \alpha/R + 2^{|A_2|} \cdot \beta/R + 2^{|A_1|+|A_2|-m}(R-\alpha-\beta)/R) \doteq 2^{77.6}$.

3.2 New Experimental Observations on the Attack

We reimplement the family of KTANTAN from its design paper [3] and examine the key schedule according to the attack in [2], different sets of neutral key bits are found and it suggests that the key schedule is much weaker than reported. Under this observation, the MITM approach brings non-marginal attacks for the full ciphers of the entire family. The previous results and our attacks (*) are summarized in Table 1.

Table 1. The B-R attack and our results

b	R	α	β	A_1	A_2	m	Time	Data	
32	254	105	118	15, 79	5, 37, 69	3	$2^{79.0}$	3 KP	[2]
48	251	107	112	11, 15, 75, 79	5, 69	1	$2^{79.7}$	2 KP	[2]
64	248	107	112	9, 73	5, 69	2	$2^{79.58}$	2 KP	[2]
32	254	111	122	3, 20, 41, 47, 63, 74	32, 39, 44, 61, 66, 75	12	$2^{73.88}$	3 KP	*
32	254	110	122	3, 20, 41, 47, 63, 74	27, 32, 39, 44, 59, 61, 66, 75	4	$2^{73.88}$	3 KP	*
32	254	109	122	3, 20, 41, 47, 63, 74	13, 27, 32, 39, 44, 59, 61, 66, 75	3	$2^{74.33}$	3 KP	*
48	254	123	122	3, 20, 41, 47, 63, 74	32, 44, 61, 66, 75	37	$2^{74.53}$	2 KP	*
48	254	111	121	3, 20, 41, 47, 63, 74	32, 39, 44, 61, 66, 75	4	$2^{73.97}$	2 KP	*
64	254	123	122	3, 20, 41, 47, 63, 74	32, 44, 61, 66, 75	44	$2^{74.53}$	2 KP	*

3.3 Low Complexity Implementation of the Attack

In the cases that the secret keys are not derived with full 80-bit entropy, the attack may become a real threat when the time is 2^6 to 2^7 times less due to the attack of this paper. For example, it is not hard to eavesdrop a small amount of ciphertext corresponding to known protocol headers and it is feasible to launch

---

[1] The authors later updated in http://eprint.iacr.org/2010/532.pdf and [1].

an attack. We implement a low complexity version of an attack to KTANTAN32 in Table 1. We assume 40 bits of A_0 are known by the attacker, the attack has $\alpha = 111, \beta = 122, A_1 = \{3, 20, 41, 47, 63, 74\}, A_2 = \{32, 39, 44, 61, 66, 75\}$ and 12 bits match. The attack successfully recovers the 40-bits in 5 hours 34 seconds on a Quad-core HP xw4600 workstation at 2.40GHz. With 40 bits known, the estimated complexity 2^{74} is reduced to 2^{34}. The experiment confirms roughly a reduction of 2^6 as encrypting 2^{26} plaintexts takes 45 seconds. As a comparison, recovering 40 bits by exhaustive search would take roughly half a month if using the same workstation.

4   More General MITM Attacks on KTANTAN Family

The effeciency of the attacks discussed in Section 3 depends crucially on the number of neutral key bits in the forward phase and backward phases. Hence a natural question is, can we improve the attack by finding more neutral key bits or more bits for matching? We show how *splice-and-cut* and *indirect-partial-matching* address this question.

4.1   The Observations and Search

In round r the computations for $f_{r,a}$ and $f_{r,b}$ are repeated 1, 2 or 3 times for 3 respective block sizes. For each evaluation[2] of $f_a(L_1)$ or $f_b(L_2)$, a single round key bit is mixed XOR-linearly into the LSB of L_1 or L_2, hence affecting the lowest 1 to 3 bits considering the shift(s). In the round that follows, only a few bits of the state get involved in the nonlinear part in computing $f_{r,a}$ and $f_{r,b}$. We observe that the round key bits remain linear in the state bits for some rounds, hence expecting more bits to be matched by IPM.

Let x_i be the state after round i for $1 \leq i \leq 254$ then x_{254} is the ciphertext and denote the plaintext x_0. For φ_{b_0,b_1} as the backward phase and φ_{f_0,f_1} as forward, the rounds between (exclusive) b_0 and f_0 are used for the *initial structure*. We search exhaustively for all feasible combinations of (f_0, f_1, b_0, b_1) and compute the complexities. IS is applicable and we set $f_0 - b_0 - 1$ to up to 20. The search shows that IS is not contributing to a better attack, hence $f_0 = b_0 + 1$. We list the best attacks in Table 2. The IPM checks between two fully determined states $M := x_{f_1}$ (after round f_1) and $M' := x_{b_1-1}$ (before round b_1).

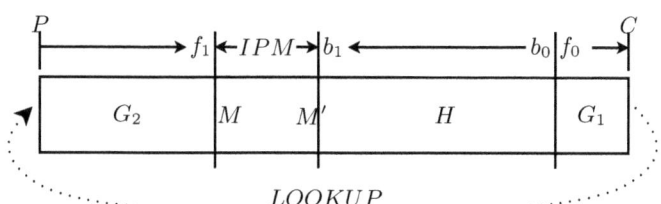

Fig. 2. Illustration of MITM attack with splice-and-cut and IPM

---

[2] Notations follow from [3].

4.2   The Attack with Splice-and-Cut and Indirect-Partial-Matching

First we construct the table $LOOKUP$ to *splice* two ends of cipher. Select a random value for x_{b_0} and compute $C := G_1(x_{b_0})$ for all 4 possible outputs of $G_1 := \varphi_{f_0,254}$. Add the chosen-ciphertext pair (C, P) to the table. Let A_1 be the key bits neutral to $H^{-1} := \varphi^{-1}_{b_1,b_0}$ and A_2 for $G := G_2 \circ LOOKUP \circ G_1 = \varphi_{1,f_1} \circ LOOKUP \circ \varphi_{f_0,254}$.

The attack goes as follows: for each guess of A_0 we try parallel guesses for A_1 and A_2, computing $M := G(x_{b_0})$ and $M' := H^{-1}(x_{b_0})$. m-bit *partial matching signature* s can be computed from both M and M', it is used as a filter of ratio 2^{-m} and the matching is done in a table. A survival key is then tested on whether $M' = \varphi_{f_1+1,b_1-1}(M)$ for K, and on other pairs of (P, C). The right key is the one that services all $\lceil 80/b \rceil$ pairs.

The m-bit partial matching signature is computed from the matching position, as Section 2 on IPM. The signature includes state bits independent or linearly dependent with the active (non-neutral) key bits in the IPM phase. The technique significantly improves the number of bits that can be matched, the matching can be extended for more rounds to have more neutral bits. For KTANTAN32, the matching position is at x_{115}. Denote x_{115} as x and let $x[i]$ be the i-th bit of x, for $0 \leq i \leq 31$. For the forward part of IPM, the following key bits are linear in the corresponding state bits, k_{27} in $x[0]$, k_{13} in $x[1]$, k_{39} in $x[3]$, k_{59} in $x[4]$ and k_{39} in $x[22]$. For the backward part, we have k_{74} in $x[26]$, k_{74} in $x[21]$, k_{74} in $x[3]$ and k_{20} in $x[2]$. Hence the matching signature is $(x[26] - k_{74}, x[22] - k_{39}, x[21] - k_{74}, x[7], x[6], x[5], x[4] - k_{59}, x[3] - k_{39} - k_{74}, x[2] - k_{20}, x[1] - k_{13}, x[0] - k_{27})$ which can be computed from both sides without knowing the value for the active key bits from that side.

Table 2. MITM attack with splice-and-cut and indirect-partial-matching

b	R	b_1	b_0	f_0	f_1	A_1	A_2	m	Time	Data
32	254	148	253	254	109	13, 27, 32, 39, 44, 59, 61, 66, 75	3, 20, 41, 47, 63, 74	11	$2^{72.93}$	4 CC
48	254	150	253	254	111	32, 39, 44, 61, 66, 75	3, 20, 41, 47, 63, 74	15	$2^{73.77}$	4 CC
64	254	151	253	254	112	32, 44, 61, 66, 75	3, 20, 41, 47, 63, 74	54	$2^{74.38}$	4 CC

5   Conclusions

We have shown corrected results for the KTANTAN key schedule for MITM attacks, and have confirmed the attack with an experiment. Moreover, we have shown some techniques from hash function MITM preimage attacks effective for improving the results on KTANTAN. In particular, *splice-and-cut* gives more neutral bits and *indirect-partial-matching* improves over *partial-matching* with much better matching. It is open to examine if better enhancements can be discovered for MITM attacks and dedicated techniques to be found for particular ciphers.

Acknowledgments. This work was supported in part by the Singapore National Research Foundation under Research Grant NRF-CRP2-2007-03.

References

1. Bogdanov, A., Rechberger, C.: A 3-Subset Meet-in-the-Middle Attack: Cryptanalysis of the Lightweight Block Cipher KTANTAN. In: Biryukov, A., Gong, G., Stinson, D.R. (eds.) SAC 2010. LNCS, vol. 6544, pp. 229–240. Springer, Heidelberg (2011)
2. Bogdanov, A., Rechberger, C.: Generalized Meet-in-the-Middle Attacks: Cryptanalysis of the Lightweight Block Cipher KTANTAN. In: Preproceedings of SAC (2010), http://homes.esat.kuleuven.be/~abogdano/talks/ktantan_sac10.pdf
3. Cannière, C.D., Dunkelman, O., Knezevic, M.: KATAN and KTANTAN - A Family of Small and Efficient Hardware-Oriented Block Ciphers. In: Clavier, C., Gaj, K. (eds.) CHES 2009. LNCS, vol. 5747, pp. 272–288. Springer, Heidelberg (2009)
4. Sasaki, Y., Aoki, K.: Preimage Attacks on 3, 4, and 5-Pass HAVAL. In: Pieprzyk, J. (ed.) ASIACRYPT 2008. LNCS, vol. 5350, pp. 253–271. Springer, Heidelberg (2008)

Toward Dynamic Attribute-Based Signcryption (Poster)

Keita Emura[1], Atsuko Miyaji[2], and Mohammad Shahriar Rahman[2]

[1] Center for Highly Dependable Embedded Systems Technology
[2] School of Information Science
Japan Advanced Institute of Science and Technology, 1-1, Asahidai, Nomi, Ishikawa, 923-1292, Japan
{k-emura,miyaji,mohammad}@jaist.ac.jp

Abstract. This paper presents an ongoing work toward the proposal of the new concept of the attribute-based cryptosystem. In SCN2010, Gagné, Narayan, and Safavi-Naini proposed attribute-based signcryption (ABSC) with threshold structure. As in ciphertext-policy attribute-based encryption (CP-ABE), an encryptor can specify the access structure of decryptors, and as in attribute-based signature (ABS), each decryptor can verify the encryptor's attributes. On the contrary to the access structure of decryptors, the access structure of the encryptor needs to be fixed in the setup phase. In this paper, we investigate ABSC with dynamic property, called dynamic ABSC (DABSC), where access structures of encryptor can be updated flexibly without re-issuing secret keys of users.

1 Introduction

1.1 Attribute-Based Signcryption (ABSC)

Recently, Gagné, Narayan, and Safavi-Naini proposed ABSC with threshold structure [3] to achieve Cost(ABS & CP-ABE) < Cost(ABS)+Cost(CP-ABE). That is, their ABSC scheme is efficient compared with encrypt-then-sign paradigm. As in Ciphertext-Policy Attribute-Based Encryption (CP-ABE) [1], an encryptor can specify the access structure of decryptors, and as in Attribute-Based Signature (ABS) [4], each decryptor can verify the encryptor's attributes. Note that, in the Gagné et al. definition, a decryptor can verify the encryptor's attribute *explicitly*. This property is preferable for the following encrypted storage system usage.

1.2 Encrypted Storage System

Encrypted storage system is a well-known application of CP-ABE. To indicate the set of common attributes of decryptors (such as affiliation, post, and so on), CP-ABE schemes can achieve a fine-grained access control without increasing the number of keys. On the contrary to the decryptor's attributes, there is no way to verify the set of attributes of encryptor if CP-ABE is applied only. To check the source of storage files, attributes of encryptor is important information. By applying the Gagné et al. ABSC, both CP-ABE and ABS properties can be

U. Parampalli and P. Hawkes (Eds.): ACISP 2011, LNCS 6812, pp. 439–443, 2011.
© Springer-Verlag Berlin Heidelberg 2011

handled for encrypted storage system usage, simultaneously. So, a decryptor can check the encryptor's attribute explicitly. However, the threshold structure (which is supported by the Gagné et al. ABSC) is not suitable for encrypted storage system usage, although it is useful for fault tolerance usage. In addition, the access structure of the encryptor is specified only once, and it cannot be changed. More precisely, the threshold value is decided in the key generation phase, and it cannot be updated without re-issuing the new key.

2   Our Approach: Dynamic ABSC

In this paper, we investigate the new concept "ABSC with dynamic property", called Dynamic ABSC (DABSC), where access structures of encryptor can be changed without re-issuing secret keys of users. As an application of DABSC, we consider authenticated fine-grained storage systems.

For example, let a teaching assistant of a lecture "Applied Cryptography" would like to store an encrypted examination data for students (who take Applied Cryptography) only. In addition, students would like to check whether a stored file was made by a teaching assistant of Applied Cryptography. Then an encryptor makes a ciphertext part associated with attributes of a decryptor (Student $\wedge$ Applied Cryptography), and also makes a signature part associated with attributes of the encryptor (Teaching Assistant $\wedge$ Applied Cryptography).

The dynamic property is suitable for the following example. We assume that the encryptor (who is a teaching assistant of Applied Cryptography) becomes a teaching assistant of a lecture "Discrete Mathematics", and the encryptor has obtained the secret key for attributes Teaching Assistant and Applied Cryptography. If the dynamic property is not handled, then key generation center (KGC) needs to re-issue the secret key of both Applied Cryptography and Teaching Assistant for handling the updated access structure of encryptor. It is quite inefficient and impractical (See Table 1).

Table 1. Computational complexity of changing predicate

	KGC	User
Non-dynamic scheme	$O(N \cdot n_e)$	$O(n_e)$
Dynamic scheme	$O(n_e)$	None

N : the number of users
n_e : the maximum number of attributes having each user

Under the dynamic property, KGC has only to issue the secret key of Discrete Mathematics, and no computation that the encryptor is required. While the above example describes the case of small number of predicates, we believe that the dynamic property gives us a very efficient and practical solution when the number of predicates grows large. Actually, the number of attributes is polynomial-size (of the security parameter k, i.e., $O(poly(k))$) and the corresponding predicates can grow exponentially (i.e., $O(2^{poly(k)})$) in large systems. That is, as an expectation, the opportunity of updating predicates also increases in such large systems. Under the dynamic property, even if the current predicate is updated, users do not have to be involved the updating procedure. This is the most benefit point of our proposal.

3 System Operations of DABSC

In the following, values are subscripted by e for encryptors, and values are subscripted by d for decryptors. Let $\mathbb{A}_e = (att_1, att_2, \ldots, att_{n_e})$ be the universe of possible attributes of encryptors, $\mathbb{A}_d = (att_1, att_2, \ldots, att_{n_d})$ be the universe of possible attributes of decryptors, and Υ_e (resp. Υ_d) be a claim-predicate over $\mathbb{A}_e$ (resp. $\mathbb{A}_d$) of encryptors (resp. decryptors). We say that an attribute set $\Gamma_e \subseteq \mathbb{A}_e$ (resp. $\Gamma_d \subseteq \mathbb{A}_d$) satisfies a claim-predicate Υ_e (resp. Υ_d) if $\Upsilon_e(\Gamma_e) = 1$ (resp. $\Upsilon_d(\Gamma_d) = 1$). In the following definition, an encryptor can select an access structure of decryptor Υ_d for each signcryption ciphertext (which follows the ciphertext-policy property of ABE). On the contrary, an access structure of encryptor Υ_e is a publicly opened. This means that legitimate encryptor who have attributes satisfying Υ_e can make a signcryption ciphertext.

Next, we modify the definitions of the Gagné et al. ABSC [3] to handle the dynamic property.

Definition 1 (Dynamic Attribute-Based Signcryption (DABSC)).

Setup: *This algorithm takes as inputs a security parameter $k \in \mathbb{N}$, and returns public parameters params and a master key msk.*

sExtract: *This algorithm takes as inputs params, msk, and a set of attributes of an encryptor $\Gamma_e \subseteq \mathbb{A}_e$, and returns signing keys $\{sk_{e,i}\}_{att_i \in \Gamma_e}$.*

uExtract: *This algorithm takes as inputs params, msk, and a set of attributes of a decryptor $\Gamma_d \subseteq \mathbb{A}_d$, and returns decryption keys $\{sk_{d,i}\}_{att_i \in \Gamma_d}$.*

BuildPredicate: *This algorithm takes as inputs params, msk, and the ℓ-th access tree T_ℓ, and returns the public value of ℓ-th access tree Υ_e^ℓ.*

Signcrypt: *This algorithm takes as inputs params, Υ_e^ℓ, $\{sk_{e,i}\}_{att_i \in \Gamma_e}$, where $\Upsilon_e^\ell(\Gamma_e) = 1$, an access structure Υ_d, and a plaintext M, and returns a ciphertext C on M. We assume that Γ_e and Υ_d are included into C.*

Unsigncrypt: *This algorithm takes as inputs params, Υ_e^ℓ, $\{sk_{d,i}\}_{att_i \in \Gamma_d}$, where $\Upsilon_d(\Gamma_d) = 1$, and C, and verifies whether the encryptor's attributes satisfy Υ_e^ℓ or not, along with Γ_e and Υ_e^ℓ. If not, then output $\bot$, and M otherwise.*

The above algorithms follow the correctness requirement: for all $(params, msk)$ $\leftarrow$ Setup(1^k), $\{sk_{e,i}\}_{att_i \in \Gamma_e}$ $\leftarrow$ sExtract$(params, msk, \Gamma_e)$, $\{sk_{d,i}\}_{att_i \in \Gamma_d}$ $\leftarrow$ uExtract$(params, msk, \Gamma_d)$, $\Upsilon_e^\ell \leftarrow$ BuildPredicate$(params, msk, T_\ell)$, and $C \leftarrow$ Signcrypt$(params, \Upsilon_e^\ell, \{sk_{e,i}\}_{att_i \in \Gamma_e}, \Upsilon_d, M)$ with $\Upsilon_e^\ell(\Gamma_e) = 1$, $M \leftarrow$ Unsigncrypt$(params, \Upsilon_e^\ell, \{sk_{d,i}\}_{att_i \in \Gamma_d}, C)$ holds when $\Upsilon_d(\Gamma_d) = 1$.

Table 2. DABSC Experiments

S-IND-DABSC-CCA2
$Adv_{\mathcal{A}}^{\text{S-IND-DABSC-CCA2}}(k) =$
$\big\| \Pr\big[(\Upsilon_d^*, T_0, State) \leftarrow \mathcal{A}(k);\ (params, msk) \leftarrow \mathsf{Setup}(1^k);$
$\quad$ Set $\mathcal{O} := \{\mathsf{sExtract}(params, msk, \cdot), \mathsf{uExtract}(params, msk, \cdot),$
$\quad$ $\mathsf{Unsigncrypt}(params, \cdot, \cdot), \mathsf{BuildPredicate}(params, msk, \cdot)\};$
$\quad (M_0^*, M_1^*, \Gamma_e^*, State) \leftarrow \mathcal{A}^{\mathcal{O}}(params, State);\ b \xleftarrow{\$} \{0,1\};$
$\quad C^* \leftarrow \mathsf{Signcrypt}(params, \Upsilon_e^\ell, \{sk_{e,i}\}_{att_i \in \Gamma_e^*}, \Upsilon_d^*, M_b^*);\ b' \leftarrow \mathcal{A}^{\mathcal{O}}(C^*, State);\ b = b'\big] - \frac{1}{2}\big\|$
S-EUF-DABSC-CMA
$Adv_{\mathcal{A}}^{\text{S-EUF-DABSC-CMA}}(k) =$
$\Pr\big[(T_e^*, T_0, State) \leftarrow \mathcal{A}(k);\ (params, msk) \leftarrow \mathsf{Setup}(1^k);$
$\quad$ Set $\mathcal{O} := \{\mathsf{sExtract}(params, msk, \cdot), \mathsf{uExtract}(params, msk, \cdot),$
$\quad \mathsf{BuildPredicate}(params, msk, \cdot), \mathsf{Signcrypt}(params, \cdot, \cdot, \cdot)\};\ (C^*, \Gamma_d^*) \leftarrow \mathcal{A}^{\mathcal{O}}(params, State);$
$\quad \mathsf{Unsigncrypt}(params, \Upsilon_e^*, \{sk_{d,i}\}_{att_i \in \Gamma_d^*}, C^*) = M^* \neq \bot;$
$\quad$ (For Γ_e where $\Upsilon_e^*(\Gamma_e) = 1$, $\mathcal{A}$ did not query either $(M, \Gamma_e, \Upsilon_d^*)$ to the
$\quad$ Signcrypt oracle or Γ_e to the sExtract oracle, where $\Upsilon_e^*(\Gamma_e) = 1) \vee (\Upsilon_e^*(\Gamma_e^*) \neq 1)\big]$

Next, we define indistinguishability against adaptive chosen-ciphertext attack property under selective attribute model (S-IND-DABSC-CCA2), existential unforgeability against chosen-message attack in the selective attribute model (S-EUF-DABSC-CMA). In the following, T (and the initial access tree T_0 also) must follow the condition that leaves of trees are appeared in $\mathbb{A}_e$. S-IND-DABSC-CCA2 guarantees that no PPT adversary $\mathcal{A}$ (which is essentially the same as the CCA adversary of CP-ABE [1]) can guess whether the actual plaintext is M_0^* or M_1^*, namely, no plaintext information is revealed from the ciphertext. Note that S-IND-DABSC-CCA2 captures collusion resistance (i.e., $\mathcal{A}$ is allowed to issue Γ_d and Γ_d' to the uExtract oracle such that $\Upsilon_d^*(\Gamma_d) \neq 1$, $\Upsilon_d^*(\Gamma_d') \neq 1$, $\Gamma_d \cup \Gamma_d' = \Gamma_d^*$, and $\Upsilon_d^*(\Gamma_d^*) = 1$) as in the conventional CP-ABE definition.

Definition 2 (S-IND-DABSC-CCA2). *A DABSC scheme is said to be S-IND-DABSC-CCA2 secure if the advantage $Adv_{\mathcal{A}}^{S\text{-}IND\text{-}DABSC\text{-}CCA2}(k)$ is negligible for any PPT adversary $\mathcal{A}$ in the S-IND-DABSC-CCA2 experiment (defined in Table 2).*

Note that we require $\Upsilon_e^\ell(\Gamma_e^*) = 1$, where Υ_e^ℓ is the public predicate in the challenge phase. In addition, for (C, Γ_d) which is an input of the unsigncryption oracle Unsigncrypt, if $C = C^*$ and $\Upsilon_d^*(\Gamma_d) = 1$, then the oracle returns $\bot$. Otherwise, it returns the result of $\mathsf{Unsigncrypt}(params, \Upsilon_e^i, \{sk_{d,i}\}_{att_i \in \Gamma_d}, C)$, where Υ_e^i is the current predicate when $\mathcal{A}$ issues the unsigncryption query.

Next, we define S-EUF-DABSC-CMA. In the definition of S-EUF-DABSC-CMA, we consider two types adversaries. S-EUF-DABSC-CMA guarantees that no (type 1) adversary $\mathcal{A}$ can make a forged ciphertext which is correctly decrypted (i.e., the Unsigncrypt algorithm outputs $M \neq \bot$) even though $\mathcal{A}$ did not issue either Γ_e to the sExtract oracle such that $\Upsilon_e(\Gamma_e) = 1$ or $(M, \Gamma_e, \Upsilon_d^*)$ to the Signcrypt oracle such that $\Upsilon_e(\Gamma_e) = 1$, and no (type 2) $\mathcal{A}$ (who can obtain all $\{sk_{e,i}\}_{att_i \in \Gamma_e}$) can make a forged ciphertext which is correctly decrypted even though $\Upsilon_e^*(\Gamma_e) \neq 1$. Type 1 adversary (which is the same as the unforgeability adversary of ABS [4]) captures collusion resistance (i.e., $\mathcal{A}$ is allowed to issue Γ_e

and Γ'_e to the sExtract oracle such that $\Upsilon^*_e(\Gamma_e) \neq 1$, $\Upsilon^*_e(\Gamma'_e) \neq 1$, $\Gamma_e \cup \Gamma'_e = \Gamma^*_e$, and $\Upsilon^*_e(\Gamma^*_e) = 1$). Type 2 adversary captures that the Unsigncrypt algorithm does not accept the ciphertext made by Γ_e such that $\Upsilon^*_e(\Gamma_e) \neq 1$ with overwhelming probability.

Definition 3 (S-EUF-DABSC-CMA). *A DABSC scheme is said to be wS-EUF-DABSC-CMA secure if the advantage $Adv_{\mathcal{A}}^{S\text{-}EUF\text{-}DABSC\text{-}CMA}(k)$ is negligible for any PPT adversary $\mathcal{A}$ in the S-EUF-DABSC-CMA experiment (defined in Table 2).*

Note that, let $\Upsilon^*_e \leftarrow \mathsf{BuildPredicate}(params, msk, T^*_e)$ be the predicate when $\mathcal{A}$ outputs the forged ciphertext.

4 Conclusion and toward the Concrete Construction of DABSC Scheme

This paper has presented an ongoing work toward the new concept, called DABSC. We define the system operations and the security requirements of DABSC. Toward the concrete construction of DABSC scheme, our methodology is described as follows. The dynamic property is achieved by applying a *bottom-up approach* construction [2], where first all secret values (assigned with leaves) are chosen, and then each parents secret is computed from bottom up. It seems that DABSC can be implemented based on appropriate CP-ABE and ABS with the bottom-up approach. It is particularly worth noting that the bottom-up approach construction itself does not require the random oracle, although the eventual dynamic ABGS [2] requires the random oracle. That is, DABSC secure in the standard model is expected.

It might be the case that the actual complexity of Signcrypt/Unsigncrypt algorithms in the dynamic scheme is worse as compared to the non-dynamic schemes since certain dynamic-property-related values may have to be included in the ciphertext. As for small system the dynamic property may not be very effective. It remains to be seen how large the number of attributes should be set as threshold between the dynamic and non-dynamic schemes.

References

1. Cheung, L., Newport, C.C.: Provably secure ciphertext policy ABE. In: Ning, P., di Vimercati, S.D.C., Syverson, P.F. (eds.) ACM Conference on Computer and Communications Security, pp. 456–465. ACM, New York (2007)
2. Emura, K., Miyaji, A., Omote, K.: A dynamic attribute-based group signature scheme and its application in an anonymous survey for the collection of attribute statistics. IPSJ Journal 50(9), 1968–1983 (2009)
3. Gagné, M., Narayan, S., Safavi-Naini, R.: Threshold attribute-based signcryption. In: Garay, J.A., De Prisco, R. (eds.) SCN 2010. LNCS, vol. 6280, pp. 154–171. Springer, Heidelberg (2010)
4. Li, J., Au, M.H., Susilo, W., Xie, D., Ren, K.: Attribute-based signature and its applications. In: Feng, D., Basin, D.A., Liu, P. (eds.) ASIACCS, pp. 13–16 (2010)

A Verifiable Distributed Oblivious Transfer Protocol

Christian L.F. Corniaux and Hossein Ghodosi

James Cook University, Townsville QLD 4811, Australia
chris.corniaux@my.jcu.edu.au, hossein.ghodosi@jcu.edu.au

Abstract. In the various distributed oblivious transfer (DOT) protocols designed in an unconditionally secure environment, a receiver contacts k out of m servers to obtain one of the n secrets held by a sender. After a protocol has been executed, the sender has no information on the choice of the receiver and the receiver has no information on the secrets she did not obtain.

These protocols are based on a semi-honest model: no mechanism prevents a group of malicious servers from disrupting the protocol such that the secret obtained by the receiver does not correspond to the chosen secret.

This paper presents ongoing work towards the definition of the first unconditionally secure verifiable DOT protocol in the presence of an active adversary who may corrupt up to $k-1$ servers. In addition to the active adversary, we also assume that the sender may (passively) corrupt up to $k-1$ servers to learn the choice of the receiver. Similarly, the receiver may (passively) corrupt up to $k-1$ servers to learn more than the chosen secret.

Our DOT protocol allows the receiver to contact $4k-3$ servers to obtain one secret, while the required security is maintained.

Keywords: Cryptographic Protocol, Privacy and Security, Distributed Oblivious Transfer, Verifiable Oblivious Transfer.

1 Introduction

In the unconditionally secure distributed oblivious transfer (DOT) schemes presented in [7,2,8,3], a sender $\mathcal{S}$ holds n secrets and a receiver $\mathcal{R}$ wishes to obtain one of them. The model encompasses a distributed environment including m servers. The sender distributes shares of the secrets to the servers and does not intervene in the rest of the protocol. The receiver selects the index of a secret, sends shares of this index to k servers and receives back k shares allowing her to reconstruct the chosen secret. The security of these protocols may be assessed thanks to the following four security conditions defined by Blundo, D'Arco, De Santis and Stinson [2,3]: correctness (C_1), receiver's privacy (C_2), sender's privacy with respect to $k-1$ servers (C_3) and sender's privacy with respect to a "greedy" receiver colluding with $k-1$ corrupted servers (C_4).

In these protocols, security condition C_1 is satisfied in a semi-honest model; No mechanism prevents a set of up to $k-1$ malicious servers from disrupting a

U. Parampalli and P. Hawkes (Eds.): ACISP 2011, LNCS 6812, pp. 444–450, 2011.

protocol such that the secret obtained by the receiver does not correspond to the chosen secret. We introduce this kind of mechanism in one of the (k, m)-DOT-$\binom{n}{1}$ protocols presented by Blundo et als. [2,3].

The two key ideas of our protocol are (1) to let the receiver distribute her requests to the servers thanks to a verifiable secret sharing scheme (VSSS) instead of a secret sharing scheme (SSS), which allows the contacted servers to verify the consistency of the shares and (2) to let the receiver collect enough shares to determine the chosen secret thanks to an error-correcting code decoding scheme.

Our protocol guarantees security conditions C_1, C_2 and C_3, despite the presence of up to $k - 1$ malicious servers among the m servers participating to the protocol. Like in Blundo et al.' one-round DOT protocols, condition C_4 is not satisfied.

This paper is organised as follows. The next section shortly describes the main three components of our protocol. Then, in Sect. 3, we describe our model. Sect. 4 is devoted to the detailed description of the protocol. Finally, in Sect. 5, the security of the protocol is briefly analysed.

2   Background

Although there have been a few DOT protocols studied in the past 15 years, e.g. [7,2,8,11,12,3], the verifiability of distributed shares in such protocols was rarely tackled. The main contribution to the subject was made by Zhong and Yang [11,12], but the setting of their proposed scheme is conditionally secure (difficulty to compute a discrete logarithm).

We present the first unconditionally secure verifiable DOT combining the DOT protocol introduced by Blundo et al. [2,3], the VSSS introduced by Gennaro, Ishai, Kushilevitz and Rabin [6] and the Reed-Solomon error-correcting code decoding scheme introduced by Gao [5].

2.1   Distributed Oblivious Transfer Protocol

The basic principles underlying DOT protocols are conceptually similar. In the original DOT protocol [7] introduced by Naor and Pinkas, as well as in its generalization [2,3] presented by Blundo et al., a sender distributes some information amongst m servers so that, by contacting k servers, a receiver is able to learn only one of the secrets held by the sender. A simplified overview of the (k, m)-DOT-$\binom{n}{1}$ protocol [2,3] we adapt in this paper may be described as follows (operations are executed in a finite field $\mathbb{F}_p$, where p is a prime number):

– In the setup phase, the sender, who holds n secrets $\omega_0, \ldots, \omega_{n-1}$ generates a sparse n-variate polynomial function $Q(x, y_1, \ldots, y_{n-1}) = \omega_0 + \sum_{i=1}^{k-1} a_i x^i + \sum_{i=1}^{n-1} (\omega_i - \omega_0) \times y_i$, where the coefficients a_i $(1 \leq i \leq k - 1)$ are numbers randomly selected in $\mathbb{F}_p$. Then, to each server S_j $(1 \leq j \leq m)$, the sender transmits the $(n - 1)$-variate polynomial function $F_j(y_1, \ldots, y_{n-1}) = Q(j, y_1, \ldots, y_{n-1})$.
– In the transfer phase, the receiver chooses the identifier ℓ of one secret and generates univariate polynomial functions Z_i $(1 \leq i \leq n - 1)$ of degree at

most $k - 1$ such that $(Z_1(0), \ldots, Z_{n-1}(0))$ is an $(n - 1)$-tuple of zeros if the receiver is interested in ω_0 (i.e., $\ell = 0$), or an $(n - 1)$-tuple of zeros and a single one in position ℓ, where $\ell \in \{1, \ldots, n-1\}$, if the receiver is interested in ω_ℓ. Then, the receiver selects a subset $\mathcal{I}_k \subset \{1, \ldots, m\}$ of k indices and sends to each server S_i $(i \in \mathcal{I}_k)$ a request $(i, Z_1(i), \ldots, Z_{n-1}(i))$. When a server S_i receives such a request, it replies with the share $F_i (Z_1(i), \ldots, Z_{n-1}(i))$. After receiving k responses, the receiver interpolates a univariate polynomial R from the k points $(i, F_i (Z_1(i), \ldots, Z_{n-1}(i)))$ and calculates the chosen secret: $\omega_\ell = R(0)$.

Moreover, the secrets are masked and the protocol is executed twice: the first run allows the receiver to obtain a masked secret and the second run allows her to obtain the corresponding mask.

2.2   Verifiable Secret Sharing Scheme

A component common to most unconditionally secure DOT protocols is the threshold SSS introduced by Shamir [10] in 1979. In this scheme, the dealer who shares a secret is honest: the m pieces he generates are built in compliance with the protocol and so, the k pieces used by the players to reconstruct the secret are genuine. In 1985, Chor, Goldwasser, Micali and Awerbuch [4] assumed that the dealer could be cheating and transmit some invalid shares to some players. They introduced the concept of VSSS to detect any deviation of the dealer from the sharing protocol. Moreover, during the reconstruction phase, some malicious players could provide honest players with incorrect shares to make honest players reconstruct an incorrect secret while they, the dishonest players, would reconstruct the original secret.

Following Chor et al.'s scheme, there have been a considerable research on VSSS. To replace Shamir's SSS, we have chosen the VSSS introduced by Gennaro, Ishai, Kushilevitz and Rabin [6] because its correctness is guaranteed, no shares are made public along its execution and its complexity is polynomial.

The setting of Gennaro et al.'s VSSS encompasses a dealer who holds a secret ω, m players, each of them receiving a (k, m)-threshold share of ω and a combiner collecting the shares from ℓ $(1 < k \leq \ell < m)$ players. Up to $t < k$ participants (dealer or players) may cheat during the execution of the protocol. In particular, cheating players may provide incorrect shares to the combiner. The protocol allows the combiner and each player to determine three sets of players: H_ω, the honest players, D_ω, the players with some incorrect shares which may be corrected by a majority of honest players and C_ω, the players with incorrect shares which cannot be corrected. Thanks to these three sets, provided that $\ell \geq 4t + 1$, the combiner is able to collect enough correct shares to determine ω.

2.3   Error-correcting Code Decoding Scheme

In Shamir's (k, m)-threshold SSS, the combiner is able to check if the ℓ collected shares $(k \leq \ell \leq m)$ were generating from a same sharing polynomial. To perform this task, the combiner applies Lagrange interpolation formula on the first k

collected shares and obtains a polynomial f of degree at most $k - 1$. Then the combiner checks that the $\ell - k$ remaining shares agree with f. If no incorrect shares are detected, the combiner calculates the secret $s = f(0)$.

If incorrect shares are detected, they could be identified and corrected, for example, by the Berlekamp-Massey algorithm [1]. However, if the main objective of error-correcting codes is to restore original codes from corrupted codes, our goal is to reconstruct the polynomial which was used to generates the codes and to evaluate this polynomial at zero. In other words, the combiner does not need to identify the incorrect shares, but needs to interpolate the sharing polynomial from the received shares. This is why a Reed-Solomon codes [9] decoding algorithm like the algorithm introduced by Gao [5] is preferred. This algorithm allows the combiner to reconstruct the sharing polynomial, in spite of up to $t \leq \frac{\ell-k}{2}$ incorrect shares.

3 Our Model

The setting of our model encompasses a sender $\mathcal{S}$ who owns n secrets $w_0, \ldots, w_{n-1}$ ($n > 1$) of $\mathbb{F}_p$ (p prime), a receiver $\mathcal{R}$ who wishes to obtain a secret w_σ ($\sigma \in \{0, \ldots n-1\}$), m servers $S_1, \ldots, S_m$ and an active adversary. The communication model and the adversary model are described below.

3.1 Communication Model

Our protocol requires the availability of private secure communication channels between the sender and the servers, the receiver and the servers and among the servers. It also requires a broadcast channel, allowing all participants to receive simultaneously information sent by one participant through this channel.

3.2 Adversary Model

Three parties may try to breach the security of our protocol:

- The sender $\mathcal{S}$, with possibly a coalition of up to $k-1$ corrupt servers, plotting against the receiver to obtain the receiver's choice σ. Servers of the coalition are passive, i.e. follow the protocol, and because they are corrupted by $\mathcal{S}$, they make any information they hold available to the sender. In addition, $\mathcal{S}$ distributes correct shares of the secrets he holds to the servers. For this part, we will consider him as honest.
- The receiver $\mathcal{R}$, colluding with up to $k-1$ corrupt servers to obtain information not only about the chosen secret w_σ, but about other secrets too. Here too, the servers of the coalition are passive and so, follow the protocol. On the contrary, $\mathcal{R}$ may not follow the protocol to obtain additional information on the secrets held by $\mathcal{S}$.
- An active adversary, with the help of a group of up to $k - 1$ malicious servers, who intends to disrupt the protocol such that the secret obtained by the receiver does not correspond to the chosen secret. In particular, when

Let $S_1, \ldots, S_m$ be m servers.

Input   The sender $\mathcal{S}$, contributes with n secrets $\omega_0, \ldots, \omega_{n-1} \in \mathbb{K} = \mathbb{F}_p$ (p prime)

The receiver $\mathcal{R}$, chooses an index $\sigma \in \{0, \ldots, n-1\}$, and contributes with $n-1$ private values $\delta_{\sigma 1}, \ldots, \delta_{\sigma(n-1)} \in \{0, 1\}$ ($\delta_{ij} = 1$ if $i = j$ and 0 otherwise)

Output $\mathcal{R}$ is either detected as a cheater and the protocol stops. Otherwise, if she follows the protocol, $\mathcal{R}$ receives ω_σ, while $\mathcal{S}$ receives nothing.

Phase 1 – Sharing of the Sender's Secrets

1. $\mathcal{S}$ generates an n-variate polynomial function $Q(x, y_1, \ldots, y_{n-1}) = P(x) + \sum_{i=1}^{n-1}(\omega_i - \omega_0)y_i$ where $P \in \mathbb{K}[X]$ is a polynomial of degree at most $k-1$ whose constant term is ω_0 and other coefficients are randomly selected in $\mathbb{K}$.
2. $\mathcal{S}$ transmits to the server S_ℓ ($\ell \in \mathcal{I}_m = \{1, \ldots, m\}$) the $(n-1)$-variate polynomial function $F_\ell(y_1, \ldots, y_{n-1}) = Q(\ell, y_1, \ldots, y_{n-1})$.

Phase 2 – Sharing of the Receiver's Secret Inputs

1. $\mathcal{R}$ chooses $\sigma \in \{0, \ldots, n-1\}$ and $\mathcal{I}_c \subset \mathcal{I}_m$, a set of $c \geq 4k-3$ servers to contact.
2. $\mathcal{R}$ generates a vector $\boldsymbol{\Theta} = (G_1, \ldots, G_{n-1})$ of $n-1$ bivariate polynomials $G_s \in \mathbb{K}[X, Y]$ of degree at most $k-1$ in X and of degree at most $k-1$ in Y, where the constant term of G_s is $\delta_{\sigma s}$ and other coefficients are randomly selected in $\mathbb{K}$. The polynomial function $G_s(x, i)$ ($i \in \mathcal{I}_c$) is denoted $f_{s,i}(x)$ and the polynomial function $G_s(i, y)$ ($i \in \mathcal{I}_c$) is denoted $g_{s,i}(y)$.
3. $\mathcal{R}$ builds c vectors $\boldsymbol{V_i} = (f_{1,i}, \ldots, f_{n-1,i})$ ($i \in \mathcal{I}_c$) from the polynomials generated in the previous step. Similarly, $\mathcal{R}$ generates c vectors $\boldsymbol{W_i} = (g_{1,i}, \ldots, g_{n-1,i})$ of polynomials ($i \in \mathcal{I}_c$).
4. $\mathcal{R}$ transmits to S_i ($i \in \mathcal{I}_c$) the pair of vectors $(\boldsymbol{V_i}, \boldsymbol{W_i})$.
5. For $s = 1, \ldots, n-1$, S_i ($i \in \mathcal{I}_c$) sends to the server S_j ($j \in \mathcal{I}_c$) a random element $r_{s,i,j} \in \mathbb{K}$.

Phase 3 – Detection of Cheaters

1. S_i ($i \in \mathcal{I}_c$) broadcasts $f_{s,i}(j) + r_{s,i,j}$ and $g_{s,i}(j) + r_{s,j,i}$, for $s = 1, \ldots, n-1$. From the broadcast values and for $s = 1, \ldots, n-1$, S_i builds three sets of servers H_s, D_s and C_s (See Sect. 2.2). If $|C_s| > k-1$, then $\mathcal{R}$ has cheated and the protocol stops.
2. S_i ($i \in \mathcal{I}_c$) determines $\mathcal{H} = \bigcap_{s=1}^{n-1}(H_s \cup D_s)$. If $c - |\mathcal{H}| \geq k$, then $\mathcal{R}$ has cheated and the protocol stops. Otherwise, the set of indices corresponding to the servers in $\mathcal{H}$ is denoted $\mathcal{I}_\mathcal{H}$.
3. S_i ($i \in \mathcal{I}_\mathcal{H}$) calculates $\boldsymbol{\Phi_i} = (Z_1(i), \ldots, Z_{n-1}(i))$. The share $Z_s(i)$ ($1 \leq s \leq n-1$) is calculated from the values $g_s(i)$ ($i \in \mathcal{I}_\mathcal{H}$) thanks to an error-correcting codes decoding scheme (See Sect. 2.3).

Phase 4 – Computation of the Shares of the Chosen Secret

Each server S_i ($i \in \mathcal{I}_\mathcal{H}$) calculates the share $\mu_i = F_i(\boldsymbol{\Phi_i})$ and sends it to $\mathcal{R}$.

Phase 5 – Reconstruction of the Chosen Secret

$\mathcal{R}$ interpolates a polynomial R of degree at most $k-1$, thanks to an error-correcting codes decoding scheme (See Sect. 2.3) and calculates $\omega_\sigma = R(0)$.

Fig. 1. A verifiable $(4k-3, m)$-DOT-$\binom{n}{1}$ protocol

they are requested to provide a share, these malicious servers may not reply or replace the requested share with any value, designated in this case as an *incorrect* share.

We assume that both the sender S and the receiver R wish to complete the protocol to allow R to obtain the chosen secret. The adversary collaborates neither with the sender S nor with the receiver R. Therefore, we assume that the set of malicious servers, the set of servers colluding with S and the set of servers colluding with R are disjoint.

In this paper, we consider static parties only; the sets of malicious and corrupt servers are in place before the protocol is executed and their contents do not change during the execution of the protocol.

In addition, along the protocol, some servers may be disqualified. We assume that a mechanism prevents the disqualified servers from keeping on participating in the protocol.

4   Proposed Protocol

In this section we present our verifiable DOT protocol (See Fig. 1), composed of five phases.

5   Security of the Protocol

To take into account the coalition of malicious servers, we replace the security condition C_1 with the following one:

C'_1. Correctness – If the receiver is not detected as cheating, she is able to determine the chosen secret once she receives information from the contacted servers, in spite of $k - 1$ malicious servers.

It remains to be seen that the proposed protocol satisfies all desirable conditions C'_1, C_2 and C_3.

Acknowledgments. We would like to thank Huaxiong Wang and the anonymous reviewers of ACISP 2011 for their helpful comments.

References

1. Berlekamp, E.: Algebraic coding theory revised 1984 edition. Aegean Park Press, Laguna Hills (1984)
2. Blundo, C., D'Arco, P., Santis, A.D., Stinson, D.R.: New results on unconditionally secure distributed oblivious transfer. In: Nyberg, K., Heys, H.M. (eds.) SAC 2002. LNCS, vol. 2595, pp. 291–309. Springer, Heidelberg (2003)
3. Blundo, C., D'Arco, P., Santis, A.D., Stinson, D.R.: On unconditionally secure distributed oblivious transfer. Journal of Cryptology 20(3), 323–373 (2007)

4. Chor, B., Goldwasser, S., Micali, S., Awerbuch, B.: Verifiable secret sharing and achieving simultaneity in the presence of faults. In: SFCS 1985: Proceedings of the 26th Annual Symposium on Foundations of Computer Science, pp. 383–395. IEEE Computer Society, Los Alamitos (1985)
5. Gao, S.: A new algorithm for decoding Reed-Solomon codes, vol. 712. Springer, Heidelberg (2003)
6. Gennaro, R., Ishai, Y., Kushilevitz, E., Rabin, T.: The round complexity of verifiable secret sharing and secure multicast. In: Proceedings of the Thirty-Third Annual ACM Symposium on Theory of Computing, pp. 580–589. ACM, New York (2001)
7. Naor, M., Pinkas, B.: Distributed oblivious transfer. In: Okamoto, T. (ed.) ASIACRYPT 2000. LNCS, vol. 1976, pp. 205–219. Springer, Heidelberg (2000)
8. Nikov, V., Nikova, S., Preneel, B., Vandewalle, J.: On unconditionally secure distributed oblivious transfer. In: Menezes, A., Sarkar, P. (eds.) INDOCRYPT 2002. LNCS, vol. 2551, pp. 395–408. Springer, Heidelberg (2002)
9. Reed, I., Solomon, G.: Polynomial codes over certain finite fields. Journal of the Society for Industrial and Applied Mathematics 8(2), 300–304 (1960)
10. Shamir, A.: How to share a secret. Communications of the ACM 22(11), 612–613 (1979)
11. Zhong, S., Richard Yang, Y.: Verifiable distributed oblivious transfer and mobile agent security. In: International Conference on Mobile Computing and Networking: Proceedings of the 2003 Joint Workshop on Foundations of Mobile Computing. ACM, New York (2003)
12. Zhong, S., Richard Yang, Y.: Verifiable distributed oblivious transfer and mobile agent security. Mobile Networks and Applications 11(2), 201–210 (2006)

Impracticality of Efficient PVSS in Real Life Security Standard (Poster)

Kun Peng

Institue for Infocomm Research, Singapore
dr.kun.peng@gmail.com

Abstract. This paper presents ongoing work toward employment of RSA encryption in PVSS. Two PVSS schemes are shown to be efficient only when very small RSA public keys like 3 are employed to encrypt the shares. However, too small RSA public keys like 3 are insecure in the PVSS schemes as they cannot apply padding to the encrypted messages. When practical larger RSA public keys are employed, the two PVSS schemes have to process extremely large integers and become intolerably inefficient.

1 Introduction

Secret sharing is an important cryptographic tool used in various secure applications. A dealer can employ it to share a secret among multiple share holders such that only certain subsets of them can cooperate to reconstruct the secret. The most popular secret sharing mechanism is t-out-of-n threshold secret sharing [10], where a dealer shares a secret s among n share holders $P_1, P_2, \ldots, P_n$ and allows any t of them to reconstruct it. It is widely employed in applications like secure computation, e-commerce and e-voting, where trust must be shared among multiple parties to strengthen security. For example, very often a private key must be shared among multiple parties such that decryption is under more strict control. As many such applications require public verifiability, very often secret sharing must be publicly verifiable. Namely, it must be publicly verified that all the shares are consistently generated from a unique secret and can be used to reconstruct it. Publicly verifiable secret sharing is usually called PVSS. The idea was firstly proposed by Feldman [5] and them developed into concrete schemes in [11,7,9,2,8]. There are two important requirements in PVSS: high efficiency and generality. Generality of PVSS requires that any secret can be shared and reconstructed efficiently no matter how it is generated or which source it is from, so that the PVSS technique can be employed in various applications.

Among the existing PVSS schemes [11,7,9,2,8], only two of them [7,2] are general and efficient. As the first PVSS solution, the PVSS scheme in [11] is not efficient and is the only one to claim a computational cost significantly higher than $O(tn)$. The dealer in [9] must know the discrete logarithm of the secret to share. So the PVSS scheme in [9] is not general and its application is limited due to hardness of the famous discrete logarithm problem. For example, when a

U. Parampalli and P. Hawkes (Eds.): ACISP 2011, LNCS 6812, pp. 451–455, 2011.

random password is chosen and shared it cannot work. Only the PVSS schemes in [7] and [2] do not limit the secret to share in any way and claim higher efficiency.

Unfortunately, high efficiency in [7] and [2] depends on a special condition: the share holders employ RSA encryption and use very small public keys like 3. As explained in Section 3, such small RSA public keys are often impractical in real life applications, especially in those applications unable to use appropriate padding of information like PVSS. As illustrated in Section 4, when larger practical RSA keys are employed, efficiency of the PVSS schemes in [7] and [2] decline to an intolerable level in real life as extremely large integers must be stored and used in computation.

2   Specification of an Important Proof in the Two PVSS Schemes

The most important operation in PVSS is to prove and verify that the same share is encrypted in a ciphertext and committed to in a commitment. In Step 4 of the PVSS protocol (in Page 9 of [7]), the dealer needs to prove

$$PROOF[B_i = BC_{(b,n_i)}(s_i) \wedge D_i(s_i) = 0 \bmod n_i] \tag{1}$$

where the parameter setting is as follows.

- $N = PQ$, $P = 2p + 1$ and $Q = 2q + 1$ where p and q primes. Although not explicitly stated in [7], P and Q should be large secret primes (or at least product of large secret primes), otherwise polynomial factorization of N makes calculation of the vth-root polynomial and breaks bindingness[1] of the commitment function $BC_{(b,v)}(s,r) = b^s r^v \bmod N$.
- b is a generator of the cyclic subgroup with order pq in Z_N^*. Although not explicitly stated in [7], p and q should be secret to guarantee hardness in factorizing N as implied by citation of [6] in [7].
- s_i is a secret share.
- n_i is an RSA modulus.

According to the definition of $BC()$ and $D_i()$ in [7], this proof in the form $PROOF[\ ]$ is actually a proof of knowledge of secret integers s_i and r to satisfy

$$B_i = b^{s_i} r^{n_i} \bmod N \tag{2}$$
$$s_i^{e_i} = C_i \bmod n_i \tag{3}$$

where C_i is an RSA ciphertext. The proof primitive $PROOF[\ ]$ is defined in Page 6 of [7] and implementation of the proof is given in the so-called Example 1 in the same page. Applying the proof method in Example 1 to proof of (2) and (3) in [7] is as follows.

---

[1] Bindingness here refers to the security property "......opening the commitment with different representations is equivalent to breaking RSA" in [7].

1. e_i is set to be 3.
2. The dealer publishes $c' = BC(s_i^2)$ and $c'' = BC(s_i^3)$.
3. The dealer runs two proof primitives $SQR_{(b,n_i)}(B_i : c')$ and $MUL_{(b,n_i)}(B_i, c' : c'')$ defined earlier in Page 6 of [7].
4. The dealer opens $(c''b^{-C_i})^{1/n_i}$.

In Section 4.1 of [2], its first PVSS protocol with fast recovery includes an operation, Step 2: "For $i = 1, ..., l$, Alice computes $E_i = s_i^3 \bmod n_i$ and executes $Proof6(s_i, S_i = g^{s_i} \bmod p \wedge E_i = s_i^3 \bmod n_i \wedge abs(s_i) < (n_i - 1)/2)$" to show that the same share s_i is committed to in S_i and encrypted in E_i where $Proof6(x, G = g^x \bmod p \wedge E = x^3 \bmod n_1 \wedge abs(x) < (n_1 - 1)/2)$ is realized in its Section 3.6 as follows.

1. Alice computes $\alpha = \frac{E-x^3}{n_1}$, $G_1 = g^{-x} \bmod N$, $G_2 = g^{-x^2} \bmod N$, $G_3 = g^{-x^3} \bmod N$ and $Z = g^{-\alpha n_1} \bmod N$.
2. Alice proves knowledge of x such that $G = g^x \bmod p$, $G_1 = g^{-x} \bmod N$, $G_2 = G_1^x \bmod N$, $G_3 = G_2^x \bmod N$ and $abs(x) < \lambda(N)/2)$ and knowledge of α such that $Z = (g^{-n_1})^{\alpha} \bmod N$.
3. The verifier checks the proofs, computes $T = g^{-E} \bmod N$ and checks that $G_3 = T/Z \bmod N$.

In Section 4.2 of [2], its second PVSS protocol with fast recovery includes an operation, Step 3: "For $i = 1, ..., l$, Alice computes $E_i = s_i^3 \bmod n_i$ and executes $Proof6(s_i, S_i = g^{s_i} \bmod p \wedge E_i = s_i^3 \bmod n_i \wedge abs(s_i) < (n_i - 1)/2)$" to show that the same share s_i is committed to in S_i and encrypted in E_i.

So the PVSS schemes in [7] and [2] are only specified in the case where the share holders use 3 as their RSA public keys. Their high efficiency is also achieved in this case.

3   RSA Key in Real-Life Cryptographic Protocols

It has been shown in [7] and [2] how their PVSS schemes work when RSA public key of the share holders is 3. However, a too small RSA public key like 3 or 5 is usually impractical in real-life cryptographic protocols. It is widely believed in the cryptographic community that smaller public keys make RSA cipher more vulnerable to attacks, especially when proper padding of message is absent. A famous example is that in a broadcasting protocol an attack can be launched if multiple receivers use the same small public key e (e.g. 3) and employ different RSA moduli $N_1, N_2, N_3, \ldots$. When a message m is broadcast, it is in the form $m^e \bmod N_1$, $m^e \bmod N_2$, $m^e \bmod N_3$ .... When multiple such ciphertexts are put together and the product of their multiplicative moduli in encryption is larger than m^e in Z, Chinese Remainder Theorem can be employed to obtain m^e in Z and then m can be extracted. The smaller e is, the fewer ciphertext is needed and the attack is easier. For example, when $e = 3$, three ciphertexts $m^e \bmod N_1$, $m^e \bmod N_2$ and $m^e \bmod N_3$ are enough for the attack.

A more formal and detailed analysis of vulnerability of very small RSA public keys is given in [4], which shows that RSA cipher with a too small public key like 3 fails "if the opponent knows two-thirds of the message, or if two messages agree over eight-ninths of their length; and we can find the factors of $N = PQ$ if we are given the high order bits of P". Even those cryptographic schemes aiming at improving efficiency of RSA cipher like [3] agree that "RSA as usually deployed uses a larger public exponent (e = 65537)". So, in authoritative security standards like the NIST standard [1], it is required that public key of RSA cipher must be no smaller than 65537.

In PVSS, as publicly verifiable encryption is necessary, secure padding of message is impossible (or at least very difficult). So the attacks exploiting small RSA public keys are serious in PVSS and the public keys of RSA cipher in PVSS should be more cautiously chosen. Therefore, in practical PVSS applications in real life, not only too small public keys like 3 or 5 should be avoided but also more strict security standard should be followed and larger RSA public keys should be adopted.

4   Intolerable Cost of the General Specification, Making the PVSS Scheme Impractical in Real Life

The specification of proof of validity of shares for general RSA keys seems to work in theory. However, in practice, it includes a very costly operation, especially when e_i is large. The operation is calculation of $s_i^2, s_i^4, \ldots, s_i^{2^{L-1}}$ where L is the bit length of e_i. In theory, they can be calculated using $L-1$ square operations $S_1 = s_i^2, S_2 = S_1^2, \ldots, S_{L-1} = S_{L-2}^2$ and then used in $SQR()$ and $MUL()$ as the secret logarithms to be proved knowledge of where $S_k = s_i^{2^k}$. In practice, the size of S_k increases rapidly and becomes intolerably large very quickly. In [7], $s_i = (f(i) \bmod v) + (2^m - \delta_i)v$ where $f()$ is the share-generating polynomial proposed by Shamir [10] and employed by most threshold PVSS schemes, $m = O(|N|)$, v is an integer decided by the dealer or a verifier and $\delta_i \in \{0, 1\}$. So 2^m and thus s_i should be hundreds of bits long in a practical secure setting. When e_i is large (e.g. no smaller than 65537), $s_i^{2^{L-1}}$ is millions or even billions of bits long. So large integers are difficult to store or process (e.g. use them to calculate S_k and c''), not to mention $s_i^2, s_i^4, \ldots, s_i^{2^{L-1}}$ are used in $SQR()$ and $MUL()$ as secret logarithms to be proved knowledge of.

Some reader may ask: cannot we calculate $S_k = S_{k-1}^2$ with a multiplicative modulus instead in Z so that S_k will not be too large? The key point is what modulus to use. As $s_i^2, s_i^4, \ldots, s_i^{2^{L-1}}$ are in the form of exponents to the base b, the multiplicative modulus must be a multiple of the order of b. However, the order of b and its multiples are secret since factorization of N must be hard as we have explained in Section 2. If the dealer knows the order of b, he can open his commitment in many different ways and commitment of the secret fails. So, the dealer cannot know the order of b or any of its multiples. Therefore, $s_i^2, s_i^4, \ldots, s_i^{2^{L-1}}$ must be calculated in Z and become extremely large when a large enough secure public key is employed in RSA cipher. In summary, intolerably high cost is inevitable in [7] if secure RSA public keys are employed. The same problem exists in [2] as well.

5   Conclusion

A proof technique in the only general and efficient PVSS schemes [7,2] is not a general soultion. They fail to achieve high efficiency in practical applications as desired.

References

1. The NIST special publication on computer security (2007), http://csrc.nist.gov/publications/nistpubs/ (sp 800-78 rev August 1, 2007)
2. Boudot, F., Traoré, J.: Efficient publicly verifiable secret sharing schemes with fast or delayed recovery. In: Varadharajan, V., Mu, Y. (eds.) ICICS 1999. LNCS, vol. 1726, pp. 87–102. Springer, Heidelberg (1999)
3. Boneh, D., Shacham, H.: Fast variants of RSA. CryptoBytes 5(1), 1–9 (2002)
4. Coppersmith, D.: Small Solutions to Polynomial Equations, and Low Exponent RSA Vulnerabilities. Journal of Cryptology 10(4), 233–260 (1997)
5. Feldman, P.: A practical scheme for non-interactive verifiable secret sharing. In: FOCS 1987, pp. 427–437 (1987)
6. Fujisaki, E., Okamoto, T.: Statistical zero knowledge protocols to prove modular polynomial relations. In: Kaliski Jr., B.S. (ed.) CRYPTO 1997. LNCS, vol. 1294, pp. 16–30. Springer, Heidelberg (1997)
7. Fujisaki, E., Okamoto, T.: A practical and provably secure scheme for publicly verifiable secret sharing and its applications. In: Nyberg, K. (ed.) EUROCRYPT 1998. LNCS, vol. 1403, pp. 32–46. Springer, Heidelberg (1998)
8. Peng, K., Bao, F.: Efficient publicly verifiable secret sharing with correctness, soundness and ZK privacy. In: Youm, H.Y., Yung, M. (eds.) WISA 2009. LNCS, vol. 5932, pp. 118–132. Springer, Heidelberg (2009)
9. Schoenmakers, B.: A simple publicly verifiable secret sharing scheme and its application to electronic voting. In: Wiener, M. (ed.) CRYPTO 1999. LNCS, vol. 1666, pp. 149–164. Springer, Heidelberg (1999)
10. Shamir, A.: How to share a secret. Communication of the ACM 22(11), 612–613 (1979)
11. Stadler, M.: Publicly verifiable secret sharing. In: Maurer, U.M. (ed.) EUROCRYPT 1996. LNCS, vol. 1070, pp. 190–199. Springer, Heidelberg (1996)

Electromagnetic Analysis Enhancement with Signal Processing Techniques (Poster)

Hongying Liu[1], Yukiyasu Tsunoo[2], and Satoshi Goto[1]

[1] Graduate School of Information, Production and Systems,
Waseda University, 8080135, Japan
[2] Information and Media Processing Laboratories, NEC Corp.,
Kawasaki, 2118666, Japan
liuhongying@fuji.waseda.jp, tsunoo@BL.jp.nec.com,
goto@waseda.jp

Abstract. This paper presents ongoing work toward Electromagnetic Analysis (EMA) with signal processing techniques. Electromagnetic emission leaks confidential data of cryptographic devices. EMA exploits such information and reveals secret keys. It has been actively studied. We present three signal processing techniques: bandpass filtering, signal companding and independent component analysis (ICA), and apply them to EMA. The effectiveness is demonstrated through the analyses of encryption algorithms on synthesized application-specific integrated circuit (ASIC). Experiments show that the performance of EMA is greatly enhanced. The number of signals needed to reveal keys has been dramatically reduced.

1 Introduction

Electromagnetic analysis (EMA) is performed with low-cost sensors to extract the secret keys from cryptographic devices. The accuracy of key detection largely depends on the quality of the EM signal, namely signal to noise ratio (SNR). In general, there are two types of noise that prevent a fast key exposure. One is the non-algorithm noise, which originates from external, intrinsic, sampling and quantization noise unintentionally, and the other is algorithm noise [1], which results from the countermeasures added intentionally. For example, signals may be displaced due to random delays inserted into the encryption algorithm, or multiple encryption algorithms may run simultaneously, referred as simultaneous algorithm noise. Several approaches have been investigated to reduce these noises. Le. et al.[2] adopt the fourth-order cumulant to decrease the non-algorithm noise. Homma et al.[3] apply the method of phase-based waveform matching to overcome signal displacement. However, because of the complexity and variations of the algorithm noise, there is few works deal with simultaneous algorithm noise.

In order to improve the quality of EM signals, unlike the previous work, we explore three signal processing techniques. Addressing the reduction of non-algorithm noise, we adopt the conventional bandpass filtering to attenuate unrelated frequency

U. Parampalli and P. Hawkes (Eds.): ACISP 2011, LNCS 6812, pp. 456–461, 2011.

components existed in EM signals. Aimed at enhancing the SNR of signal directly, signal companding is used to enlarge the exploitable part of the signal for EMA. Additionally in view of the simultaneous algorithm noise, we propose the approach of difference ICA to separate the uncorrelated encryption from mixed encryptions. The aim of the above techniques is to enhance the performance of EMA.

The remainder of this paper is organized as follows. Section 2 describes some related preliminaries. Section 3 presents the application of three signal processing techniques in detail. Section 4 draws conclusions and suggests future work.

2 Preliminaries

The effectiveness of EMA is assessed by the number of signal needed to perform a successful attack. A more specific metric is success rate, which expresses the number of correct key guess among all the key bytes. In our work, we test the proposed three techniques against implementations of AES and Camellia on Side-channel Attack Standard Evaluation Board-R (SASEBO-R) [4]. Experimental environment is shown in Fig.1.The cryptographic cores use $0.13\mu m$ TSMC standard library of CMOS process technology. From AES1 to AES4, the S-boxes are based on Look-up table, (Positive Polarity Reed Muler 1-stage) PPRM1, PPRM3 and the multiplicative inverse circuit with a composite field respectively.AES0 is similar to AES4 but with support of decryption.

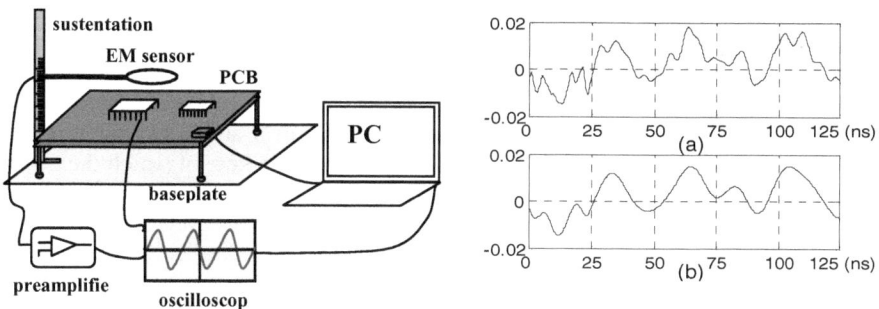

Fig. 1. Illustration of experimental environment   **Fig.2.** (a)EM signal without bandpss filtering
(b)EM signal with bandpass filtering

Output of the encryption function "add round key" in the final round of AES is chosen as a target to analyze. Encryption proceeds with 10000 random plaintexts and a fixed but random 16-byte key (the final round): 28 AF CE 9F 5A FF C8 F1 E0 54 B3 52 B0 CE 43 0E. Each measurement is repeated for 30 times. Then after signal processing, which is presented in the following section, EMA based on Hamming Distance model [5] is performed.

3 EMA with Signal Processing

Bandpass filtering. This is a technique that passes frequencies within a certain range and rejects frequencies outside that range. It is described by Eq.1.

$$Y[t] = \sum_{i=0}^{N} b_i X[t-i] \tag{1}$$

where $X[t]$ is the input signal, $Y[t]$ is the output signal, b_i are the coefficients of a filter, N is the order of filter. The encryption runs at 24MHz.The pass band of the filter is set from 0Hz to 40MHz. Signals with filtering and without filtering are shown in Fig.2. The signals become smoother, which leads to an enhanced success rate. All the key bytes are revealed at 2905 signals with filtering, while 3614 signals are needed without filtering. Through extensive experiments on setting of the pass bands, we find that low pass filtering is effective for implementations at various frequencies, such as 12MHz, 20MHz, 28MHz, etc. This indicates that the low frequency components play significant role for key detection. Further work is still under way.

Signal companding. It is a non-linear transform that includes the compressing function and expanding function, which is widely used in digital communication systems. The expanding function of μ-law algorithm is given by Eq.2, where V is the maximum value of input signal X, U is an adjustable parameter.

$$Y = (e^{X \log(1+U)/V} - 1)\frac{V}{U}\operatorname{sgn}(X) \tag{2}$$

EM signals processed with μ-law expanding function is shown in Fig.3. The amplitudes with high peaks are enlarged and the amplitudes with low peaks are remained almost unchanged, which yields a higher SNR. Thereby the performance of EMA is enhanced. The number of signals needed to recover all the key bytes is listed and compared in Table 1. The slowest key guess is the 10th key, of which the number of needed signals has been decreased from 3614 to 2981.

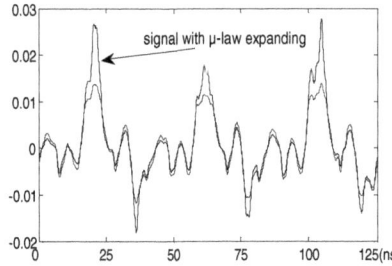 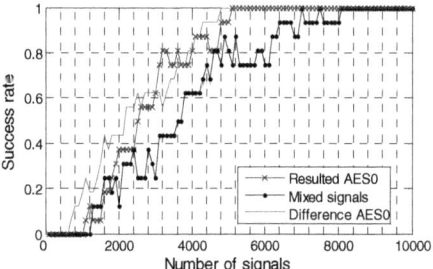

Fig.3. Signals processed with μ-law expanding

Fig.4. Success rates with three groups of signals

Table 1. The number of needed signals for revealing each key byte of AES0

No.	K1	K 2	K 3	K 4	K 5	K 6	K 7	K 8
original	3,258	2,075	2,614	1,607	2,621	3,012	2,893	2,153
expanding	2,103	2,043	2,421	1,562	2,218	3,007	2,899	2,156
No.	K 9	K 10	K 11	K 12	K 13	K 14	K 15	K16
original	1,531	3,614	3,009	3,325	2,736	2,134	2,618	2,748
expanding	1,507	2,981	2,914	2,819	2,041	1,892	2,807	2,533

Difference ICA. Multiple encryption modules may run simultaneously. This results in a slower key detection. Aimed at separating each encryption signal from mixed signals, FastICA[6] which has good performance for blind source separation (BSS), is used as basic algorithm in our work. Based on it, we propose so-called difference ICA approach, which is computing the difference between the mixed signals and the separated signal, to process various cases of mixed encryptions.

✦   Experiment1: a mixture with 2 encryption sources.
AES0 and Camellia on the ASIC execute simultaneously. Two mixed signals which are shown in Fig.5 (a)(b), with different plaintext and same key are input to FastICA algorithm. This leads to two resulted signals shown in Fig.5(d)(f) respectively. The individual executions of AES0 and Camellia are supposed to be the source signals, which are plotted in Fig.5(c)(e) respectively.

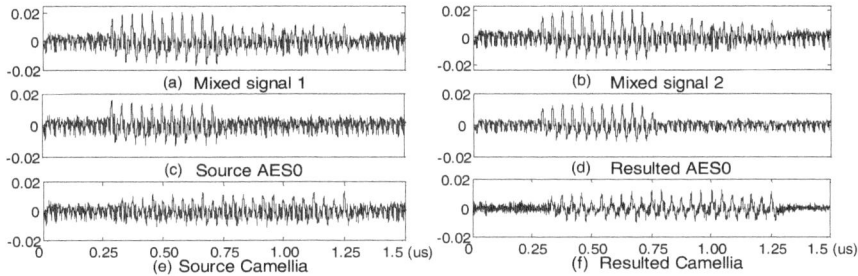

Fig. 5. The signals of two encryption source: AES0 and Camellia

Then we perform EMA with (1) mixed signals; (2) resulted AES0; (3) difference AES0, which is the signal difference of mixed signal and resulted Camellia. Success rates are compared and shown in Fig.4. The fastest key detection is with the difference AES0. All the key bytes are revealed within 4845 signals. It indicates the effectiveness of difference ICA approach.

✦   Experiment2: a mixture with 3 encryption sources.
In this case, only Camellia is separated with FastICA algorithm. Because any one of the AES executions on ASIC has a linear relation with Hamming Distance, the relation between different AES is not independent. Then according to difference ICA approach, we substrate the resulted Camellia from the mixed signals and conduct EMA. The number of signals to reveal all the key bytes and the maximal correlation

coefficient are listed in Table 2. The number of signals has been reduced 41.1% at least with the separation of Camellia in all the above cases.

✦   Experiment3: a mixture with more than 3 encryption sources.
Five encryptions, namely AES1-AES4 and Camellia are processed by difference ICA. The number of signals used to reveal all the key bytes has been reduced 47.8%, which is listed in the last line of Table 2. The evolution of the second key byte "AF" is shown in Fig.6. Only 4327 signals are needed for the appearing of correct key with the differential signals.

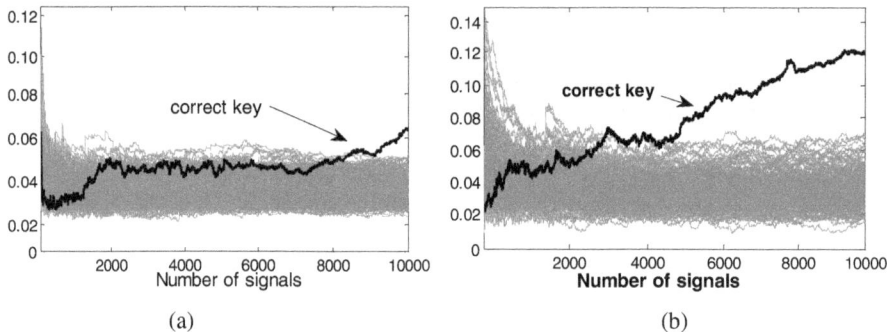

(a)   (b)

Fig.6. Correct key evolution for the second key byte: "AF". (a)with mixed signals,(b)with differential signal.

All the above three groups of experiments demonstrate the successful application of the proposed difference ICA approach.

Table 2. Number of needed signals and correlations for each mixed encryption

Mixed type	Mixed signal		Differential signal		Reduction
	No.	Corr.	No.	Corr.	rate
AES1,2, C	Fails	0.0422	5,371	0.0996	46.3%
AES1,3, C	7,012	0.0627	4,126	0.1098	41.1%
AES1,4, C	6,411	0.0703	3,679	0.1327	42.6%
AES2,3, C	Fails	0.0419	5,301	0.0921	47.0%
AES2,4, C	9,835	0.0580	5,527	0.0908	43.8%
AES3,4, C	7,164	0.0695	4,175	0.1162	41.7%
AES1-4, C	8,291	0.0613	4,327	0.1204	47.8%

4   Conclusions

The main contribution of this work is that we propose three signal processing techniques and successfully apply them in EMA: bandpass filtering, signal companding and difference ICA. This is confirmed by the experiments of EMA against the implementations of AES and Camellia on ASIC. Several conclusions are elicited. Bandpass filtering is a general processing technique, which attenuates the inference from multiple frequency components. Signal companding is useful for improving the SNR of the signals directly. Difference ICA is particularly effective to separate uncorrelated signals, which is fit for the mixed algorithm implementations. With difference ICA, the countermeasure of simultaneous algorithm noise is greatly weakened. These results may also provide enlightment for the design of countermeasures.

In the future, more advanced signal processing techniques will be investigated and studied. They will be applied to the evaluation of other countermeasures in order to improve the security of cryptographic devices.

Acknowledgments. This work was supported by Waseda University "Global COE program" of MEXT in Japan.

References

1. Messerges, T.S., Dabbish, E.A., Sloan, R.H.: Examining Smart-card security under the Threat of Power Analysis Attacks. IEEE Transactions on Computer 51(5), 541–552 (2002)
2. Le, T.H., Servière, C., Cledière, J., Lacoume, J.-L.: Noise reduction in the side channel attack using fourth order cumulants. IEEE Trans. Inf. Forensic Security 2(4), 710–720 (2007)
3. Homma, N., Nagashima, S., Imai, Y., et al.: A high-resolution phase-based waveform matching and its application to side-channel attacks. IEICE Transactions on Fundamentals E91-A(1) (2008)
4. Research Center for Information Security (RCIS) of AIST: Side-channel Attack Standard Evaluation Board (SASEBO),
   `http://www.rcis.aist.go.jp/special/SASEBO/index-en.html`
5. Brier, E., Clavier, C., Olivier, F.: Correlation Power Analysis with a Leakage Model. In: Joye, M., Quisquater, J.-J. (eds.) CHES 2004. LNCS, vol. 3156, pp. 16–29. Springer, Heidelberg (2004)
6. Hyvärinen, A.: Fast and robust fixed-point algorithms for independent component analysis. IEEE Transactions on Neural Networks 10(3), 626–634 (1999)

Erratum: Compliance or Security, What Cost? (Poster)

Craig Wright

Springer-Verlag, Computer Science Editorial, Tiergartenstr. 17,
69121 Heidelberg, Germany
{cwrigh20}@postoffice.csu.edu.au

U. Parampalli and P. Hawkes (Eds.): ACISP 2011, LNCS 6812, pp. 412–416, 2011.
© Springer-Verlag Berlin Heidelberg 2011

DOI 10.1007/978-3-642-22497-3_36

In the original version the author affiliation was wrongly added in this paper. It should read as: Charles Sturt University, Australia

The original online version for this chapter can be found at
http://dx.doi.org/10.1007/978-3-642-22497-3_27

Author Index

Akram, Raja Naeem   208
Al-Hamdan, Ali   75
Au, Man Ho   172

Bagheri, Nasour   428
Bartlett, Harry   75
Bogdanov, Andrey   106

Carlet, Claude   1
Chatterjee, Sanjit   353
Chen, Jiazhe   16
Chu, Cheng Kang   337
Corniaux, Christian L.F.   444

Dawson, Ed   75
Du, Yusong   47

Emura, Keita   439

Fan, Jia   371
Freire, Eduarda S.V.   292
Fujioka, Atsushi   319

Gauravaram, Praveen   428
Ghodosi, Hossein   444
Goto, Satoshi   456
Guo, Jian   433

Hatano, Tetsuya   189

Jia, Dingding   310
Jia, Keting   16

Knudsen, Lars R.   428

Lagishetty, Shashank   276
Li, Bao   310
Lin, Dongdai   34
Ling, San   433
Liu, Hongying   456
Liu, Joseph K.   337
Liu, Meicheng   34
Liu, Yamin   310
Long, Benjamin W.   226
Lu, Xianhui   310

Markantonakis, Konstantinos   208
Mayes, Keith   208
Menezes, Alfred   353
Minematsu, Kazuhiko   89
Miyaji, Atsuko   189, 439
Mouha, Nicky   120
Mu, Yi   172

Nguyen, Phuong Ha   61

Ohtahara, Chiaki   417
Okada, Keita   417
Okamoto, Yoshiaki   319

Paterson, Kenneth G.   292
Pei, Dingyi   34
Peng, Kun   451
Pieprzyk, Josef   407
Preneel, Bart   120, 423

Rahman, Mohammad Shahriar   439
Rechberger, Christian   433

Sabbu, Pruthvi   276
Saito, Taiichi   319
Šarinay, Juraj   142
Sasaki, Yu   417
Sato, Takashi   189
Sepahi, Reza   407
Shibutani, Kyoji   106
Shigeri, Maki   89
Shimoyama, Takeshi   417
Simpson, Leonie   75
Srinathan, Kannan   276
Steinfeld, Ron   407
Sun, Yue   120
Susilo, Willy   172
Suzaki, Tomoyasu   89

Takagi, Tsuyoshi   241
Tang, Qiang   389
Tang, Xiaohu   371
Tartary, Christophe   259
Teo, Sui-Guan   75
Tsunoo, Yukiyasu   456

Ustaoglu, Berkant   353

Wang, Huaxiong   61, 259, 433
Wang, Meiqin   120
Wang, Xiaoyun   16, 157
Watanabe, Dai   423
Wei, Lei   428, 433
Wong, Kenneth Koon-Ho   75
Wright, Craig   412, E1
Wu, Hongjun   61, 433

Yoshida, Hirotaka   423
Yu, Hongbo   16, 157

Zhang, Fangguo   47
Zhang, Mingwu   241
Zhang, Yun   259
Zheng, Yuliang   371
Zhou, Jianying   337